# Lecture Notes in Computer Science 12532

More information about this subseries at http://www.springer.com/series/7407

Haiqin Yang · Kitsuchart Pasupa ·
Andrew Chi-Sing Leung ·
James T. Kwok · Jonathan H. Chan ·
Irwin King (Eds.)

# Neural Information Processing

## 27th International Conference, ICONIP 2020
## Bangkok, Thailand, November 23–27, 2020
## Proceedings, Part I

*Editors*
Haiqin Yang
Department of AI
Ping An Life
Shenzhen, China

Andrew Chi-Sing Leung
City University of Hong Kong
Kowloon, Hong Kong

Jonathan H. Chan
School of Information Technology
King Mongkut's University
of Technology Thonburi
Bangkok, Thailand

Kitsuchart Pasupa
Faculty of Information Technology
King Mongkut's Institute
of Technology Ladkrabang
Bangkok, Thailand

James T. Kwok
Department of Computer Science
and Engineering
Hong Kong University of Science
and Technology
Hong Kong, Hong Kong

Irwin King
The Chinese University of Hong Kong
New Territories, Hong Kong

ISSN 0302-9743 ISSN 1611-3349 (electronic)
Lecture Notes in Computer Science
ISBN 978-3-030-63829-0 ISBN 978-3-030-63830-6 (eBook)
https://doi.org/10.1007/978-3-030-63830-6

LNCS Sublibrary: SL1 – Theoretical Computer Science and General Issues



This Springer imprint is published by the registered company Springer Nature Switzerland AG
The registered company address is: Gewerbestrasse 11, 6330 Cham, Switzerland

# Preface

This book is a part of the five-volume proceedings of the 27th International Conference on Neural Information Processing (ICONIP 2020), held during November 18–22, 2020. The conference aims to provide a leading international forum for researchers, scientists, and industry professionals who are working in neuroscience, neural networks, deep learning, and related fields to share their new ideas, progresses, and achievements. Due to the outbreak of COVID-19, this year's conference, which was supposed to be held in Bangkok, Thailand, was organized as fully virtual conference.

The research program of this year's edition consists of four main categories, Theory and Algorithms, Computational and Cognitive Neurosciences, Human-Centered Computing, and Applications, for refereed research papers with nine special sessions and one workshop. The research tracks attracted submissions from 1,083 distinct authors from 44 countries. All the submissions were rigorously reviewed by the conference Program Committee (PC) comprising 84 senior PC members and 367 PC members. A total of 1,351 reviews were provided, with each submission receiving at least 2 reviews, and some papers receiving 3 or more reviews. This year, we also provided rebuttals for authors to address the errors that exist in the review comments. Meta-reviews were provided with consideration of both authors' rebuttal and reviewers' comments. Finally, we accepted 187 (30.25%) of the 618 full papers that were sent out for review in three volumes of Springer's series of *Lecture Notes in Computer Science* (LNCS) and 189 (30.58%) of the 618 in two volumes of Springer's series of *Communications in Computer and Information Science* (CCIS).

We would like to take this opportunity to thank all the authors for submitting their papers to our conference, and the senior PC members, PC members, as well as all the Organizing Committee members for their hard work. We hope you enjoyed the research program at the conference.

November 2020

Haiqin Yang
Kitsuchart Pasupa

# Organization

## Honorary Chairs

| | |
|---|---|
| Jonathan Chan | King Mongkut's University of Technology Thonburi, Thailand |
| Irwin King | Chinese University of Hong Kong, Hong Kong |

## General Chairs

| | |
|---|---|
| Andrew Chi-Sing Leung | City University of Hong Kong, Hong Kong |
| James T. Kwok | Hong Kong University of Science and Technology, Hong Kong |

## Program Chairs

| | |
|---|---|
| Haiqin Yang | Ping An Life, China |
| Kitsuchart Pasupa | King Mongkut's Institute of Technology Ladkrabang, Thailand |

## Local Arrangements Chair

| | |
|---|---|
| Vithida Chongsuphajaisiddhi | King Mongkut University of Technology Thonburi, Thailand |

## Finance Chairs

| | |
|---|---|
| Vajirasak Vanijja | King Mongkut's University of Technology Thonburi, Thailand |
| Seiichi Ozawa | Kobe University, Japan |

## Special Sessions Chairs

| | |
|---|---|
| Kaizhu Huang | Xi'an Jiaotong-Liverpool University, China |
| Raymond Chi-Wing Wong | Hong Kong University of Science and Technology, Hong Kong |

## Tutorial Chairs

| | |
|---|---|
| Zenglin Xu | Harbin Institute of Technology, China |
| Jing Li | Hong Kong Polytechnic University, Hong Kong |

## Proceedings Chairs

| | |
|---|---|
| Xinyi Le | Shanghai Jiao Tong University, China |
| Jinchang Ren | University of Strathclyde, UK |

## Publicity Chairs

| | |
|---|---|
| Zeng-Guang Hou | Chinese Academy of Sciences, China |
| Ricky Ka-Chun Wong | City University of Hong Kong, Hong Kong |

## Senior Program Committee

| | |
|---|---|
| Sabri Arik | Istanbul University, Turkey |
| Davide Bacciu | University of Pisa, Italy |
| Yi Cai | South China University of Technology, China |
| Zehong Cao | University of Tasmania, Australia |
| Jonathan Chan | King Mongkut's University of Technology Thonburi, Thailand |
| Yi-Ping Phoebe Chen | La Trobe University, Australia |
| Xiaojun Chen | Shenzhen University, China |
| Wei Neng Chen | South China University of Technology, China |
| Yiran Chen | Duke University, USA |
| Yiu-ming Cheung | Hong Kong Baptist University, Hong Kong |
| Sonya Coleman | Ulster University, UK |
| Daoyi Dong | University of New South Wales, Australia |
| Leonardo Franco | University of Malaga, Spain |
| Jun Fu | Northeastern University, China |
| Xin Geng | Southeast University, China |
| Ping Guo | Beijing Normal University, China |
| Pedro Antonio Gutiérrez | Universidad de Córdoba, Spain |
| Wei He | University of Science and Technology Beijing, China |
| Akira Hirose | The University of Tokyo, Japan |
| Zengguang Hou | Chinese Academy of Sciences, China |
| Kaizhu Huang | Xi'an Jiaotong-Liverpool University, China |
| Kazushi Ikeda | Nara Institute of Science and Technology, Japan |
| Gwanggil Jeon | Incheon National University, South Korea |
| Min Jiang | Xiamen University, China |
| Abbas Khosravi | Deakin University, Australia |
| Wai Lam | Chinese University of Hong Kong, Hong Kong |
| Chi Sing Leung | City University of Hong Kong, Hong Kong |
| Kan Li | Beijing Institute of Technology, China |
| Xi Li | Zhejiang University, China |
| Jing Li | Hong Kong Polytechnic University, Hong Kong |
| Shuai Li | University of Cambridge, UK |
| Zhiyong Liu | Chinese Academy of Sciences, China |
| Zhigang Liu | Southwest Jiaotong University, China |

| | |
|---|---|
| Wei Liu | Tencent, China |
| Jun Liu | Xi'an Jiaotong University, China |
| Jiamou Liu | The University of Auckland, New Zealand |
| Lingjia Liu | Virginia Tech, USA |
| Jose A. Lozano | UPV/EHU, Spain |
| Bao-liang Lu | Shanghai Jiao Tong University, China |
| Jiancheng Lv | Sichuan University, China |
| Marley M. B. R. Vellasco | PUC of Rio de Janeiro, Brazil |
| Hiroshi Mamitsuka | Kyoto University, Japan |
| Leandro Minku | University of Birmingham, UK |
| Chaoxu Mu | Tianjin University, China |
| Wolfgang Nejdl | L3S Research Center, Germany |
| Quoc Viet Hung Nguyen | Griffith University, Australia |
| Takashi Omori | Tamagawa University, Japan |
| Seiichi Ozawa | Kobe University, Japan |
| Weike Pan | Shenzhen University, China |
| Jessie Ju Hyun Park | Yeungnam University, Japan |
| Kitsuchart Pasupa | King Mongkut's Institute of Technology Ladkrabang, Thailand |
| Abdul Rauf | Research Institute of Sweden, Sweden |
| Imran Razzak | Deakin University, Australia |
| Jinchang Ren | University of Strathclyde, UK |
| Hayaru Shouno | The University of Electro-Communications, Japan |
| Ponnuthurai Suganthan | Nanyang Technological University, Singapore |
| Yang Tang | East China University of Science and Technology, China |
| Jiliang Tang | Michigan State University, USA |
| Ivor Tsang | University of Technology Sydney, Australia |
| Peerapon Vateekul | Chulalongkorn University, Thailand |
| Brijesh Verma | Central Queensland University, Australia |
| Li-Po Wang | Nanyang Technological University, Singapore |
| Kok Wai Wong | Murdoch University, Australia |
| Ka-Chun Wong | City University of Hong Kong, Hong Kong |
| Raymond Chi-Wing Wong | Hong Kong University of Science and Technology, Hong Kong |
| Long Phil Xia | Peking University, Shenzhen Graduate School, China |
| Xin Xin | Beijing Institute of Technology, China |
| Guandong Xu | University of Technology Sydney, Australia |
| Bo Xu | Chinese Academy of Sciences, China |
| Zenglin Xu | Harbin Institute of Technology, China |
| Rui Yan | Peking University, China |
| Xiaoran Yan | Indiana University Bloomington, USA |
| Haiqin Yang | Ping An Life, China |
| Qinmin Yang | Zhejiang University, China |
| Zhirong Yang | Norwegian University of Science and Technology, Norway |

## Program Committee

| | |
|---|---|
| Chuan Chen | Sun Yat-sen University, China |
| Liang Chen | Sun Yat-sen University, China |
| Zhuangbin Chen | Chinese University of Hong Kong, Hong Kong |
| Junyi Chen | City University of Hong Kong, Hong Kong |
| Xingjian Chen | City University of Hong Kong, Hong Kong |
| Lisi Chen | Hong Kong Baptist University, Hong Kong |
| Fan Chen | Duke University, USA |
| Xiang Chen | George Mason University, USA |
| Long Cheng | Chinese Academy of Sciences, China |
| Aneesh Chivukula | University of Technology Sydney, Australia |
| Sung Bae Cho | Yonsei University, South Korea |
| Sonya Coleman | Ulster University, UK |
| Fengyu Cong | Dalian University of Technology, China |
| Jose Alfredo Ferreira Costa | Federal University of Rio Grande do Norte, Brazil |
| Ruxandra Liana Costea | Polytechnic University of Bucharest, Romania |
| Jean-Francois Couchot | University of Franche-Comté, France |
| Raphaël Couturier | University Bourgogne Franche-Comté, France |
| Zhenyu Cui | University of the Chinese Academy of Sciences, China |
| Debasmit Das | Qualcomm, USA |
| Justin Dauwels | Nanyang Technological University, Singapore |
| Xiaodan Deng | Beijing Normal University, China |
| Zhaohong Deng | Jiangnan University, China |
| Mingcong Deng | Tokyo University, Japan |
| Nat Dilokthanakul | Vidyasirimedhi Institute of Science and Technology, Thailand |
| Hai Dong | RMIT University, Australia |
| Qiulei Dong | Chinese Academy of Sciences, China |
| Shichao Dong | Shenzhen Zhiyan Technology Co., Ltd., China |
| Kenji Doya | Okinawa Institute of Science and Technology, Japan |
| Yiqun Duan | University of Sydney, Australia |
| Aritra Dutta | King Abdullah University of Science and Technology, Saudi Arabia |
| Mark Elshaw | Coventry University, UK |
| Issam Falih | Paris 13 University, France |
| Ozlem Faydasicok | Istanbul University, Turkey |
| Zunlei Feng | Zhejiang University, China |
| Leonardo Franco | University of Malaga, Spain |
| Fulvio Frati | Università degli Studi di Milano, Italy |
| Chun Che Fung | Murdoch University, Australia |
| Wai-Keung Fung | Robert Gordon University, UK |
| Claudio Gallicchio | University of Pisa, Italy |
| Yongsheng Gao | Griffith University, Australia |
| Cuiyun Gao | Harbin Institute of Technology, China |
| Hejia Gao | University of Science and Technology Beijing, China |
| Yunjun Gao | Zhejiang University, China |

| | |
|---|---|
| Xin Gao | King Abdullah University of Science and Technology, Saudi Arabia |
| Yuan Gao | Uppsala University, Sweden |
| Yuejiao Gong | South China University of Technology, China |
| Xiaotong Gu | University of Tasmania, Australia |
| Shenshen Gu | Shanghai University, China |
| Cheng Guo | Chinese Academy of Sciences, China |
| Zhishan Guo | University of Central Florida, USA |
| Akshansh Gupta | Central Electronics Engineering Research Institute, India |
| Pedro Antonio Gutiérrez | University of Córdoba, Spain |
| Christophe Guyeux | University Bourgogne Franche-Comté, France |
| Masafumi Hagiwara | Keio University, Japan |
| Ali Haidar | University of New South Wales, Australia |
| Ibrahim Hameed | Norwegian University of Science and Technology, Norway |
| Yiyan Han | Huazhong University of Science and Technology, China |
| Zhiwei Han | Southwest Jiaotong University, China |
| Xiaoyun Han | Sun Yat-sen University, China |
| Cheol Han | Korea University, South Korea |
| Takako Hashimoto | Chiba University of Commerce, Japan |
| Kun He | Shenzhen University, China |
| Xing He | Southwest University, China |
| Xiuyu He | University of Science and Technology Beijing, China |
| Wei He | University of Science and Technology Beijing, China |
| Katsuhiro Honda | Osaka Prefecture University, Japan |
| Yao Hu | Alibaba Group, China |
| Binbin Hu | Ant Group, China |
| Jin Hu | Chongqing Jiaotong University, China |
| Jinglu Hu | Waseda University, Japan |
| Shuyue Hu | National University of Singapore, Singapore |
| Qingbao Huang | Guangxi University, China |
| He Huang | Soochow University, China |
| Kaizhu Huang | Xi'an Jiaotong-Liverpool University, China |
| Chih-chieh Hung | National Chung Hsing University, Taiwan |
| Mohamed Ibn Khedher | IRT SystemX, France |
| Kazushi Ikeda | Nara Institute of Science and Technology, Japan |
| Teijiro Isokawa | University of Hyogo, Japan |
| Fuad Jamour | University of California, Riverside, USA |
| Jin-Tsong Jeng | National Formosa University, Taiwan |
| Sungmoon Jeong | Kyungpook National University, South Korea |
| Yizhang Jiang | Jiangnan University, China |
| Wenhao Jiang | Tencent, China |
| Yilun Jin | Hong Kong University of Science and Technology, Hong Kong |

| | |
|---|---|
| Wei Jin | Michigan State University, USA |
| Hamid Karimi | Michigan State University, USA |
| Dermot Kerr | Ulster University, UK |
| Tariq Khan | Deakin University, Australia |
| Rhee Man Kil | Korea Advanced Institute of Science and Technology, South Korea |
| Sangwook Kim | Kobe University, Japan |
| Sangwook Kim | Kobe University, Japan |
| DaeEun Kim | Yonsei University, South Korea |
| Jin Kyu Kim | Facebook, Inc., USA |
| Mutsumi Kimura | Ryukoku University, Japan |
| Yasuharu Koike | Tokyo Institute of Technology, Japan |
| Ven Jyn Kok | National University of Malaysia, Malaysia |
| Aneesh Krishna | Curtin University, Australia |
| Shuichi Kurogi | Kyushu Institute of Technology, Japan |
| Yoshimitsu Kuroki | National Institute of Technology, Kurume College, Japan |
| Susumu Kuroyanagi | Nagoya Institute of Technology, Japan |
| Weng Kin Lai | Tunku Abdul Rahman University College, Malaysia |
| Wai Lam | Chinese University of Hong Kong, Hong Kong |
| Kittichai Lavangnananda | King Mongkut's University of Technology Thonburi, Thailand |
| Xinyi Le | Shanghai Jiao Tong University, China |
| Teerapong Leelanupab | King Mongkut's Institute of Technology Ladkrabang, Thailand |
| Man Fai Leung | City University of Hong Kong, Hong Kong |
| Gang Li | Deakin University, Australia |
| Qian Li | University of Technology Sydney, Australia |
| Jing Li | University of Technology Sydney, Australia |
| JiaHe Li | Beijing Institute of Technology, China |
| Jian Li | Huawei Noah's Ark Lab, China |
| Xiangtao Li | Jilin University, China |
| Tao Li | Peking University, China |
| Chengdong Li | Shandong Jianzhu University, China |
| Na Li | Tencent, China |
| Baoquan Li | Tianjin Polytechnic University, China |
| Yiming Li | Tsinghua University, China |
| Yuankai Li | University of Science and Technology of China, China |
| Yang Li | Zhejiang University, China |
| Mengmeng Li | Zhengzhou University, China |
| Yaxin Li | Michigan State University, USA |
| Xiao Liang | Nankai University, China |
| Hualou Liang | Drexel University, USA |
| Hao Liao | Shenzhen University, China |
| Ming Liao | Chinese University of Hong Kong, Hong Kong |
| Alan Liew | Griffith University, Australia |

| | |
|---|---|
| Chengchuang Lin | South China Normal University, China |
| Xinshi Lin | Chinese University of Hong Kong, Hong Kong |
| Jiecong Lin | City University of Hong Kong, Hong Kong |
| Shu Liu | The Australian National University, Australia |
| Xinping Liu | University of Tasmania, Australia |
| Shaowu Liu | University of Technology Sydney, Australia |
| Weifeng Liu | China University of Petroleum, China |
| Zhiyong Liu | Chinese Academy of Sciences, China |
| Junhao Liu | Chinese Academy of Sciences, China |
| Shenglan Liu | Dalian University of Technology, China |
| Xin Liu | Huaqiao University, China |
| Xiaoyang Liu | Huazhong University of Science and Technology, China |
| Weiqiang Liu | Nanjing University of Aeronautics and Astronautics, China |
| Qingshan Liu | Southeast University, China |
| Wenqiang Liu | Southwest Jiaotong University, China |
| Hongtao Liu | Tianjin University, China |
| Yong Liu | Zhejiang University, China |
| Linjing Liu | City University of Hong Kong, Hong Kong |
| Zongying Liu | King Mongkut's Institute of Technology Ladkrabang, Thailand |
| Xiaorui Liu | Michigan State University, USA |
| Huawen Liu | The University of Texas at San Antonio, USA |
| Zhaoyang Liu | Chinese Academy of Sciences, China |
| Sirasit Lochanachit | King Mongkut's Institute of Technology Ladkrabang, Thailand |
| Xuequan Lu | Deakin University, Australia |
| Wenlian Lu | Fudan University, China |
| Ju Lu | Shandong University, China |
| Hongtao Lu | Shanghai Jiao Tong University, China |
| Huayifu Lv | Beijing Normal University, China |
| Qianli Ma | South China University of Technology, China |
| Mohammed Mahmoud | Beijing Institute of Technology, China |
| Rammohan Mallipeddi | Kyungpook National University, South Korea |
| Jiachen Mao | Duke University, USA |
| Ali Marjaninejad | University of Southern California, USA |
| Sanparith Marukatat | National Electronics and Computer Technology Center, Thailand |
| Tomas Henrique Maul | University of Nottingham Malaysia, Malaysia |
| Phayung Meesad | King Mongkut's University of Technology North Bangkok, Thailand |
| Fozia Mehboob | Research Institute of Sweden, Sweden |
| Wenjuan Mei | University of Electronic Science and Technology of China, China |
| Daisuke Miyamoto | The University of Tokyo, Japan |

| | |
|---|---|
| Kazuteru Miyazaki | National Institution for Academic Degrees and Quality Enhancement of Higher Education, Japan |
| Bonaventure Molokwu | University of Windsor, Canada |
| Hiromu Monai | Ochanomizu University, Japan |
| J. Manuel Moreno | Universitat Politècnica de Catalunya, Spain |
| Francisco J. Moreno-Barea | University of Malaga, Spain |
| Chen Mou | Nanjing University of Aeronautics and Astronautics, China |
| Ahmed Muqeem Sheri | National University of Sciences and Technology, Pakistan |
| Usman Naseem | University of Technology Sydney, Australia |
| Mehdi Neshat | The University of Adelaide, Australia |
| Quoc Viet Hung Nguyen | Griffith University, Australia |
| Thanh Toan Nguyen | Griffith University, Australia |
| Dang Nguyen | University of Canberra, Australia |
| Thanh Tam Nguyen | Ecole Polytechnique Federale de Lausanne, France |
| Giang Nguyen | Korea Advanced Institute of Science and Technology, South Korea |
| Haruhiko Nishimura | University of Hyogo, Japan |
| Stavros Ntalampiras | University of Milan, Italy |
| Anupiya Nugaliyadde | Murdoch University, Australia |
| Toshiaki Omori | Kobe University, Japan |
| Yuangang Pan | University of Technology Sydney, Australia |
| Weike Pan | Shenzhen University, China |
| Teerapong Panboonyuen | Chulalongkorn University, Thailand |
| Paul S. Pang | Federal University Australia, Australia |
| Lie Meng Pang | Southern University of Science and Technology, China |
| Hyeyoung Park | Kyungpook National University, South Korea |
| Kitsuchart Pasupa | King Mongkut's Institute of Technology Ladkrabang, Thailand |
| Yong Peng | Hangzhou Dianzi University, China |
| Olutomilayo Petinrin | City University of Hong Kong, Hong Kong |
| Geong Sen Poh | National University of Singapore, Singapore |
| Mahardhika Pratama | Nanyang Technological University, Singapore |
| Emanuele Principi | Università Politecnica delle Marche, Italy |
| Yiyan Qi | Xi'an Jiaotong University, China |
| Saifur Rahaman | International Islamic University Chittagong, Bangladesh |
| Muhammad Ramzan | Saudi Electronic University, Saudi Arabia |
| Yazhou Ren | University of Electronic Science and Technology of China, China |
| Pengjie Ren | University of Amsterdam, The Netherlands |
| Colin Samplawski | University of Massachusetts Amherst, USA |
| Yu Sang | Liaoning Technical University, China |
| Gerald Schaefer | Loughborough University, UK |

| | |
|---|---|
| Nhi N.Y. Vo | University of Technology Sydney, Australia |
| Hiroaki Wagatsuma | Kyushu Institute of Technology, Japan |
| Nobuhiko Wagatsuma | Tokyo Denki University, Japan |
| Yuanyu Wan | Nanjing University, China |
| Feng Wan | University of Macau, Macau |
| Dianhui Wang | La Trobe University, Australia |
| Lei Wang | Beihang University, China |
| Meng Wang | Beijing Institute of Technology, China |
| Sheng Wang | Henan University, China |
| Meng Wang | Southeast University, China |
| Chang-Dong Wang | Sun Yat-sen University, China |
| Qiufeng Wang | Xi'an Jiaotong-Liverpool University, China |
| Zhenhua Wang | Zhejiang University of Technology, China |
| Yue Wang | Chinese University of Hong Kong, Hong Kong |
| Jiasen Wang | City University of Hong Kong, Hong Kong |
| Jin Wang | Hanyang University, South Korea |
| Wentao Wang | Michigan State University, USA |
| Yiqi Wang | Michigan State University, USA |
| Peerasak Wangsom | CAT Telecom PCL, Thailand |
| Bunthit Watanapa | King Mongkut's University of Technology Thonburi, Thailand |
| Qinglai Wei | Chinese Academy of Sciences, China |
| Yimin Wen | Guilin University of Electronic Technology, China |
| Guanghui Wen | Southeast University, China |
| Ka-Chun Wong | City University of Hong Kong, Hong Kong |
| Kuntpong Woraratpanya | King Mongkut's Institute of Technology Ladkrabang, Thailand |
| Dongrui Wu | Huazhong University of Science and Technology, China |
| Qiujie Wu | Huazhong University of Science and Technology, China |
| Zhengguang Wu | Zhejiang University, China |
| Weibin Wu | Chinese University of Hong Kong, Hong Kong |
| Long Phil Xia | Peking University, Shenzhen Graduate School, China |
| Tao Xiang | Chongqing University, China |
| Jiaming Xu | Chinese Academy of Sciences, China |
| Bin Xu | Northwestern Polytechnical University, China |
| Qing Xu | Tianjin University, China |
| Xingchen Xu | Fermilab, USA |
| Hui Xue | Southeast University, China |
| Nobuhiko Yamaguchi | Saga University, Japan |
| Toshiyuki Yamane | IBM Research, Japan |
| Xiaoran Yan | Indiana University, USA |
| Shankai Yan | National Institutes of Health, USA |
| Jinfu Yang | Beijing University of Technology, China |
| Xu Yang | Chinese Academy of Sciences, China |

| | |
|---|---|
| Jinghui Zhong | South China University of Technology, China |
| Junping Zhong | Southwest Jiaotong University, China |
| Xiaojun Zhou | Central South University, China |
| Hao Zhou | Harbin Engineering University, China |
| Yingjiang Zhou | Nanjing University of Posts and Telecommunications, China |
| Deyu Zhou | Southeast University, China |
| Zili Zhou | The University of Manchester, UK |

# Contents – Part I

# Human-Computer Interaction

# A Genetic Feature Selection Based Two-Stream Neural Network for Anger Veracity Recognition

Chaoxing Huang(✉), Xuanying Zhu, and Tom Gedeon

Research School of Computer Science, Australian National University, ACT, Canberra 2601, Australia
{chaoxing.huang,xuanying.zhu,tom.gedeon}@anu.edu.au

**Abstract.** People can manipulate emotion expressions when interacting with others. For example, acted anger can be expressed when the stimulus is not genuinely angry with an aim to manipulate the observer. In this paper, we aim to examine if the veracity of anger can be recognized from observers' pupillary data with computational approaches. We use Genetic-based Feature Selection (GFS) methods to select time-series pupillary features of observers who see acted and genuine anger as video stimuli. We then use the selected features to train a simple fully connected neural network and a two-stream neural network. Our results show that the two-stream architecture is able to achieve a promising recognition result with an accuracy of 93.6% when the pupillary responses from both eyes are available. It also shows that genetic algorithm based feature selection method can effectively improve the classification accuracy by 3.1%. We hope our work could help current research such as human machine interaction and psychology studies that require emotion recognition.

**Keywords:** Anger veracity · Two-stream architecture · Neural network · Genetic algorithm

## 1 Introduction

The veracity of emotions plays an essential role in human interaction. It influences people's view towards others after the observer observes a certain emotion [1]. In reality, human beings are sometimes very poor at telling whether a person's emotion is genuine or posed, especially in the scenario that humans are usually asked to use verbal information to make the prediction [17]. This kind of mistake may negatively affect some current research like psychological studies which includes emotion observation [18]. Thus, it is worth looking into the problem of using computational algorithms to take the physiological responses of humans to aid the recognition. Also, in human-machine interaction, it is important to let the machine know whether a human's emotion is disguised or genuine if the interaction involves emotion [3]. There has been work using physiological

This work was done when Chaoxing was at ANU.


H. Yang et al. (Eds.): ICONIP 2020, LNCS 12532, pp. 3–11, 2020.
https://doi.org/10.1007/978-3-030-63830-6_1

signals of observers who are exposed to emotional stimuli to interpret the emotion of the stimuli. In [2] and [3], a classifier is trained to identify if a person's smile or anger is genuine or posed. Meanwhile, a human thermal data based algorithm is proposed to analyse human's stress in [4].

Neural Networks (NNs) are able to learn their parameters automatically via back-propagation and can be used to map physiological data to emotion veracity. However, since human beings interact with the environment, it is likely that physiological data being collected by sensors are noisy. Noisy features can dampen the learning process of the NN on the data-set, since the model needs to learn the underlying pattern of the noise. On the other hand, the model can overfit the dataset when the training time has to be escalated. Therefore, it is crucial to look into the problem of selecting useful features from the physiological data-set. Anomaly detection has a profound studied history. One of the most classic methods is the generative model learning approach [10]. However, this method requires a cumbersome learning process and relies heavily on distribution assumptions. Other works have also been done to detect noisy and fraud features [6,7], and genetic algorithm is used in [5] to select features without much human intervention. Since the noise in physiological data is usually not obvious to non-expert humans and has long temporal sequences, the evolutionary based genetic algorithm becomes a reasonable way for avoiding intractable manual selection. In the original work of [3], it is shown that using pupillary data, the model can recognise anger veracity with an accuracy of 95%, which is a significant improvement over verbal data. However, the collected data may contain environment-affected noise and the sensors occasionally fail to collect physiological data at some time stages. When we take the time-series information into consideration, not all the recorded data from the sensor plays essential roles in classification due to this noisiness and redundancy, which thus requires feature selection, and the genetic algorithm provides a way to achieve this. Therefore, we study the effect of genetic-based feature selection (GFS) [13] on anger veracity recognition in this paper.

The contribution of this paper is two-fold:

- We adopt a two-stream neural-network to effectively use the physiological data (pupil diameters) from the two eyes of humans to predict the anger veracity of the emotion stimuli.
- We adopt the genetic-based feature selection method to select useful features from the noisy temporal data due to environment noise and occasional sensor failures (e.g. eye-blinks) and thus to enhance the recognition performance.

In this paper, we first tune a baseline NN with one hidden layer by taking the pupillary data from one eye as input. Then we apply GFS to the time-series data and verify our proposed two-stream model can handle the binocular pupillary information. The rest of this paper is organised as follows: Sect. 2 introduces the NN architecture, and the GFS pipeline. Section 3 is about the experiments and results. Discussions are also provided in this section. Section 4 includes future work and concludes this paper.

# 2 Method

## 2.1 Dataset

We use Chen et al.'s anger dataset [3]. The dataset was collected by displaying 20 video segments to 22 different persons (observers). The observers watched the presenters' anger expression in the video and the pupillary response of the observers were collected by an eye-tracking sensor. A sample in the dataset means the pupillary data of a person which was collected when the person watched a video, and each of them is labelled with "Genuine" or "Posed", meaning a genuine anger expression or a posed anger expression is observed. The videos have various lengths and the recorded data sequence length of each samples varies from 60 time-steps to 186 time-steps. The sensor recording rate for each of the samples 60 Hz. The dataset contains the pupillary response from each observer's two eyes at each time-step as well as the mean statistics.

## 2.2 Network Architecture

**Baseline Architecture.** In the baseline model, we adopt a simple fully-connected neural network architecture with one hidden layer with $n$ hidden neurons. There are three potential choices of activation function in our NN, which are Sigmoid, Tanh and ReLU. We will investigate the effect of different choices of $n$ and activation function type in the experiment part. Since this is a binary classification, we choose cross-entropy loss as the loss function.

## 2.3 Two-Stream Architecture

Inspired by the two-stream architecture in video recognition [11,12], we adopt a two-stream fully connected architecture in our classification task, which is shown in Fig. 1. For every stream, the sub-stream network is the baseline model and the feature vector from the two streams are fused together to a one-layer fully connected layer for final prediction. There are two potential kinds of input to the network. The first scenario is, the first stream takes the pupillary temporal data from the left eye and the second stream takes the pupillary data from the right eye. The second scenario is, the first stream takes the pupillary temporal data from the left (right) eye and the second stream takes the pupillary diameter differences at each time step from the left (right) eye. The pupillary difference for each time step is the data at the current time step minus the data at the previous step.

## 2.4 Data Pre-processing and Feature Selection

**Data Pre-processing.** The data in the dataset is temporal, with values from both eyes at each time-step. We regard every sequence of time-series data of each sample as an input vector to the neural network. To deal with the length varying issue, we use zero padding to pad every feature vector to the same length $(186 \times 1)$.

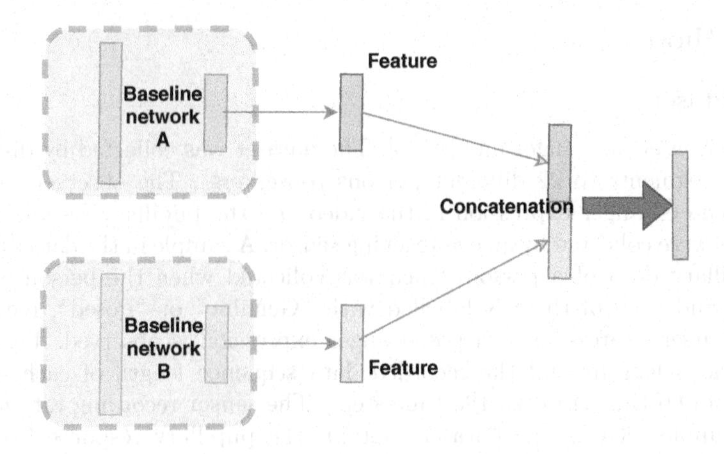


**Fig. 1.** Two-stream network architecture

**Genetic-Based Feature Selection.** The feature selection mask is indicated by a binary vector with the length of the feature-vector (0 for omitting a feature and 1 for keeping a feature). In a genetic algorithm, the selection mask is regarded as the chromosome. We first initialize the population size as $n+1$, and we adopt a neural network to compute the validation classification accuracy as the fitness value. Note that for different chromosomes, the input size of the neural network is different, and thus we are not only doing a feature selection but also conducting a network architecture selection. We adopt a tournament-based reproduction [14], in which we create $\frac{n}{2}$ sets of tournament-group, and we randomly choose a fix size of members from the current generation to form the tournament-groups as the population pool for generating off-spring. Note that we actually repeat $\frac{n}{2}$ times of tournament group creation, which means one chromosome can appear in different tournament groups. In each tournament-group, two parents are selected by using the selection probability which is obtained by its normalized fitness value in the population (proportional selection). The crossover generates two off-springs by a one-point crossing. Therefore, the tournament-reproduction can generates $n$ off-springs, while the selected one is the chromosome with the highest fitness value in the current generation. Every generated off-spring goes through a mutation process to increase the gene diversity. To sum up, the population of each generation retains at $n+1$ while the parents' selection in every generation's reproduction need to go through a fierce tournament competition. The pipeline is shown in Fig. 2.

## 3 Experiments and Discussions

### 3.1 Experiment Settings

We first shuffle the data-set and randomly split out 80% of the data as training patterns. The rest of the data are for testing. We use Python3.6 and Pytorch

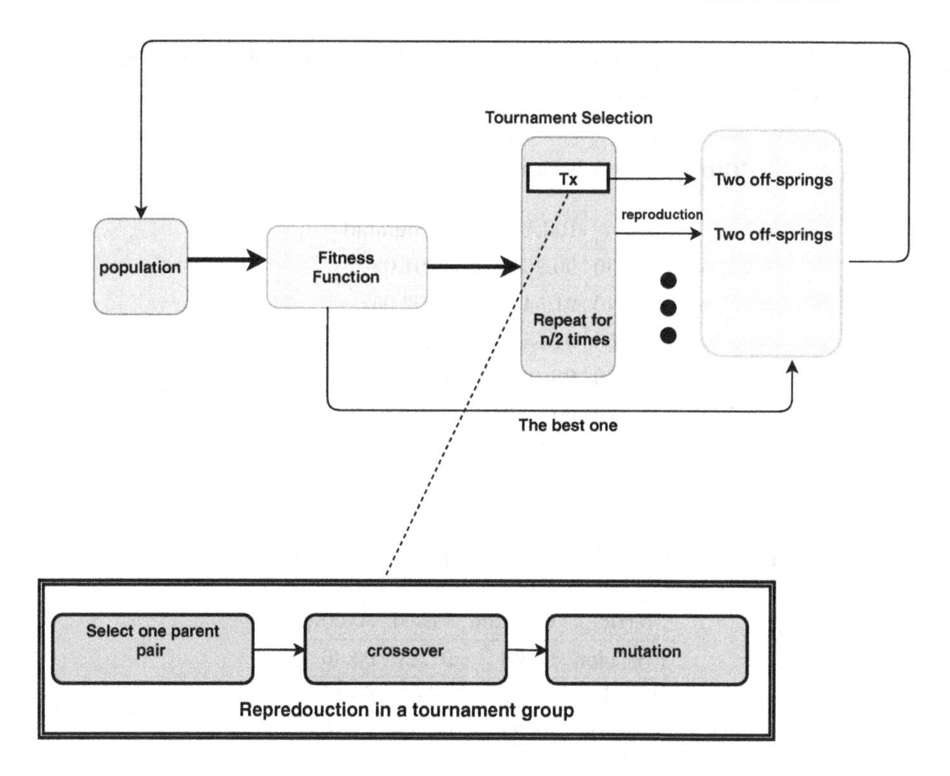


**Fig. 2.** Genetic-algorithm pipeline

1.0 [9] to implement the experiment and the environment is Windows 10. Since the training set is small, we only use an i7-8750H CPU for computation and we adopt batch gradient descent with an Adam [8] optimizer. For the optimizer, the hyper-parameters are: $\beta_1 = 0.9$ and $\beta_2 = 0.999$. The learning rate is set to be 1e–4 and a weight-decay of 1e–5 is used to prevent over-fitting. The iteration number of training is set to be 1000 epochs. The random seed numbers are set as 1000, 2000, 3000, 4000 and 5000. We will denote them as sets 1 to 5 respectively in this context. All the results are reported based on five-fold cross-validation by using set 1 to set 5.

### 3.2 Baseline Model

We train the model under different settings of hidden neuron number $n$ and different activation function types. We take the left eye pupillary data as input. The results, which are the average accuracy of the 5 runs, are shown in Table 1. As is clearly shown, the performance of using ReLU activation function with 60 neurons shows significant advantage over those of Tanh and Sigmoid since it avoids the problem of gradient diminishing and overfitting. In the rest of the experiments, we will use the model using ReLU with 60 hidden units as the baseline. We also present the precision, recall and F1 score of the baseline model after five runs in Table 2.

**Table 1.** Test accuracy of the baseline model (%)

| $n$ | ReLU | Tanh | Sigmoid |
|---|---|---|---|
| 30 | 90.26 | 91.03 | 91.03 |
| 40 | 91.54 | 91.79 | 92.05 |
| 50 | 91.54 | 90.30 | 92.05 |
| 60 | **93.08** | 91.54 | 91.80 |
| 70 | 90.52 | 91.27 | 91.80 |
| 80 | 91.54 | 91.28 | 92.05 |

**Table 2.** Precision, recall and F1 score of baseline model

| Metric | Genuine | Posed | Average |
|---|---|---|---|
| Precision | 95.68 | 91.23 | 93.46 |
| Recall | 88.57 | 97.65 | 93.11 |
| F1 score | 91.99 | 94.33 | 93.16 |

### 3.3 Experiments on GFS and Two-Stream Architecture

**Classification Without Feature-Selection.** We first conduct experiments on training the fully-connected classifier without applying GFS on the temporal features. We compare the 5-set averaged results of different input settings of both single-stream and two-stream models, and the results are shown in Table 3. The hidden layer number is still 60. We also present the precision, recall as well as the F1 score of the best model here. The results are shown in Table 4. We can also notice that using information from two eyes achieves a better performance than only using information from either single eye, and this is aligned with our human intuition. It should also be noted that taking the diameter differences into consideration does not provide an improvement, this may be due to the fact that the diameter change in every time-step is very small and does not provide much significant temporal information.

**Effectiveness of GFS.** We study the effect of applying Genetic feature selection to the model. During the feature selection stage, 80% of the training data is used for training while 20% of the training data is used for validation to compute the fitness value. After selecting the best chromosome, we train our model on the entire training set. For every tournament group, the member number is set to be 9. The generation number is set to be 10. The population number is set to

**Table 3.** Test accuracy of different model (%)

| Input | Test accuracy |
|---|---|
| Double-eyes (two-stream) | **93.58** |
| Left-eye (single-stream) | 93.08 |
| Right-eye (single-stream) | 92.31 |
| Left-eye+Left-differences (two-stream) | 91.03 |
| Right-eye+Right-differences (two-stream) | 91.03 |

**Table 4.** Precision, recall and F1 score of double-eye two-stream model

| Metric | Genuine | Posed | Average |
|---|---|---|---|
| Precision | 96.88 | 91.30 | 94.09 |
| Recall | 88.57 | 97.67 | 93.12 |
| F1 score | 92.54 | 94.38 | 93.39 |

be 21 and the mutation rate is 0.001. In the two-stream case, both the streams use the same feature selection mask. We conduct the GFS experiment on the double-eye two stream model as well as the left-eye single stream model. The results are shown in Table 5.

**Table 5.** Performance of applying GFS

| Two-stream | Genuine | Posed | Average | Accuracy:96.15% |
|---|---|---|---|---|
| Precision | 97.06 | 95.45 | 96.26 | |
| Recall | 94.29 | 97.67 | 95.98 | |
| F1 score | 95.65 | 96.54 | 96.10 | |
| One-stream | Genuine | Posed | Average | Accuracy:94.82% |
| Precision | 94.29 | 95.35 | 94.82 | |
| Recall | 94.29 | 95.35 | 94.82 | |
| F1 score | 94.29 | 95.35 | 94.82 | |

As we can see from Table 3 and 5, by using the GFS method, the single-stream model with left eye input can achieve a better performance by applying feature selection, the single-stream model using feature selection can even surpass the performance of the double-eye two stream model that does not have feature selection. Moreover, the double-eye two stream model with feature-selection achieves the best performance among all. The result reveals an underlying drawback, that using zero padding to fill up the length of those "short" vector create feature redundancy, and feature-selection can improve the recognition performance.

### 3.4 Discussion

**Why Does the GFS Improve the Recognition Performance?** Perhaps the most plausible explanation is that the GFS helps to eliminate some of the noisy pupillary data during collection. In the original paper using the dataset, it is mentioned that the eye-blinks of the humans exist, and which is prone to affect some of the collected pupillary diameter values. In fact, we have found that many of the feature selection masks have many zero entries before the zero padding stage, which means those actual collected data are removed.

## 4 Conclusion and Future Work

This paper examines applying two-stream neural networks with genetic-based feature selection on anger veracity recognition, to tackle the problem of noisy data collection and sensor failure during physiological data collection. From the experimental results, it can be concluded that the two-stream architecture can effectively handle the data from both eyes of humans and it is crucial to take these binocular physiological reactions into consideration when doing anger veracity recognition. It can also be concluded that applying genetic-based feature selection can effectively improve the model performance and remove redundant or noisy features.

In our work, we use the same feature mask for selecting features from both eyes. One might question that the data behaviour from two eyes may be different and applying an identical mask overlooks this possibility to a certain extent. Indeed, the asymmetric reaction of both eyes can affect the data collection of the sensor differently, and it is worth looking into the problems of finding two feature selection masks that can reflect the linkage and difference between the two eyes.

As for the model that uses time-series data, our fully-connected model requires zero padding to deal with varying length data, which creates redundancy and reduces flexibility. Therefore, it is worth looking into the method of applying RNN/LSTM [15] or Transformer models [16] in the future.

**Acknowledgments.** We would like to thank those contributors who collected and provided the dataset. We would also like to thank the Human Centred Computing team at ANU for providing useful advice for this paper.

## References

1. Knutson, B.: Facial expressions of emotion influence interpersonal trait inferences. J. Nonverbal Behav. **20**(3), 165–182 (1996)
2. Hossain, M.Z., Gedeon, T.: Classifying posed and real smiles from observers' peripheral physiology. In: Proceedings of the 11th EAI International Conference on Pervasive Computing Technologies for Healthcare. pp. 460–463 (2017)
3. Chen, L., Gedeon, T., Hossain, M.Z., Caldwell, S.: Are you really angry? detecting emotion veracity as a proposed tool for interaction. In: Proceedings of the 29th Australian Conference on Computer-Human Interaction. pp. 412–416 (2017)

4. Irani, R., Nasrollahi, K., Dhall, A., Moeslund, T.B., Gedeon, T.: Thermal super-pixels for bimodal stress recognition. In: 2016 Sixth International Conference on Image Processing Theory, Tools and Applications (IPTA). pp. 1–6. IEEE (2016)
5. Tan, F., Fu, X., Zhang, Y., Bourgeois, A.G.: A genetic algorithm-based method for feature subset selection. Soft. Comput. **12**(2), 111–120 (2008)
6. Hawkins, S., He, H., Williams, G., Baxter, R.: Outlier detection using replicator neural networks. In: Kambayashi, Y., Winiwarter, W., Arikawa, M. (eds.) DaWaK 2002. LNCS, vol. 2454, pp. 170–180. Springer, Heidelberg (2002). https://doi.org/10.1007/3-540-46145-0_17
7. Gedeon, T.D., Wong, P.M., Harris, D.: Balancing bias and variance: network topology and pattern set reduction techniques. In: Mira, J., Sandoval, F. (eds.) IWANN 1995. LNCS, vol. 930, pp. 551–558. Springer, Heidelberg (1995). https://doi.org/10.1007/3-540-59497-3_222
8. Kingma, D.P., Ba, J.: Adam: A method for stochastic optimization. arXiv preprint arXiv:1412.6980 (2014)
9. Paszke, A., et al.: Pytorch: an imperative style, high-performance deep learning library. In: Advances in Neural Information Processing Systems. pp. 8024–8035 (2019)
10. Svensen, M., Bishop, C.M.: Pattern recognition and machine learning (2007)
11. Feichtenhofer, C., Pinz, A., Zisserman, A.: Convolutional two-stream network fusion for video actionrecognition. In: Proceedings of the IEEE Conference on Computer Vision and Pattern Recognition. pp. 1933–1941 (2016)
12. Simonyan, K., Zisserman, A.: Two-stream convolutional networks for action recognition in videos. In: Advances in Neural Information Processing Systems. pp. 568–576 (2014)
13. Vafaie, H., De Jong, K.A.: Genetic algorithms as a tool for feature selection in machine learning. In: ICTAI. pp. 200–203 (1992)
14. Yang, J., Soh, C.K.: Structural optimization by genetic algorithms with tournament selection. J. Comput. Civil Eng. **11**(3), 195–200 (1997)
15. Gers, F.A., Schmidhuber, J., Cummins, F.: Learning to forget: continual prediction with lstm (1999)
16. Vaswani, A., et al.: Attention is all you need. In: Advances in Neural Information Processing Systems. pp. 5998–6008 (2017)
17. Qin, Z., Gedeon, T., Caldwell, S.: Neural networks assist crowd predictions in discerning the veracity of emotional expressions. In: Cheng, L., Leung, A.C.S., Ozawa, S. (eds.) ICONIP 2018. LNCS, vol. 11306, pp. 205–216. Springer, Cham (2018). https://doi.org/10.1007/978-3-030-04224-0_18
18. Strongman, K.T.: The psychology of emotion: Theories of emotion in perspective. John Wiley and Sons, New York (1996)

# An Efficient Joint Training Framework for Robust Small-Footprint Keyword Spotting

Yue Gu[1], Zhihao Du[2], Hui Zhang[1], and Xueliang Zhang[1](✉)

[1] Inner Mongolia Key Laboratory of Mongolian Information Processing Technology, Inner Mongolia University, Hohhot, China
427gy@sina.com, alzhu.san@163.com, cszxl@imu.edu.cn
[2] School of Computer Science and Technology, Harbin Institute of Technology, Harbin, China
duzhihao@hit.edu.cn

**Abstract.** In real-world applications, robustness against noise is crucial for small-footprint keyword spotting (KWS) systems which are deployed on resource-limited devices. To improve the noise robustness, a reasonable approach is employing a speech enhancement model to enhance the noisy speeches first. However, current enhancement models need a lot of parameters and computation, which do not satisfy the small-footprint requirement. In this paper, we design a lightweight enhancement model, which consists of the convolutional layers for feature extracting, recurrent layers for temporal modeling and deconvolutional layers for feature recovering. To reduce the mismatch between the enhanced features and KWS system desired ones, we further propose an efficient joint training framework, in which the enhancement model and KWS system are concatenated and jointly fine-tuned through a trainable feature transformation block. With the joint training, linguistic information can back-propagate from the KWS system to the enhancement model and guide its training. Our experimental results show that the proposed small-footprint enhancement model significantly improves the noise robustness of KWS systems without much increasing model or computation complexity. Moreover, the recognition performance can be further improved through the proposed joint training framework.

**Keywords:** Small footprint · Robust KWS · Speech enhancement

## 1 Introduction

Keyword spotting (KWS) aims at detecting predefined keywords from a continuous audio stream, which is an important technique for human-computer interaction. A typical application of KWS is detecting the wake-up word on the mobile devices. Since KWS systems usually run in "always-on" mode, their model and computation complexity should be low enough, especially in the small-footprint embedded devices which have limited memory and computation resources.

This research was supported in part by the China National Nature Science Foundation (No. 61876214, No. 61866030).


H. Yang et al. (Eds.): ICONIP 2020, LNCS 12532, pp. 12–23, 2020.
https://doi.org/10.1007/978-3-030-63830-6_2

In a quiet environment, small-footprint KWS systems have achieved a high accuracy, such as the residual learning based KWS model [12] on Google Speech Commands dataset [16]. Unfortunately, in noisy environments, their performance degrades significantly. To improve the robustness against noise, there are two optional approaches. One is multi-conditional training [6,7,9], in which the KWS models are trained with speeches in different acoustic scenes, simultaneously. However, the performance of multi-condition training is still worse than desired, since the KWS system needs to model various acoustic scenes with the limited parameters. Another reasonable approach is involving a speech enhancement model as the front-end processing of KWS systems. Thanks to the deep learning techniques, speech enhancement has achieved a huge progress [13], and improved the noise robustness of automatic speech recognition (ASR) systems [1,8]. In the context of KWS, a text-dependent enhancement model is developed in [17], which does not fit the small-footprint device due to the large model size.

In this paper, we propose a small-footprint enhancement model with limited parameters and computation complexity. Specifically, the proposed model consists of a convolutional encoder for feature extraction, recurrent layers for temporal modeling, and a deconvolutional decoder for feature reconstruction. We also adopt the deep residual learning by adding skip connections between the corresponding encoder and decoder layers. Since speech enhancement and KWS can benefit from each other, we further propose an efficient joint training framework, in which the enhancement model and KWS system are concatenated and jointly fine-tuned by using a trainable feature transformation block. Through the joint training framework, semantic information can back-propagate from the KWS system to the enhancement model and guide its training process. In addition, we also change the enhancement domain from the power spectrum to Mel spectrogram resulting in less computation without any performance degradation.

## 2 System Description

The overall system is illustrated in Fig. 1. There are three components in the proposed system, i.e., speech enhancement model, feature transformation block and KWS model. The speech enhancement model is trained to predict the ideal ratio masks (IRMs) [5], and the enhanced time-frequency (T-F) features are obtained by point-wisely multiplying the noisy T-F features with the predicted masks. Then, the enhanced T-F features are transformed to the Mel-frequency cepstral coefficients (MFCCs) by the feature transformation block. Given the enhanced MFCCs as input, the KWS system is trained to predict the posterior probability of keywords. More details can be found in our released code[1].

### 2.1 Masking-Based Speech Enhancement Method

In the proposed system, we employ the masking-based enhancement method, which has successfully improved the speech perceptive quality [14] and the noise

[1] Available at https://github.com/ZhihaoDU/du2020kws.

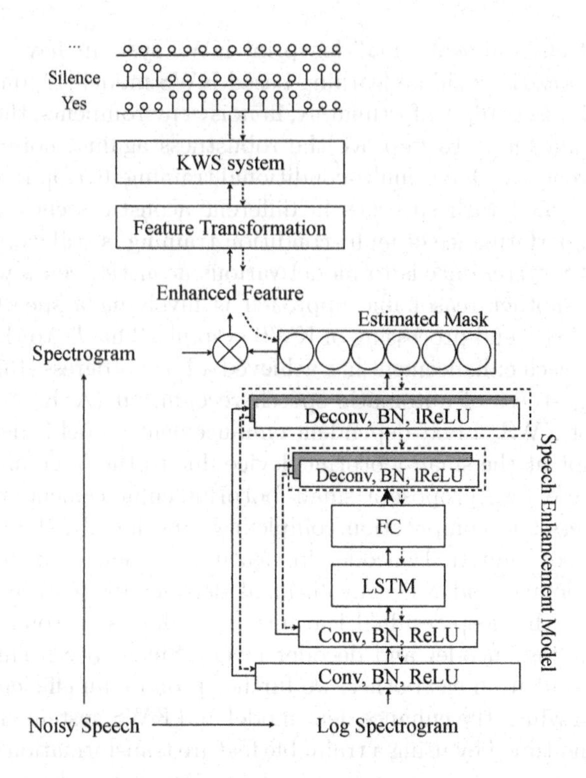


**Fig. 1.** Schematic diagram of the proposed system. Solid and dotted arrows indicate the directions of forward and backward passes, respectively.

robustness of ASR [15] . The loss function of masking-based method is defined as follows:

$$\mathcal{L}_{\mathbf{MSE}} = \frac{1}{T}\frac{1}{F}\sum_{t=1}^{T}\sum_{f=1}^{F}\|M(t,f) - \hat{M}(t,f)\|_2^2 \tag{1}$$

where $M(t, f)$ and $\hat{M}(t, f)$ are the ideal and predicted T-F masks at time $t$ and frequency $f$, respectively. $T$ and $F$ represent the total numbers of frames and frequency bins, respectively. The ideal mask $M$ is defined as follows:

$$M(t,f) = \sqrt{\frac{|S(t,f)|^2}{|S(t,f)|^2 + |N(t,f)|^2}} \tag{2}$$

where $S$ and $N$ represent the T-F features of the clean speeches and background noises, respectively.

In the test stage, IRMs are predicted by the enhancement model according to the noisy T-F features. Then, the enhanced features are obtained as follows:

$$\hat{S} = Y \otimes \hat{M} \tag{3}$$

where $\hat{M}$ is the mask predicted by the enhancement model. $Y$ represents the T-F features of noisy speeches. $\otimes$ means the element-wise multiplication.

To match the small-footprint requirement, we design a novel convolution recurrent network (CRN) [10] with limited parameters and computation. The architecture of CRN is shown in the lower part of the Fig. 1. There are three components in the CRN, i.e., the convolutional encoder, deconvolutional decoder and the recurrent layers with long short term memory (LSTM) cells followed by a linear projection layer. Skip connections are added to the corresponding layers between the encoder and decoder. Batch normalization [2] and rectified linear units (ReLUs) [4] are employed in the convolutional layers, and the leaky ReLUs are used in the deconvolutional layers. The sigmoid non-linearity function is employed for the output layer. We propose two modifications to make the CRN more suitable for the small-footprint KWS purpose. First, to reduce the computation complexity, the convolutional layers in our CRN have the strides on both time and frequency axes, while the original CRN only has strides on the frequency axis. Second, we employ the leaky ReLUs in the decoding layers instead of ReLUs, which guarantee the nonzero gradients everywhere and benefit the training process.

In general, IRMs can be defined in different T-F feature domains. Although the power spectrum is a common choice in the speech enhancement community, it is not the case for KWS purpose. In small-footprint KWS systems, the Mel-Frequency cepstral coefficients (MFCCs) are employed as the input features. MFCCs are extracted by integrating the frequency bins in the power spectrum through the Mel filter-bank. It means that many details contained in the power spectrum are filtered out. Therefore, it is inefficient and unnecessary to perform enhancement on the power spectrum domain. On the contrary, Mel spectrogram is a better choice, which can be transformed to MFCC by multiplying with the discrete cosine coefficients. By changing the enhancement domain from power spectrum to Mel spectrogram, the model and computation complexity can be further reduced, which is more friendly to resource-limited devices.

### 2.2 Feature Transformation Block

The input of KWS system is MFCC, while the output of enhancement model is Mel spectrogram. To transform the enhanced Mel spectrogram to MFCC, we design the Mel feature transformation block (Mel-FTB) as shown in Fig. 2 (a). In Mel-FTB, the Mel spectrogram is first compressed by element-wise logarithm, and then multiplied with the discrete cosine transformation (DCT) coefficient matrix. In addition, we also design a power feature transformation block (Pow-FTB) for enhancement models that are trained on the power spectrum domain. In Pow-FTB, the power spectrum is first multiplied with the Mel filter-bank, then it is compressed by element-wise logarithm, finally it is multiplied with the DCT coefficient matrix. The transformation process of Pow-FTB is given in Fig. 2 (b). By using the feature transformation blocks, enhancement models and KWS systems can be concatenated and jointly fine-tuned with the back-propagation algorithm. Note that all operations in the feature transformation

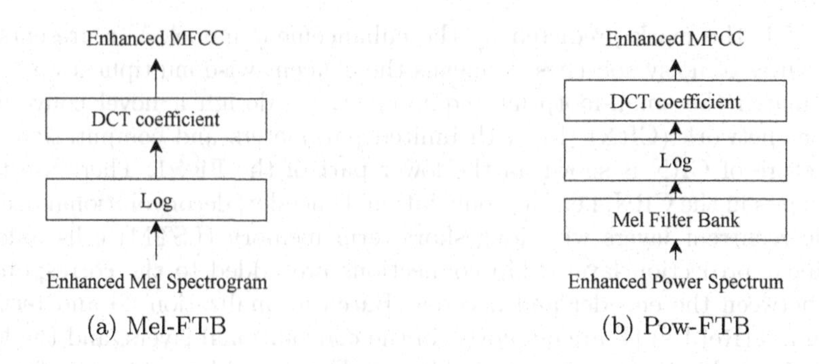


(a) Mel-FTB (b) Pow-FTB

**Fig. 2.** The (a) Mel and (b) power feature transformation blocks.

blocks can be implemented directly through a modern machine learning framework, such as PyTorch and TensorFlow.

### 2.3 Keyword Spotting System

We employ the fully convolutional model `cnn-trad-pool2` [11] as our KWS system, which is a modified version of `cnn-trad-fpool3` [7]. In `cnn-trad-pool2`, the size and stride of the first max-pooling layer are set to 2, and the hidden linear layers are dropped. These modifications lead to higher recognition accuracy. The details of `cnn-trad-pool2` are shown in Table 1.

## 3 Experiments and Results

### 3.1 Experimental Settings

We evaluate the proposed system on Google's Speech Commands dataset [16], which contains 105,829 one-second long utterances of 35 command words recorded by thousands of speakers. We train the KWS system to detect 10 predefined keywords, and the other 25 non-keywords are treated as unknown tokens. We split the dataset into three parts for training, validation, and test set with the ratio of 8:1:1. To simulate different acoustic scenes, six background noise records are employed to create noisy utterances, which include stationary noises (pink and white) and daily environmental sounds (such as dishes and bikes) [16]. Noisy utterances are obtained by mixing up with 6 noises at the signal-to-noise ratios (SNRs) of $-3, 0, 3, 6$ dB. There are about 812,000 and 97,000 noisy utterances in the training and validation sets, respectively. In the test set, there are 210,000 noisy utterances with the keywords and non-keywords ratio of 1.3:1. To evaluate the generalization ability on untrained noises, we also build another noise-unmatched set by mixing up the clean utterances with 100 non-speech sounds [2].

[2] http://web.cse.ohio-state.edu/pnl/corpus/HuNonspeech/HuCorpus.html.

**Table 1.** The network architecture of **cnn-trad-pool2**. $T$ denotes the total number of frames in the MFCCs

| Layer Name | Input Size | Parameter |
|---|---|---|
| reshape_1 | $T \times 40$ | - |
| conv_1 | $1 \times T \times 40$ | $(20, 8), (1, 1), 64$ |
| pool_1 | $64 \times (T - 19) \times 33$ | $(2, 2), (2, 2)$ |
| conv_2 | $64 \times T/2 - 9 \times 16$ | $(10, 4), (1, 1), 64$ |
| reshape_2 | $64 \times T/2 - 18 \times 13$ | - |
| Linear | $64 * (T/2 - 18) * 13$ | 12 |
| Output | 12 | |

The noise-unmatched test set contains nearly 3600,000 noisy utterances. All utterances are sampled to 16,000 Hz. We use Hanning window to extract T-F features with the window length of 30 ms and the hop length of 10 ms. The Mel filter-bank is calculated with the low frequency 20 Hz and the high frequency of 4,000 Hz. 40-dimension DCT coefficients are used to extract MFCC.

The utterance-level classification accuracy is the main metric, which is simply measured as the fraction of classification decisions that correct. We also plot receiver operating characteristic (ROC) curves, in which the $x$ and $y$ axes represent the false alarm rate (FAR) and false reject rate (FRR), respectively. Generally, methods with less area under the curve (AUC) are better. In addition, the equal error rate (EER) is also used to evaluate the recognition performance.

All models are trained with the Adam optimizer [3] and the batch size of 256 on the utterance level. We set the learning rate to 0.0001. The mean squared error (MSE) and cross entropy (CE) are employed as the objective functions of the enhancement model and KWS system, respectively. The best models are selected by classification accuracy on the validation set.

For different enhancement domains, i.e. power spectrum and Mel spectrogram, we design different small-footprint enhancement models. The full-size models trained on the power spectrum and Mel spectrogram are denoted as `PowCRN32` and `MelCRN32`, respectively. In addition, we also design a narrow version of these models, which are denoted as `PowCRN16` and `MelCRN16`, respectively. The details of model architecture are given in Table 2. We compare the proposed small-footprint enhancement model with a widely used LSTM model [17], which consists of two hidden layers with 384 bidirectional LSTM cells followed by a linear projection layer with 241 units. We refer this enhancement model as `BiLSTM`. In Table 3, we compare the number of parameters and multiply-add operations for each model.

**Table 2.** The model architectures of small-footprint enhancement models. $T$ denotes the number of time frames in the T-F features. $(f, h)$ is set to $(16, 32)$ and $(32, 64)$ for the narrow and full-size models, respectively. For convolutional and deconvolutional layers, the parameter indicates kernel size, stride and filter number. $h$ represents the number of LSTM cells.

| Layer Name | Input Size | Parameter | Output Size |
|---|---|---|---|
| PowCRN | | | |
| reshape_1 | $T \times 241$ | - | $1 \times T \times 241$ |
| conv2d_1 | $1 \times T \times 241$ | $8, 4, f$ | $f \times T/4 \times 60$ |
| conv2d_2 | $f \times T/4 \times 60$ | $8, 4, f$ | $f \times T/16 \times 15$ |
| reshape_1 | $f \times T/16 \times 15$ | - | $T/16 \times 15f$ |
| LSTM | $T/16 \times 15f$ | $h$ | $T/16 \times h$ |
| FC | $T/16 \times h$ | $15f$ | $T/16 \times 15f$ |
| reshape_2 | $T/16 \times 15f$ | - | $2f \times T/16 \times 15$ |
| deconv2d_2 | $2f \times T/16 \times 15$ | $8, 4, f$ | $2f \times T/4 \times 60$ |
| deconv2d_1 | $2f \times T/4 \times 60$ | $9, 4, f$ | $f \times T \times 241$ |
| conv2d_out | $f \times T \times 241$ | $3, 1, 1$ | $1 \times T \times 241$ |
| reshape_3 | $1 \times T \times 241$ | - | $T \times 241$ |
| MelCRN | | | |
| reshape_1 | $T \times 40$ | - | $1 \times T \times 40$ |
| conv2d_1 | $1 \times T \times 40$ | $4, 2, f$ | $f \times T/2 \times 20$ |
| conv2d_2 | $f \times T/2 \times 20$ | $4, 2, 2f$ | $2f \times T/4 \times 10$ |
| conv2d_3 | $2f \times T/4 \times 10$ | $(3, 4), (1, 2), 4f$ | $4f \times T/4 \times 5$ |
| reshape_1 | $4f \times T/4 \times 5$ | - | $T/4 \times 20f$ |
| LSTM | $T/4 \times 20f$ | $h$ | $T/4 \times h$ |
| FC | $T/4 \times h$ | $20f$ | $T/4 \times 20f$ |
| reshape_2 | $T/4 \times 20f$ | - | $8f \times T/4 \times 5$ |
| deconv2d_3 | $8f \times T/4 \times 5$ | $(3, 4), (1, 2), 2f$ | $4f \times T/4 \times 10$ |
| deconv2d_2 | $4f \times T/4 \times 10$ | $4, 2, f$ | $2f \times T/2 \times 20$ |
| deconv2d_1 | $2f \times T/2 \times 20$ | $4, 2, f$ | $f \times T \times 40$ |
| conv2d_out | $f \times T \times 40$ | $3, 1, 1$ | $1 \times T \times 40$ |
| reshape_3 | $1 \times T \times 40$ | - | $T \times 40$ |

### 3.2 Results

**Model Comparison.** Table 4 shows the comparisons on different models. From the table, we can see that the recognition performance can be significantly improved by involving a speech enhancement model into the KWS system with respect to classification accuracy, AUC and EER. Although `BiLSTM` achieves the lowest AUC and EER with the largest model size and computation complexity, it does not match the small-footprint requirement, which cannot be deployed on

**Table 3.** The number of parameters and multiply-add operations in the KWS and enhancement models.

| Model Name | Parameters | Multiplies |
|---|---|---|
| cnn-trad-pool2 | 493.7K | 95.87M |
| BiLSTM | 5661.0K | 432.7M |
| PowCRN32 | 724.0K | 280.1M |
| PowCRN16 | 182.3K | 73.0M |
| MelCRN32 | 881.3K | 115.1M |
| MelCRN16 | 221.5K | 29.2M |

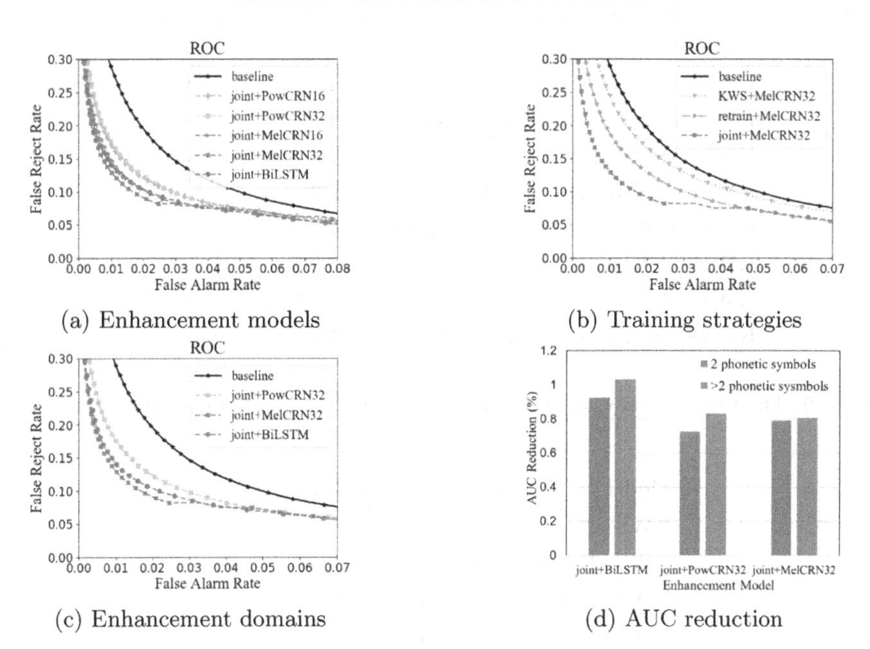


**Fig. 3.** Comparison on ROC curves from the perspective of (a) different enhancement models, (b) training strategy, (c) feature domain, and (d) AUC reduction against phonetic symbol length.

resource-limited devices. On the contrary, the proposed small-footprint CRNs achieve comparable or even better classification accuracy, AUC and EER with much less parameters and computation complexity. To our surprise, even the narrow version of MelCRN (`MelCRN16`) can achieve higher classification accuracy than `BiLSTM`.

In Fig. 3 (a), we plot the ROC curves of different models. From the figure, we find that although `BiLSTM` obtains lower AUC and EER, its false rejection rate is higher than `MelCRN32` and `MelCRN16` when the false alarm rate is fixed on a low level ($< 2\%$). In real-life applications, a KWS system should have a low

**Table 4.** The test accuracy (ACC), EER and AUR of each model under matched noise condition.

| Model | ACC (%) | AUC (%) | EER (%) |
|---|---|---|---|
| KWS | 80.89 | 1.99 | 7.28 |
| KWS+BiLSTM | 87.64 | 1.30 | 6.66 |
| KWS+BiLSTM+retrain | 90.18 | 1.17 | **5.92** |
| KWS+BiLSTM+joint | 91.64 | **1.01** | 6.15 |
| KWS+PowCRN32 | 86.42 | 1.52 | 6.67 |
| KWS+PowCRN32+retrain | 87.69 | 1.53 | 6.63 |
| KWS+PowCRN32+joint | 91.07 | 1.20 | 6.27 |
| KWS+PowCRN16 | 86.20 | 1.61 | 6.73 |
| KWS+PowCRN16+retrain | 87.01 | 1.67 | 6.88 |
| KWS+PowCRN16+joint | 90.68 | 1.22 | 6.50 |
| KWS+MelCRN32 | 87.59 | 1.59 | 6.97 |
| KWS+MelCRN32+retrain | 89.17 | 1.35 | 6.10 |
| KWS+MelCRN32+joint | **93.17** | 1.19 | 6.20 |
| KWS+MelCRN16 | 86.87 | 1.64 | 7.00 |
| KWS+MelCRN16+retrain | 88.20 | 1.42 | 6.49 |
| KWS+MelCRN16+joint | 92.56 | 1.28 | 6.39 |

**Table 5.** The test accuracy of joint-trianing models under unmatched noise condition.

| Model | Test accuracy(%) |
|---|---|
| KWS | 68.81 |
| KWS+BiLSTM+joint | 73.74 |
| KWS+PowCRN32+joint | 75.19 |
| KWS+PowCRN16+joint | 72.49 |
| KWS+MelCRN32+joint | **78.12** |
| KWS+MelCRN16+joint | 75.67 |

false alarm rate to avoid undesired wake-up operation. From this respective, the proposed small-footprint CRNs has better ROC curves.

**The Impact of Joint Training.** In Table 4, we mark the systems with and without joint training as "KWS+⋆+joint" and "KWS+⋆". From the table, we can see that, no matter what enhancement model is employed, the proposed joint training strategy can much improve the recognition performance in terms of test accuracy, AUC and EER. This is because the joint training strategy reduces the mismatch between the enhanced and desired features of KWS system.

In addition, we also compare our joint training with the retraining strategy in Fig. 3 (b), which is widely used in ASR community. In the retraining strategy, the KWS model is trained with the enhanced MFCCs rather than the noisy ones. We mark the retraining based models with "KWS+$\star$+retrain" in Table 4. From the table, we can see that the retraining strategy also reduces the mismatch between the outputs of enhancement model and the desired inputs of KWS model, resulting in better recognition performance than the models without retraining and joint training. However, the proposed joint training framework much outperforms the retraining strategy in terms of classification accuracy, AUC and EER, because the semantic information can back-propagate from the KWS system to the enhancement model and guide the training process.

**Comparisons on Enhancement Domains.** We also compare different enhancement domains in Table 4. From the table, we can see that although the model sizes of PowCRNs and MelCRNs are similar, MelCRNs achieve much better recognition performance than PowCRNs, no matter the full-size or narrow versions. In addition, since the dimension of Mel spectrogram is much less than the power spectrum, the number of multiply-add operations is significantly reduced, resulting in much lower computation complexity. Therefore, Mel spectrogram is more suitable for small-footprint KWS systems.

In Fig. 3 (c), we find that MelCRN32 achieves a lower false rejection rate than PowCRN32 and BiLSTM, when the false alarm rate is set to a low level. This is an important character for real-life applications which always place demand for low false alarm rate. It should be noticed that this advantage is also retained by the narrow models.

Since keywords have different numbers of phonetic symbols, we wonder how the enhancement models are sensitive to the number of phonetic symbols. We split the dataset into two sets, i.e., the keywords with 2 and more phonetic symbols. Figure 3 (d) shows AUC reduction for keywords with different numbers of phonetic symbols, where the models with less reduction are better. From the figure, we can see that the Mel spectrogram based method is less sensitive to the number of phonetic symbols in the keywords, and achieves less AUC reduction than the power spectrum based models.

**Noise Generalization Ability.** From the above analysis, we can see that the proposed MelCRNs obtain the best recognition performance under matched noise conditions. To evaluate the noise generalization ability, we provide the classification accuracy under the unmatched noise conditions in Table 5. The proposed MelCRNs achieve higher accuracy on untrained noises than `BiLSTM` and PowCRNs. Note that these results are consist with the matched noise conditions. This indicates that the proposed small-footprint MelCRNs have better noise generalization ability.

## 4 Conclusions

In this paper, we proposed a lightweight convolutional recurrent network (CRN) as the enhancement model for small-footprint KWS systems, which consists of a convolutional encoder for feature extraction, recurrent layers for temporal modeling and a deconvolutional decoder for feature reconstruction. To reduce the mismatch between the enhanced features and the desired inputs of KWS system, we propose an efficient joint training framework, in which a feature transformation block is proposed to concatenate and jointly fine-tune the enhancement and KWS models. Through the joint training framework, the proposed CRNs improve the noise robustness of KWS systems significantly in terms of classification accuracy, AUC and EER. Meanwhile, the parameter and computation complexity of our CRNs is much lower than the commonly used BiLSTM model. In addition, our models also have better generalization ability on untrained noise conditions and lower sensitivity to the phonetic symbol numbers of the keywords.

## References

1. Du, J., Wang, Q., Gao, T., Xu, Y., Dai, L.R., Lee, C.H.: Robust speech recognition with speech enhanced deep neural networks. In: INTERSPEECH (2014)
2. Ioffe, S., Szegedy, C.: Batch normalization: Accelerating deep network training by reducing internal covariate shift. ICML **37**, 448–456 (2015)
3. Kingma, D.P., Ba, J.: Adam: A method for stochastic optimization. In: ICLR (Poster) (2015)
4. Nair, V., Hinton, G.E.: Rectified linear units improve restricted boltzmann machines. In: ICML. pp. 807–814 (2010)
5. Narayanan, A., Wang, D.: Ideal ratio mask estimation using deep neural networks for robust speech recognition. In: ICASSP. pp. 7092–7096 (2013)
6. Prabhavalkar, R., Alvarez, R., Parada, C., Nakkiran, P., Sainath, T.N.: Automatic gain control and multi-style training for robust small-footprint keyword spotting with deep neural networks. In: ICASSP. pp. 4704–4708 (2015)
7. Sainath, T., Parada, C.: Convolutional neural networks for small-footprint keyword spotting. In: INTERSPEECH (2015)
8. Seltzer, M.L., Yu, D., Wang, Y.: An investigation of deep neural networks for noise robust speech recognition. In: ICASSP. pp. 7398–7402 (2013)
9. Shan, C., Zhang, J., Wang, Y., Xie, L.: Attention-based end-to-end models for small-footprint keyword spotting. pp. 2037–2041 (2018)
10. Tan, K., Wang, D.: A convolutional recurrent neural network for real-time speech enhancement. In: INTERSPEECH. pp. 3229–3233 (2018)
11. Tang, R., Lin, J.: Honk: A pytorch reimplementation of convolutional neural networks for keyword spotting. arXiv preprint arXiv:1710.06554 (2017)
12. Tang, R., Lin, J.: Deep residual learning for small-footprint keyword spotting. In: ICASSP. pp. 5484–5488 (2018)

13. Wang, D., Chen, J.: Supervised speech separation based on deep learning: an overview. IEEE/ACM Trans. Audio, Speech, and Language Process. **26**(10), 1702–1726 (2018)
14. Wang, Y., Narayanan, A., Wang, D.: On training targets for supervised speech separation. IEEE/ACM Trans. Audio, Speech, and Language Process. **22**(12), 1849–1858 (2014)
15. Wang, Z.Q., Wang, D.: A joint training framework for robust automatic speech recognition. IEEE/ACM Trans. Audio, Speech, and Language Process. **24**(4), 796–806 (2016)
16. Warden, P.: Speech commands: A dataset for limited-vocabulary speech recognition. arXiv preprint arXiv:1804.03209 (2018)
17. Yu, M., et al.: Text-dependent speech enhancement for small-footprint robust keyword detection. In: Interspeech. pp. 2613–2617 (2018)

# Hierarchical Interactive Matching Network for Multi-turn Response Selection in Retrieval-Based Chatbots

Ting Yang[1], Ruifang He[1], Longbiao Wang[1](✉), Xiangyu Zhao[1], and Jiangwu Dang[2]

[1] Tianjin Key Laboratory of Cognitive Computing and Application, College of Intelligence and Computing, Tianjin University, Tianjin, China
16622898776@163.com, {rfhe,longbiao_wang}@tju.edu.cn, zhaoxiangyu009@163.com

[2] Japan Advanced Institute of Science and Technology, Ishikawa, Japan
jdang@jaist.ac.jp

**Abstract.** We study multi-turn response selection in open domain dialogue systems, where the best-matched response is selected according to a conversation context. The widely used sequential matching models match a response candidate with each utterance in the conversation context through a representation-interaction-aggregation framework, but do not pay enough attention to the inter-utterance dependencies at the representation stage and global information guidance at the interaction stage. They may lead to the result that the matching features of utterance-response pairs may be one-sided or even noisy. In this paper, we propose a hierarchical interactive matching network (HIMN) to model both aspects in a unified framework. In HIMN, we model the dependencies between adjacency utterances in the context with multi-level attention mechanism. Then a two-level hierarchical interactive matching is exploited to introduce the global context information to assist in distilling important matching features of each utterance-response pair at the interaction stage. Finally, the two-level matching features are merged through gate mechanism. Empirical results on both Douban Corpus and Ecommerce Corpus show that HIMN can significantly outperform the competitive baseline models for multi-turn response selection.

**Keywords:** Dialogue systems · Response selection · Hierarchical matching

## 1 Introduction

Building human-machine dialogue systems is a core problem in Artificial Intelligence. In recent years, there has been a growing interest for open domain dialogue systems which aim to realize natural communication with people under a wide range of topics. Generally, the methods of establishing open domain dialogue systems are two types. One is to learn a response generation model within an encoder-decoder framework [13], and the other is to select the best-matched response from the pre-built massive repository [7,17]. Compared with


H. Yang et al. (Eds.): ICONIP 2020, LNCS 12532, pp. 24–35, 2020.
https://doi.org/10.1007/978-3-030-63830-6_3

the generation-based method, retrieving response from the repository can ensure informativeness and stability to a certain extent. And it is practical in a production setting [6].

Early retrieval-based conversation systems [8,15] focus on utilizing a message to select the response, but conversations in real scenarios are always complicated which may last for several turns. Hence more recently, researchers have taken conversation history into consideration, where a response is selected given the conversation context consisting of a sequence of utterances. For this task, the key is to calculate the matching score of a conversation context and a response candidate. Existing methods, such as SMN [17], DUA [20] and IOI [11], have achieved impressive performance within representation-interaction-aggregation framework where the response interacts with each utterance in the context based on their representations and then matching features of each utterance-response pair are aggregated as context-response matching score. Obviously, only after the matching features of each utterance-response pair is robust can high quality of the aggregated matching features be guaranteed. However, at the representation stage, the utterances are modeled independently from each other without making full use of inter-utterance dependencies. Most of them just aggregate matching features of utterance-response pairs in a chronological order to indirectly model the relationships among utterances in the context. Furthermore, at the interaction stage, they capture the important matching signals in the utterance through the supervision of the response without the guidance of the global information, which may lead to the result that the current matching features may be one-sided or even noisy.

In this paper, we propose a hierarchical interactive matching network (HIMN) to tackle the above problems. For one thing, a conversation is made up of a series of two consecutive utterances, which have the flowing characteristics: adjacent; produced by different interlocutors; ordered as a first part and second part; typed so that a particular first part requires a particular second part and are defined as adjacency pairs [10]. We postulate that modeling such adjacency dependencies can contribute to a deep context representation and is beneficial for matching. And for another, we attempt to introduce an over-all view of context, global information, to guide local utterance-response pairs to gain good matching features. Therefore, the multi-level attention mechanism is leveraged to represent the context instead of a flat model. We use interleaved attention to model the adjacency dependencies between utterances. And utilize recurrent attention to get representation of the global context based on adjacency pairs representations, considering the chronological order of adjacency pairs. With the rich representations, we perform two-level hierarchical global interactive matching. At the lower level, we make the global context attend to each adjacency pair at first. Thus, we can emphasize the global context-relevant part of the adjacency pair representations. After that, each adjacency pair matches with the response. At the upper level, the response directly matches with the global context. Convolutional neural networks are used to distill matching features at two levels. Finally, the two-level matching features are merged through gate mechanism to obtain the final context-response matching features, which are then fused into one matching score via a single-layer perceptron. Our contributions in this paper are listed as follows:

- We propose a novel matching network for multi-turn response selection which explicitly models inter-utterance dependencies in the context , especially the dependencies between adjacency pairs.
- We perform hierarchical interactive matching under the guidance of global information, which avoids distilling partial one-sided matching features in the context.
- The experimental results on Douban Corpus and Ecommerce Corpus significantly verify the effectiveness of the proposed model.

## 2 Related Work

With the increase in availability of open-sourced conversation datasets, a large number of data-driven approaches have been proposed for open domain dialogue systems, which has attracted widespread attention in academia and industry in recent years. These approaches are either generation-based [13,19] or retrieval-based [11,16]. The generation-based approaches treat generation task as statistical machine translation, where the goal is to synthesize a response given the previous dialogue utterances, while the retrieval-based approaches aim to select a response that best matches with the input.

Many previous works study single-turn response selection [8,15], where the response candidate matches with a single message. Recently multi-turn response selection has drown much attention. The matching process is more difficult since there are more context information and constrains to consider, which brings new challenges for researchers in this area. [7] publish the Ubuntu dialogue corpus and propose a dual LSTM model to convert the entire context and the response into fixed-length vectors respectively, and then calculate the matching score by using a bilinear function. [21] integrate context information from both word sequence view and utterance sequence view. [18] reformulate the the current message by adding one or more utterances from the context. [17] propose a sequential matching network to match a response candidate with each utterance in the context at word level and segment level by the word embedding and hidden states of GRU. [20] model the relationships between the last utterance and the rest utterances and use self-matching attention to distill the import matching features. [22] construct utterance semantic representation at multi-granularity by exploiting stacked self-attention mechanism and they implicitly model the relationships between utterances through 3D convolution. [11] present an interaction-over-interaction network to accumulate useful matching information in an iterative way. Different form the previous methods, we explicitly model the dependencies of adjacency pairs in the context and leverage global information at two level to help distill important matching features.

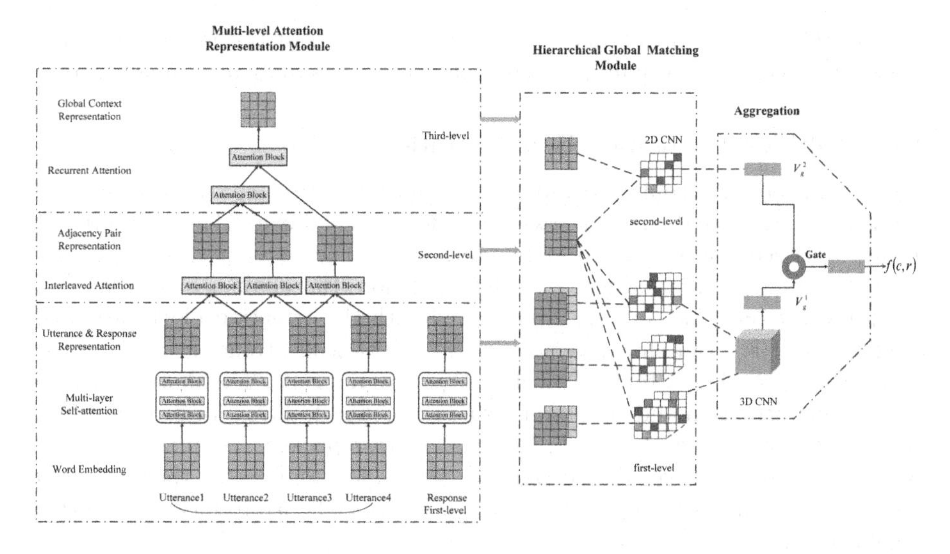


**Fig. 1.** The HIMN framework. Matrices of the same color are of the same type, especially the yellow matrix is the representation obtained by attending global context representation with each adjacency pair representation.(Color figure online)

# 3 Hierarchical Interactive Matching Network

## 3.1 Task Description

Given a data set $D = \{(c, y, r)_i\}_{i=1}^{N}$, where $c = \{u_1, u_2, ..., u_n\}$ is a conversation context consisting of $n$ utterances, $u_i = \{w_{u_i,k}\}_{k=1}^{n_{u_i}}$ represents the $i$-th utterance with $n_{u_i}$ words. $r = \{w_{r,j}\}_{j=1}^{n_r}$ is a response candidate with $n_r$ words and $y \in \{0, 1\}$ denotes a label. $y = 1$ indicates that the response $r$ is appropriate for the given conversation context $c$, otherwise $y = 0$. The goal is to learn a matching model $f = (\cdot, \cdot)$ with D. For a new context-response pair $(c, r)$, $f = (c, r)$ returns the matching score of $c$ and $r$.

## 3.2 Model Overview

As illustrated in Fig. 1, HIMN consists of three parts: a multi-level attention representation module, a two-level hierarchical interactive matching module and an aggregation module. The multi-level attention representation module adopts multi-layer self-attention, interleaved attention, and recurrent attention to obtain deep utterances representations, adjacency pairs representations and global context representations respectively. The multi-layer self-attention is also applied to represent the response. For the two-level hierarchical interactive matching module, at the lower level, adjacency pairs sequentially matches with the response under the guidance of global context information. At the upper level, the response directly matches with global context. The convolution neural network (CNN) is used to obtain the matching features at two levels, which are merged through gate mechanism as the final context-response matching features.

### 3.3 Multi-level Attention Representation

Given a conversation context $c$ and a response candidate $r$, $u_i$ is an utterance in $c$. Each word in $u_i$ and $r$ is converted to its word embedding. Then we represent $u_i$ and $r$ as sequences of word embeddings $E_{u_i} = \left[e_{u_i,1}, e_{u_i,2}, ..., e_{u_i,n_{u_i}}\right]$ and $E_r = [e_{r,1}, e_{r,2}, ..., e_{r,n_r}]$, where $n_{u_i}$ and $n_r$ denote the number of words in $u_i$ and $r$ respectively. $E_{u_i}$ and $E_r$ are encoded with the attention block proposed in [22], which is a simplified version of the multi-head attention block in the transformer [12]. The attention block has two components. The first is a scaled dot-product attention and the second is a fully connected feed-forward network. We use $Q \in \mathbb{R}^{n_Q \times d}$, $K \in \mathbb{R}^{n_K \times d}$, $V \in \mathbb{R}^{n_V \times d}$ to represent the query, key, and value embedding matrices respectively, where $n_Q$, $n_K$, and $n_V$ are the number of words and $d$ means $d$-dimension embedding. The scaled dot-production attention is defined as:

$$\text{Att}(Q, K, V) = \text{softmax}\left(\frac{QK^T}{\sqrt{d}}\right) V \tag{1}$$

where usually $K = V$ and $Q$ is represented by similar entries of $V$. Then the fully connected feed-forward network is formulated as:

$$\text{FFN}(x) = \max(0, xW_1 + b_1) W_2 + b_2 \tag{2}$$

where $x \in \mathbb{R}^{n_Q \times d}$ and $W_1, W_2, b_1, b_2$ are parameters. A residual connection [2] is employed around each of the above two components, followed by layer normalization operation [1]. We refer the whole module as AttentionBlock$(Q, K, V)$.

At the first representation level, $u_i$ and $r$ pass through the multi-layer self-attention. $U_i^l$ denotes the the output of $l^{th}$ layer, which takes the output of $l-1^{th}$ layer as input and $R^l$ is expressed in the same way.

$$U_i^l = \text{AttentionBlock}\left(U_i^{l-1}, U_i^{l-1}, U_i^{l-1}\right) \tag{3}$$

$$R^l = \text{AttentionBlock}\left(R^{l-1}, R^{l-1}, R^{l-1}\right) \tag{4}$$

where $U_i^0 = E_{u_i}$, $R^0 = E_r$, $l \in \{1, 2, ..., L\}$ and $L$ is the number of layers of the attention block. The outputs of all layers are then combined linearly to obtain the deep enhanced representations of $u_i$ and $r$, namely $\tilde{U}_i$ and $\tilde{R}$.

$$\tilde{U}_i = \sum_{l=1}^{L} \alpha_l U_i^l, \tilde{R} = \sum_{l=1}^{L} \alpha_l R^l \tag{5}$$

where $\alpha_l$ are the softmax-normalized weights shared between utterances and the response.

At the second level, the interleaved attention is employed to model high dependencies between the two utterances in adjacency pairs, which takes each word in the second utterance to attend to the words in the first utterance.

$$P_j = \text{AttentionBlock}\left(\tilde{U}_{j+1}, \tilde{U}_j, \tilde{U}_j\right), j \in \{1, 2, ..., n-1\} \tag{6}$$

$P_j$ represents the $j$-th adjacency pair, where $\tilde{U}_j \in \mathbb{R}^{n_{u_j} \times d}$ , $P_j \in \mathbb{R}^{n_{p_j} \times d}$ and $n_{p_j} = n_{u_{j+1}}$. $n$ means the number of utterances in the conversation context and $n_{u_{j+1}}$ the the number of words in the $u_{j+1}$.

At the third level, the global context representation is learned following the chronological order of adjacency pairs in the context with recurrent attention. We get the set of adjacency pairs representations $P = \{P_1, P_2, ..., P_{n-1}\}$ at the second level.

$$G_{n-1} = \text{AttentionBlock}\left(P_{n-1}, G_{n-2}, G_{n-2}\right) \tag{7}$$

where we set $G_1 = P_1$, $C = G_{n-1} \in \mathbb{R}^{n_{p_{n-1}} \times d}$ and $C$ denotes the global context.

### 3.4 Two-Level Hierarchical Interactive Matching

At the first matching level, given the adjacency pair $P_j$ and the response $R$, three kinds of similarity matrices are constructed: self-matching matrix $M^j_{self}$, cross-matching matrix $M^j_{cross}$ and global-matching matrix $M^j_{global}$. $M^j_{self}$ are formulated as:

$$M^j_{self} = \frac{P_j R^T}{\sqrt{d}} \tag{8}$$

Then we construct $M^j_{cross}$ by making $P_j$ and $R$ attend to each other, which are defined as:

$$\hat{P}_j = \text{AttentionBlock}\left(P_j, R, R\right), \hat{R} = \text{AttentionBlock}\left(R, P_j, P_j\right)$$

$$M^j_{cross} = \frac{\hat{P}_j \hat{R}^T}{\sqrt{d}} \tag{9}$$

In order to introduce global guidance when distilling important information of adjacency pairs at the lower matching level, we make $P_j$ attend to $C$ to obtain $\tilde{P}_j$, which is represented by the yellow matrix in Fig. 1.

$$\tilde{P}_j = \text{AttentionBlock}\left(P_j, C, C\right) \tag{10}$$

where $\tilde{P}_j \in \mathbb{R}^{n_{p_j} \times d}$ has the same shape as $P_j$. $M^j_{global}$ is then constructed as:

$$M^j_{global} = \frac{\tilde{P}_j R^T}{\sqrt{d}} \tag{11}$$

where $M^j_{self} \in \mathbb{R}^{n_{p_j} \times n_r}$, $M^j_{cross} \in \mathbb{R}^{n_{p_j} \times n_r}$, $M^j_{global} \in \mathbb{R}^{n_{p_j} \times n_r}$ and $d$ is the dimension of word embedding.

At the upper level, the response candidate $R$ directly interacts with the global context $C$. The similarity matrices can be calculated as:

$$M'_1 = \frac{CR^T}{\sqrt{d}}, \tag{12}$$

where $M'_1 \in \mathbb{R}^{n_{p_{n-1}} \times n_r}$ and $M'_2 \in \mathbb{R}^{n_{p_{n-1}} \times n_r}$. The two matching matrices are concatenated as $M' \in \mathbb{R}^{n_{p_{n-1}} \times n_r \times 2}$:

$$M' = M'_1 \oplus M'_2 \tag{13}$$

### 3.5 Aggregation

The three matching matrices of all adjacency pairs at the first matching level are aggregated into a 4D-cube.

$$M = \{M_{j,k,t}\}_{n-1 \times n_{p_j} \times n_r}$$

$$M_{j,k,t} = M^j_{self}[k,t] \oplus M^j_{cross}[k,t] \oplus M^j_{global}[k,t] \tag{14}$$

where $M \in \mathbb{R}^{n-1 \times n_{p_j} \times n_r \times 3}$, $M_{j,k,t} \in \mathbb{R}^3$ represents that each pixel in $M$ has three channels and $\oplus$ means concatenation operation. Similar to [22], we leverage two layers of 3D convolution with max pooling followed by the batch normalization layer [3], which is conducive to preventing vanishing or exploding of gradients, to distill important matching information in $M$. The output of 3D CNN is flattened and mapped to $d_g$-dimentional matching features $V_{g_1} \in \mathbb{R}^{d_g}$.

The matching matrices at the second level $M'$ are processed by a 2D convolution with max pooling operations followed by the batch normalization layer to compute the matching features $V_{g_2} \in \mathbb{R}^{d_g}$.

A gate is proposed to integrate two levels of matching features. The importance of the two-level matching vectors is dynamically determined by the network. The gate is given by:

$$z = \sigma\left(W_{g_1} V_{g_1} + W_{g_2} V_{g_2}\right) \tag{15}$$

where $\sigma$ is a sigmoid function and $W_{g_1} \in \mathbb{R}^{d_g \times d_g}$ , $W_{g_2} \in \mathbb{R}^{d_g \times d_g}$ are learnable parameters.The final context-response matching vector $V$ is calculated as:

$$V = z \odot V_{g_1} + (1 - z) \odot V_{g_2} \tag{16}$$

where $\odot$ is an element-wise multiplication operation. The matching score $f(c,r)$ of the conversation context and the response candidate is calculated by a single perception.

$$f(c,r) = \sigma\left(W_v V + b_v\right) \tag{17}$$

where $W_v$, $b_v$ are weights and $\sigma$ is the sigmoid function. Finally, the following loss function is used to optimize the training process.

$$-\sum_{i=1}^{N} \left[y_i \log f(c_i, r_i) + (1 - y_i) \log\left(1 - f(c_i, r_i)\right)\right] \tag{18}$$

## 4 Experiments

### 4.1 Dataset

The first data we use is the Douban Corpus [17], which is composed of dyadic dialogues crawled from Douban group[1]. The data set consists of 1 million context-response pairs for training, 50 thousand pairs for validation, and 10 thousand pairs

---

[1] https://www.douban.com/group/.

**Table 1.** Statistics of the two data sets.

| | Douban corpus | | | E-commerce corpus | | |
|---|---|---|---|---|---|---|
| | Train | Valid | Test | Train | Valid | Test |
| # context-response pairs | 1M | 50K | 10K | 1M | 10K | 10K |
| # candidates per context | 2 | 2 | 10 | 2 | 2 | 10 |
| Av turns/conversation | 6.69 | 6.75 | 6.45 | 5.51 | 5.48 | 5.64 |
| Av length of utterance | 18.56 | 18.50 | 20.74 | 7.02 | 6.99 | 7.11 |

for testing. For each dialogue in the testing data set, the last turn is taken as a message to retrieve 10 response candidates from the pre-built index with Apache Lucene and three people are recruited to judge if a response candidate is positive.

In addition to the Douban Corpus, we also conduct experiments with E-commerce Corpus [20]. The E-commerce Corpus collects real conversations between customers and customer service staff from Taobao[2], which is the largest e-commerce platform. There are 1 million context-response pairs for training, 10 thousand pairs for validation and 10 thousand pairs for testing. The ratio of the positive and the negative is 1:1 in the training set and validation set, and 1:9 in the testing set. More statistics of the two data sets are shown in Table 1.

### 4.2 Evaluation Metric

On the E-commerce Corpus, we employ recall at position $k$ in $n$ candidates $\left(R_n@k = \frac{\sum_{i=1}^{k} y_i}{\sum_{i=1}^{n} y_i}\right)$, where $k \in \{1, 2, 5\}$ and $y_i$ denotes the binary label for each response candidate, as evaluation metrics according to the previous works [7,20]. In addition to $R_n@k$, we follow by [17] and adopt mean average precision (MAP), mean reciprocal rank (MRR), and precision at position $P@1$ as evaluation metrics on the Douban Corpus.

### 4.3 Baseline Models

We compare our proposed model with the following baseline models:

**Basic deep Matching Models:** These models concatenate all the utterances in a context as a long document and encode the long text and a response candidate into two vectors by RNN [7], CNN [4], LSTM [4], BiLSTM [4] and advanced models, such as MV-LSTM [14], Match-LSTM [16].

**Multi-View** [21]**:** This model represents the conversation context and the response in two views: word sequence view and utterance sequence view, using RNN and CNN respectively.

**DL2R** [18]**:** This model reformulates query and selects one or more utterances from dialogue history to enhance the representation of the current query. Then returns a response that is relevant to the reformulated query according to a DNN-based ranker.

[2] https://www.taobao.com/.

**SMN** [17]**:** This model allows a response candidate to interact with each utterance in the context at the word and segment level. Then distills the important matching features from each utterance-response pair with CNN. Finally, the matching features are accumulated through RNN to obtain the matching score.

**DUA** [20]**:** This model concatenates the last utterance with the preceding utterances, and then a self-attention mechanism is introduced to mine important information in each refined utterance.

**DAM** [22]**:** This model constructs representations of utterances in the context and the response at multi-granularity with stacked self-attention.

**IoI** [11]**:** This model stacks multiple interaction blocks to iteratively accumulate vital matching signals from utterance-response pairs.

### 4.4 Experiment Settings

Our model is implemented with tensorflow[3] framework. The dimension of word embedding, which is initialized by the Word2Vec [9] pretrained on the training set, is set to 200 and the batch size is 128. In HIMN, the multiple layers self-attention is composed of 6 identical layers. For the lower level in the matching module, the window size of 2D convolution and max pooling is set as $(3, 3)$ finally. The number of feature maps is 8. At the higher matching level, we set the window size of two-layer 3D convolution and max pooling kernels as $(3, 3, 3)$. Both convolutional layers have 16 filters. We pad zeros if the number of words in an utterance (response) is less than 50, otherwise we keep only the last 50 words. We employ Adam [5] for optimization with a learning rate of 0.0005, which is gradually decreased during training.

### 4.5 Experiment Results

Table 2 shows the evaluation results on the two datasets. HIMN significantly outperforms all the baseline models in terms of most of the metrics on both datasets. Our proposed model extends from DAM [22], and it outperforms the DAM by a large margin of 3.9% in terms of MAP and 4.2% in terms of MRR on the Douban Conversation Corpus and 6.1% in terms of $R_{10}@1$ on the Eommerce Conversation Corpus. Compared with the best performing baseline on the two datasets, our model achieves 3.2% absolute improvement on $R_{10}@2$ on the Douban Corpus and 2.4% absolute improvement on $R_{10}@1$ on the Eommerce Corpus. The improvement verifies effectiveness of considering inter-utterance dependencies in the context, especially the dependence of adjacency pair, and the importance of introducing global information supervision at the interaction stage.

### 4.6 Discussions

**Ablation Study.** As it is shown in Table 3, we conduct an ablation study on the Douban Corpus, where we aim to examine the effect of each part in our

[3] https://github.com/tensorflow/tensorflow.

**Table 2.** Evaluation results for baselines and our proposed model on Douban Corpus and E-commerce Corpus.

| | Douban corpus | | | | | | E-commerce corpus | | |
|---|---|---|---|---|---|---|---|---|---|
| | MAP | MRR | P@1 | $R_{10}$@1 | $R_{10}$@2 | $R_{10}$@5 | $R_{10}$@1 | $R_{10}$@2 | $R_{10}$@5 |
| RNN | 0.390 | 0.422 | 0.208 | 0.118 | 0.223 | 0.589 | 0.325 | 0.463 | 0.775 |
| CNN | 0.417 | 0.440 | 0.226 | 0.121 | 0.252 | 0.647 | 0.328 | 0.515 | 0.792 |
| BiLSTM | 0.479 | 0.514 | 0.313 | 0.184 | 0.330 | 0.716 | 0.355 | 0.525 | 0.825 |
| LSTM | 0.485 | 0.527 | 0.320 | 0.187 | 0.343 | 0.720 | 0.365 | 0.536 | 0.828 |
| DL2R | 0.488 | 0.527 | 0.330 | 0.193 | 0.342 | 0.705 | 0.399 | 0.571 | 0.842 |
| MV-LSTM | 0.498 | 0.538 | 0.348 | 0.202 | 0.351 | 0.710 | 0.412 | 0.591 | 0.857 |
| Match-LSTM | 0.500 | 0.537 | 0.345 | 0.202 | 0.348 | 0.720 | 0.410 | 0.590 | 0.858 |
| Multi-view | 0.505 | 0.543 | 0.342 | 0.202 | 0.350 | 0.729 | 0.421 | 0.601 | 0.861 |
| SMN | 0.529 | 0.569 | 0.397 | 0.233 | 0.396 | 0.724 | 0.453 | 0.654 | 0.886 |
| DUA | 0.551 | 0.599 | 0.421 | 0.243 | 0.421 | 0.780 | 0.501 | 0.700 | 0.921 |
| DAM | 0.550 | 0.601 | 0.427 | 0.254 | 0.410 | 0.757 | 0.526 | 0.727 | 0.933 |
| IoI | 0.573 | 0.621 | 0.444 | 0.269 | 0.451 | 0.786 | 0.563 | 0.768 | 0.950 |
| HIMN | **0.589** | **0.643** | **0.465** | **0.286** | **0.483** | **0.797** | **0.587** | **0.770** | **0.964** |

proposed model. Firstly, we use the complete HIMN as the baseline. Then, we gradually remove its modules as follows:

- Removing recurrent attention part (denoted as **HIMN-Interleaved**) makes the performance drop dramatically. This indicates the supervision form both global information and responses can indeed assist in distilling important matching part in the context. We think the main reason is that the introduction of global guidance can make the model pay more attention to the matching parts related to the entire contex.
- The performance drops slightly when removing the interleaved attention part (denoted as **HIMN-Global**), proving that taking into account the adjacency dependencies between utterances is useful for matching. The next utterance is usually a response to the previous utterance. In addition, HIMN-Global has better performance than HIMN-Interleaved, demonstrating that global guidance is more effective than adjacency dependencies in the matching process.
- Compared with HIMN-base, removing interleaved attention and recurrent attention parts (denoted as **HIMN-Single**) leads to performance decay a large margin, which shows that considering the adjacency pairs dependencies and global information guidance are indeed help to improve the selection performance.

**Impact of Self-attention Layers.** Figure 2 illustrates how HIMN performs as the number of self-attention layers of multi-level attention representation module changes on the test set of the Douban Corpus. From the figure, we can observe that there is significantly improvement in the first few layers. The performance of the model reaches the highest when the number of self-attention layers was 6, which outperforms IoI [11] by a margin of 2.2% in term of MRR and

**Table 3.** Evaluation results of the ablation study on Douban Corpus.

| | Douban corpus | | | | | |
|---|---|---|---|---|---|---|
| | MAP | MRR | P@1 | $R_{10}$@1 | $R_{10}$@2 | $R_{10}$@5 |
| HIMN-Single | 0.557 | 0.604 | 0.421 | 0.249 | 0.427 | 0.789 |
| HIMN-Interleaved | 0.561 | 0.614 | 0.436 | 0.264 | 0.436 | 0.772 |
| HIMN-Global | 0.574 | 0.624 | 0.442 | 0.268 | 0.463 | 0.787 |
| HIMN-Base | 0.589 | 0.643 | 0.465 | 0.286 | 0.483 | 0.797 |

1.6% in terms of MAP. After that it begins to gradually decrease. As the result demonstrated, incorporating the representations of the appropriate number of self-attention layer can indeed improve the matching performance of context and the response candidate.

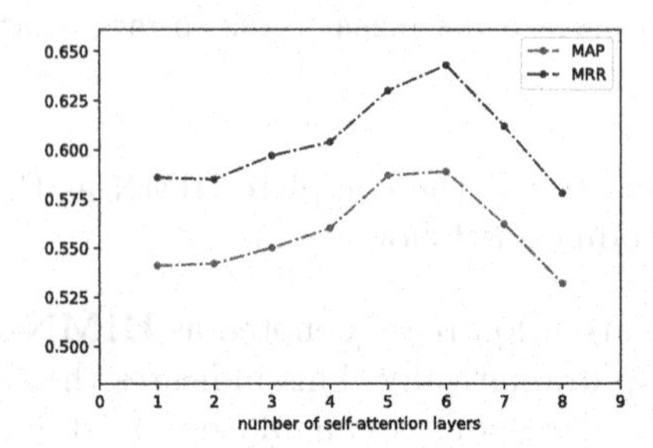


**Fig. 2.** Performance of HIMN for different numbers of self-attention layers.

## 5 Conclusion

In this paper, we propose a hierarchical interactive matching network for multi-turn response selection. This model considers adjacency dependencies between utterances in the context and obtains global information through multi-level attention representation. An empirical study on two multi-turn dialogue datasets demonstrates that our proposed model significantly outperforms the baseline models. In the future, we plan to integrate our model with pretrained models to study if the performance can be further improved.

**Acknowledgments.** This work was supported in part by the National Natural Science Foundation of China under Grant 61771333, the Tianjin Municipal Science and Technology Project under Grant 18ZXZNGX00330.

## References

1. Ba, J., Kiros, J.R., Hinton, G.E.: Layer normalization. arXiv abs/1607.06450 (2016)

2. He, K., Zhang, X., Ren, S., Sun, J.: Deep residual learning for image recognition. In: IEEE Conference on Computer Vision and Pattern Recognition pp. 770–778 (2015)
3. Ioffe, S., Szegedy, C.: Batch normalization: accelerating deep network training by reducing internal covariate shift. arXiv abs/1502.03167 (2015)
4. Kadlec, R., Schmid, M., Kleindienst, J.: Improved deep learning baselines for ubuntu corpus dialogs. arXiv abs/1510.03753 (2015)
5. Kingma, D., Ba, J.: Adam: A method for stochastic optimization. In: ICLR (2014)
6. Kyle, S., Lili, Y., Christopher, F., WohlwendJeremy, Tao, L.: Building a production model for retrieval-based chatbots. In: Proceedings of the First Workshop on NLP for Conversational AI, pp. 32–41 (2019)
7. Lowe, R., Pow, N., Serban, I., Pineau, J.: The Ubuntu dialogue corpus: a large dataset for research in unstructured multi-turn dialogue systems. In: Proceedings of the 16th Annual Meeting of the Special Interest Group on Discourse and Dialogue. pp. 285–294 (2015)
8. Lu, Z., Li, H.: A deep architecture for matching short texts. In: International Conference on Neural Information Processing Systems, pp. 1367–1375 (2013)
9. Mikolov, T., Chen, K., Corrado, G., Dean, J.: Efficient estimation of word representations in vector space (2013)
10. Schegloff, E.A., Sacks, H.: Opening up closings. J. Int. Assoc. Semiotic Stud. **8**(4), (1973)
11. Tao, C., Wu, W., Xu, C., Hu, W., Zhao, D., Yan, R.: One time of interaction may not be enough:go deep with an interaction-over-interaction network for response selection in dialogues. In: Proceedings of the 57th Annual Meeting of the Association for Computational Linguistics. pp. 1–11 (2019)
12. Vaswani, A., et al.: Attention is all you need. In: International Conference on Neural Information Processing Systems (2017)
13. Vinyals, O., Le, Q.V.: A neural conversational model. arXiv abs/1506.05869 (2015)
14. Wan, S., Lan, Y., Xu, J., Guo, J., Pang, L., Cheng, X.: Match-srnn: Modeling the recursive matching structure with spatial RNN. In: IJCAI (2016)
15. Wang, M., Lu, Z., Li, H., Liu, Q.: Syntax-based deep matching of short texts. In: IJCAI, pp. 1354–1361 (2015)
16. Wang, S., Jiang, J.: Learning natural language inference with LSTM. In: NAACL, pp. 1442–1451 (2016)
17. Wu, Y., Wu, W., Xing, C., Zhou, M., Li, Z.: Sequential matching network: a new architecture for multi-turn response selection in retrieval-based chatbots. In: ACL, pp. 496–505 (2017)
18. Yan, R., Song, Y., Wu, H.: Learning to respond with deep neural networks for retrieval-based human-computer conversation system. In: SIGIR 2016 (2016)
19. Zhang, H., Lan, Y., Pang, L., Guo, J., Cheng, X.: ReCoSa: detecting the relevant contexts with self-attention for multi-turn dialogue generation. In: ACL, pp. 3721–3730 (2019)
20. Zhang, Z., Li, J., Zhu, P., Zhao, H., Liu, G.: Modeling multi-turn conversation with deep utterance aggregation. In: Proceedings of the 27th International Conference on Computational Linguistics. pp. 3740–3752 (2018)
21. Zhou, X., et al.: Multi-view response selection for human-computer conversation. In: EMNLP, pp. 372–381 (2016)
22. Zhou, X., et al.: Multi-turn response selection for chatbots with deep attention matching network. In: Proceedings of the 56th Annual Meeting of the Association for Computational Linguistics, pp. 1118–1127 (2018)

# Investigation of Effectively Synthesizing Code-Switched Speech Using Highly Imbalanced Mix-Lingual Data

Shaotong Guo[1], Longbiao Wang[1](✉), Sheng Li[4], Ju Zhang[3](✉), Cheng Gong[1], Yuguang Wang[3], Jianwu Dang[2], and Kiyoshi Honda[1]

[1] Tianjin Key Laboratory of Cognitive Computing and Application, College of Intelligence and Computing, Tianjin University, Tianjin, China
{shaotong_guo,longbiao_wang}@tju.edu.cn

[2] Japan Advanced Institute of Science and Technology, Ishikawa, Japan
juzhang@huiyan-tech.com

[3] Huiyan Technology (Tianjin) Co., Ltd., Tianjin, China

[4] National Institute of Information and Communications Technology, Kyoto, Japan

**Abstract.** End-to-end text-to-speech (TTS) can synthesize monolingual speech with high naturalness and intelligibility. Recently, the end-to-end model has also been used in code-switching (CS) TTS and performs well on naturalness, intelligibility and speaker consistency. However, existing systems rely on skillful bilingual speakers to build a CS mix-lingual data set with a high Language-Mix-Ratio (LMR), while simply mixing monolingual data sets results in accent problems. To reduce the cost of recording and maintain the speaker consistency, in this paper, we investigate an effective method to use a low LMR imbalanced mix-lingual data set. Experiments show that it is possible to construct a CS TTS system with a low LMR imbalanced mix-lingual data set with diverse input text presentations, meanwhile produce acceptable synthetic CS speech with more than 4.0 Mean Opinion Score (MOS). We also find that the result will be improved if the mix-lingual data set is augmented with monolingual English data.

**Keywords:** Speech synthesis · Code-switching · Text representation · Imbalanced mix-lingual data

## 1 Introduction

Speech synthesis aims to generate human-like speech with high naturalness and intelligibility. The recent studies on the end-to-end text-to-speech system allow us to produce a human-like synthesized speech with an integrated end-to-end text-to-speech (TTS) model, as seen as Tacotron-2 [1] and DeepVoice [2] that can synthesize state-of-the-art monolingual speech. However, synthesizing at least two languages simultaneously from one model also is a practical need in human-machine interaction and other scenarios. The code-switching (CS)[1] is a more

[1] Code-switching is also known as code-mixing. In this paper, we use code-switching.


H. Yang et al. (Eds.): ICONIP 2020, LNCS 12532, pp. 36–47, 2020.
https://doi.org/10.1007/978-3-030-63830-6_4

common practice in daily communication compared to multi-lingual communications. The CS is done by embedding words or phrases of a non-native language into the native utterance. Therefore, it is desirable to enable the TTS system to synthesize CS speech with high naturalness, intelligibility, and speaker consistency.

Existing CS TTS systems rely on skillful bilingual speakers to build a bilingual data set with high cost or a CS mix-lingual data set with high Language-Mix-Ratio (LMR), while simply mixing monolingual data sets results in accent problems. It is necessary to investigate methods to reduce the cost of data recording while maintaining the speaker consistency.

Another problem for CS TTS is which input text representation is better for the final result, especially for distinct languages like Mandarin and English. In the monolingual TTS system, various input representations could be chosen, but in the CS TTS system, the model has to learn at least two linguistic features at the same time. [3] evaluated different input representations on monolingual data sets in terms of the numbers of input tokens. However, they did not investigate different input text representations for different languages.

In this paper, we build an imbalanced mix-lingual data set with the low LMR and embedded with simple English words. Then we investigate the combination of different text representations to model Mandarin and English text in the end-to-end CS TTS system on this imbalanced data set.

The contribution of this paper is two-fold:

1. We investigate an effective method to synthesize acceptable CS speech on an imbalanced mix-lingual data set, to reduce the cost of data recording, meanwhile maintaining the speaker consistency.
2. We also discover that diverse text representations and augmentation with English monolingual data for the imbalanced mix-lingual data set can achieve better results.

The rest of the paper is organized as follows. Section 2 describes the related work of this study. Framework and data descriptions are in Sect. 3 and Sect. 4, and experiments in Sect. 5. Finally, we conclude the work in Sect. 6.

## 2 Related Work

### 2.1 Data Sets for the CS TTS

There are three kinds of data set for CS model training listed as follows.

**Bilingual Data Set from a Bilingual Speaker:** [4] implemented a polyglot TTS system using a multi-lingual database recorded by a multi-lingual speaker, which required four languages lacked in foreign accents. A bilingual speech corpus recorded by a professional female speaker was used in [5]. [6] described a bilingual Chinese and English parametric neural TTS system trained on an imbalanced data set from a bilingual speaker who is a Chinese native speaker with average level English skills. However, bilingual speakers with professional skills of two languages are hard to find, and the data sets recorded by bilingual speakers are expensive to collect.

**Monolingual Data Set from Native Speakers:** [7] implement a CS TTS system for Hindi-English and German-English using monolingual data. However, there are different performance between English and German. [8] built an end-to-end CS TTS system with a mixture of monolingual data sets recorded by different speakers. However, the system performed poorly on English text in the synthesis with the Mandarin speaker's voice. [9] also used the same type of data set. However, this method can not disentangle speaker identity from the language very well, which makes it difficult to keep speaker consistency.

**Mix-lingual Data Set:** This is a feasible method to resolve these problems. [10] applied a mix-lingual data set recorded by a bilingual speaker. However, the data set is with a high LMR, which is the ratio of the number of words from the embedded language to the number of words in the dominant language. However, the high LMR also means a high requirement of a bilingual skill for speakers as well as a relatively high cost.

### 2.2 Text Representation for CS TTS

Input text representation is an essential factor because the model needs to learn different linguistic features simultaneously. A direct method is mapping acoustically similar phonemes from the foreign language into native language [11], but mismatches often occur when there is no one-to-one mapping. An statistical parametric speech synthesis (SPSS) based TTS system [12] and a cross-lingual multi-speaker TTS system [13] mainly uses IPA [14] as input representations to cover pronunciation of different languages. Similarly, there is no strict one-to-one mapping between the IPA symbols of two distinct languages. [15] use allophones, a kind of conceptual unit smaller than phoneme units, to provide more chances of unit sharing between Mandarin and English. However, this method will make the text processing front-end overly complex.

An alternative method is using different text representations for different languages. [16] used the radical, a kind of basic unit in Mandarin writing system, for Mandarin and kana for Japanese in the ASR system to fill the gap between two languages. This study suggests that different representations can be employed for different languages. Diverse representations were used in [17,18] to generate linguistic features respectively and combine them into a uniform or universal input features. [19] used different representations, however, auxiliary information was used. End-to-end models simplify input representations because the encoder can learn linguistic features automatically. [20] used diverse representations to build a cross-lingual TTS system. [8] used pinyin for Mandarin and alphabet for English, and [9] used pinyin and CMU phonemes to build CS TTS systems. However, The amount of Chinese and English of the data set used in these methods is relatively balanced.

# 3 Proposed Method

## 3.1 General Framework

The general framework is shown in Fig. 1. The CS text is first fed into a CS front-end to transform input text into diverse representations. In the synthesis module, the derived output from CS front-end is transformed into waveforms. We describe more details in the following subsections.

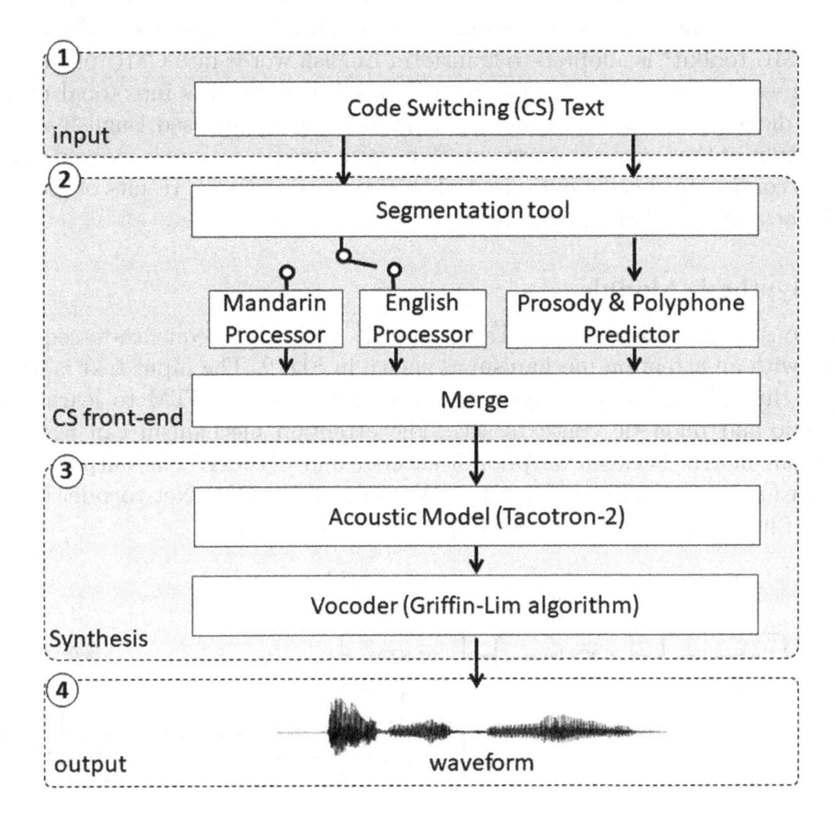


**Fig. 1.** A general framework of the proposed method (In the CS front-end, the diverse representations from Mandarin processor and English processor will be merged according to specified options.)

### 3.2 CS Front-End

Our front-end to process input CS text is shown in Fig. 1. The input text is processed by a segmentation tool to segment text and generate part-of-speech tagging. The result is fed into the polyphone predictor and the prosody predictor in parallel, and outputs the predicted results, respectively. An LSTM-based polyphone predictor is used to predict the pronunciations of polyphones in the text, and a conditional random field (CRF) based prosody predictor to obtain prosody information.

A discriminator is used to distinguish Mandarin characters and English words from the segmentation results and then sends them into the Mandarin processor and English processor, respectively. For the Mandarin processor, a dictionary is used to produce tonal pinyin from Mandarin characters. For the English processor, four representations can be produced. A modified CMU phonemes dictionary from CMU toolkit[2] is adopted to transform English words into CMU phonemes. English words can also be transformed from CMU phonemes into tonal pinyin with a dictionary. It also outputs the alphabet and uppercased English words. The output of these diverse representations from two processors is merged differently according to specific options, and then merged with the results of prosody predictor and polyphone predictor.

### 3.3 Synthesis Module

Our acoustic model is based on Tacotron-2 [1], which is a sequence-to-sequence model with an attention mechanism as shown in Fig. 2. The input text is firstly passed through three CNN layers and then fed into a BLSTM to learn both linguistic and phonetic contexts [21]. The attention mechanism can learn an alignment matrix between outputs of encoder and decoder. The output of the decoder forms 80-dim log-mel spectra. We replaced the WaveNet vocoder by the Griffin-Lim algorithm [22] to speed up the work process.

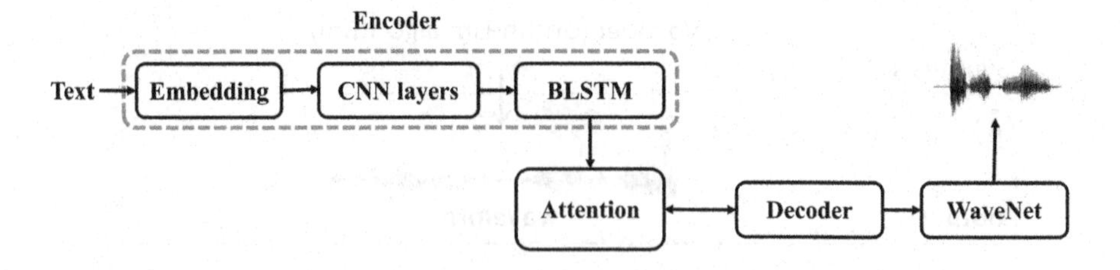


**Fig. 2.** Overview of the Structure of Tacotron-2

For the input text, we use characters as modeling units for Mandarin and English. Both Chinese and English characters are universally defined into a compact unit set, including twenty-six lowercase letters, twenty-six capitalized letters, ten numbers, space, special symbols, and several punctuation marks.

[2] http://www.speech.cs.cmu.edu/cgi-bin/cmudict.

## 4 Data Description

To ensure the reduction of costs of recording data set and meanwhile the speaker consistency, we build a mix-lingual data set with a low LMR in particular. In this data set, the Mandarin is the dominant language, while about 15% – 20% sentences are embedded with short syllable English words. Every CS sentence contains one or two CS points. In daily communication, non-native people can handle short syllable English words, even their English skills are not excellent. Thus, in our mix-lingual data set, all the English words are acronyms and short syllable words, which does not require a speaker with native English skills, meanwhile can minimize the labor because of the low LMR.

Our mix-lingual data set is uttered by three native Mandarin speakers, including two males and a female. All of them can speak simple English words like 'NBA', 'nice' and 'OK' fluently. The text materials are from three different text files. Each text file contains 15% – 20% CS sentences, and each CS sentence has one or two CS points. The majority of the embedded English words are acronyms and short words. The total time of the three data sets was about 12 h. We name these three data set as the internal-1, internal-2, and internal-3. The statistical information of these data sets is listed in Table 1. The LMR of these three data sets is between 0.72% and 1.18%, which is much smaller than 9.42% – 23.92% from another English-Hindi mix-lingual data [10].

**Table 1.** Statistics of mix-lingual data set (The En is the ratio of the number of characters from the English words over the number of the total characters.)

| Data set | Gender | #Sent | CS(%) | LMR(%) | En(%) |
|---|---|---|---|---|---|
| Internal-1 | Female | 2532 | 15.7 | 0.76 | 1.97 |
| Internal-2 | Male | 4048 | 24.41 | 1.18 | 4.63 |
| Internal-3 | Male | 6500 | 15.35 | 0.72 | 1.91 |

Since pre-training a model with a large amount of data from rich-resource language can enhance the performance of the TTS model for a low-resource language [23]. We also employ monolingual Mandarin data from Data Baker Technology[3] from a female to pre-train the model. A subset of the LJ Speech dataset,[4] an open-sourced female English data, also been used to investigate the effectiveness of data augmentation with the monolingual English data. The duration of the subset is about 10 h.

Our test data are 150 CS sentences crowed from the Internet, including news, talking show, and reading books. Most of the sentences contain acronyms and short words, and other sentences contain long English words and phrases.

[3] https://www.data-baker.com/hc_znv_1_en.html.
[4] https://keithito.com/LJ-Speech-Dataset/.

# 5 Experiments

We pre-train our model with a 10 h Mandarin female data set, and fine-tune them with mix-lingual data sets. A pure English data set also been used in our experiments aiming to improve the performance of code-switching synthesizing. The results of the combination of the representations mentioned in Sect. 3.2 were evaluated and analyzed with the MOS score and AB-test.

## 5.1 Input Representations

For the Mandarin representation we use tonal pinyin. All the pinyin symbols are lowercase ending with tone markers 1,2,3 and 4 for cardinal tones and 5 for none-tone.

Four methods were attempted to describe English representations. The alphabet is the default input representation, and the second is the uppercase for all English characters, assuming that this method may help the model distinguish the Mandarin text and English text better. Thirdly, because of the low LMR, all the English words transformed into pinyin with phonemes mapping method, with tone 1 for ending all the words. Finally, a CMU dictionary is used to transform English words into phonemes, with all phonemes in upper-case. Examples of the representations are shown in Table 2.

**Table 2.** An example of representation

| Text | Representation | Example |
|---|---|---|
| English (asking) | Alphabet (AP) | asking@ |
| | Uppercase (UP) | ASKING@ |
| | Pinyin (PY) | ask$ in1 |
| | CMU-phonemes (PH) | AE S K$ IN NG@ |
| Mandarin (ben:book) | tonal pinyin (PY) | ben3 |

To synthesize more expressive speech, prosody information was added into sentences. For the whole sentence, symbols '#1', '#2', '#3' and '#4' were used to indicate segmentation, pause, silence, and end of the sentence, respectively. '$' is to indicate a syllable in CMU phonemes and pinyin transformed from the CMU, and '@' is for the end-of-word symbol in all four English representations. There is a '#1' after every English word.

These representations from Mandarin and English were investigated as to which generates better results on the imbalanced data set. For diverse representations, we combine them in three ways: tonal pinyin and alphabet, and tonal pinyin and uppercase, tonal pinyin and CMU phonemes. For the same representation, we use tonal pinyin for both Mandarin and English. We also design a contrast experiment that pure English data used in the training stage to investigate the effect of English data on the final results.

### 5.2 Experimental Setup

Because the dominant language in the mix-lingual data-set is Mandarin, a pure Mandarin data set was used to pre-train our model for 100k steps. With this pre-trained model, some linguistic features could be learned from Mandarin text, which may help to learn English phoneme features and initialize decoder parameters from acoustic features [24]. Three mix-lingual data sets are processed to fine-tune the model one by one to ensure speaker consistency. The female mix-lingual data set (internal-1) was used first, the smaller male mix-lingual data set (internal-2) to follow, and the larger male mix-lingual (internal-3) used at last. However, to investigate the effect of English data on the final results, we use a pure English data before internal-3 in the contrast experiments. Every data set trained for 20k steps after pre-training.

We train our model on one GTX-1080, with the batch-size set as 16. The learning rate starts at 0.001. For the pre-trained model, the learning rate decays starts at 50k steps and halves for every 10k. Inspired by [25], the learning rate will be reset when fine-tuning new data, starts decay after 10k steps training, and halves for every 2k steps. Other hyper-parameters are the same as the Tacotron-2 default, but the LSTM units of decoder were set as 512 to keep the model sizes compressed.

### 5.3 Experimental Results

We use four types of text representations and augmented them by a pure English data. They are listed as follows.

- PY-AP: Tonal pinyin for Mandarin and the alphabet for English as the baseline.
- PY-UP: Tonal pinyin for Mandarin and the upper-cased alphabet for English.
- PY-PH: Tonal pinyin for Mandarin, and the CMU phonemes for English.
- PY-PY: Tonal pinyin both for Mandarin and English.
- AUG: A pure English data set were used in the training stage to improve the intelligibility of CS speech.

**Evaluation 1:** We use the MOS score to evaluate the results of different input text representations[5]. All the samples were presented at the 16k Hz sampling rate. The raters knew that the results were derived from different experiments. At least ten raters were involved in each experiment. The raters were asked to score the naturalness and intelligibility on a scale from 1 to 5, 1 means the poorest results to understand, and 5 is the best result, with 0.5 points increments for every level.

The results of the MOS score are shown in Fig. 3. In the baseline PY-AP, the model fails to synthesize correct English words or acronyms in most cases. The reason is that the difference between these two representations is not significant, and the data is not enough. The model can not distinguish text from two

[5] Some samples are available in "https://pandagst.github.io/".

languages and learn English linguistic features very well. Still, the results of the baseline are better than the method PY-UP. This is different from our hypothesis, and we will search for the reason for future work. Besides, the pronunciation of English words was influenced by Chinese pronunciation firmly in these two methods.

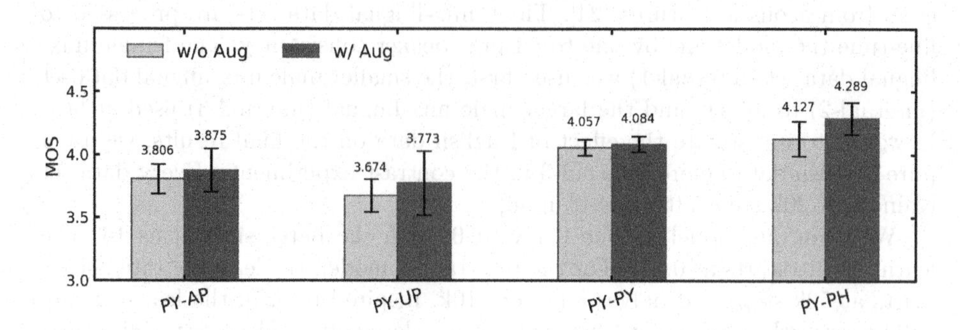


**Fig. 3.** MOS score ratings with 95% confidence intervals of diverse representations

The method PY-PH is much better than the previous two. The model can not only synthesize acronyms and short syllable words, but most of the longer words or even phrases can also be synthesized. However, the model fails to synthesize very long words and phrases wholly and correctly. The reason is that all the English words in the training data are simple and short. The model can learn the pronunciation of a few phonemes sequence, but can not handle the joint pronunciation in long English words. However, the method PY-PY performs worse than PY-PH, this is because of many mismatches between pinyin and CMU phonemes, and pinyin cannot represent all of CMU phonemes correctly. The results of AUG show better performance in long words and phrases than the method without data augmentation.

**Evaluation 2:** The AB-test is used between with and without data augmentation among PY-AP, PY-UP, PY-PH and PY-PY to investigate whether pure English monolingual data effects the final result. The AB-test is also used between with and without pure mandarin pre-train. For each test case, a paired samples with the same text were played in a sound-proof studio, and ten bilingual listeners joined the listening test. They give answers whether they prefer A or B according to the quality of CS speech. 'NP' will be selected when there is no significant difference between paired samples.

The results from the AB-test in Fig. 4 show that the listeners prefer the methods with AUG in comparison to those without AUG. This result indicates that the performance excels when the English data is adopted into the training data. However, for PY-PH, about 68.67% were selected as 'NP' because the method with this representation gives better results than other methods, and listeners cannot find a significant difference between the results with and without data augmentation. The experiment also shows that the influence of pre-training

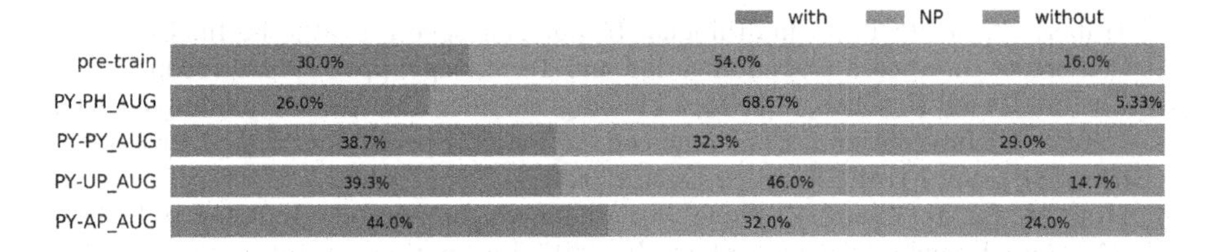


**Fig. 4.** AB-test result

with pure Mandarin data is not substantial. This is because Mandarin is the dominant language in the mix-lingual data set (almost 10 h), which is capable enough to train a TTS model [24].

## 6 Conclusion

In this paper, a method for CS TTS to reduce the cost of data recording meanwhile maintain the speaker consistency was proposed, where the CS model was trained by using an imbalanced mix-lingual data set recorded by native speakers. The combination of different text representations was conducted in the experiments. The results showed that the high quality CS speech could be synthesized by the representations using tonal pinyin for Mandarin and CMU phonemes for English, even if the training data only contains a small quantity of CS sentences. The CS result will be better when there is a data augmentation with pure English data also being demonstrated. In the future, we will study how to use the alphabet for English to simplify the front-end and train the CS model when training data is deficient and imbalanced.

**Acknowledgments.** This work was supported in part by the National Natural Science Foundation of China under Grant 61771333, the Tianjin Municipal Science and Technology Project under Grant 18ZXZNGX00330 and JSPS KAKENHI Grant No. 19K24376 and NICT international fund 2020 "Bridging Eurasia: Multilingual Speech Recognition along the Silk Road", Japan.

## References

1. Shen, J., et al.: Natural TTS synthesis by conditioningwavenet on MEL spectrogram predictions. International Conference on Acoustics. Speech and Signal Processing (ICASSP), pp. 4779–4783. IEEE, Calgary (2018)
2. Ping, W., Peng, K., Gibiansky, A., Arik, S.O., Kannan, A., Narang, S., Jonathan, R., Miller, J.: Deep voice 3: Scaling text-to-speech with convolutional sequence learning. In: 6th International Conference on Learning Representations (ICLR), Vancouver (2018)
3. Zhang, Y., et al.: Learning to speak fluently in a foreign language: multilingual speech synthesis and cross-language voice cloning. In: 20th International Speech Communication Association (INTERSPEECH), pp. 2080–2084. ISCA, Graz (2019)

4. Traber, C., et al.: From multilingual to polyglot speech synthesis. In: European Conference on Speech Communication and Technology, pp. 835–839 (1999)
5. Chu, M., Peng, H., Zhao, Y., Niu, Z., Chang, E.: Microsoft Mulan - a bilingual TTS system. In: International Conference on Acoustics, Speech, and Signal Processing (ICASSP), pp. I-I. IEEE, Hong Kong (2003)
6. Ming, H., Lu, Y., Zhang, Z., Dong, M.: A light-weight methodof building an LSTM-RNN-based bilingual TTS system. In: 2017 International Conference on Asian Language Processing (IALP), pp. 201–205. IEEE, Singapore (2017)
7. Sitaram, S., Rallabandi, S.K., Rijhwani, S., Black, A.W.: Experiments with cross-lingual systems for synthesis of code-mixed text. In: SSW, pp. 76–81 (2016)
8. Cao, Y., et al.: End-to-end code-switched tts with mix of monolingual recordings. In: 2019 IEEE International Conference on Acoustics. Speech and Signal Processing (ICASSP), pp. 6935–6939. IEEE, Brighton (2019)
9. Xue, L., Song, W., Xu, G., Xie, L., Wu, Z.: Building a mixed-lingual neural tts system with only monolingual data. In: 20th International Speech Communication Association (INTERSPEECH), pp. 2060–2064. ISCA, Graz (2019)
10. Chandu, K.R., Rallabandi, S.K., Sitaram, S., Black, A.W.: Speech synthesis for mixed-language navigation instructions. In: 18th International Speech Communication Association (INTERSPEECH), pp. 57–61. ISCA, Stockholm (2017)
11. Campbell, N.: Talking foreign-concatenative speech synthesis and the language barrier. In: 7th European Conference on Speech Communication and Technology (EUROSPEECH), pp. 337–340. ISCA, Aalborg (2001)
12. Zen, H., Braunschweiler, N., Buchholz, S., Gales, M.J., Knill, K., Krstulovic, S., Latorre, J.: Statistical parametric speech synthesis based on speaker and language factorization. In: IEEE Transactions on Audio, Speech, and Language Processing, vol. 20, no. 6, pp. 1713–1724. IEEE (2012)
13. Chen, M., et al.: Cross-lingual, multi-speaker text-to-speech synthesis using neural speaker embedding. In: 20th International Speech Communication Association (INTERSPEECH), pp. 2105–2109. ISCA, Graz (2019)
14. Spa, J.: Handbook of the international phonetic association. a guide to the use of the international phonetic alphabet. Word-J. Int. Ling. Assoc. **53**(3), 421–424 (2002)
15. Qian, Y., Cao, H., Soong, F.K.: HMM-based mixed-language (Mandarin-English) speech synthesis. In: Proceedings of the 2008 6th International Symposium on Chinese Spoken Language Processing, pp. 1–4. IEEE, Kunming (2008)
16. Li, S., Lu, X., Ding, C., Shen, P., Kawahara, T.: investigating radical-based end-to-end speech recognition systems for chinese dialects and japanese. In: 20th International Speech Communication Association (INTERSPEECH), pp. 2200–2204. ISCA, Graz (2019)
17. Li, B., Zen, H.: Multi-language multi-speaker acoustic modeling for lstm-rnn based statistical parametric speech synthesis. In: 17th International Speech Communication Association (INTERSPEECH), pp. 2468–2472. ISCA, San Francisco (2016)
18. Yu, Q., Liu, P., Wu, Z., Ang, S.K., Meng, H., Cai, L.: Learning cross-lingual information with multilingual BLSTM for speech synthesis of low-resource languages. IEEE International Conference on Acoustics. Speech and Signal Processing (ICASSP), pp. 5545–5549. IEEE, Shanghai (2016)
19. Sitaram, S., Black, A.W.: Speech synthesis of code-mixed text. In: 10th International Conference on Language Resources and Evaluation (LREC), pp. 3422–3428. ELRA, Portoroz (2016)

20. Chen, Y., Tu, T., Yeh, C., Lee, H.Y.: End-to-end text-to-speech for low-resource languages by cross-lingual transfer learning. In: 20th International Speech Communication Association (INTERSPEECH), pp. 2075–2079. ISCA, Graz (2019)
21. Mametani, K., Kato, T., Yamamoto, S.: Investigating context features hidden in End-to-End TTS. In: IEEE International Conference on Acoustics. Speech and Signal Processing (ICASSP), pp. 6920–6924. IEEE, Brighton (2019)
22. Griffin, D.W., Lim, J.S.: Signal estimation from modified short-time fourier transform. In: IEEE International Conference on Acoustics. Speech, and Signal Processing, pp. 804–807. IEEE, Boston (1983)
23. Lee, Y., Shon, S., Kim, T.: Learning pronunciation from a foreign language in speech synthesis networks. In: arXiv preprint arXiv:1811.09364, (2018)
24. Chung, Y.A., Wang, Y., Hsu, W.N., Zhang, Y., Skerry-Ryan, R.J.: Semi-supervised training for improving data efficiency in end-to-end speech synthesis. In: 2019 IEEE International Conference on Acoustics. Speech and Signal Processing (ICASSP), pp. 6940–6944. IEEE, Brighton (2019)
25. Li, B., Zhang, Y., Sainath, T., Wu, Y., Chan, W.: Bytes are all you need: End-to-end multilingual speech recognition and synthesis with bytes. In: 2019 IEEE International Conference on Acoustics. Speech and Signal Processing (ICASSP), pp. 5621–5625. IEEE, Brighton (2019)

# Image Processing and Computer Vision

# A Feature Fusion Network for Multi-modal Mesoscale Eddy Detection

Zhenlin Fan, Guoqiang Zhong(✉), and Haitao Li

Department of Computer Science and Technology, Ocean University of China, Qingdao 266100, China
916056589@qq.com, {gqzhong,lihaitao}@ouc.edu.cn

**Abstract.** As a marine phenomenon, mesoscale eddies have important impacts on global climate and ocean circulation. Many researchers have devoted themselves to the field of mesoscale eddy detection. In recent years, some methods based on deep learning for mesoscale eddy detection have been proposed. However, a major disadvantage of these methods is that only single-modal data are used. In this paper, we construct a multi-modal dataset containing sea surface height (SSH), sea surface temperature (SST) and velocity of flow(VoF), which are useful for mesoscale eddy detection. Moreover, we propose a feature fusion network named FusionNet, which consists of a downsampling stage and an upsampling stage. We take ResNet as backbone of the downsampling stage, and achieve multi-scale feature maps fusion via vertical connections. Additionally, dilated convolutions are applied in the FusionNet to aggregate multi-scale contextual information. Experimental results on the constructed multi-modal mesoscale eddy dataset demonstrated the superiority of FusionNet over previous deep models for mesoscale eddy detection.

**Keywords:** Mesoscale eddy detection · FusionNet · Multi-modal data

## 1 Introduction

Nowadays, deep learning [1] has been a popular approach to solve many practical problems, such as image classification, object recognition and semantic segmentation [2–4]. Among them, semantic segmentation has made significant progress since convolutional networks are applied in this area [5–8]. Specifically, the semantic segmentation methods generally work on the pixel level, in other words, the methods assign each pixel in the image to a class. Motivated by the idea of semantic segmentation, we propose a semantic segmentation method to tackle the mesoscale eddy detection problem.

In the ocean, mesoscale eddies are known as weather-type ocean eddies, the radius of which is 100 to 300 km and the life span of which is 2 to 10 months. There are two types of mesoscale eddies: cyclonic eddies (counterclockwise rotation in the northern hemisphere) and anti-cyclonic eddies (counterclockwise rotation in the southern hemisphere). Mesoscale eddies play an important role in


H. Yang et al. (Eds.): ICONIP 2020, LNCS 12532, pp. 51–61, 2020.
https://doi.org/10.1007/978-3-030-63830-6_5

global climate and marine ecosystem, so that have been widely studied for many years. In the early days, mesoscale eddies are directly identified by experts, resulting in a lot of time wasting and low accuracy. With the continuous improvement of observation accuracy and resolution of the satellite data, mesoscale eddy detection has entered a new stage. Many methods using mathematical or physical knowledge to characterize the structure of mesoscale eddies have been proposed, but the performance of these methods are still in low accuracy [9–11]. The popularity of deep learning provides another way for mesoscale eddy detection. Hence, some deep learning models have been put forward for mesoscale eddy detection. However, the architectures of these networks are simple. Furthermore, only single-modal data are used to detect mesoscale eddies, which lack the data of other modals beneficial to mesoscale eddy detection.

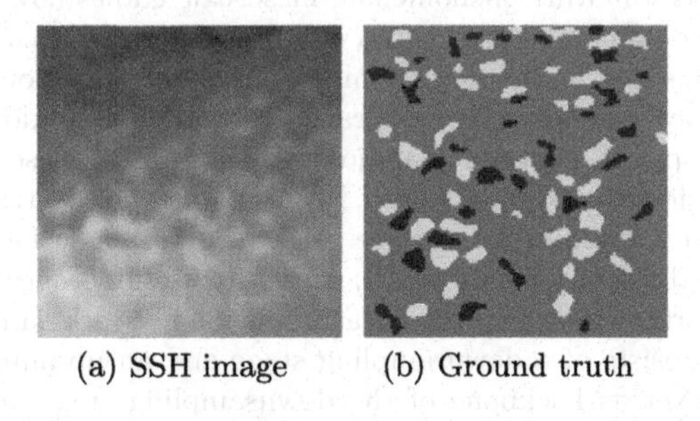

(a) SSH image (b) Ground truth

**Fig. 1.** The examples of our constructed dataset. (a) The SSH image; and (b) The ground truth based on the SSH image.

To address the above challenges of mesoscale eddy detection, in this work, we propose a feature fusion network named FusionNet to fuse multi-modal data and detect mesoscale eddies. Concretely, we construct a multi-modal dataset composed of sea surface height (SSH), sea surface temperature (SST) and velocity of flow(VoF) to overcome the single-modal data problem. Considering that most deep learning approaches only use SSH to detect mesoscale eddies, we invite experts to annotate the SSH images as ground truth of our dataset. The examples of our dataset are shown in Fig. 1, where the SSH image is shown in Fig. 1(a) and the ground truth is shown in Fig. 1(b). In the ground truth image, experts define the yellow areas as the anti-cyclonic eddies, the dark blue areas as the cyclonic eddies, and the light blue areas as the area without mesoscale eddies. Furthermore, we have designed a feature fusion network called FusionNet for mesoscale eddy detection, which is composed of a downsampling stage and an upsampling stage. In the downsampling stage, we use ResNet-101 as the backbone to get multi-scale feature maps with rich semantic information. In the upsampling stage, the feature maps with rich semantic information are fused with the feature maps of high resolution by vertical connections, finally obtaining the feature maps with both rich semantic and contextual information. Moreover, we apply dilated convolutions to FusionNet to attain multi-scale contextual information. Therefore, our method tackles the current problems of mesoscale eddy detection from both datasets and network architectures, getting improvement of the current mesoscale eddy detection results.

## 2 Related Work

In this section, we introduce existing mesoscale eddy detection algorithms from two aspects: non-deep learning algorithms and deep learning algorithms.

### 2.1 Non-deep Learning Algorithms

In the early days, mesoscale eddies were detected by experts' subjective annotation, leading to inefficient and inaccurate detection. To obtain better mesoscale eddy detection results, some automatic mesoscale eddy detection algorithms began to appear [12]. Thonet *et al.* used the thermal difference of ocean currents to establish a mesoscale eddy extraction technique [13]. Peckinpaugh and Holyer utilized the Hough transformed circle identification method [14] to perform edge detection on the remote sensing images [9]. However, the edge curves of the mesoscale eddies are not standard circles, so that this algorithm is not particularly rigorous. To tackle this problem, Ji *et al.* designed a mesoscale eddy detection algorithm based on the ellipse identification method [15]. Besides, many algorithms making use of sea surface height or sea surface temperature information to detect mesoscale eddies have also been proposed [11,16–19].

However, these non-deep learning algorithms are rough in the selection of used data and algorithms, resulting in low accuracy of mesoscale eddy detection.

### 2.2 Deep Learning Algorithms

Due to the great success of deep learning in recent years, scientists began to solve mesoscale eddy detection problems from the perspective of deep learning. So far, not many deep learning algorithms for mesoscale eddy detection have been proposed. Lguensat *et al.* designed the EddyNet to segment mesoscale eddies based on SSH images [20]. Although this method applies deep learning to mesoscale eddy detection, the architecture of EddyNet based on U-Net [21] is simple. In addition, they only chose SSH, not taking data of other modals that are beneficial for mesoscale eddy detection into account. To deepen the network, Xu *et al.* took advantage of the pyramid scene parsing network (PSPNet) [22] to segment mesoscale eddies [23]. However, only SSH was used. With the improvement of resolution of synthetic aperture radar (SAR) images, Du *et al.* designed the DeepEddy to classify SAR images [24]. They took the principal component analysis network (PCANet) [25] as the backbone for DeepEddy, and applied spatial pyramid pooling (SPP) [26] to this network. Nevertheless, this algorithm just achieves a two-class classification task, not segmenting the shape of mesoscale eddies from SAR images.

Consequently, the main problems of existing deep learning algorithms for mesoscale eddy detection are that the networks are simple and the used data are in single-modal. Therefore, the proposed FusionNet based on multi-modal data overcomes these difficulties, achieving better performance than other existing deep learning algorithms for mesoscale eddy detection.

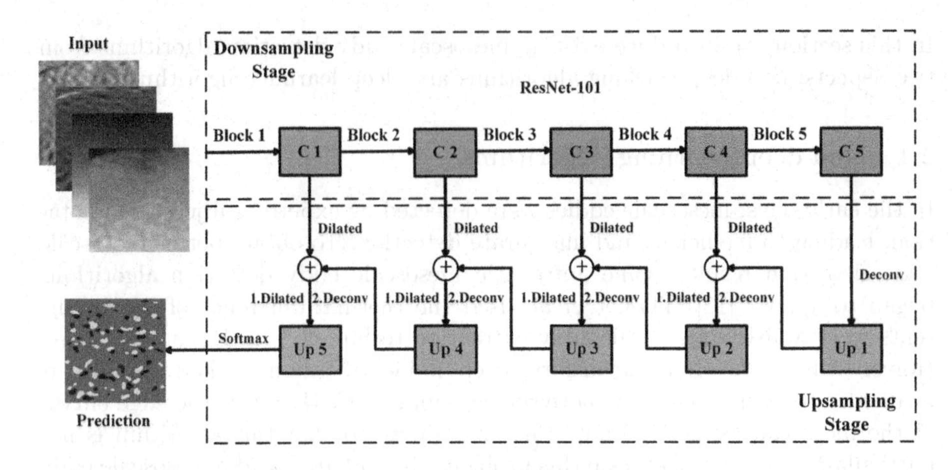


**Fig. 2.** The structure of FusionNet.

## 3 Methodology

In this section, we firstly introduce the structure of FusionNet in detail. As shown in Fig. 2, the FusionNet contains a downsampling stage and an upsampling stage, achieving multi-scale feature maps fusion from these two stages via vertical connections. Then, we introduce the loss function for the training of FusionNet.

### 3.1 FusionNet

**The Downsampling Stage.** Due to the great success of convolutional networks in feature extraction, we choose ResNet-101 as the backbone for the downsampling stage. As a deep network, ResNet-101 has been successfully used to solve many problems. ResNet mainly contains 5 convolutional blocks, each of which consists of a series of convolutional layers. The size of feature maps is halved when performing a new convolutional block.

In the downsampling stage, we specify the output of each convolutional block as C1, C2, C3, C4 and C5, respectively. As convolutional blocks are performed continually, the semantic information of the feature maps become richer and richer. However, the resolution of the feature maps decreases, causing the loss of contextual information of the feature maps. Therefore, we come up with an idea that fuses the feature maps of rich contextual information with the feature maps of rich semantic information, which is described in the next subsection.

In summary, we can get multi-scale feature maps based on C1, C2, C3, C4 and C5 in the downsampling stage. These multi-scale feature maps will be fused with the feature maps from the upsampling stage to obtain feature maps with rich semantic and contextual information.

**The Upsampling Stage and Vertical Connections.** After going through the downsampling stage, C5 gets rich semantic information and loses a lot of contextual information. Therefore, Up1 is obtained by upsampling C5 with a factor 2 for recovering lost contextual information. However, the lost contextual information can not be recovered fully. Thus, the fusion of Up1 and C4 with the same size via vertical connection provides a solution to this problem.

From Fig. 2, we can see the detail of vertical connections. Firstly, C4 is performed a $3 \times 3$ dilated convolution with a rate of 4 to change channels. Besides, dilated convolution is used to gain more contextual information. Then, we add the Up1 to the feature maps obtained by performing dilated convolution on C4 element by element. Finally, the feature maps obtained by addition are performed a $3 \times 3$ dilated convolution with a rate of 4 again to get contextual information.

In this manner of vertical connections, we fuse the feature maps from downsampling stage with the feature maps from upsampling stage continuously. We end the fusion when we get the Up5. We apply a softmax layer to Up5 with rich semantic and contextual information to classify each pixel of the feature maps, segmenting mesoscale eddies in high accuracy.

### 3.2 The Loss Function

The loss function used for training the proposed FusionNet combines the dice loss function and the cross-entropy loss function, which is defined as:

$$L(P, G) = -log(1 - DL(P, G)) + L_{log}(P, G), \tag{1}$$

where $DL(P, G)$ is the dice loss function, and $L_{log}(P, G)$ is the cross-entropy loss function.

As aforementioned, we use the idea of semantic segmentation to detect mesoscale eddies. As a loss function widely used in semantic segmentation problems, the dice loss function is chosen as the part of our loss function. In the dice loss function, there is a dice coefficient that defines the similarity of two samples. The formula of the dice coefficient is defined as:

$$DC(P, G) = \frac{2|P \bigcap G|}{|P| + |G|}, \tag{2}$$

where $P$ and $G$ refer to the prediction and the ground truth, respectively. $|P|$ and $|G|$ represent the sums of elements in $P$ and $G$, respectively. According to the formula of the dice coefficient, we know that the degree of similarity between two samples increases as the dice coefficient increases. Therefore, the dice loss function is defined as:

$$DL(P,G) = 1 - DC(P,G) = 1 - \frac{2|P \bigcap G|}{|P| + |G|}. \quad (3)$$

However, it is difficult to train our network based on the dice loss function alone. Our proposed mesoscale eddy detection method, which classify each pixel, is actually a multi-class classification problem. Hence, the cross-entropy loss function $L_{log}(P,G)$ is added to the loss function to optimize our network.

At last, the loss function combining the dice loss function and the cross-entropy loss function is used to optimize our network, and deliver good results for mesoscale eddy detection.

# 4 Experiments

In this section, we introduce the conducted experiments in detail. In Sect. 4.1, the building of our multi-modal dataset is described in detail. In Sect. 4.2, we compare the proposed FusionNet with the state-of-the-are mesoscale eddy detection approachs.

## 4.1 The Multi-modal Dataset

In order to solve the problem of insufficient datasets in mesoscale eddy detection, a multi-modal dataset was collected. In addition to the SSH widely used in current mesoscale eddy detection methods, the SST and the VoF were added to the multi-modal dataset. In the multi-modal dataset, the SSH and SST occupy one channel respectively. However, there are two channels occupied by the VoF, because the VoF contains both the east/west direction and north/south direction. The SSH, SST and VoF come from the same time (2000–2009) and the same sea area. We randomly selected 128 × 128 patches from the data of these three modals as the input data of mesoscale eddy detection models, each channel of which was from the same time and the same sea area. Besides, we invited experts to annotate the SSH images for convenient comparison, since existing deep models generally use SSH to detect mesoscale eddies.

To the end, 512 patches and 256 patches were selected as training and test sets of our multi-modal mesoscale eddy detection dataset, respectively. Although our collected dataset is small, we can see that the small amount of data are enough to prove the superiority of our method based on the experimental results in Sect. 4.2.

## 4.2 Experimental Results

**Comparison on Different Modals of Data.** To test the effectiveness of the collected multi-modal dataset for mesoscale eddy detection, we trained our network on different modals of data of training set, i.e. the SSH, SST, VoF and multi-modal data. Then, we tested it on the corresponding four modals of data of

**Table 1.** Accuracy using our proposed FusionNet on different modals of data of test set. The best result is highlighted in boldface.

| Data | Accuracy |
|---|---|
| SSH | 98.09% |
| SST | 93.29% |
| VoF | 97.70% |
| Multi-modal data | **98.47%** |

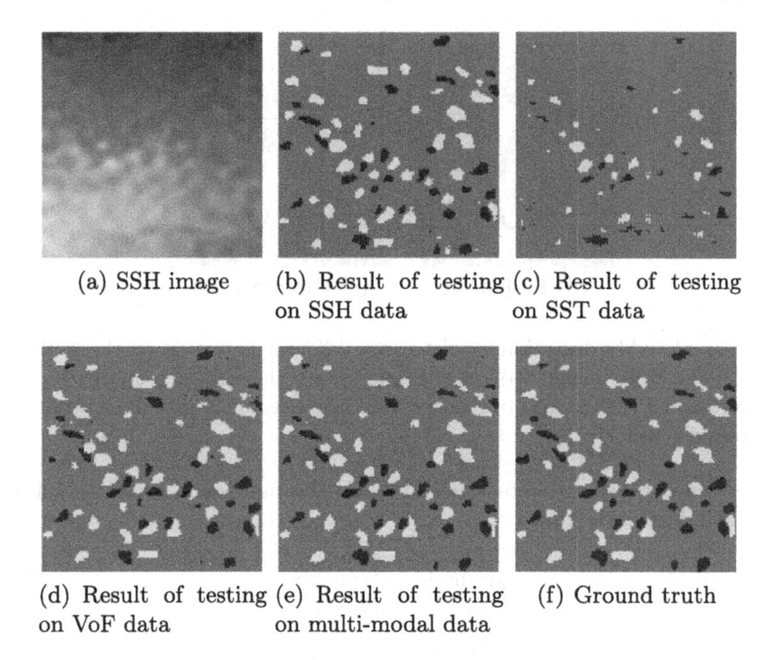

(a) SSH image (b) Result of testing on SSH data (c) Result of testing on SST data

(d) Result of testing on VoF data (e) Result of testing on multi-modal data (f) Ground truth

**Fig. 3.** Mesoscale eddy detection results using our proposed FusionNet on different modals of data of test set. (a) The SSH image. (b) (c) (d) (e) The mesoscale eddy detection results of testing on the SSH, SST, VoF and the multi-modal data, respectively. (f) The ground truth.

test set. The accuracy using our proposed FusionNet on different modals of data of test set is shown in Table 1. From Table 1, we can see that the result based on our multi-modal data is superior to that based on other three single-modal data.

We show mesoscale eddy detection results using our proposed FusionNet on different modals of data of test set in Fig. 3. From Fig. 3, we can see that the results based on the multi-modal data is closest to the ground truth. Hence, compared to other single-modal data, the collected multi-modal data can improve the performance of mesoscale eddy detection.

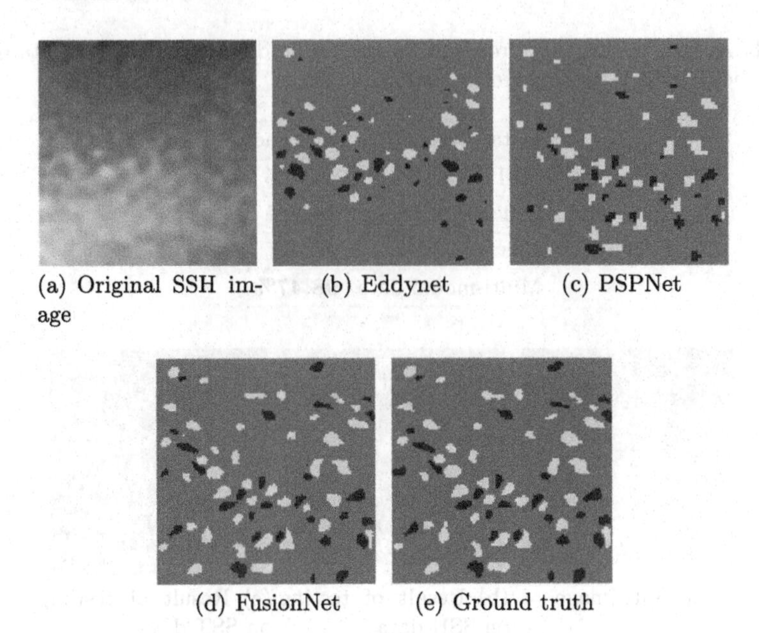

(a) Original SSH image (b) Eddynet (c) PSPNet

(d) FusionNet (e) Ground truth

**Fig. 4.** Mesoscale eddy detection results using different networks on the multi-modal test set. (a) The SSH image. (b) (c) (d) The mesoscale eddy detection results using the EddyNet and PSPNet and our proposed FusionNet on the multi-modal test set. (e) The ground truth.

**Table 2.** Accuracy using different networks on the multi-modal test set. The best result is highlighted in boldface.

| Method | Accuracy |
|---|---|
| Eddynet | 93.77% |
| PSPNet | 96.25% |
| FusionNet | **98.47%** |

**Comparison on Different Networks.** Among the existing mesoscale eddy detection methods, the EddyNet and PSPNet use semantic segmentation to detect mesoscale eddies. However, the DeepEddy can only achieve classification of the SAR images, so that we only chose the EddyNet and PSPNet to compare with our proposed FusionNet. We show the accuracy using different networks based on the multi-modal test set in Table 2. Table 2 shows that our proposed FusionNet gains accuracy of 98.47% for mesoscale eddy detection, which is 4.7% higher than that of EddyNet and 2.22% higher than that of PSPNet.

Figure 4 shows the mesoscale eddy detection results using different networks based on the multi-modal test set, from which we can see that the results using the FusionNet is closer to the ground truth than that using the EddyNet and PSPNet.

**Table 3.** Accuracy using different networks on the four modals of data of test set. The best result is highlighted in boldface.

| Method | SSH | SST | VoF | Multi-modal data |
|---|---|---|---|---|
| Eddynet | 93.66% | 89.59% | 93.55% | 93.77% |
| PSPNet | 96.15% | 89.80% | 95.77% | 96.25% |
| FusionNet | 98.09% | 93.29% | 97.70% | **98.47%** |

Last but not the least, in order to demonstrate that the result using the proposed FusionNet based on the multi-modal dataset is the best, Table 3 shows the results using different networks on the four modals of data of test set, i.e., the SSH, SST, VoF and multi-modal data. From Table 3, we can see that no matter what network, the multi-modal data attains the best results and no matter what data, our proposed FusionNet attains the best results.

## 5 Conclusion

In this paper, we construct a multi-modal dataset for mesoscale eddy detection, which is composed of the SSH, SST and VoF data. Moreover, we propose a feature fusion network named FusionNet, which consists of a downsampling stage and an upsampling stage. In the FusionNet, We fuse the feature maps from downsampling stage with the feature maps from upsampling stage via vertical connections, obtaining the feature maps with rich semantic and contextual information. Additionally, dilated convolutions are used in the FusionNet to aggregate multi-scale contextual information. Consequently, we demonstrate the effectiveness of the proposed FusionNet for mesoscale eddy detection with extensive experiments.

**Acknowledgments.** This work was supported by the Major Project for New Generation of AI under Grant No. 2018AAA0100400, the National Natural Science Foundation of China (NSFC) under Grant No. 41706010, the Joint Fund of the Equipments PreResearch and Ministry of Education of China under Grand No. 6141A020337, the Open Project Program of Key Laboratory of Marine Hazards Forecasting, Ministry of Natural Resources, under Grant No. LOMF1802, and the Fundamental Research Funds for the Central Universities of China.

## References

1. LeCun, Y., Bengio, Y., Hinton, G.: Deep learning. Nature **521**(7553), 436–444 (2015)
2. Cireşan, D., Meier, U., Schmidhuber, J.: Multi-column deep neural networks for image classification. In: CVPR, pp. 3642–3649 (2012)
3. Girshick, R., Donahue, J., Darrell, T., Malik, J.: In: Rich feature hierarchies for accurate object detection and semantic segmentation. In: CVPR, pp. 580–587 (2014)

4. Long, J., Shelhamer, E., Darrell, T.: Fully convolutional networks for semantic segmentation. In: CVPR, pp. 3431–3440 (2015)
5. Chen, L., Papandreou, G., Kokkinos, I., Murphy, K., Yuille, A.L.: Semantic image segmentation with deep convolutional nets and fully connected crfs. In: ICLR (2015)
6. Chen, L.-C., Papandreou, G., Kokkinos, I., Murphy, K., Yuille, A.L.: Deeplab: semantic image segmentation with deep convolutional nets, atrous convolution, and fully connected crfs. IEEE Trans. Pattern Anal. Mach. Intell. **40**(4), 834–848 (2017)
7. Chen, L.-C., Papandreou, G., Schroff, F., Adam, H.: Rethinking atrous convolution for semantic image segmentation. arXiv preprint arXiv:1706.05587 (2017)
8. Chen, L.C., Zhu, Y., Papandreou, G., Schroff, F. and Adam, H.: Encoder-decoder with atrous separable convolution for semantic image segmentation. In: Proceedings of the European conference on computer vision (ECCV) pp. 801–818 (2018)
9. Peckinpaugh, S.H., Holyer, R.J.: Circle detection for extracting eddy size and position from satellite imagery of the ocean. IEEE Trans. Geosci. Remote Sens. **32**(2), 267–273 (1994)
10. Chelton, D.B., Schlax, M.G.: Global observations of oceanic rossby waves. Science **272**(5259), 234–238 (1996)
11. Fernandes, A., Nascimento, S.: Automatic water eddy detection in sst maps using random ellipse fitting and vectorial fields for image segmentation. In: Todorovski, L., Lavrac, N., Jantke, K.P., (eds.) Discovery Science. DS 2006. Lecture Notes in Computer Science, vol 4265. Springer, Berlin, Heidelberg (2006) https://doi.org/10.1007/11893318_11
12. Nichol, D.G.: Autonomous extraction of an eddy-like structure from infrared images of the ocean. IEEE Trans. Geosci. Remote Sens., (1), 28–34 (1987)
13. Thonet, H., Lemonnier, B., Delmas, R.: Automatic segmentation of oceanic eddies on AVHRR thermal infrared sea surface images. In: Challenges of Our Changing Global Environment. Conference Proceedings. OCEANS 1995 MTS/IEEE, vol. 2, pp. 1122–1127. IEEE (1995)
14. Illingworth, J., Kittler, J.: A survey of the hough transform. Computer Vision, Graphics, and Image Process. **44**(1), 87–116 (1988)
15. Ji, G., Chen, X., Huo, Y., Jia, T.: A automatic detection method for mesoscale eddies in ocean remote sensing image. Ocean and Lake **33**(2), 139–144 (2002)
16. Gairola, R., Basu, S., Pandey, P.: Eddy detection over southern indian ocean using topex/poseidon altimeter data. Mar. Geodesy **24**(2), 107–121 (2001)
17. Chelton, D.B., Schlax, M.G., Samelson, R.M.: Global observations of nonlinear mesoscale eddies. Prog. Oceanogr. **91**(2), 167–216 (2011)
18. Faghmous, J.H.: Eddyscan: a physically consistent ocean eddy monitoring application. In: 2012 Conference on Intelligent Data Understanding, pp. 96–103. IEEE (2012)
19. Chen, W., Zhang, C., Zhao, S., Li, H.: Automatic mesoscale eddy extraction from satellite remote sensing sea surface height image. Opt. Preci. Eng. **21**(10), 2704–2712 (2013)
20. Lguensat, R., Sun, M., Fablet, R., Tandeo, P., Mason, E., Chen, G.: EddyNet: A deep neural network for pixel-wise classification of oceanic eddies. In: IGARSS 2018-2018 IEEE International Geoscience and Remote Sensing Symposium. pp. 1764–1767. IEEE (2018)

21. Ronneberger, O., Fischer, P., Brox, T.: U-net: convolutional networks for biomedical image segmentation. In: Navab, N., Hornegger, J., Wells, W.M., Frangi, A.F. (eds.) MICCAI 2015. LNCS, vol. 9351, pp. 234–241. Springer, Cham (2015). https://doi.org/10.1007/978-3-319-24574-4_28
22. Zhao, H., Shi, J., Qi, X., Wang, X., Jia, J.: Pyramid scene parsing network. In Proceedings of the IEEE Conference on Computer Vision and Pattern Recognition. pp. 2881–2890 (2017)
23. Xu, G., Cheng, C., Yang, W., Xie, W., Kong, L., Hang, R., Ma, F., Dong, C., Yang, J.: Oceanic eddy identification using an ai scheme. Remote Sens. **11**(11), 1349 (2019)
24. Du, Y., Song, W., He, Q., Huang, D., Liotta, A., Su, C.: Deep learning with multi-scale feature fusion in remote sensing for automatic oceanic eddy detection. Inf. Fusion **49**, 89–99 (2019)
25. Chan, T.-H., Jia, K., Gao, S., Lu, J., Zeng, Z., Ma, Y.: Pcanet: A simple deep learning baseline for image classification? IEEE Trans. Image Process. **24**(12), 5017–5032 (2015)
26. He, K., Zhang, X., Ren, S., Sun, J.: Spatial pyramid pooling in deep convolutional networks for visual recognition. IEEE Trans. Pattern Anal. Mach. Intell. **37**(9), 1904–1916 (2015)

# A Hybrid Self-Attention Model for Pedestrians Detection

Yuan Wang, Chao Zhu(✉), and Xu-Cheng Yin

School of Computer and Communication Engineering,
University of Science and Technology Beijing, Beijing, China
LHMY599@163.com, {chaozhu,xuchengyin}@ustb.edu.cn

**Abstract.** In recent years, with the research enthusiasm of deep learning, pedestrian detection has made significant progress. However, the performance of state-of-the-art algorithms are still limited due to the high complexity of the detection scene. Therefore, in order to better distinguish between pedestrians and background, we propose a novel hybrid attention module which is capable of obtaining inter-dependencies between features from both channel and spatial dimensions through local convolution and dual-pass pooling, and guiding the network to focus on better pedestrians' feature representation while suppressing background noise. Further, we complement the information of channel attention and spatial attention through an effective fusion mechanism. To validate the effectiveness of the proposed hybrid attention module, we embed it into a representative pedestrian detection framework named Center and Scale Prediction (CSP) based detector. The experimental results on the Caltech Pedestrians Benchmark, one of the largest pedestrian detection datasets, show that the proposed method outperform not only the baseline framework but also several state-of-the-arts.

**Keywords:** Pedestrian detection · Channel attention · Spatial attention · Fusion mechanism

## 1 Introduction

Pedestrian detection has been an important research hotspot in the field of computer vision for many years and has attracted widespread attention in both academia and industry. In recent years, due to the great success of deep learning in general object detection, many effective algorithms [1–3] have been adjusted and transplanted into pedestrian detection, which greatly improves the performance of pedestrian detection. However, the performance of the state-of-the-art pedestrian detection algorithm still does not reach human standards [4]. Due to the complexity of the pedestrian detection environment in real scenes, it is difficult to distinguish between pedestrians and backgrounds, resulting in many network detection performance limitations. Therefore, the way to enhance the capability of pedestrians' feature representation is the key to further improving the performance of pedestrian detection.


H. Yang et al. (Eds.): ICONIP 2020, LNCS 12532, pp. 62–74, 2020.
https://doi.org/10.1007/978-3-030-63830-6_6

To achieve this, [5] reduces false positive samples from backgrounds by fusing multiple parallel networks, but it is very time-consuming. [6] gradually locates pedestrians for better classification through multi-step predictions that raise the IOU threshold multiple times, but it does not perform well at low IOU thresholds. [7,8] proposes two new loss functions to guide the network to optimize pedestrian feature representation, but they require complex hyperparameter adjustments.

Recently, attention mechanism is applied to pedestrian detection with excellent feature enhancement and discrimination ability. [9] proposes a channel attention model to enhance the feature representation of the visible part of the human body through different feature channels. [10] integrates bounding-box level segmentation information of pedestrian visible areas into detection branches to guide detector to better distinguish background and pedestrian. However, these models require additional data sets for pre-training or extra information for supervision. Also, few methods in the literature of pedestrian detection use both channel and spatial attention to guide the network with multi-dimensional information fusion for better pedestrian detection performance.

Therefore, in this paper, we propose a novel hybrid attention module with both channel and spatial attention to boost pedestrian detection. Different from common attention models to obtain global dependencies, our channel attention and spatial attention use one-dimensional convolution and stacked dilated convolution to obtain local dependencies. Such an operation can not only avoid the interference of obtaining long-distance irrelevant information, but also improve the efficiency of attention map calculation. These two attention mechanism can guide the network to focus on better pedestrian feature learning while suppressing background noise after an effective fusion strategy. Our attention model requires neither additional guidance information nor additional database, which makes our model easier to train and optimize. To validate the effectiveness of the proposed method for pedestrian detection, we embed our attention model into a representative pedestrian detection framework named Center and Scale Prediction (CSP) based detector [11], which is a detector that achieves state-of-the-art performance for pedestrian detection. The Caltech Pedestrian Benchmark, one of the largest pedestrian detection databases, is adopted to conduct experimental evaluation and comparison of the proposed method against other state-of-the-arts.

In summary, our main contributions are as follows: (1) We propose a novel hybrid self-attention model to obtain local dependencies through channel attention and spatial attention to boost pedestrian detection. (2) The proposed hybrid attention model is based on self-attention information acquisition and thus requires no additional supervision information or pre-trained databases, which makes our model easier to train and optimize. (3) The experimental results demonstrate that being embedded into a representative baseline (CSP) framework, our proposed method can achieve superior detection performance than not only the baseline detector but also several state-of-the-arts.

## 2 Related Work

### 2.1 Pedestrian Detection

In recent years, Faster-RCNN [1] is widely used as backbone network in deep learning pedestrians detection algorithms because it's high detection accuracy. RPN+BF [12] adopt the RPN sub-network in Faster-RCNN to generate proposals and then use cascaded boosted forest to refine the proposals in RPN. MS-CNN [13] propose multi-scale detection under different feature maps to enhance pedestrian' feature representation capabilities of Faster-RCNN at multiple scales. SDS-RCNN [14] uses bounding-box level semantic segmentation as additional supervision information to enhance the classification ability of Faster-RCNN. SSD [3] is a another backbone for pedestrian detection because of its high speed. ALFNet [6] use multi-step prediction SSD framework to gradually locate the pedestrian for better accuracy. In order for the network to adaptively enhance pedestrian representation capabilities, OR-CNN [7] and RepLoss [8] proposes two novel loss function to ensure the bounding-box distance between proposals and ground truths as short as possible. Adaptive NMS [15] and CSID [16] shows the new NMS strategy can reduce the possibility of targets filtered by NMS. CSP [11] is an anchor-free detector, it doesn't need tedious design of anchor boxes. CSP can obtain sufficient high-level semantic information by fusing feature maps of different resolutions in the backbone network. CSP reaches the new state-of-art performance of pedestrian detection at that time and thus is chosen as the baseline framework in this work.

Recently, the application of attention mechanism has become a new perspective to further improve the performance of pedestrian detection. Faster-RCNN+ATT [9] designs three subnet with attention methods plugged into Faster-RCNN to guide the network to focus on the visible parts of pedestrians. But Faster-RCNN+ATT requires additional datasets for pre-training, which does not meet the end-to-end manner. MGAN [10] uses two branches to predict the visible part and the whole body of the pedestrian, respectively, then uses the segmentation of the visible part bounding-boxes as a spatial attention mask to multiply into the whole body branch to enhance the learning of advanced semantic features of pedestrians. SSA-CNN [17] share the similar ideas with MGAN [10], but SSA-CNN uses the whole-body segmentation box as a guide and directly cascades the spatial attention mask and detection branch. These methods all require additional supervision information, which increases the difficulty of network optimization. Differently, our goal is to propose a self-attention module that is constructed directly from the inter-dependencies between channels and pixels without additional supervision information.

### 2.2 Attention Mechanism

Attention mechanism can be divided according to weighted approach, channel attention is given different weights to different feature channels; spatial attention is given different weights according to the importance between pixels in feature maps.

For Channel attention, SENet [18] integrates information between channel levels through global average pooling, then uses two fully connected layers for transformation operations, and finally uses the obtained vector to re-weight the original feature map. Similar to SENet, GENet [19] defines a gather operator with parameters using depthwise separable convolution to learn the context information and a excite operator to adjust the importance of different channels. LCT [20] uses group normalization and $1 * 1$ convolution for normalization operator and transform operator. SKNet [21] consists of multiple branches using different convolution kernel sizes and softmax fusion method, which can dynamically adjust the receptive field according to the target size. Different from them, we apply 1D convolution for channel attention, which can reduce the parameters in information transform. And we adopt max pooling and average pooling as two path to sum together for information aggregation and complementarity.

For spatial attention, STN [22] helps the network to correct image distortion by letting the network learn to calculate the spatial mapping of input map to output map. RAN [23] weights high-level semantic feature maps to low-level detailed feature maps in order to enrich the feature representation ability for lower layer. Non-Local Block [24] calculates the weighted sum of all position on the feature map as the response of a position to obtain the long-distance dependency. Although Non-Local Block doing pretty well on video classification task, but the calculation of NL is very time consuming and the memory consumption is very large. So Several algorithms [25–27] aims to reduce the computational complexity through different computational decomposition methods. Specifically, we stack multiple efficient dilation convolution with different dilation rates for spatial attention, which can well balance the calculation speed and accuracy for spatial attention.

The fusion mechanism of spatial attention and channel attention is an important part of designing an attention mechanism. CBAM [28] uses a serial structure to connect the two in the order of CA (channel attention) and SA (spatial attention). GCNet [29] also adopt a serial structure, but GCNet lets SA replace the pooling operations in CA, reducing the spatial information lost by channel attention when integrating channel information. DANet [30] treats CA and SA as two parallel branches, then fuses them through convolution layers and element-wise summation. In our work, we design CA and SA as independent modules for efficiency and then connect them in series.

## 3 Proposed Method

### 3.1 Revisiting the CSP Detector

The CSP detector is based on the idea of anchor-free detection, which can abandon the complexity of the anchor boxes and sliding windows design. It regards pedestrian detection as a high-level semantic feature detection task, and uses the pedestrian center point and height as abstract high-level semantic information features.

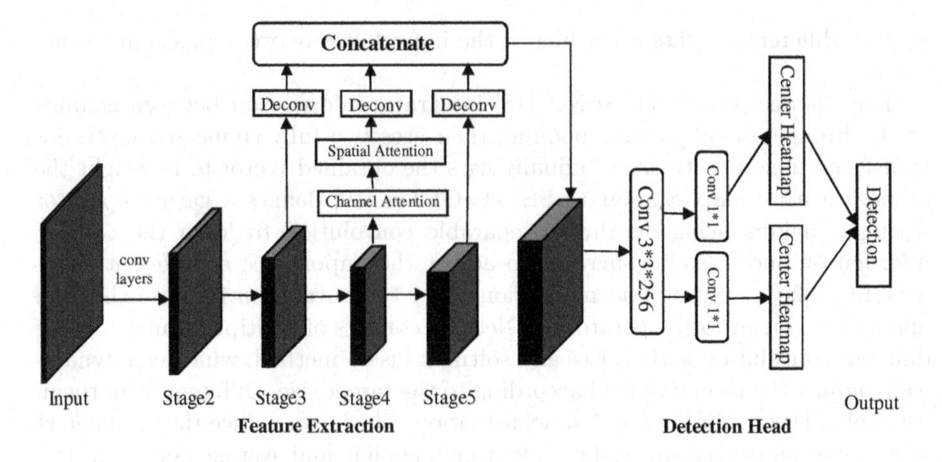


**Fig. 1.** The architecture after embedding our hybrid attention model into CSP.

The CSP framework consists of two modules: feature extraction module and detection head. For feature extraction, CSP uses ResNet-50 and MobileNet as backbone networks. Taking ResNet-50 as an example, its convolutional layer can be divided into 5 stages with down-sampling rates of the input image of 2, 4, 8, 16, and 16, respectively. The fifth stage uses dilated convolution to keep the size of the feature map unchanged and increase the receptive field of the convolution layers. Then the CSP uses deconvolution to up-sample the feature maps of the stage 3, 4, and 5 to the same size as stage 2, and concatenate them to feed into the detection head. For detection head, CSP reduces the channel dimension through a $3*3$ convolution layer, and then uses the two $1*1$ convolution layers to generate two branches of the heat map that predict the center point and scale, respectively. Although CSP achieves the state-of-the-art performance of pedestrian detection at that time, we demonstrate that it still can achieve better detection results when enhanced with more discriminative information on pedestrians and backgrounds through our hybrid self-attention module, as shown in Fig. 1. Note that our attention module is added after stage 4 of the CSP backbone network, because stage 4 has enough information for learning channel discrimination and spatial context.

### 3.2 Channel Attention

Channel attention is designed to help the network acquire inter-dependencies between channels. The first step for Channel attention is compresses spatial dimensions to simplify the calculation of global context information integration for spatial. However, unlike global average pooling adopted in SENet [18] that preserves background information, global maximum pooling can extract pedestrian texture information that is helpful for detection. So we use both global average pooling and global maximum pooling for global context information integration. Different from SENet [18], which uses two fully connected layers

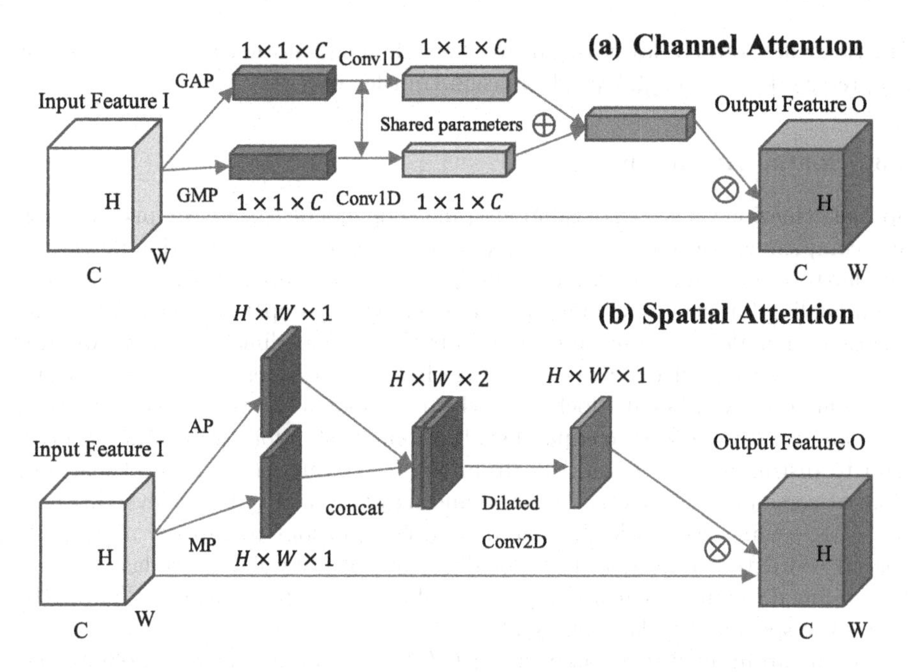


**Fig. 2.** The architecture of our channel attention and spatial attention.

to obtain the internal dependencies between all channels, we believe that this relationship can be calculated only locally in deep networks because of the huge differences between non local channels in deep layers. So we just use a single 1D convolution layer with a larger kernel size to capture this relationship. This can make the design of the attention module more concise, while reducing the amounts of parameters. Finally, we rescale the input features by obtaining the weighted vector from the previous step. Figure 2(a) shows the specific processing flow of our channel attention module.

Giving an input feature map as $I \in R^{(H\times W\times C)}$, we use two branches to obtain global context information from spatial by global average pooling (GAP) and global maximum pooling (GMP), then we generate two one-dimensional vectors: $I_{GAP} \in R^{(1\times1\times C)}$ and $I_{GMP} \in R^{(1\times1\times C)}$. We use these two vectors to pass a same 1D convolution layer with kernel size of $9 * 9(Conv1D^9)$ meanwhile keep the dimensions unchanged. This shared convolution layer can reduce the amounts of parameters and the possibility of overfitting. Then we add the two vectors by element-wise summation and go through the sigmoid activation function, and finally multiply the merged attention vector with the input feature map to get the final output $O \in R^{(H\times W\times C)}$. This process can be expressed by the following formula:

$$O = I \bigotimes (S(Conv1D^9(I_{GAP}) \bigoplus Conv1D^9(I_{GMP}))) \tag{1}$$

where $S$ means sigmoid activation function, $\bigotimes$ and $\bigoplus$ represent multiply per channel and element-wise summation. Note that we do not use activation functions in Conv1D layer to ensure that channel information is not lost or suppressed during the global information transfer process.

### 3.3 Spatial Attention

Spatial attention is used to obtain long-range dependencies between pixels, it can also complement the channel attention information to help pedestrian detection. Similar to the channel attention we designed, we do average pooling and maximum pooling for each pixel along the channel axis, this operation can obtain the abstraction of global context information in the channel dimension, then we concatenate the two pooled feature maps together for subsequent convolution layers. Different from Non-Local Block [24], which calculates the relationship between any two pixels, we intend to expand spatial contextual information and receptive field to distinguish the pedestrian from the background in high-level features. Dilated convolution is a very intuitive and lightweight way to achieve this purpose. Meanwhile, we stack multiple dilated convolutions to capture multi-scale spatial contextual information under different receptive fields. At last, we use the obtained weighted feature map to recalibrate the input features. Figure 2(b) shows the specific details of our spatial attention module.

Giving an input feature map as $I \in R^{(H \times W \times C)}$, we use average pooling and maximum pooling along the channel axis to obtain two feature maps $I_{AP} \in R^{(H \times W \times 1)}$ and $I_{MP} \in R^{(H \times W \times 1)}$ containing global spatial information. Then the two feature maps are concatenated together and pass 3 layers of dilated convolution, the filters are all set to 1, the kernel sizes are all set to 3 * 3 and the dilated rates are 1, 2, and $5 (DilaConv2D^{1,2,5})$, respectively. Finally we compute the output feature map $O \in R^{(H \times W \times C)}$ by multiplying the obtained attention map and the input feature map. Spatial attention can be calculated by the following formula:

$$O = I \bigotimes (S(DilaConv2D^{1,2,5}(I_{AP}, I_{MP}))) \tag{2}$$

where [,] means the concatenation of two feature maps, $\bigotimes$ represent multiply per pixel.

### 3.4 Hybrid Attention Fusion Strategy

Fusion of channel attention and spatial attention helps aggregate and complement global context information in different dimensions. Since our channel attention module and spatial attention module are two independent embeddable models with the same input and output, a very efficient fusion strategy is to connect two modules in the order of either channel attention first or spatial attention first. Another way of fusion is parallel structures, where channel attention and spatial attention modules are processed separately and added by element-wise summation. The experimental evaluation demonstrates that the first fusion strategy is better than the second one.

# 4 Experiments

## 4.1 Dataset and Evaluation Metrics

The Caltech Pedestrians Benchmark [31] is one of the largest and most widely studied pedestrian detection datasets. It contains a 10-h of 30HZ 640 * 480 vehicle driving video through regular traffic in an urban environment. This dataset has a total of 350,000 bounding boxes for about 2300 unique pedestrians. For the training set, we extract a picture from set00-05 every 3 frames (42,782 images in total). For the test set, we extract a picture from set06-10 every 30 frames (4024 images in total). Note that during training and testing phases we use the Caltech new annotations provided by [4].

We follow the average log missing rate(MR) of false positive per image(FPPI) between $10^{-2}$ and $10^{0}$ as the evaluation standard. Particularly, the pedestrians from three subsets of Reasonable, All and Heavy Occlusion are used for evaluation, and they are set as follows: The pedestrians height range for the three subsets are [50, inf], [20, inf], [50, inf]. (inf means infinite). The visible portion of pedestrians are [0.65, inf], [0.2, inf], and [0.2, 0.65] respectively.

## 4.2 Ablation Study

**Channel Attention.** The key to the design of channel attention is the size of the convolution kernel in the one-dimensional convolution, because it represents how wide the correlation is obtained between channels. Table 1 shows the influence of different convolution kernel sizes on channel attention performance. The first row of Table 1 is the original MR on the Reasonable set in the CSP [11] paper, while the second row is our own reproduced result by using its opensource code[1]. It's worth noting that our reproduced results are always worse than the original results in the CSP [11] paper by about 0.5 % on the Reasonable set. Therefore in the subsequent experiments, we will take our reproduced results as baseline for fair comparisons. From the results in Table 1, we can observe that as the size of the convolution kernel increases, a larger range of channel interdependencies can be obtained, and the performance of channel attention continues to get improved. In addition, different kernel sizes have improved pedestrian detection performance, validating the design rationality of our channel attention module.

**Spatial Attention.** The setting of the dilated rates affects the scope of contextual information acquisition in spatial. As shown in Table 2, we set up three combinations of dilated rates to verify its influence on spatial attention. We first set the dilated rates to 1-2-1, which slightly improves detection performance. Then we further expand the dilated rates to 1-2-5 and 5-2-1 in a way of increasing and decreasing. These two ways prove that getting more contextual information and a larger receptive field through a larger dilated rates setting is crucial to

[1] https://github.com/liuwei16/CSP.

**Table 1.** Comparison of different convolution kernel sizes for channel attention.

| Algorithm | Reasonable set | |
|---|---|---|
| | $MR(\%)$ | $\Delta MR(\%)$ |
| CSP [11] | 4.54 | – |
| CSP (reproduced) | 5.02 | – |
| CSP+CA (k5) | 4.94 | +0.08 |
| CSP+CA (k7) | 4.85 | +0.17 |
| CSP+CA (k9) | 4.37 | +0.65 |

**Table 2.** Comparison of different dilated rate settings for spatial attention.

| Algorithm | Reasonable set | |
|---|---|---|
| | $MR(\%)$ | $\Delta MR(\%)$ |
| CSP [11] | 4.54 | – |
| CSP (reproduced) | 5.02 | – |
| CSP+SA (dr 1-2-1) | 4.84 | +0.18 |
| CSP+SA (dr 1-2-5) | 4.42 | +0.60 |
| CSP+SA (dr 5-2-1) | 4.43 | +0.59 |

improve the performance of spatial attention. However, setting the dilated rates either in increasing way or in decreasing way has little affect on the results.

**Fusion Strategies.** After fixing the size of the one-dimensional convolution kernel to 9 and the dilated rates setting to 1-2-5, we further study the effects of different fusion strategies for integrating channel attention and spatial attention. As seen the results in Table 3, the sequential combination is significantly better than the parallel combination. This may be due to the fact that channel attention and spatial attention give weights to features in different dimensions, and adding them directly will cause chaos in the weight distribution. For the sequential fusion structure, we find that putting channel attention first is better than putting spatial attention first, thus in subsequent experiments, we will apply sequential fusion strategy in the order of first channel attention and then spatial attention. Moreover, we can see that the results of sequential fusion are better than the result of either channel attention or spatial attention alone, proving that the proposed hybrid attention model can provide complementary information of two single attention module and further improve pedestrian detection performance.

### 4.3 Comparison with State of the Arts

Finally, we compare the proposed method with several state-of-the-art pedestrian detection approaches on three Caltech subsets of Reasonable, All, and

**Table 3.** Comparison of different hybrid attention fusion strategies.

| Algorithm | Reasonable set | |
|---|---|---|
| | $MR(\%)$ | $\Delta MR(\%)$ |
| CSP [11] | 4.54 | – |
| CSP (reproduced) | 5.02 | – |
| CSP+CA+SA | 3.84 | +1.18 |
| CSP+SA+CA | 4.01 | +1.01 |
| CSP+CASA (add) | 4.75 | +0.27 |

**Table 4.** Comparison with state of the arts (the best results are in bold).

| Algorithm | $MR\%$ | | |
|---|---|---|---|
| | Reasonable | All | Heavt Occlusion |
| CSP [11] | 4.5 | 56.9 | 45.8 |
| CSP (reproduced) | 5.0 | 57.9 | 49.1 |
| RPN+BF [12] | 7.3 | – | 54.6 |
| ALFNet [6] | 6.1 | 59.1 | 51.0 |
| RepLoss [8] | 5.0 | 59.0 | 47.9 |
| OR-CNN (city) [7] | 4.1 | 58.8 | **45.0** |
| Ours | **3.8** | **56.8** | 46.4 |

Heavy Occlusion. Note that we only compare algorithms trained and tested with the Caltech new annotations [4], the results are shown in Table 4. On Reasonable subset, compared with our reproduced result, we have significantly improved the performance by 24% after embedding our hybrid attention module. Even compared with the original results in [11], we still achieve a 15% improvement.

The performances of the proposed method are also better than other state-of-the-art algorithms, even better than OR-CNN [7] for 4.1% MR, which is a method for pre-training with additional Citypersons dataset [32]. For All subset, our hybrid attention module also achieved the best results, showing that it can guide the network to enhance detection performance in various sizes of pedestrians. Considering that our reproduced results on the Heavy Occlusion subset are far from the results in the CSP paper, we achieved slightly lower results after embedding the proposed attention module. However, our results are still competitive with other state-of-the-arts. Since we have significantly improved the detection performance on heavy occlusion pedestrians compared with the reproduced baseline result, it has been revealed that our method can pay more attention to the feature representation of pedestrians' visible parts.

## 5 Conclusion

In this paper, we propose a novel hybrid attention model for pedestrian detection from both channel attention and spatial attention aspects to obtain feature inter-dependencies in different dimensions. Our model can provide the detection network with more discriminative guidance information for pedestrians and backgrounds to enhance the capabilities of pedestrians' feature representation. By embedding the proposed attention module into the CSP baseline framework, the detection performance has been further improved on the standard Caltech pedestrian detection benchmark. The ablation study demonstrates the effectiveness of the proposed channel attention, spatial attention, and their hybrid fusion strategy. Our model also achieves superior performances than several state-of-the-art algorithms on the subsets named Reasonable, All, and Heavy Occlusion of the Caltech benchmark.

**Acknowledgment.** This work was supported by National Natural Science Foundation of China under Grant 61703039, and Beijing Natural Science Foundation under Grant 4174095.

## References

1. Ren, S., He, K., Girshick, R., Sun, J.: Faster r-cnn: towards real-time object detection with region proposal networks. In: Advances in neural information processing systems. pp. 91–99. (2015)
2. Redmon, J., Farhadi, A.: YOLO9000: better, faster, stronger. In: Proceedings of the IEEE Conference on Computer Vision and Pattern Recognition. pp. 7263–7271 (2017)
3. Liu, W., Anguelov, D., Erhan, D., et al.: SSD: Single shot multibox detector. In: European Conference on Computer Vision, pp. 21–37. Springer, Cham (2016)
4. Zhang, S., Benenson, R., Omran, M., Hosang, J., Schiele, B.: How far are we from solving pedestrian detection? In: Proceedings of the IEEE Conference on Computer Vision and Pattern Recognition. pp. 1259–1267 (2016)
5. Du, X., El-Khamy, M., Lee, J., et al.: Fused DNN: a deep neural network fusion approach to fast and robust pedestrian detection. In: 2017 IEEE Winter Conference on Applications of Computer Vision (WACV). IEEE 953–961 (2017)
6. Liu, W., Liao, S., Hu, W., Liang, X., Chen, X.: Learning efficient single-stage pedestrian detectors by asymptotic localization fitting. In: Proceedings of the European Conference on Computer Vision (ECCV). pp. 618–634 (2018)
7. Zhang, S., Wen, L., Bian, X., Lei, Z., Li, S.Z.: Occlusion-aware R-CNN: detecting pedestrians in a crowd. In: Proceedings of the European Conference on Computer Vision (ECCV). pp. 637–653 (2018)
8. Wang, X., Xiao, T., Jiang, Y., Shao, S., Sun, J., Shen, C.: Repulsion loss: detecting pedestrians in a crowd. In: Proceedings of the IEEE Conference on Computer Vision and Pattern Recognition pp. 7774–7783 (2018)
9. Zhang, S., Yang, J., Schiele, B.: Occluded pedestrian detection through guided attention in cnns. In: Proceedings of the IEEE Conference on Computer Vision and Pattern Recognition. pp. 6995–7003 (2018)

10. Pang, Y., Xie, J., Khan, M.H., Anwer, R.M., Khan, F.S., Shao, L.: Mask-guided attention network for occluded pedestrian detection. In: Proceedings of the IEEE International Conference on Computer Vision. pp. 4967–4975 (2019)
11. Liu, W., Liao, S., Ren, W., Hu, W., Yu, Y.: High-level semantic feature detection: a new perspective for pedestrian detection. In: Proceedings of the IEEE Conference on Computer Vision and Pattern Recognition pp. 5187–5196 (2019)
12. Zhang, L., Lin, L., Liang, X., et al.: Is Faster R-CNN doing well for pedestrian detection?. In: European Conference on Computer Vision, pp. 443–457. Springer, Cham (2016)
13. Cai, Z., Fan, Q., Feris, R.S., et al.: A unified multi-scale deep convolutional neural network for fast object detection. In: European Conference on Computer Vision, pp. 354–370. Springer, Cham (2016)
14. Brazil, G., Yin, X., Liu, X.: Illuminating pedestrians via simultaneous detection and segmentation. In: Proceedings of the IEEE International Conference on Computer Vision. pp. 4950–4959 (2017)
15. Liu, S., Huang, D., Wang, Y.: Adaptive nms: refining pedestrian detection in a crowd. In: Proceedings of the IEEE Conference on Computer Vision and Pattern Recognition pp. 6459-6468 (2019)
16. Zhang, J., Lin, L., Chen, Y., et al.: CSID: center, scale, identity and density-aware pedestrian detection in a crowd[OL]. arXiv preprint arXiv:1910.09188 (2019)
17. Zhou, C., Wu, M., Lam, S.K.: SSA-CNN: Semantic self-attention CNN for pedestrian detection[OL]. arXiv preprint arXiv:1902.09080 (2019)
18. Hu, J., Shen, L., Sun, G.: Squeeze-and-excitation networks. In: Proceedings of the IEEE Conference on Computer Vision And Pattern Recognition pp. 7132–7141 (2018)
19. Hu, J., Shen, L., Albanie, S., Sun, G., Vedaldi, A.: Gather-excite: exploiting feature context in convolutional neural networks. In Advances in Neural Information Processing Systems. pp. 9401–9411 (2018)
20. Dongsheng, R., Jun, W., Nenggan, Z.: Linear context transform block. arXiv preprint arXiv:1909.03834 (2019)
21. Li, X., Wang, W., Hu, X., Yang, J.: Selective kernel networks. In Proceedings of the IEEE Conference on Computer Vision and Pattern Recognition. pp. 510–519 (2019)
22. Jaderberg, M., Simonyan, K., Zisserman, A.: Spatial transformer networks. In: Advances in Neural Information Processing Systems. pp. 2017–2025 (2015)
23. Wang, F., Jiang, M., Qian, C., et al.: Residual attention network for image classification. In: Proceedings of the IEEE Conference on Computer Vision and Pattern Recognition. pp. 3156–3164 (2017)
24. Wang, X., Girshick, R., Gupta, A., He, K.: Non-local neural networks. In: Proceedings of the IEEE Conference on Computer Vision and Pattern Recognition. pp. 7794–7803 (2018)
25. Huang, Z., Wang, X., Huang, L., Huang, C., Wei, Y., Liu, W.: Ccnet: criss-cross attention for semantic segmentation. In: Proceedings of the IEEE International Conference on Computer Vision. pp. 603–612 (2019)
26. Yue, K., Sun, M., Yuan, Y., et al.: Compact generalized non-local network. In: Advances in Neural Information Processing Systems. pp. 6510–6519 (2018)
27. Chen, Y., Kalantidis, Y., Li, J., et al.: $A^2$-nets: Double attention networks. In: Advances in Neural Information Processing Systems. pp. 352–361 (2018)
28. Woo, S., Park, J., Lee, J.Y., et al.: CBAM: convolutional block attention module. In: Proceedings of the European Conference on Computer Vision (ECCV). pp. 3–19 (2018)

29. Cao, Y., Xu, J., Lin, S., et al.: GCNet: Non-local networks meet squeeze-excitation networks and beyond. In: Proceedings of the IEEE International Conference on Computer Vision Workshops (2019)
30. Fu, J., Liu, J., Tian, H., et al.: Dual attention network for scene segmentation. In: Proceedings of the IEEE Conference on Computer Vision and Pattern Recognition. pp. 3146–3154 (2019)
31. Dollar, P., Wojek, C., Schiele, B., et al.: Pedestrian detection: an evaluation of the state of the art. IEEE Trans. Pattern Anal. Mach. Intell. **34**(4), 743–761 (2011)
32. Zhang, S., Benenson, R., Schiele, B.: Citypersons: a diverse dataset for pedestrian detection. In: Proceedings of the IEEE Conference on Computer Vision and Pattern Recognition. pp. 3213–3221 (2017)

# DF-PLSTM-FCN: A Method for Unmanned Driving Based on Dual-Fusions and Parallel LSTM-FCN

Meng Wei[1,2], Yuchen Fu[1,2](✉), Shan Zhong[2], and Zicheng Li[1,2]

[1] School of Computer Science and Technology, Soochow University, Suzhou 215000, Jiangsu, China
yuchenfu@cslg.edu.cn

[2] School of Computer Science and Engineering, Changshu Institute of Technology, Changshu 215500, Jiangsu, China

**Abstract.** Learning algorithms are increasingly being applied to behavioral decision systems for unmanned vehicles. In multi-source road environments, it is one of the key technologies to solve the decision-making problem of driverless vehicles. This paper proposes a parallel network, called DF-PLSTM-FCN, which is composed of LSTM-FCN-variant and LSTM-FCN. As an end-to-end model, it will jointly learn a mapping from the visual state and previous driving data of the vehicle to the specific behavior. Different from LSTM-FCN, LSTM-FCN-variant provides more discernible features for the current vehicle by introducing dual feature fusions. Furthermore, decision fusion is adopted to fuse the decisions made by LSTM-FCN-variant and LSTM-FCN. The parallel network structure with dual fusion on both features and decisions can take advantage of the two different networks to improve the prediction for the decision, without the significant increase in computation. Compared with other deep-learning-based models, our experiment presents competitive results on the large-scale driving dataset BDDV.

**Keywords:** Unmanned driving · Decision-making · Feature fusion · Decision fusion

## 1 Introduction

Behavioral decision of unmanned vehicles is a key field of unmanned driving, aiming to predict the future's movement based on the current position, speed, driving direction and surrounding environment, combined with the driving data of the vehicle. The current decision-making research has yielded some satisfactory results. However, these solutions are implemented under specially calibrated drives or correspondingly simplified conditions, as a result, they are not suitable for complex real traffic scenarios.

With the development of large public image repositories (such as ImageNet [15]) and high-performance computing systems (such as GPUs or large-scale distributed clusters [20]), deep learning has also achieved great success in


H. Yang et al. (Eds.): ICONIP 2020, LNCS 12532, pp. 75–87, 2020.
https://doi.org/10.1007/978-3-030-63830-6_7

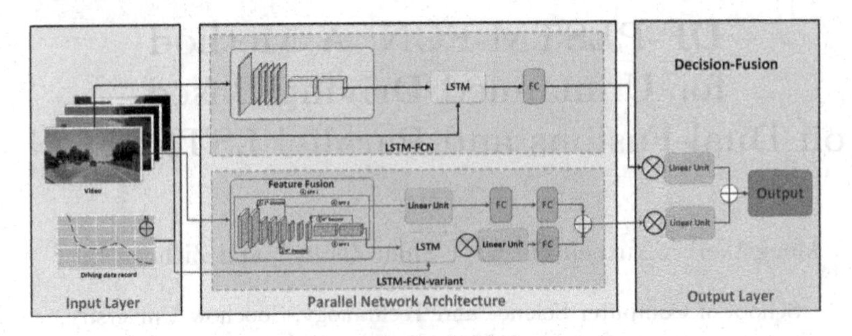


**Fig. 1.** Overall Network Architecture. The DF-PLSTM-FCN includes LSTM-FCN and LSTM-FCN-variant two networks. Among them, FCN and the FCN-variant is shown in Fig. 2. Dual feature fusion technology is used in FCN-variant. Decision fusion technology is used to fuse the output of the two networks

the large-scale image and video recognition [14]. Convolutional neural networks (CNN) is regarded as the most popular method in deep learning due to its ability in extracting features. Some researchers have attempted to apply CNN to driverless decision-making systems, e.g. [1,11]. Most semantic segmentation frameworks follow the design of a fully convolutional network [13]. Because it can not only accept images with any size as input but also achieve pixel-level classification by deconvolution. Long Short-Term Memory (LSTM) network is proposed by Hochreite [8] et al. which is a circular memory network with the advantage of processing temporal information. Inspired by the ability of CNN, FCN and LSTM in feature extraction and temporal information capturing, we propose a driving behavior decision-making model by combining LSTM-FCN and its variant. Moreover, the model is trained by using a large-scale driverless dataset so that it can be applied in more complex real-world scenarios.

The contributions of our work can be summarized as:

- A parallel network structure composed of LSTM-FCN and LSTM-FCN-variant is designed, to learn more distinguishable features of the environment information and make better decision-making behaviors.
- Dual feature fusion is applied, one is to fuse the feature descriptors of different depths for FCN-variant, and the other is to fuse those generated from SPPs in FCN-variant. The two fusions are achieved by deconvolution operation and the adaptive linear model, respectively.
- To make the decision more accurate, decision fusion, also achieved by using adaptive linear models, is implemented in both LSTM-FCN-variant and LSTM-FCN to weigh the executing probability for each action.

## 2 Related Work

A common method for solving decision-making in autonomous driving is trajectory optimization. Xu et al. [19] obtained a group of paths by measuring the distance between the vehicle and the obstacles, with high efficiency. However, the proposed method represented all the obstacles in a uniformly predefined approach regardless it is static or dynamic, making it hard to be applied to the domains with real-time changes. Tran and Diehl [4] proposed a convex optimization method for unmanned vehicles with a static environment, where a model for approximating the appearance of obstacles in simulation is designed. Although the obstacles generated from probability estimation will affect the decision-making process of unmanned vehicles, it still cannot work well for the problems with nonlinear hypothetical dynamic models or discrete decision making.

By taking environmental dynamics into consideration, Gu and Dolan [6] adopted dynamic programming to generate the trajectories. Even if it gained much success in the simulated experiment, its performance in real-world applications can not be guaranteed. Kuderer [10] used inverse reinforcement learning to learn driving style from trajectory presentations, where the autonomous vehicle trajectories are generated by trajectory optimization. But the resulting trajectories after optimization do not exactly meet the optimality criteria. Felipe et al. [3] proposed conditional imitation learning, an approach to learning the driving behavior from expert demonstrations with low-level controls and high-level commands. However, it is only suitable for simple simulation environments.

To make real-time decisions, Cunningham [4] proposed multipolicy decision-making (MPDM) method, where the controlled vehicles and other vehicles are modeled using interactive forward simulations. Galceran et al. [5] proposed a comprehensive inference and decision-making method for autonomous driving to capture the behavior and intention of the vehicle. Specifically, they estimated the potential policy distribution according to the historical data of the vehicles, and then sampled from the distribution to get the decision with the highest probability. Though the decision-making approach can choose between a set of possible policies, it will result in computational limitations as the number of policy assignments increase. However, their results are also limited to simulations.

Although previous works have achieved many outcomes, they are often implemented in simulated or simplified environments, hindering further application in real or complex environments. To learn a more common driving model, this paper proposes the DF-PLSTM-FCN model. It is a parallel structure consists of two networks LSTM-FCN and LSTM-FCN-variant, to learn the mapping from the current state (the visual observation along the road and the previous driving data of the vehicle) to the driving behavior of the next moment. Experimental evaluations are implemented on Berkeley DeepDrive Video (BDDV) which is a real driving video dataset composed of a variety of driving scenarios (e.g., several major cities, highways, towns, and rural areas in the United States, etc.).

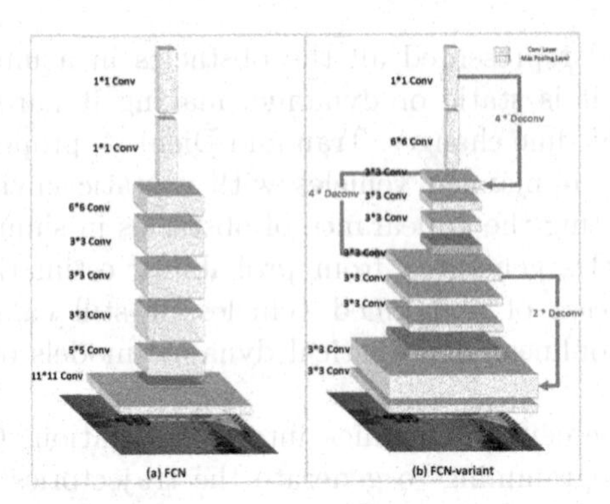

**Fig. 2.** Parallel FCN Model. FCN contains 7 convolutional layers and a pooling layer. FCN-variant consists of 10 convolution layers and 2 pooling layers. The last two convolutional layers serve as fully connected layers

# 3 DF-PLSTM-FCN

## 3.1 Driving Model

DF-PLSTM-FCN focuses on learning a general driving model from a large-scale driving dataset. The action set is shown as:

$$A = \{straight, stop, turn - left, turn - right\}. \quad (1)$$

Given the visual state along the road (the images captured by the front-positioned camera) and the driving data of the controlled vehicle (i.e., the velocity, acceleration velocity), the model can be used to estimate the probability of executing any candidate action at the next moment, as shown in:

$$\mathrm{F}(s, \mathrm{a}) : S \times \mathrm{A} \rightarrow \mathrm{R}, \quad (2)$$

where $S$ is the state space, and $s \in S$ represents any candidate state, i.e., the visual state captured by the front-positioned camera. $A$ is the action set, and $a \in A$ represents any candidate action in the action set.

## 3.2 Network Structure

The structure of DF-PLSTM-FCN is shown as Fig. 1, which consists of two parallel networks, LSTM-FCN and LSTM-FCN-variant. The two networks can run in parallel to speed the learning of the network considerably. LSTM-FCN, which is combining a fully convolutional network (FCN) and long short-term memory (LSTM), was proposed by Xu et al. [18]. It can be regarded as a variant of AlexNet [9] by removing POOL2 and POOL5 layers and replacing fully connected layers with convolutional layers. LSTM is used to process the sequential driving data of the controlled vehicle (velocity, angular velocity), while FCN is adopted to extract the features of the images obtained from the front-positioned camera. The FCN network is shown in the Fig. 2(a). Through such a combination, LSTM-FCN can provide a more complete state for decision-making.

The relatively small size of the convolutional kernel is proved to have a significant improvement in the precision for object recognition [17]. This may be resulted from that the small size of the convolutional kernel can keep more local and detailed information. Inspired by this phenomenon, we provide a variant of LSTM-FCN, namely, LSTM-FCN-variant, by updating the size of kernels to $3 \times 3$ in all convolutional layer except for the last two. Besides, the depth of LSTM-FCN-variant is also increased. The network structure is shown as the Fig. 2(b). As a result, the representation ability of the network can be strengthened without a significant increase in parameters. The feature fusion between neighbor layers is also introduced to make the final feature more robust, i.e., containing not only the local information but also the semantic information.

Four linear models are also added to DF-PLSTM-FCN to achieve the fusions both on feature extract and decision-making, and the specific details will be described in the following subsection.

### 3.3 Feature Fusion

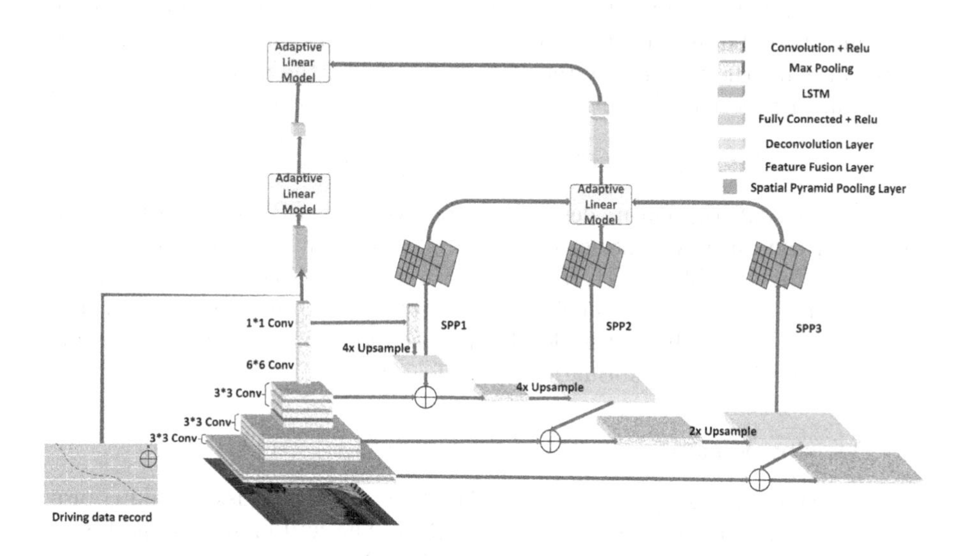


**Fig. 3.** LSTM-FCN-variant. Dual feature fusion is applied. The first one is to fuse the results of deconvolution from different depth. And the other is to fuse the features from spp by adaptive linear model

To capture features better, we adopt dual fusion, where the first is for fusing the feature descriptors obtained from different layers, and the other is to fuse the final features which are generated from SPP [7]. As shown in Fig. 3, the LSTM-FCN-variant network is composed of dual feature fusion.

Image upsampling can improve the accuracy of image recognition [21], and it is helpful for recovering the lost valuable information due to downsampling. Here we use deconvolution to achieve upsampling as the first fusion. As shown in Fig. 3, feature descriptors generated from the 10th, 8th, and 5th layers are deconvoluted and fused with the ones from the 8th, 5th, and 2nd layers, respectively. The first fusion function is shown as:

$$\mathrm{f}: a \times X_{\mathrm{m}} + X_n \rightarrow y_r, \tag{3}$$

where $\mathrm{X_m} \in R^{HWD}$ represents the feature descriptor generated from the m-th convolution layer, and $\mathrm{X}_n \in R^{HWD}$ denotes the feature descriptor from the n-th convolution layer, with $H$, $W$ and $D$ denoting the length, width and channel of the feature descriptor respectively. $a$ denotes a deconvolution kernel that will perform an up-sampling calculation on $X_m$. The variable $y_r, 1 \leq r \leq 3$ represents the $r-th$ fused feature by summing the up-sampling result $aX_m$ and the original feature descriptor $X_n$.

After deconvolution, there will be three fused feature descriptors with different sizes. To uniform them, SPP is introduced to pool on them. Afterward, the adaptive linear model is introduced to weigh the three fused feature descriptors. This layer is called the buffer layer. Special attention is paid to the buffer layer in our architecture, because it fuses the features generated from SPP further. Such fusion can make the features better without increasing the number of channels.

The linear function for the second fusion can be denoted as:

$$Y_{i,j,d}^{sum} = w_1 X_{i,j,d}^1 + w_2 X_{i,j,d}^2 + w_3 X_{i,j,d}^3, \tag{4}$$

where $X$ represents the feature descriptor generated from SPP, and $Y$ is the fused result from the linear weighting. $w_r, 1 \leq r \leq 3$ is the $r-th$ weighted feature descriptor. The variables $i \in [1, H]$, $j \in [1, W]$, and $d \in [1, D]$ denote the indexs of the height, width and channel of the feature descriptor, respectively. The buffer layer enhances the feature-level expression of the discernible descriptors and also strengthens the information of the feature.

### 3.4 Decision Fusion

Decision fusion strategy provides greater flexibility for the decision-making process and enhances its generalization capabilities. The well-known method is to set equivalent weights for all the classifiers [2,12]. However, it is unreasonable to assign equal weights to different feature classifiers if they play different roles in the final decision-making. As for DF-PLSTM-FCN, FCN tends to learn semantic features, and FCN-variant may be more likely to learn compound ones including not only local but also semantic features due to the fusions. The different networks in DF-PLSTM-FCN have significant differences in handling different driving environments. They also may play different advantages in different cases, which helps to improve the robustness of the training model. Therefore, we

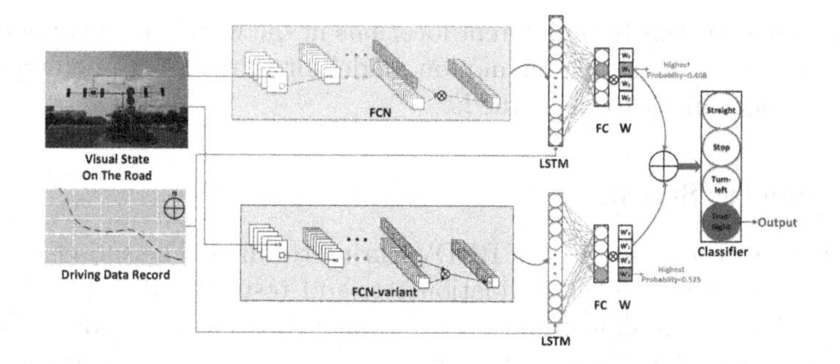


**Fig. 4.** Decision Fusion Process. The networks LSTM-FCN and LSTM-FCN-variant in DF-PLSTM-FCN correspond to two classifiers. The action with the largest weighted probability will be selected as the final optimal action

also weight the output of the two parallel networks. The larger the weight is, the more important the corresponding network will play and vice versa. The decision fusion structure is shown in Fig. 4, where two network classifiers are assigned with the weights learned from training, and thus the fusion result is the combining of two sets of probability distributions. The action whose weighted probability is largest will be selected as the final optimal action:

$$a_{pre} = \arg\max_{a_i}\{l_i^1 F_i^1(s, a_i), l_i^2 F_i^2(s, a_i), 1 \le i \le 4\}, \tag{5}$$

where $l_i^1$ and $F_i^1$, $1 \le i \le 4$ are the probabilities and weights learned from LSTM-FCN and its corresponding adaptive linear model, and $l_i^2$ and $F_i^2$, $1 \le i \le 4$ denote the ones from LSTM-FCN-variant.

### 3.5 Model Evaluation

Given an action sequence $a_1, a_2, \cdots a_t$ conditioned on state sequence $s_1, s_2, \cdots s_t$, then the probability of executing $a_{t+1}$ can be represented as:

$$p(s_t, a_t) = max\{l_1^1 F_1^1(s_t, a_t), l_i^2 F_i^2(s_t, a_t)\}. \tag{6}$$

To train the network, the loss function is designed as the cross-entropy loss between the probability of the model prediction, $p(s_t, a_t)$, and the probability of the actual action, shown as:

$$perplexity = \exp\left\{-\frac{1}{t}\sum_{i=1}^{t}\log p(s_i, a_i)\right\}. \tag{7}$$

## 4 Experiment

### 4.1 Dataset

The Berkeley DeepDrive Video dataset (BDDV) [18] is a dataset comprised of real driving videos and GPS/IMU data. It contains over 10k hours of driving dash-cam video streams from different locations in the world. There are multiple driving scenarios. A more generic motion model for vehicles with heterogeneous actuators is more likely to be learned.

### 4.2 Parameter Setting

DF-PLSTM-FCN uses a subset of BDDV during training. The subset is divided into three parts, training set, validation set, and test set. After preprocessing data, a total of 2.9 million frames are retained. Stochastic Gradient Descent (SGD) is used to update the model parameters in training. The learning rate, the momentum and the batch size are initialized as $10^{-4}$, 0.99, and 2, respectively. The learning rate is decayed by 0.5 whenever the training loss plateaus.

**Table 1.** Comparative experiment

| Configuration | Temporal | Perplexity | Accuracy |
|---|---|---|---|
| LSTM-FCN | LSTM | 0.452 | 83.08% |
| LSTM-FCN-Adaptive-linear-model | LSTM | 0.442 | 83.43% |
| LSTM-FCN-Adaptive-linear-model-Feature-Fusion | LSTM | 0.448 | 83.60% |
| LSTM-VGG-19 | LSTM | 0.468 | 82.65% |
| LSTM-FCN-variant' | LSTM | 0.415 | 84.40% |
| LSTM-FCN-variant'-Adaptive-linear-model | LSTM | 0.409 | 84.76% |
| LSTM-FCN-variant | LSTM | 0.408 | 84.80% |
| DF-PLSTM-FCN | LSTM | 0.421 | 85.02% |

### 4.3 Experiment Analysis

We carried out several groups of ablation experiments to study the importance of different components in the model. Table 1 records the comparison of experiments between the DF-PLSTM-FCN and the variants of the method.

**Kernel Size.** In unmanned driving, the detailed information of the environmental observation may have an important effect on decision making. Most of the kernels in FCN take a relatively large size, which may result in a loss of detailed information. To verify, we firstly replaced the feature extraction layers of FCN with VGG-19 and generated a model named LSTM-VGG-19. As shown in Table 1, the experimental performance of LSTM-VGG-19 is nearly equal to that of the LSTM-FCN network. However, since it takes the kernels with small size, it has a lower computation complexity. Afterward, we compared the networks among LSTM-VGG-19 and LSTM-FCN, and LSTM-FCN-variant', where LSTM-FCN-variant' is a variant of LSTM-FCN-variant by removing the feature fusion between different convolution layers. Compared with LSTM-VGG-19, LSTM-FCN-variant' has an approximate 1.7% improvement on accuracy. The perplexity of LSTM-VGG-19 was decreased by approximately 0.53. Compared with LSTM-FCN, the accuracy of LSTM-FCN-variant' is higher 1.32%, while the perplexity is lower 0.037. The above experiments demonstrate the small kernels had a positive effect on final decision making.

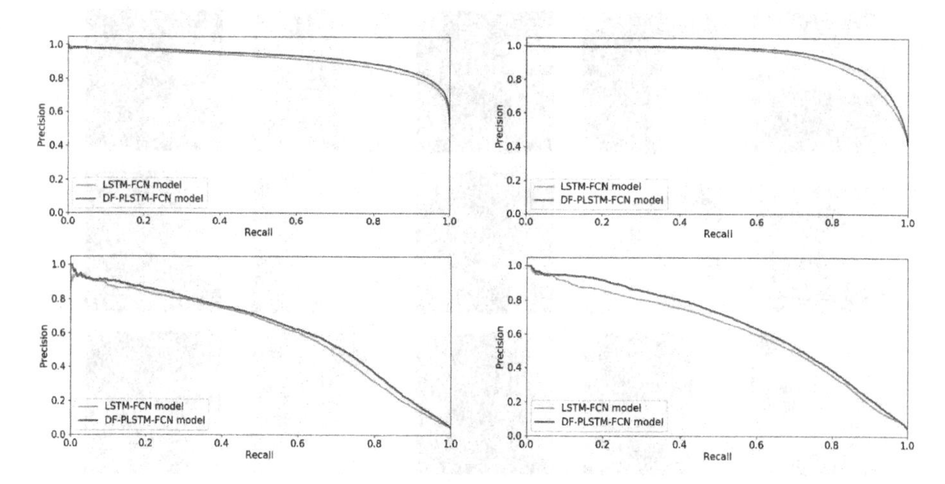


**Fig. 5.** PR-Curve. The red curve represents the DF-PLSTM-FCN model, and the orange curve represents the LSTM-FCN model. The four pictures, in turn, represent the changes of the P-R curves predicted by the two models on four actions straight, stop, turn left, and turn right (color figure online)

**Dual Feature Fusion.** The second part is on the feature fusion algorithm. Firstly, we add adaptive linear models to LSTM-FCN-variant'. Compared with the LSTM-FCN-variant', the addition of the adaptive linear model improves its accuracy by 0.36%, while the perplexity decreases by 0.06 accordingly. Therefore, we can inference that the adaptive linear model is an effective component to enhance the network learning feature descriptor. LSTM-FCN-Adaptive-linear-model in Table 1 also verifies this point. At the same time, the accuracy of LSTM-FCN-variant'-Adaptive-linear-model is about 1.33% higher than that of LSTM-FCN-Adaptive-linear-model, which still shows that LSTM-FCN-variant' performs better than the LSTM-FCN. After that, feature fusion layers have been added on the basis of LSTM-FCN-variant', that is, the LSTM-FCN-variant described in Fig. 3. It can be seen from the data that the insertion of the feature fusion layers improves the accuracy of LSTM-FCN-Adaptive-linear-model by about 0.17%. The LSTM-FCN-variant'-Adaptive-linear-model is improved by 0.04%. Based on the results, we can speculate that feature fusion technology through deconvolution is helpful for the network to recover the lost information and improve the accuracy of the model prediction.

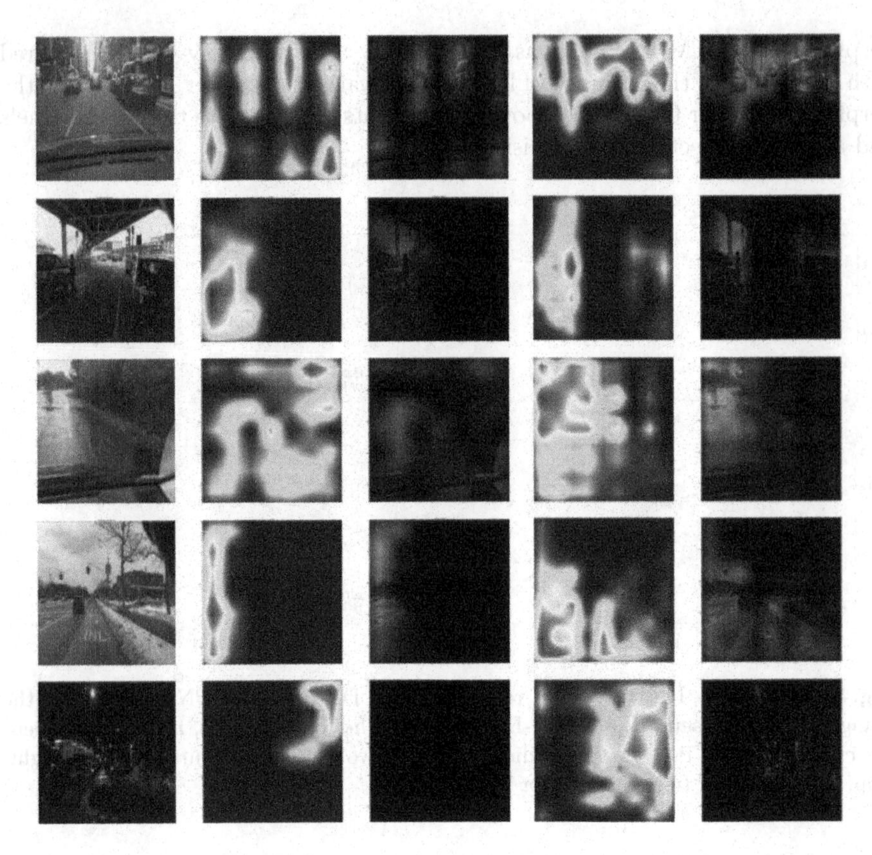

**Fig. 6.** Network Visualization Graphs. From left to right are Original graph, Grad-CAM by LSTM-FCN, Guided Backpropagation by LSTM-FCN, Grad-CAM by DF-PLSTM-FCN and Guided Backpropagation by DF-PLSTM-FCN

**Decision Fusion.** To further improve the accuracy of predict result, we also tried to use decision fusion technology. It combines the two networks in the second part of the experiment through adaptive linear models, which is the DF-PLSTM-FCN model finally proposed in the paper. The addition of decision fusion improves the model slightly, with an accuracy rate of about 85%.

### 4.4 Evaluation Index

Vehicles may perform multiple actions while driving. Therefore, behavioral decision-making is a multi-classification problem. And the movement of going straight is obviously more common than that of turning. Therefore, the data classification in the dataset is unbalanced. In view of this phenomenon, PR curve was selected to evaluate the prediction performance of DF-PLSTM-FCN and LSTM-FCN. As shown in Fig. 5. It can be seen from the figure that the red curve performs significantly better than that of orange. In the real world, going straight is indeed more common than turning. This is the reason why the prediction results of training samples are highly concentrated on going straight and stopping. Besides, it can be seen from our figure that the AP value of DF-PLSTM-FCN is significantly higher than that of LSTM-FCN.

Gradient-weighted Class Activation Mapping (Grad-cam) [16] is a visual interpretation technique designed for decision-making by convolutional neural networks. We use it to visualize the convolution results of DF-PLSTM-FCN and LSTM-FCN. The key monitoring domains which play an important role in model prediction are generated as the heat maps on the original image. As shown in Fig. 6, compared with the LSTM-FCN network, DF-PLSTM-FCN obtained better results in the recognition of objects on the road.

## 5 Conclusion

Given the road observation images and vehicle driving data, this paper proposes a dual-fusion-based parallel LSTM-FCN network, named DF-PLSTM-FCN, to make end-to-end decisions. Compared with LSTM-FCN, LSTM-FCN-variant is deepened, the representation ability of the subnetwork is enhanced. Fortunately, the parameters are not increased due to the adoption of the small size kernels.

To obtain better features, we use dual feature fusion in the extraction of features. The first fusion is to fuse the feature descriptors in different layers, while the second is to further fuse the features generated from SPP. The experiment demonstrates both fusions are contributed to the improvement of precision. We also take a decision fusion method, an adaptive linear model, to merge the probability distribution results of the parallel networks. It helps the network to make optimal decisions in complex traffic environments.

In the real unmanned driving scenarios, on-line decision making is required. The proposed model may be limited in coping with the unknown environment. As a well-known method for on-line learning, reinforcement learning handle the uncertainty of the environment better. Therefore, future work will focus on building a dynamics model during the learning of decision-making, then we can adopt model-based reinforcement learning methods for online learning.

## References


1. Caltagirone, L., Bellone, M., Svensson, L., Wahde, M.: Lidar-camera fusion for road detection using fully convolutional neural networks. Robot. Auton. Syst. **111**, 125–131 (2019)
2. Chen, C., Jafari, R., Kehtarnavaz, N.: Action recognition from depth sequences using depth motion maps-based local binary patterns. In: 2015 IEEE Winter Conference on Applications of Computer Vision, pp. 1092–1099. IEEE (2015)

3. Codevilla, F., Miiller, M., López, A., Koltun, V., Dosovitskiy, A.: End-to-end driving via conditional imitation learning. In: 2018 IEEE International Conference on Robotics and Automation (ICRA), pp. 1–9. IEEE (2018)
4. Cunningham, A.G., Galceran, E., Eustice, R.M., Olson, E.: MPDM: multipolicy decision-making in dynamic, uncertain environments for autonomous driving. In: 2015 IEEE International Conference on Robotics and Automation (ICRA), pp. 1670–1677. IEEE (2015)
5. Galceran, E., Cunningham, A.G., Eustice, R.M., Olson, E.: Multipolicy decision-making for autonomous driving via changepoint-based behavior prediction: theory and experiment. Auton. Robots **41**(6), 1367–1382 (2017). https://doi.org/10.1007/s10514-017-9619-z
6. Gu, T., Dolan, J.M.: On-road motion planning for autonomous vehicles. In: Su, C.-Y., Rakheja, S., Liu, H. (eds.) ICIRA 2012. LNCS (LNAI), vol. 7508, pp. 588–597. Springer, Heidelberg (2012). https://doi.org/10.1007/978-3-642-33503-7_57
7. He, K., Zhang, X., Ren, S., Sun, J.: Spatial pyramid pooling in deep convolutional networks for visual recognition. IEEE Trans. Pattern Anal. Mach. Intell. **37**(9), 1904–1916 (2015)
8. Hochreiter, S., Schmidhuber, J.: Long short-term memory. Neural Comput. **9**(8), 1735–1780 (1997)
9. Krizhevsky, A., Sutskever, I., Hinton, G.E.: Imagenet classification with deep convolutional neural networks. In: Advances in Neural Information Processing Systems, pp. 1097–1105 (2012)
10. Kuderer, M., Gulati, S., Burgard, W.: Learning driving styles for autonomous vehicles from demonstration. In: 2015 IEEE International Conference on Robotics and Automation (ICRA), pp. 2641–2646. IEEE (2015)
11. Li, L., Ota, K., Dong, M.: Humanlike driving: empirical decision-making system for autonomous vehicles. IEEE Trans. Veh. Technol. **67**(8), 6814–6823 (2018)
12. Li, W., Chen, C., Su, H., Du, Q.: Local binary patterns and extreme learning machine for hyperspectral imagery classification. IEEE Trans. Geosci. Remote Sens. **53**(7), 3681–3693 (2015)
13. Long, J., Shelhamer, E., Darrell, T.: Fully convolutional networks for semantic segmentation. In: Proceedings of the IEEE Conference on Computer Vision and Pattern Recognition, pp. 3431–3440 (2015)
14. Qassim, H., Verma, A., Feinzimer, D.: Compressed residual-VGG16 CNN model for big data places image recognition. In: 2018 IEEE 8th Annual Computing and Communication Workshop and Conference (CCWC), pp. 169–175. IEEE (2018)
15. Russakovsky, O., et al.: Imagenet large scale visual recognition challenge. Int. J. Comput. Vis. **115**(3), 211–252 (2015). https://doi.org/10.1007/s11263-015-0816-y
16. Selvaraju, R.R., Cogswell, M., Das, A., Vedantam, R., Parikh, D., Batra, D.: Grad-cam: visual explanations from deep networks via gradient-based localization. In: Proceedings of the IEEE International Conference on Computer Vision, pp. 618–626 (2017)
17. Simonyan, K., Zisserman, A.: Very deep convolutional networks for large-scale image recognition. arXiv preprint arXiv:1409.1556 (2014)
18. Xu, H., Gao, Y., Yu, F., Darrell, T.: End-to-end learning of driving models from large-scale video datasets. In: Proceedings of the IEEE Conference on Computer Vision and Pattern Recognition, pp. 2174–2182 (2017)
19. Xu, W., Wei, J., Dolan, J.M., Zhao, H., Zha, H.: A real-time motion planner with trajectory optimization for autonomous vehicles. In: 2012 IEEE International Conference on Robotics and Automation, pp. 2061–2067. IEEE (2012)

20. Zhang, Q., Yang, L.T., Chen, Z., Li, P.: A survey on deep learning for big data. Inf. Fusion **42**, 146–157 (2018)
21. Zhang, Z., Zhang, X., Peng, C., Xue, X., Sun, J.: Exfuse: enhancing feature fusion for semantic segmentation. In: Proceedings of the European Conference on Computer Vision (ECCV), pp. 269–284 (2018)

# A Modified Joint Geometrical and Statistical Alignment Approach for Low-Resolution Face Recognition

Rakesh Kumar Sanodiya[1], Pranav Kumar[2], Mrinalini Tiwari[2], Leehter Yao[1](✉), and Jimson Mathew[2]

[1] National Taipei University of Technology, Taipei 10608, Taiwan
rakesh.pcs16@gmail.com, ltyao@ntut.edu.tw
[2] Indian Institute of Technology Patna, Patna, India
pranavkmr79@gmail.com, mrinalini.cse0206@gmail.com, jimson@iitp.ac.in

**Abstract.** Domain Adaptation (DA) or Transfer Learning (TL) makes use of the already available source domain information for training the target domain classifier. Traditional ML algorithms require abundant amount of labeled data for training the model, and also they assume that both training and testing data follow similar distributions. However, in a real-world scenario, this does not always work. The scarcity of labeled data in the target domain is a big issue. Also, the source and the target domains have distinct data distributions. So, lessening the gap between the distributions of the two domains is very important so that a model that is trained using source domain information can be deployed to classify the target domain information efficiently. The already existing domain adaptation technique tries to reduce this distribution interval statistically and geometrically to an extent. Nevertheless, it requires some important components such as Laplacian regularization and maximizing source domain variance. Hence, we propose a Modified Joint Geometrical and Statistical Alignment (MJGSA) approach for Low-Resolution Face Recognition that enhances the previous transfer learning methods by incorporating all the necessary objectives that are useful for diminishing the distribution gap between the domains. Rigorous experiments on several real-world datasets verify that our proposed MJGSA approach surpasses other state-of-the-art existing methods.

**Keywords:** Unsupervised learning · Domain adaptation · Low-resolution face recognition · Dimensionality reduction · Transfer learning · Classification

## 1 Introduction

Nowadays, Face recognition has become the trending topic for research in the field of machine learning. This is because it is widely applicable for many real world applications such as identify and track criminals, find missing children and


H. Yang et al. (Eds.): ICONIP 2020, LNCS 12532, pp. 88–100, 2020.
https://doi.org/10.1007/978-3-030-63830-6_8

disoriented adults etc. Most of the face recognition systems attains good recognition accuracy under High-resolution face images. But practically, the performance of these systems degrades due to the negative impact of Low-Resolution (LR) Images. Hence, enhancing the performance of LR face recognition systems is equally important.

Conventional machine learning approaches follow the assumption that both training and test data belong to same set of distributions [8]. However, practically there is a large distribution gap between both the source and the target domain images because of various factors such as different illumination, poses, backgrounds, low-resolutions etc. Also, the source domain contains a sufficient amount of labeled images for the training purpose, while the target domain has either less or no labeled images. So, it is not possible to use the classifier trained with the source domain images to classify the target domain unlabeled images efficiently. Domain adaptation is a subfield of machine learning which minimizes the distribution gap between the two domains by considering many of the objectives like minimizing the marginal and conditional distributions, maximizing the target domain variance, subspace alignment, etc. This technique has a huge application in cross-domain image classification which also focuses on enhancing the efficiency of Low-Resolution Face Recognition Systems.

Domain adaptation (DA) [10,11] utilizes already available labeled source data information for hiking the performance of diverse but interrelated target domain applications. The DA algorithms can be categorized into two broad types: a) Semi-supervised DA [1,6] and b) Unsupervised DA [9]. Semi-supervised DA algorithms have fewer labeled data, and Unsupervised DA algorithms have no labeled data for the target domain. In this paper, we deal with the Unsupervised domain adaptation as in real world, getting labelled data is quite a tedious task. Also, the approaches for domain adaptation are broadly classified as: (i) Instance-based DA [7] approaches, which utilizes re-weighting of data samples of source domain that can be used further for training the classifier of the target domain, and (ii) Feature based DA [11] approaches, which aims to develop a common feature subspace between the two domains.

The already existing DA approaches try to reduce the distribution gap between the two domains to an extent by covering many important components like Target Variance maximization, Marginal and Conditional distribution minimization, subspace alignment, etc. Still, they lack in covering some of the important components like Laplacian regularization and Source domain variance maximization. We have proposed a novel Modified Joint Geometrical and Statistical Approach (MJGSA) for Low-resolution Face recognition, which has given equal importance to all the components (listed in Subsect. 3.2) needed to reduce the distribution discrepancy between the domains. The significant contributions of our novel MJGSA method are listed as under:

1) Our proposed approach tries to minimize the domain shift by taking into account both geometrical and statistical properties of the domains.
2) Our proposed MJGSA approach diminishes the distribution discrepancy between the domains by taking into account several components like

Minimization of source domain variance, Minimization of target domain variance, Distribution divergence minimization, Manifold geometry preservation, Subspace divergence minimization.

3) Our proposed MJGSA algorithm is working very well on six standard face datasets. For indicating the superiority of our proposed approach, we have compared it with existing DA approaches and have shown that our method successfully outperforms all the other ones.

## 2 Related Work

Plenty of works have been done in the field of Domain adaptation. Here we will discuss the techniques that are related to our proposed approach.

A popular technique, MMD (Maximum Mean Discrepancy) [4] aims to diminish the distribution gap by learning the common feature space between the domains. Transfer Component Analysis (TCA) [5] extends this idea by determining the common feature space in an RKHS (Reproducing Kernel Hilbert Space) using the MMD approach. These common features are then utilized by the machine learning algorithms for getting better classification results.

Another unsupervised technique, Joint Distribution Adaptation (JDA) [3], further extends TCA by taking into consideration both the marginal and conditional distribution gap between the domains. In this approach, Principal Component Analysis (PCA) and MMD bound together to minimize the joint distribution gap.

Further, a framework for visual domain adaptation is introduced, i.e., JGSA (Joint Geometrical and Statistical Adaptation) [11], which aims to diminish the distribution gap both statistically and geometrically to an extent. This technique aims to determine coupled projections for both the source and the target domains. Existing methods fail to take into account some of the issues like preserving the source domain samples information and maximization of source domain variance. Hence we proposed a novel method MJGSA, which enhances JGSA by incorporating these objectives in the newly formulated objective function that finally diminishes the distribution discrepancy between the domains.

## 3 Framework for Visual Domain Adaptation

Our proposed method and related concepts are described in this section.

### 3.1 Problem Description

In our research, we focus on Unsupervised DA technique in which source domain has abundant labelled data while target domain has no labelled data at all. $\mathcal{V}_s = \{(x_i, y_i)\}_{i=1}^{n_s}$, $x_i \in \mathrm{IR}^{\mathrm{d}}$ represents the source domain and $\mathcal{V}_t = \{(x_j)\}_{j=1}^{n_t}$, $x_j \in \mathrm{IR}^{\mathrm{d}}$ represents the target domain where d represents data dimensions. $n_s$ represents the number of labelled data samples while $n_t$ represents the number of unlabelled data samples. Source domain labels and target domain labels are depicted by $\mathcal{Y}_s$ and $\mathcal{Y}_t$. Combined data of both the domains is represented by $\mathcal{V}$. For experimental purpose, there is a pre-assumption that the domains have similar feature spaces and label spaces i.e. $\mathcal{V}_s = \mathcal{V}_t$ and $\mathcal{Y}_s = \mathcal{Y}_t$.

### 3.2 Model Formulation

The proposed approach MJGSA for low-resolution face recognition aims to lessen the distribution gap between the domains, both statistically and geometrically, by taking into account all the necessary components in the defined objective function. The main focus of MJGSA approach is to find two coupled mappings, E (for source domain) and F (for target domain) for projecting both the domains feature spaces onto a common subspace. The description of the different components used in our objective function is as follows:

**Maximization of Source Domain Variance (SDV).** As the necessary amount of data is available for the source domain, we can maximize the source domain variance by projecting it into a new subspace. We have utilized Principal Component Analysis (PCA) [2] for maximizing the SDV with the help of a source domain projection matrix $E$. This can be done using the following formula:

$$\underset{E}{\text{Max}}\ E^T J_s E \tag{1}$$

where $J_s = \mathcal{V}_s^T H_t \mathcal{V}_s$, $H_s = I_s - \frac{1}{n_s} 1_s (1_s)^T$, $I_s$, and $1_s \in \mathrm{IR}^{n_s}$ are the scatter matrix, the centering matrix of the source domain, identity matrix of size $n_s$, and the column vector with all ones, respectively.

**Maximization of Target Domain Variance (TDV).** The target domain variance can be maximized when the data samples will get cast onto a new subspace by utilizing PCA as the dimensionality reduction technique. With the help of target domain projection vector matrix $F$, the TDV can be maximized as follows:

$$\underset{F}{\text{Max}}\ F^T J_t F \tag{2}$$

where $J_t = \mathcal{V}_t^T H_t \mathcal{V}_t$ and $H_t = I_t - \frac{1}{n_t} 1_t (1_t)^T$, $I_t$, and $1_t \in \mathrm{IR}^{n_t}$ are the scatter matrix, the centering matrix of the target domain, identity matrix of size $n_t$, and the column vector with all ones, respectively.

**Minimization of Distribution Divergence (DD).** As proposed in the JGSA [11] technique, we also try to diminish both the marginal and conditional distribution variance.

The statistical probability distributions of the two domains are compared using the Maximum Mean Discrepancy (MMD) [3] approach. The distribution distance between the domains is computed to reduce the Marginal distribution gap, as shown by the equation:

$$\underset{E,F}{\text{Min}} \left\| \frac{1}{n_s} \sum_{x_i \in \mathcal{V}_s} E^T x_i - \frac{1}{n_t} \sum_{x_j \in \mathcal{V}_t} F^T x_j \right\|_f^2 \tag{3}$$

The marginal distribution is not sufficient to minimize the distribution discrepancy if the data is class-wise distributed. We have to take into account the conditional distribution also. Long et al. [3] proposed a technique to predict pseudo labels, by applying base classifiers onto the labelled source domain data and then iteratively refine them. The conditional distribution shift is reduced by the formula:

$$\underset{E,F}{\text{Min}} \sum_{i=1}^{C} \left\| \frac{1}{n_s^i} \sum_{x_i \in \mathcal{V}_s^i} E^T x_i - \frac{1}{n_t^i} \sum_{x_j \in \mathcal{V}_t^i} F^T x_j \right\|_f^2 \tag{4}$$

For effective domain adaptation, we have to diminish both the marginal as well as conditional shift between the two domains. Equation (3) and Eq. (4) constitutes together to form the final joint distribution minimization term as follows:

$$\underset{E,F}{\text{Min}} Tr\left( \begin{bmatrix} E^T & F^T \end{bmatrix} \begin{bmatrix} G_s & G_{st} \\ G_{ts} & G_t \end{bmatrix} \begin{bmatrix} E \\ F \end{bmatrix} \right) \tag{5}$$

The calculation of $G_s$, $G_t$, $G_{st}$ and $G_{ts}$ can be in same manner as proposed by JGSA [11] method.

**Geometrical Diffusion on Manifolds (GDM).** The prior knowledge of the geometry of unlabeled data from both the domains can be considered as the samples that are closer in the higher manifold are likely to be nearer in low dimensional manifold as well. Hence, by utilizing the data samples $\mathcal{V}$, a K-nearest neighbour graph is constructed to model the relationship between nearby data samples. A new component, i.e., Laplacian regularization term R is incorporated in the objective function for taking into account the exact geometry of manifolds.

The similarity between any two data samples $x_i$ and $x_j$ can be shown by the Weight matrix ($W$) as:

$$W_{ij} = \begin{cases} 1, & \text{if } x_i \in nn_k(x_j) \mid x_j \in nn_k(x_i) \\ 0, & \text{otherwise} \end{cases}$$

where, $nn_k(x_j)$ and $nn_k(x_i)$ depicts the set of K nearest neighbours of $x_j$ and $x_i$ respectively. The Laplacian matrix is computed as $L = (D - W)$, where the sum of columns of $W$ are the diagonal matrix D entries i.e. $D_{ii} = \sum_j W_{ij}$. Finally, the regularization term can be derived as:

$$R = \sum_{i,j} (x_i - x_j)^2 W_{ij} = 2\mathcal{V}L\mathcal{V}^T \tag{6}$$

**Minimization of Subspace Divergence (SD).** The divergence between the distribution of domains (source and target) can also be reduced by bringing their feature subspaces closer as well. The projection vectors E and F can be optimized without projecting the two subspaces onto a new matrix. This can be depicted as follows:

$$\underset{E,F}{\text{Min}} \|E - F\|_f^2 \tag{7}$$

The shared as well as the domain-specific features are utilized, as proposed in the JGSA [11] method.

### 3.3 New Objective Function

Here, the objective function for our proposed MJGSA framework that is formulated by incorporating all the components listed in Subsect. 3.2, is described. It can be represented as:

$$\text{Max} \frac{\{(\mu\{\text{TDV}\} + \beta\{\text{SDV}\})\}}{(\alpha\{\text{SD}\} + \lambda\{\text{GDM}\} + \{\text{DD}\})}$$

where $\mu$, $\beta$, $\alpha$, and $\lambda$ are various trade-off parameters that balance the significance of every term. We can represent the objective function as:

$$\underset{E,F}{\text{Max}} \frac{\text{Tr}\left(\begin{bmatrix} E^T & F^T \end{bmatrix} \begin{bmatrix} \beta J_s & 0 \\ 0 & \mu J_t \end{bmatrix} \begin{bmatrix} E \\ F \end{bmatrix}\right)}{\text{Tr}\left(\begin{bmatrix} E^T & F^T \end{bmatrix} \begin{bmatrix} G_s + \alpha I + \eta R & G_{st} - \alpha I + \eta R \\ G_{ts} - \alpha I + \eta R & G_t + (\alpha + \mu) I + \eta R \end{bmatrix} \begin{bmatrix} E \\ F \end{bmatrix}\right)}$$

where $I \in \mathbb{R}^{d*d}$ is the identity matrix.

For optimizing the objective function, the same steps proposed by JGSA [11] algorithm are followed. After rewriting $\begin{bmatrix} E^T & F^T \end{bmatrix}$ as $Q^T$, and reformulating the above equations by applying the Lagrange function $\delta$, we will end up to the generalized eigenvalue decomposition problem whose solutions are the matrices $\delta$ and Q. The diagonal matrix $\delta$ comprises of the top k eigenvalues, and Q comprises of the corresponding eigenvectors. Thus, E and F can be computed from Q. The pseudo-code of our proposed approach MJGSA is depicted in Algorithm 1.

## 4 Experiments

In this section, we have used the six benchmark face databases and applied the proposed method to check the effectiveness of it. We have compared the results of the proposed approach with three state-of-the-art DA techniques (i.e., Joint Geometrical and Statistical Alignment (JGSA) [11], Joint distribution analysis (JDA) [3] and Transfer component analysis (TCA) [5])

**Algorithm 1:** Proposed MJGSA for visual domain adaptation

**Input** : Source domain data ($\mathcal{V}_s$), Target domain data ($\mathcal{V}_t$), Source labels ($\mathcal{Y}_s$), Trade-off parameters ($\mu$, $\beta$, $\alpha$, $\lambda$, *graph*) and maximum number of iterations ($I_{max}$)

**Output**: Subspaces $E$ and $F$, Transformation matrix Q.

1 Build $J_s$, $J_t$.
2 Initialize i = 0.
3 Compute the initial Pseudo label from $\mathcal{V}_s, \mathcal{Y}_s$ and $\mathcal{V}_t$.
4 **while** $i < I_{max}$ **do**
5 Construct MMD matrix $G_s$, $G_{st}$, $G_{ts}$ and $G_t$ by using Pseudo label;
6 Compute the value of E and F by solving the eigenvalue decomposition function;
7 Project the source and target data to their respective subspaces as $\mathcal{Z}_s = E^T \mathcal{V}_s$ and $\mathcal{Z}_t = F^T \mathcal{V}_t$ ;
8 Train the classifier with $\mathcal{Z}_s$, $\mathcal{Y}_s$, and $\mathcal{Z}_T$ and compute Pseudo label of target domain;
9 **end**
10 Obtain the final accuracy.

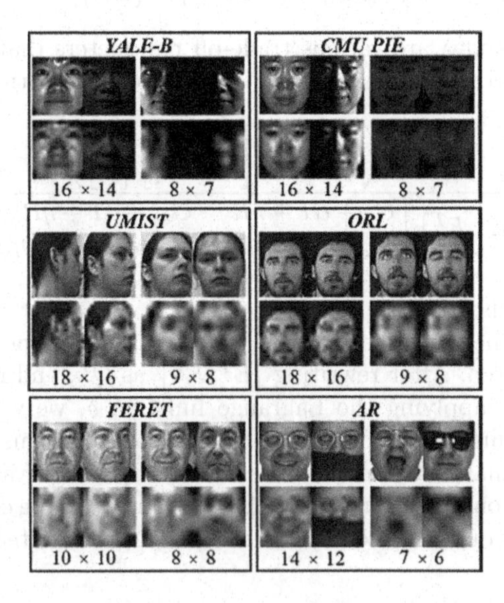


**Fig. 1.** Some sample images of the benchmark face databases. (1st row depicts HR images, while 2nd row depicts LR images.)

### 4.1 Benchmark Datasets

In the experiments, we have used six benchmark face databases, namely AR, CMU-PIE, FERET, ORL, UMIST, and YALE-B, for accessing the performance of our proposed MJGSA method. Figure 1 shows some sample images selected from the above-mentioned databases. The descriptions of the databases are as follows:

AR database consists of images of 116 distinct subjects (total 3288 images) with different expressions, poses, and illumination conditions. For experimental purposes, 100 individuals out of 116 subjects are randomly chosen at a time, and each one has 26 face images with the frontal pose, illumination conditions, and expressions. FERET database consists of images of 1199 distinct subjects (total 14,126 images) with different expressions, poses, and illumination conditions. For experimental purposes, 200 individuals out of 1199 subjects are randomly chosen at a time, and each one has 7 face images with the frontal pose, illumination conditions, and expressions.

The ORL database comprises of images of 40 subjects (total of 400 images) with distinct postures and expressions. The UMIST Face database consists of images of 20 subjects (total of 564 images) with different postures and angles. The CMU-PIE database consists of images of 68 subjects (total 41,368 images) with 43 distinct illumination conditions, 13 distinct poses, and 4 distinct expressions. The Extended YALE-B database consists of images of 38 subjects (total 16,128 images) with 9 distinct poses and 64 distinct illumination conditions.

For each of the face databases, we randomly choose one-half images of each subject as the source domain, and the other half images of each subject as the target domain. Images of each face database are not of low resolution. The high-resolution images of the above-mentioned databases are smoothly downsampled to low-resolution sizes. Images of the AR Face database are downsampled to the sizes $14 \times 12$ and $7 \times 6$, while FERET database images are downsampled to the sizes $10 \times 10$ and $8 \times 8$. Images of CMU-PIE and YALE-B databases are downsampled to the sizes $8 \times 7$ and $16 \times 14$, while ORL and UMIST face database images are downsampled to the sizes $9 \times 8$ and $18 \times 16$.

### 4.2 Experimental Results

The recognition accuracy of our proposed MJGSA technique is compared with the three state-of-the-art techniques on six face databases, as depicted in Table 1. From the table, we can see that the proposed method constantly achieves the best recognition accuracy as compared to TCA, JDA, and JGSA on six face databases [12]. For the resolution $8 \times 7$ and $16 \times 14$ of the YALE-B image database, accuracy of the proposed method is around 10% higher than the other approaches. The recognition rate of the proposed algorithm on the CMU-PIE face database is around 97%, which shows the efficiency of the proposed algorithm. Again on AR image database, accuracy of the proposed algorithm is around 10% more than the other state-of-art approaches. Accuracy of the proposed algorithm is

**Table 1.** Accuracy (%) comparison of MJGSA with three state-of-the-art methods for two resolutions of each of six face databases

| Face database | Resolution | TCA | JDA | JGSA | JGSALR (Proposed) |
|---|---|---|---|---|---|
| YALE-B | $8 \times 7$ | 57.1 | 57.74 | 43.22 | 67.3 |
| | $16 \times 14$ | 63.72 | 66.27 | 68.5 | 80.62 |
| UMIST | $9 \times 8$ | 89.6 | 90.0 | 91.8 | 93.1 |
| | $18 \times 16$ | 90.1 | 91.0 | 92 | 94 |
| ORL | $9 \times 8$ | 86.0 | 90.6 | 86.4 | 92.0 |
| | $18 \times 16$ | 88.0 | 91.0 | 90.5 | 91.0 |
| FERET | $8 \times 8$ | 43.01 | 70.23 | 45.0 | 80.4 |
| | $10 \times 10$ | 43.5 | 72.00 | 48.0 | 84.0 |
| CMU-PIE | $8 \times 7$ | 86.0 | 93.0 | 78.0 | 95.0 |
| | $16 \times 14$ | 91.00 | 94.0 | 87.6 | 96.8 |
| AR | $7 \times 6$ | 36.8 | 55.1 | 35.0 | 65.3 |
| | $14 \times 12$ | 41.2 | 58.6 | 39.0 | 69.2 |

also more than the accuracy of TCA, JDA, and JGSA on UMIST, ORL, and FERET face databases.

The higher performance mainly achieved due to incorporating Laplacian regularization in order to preserve the source domain discriminative information and using the concept of source domain variance maximization.

### 4.3 Experimental Analysis

Here, the variation in the recognition accuracy of our proposed MJGSA is analyzed by varying the value of k(Dim), dimensionality of feature of different subspace. For this experiment, we have taken low-resolution images of all six face databases. We have set the size of YALE-B and CMU-PIE to $8 \times 7$, that of FERET to $8 \times 8$, that of UMIST and ORL to $9 \times 8$, and similarly the resolution $7 \times 6$ for AR face database. As shown in Fig. 2, the best performance can be achieved by the proposed MJGSA method when the highest feature dimension matching is done.

Also, the accuracy results of the proposed MJGSA are compared with other state-of-the-art methods by two resolution levels of all six face databases. We have chosen resolution for CMU-PIE face and YALE-B databases as $8 \times 7$ and $16 \times 14$, Resolution of $8 \times 8$ and $10 \times 10$ are selected for the FERET database, at the same time resolution of $9 \times 8$ and $18 \times 16$ for the ORL and the UMIST databases, at the end resolution of $7 \times 6$ and $14 \times 12$ for the AR face database. Figure 4 illustrates the comparison of results for two resolution levels of different methods. Here, we can see that the proposed method shows better performance than other methods in all cases.

### 4.4 Parameter Sensitivity Test

In the proposed method, performance got effected due to five vital parameters, namely $\alpha$, $\beta$, $\mu$, $\lambda$, and graph. For the testing purpose, at first, we have taken face database YALE-B as test database and repeated the experiment five times to find the effect of these five parameters. We have set the size of the image to $16 \times 14$. At first, we have set the value of $\beta = .0001$, $\mu = .0001$, $\lambda = .001$ and graph $= 3$ and then vary the value of $\alpha$ from 0.5 to 2.5 with an interval of 0.5. We notice that at first, with an increase of $\alpha$, accuracy gets increased and then decreases rapidly, as illustrated in Fig. 3a. We have set the value of $\alpha = 1.5$, $\mu = .0001$, $\lambda = .001$ and graph $= 3$ and then vary the value of $\beta$ from 1 to .0001 by a factor of $10^{-1}$ in each experiment. Here we see that performance of the proposed algorithm get increased with a decrease in the value of $\beta$, as shown in Fig. 3b. For finding the effect of $\mu$ on the performance of the proposed method, we have set the value of $\alpha = 1.5$, $\beta = .0001$, $\lambda = .001$ and graph $= 3$ and then vary the value of $\mu$ from 1 to .0001 by a factor of $10^{-1}$ in each experiment. Again

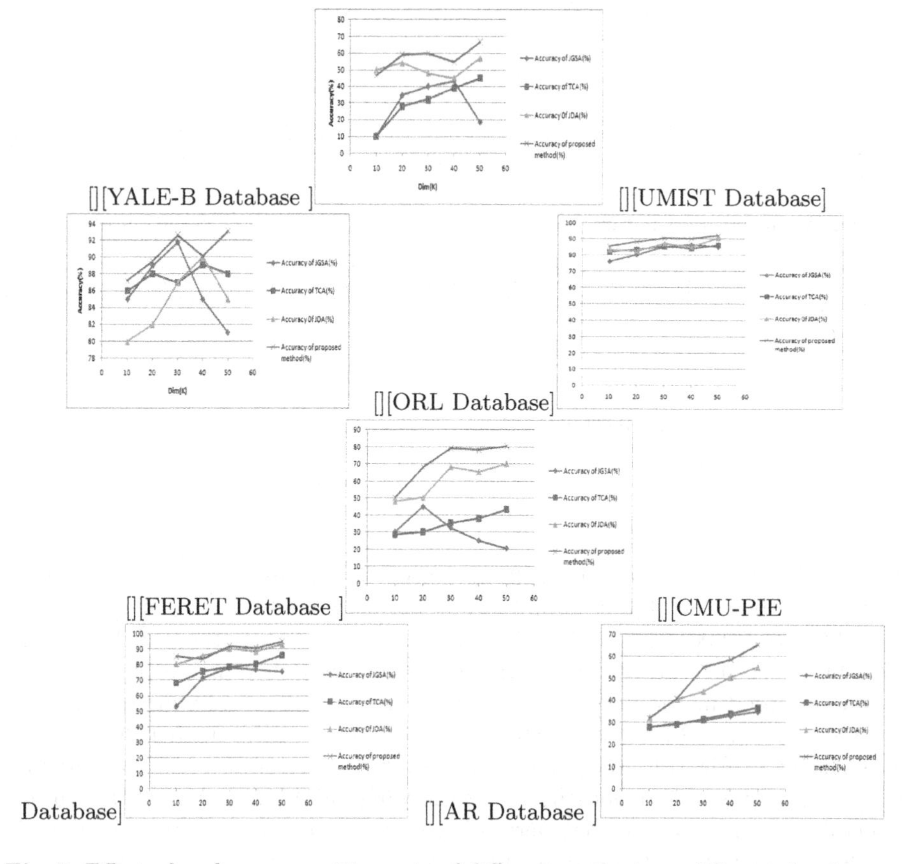


**Fig. 2.** Effect of rank on recognition rate of different methods on different databases.

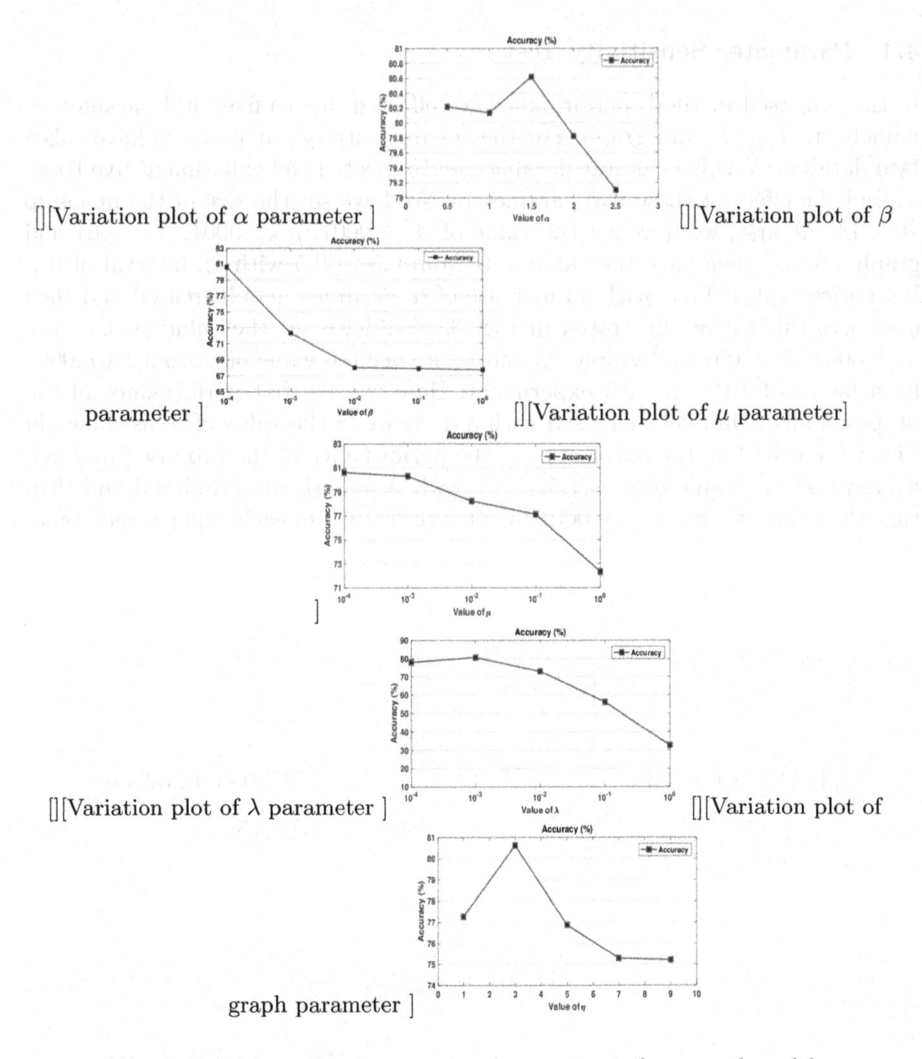


**Fig. 3.** Variations of parameters on the accuracy of proposed model

we find that the performance of the proposed algorithm gets increased with a decrease in the value of $\mu$, as shown in Fig. 3c. We have set the value of $\alpha = 1.5$, $\beta = .0001$, $\mu = .0001$ and graph $= 3$ and then vary the value of $\lambda$ from 1 to .0001 by a factor of $10^{-1}$ each time to illustrate the effect of $\lambda$ on the performance of our proposed method. We notice that at first, accuracy gets increased with a decrease in $\lambda$, and after $\lambda = .001$ it gets decreased with a further decrease in $\lambda$ as shown in Fig. 3d. At last, we have set the value of $\alpha = 1.5$, $\beta = .0001$, $\mu = .0001$ and $\lambda = .001$ and then vary the value of graph from 1 to 9 with an interval of 2. Here we see that performance at first increases with the value of graph and then starts to decrease after the value of graph $= 3$, as illustrated in Fig. 3e.

[][YALE-B (left), UMIST (right)] [][ORL (left), FERET (right)] [][CMU-PIE (left), AR (right)]

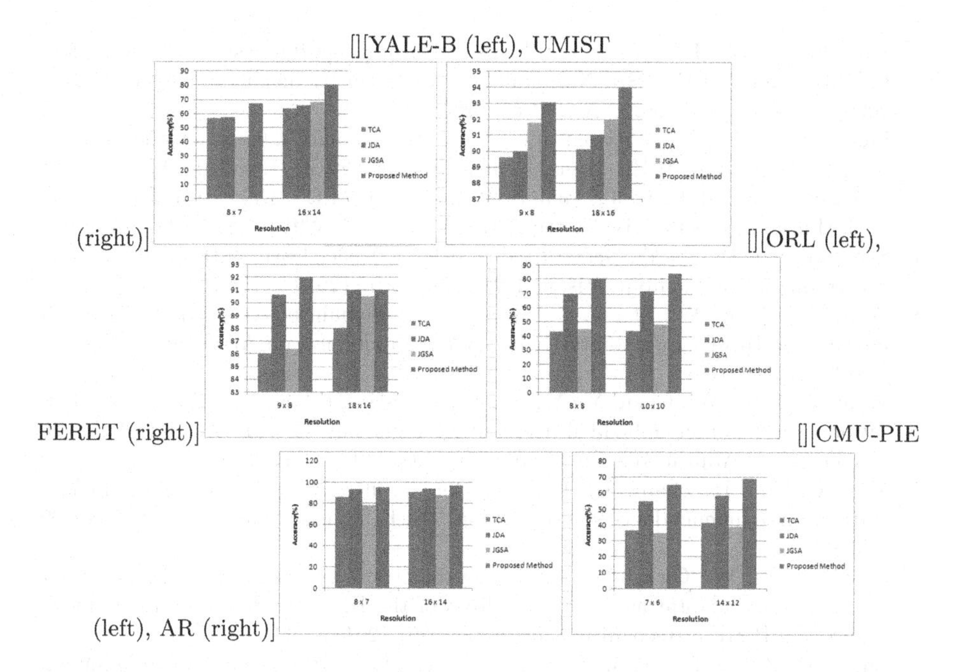


**Fig. 4.** Accuracy of different techniques with two distinct resolutions on six face databases.

## 5 Conclusion

In this paper, we have proposed a novel approach, MJGSA for low resolution face recognition. This approach tries to diminish the distribution discrepancy between the domains both statistically and geometrically. Our proposed method enhances JGSA approach by incorporating new objectives like source domain variance maximization and laplacian regularization into the proposed fitness function. The effectiveness of the proposed approach is verified by comparing it with existing DA approaches. The experimental results of our proposed approach on six different databases have indicated the superiority of our approach over other methods.

## References

1. Donahue, J., Hoffman, J., Rodner, E., Saenko, K., Darrell, T.: Semi-supervised domain adaptation with instance constraints. In: Proceedings of the IEEE Conference on Computer Vision and Pattern Recognition, pp. 668–675 (2013)
2. Duchene, J., Leclercq, S.: An optimal transformation for discriminant and principal component analysis. IEEE Trans. Pattern Anal. Mach. Intell. **10**(6), 978–983 (1988)
3. Long, M., Wang, J., Ding, G., Sun, J., Yu, P.S.: Transfer feature learning with joint distribution adaptation. In: Proceedings of the IEEE International Conference on Computer Vision, pp. 2200–2207 (2013)

4. Pan, S.J., Kwok, J.T., Yang, Q.: Transfer learning via dimensionality reduction. In: Proceedings of the 23rd National Conference on Artificial Intelligence, AAAI 2008, vol. 2, pp. 677–682. AAAI Press (2008)
5. Pan, S.J., Tsang, I.W., Kwok, J.T., Yang, Q.: Domain adaptation via transfer component analysis. IEEE Trans. Neural Networks **22**(2), 199–210 (2011)
6. Sanodiya, R.K., Mathew, J., Saha, S., Thalakottur, M.D.: A new transfer learning algorithm in semi-supervised setting. IEEE Access **7**, 42956–42967 (2019)
7. Shao, M., Kit, D., Fu, Y.: Generalized transfer subspace learning through low-rank constraint. Int. J. Comput. Vision **109**(1–2), 74–93 (2014)
8. Wang, H., Wang, W., Zhang, C., Xu, F.: Cross-domain metric learning based on information theory. In: Twenty-Eighth AAAI Conference on Artificial Intelligence (2014)
9. Wang, J., Feng, W., Chen, Y., Yu, H., Huang, M., Yu, P.S.: Visual domain adaptation with manifold embedded distribution alignment. In: 2018 ACM Multimedia Conference on Multimedia Conference, pp. 402–410. ACM (2018)
10. Yan, K., Kou, L., Zhang, D.: Learning domain-invariant subspace using domain features and independence maximization. IEEE Trans. Cybern. **48**(1), 288–299 (2018)
11. Zhang, J., Li, W., Ogunbona, P.: Joint geometrical and statistical alignment for visual domain adaptation. In: Proceedings of the IEEE Conference on Computer Vision and Pattern Recognition, pp. 1859–1867 (2017)
12. Zheng, D., Zhang, K., Lu, J., Jing, J., Xiong, Z.: Active discriminative cross-domain alignment for low-resolution face recognition. IEEE Access **8**, 97503–97515 (2020)

# A Part Fusion Model for Action Recognition in Still Images

Wei Wu(✉) and Jiale Yu

College of Computer, Inner Mongolia University, Huhhot, China
cswuwei@imu.edu.cn

**Abstract.** Recognizing actions from still images is very challenging due to the lack of human movement information and the variance of background in action categories. The existing methods in action recognition focus on extracting scene context information, modeling the human-object pair and its interactions or using human body information, of which part-based methods are one of the successful methods, which extract rich semantic information from the human body parts or pose. However, most of part-based models need to rely on expensive part annotations to improve the recognition accuracy. Different from these methods, in this paper, we propose a part fusion model which can effectively combine the discriminative parts information without extra annotations for action recognition. In our model, a guided attention module is used to further extract the more discriminative information in the image. And the part-level features are trained with the different weighted loss, which is mainly based on different object and background parts' characteristics. In order to further enhance the model performance, part-level features are fused to form new image-level features in the global supervision learning. This method achieves state-of-the-art result on the PPMI dataset and significant competitive performance on the Stanford-40 dataset, which demonstrates the effectiveness of our method for characterizing actions.

**Keywords:** Action recognition · The part-based method · Attention mechanism

## 1 Introduction

Although image-based action recognition has a wide range of applications in computer vision, it is still very challenging. Unlike the video-based action recognition [1–3] which is largely based on the spatial–temporal features, action recognition from still images is unable to rely on the motions, which greatly increases the difficulties in handling the difference in the human pose and background. Because of the lack of temporal-stream information, image-based methods exploit cues such as interactive objects, part appearance, or contextual information to improve recognition performances. Since human body parts provide


H. Yang et al. (Eds.): ICONIP 2020, LNCS 12532, pp. 101–112, 2020.
https://doi.org/10.1007/978-3-030-63830-6_9

rich information for recognizing human actions, part-based methods are one of the most effective approaches, which extracts appearance features from body parts or poses. Gkioxari et al. [4] used the sliding windows to detect the human body parts, and then combined them with the target person bounding box to jointly train the network. In [5], Zhao et al. proposed to use the bounding boxes to detect semantic body parts, and arrange their features in spatial order to find the differences in behavior classes. Deformable part-based models (DPM) [6] have demonstrated excellent results for object detection. The Poselets model [7] further develops DPM, which employs key points to detect some specific body parts, and achieves improved performance in action recognition vision tasks. However, a significant drawback of most of these methods [4,5,7] is the highly reliance on part annotations.

Contrastingly, we propose a part fusion network without external annotations for image-based action recognition. Overall, our model composes of the guided attention module and a two-level classification network. The attention module guided by image-level classification loss can enhance the localization ability of our proposed network and further improve the discriminative capability of our model. Most existing part-based methods to combine original image global information with the parts' information are concatenating their features and then use the classifier to predict. In order to utilize the image-level and part-level features, we combine part-level loss and image-level loss together to train our model. In the two-level classifier, we divide part proposal candidates generated by the Selective Search Windows (SSW) method [8] into three different parts: easy, hard and background parts. According to the difference in recognition difficulty, different weighting values are assigned to different parts in the part-level loss. On this basis, we can pay more attention to harder parts and reduce the harmful impact of background. Different from the existing approaches, the part-level features are fused to form a new global image-level feature, instead of concatenating the part-level features and the original image global features.

We evaluate our proposed model on two challenging action dataset: (1) PPMI [9] (2) Stanford-40 [10]. With extensive experiments, we obtain the state-of-the-art mAP (mean average precision) result of 86.46% in the PPMI and competitive mAP result of 88.73% in the Stanford-40, respectively. The experimental results demonstrate the effectiveness of our method compared to the other approaches.

The paper is organized as follows. We introduce our proposed method in detail in Sect. 2 and we show experimental results in two action datasets as well as ablation analysis in Sect. 3. Finally, we conclude our work in Sect. 4.

## 2 Method

In this section, we firstly introduce our proposed model, which mainly includes two major parts: the guided attention module and two-level classification network. The Fig. 1 shows the overall architecture of our model. For an input image $I \in \mathbb{R}^{H \times W \times 3}$, we exploit the SSW to generate the part proposal candidates. An enhanced feature map is obtained by the ConvNet with the guided attention module. The features of the part proposal candidates generated by the RoI Pooling layer are sent to a two-level classifier. In this two-level classifier, we exploit the weighted part-level loss to train the extracted part features with image-level annotations and then fuse these extracted features to form a new feature. The final loss of the whole network is composed of three parts. The rest of this section introduces the above components in detail.

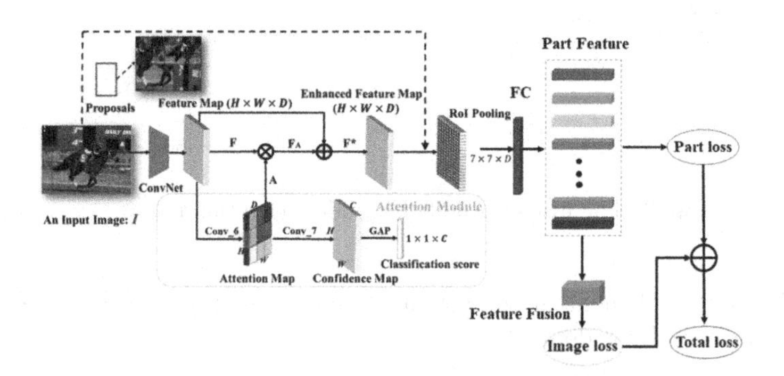


**Fig. 1.** The overall structure of our model.

### 2.1 The Guided Attention Module

We firstly introduce the overall framework of the guided attention module, which discriminatively selects the most informative locations within the background. Let $\mathrm{F} \in \mathbb{R}^{\mathrm{W \times H \times D}}$ denotes the feature map of an input image $I$ fed into the ConvNet, in which W, H, and D represent the width, height, and channel size of the feature map, respectively. The feature map F is then convolved by a $1 \times 1$ convolution layer to obtain a spatial attention weight map A. The attention map A is taken as a spatial regularizer, which effectively enhances the information of discriminative regions and suppresses the unimportant information for feature F. After that, the attention map is element-wise multiplied with F to generate an attentive feature $F_A$. Inspired by the residual learning of ResNet [11], $F_A$ is added to F to get an enhanced feature $F^*$ to improve the capability of feature representation. Figure 2 shows the guided attention mechanism implemented in our proposed model.

Formally, the attention module includes a convolution layer, a non-linear activation layer and a batch normalization. Given the score map Z is obtained by the 1×1 convolution layer, following is the practice of attention mechanism, and the score map is further processed by a non-linear Softmax which is given as follows:

$$A_{i,j} = \frac{\exp(Z_{i,j})}{\sum_i^w \sum_j^h \exp(Z_{i,j})} \tag{1}$$

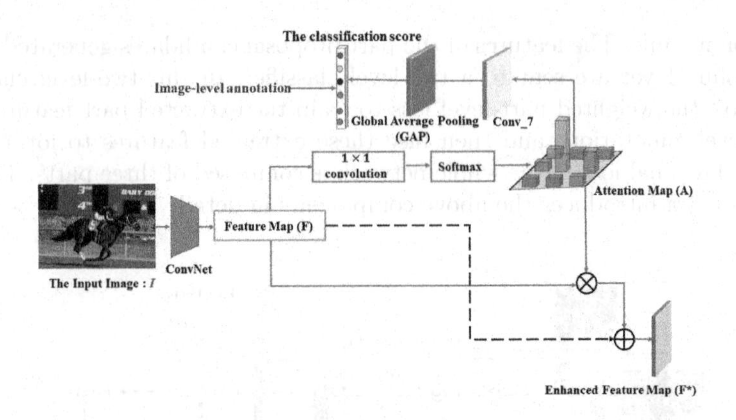


**Fig. 2.** The framework of the guided attention module.

where $A_{i,j}$ is the element of the attention map A at position $(i, j)$. The final enhanced feature $F^*$ is defined as:

$$F^*_{i,j} = (1 + A_{i,j}) \odot F_{i,j} \tag{2}$$

The convolution attention map is class-agnostic. In order to get more discriminative foreground information, we utilize the classification to guide the learning of the attention weighting. Although the spatial attention focuses on "where", which is an informative location, the channel-wise attention also obtains a more discriminative region for the target by suppressing the rest of noise interference. To achieve it, we modify the original spatial attention model to both channel and spatial-wise attention. More specially, the attention map is diverted from $A \in \mathbb{R}^{H \times W}$ to $A \in \mathbb{R}^{H \times W \times D}$. Therefore, the attention map is defined as:

$$A_{i,j,d} = \frac{Z_{i,j,d}}{1 + \exp(-Z_{i,j,d})} \tag{3}$$

where $d$ represents the $d$-th channel. Correspondingly, an enhanced feature map is finally calculated as:

$$F^*_{i,j,d} = (1 + A_{i,j,d}) \odot F_{i,j,d} \tag{4}$$

In order to realize the learning of attention weighting values, the attention map A is sent to the conv_7 and a global average pooling (GAP) layer to obtain the final classification prediction. This attention module is trained by the standard cross-entropy classification loss.

### 2.2 Two-Level Classification Networks

The part-level and image-level loss are combined to train the extracted features of proposal regions and fusion features with image-level annotations. For an input image $I$ and the proposal candidates $R = (R_1, R_2, ..., R_N)$, the features of proposals $\varphi(F^*; \mathrm{R})$ are extracted through the front ConvNet and RoI Pooling, which are fed to a FC layer with the softmax layer to produce the feature matrix $X$:

$$X = f(\varphi(F^*; \mathrm{R})) \tag{5}$$

where $f$ denotes a softmax function, and $\varphi$ is the series of operations of the model. $X \in \mathbb{R}^{C \times N}$ is a feature matrix, where C and N represent the number of image classes and region proposals, respectively. $X_i$ denotes a part feature $\varphi(F^*; \mathrm{R}_i)$. Next, the two-level loss is introduced in detail.

**Part-Level Loss.** The proposed parts R come from any position of an input image $I$, which results in containing or overlapping with foreground objects, and even including only the background. Because of the lack of the part-level annotations, we only use their corresponding image labels to train the model directly. The cross-entropy loss function is used to train proposal candidates with image label $Y$. The part-level loss $\mathcal{L}_{part-level}$ can be formulated as follows:

$$\mathcal{L}_{part-level} = -\frac{1}{N} \sum_{1 \leqslant i \leqslant N} \log X_{Y;i} \tag{6}$$

However, each proposal candidate has different overlaps with the objects. According to the degree of the difficulty of identifying part features, we divide the region proposals into three types: easy, hard and background parts. Although easy parts can give high confident predictions on the real image label in early epochs, but as the training epochs increase, they cannot contribute more useful information. And hard and background parts only provide relatively limited confidence predictions for action categories. Moreover, the loss of easy parts becomes very small after several epochs, but the total loss of large amount of easy parts still accounts for a large part of the whole loss. Therefore, it will lead to meaningless gradient descent, and even further affect the network optimization. In order to overcome the shortcoming of the above cross-entropy part-loss $\mathcal{L}_{part-level}$, the weighting mechanism is introduced to train three different parts. A part-level loss is designed to assigned automatically different weightings to three kinds of parts based on their characteristics. To be more especially, the weighting value $(1-X_{Y;i})^{\beta}$ is added on the basis of the cross-entropy part loss. Formally, the weighted part-level loss $\mathcal{L}_{weighted-part-level}$ is given by:

$$\mathcal{L}_{weighted-part-level} = -\frac{1}{N} \sum_{1 \leqslant i \leqslant N} (1 - X_{Y;i})^{\beta} \log X_{Y;i} \tag{7}$$

where $\beta$ is a hyper-parameter to control the ratio of weights. Less weight is assigned to the easy-to-classify parts. As shown in Fig. 3, the weighted part-level loss on the PPMI dataset makes our model to pay more attention to harder parts in later epochs and reduce the negative influence of easy parts.

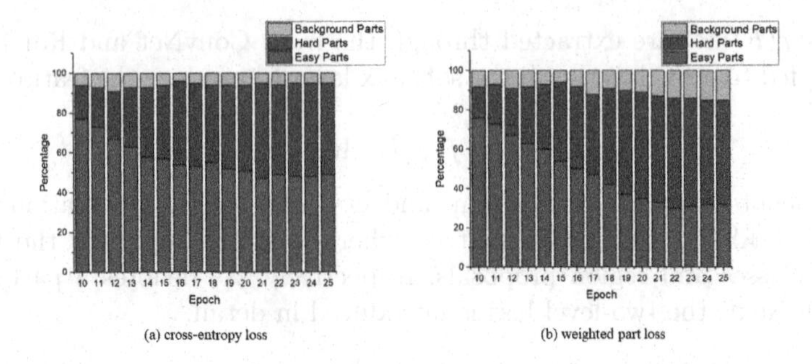


**Fig. 3.** Percentage of losses contributed by three parts on PPMI. a and b show that percentage of losses are calculated by the cross-entropy loss and the weighted part loss respectively.

**Image-Level Loss.** Because no external annotations are used, it is very difficult to distinguish object and background parts in the proposal candidates. Therefore, we only use the image-level labels to train background and object parts. Although easy parts and hard parts are more distinguishable than background in classification, the existence of a certain amount background parts still have a negative effect on the feature fusion. In order to further improve our model performance, we intend to reduce the impact of harmful background parts in the fusion features. Therefore, the threshold value $\theta \in [0, 1]$ is set to remove the harmful background. The fusion feature is defined as follows:

$$\mathrm{F}^{'} = \frac{1}{n} \sum_{j \leqslant n} X_{Y;j} \tag{8}$$

where n ($n \leqslant N$) denotes the number of parts which satisfy $\max(X_{Y;j}) > \theta$. A fusion feature $\mathrm{F}^{'}$ is generated to utilize the informative easy and harder parts. We replace the softmax with a new image loss function to train the fusion features $\mathrm{F}^{'}$. The loss function $\mathcal{L}_{image-level}$ is shown as:

$$\mathcal{L}_{image-level} = -\log \mathrm{F}_{Y}^{'} \tag{9}$$

where $\mathrm{F}_{Y}^{'}$ is y-th element of $\mathrm{F}^{'}$, which is the prediction probability of the corresponding image-level annotation. Therefore, the final loss of the overall network $\mathcal{L}$ is given as:

$$\begin{aligned}\mathcal{L} &= \mathcal{L}_{image-level} + \alpha \mathcal{L}_{weighted-part-level}(R) + \mathcal{L}_{GAM} \\ &= -\log \mathrm{F}_{Y}^{'} - \frac{\alpha}{N} \sum_{1 \leqslant i \leqslant N} (1 - X_{Y;i})^{\beta} \log X_{Y;i} + \mathcal{L}_{GAM}\end{aligned} \tag{10}$$

where $\alpha$ represents the balance factor of two-level loss, $\beta$ is the weight of part-level loss, and $\mathcal{L}_{GAM}$ represents the classification loss of the guided attention module.

# 3 Experiments

## 3.1 Experimental Setup

**Implementation Details.** In this experiment, we use VGG-16 [12] pre-trained on ImageNet [13] to construct the proposed model for fair comparison. For constructing our model, we add several new layers to build up a guided attention module, and remove the layers (after the Conv5_3) to add a RoI Pooling layer. We exploit the Selective Search Windows method (SSW) to generated object proposals.

Our model is trained using SGD on a single NVIDIA Tesla P40 GPU with momentum 0.9 and a mini-batch size of 8. The initial learning rate is set to 0.0001, which decays by 0.1 for every 10 epochs. Each input image is resized to $224 \times 224$. Three kinds of data augmentation are employed in the training and testing stage: three image scales, i.e., $\{450, 550, 650\}$, center crops and random flips. We use the official split of training and testing images for evaluation on two datasets. We take the Mean of AP (mAP) and average precision (AP) as the evaluation metrics to evaluate the method on the test set. We follow the above experimental settings to train on each dataset for 40 epochs and use the deep learning framework of PyTorch to implement our method.

## 3.2 Comparison with State-of-the-Art Method

**PPMI Dataset.** Table 1 shows the mean AP results of our approach and comparison with other methods. It can be seen that our approach achieves a mean AP of 86.46% on the testing set, which outperforms the traditional and the latest deep learning methods by a large margin. In contrast to the traditional methods [7,14,15], our model could extract efficiently discriminative image features to improve the accuracy of action recognition by deep learning. Qi et al. [16] proposed a novel action recognition model by integrating the pose hints to the CNN and obtained a mean AP of 82.2%. Yan et al. [17] encoded the features of image patches generated by the EdgeBox with the VLAD encoding scheme for action recognition and obtained a mean AP of 81.3%. Yet our method outperforms all of the two methods by 4.26% and 5.16%, respectively, which shows our method could fully obtain the human body part or pose features to improve recognition accuracy.

We also analyze the performance of our methods in different action classes. Figure 4 shows the average precision (AP) on every action compared with the other approaches. It is shown that our model achieves the best performance on 24 categories. Our method performs differently in all action categories, ranging from 98% on "playing violin" to 71% on "with recorder". It also can be observed that the overall accuracy of playing instrument is higher than that of with instrument. This is because the changes in the body parts of playing instrument are significantly easier to identify than with instrument.

**Table 1.** Comparison of the mean precision results using the proposed approach and the other methods based on the PPMI dataset.

| Method | Poselets [7] | SPM [14] | Grouplet [15] | Yao [9] | Simonyan [12] | Yan | Qi | Ours |
|---|---|---|---|---|---|---|---|---|
| mAP(%) | 23.40 | 40.00 | 42.00 | 48.00 | 74.35 | 81.3 | 82.2 | 86.46 |

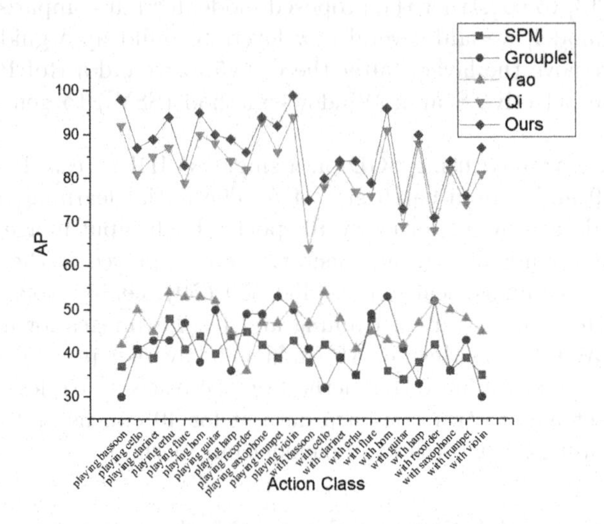


**Fig. 4.** Average precision (AP) at each class on the PPMI dataset.

**Stanford-40 Dataset.** The Stanford-40 dataset contains 9532 images in total corresponding to 40 different types of daily human activities and the number of images for each class ranges from 180 to 300. We evaluate our model on the testing set and it obtains a mean AP of 88.73%, with performance varying from 73.5% for "taking photos" to 98% for "riding a horse". The comparisons with previously published methods are shown in Table 2, which demonstrates that our method is quite competitive. Although Multi-branch attention network obtains the best performance, it need to exploit the manually annotated bounding-boxes to accurately extract the target person for classification. Compared with [12,20], no part-level annotations like bounding-boxes are used in our method.

### 3.3 Ablation Studies

**Loss Analysis.** Our proposed method mainly consists of the classification loss of GAM and two-level loss. We do extensive ablation studies on the PPMI and Stanford-40 dataset to verify the effectiveness of different losses. As shown in Table 3, it can be seen that all settings outperform the baseline VGG-16 with a large margin, which shows the effectiveness of our model. The classification loss of GAM achieves accuracies with 2.05% and 1.73% improvement on PPMI and Stanford-40 respectively than the baseline, while the image-level loss brings 3.15% and 2.41% performance gains on two datasets. However, the weighted part

**Table 2.** Comparison of the mean precision results using the proposed approach and the other methods based on the Stanford-40 dataset.

| Method | Network | mAP(%) |
|---|---|---|
| Simonyan [12] | 16 and 19 layer networks | 72.40 |
| Qi [16] | VGG-16 | 80.69 |
| TDP [5] | VGG-16 | 80.80 |
| AttSPP-net [18] | VGG-19 | 81.62 |
| Action Mask [19] | VGG-16 | 82.60 |
| Deep VLAD spatial pyramids [17] | VGG-16 | 88.50 |
| Multi-branch attention network [20] | VGG-16 | 90.70 |
| Ours | VGG-16 | 88.73 |

loss achieves higher accuracies with 6.41% and 5.75% improvement on PPMI and Stanford-40 than the baseline. It indicates that three proposed losses are essential for our model, especially the weighted part loss. Compared with using only one loss, their combination performs better than a single loss function. However, it can be found that the combination of image-level loss and classification loss of GAM outperforms worse than the weighted part loss. It means that the weighted loss plays an important role in the whole model. We use the cross-entropy loss (CE Loss) instead of the weighted part-level loss, and find that the accuracies on two datasets descend. This demonstrates that the proposed weighted part-level loss is effective and its performance is better than that of the cross-entropy loss. When the part-level loss using the cross-entropy loss, we also find that the performance of the combinations with part-level loss is worse than that of part-level loss using the weighted part loss.

**Table 3.** Ablation study of using different losses on the PPMI and Stanford-40 dataset.

| Part-level loss | Image-level loss | Loss of GAM | PPMI | Stanford-40 |
|---|---|---|---|---|
| VGG-16(baseline) | – | – | 76.48% | 78.82% |
| – | – | CE Loss | 78.53% | 80.55% |
| Weighted Part Loss | – | – | 82.89% | 84.57% |
| – | CE Loss | – | 79.63% | 81.23% |
| – | CE Loss | CE Loss | 80.17% | 82.62% |
| CE Loss | CE Loss | – | 80.79% | 83.58% |
| CE Loss | – | CE Loss | 79.54% | 80.83% |
| Weighted Part Loss | CE Loss | – | 85.37% | 86.64% |
| Weighted Part Loss | – | CE Loss | 83.43% | 85.37% |
| CE Loss | CE Loss | CE Loss | 82.25% | 84.35% |
| Weighted Part Loss | CE Loss | CE Loss | 86.46% | 88.73% |

**Parameter Analysis.** This subsection mainly introduces the impact on four hyper-parameters: part proposals N, weighting parameter $\beta$, threshold value $\theta$ as well as balance factor $\alpha$. When one parameter is evaluated, the other parameters keep unchanged. Since region proposals are generated by the SSW, we select 5 different numbers of proposal candidates N to evaluate the model' performance on the PPMI and Stanford-40 dataset. As shown in Fig. 5a, with the rising of N, the accuracies on two datasets also increase. When $N = 1000$, the model obtains the best results on two datasets. The weighting value is used to assign different weighting values to three different parts in the part-level loss function. The influence of different values on two action datasets is illustrated in Fig. 5b. When we choose $\beta = 2$ on Stanford-40, $\beta = 3$ on PPMI, the model obtains good performance. When $\beta$ is becoming large, the part-level loss will assign more weighting value to hard parts. However, the larger $\beta$ may deteriorate the performance of the model. It means that too much attention on harder parts brings the negative effects.

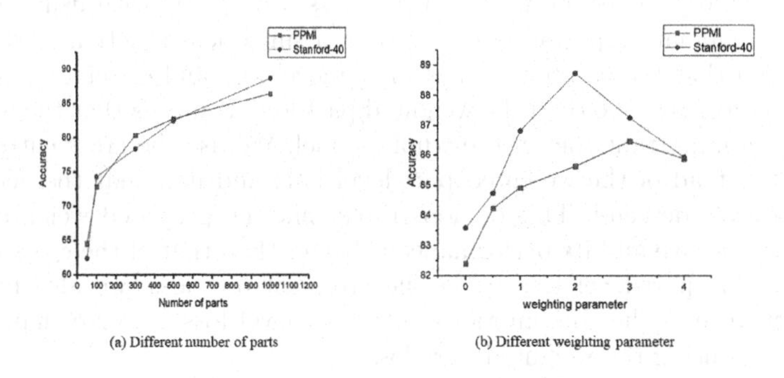


**Fig. 5.** (a) demonstrates the influence of different number of part proposals, and (b) shows the impact of weighting values.

In order to reduce the impact of harmful background parts, threshold value $\theta$ is proposed to select part features. As shown in Fig. 6a, $\theta = 0$ denotes that no selection, and the accuracies are about 83.75% and 85.58% on the PPMI and Stanford-40, respectively. In this case, the model may be negatively affected by background parts. When $\theta$ is between 0.05 and 0.1, our model achieves good performance. However, too large $\theta$ results in the decline of the model performance. This reason is that in addition to background parts, some useful information has also been removed. Thence, a suitable value 0.1 is selected in the experiment. The parameter $\alpha$ is a balance factor, which enables to allocate different weighting values to two-level losses and thus adjust the rates of part and global information. The accuracies of the PPMI and Stanford-40 are shown in Fig. 6b. When $\alpha = 0$, we only use the image-level loss to train the model, and it obtains the mAP results of 83.72% and 84.60%, respectively. With the increase of $\alpha$, the accuracies are also promoting. It can be seen that when $\alpha$ is between 8 and 12, the model performs well. In this instance, image-level and part-level loss keep the better state of balance. As shown in Fig. 6b, it can be observed that $\alpha$ is set to 10 and 12 on two datasets respectively, the model obtains the best performance. But with the increase of $\alpha$, the performance becomes worse. In this situation, the global information plays a leading role compared with part information.

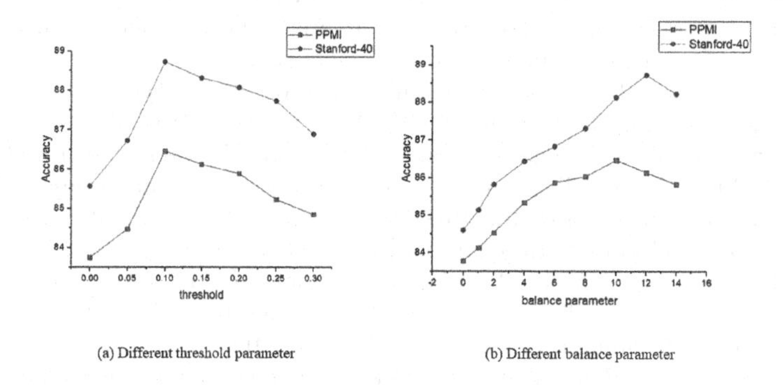


(a) Different threshold parameter

(b) Different balance parameter

**Fig. 6.** (a) represents the influence of threshold value in part fusion process, and (b) shows the effect of balance factor.

# 4 Conclusion

In the paper, we propose a part fusion framework to deal with image-based action recognition. In our model, a two-level loss is designed to distinguish different kinds of parts without external annotations. Additionally, the guided attention branch is added for better locating distinctive regions. We evaluate the proposed method on two image-based action datasets and the results demonstrate that our approach achieves the outstanding performance on the PPMI dataset than other state-of-the-art methods, and obtains the competitive performance on the Stanford-40. Through extensive experiments, we also verify the effectiveness of the two-level loss and the guided attention module.

**Acknowledgement.** This work is supported by National Natural Science Foundation of China (No. 61763035) and Inner Mongolia Natural Science Foundation (No. 2020MS06006).

# References

1. Hao, W., Zhang, Z.: Spatiotemporal distilled dense-connectivity network for video action recognition. Pattern Recogn. **2**, 13–24 (2019)
2. Dai, M., Srivastava, A.: Video-based action recognition using dimension reduction of deep covariance trajectories. In: 2019 IEEE/CVF Conference on Computer Vision and Pattern Recognition Workshops (CVPRW), pp. 611–620 (2019)
3. Yang, Z., Li, Y., Yang, J., Luo, J.: Action recognition with spatio-temporal visual attention on skeleton image sequences. IEEE Trans. Circuits Syst. Video Technol. **29**(8), 2405–2415 (2019)

4. Gkioxari, G., Girshick, R., Malik, J.: Actions and attributes from wholes and parts. In: 2015 IEEE International Conference on Computer Vision (ICCV), pp. 2470–2478 (2015)
5. Zhao, Z., Ma, H., Chen, X.: Semantic parts based top-down pyramid for action recognition. Pattern Recogn. Lett. **84**, 134–141 (2016)
6. Felzenszwalb, P.F., Girshick, R.B., McAllester, D.: Cascade object detection with deformable part models. In: 2010 IEEE Computer Society Conference on Computer Vision and Pattern Recognition, pp. 2241–2248 (2010)
7. Maji, S., Bourdev, L., Malik, J.: Action recognition from a distributed representation of pose and appearance. In: CVPR 2011, pp. 3177–3184 (2011)
8. Kulkarni, A., Callan, J.: Selective search. ACM Trans. Inf. Syst. **33**(4), 1–33 (2015)
9. Yao, B., Li, F.F.: Modeling mutual context of object and human pose in human-object interaction activities. In: 2010 IEEE Computer Society Conference on Computer Vision and Pattern Recognition, pp. 17–24 (2010)
10. Yao, B., Jiang, X., Khosla, A., Lin, A.L., Guibas, L., Li, F.-F.: Human action recognition by learning bases of action attributes and parts. In: 2011 International Conference on Computer Vision, pp. 1331–1338 (2011)
11. He, K., Zhang, X., Ren, S., Sun, J.: Deep residual learning for image recognition. In: 2016 IEEE Conference on Computer Vision and Pattern Recognition (CVPR), pp. 770–778 (2010)
12. Simonyan, K., Zisserman, A.: Very deep convolutional networks for large-scale image recognition. In: Proceedings of the International Conference on Learning Representations (2015)
13. Deng, J., Dong, W., Socher, R., Li, L., Li, K., Li, F.-F.: ImageNet: a large-scale hierarchical image database. In: 2009 IEEE Conference on Computer Vision and Pattern Recognition, pp. 248–255 (2009)
14. Lazebnik, S., Schmid, C., Ponce, J.: Beyond bags of features: spatial pyramid matching for recognizing natural scene categories. In: 2006 IEEE Computer Society Conference on Computer Vision and Pattern Recognition, pp. 2169–2178 (2010)
15. Yao, B., Li, F.-F.: Grouplet: a structured image representation for recognizing human and object interactions. In: 2010 IEEE Computer Society Conference on Computer Vision and Pattern Recognition, pp. 9–16 (2010)
16. Qi, T.Q., Xu, Y., Quan, Y.H., Wang, Y.D., Ling, H.B.: Image-based action recognition using hint-enhanced deep neural networks. Neurocomputing **267**, 475–488 (2017)
17. Yan, S.Y., Smith, J.S., Zhang, B.: Action recognition from still images based on deep VLAD spatial pyramids. Signal Process. Image Commun. **54**, 118–129 (2017)
18. Feng, W., Zhang, X., Huang, X., Luo, Z.: Attention focused spatial pyramid pooling for boxless action recognition in still images. In: Lintas, A., Rovetta, S., Verschure, P.F.M.J., Villa, A.E.P. (eds.) ICANN 2017. LNCS, vol. 10614, pp. 574–581. Springer, Cham (2017). https://doi.org/10.1007/978-3-319-68612-7_65
19. Zhang, Y., Li, C., Wu, J., Cai, J., Do, M.N., Lu, J.: Action recognition in still images with minimum annotation efforts. IEEE Trans. Image Process. **25**(11), 5479–5490 (2016)
20. Yan, S., Smith, J.S., Lu, W., Zhang, B.: Multibranch attention networks for action recognition in still images. IEEE Trans. Cogn. Dev. Syst. **10**(4), 1116–1125 (2018)

# An Empirical Study of Deep Neural Networks for Glioma Detection from MRI Sequences

Matthieu Coupet[1], Thierry Urruty[1,2(✉)], Teerapong Leelanupab[5], Mathieu Naudin[2,3], Pascal Bourdon[1,2], Christine Fernandez-Maloigne[1,2], and Rémy Guillevin[2,3,4]

[1] XLIM Laboratory, University of Poitiers, UMR CNRS 7252, Poitiers, France
thierry.urruty@univ-poitiers.fr
[2] I3M, Common Laboratory CNRS-Siemens, University and Hospital of Poitiers, Poitiers, France
[3] Poitiers University Hospital, CHU, Poitiers, France
[4] DACTIM-MIS/LMA Laboratory University of Poitiers, UMR CNRS 7348, Poitiers, France
[5] Faculty of Information Technology, King Mongkut's Institute of Technology Ladkrabang (KMITL), Bangkok 10520, Thailand

**Abstract.** Gliomas are the most common central nervous system tumors. They represent 1.3% of cancers and are the 15th most common cancer for men and women. For the diagnosis of such pathology, doctors commonly use Magnetic Resonance Imaging (MRI) with different sequences. In this work, we propose a global framework using convolutional neural networks to create an intelligent assistant system for neurologists to diagnose the brain gliomas. Within this framework, we study the performance of different neural networks on four MRI modalities. This work allows us to highlight the most specific MRI sequences so that the presence of gliomas in brain tissue can be classified. We also visually analyze extracted features from the different modalities and networks with an aim to improve the interpretability and analysis of the performance obtained. We apply our study on the MRI sequences that are obtained from BraTS datasets.

**Keywords:** Glioma · Convolutional neural network · Brain · Visualization · MRI · Transfer learning

## 1 Introduction

In 2019, an estimated one in ten people died from cancer. This fact makes cancer one of the major health problems worldwide, with nearly 10 million people dying every year [32]. Cancer is therefore the second leading cause of death in the world behind cardiovascular diseases. Thus, it is urgent to make progress in the fight


H. Yang et al. (Eds.): ICONIP 2020, LNCS 12532, pp. 113–125, 2020.
https://doi.org/10.1007/978-3-030-63830-6_10

against this disease. Among cancer diseases, gliomas make up approximately 30% of all brain and central nervous system tumors and 80% of all malignant brain tumors [12]. Gliomas are pathologies of the glia, the brain's support tissue. They originate from the glial cells of the spine and can be classified according to the cell type, grade and localization [18].

In medicine, as in health research, the use of different types of imaging is a common tool for diagnosis of various diseases. The use of Magnetic Resonance Imaging (MRI) is particularly helpful for brain glioma diagnosis, but also precious for biomedical engineering research. MRI is interesting due to its non-invasiveness for obtaining highly resolved images, as well as information that would not be recoverable other than by an extremely invasive examination like a brain biopsy. All the information collected from MRI usually help neurological experts, in particular for the diagnosis and the research study of most cancer pathologies, such as on the detection, classification and segmentation of tumors including gliomas.

Recent developments in deep learning showed great potential [21,34] for the health field in medical diagnosis and particularly in the analysis of different medical imaging technologies [23]. The use of deep learning on digital medical imagery can be very beneficial for targeted and automated aid for doctors. Indeed, when a fairly effective network is set up, diagnostic assistance can support the doctors in their work. Some networks, for example, have already been set up to help with glioma segmentation [15,30].

Numerous available neural networks make it difficult to decide which one will perform best. Therefore, the contribution of this paper is to propose a framework that studies the performance, for a specific task, of the most effective neural networks in the literature. Our framework also includes the study of the pertinence of most common modalities of MRI acquisition for the task. Those modalities represent a huge amount of data, and our results show the importance of selecting and combining such modalities. We also propose to investigate the features extracted by the neural networks, which improve the interpretability of the deep learning approaches and allow a better understanding of their performance. Our dedicated task in this paper is the detection of brain tumors, but note that this framework can be applied to other computer-aided medical tasks.

The remainder of this paper is organized as follows. In Sect. 2, we outline the literature of deep learning approaches for computer-aided medical tasks. Section 3 presents the material and methods, including the presentation of our general framework. Results of our exhaustive experiments are explained in Sect. 4 which also proposes an analysis of the interpretability of the deep neural network. In Sect. 5, we conclude and give perspectives of this work.

## 2 State of the Art

Before the rise of deep learning, Support Vector Machine (SVM) was a common method for the analysis of images by magnetic resonance. Bahadure *et al.* [1] showed that SVM could somewhat effectively detect and extract the characteristics of brain tumors. In addition, K-Nearest Neighbor (KNN) was applied for the segmentation of parts of the brain recognized as being abnormal [17]. Less generally, the use of Wavelet-Entropy and Naive Bayes classifiers for characteristic extraction and detection of brain pathologies gave interesting performance [40].

A neural network is a well-known Machine Learning (ML) approach, which nowadays receives increasing attention due to the success of deep learning and computer technological advances. Deep learning is a very large neural network comprising many layers and nodes in each layer. Although deep learning provides a higher performance for dedicated tasks, deep refers to bigger and more sophisticated models requiring huge amount of data to train.

There are many fields of application in deep learning on MRI. Examples of possible applications are image acquisition and reconstruction [36,39], image restoration (e.g., denoising and artifact detection) [5,6], and the increase of image resolution [2]. Among those, the applications of interest for us are the classification of image data [22], segmentation [19,25], and diagnostic assistance and disease prediction [24]. Many articles highlighted the importance of MRI modalities in deep learning. However, so far very few articles have presented the importance of each modality, especially those we examine here for their use for the classification between healthy tissue and a brain with a tumor. For example, Feng *et al.* [11] worked on a self-adaptive neural network based on missing modalities for the lesion segmentation of multiple sclerosis. The self-adaptive neural network plays an essential role here when certain MRI sequences could not be acquired correctly by a medical exam.

In this study, we use four well-known pre-trained convolutional neural networks with a transfer learning method. The selected networks were used for the participation in the ILSVRC (ImageNet Large Scale Visual Recognition Challenge) competition. These networks include VGG19 [35], InceptionV3 [37], Resnet50 [13] and DenseNet [14]. These pre-trained neural networks are widely used in many fields of application. They have strong synaptic weights from the acquired knowledge. Apart from the prediction of the training dataset of ILSVRC [33], such already-built models can be employed to improve baseline performance, reduce model-development time, and in particular optimally utilize existing knowledge to solve a different target task. This design methodology is so-called *transfer learning*. In the medical field, for example, Resnet50 was used to leverage the formation of brain tumors to detect abnormal methylation in the brain tissue [20]. For Densenet, its dense connectivity was applied to improve MRI image resolution [8]. Finally and very recently, the Resnet50 and InceptionV3 networks were employed for the detection of Covid19 on X-ray chest radiography [28].

3D neural network models have also shown their interest. They are specifically designed for learning dense volumetric segmentation from sparse annotation [9], and today widely used in the ML field of medical images. We can find its use for various applications, such as cell counting [10], segmentation of the pancreas [29], or, as what interests us, the segmentation of brain tumor [7].

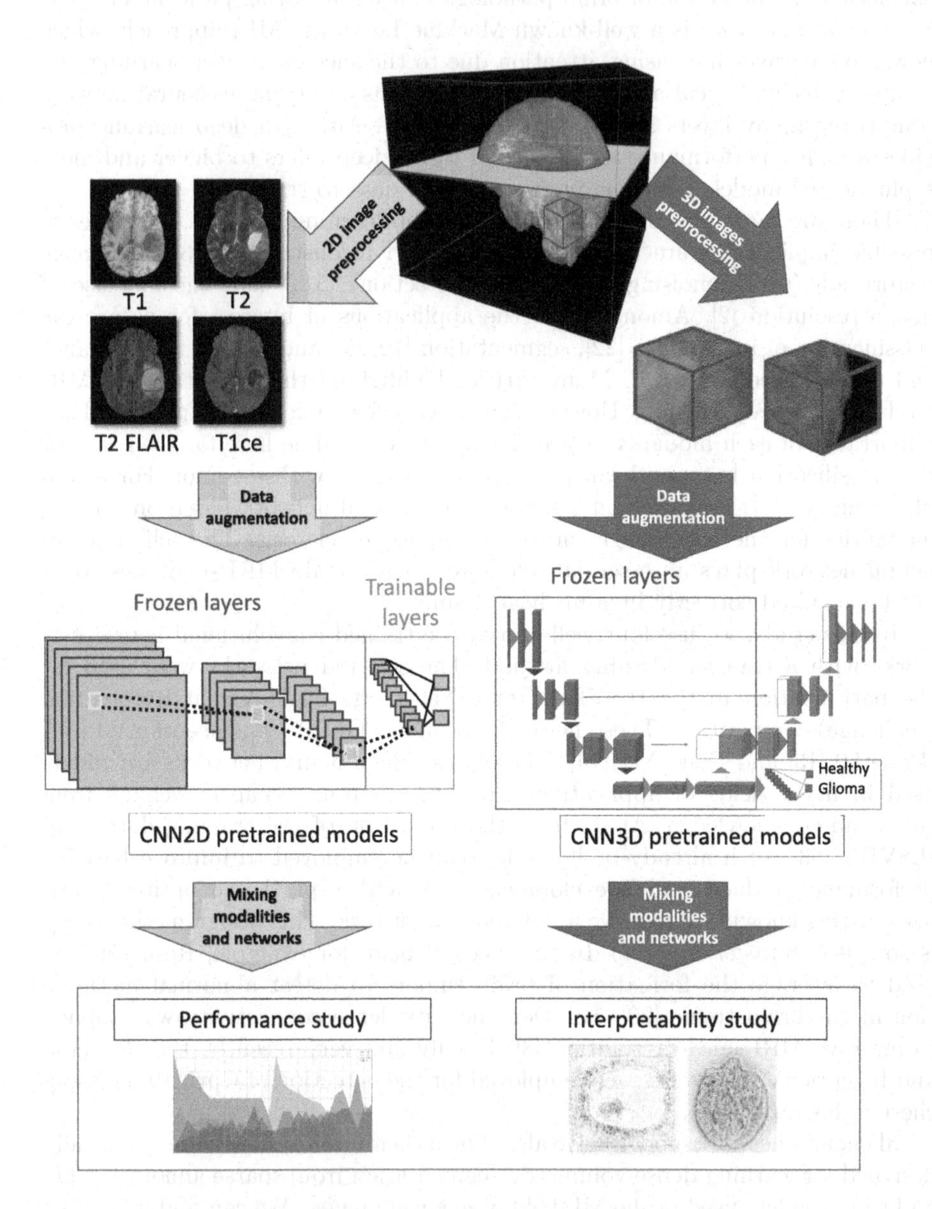


**Fig. 1.** Overall data processing workflow of our approach

# 3 Materials and Methods

Figure 1 shows the general workflow of our approach. Starting from an MRI database, we extract 2D slices and 3D cubes for each modality that undergo an usual pre-processing step, including skull stripping and normalization of the images. Then, we apply a data augmentation methodology to the 2D slices and the 3D cubes, essential to feed the 2D and 3D pre-trained neural networks. We use a transfer learning approach to train the models for our specific classification task.

Finally, we analyse our exhaustive experiments with a performance study on the combinations of two different modalities - network couples. In parallel, the extraction and analysis of deep features can be performed for an interpretability analysis. Our investigation of the features extracted by multi-modality network gives a better understanding of their performance for our task.

## 3.1 Material

For our research, the use of medical MRI for glioma detection is very common. Different tumor tissues, such as necrotic core, active edge and edema, can be imaged using different MRI modalities. For example, the T2 FLAIR (Fluid-attenuated inversion recovery) sequence can be used for the detection of edema by doctors while the T1ce sequence, a T1 sequence coupled with injections of gadolinium as a contrast agent, can be used for the detection of the tumor core, often necrotic tissue.

To do this work, we selected a database with different modalities, i.e., the BraTS database [3,4,26]. The BraTS challenge 2018 database (Multimodal Brain Tumor Segmentation) is a database produced by CBICA (Center for Biomedical Image Computing and Analysis) in partnership with MICCAI (Medical Image Computing and Computer-Assisted Intervention Society). This database presents MRI brain data in NIFTI file in four different sequences as illustrated in Fig. 2: longitudinal relaxation (T1), longitudinal relaxation with addition of contrast medium (T1-Contrast Enhanced, T1ce), transverse relaxation (T2), and transverse relaxation with inversion recovery (attenuation of fluids (T2 FLAIR).

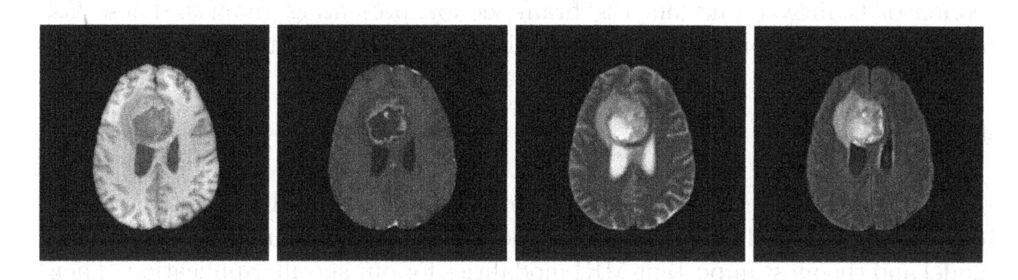

**Fig. 2.** Transverse section of a patient, who suffers from a high grade glioma in T1 sequence (a), T1ce sequence (b), T2 sequence (c) and T2-FLAIR (d).

These data present MRI images of brains with different grades of gliomas and are produced by 19 different institutions, segmented by hand using the same procedure by approved and experienced neuro-radiologists. These data therefore include 210 patients with high grade gliomas and 74 patients with low grade gliomas. For glioma diagnosis, the neural networks can utilize this data to train and test their classifiers to predict at least two different outputs[1]: healthy tissues and abnormal tissues of brain with glioma. The data undergo usual MRI pre-processing including normalization and preparation for training as data augmentation.

### 3.2 Model Implementation

In the following section, we focus on the implementation of our different CNN models. Each model used in our study has been loaded with the synaptic weight trained on ImageNet [33]. ImageNet is a database organized according to a hierarchy of words used on WordNet, a lexical database developed by linguists in the cognitive science laboratory at Princeton University.

The used networks require tremendous computing power and time to achieve the performance they obtained in the ILSVRC competition. In fact, these networks have between 17 and 31 million trainable parameters. Hoping to adjust each of these parameters on a small MRI dataset would be illusory. The aim of *transfer learning* is to keep all the synaptic weights already learned from the networks on ImageNet images and to freeze these layers to make the parameters non-entrainable. The idea behind this approach is that MRI images have characteristics (e.g., certain shapes, shades of colors, etc.) that would allow them to be differentiated in the same way as they were done for the image classification of the ILSVRC competition. Transfer learning approaches has shown great potential for specific task which has little annotated data [38].

The output of those models, also called bottleneck, is removed and replaced to adapt to our task. For each model, a flatten layer is added to standardize the data into a single vector. This layer is followed by a small fully connected network that contains a mix of fully connected and dropout layers. Finally, the output of each network is linked to the number of classes being sought. In this study, two neurons are used as output for the classification between the brain section of healthy tissue and the brain section presenting gliomas. These last layers are the only ones being trained with the MRI annotated dataset.

## 4 Results

First, we compared each 2D model previously chosen for each of the four modalities. We synthesized our exhaustive experiments to bring out the most efficient model and the most important MRI modalities for our specific application. Then,

[1] Multi-class classification can also be applied to this dataset for different grades of gliomas.

we compared 2D CNN with the UNET 3D network, also widely used in segmentation for medical imagery. Finally, extraction and analysis of the deep features were carried out in order to improve the interpretability of the data.

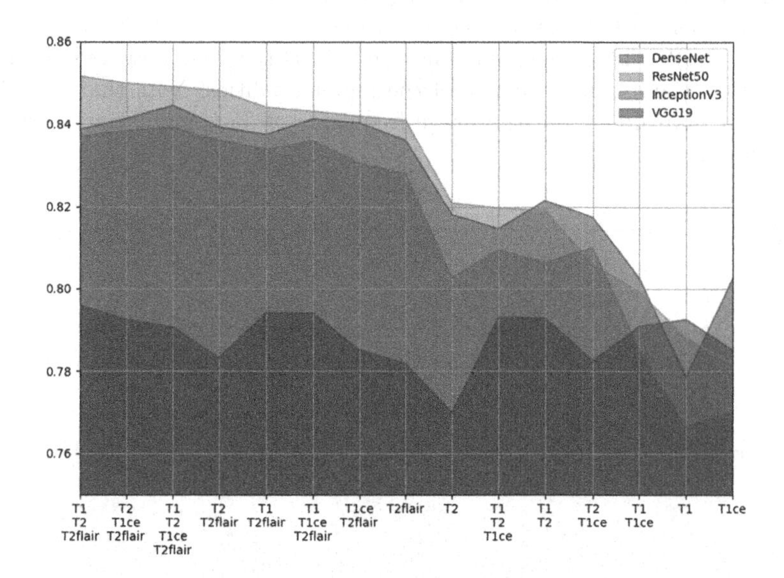


**Fig. 3.** Classification accuracy with respect to combinations of modalities and CNN models (Color figure online)

## 4.1 Mixing All

The goal of our study is to test all combinations of MRI modalities coupled with neural network models. This will determine whether a predominant modality-network couple emerges for determining the presence of gliomas in a patient brain.

In our first experiment, we aim to test different neural network models in all combinations of MRI modalities. In Fig. 3, these different combinations are sorted in a descending order according to their performance with the best network (i.e., ResNet). The classification accuracy for mixed modalities ranges from 0.85 to 0.78. We can also observe a certain preferential order towards the use of specific neural networks. Indeed, the ResNet50 model (orange) seems to be more effective than DenseNet (blue) and InceptionV3 (green) models. Finally VGG19 model (red) seems to be the least accurate model for this task of glioma detection. Figure 3 also shows a predominance of combinations including the T2 FLAIR modality with a high classification accuracy for most models (above 84%). Even T2 Flair alone performs well and the accuracy loss is clear as soon as T2 FLAIR modality disappears from the combination.

A second experiment proposes to test all possible network-modality combinations. Thus, at the input of a model with a dense layer 2, we used the prediction of each type of model for each of the modalities. Thereby, we have four different models and four different modalities, i.e., a total of 16 different possible inputs. We then tested the best combinations of four out of the 16, resulting in $16!/(16-4)! \cdot 4! = 1820$ different combinations to be observed. Figure 4 demonstrates the usage frequency of the four modalities. Note that, for visual purpose, we sum up by bin the usage frequency of 35 combinations sorted by their performance.

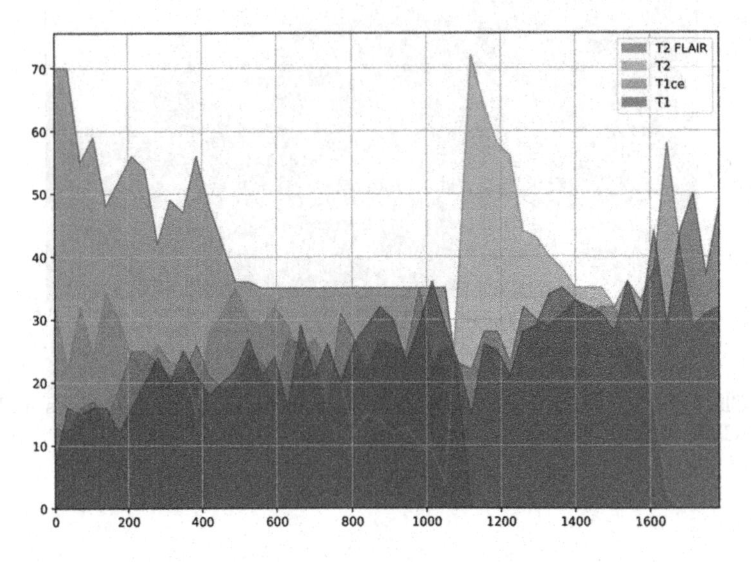


**Fig. 4.** Usage frequency of each modality according to accuracy of 4 modalities out of 16

As already noticed in Fig. 3, we have a clear predominance of the T2 FLAIR modality (in blue) for the most important accuracy (ranked first). When this modality is not used any more (rank above 1100), the T2 modality (orange) compensates for the loss of information that the T2 FLAIR modality provided. Finally, when the T2 FLAIR and T2 modes are no longer active, T1ce takes over and then finally T1 does. Note that best 10 combinations include ResNet model twice and T2 FLAIR modality two or three times. So, we can determine a clear ranking of the importance of the different modalities for our glioma detection task. The T2 FLAIR modality is the most suitable, followed by the T2 modality, then T1ce and finally T1.

Note that we have recently tested on the BraTS-2020 database, which contains 30% of different patients in comparison with BraTS-2018. The performance is very similar with a slight increase of accuracy (+2%).

### 4.2 Comparison with UNET3D

As already mentioned in the state-of-art section, UNET-3D [31] has become popular when dealing with 3D medical imaging. UNET-3D is a network specialized in segmentation task for medical imaging; therefore, it can be useful and easily adapt for a detection task. UNET-3D model can be decomposed into two parts. The first part is a contracting path, which contains a set of convolutional blocks (containing 3D convolutional layers with ReLu activation) and pooling layers from 3D cubes to a small feature map. The second part of the model is an expansive path, which contains deconvolutional layers and upscaling layers. At the end of this second part, a segmentation mask classifies each voxel with respect to the desired task.

Similar to a procedure of training the 2D pretrained models, the procedure of training UNET-3D model was applied by using all possible combinations of MRI sequences. We use the deep features extracted after the convolutional part, to which we added a dense fully connected network of 64 perceptrons, followed by a dense layer for our detection task. The pre-learned parameters of the basic model were maintained and the fully connected layers were trained with the previously described dataset.

In this study, all selected 2D CNNs are compared with UNET-3D. The main goal of this comparison is to see whether the used 2D networks are obsolete or not when comparing them with the networks dedicated to the specific tasks of the study, such as UNET-3D. To do this, each brain in the database was cut into cubes of $32 \times 32 \times 32$ voxels. For the training set, cubes containing more than 20% of black voxels were excluded due to the lack of information. Besides, if the cubes contain more than 15% of voxels belonging to the glioma, the image cube will then be labeled as a diseased brain, and vice versa if the cubes contain less than 1% of voxels presenting a glioma, the cube will then be considered healthy. Furthermore, if the cubes are between those two percentages (1% to 15%), the cube will be labeled as a diseased brain but are not used for training.

We observe that the UNET-3D model has, on average, similar effectiveness and accuracy (around 83%) to a mix of 2D CNN models, and outperforms every single 2D model. UNET-3D obtains its maximum accuracy (87%) for the combination of the modalities T1, T2, and T2-FLAIR and can drop to 77% using T1ce alone. To explain this, we hypothesize that the information presents in the cubes are much richer than that in 2D images. We can especially see a drop in the performance of 2D neural networks for tangent image slices of a glioma, which is filtered out in the case of a $32 \times 32 \times 32$ voxel cube.

We therefore observe similar performances between the two approaches with either 2D or 3D models. The choice of the type of neural networks can therefore be free depending on the approaches as desired by the experimenters. As a reminder, the overall goal of the approach can be done for different databases and different diseases. Thus, if the study was carried out, for example, only on 2D images (e.g., the classification of melanoma) where a 3D approach was not possible, the 2D approach would remain just as effective.

### 4.3 Deep Feature Extraction and Interpretation Analysis

Another important aspect not covered so far is to inspect whether the model appears to perform in accordance with acceptance criteria of radiological diagnosis or not. We assume that if the model is successful in performing for full-image classification, the prediction rules will focus their attention on either glioma-induced hyperintensities or mass effects. With this prospect, we tested different saliency methods commonly used for model interpretability. Such methods are able to produce a visual feedback under the form of heatmaps that highlight the importance of each individual pixel for class predictions.

Among numerous suggestions on how to achieve explainable artificial intelligence, we decided to put our efforts on Layer-Wise Relevance Propagation (LRP) [27] and Deep Taylor decomposition [16] methods. We used those approaches to highlight image areas that are the most relevant for prediction depending on MRI modalities. We chose the ResNet50 network, which achieves the best performance in our study. The results of this analysis are showed in Fig. 5. An axial section of the brain of a high-grade glioma patient with an apparent necrotic core was selected and compared to a similar section of a healthy patient.

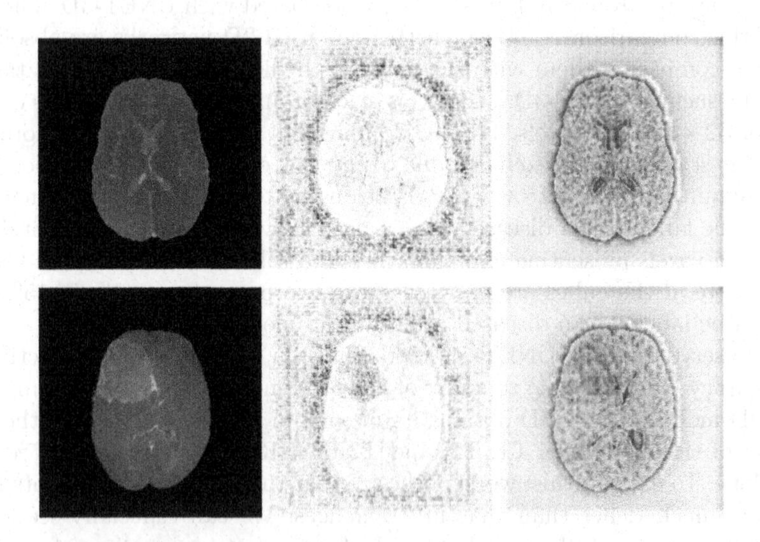

**Fig. 5.** Deep features visualization for a better interpretation. Healthy brain vs glioma, in order of columns: RGB recomposition from T2 FLAIR, T2 and T1ce sequences, LRP-Epsilon [27] and Deep Taylor [16]

While each modality has its importance in the detection of the different histological parts of the glioma, we focus solely on T2 FLAIR, T2, and T1ce modalities. From those three modalities, we recompose RGB images that serve as input of this analysis. As suggested, the CNN model focuses mainly on brain abnormalities such as brain structure modification or T2 FLAIR edema, as the radiologists commonly do. In Fig. 5, the lesions are highlighted by the LRP-Epsilon map for HGG. For the healthy subject, a diffuse noise within the brain parenchyma is present on both maps, which we assume is a hint to classify a patient's brain as healthy. Demonstrated to our medical expert team for their feedback, this visualization tool helps them to better understand and interact with the outputs of the CNN models. They also agreed and highlighted the importance of a pertinent choice of modality-network couple for a specific task.

## 5 Conclusion

In this paper, we proposed a framework using CNN models to create an intelligent computer-aided system for neurologists to diagnose the brain gliomas and conducted an empirical study of the performance of deep neural networks for glioma detection from MRI sequences. All combinations of MRI modalities coupled with 2D and 3D CNN models were examined. Our results highlighted the importance of mixing modalities to improve the overall performance. Selecting a well-adapted CNN model for a specific medical imaging application (e.g., glioma detection) is of importance. Moreover, we proposed a visual analysis and interpretation of the multimodal features extracted from the different models. Thus, medical experts can have a better understanding of obtained results. Our future work will focus on improving the interactivity between the visualization tool and medical experts to enhance the learning process of our framework.

## References


1. Bahadure, N.B., Ray, A.K., Thethi, H.P.: Image analysis for MRI based brain tumor detection and feature extraction using biologically inspired BWT and SVM. Int. J. Biomed. Imaging **2017**, 12 p. (2017). Article ID 9749108. https://doi.org/10.1155/2017/9749108
2. Bahrami, K., Shi, F., Rekik, I., Gao, Y., Shen, D.: 7T-guided super-resolution of 3T MRI. Med. Phys. **44**(5), 1661–1677 (2017)
3. Bakas, S., et al.: Advancing the cancer genome atlas glioma MRI collections with expert segmentation labels and radiomic features. Sci. Data **4**, 170117 (2017)
4. Bakas, S., et al.: Identifying the best machine learning algorithms for brain tumor segmentation, progression assessment, and overall survival prediction in the brats challenge. arXiv preprint arXiv:1811.02629 (2018)
5. Benou, A., Veksler, R., Friedman, A., Raviv, T.R.: Ensemble of expert deep neural networks for spatio-temporal denoising of contrast-enhanced MRI sequences. Med. Image Anal. **42**, 145–159 (2017)
6. Bermudez, C., Plassard, A.J., Davis, L.T., Newton, A.T., Resnick, S.M., Landman, B.A.: Learning implicit brain MRI manifolds with deep learning. In: Medical Imaging 2018: Image Processing, vol. 10574, p. 105741L. International Society for Optics and Photonics (2018)
7. Chen, W., Liu, B., Peng, S., Sun, J., Qiao, X.: S3D-UNet: separable 3D U-Net for brain tumor segmentation. In: Crimi, A., Bakas, S., Kuijf, H., Keyvan, F., Reyes, M., van Walsum, T. (eds.) BrainLes 2018. LNCS, vol. 11384, pp. 358–368. Springer, Cham (2019). https://doi.org/10.1007/978-3-030-11726-9_32

8. Chen, Y., Shi, F., Christodoulou, A.G., Xie, Y., Zhou, Z., Li, D.: Efficient and accurate MRI super-resolution using a generative adversarial network and 3D multi-level densely connected network. In: Frangi, A.F., Schnabel, J.A., Davatzikos, C., Alberola-López, C., Fichtinger, G. (eds.) MICCAI 2018. LNCS, vol. 11070, pp. 91–99. Springer, Cham (2018). https://doi.org/10.1007/978-3-030-00928-1_11
9. Çiçek, Ö., Abdulkadir, A., Lienkamp, S.S., Brox, T., Ronneberger, O.: 3D U-Net: learning dense volumetric segmentation from sparse annotation. In: Ourselin, S., Joskowicz, L., Sabuncu, M.R., Unal, G., Wells, W. (eds.) MICCAI 2016. LNCS, vol. 9901, pp. 424–432. Springer, Cham (2016). https://doi.org/10.1007/978-3-319-46723-8_49
10. Falk, T., et al.: U-Net: deep learning for cell counting, detection, and morphometry. Nat. Methods **16**(1), 67–70 (2019)
11. Feng, Y., Pan, H., Meyer, C., Feng, X.: A self-adaptive network for multiple sclerosis lesion segmentation from multi-contrast MRI with various imaging protocols. arXiv preprint arXiv:1811.07491 (2018)
12. Goodenberger, M.L., Jenkins, R.B.: Genetics of adult glioma. Cancer Genet. **205**(12), 613–621 (2012)
13. He, K., Zhang, X., Ren, S., Sun, J.: Deep residual learning for image recognition. In: Proceedings of the IEEE Conference on Computer Vision and Pattern Recognition, pp. 770–778 (2016)
14. Huang, G., Liu, Z., Van Der Maaten, L., Weinberger, K.Q.: Densely connected convolutional networks. In: Proceedings of the IEEE Conference on Computer Vision and Pattern Recognition, pp. 4700–4708 (2017)
15. Işın, A., Direkoğlu, C., Şah, M.: Review of MRI-based brain tumor image segmentation using deep learning methods. Procedia Comput. Sci. **102**, 317–324 (2016)
16. Kauffmann, J., Müller, K., Montavon, G.: Towards explaining anomalies: a deep taylor decomposition of one-class models. Pattern Recognit. **101**, 107198 (2020). https://doi.org/10.1016/j.patcog.2020.107198
17. Khalid, N.E.A., Ibrahim, S., Haniff, P.: MRI brain abnormalities segmentation using k-nearest neighbors (K-NN). Int. J. Comput. Sci. Eng. **3**(2), 980–990 (2011)
18. Kleihues, P., Soylemezoglu, F., Schäuble, B., Scheithauer, B.W., Burger, P.C.: Histopathology, classification, and grading of gliomas. Glia **15**(3), 211–221 (1995)
19. Kline, T.L., et al.: Performance of an artificial multi-observer deep neural network for fully automated segmentation of polycystic kidneys. J. Digit. Imaging **30**(4), 442–448 (2017)
20. Korfiatis, P., Kline, T.L., Lachance, D.H., Parney, I.F., Buckner, J.C., Erickson, B.J.: Residual deep convolutional neural network predicts MGMT methylation status. J. Digit. Imaging **30**(5), 622–628 (2017)
21. LeCun, Y., Bengio, Y., Hinton, G.: Deep learning. Nature **521**(7553), 436 (2015)
22. Li, H., Parikh, N.A., He, L.: A novel transfer learning approach to enhance deep neural network classification of brain functional connectomes. Front. Neurosci. **12**, 491 (2018)
23. Li, Y., Shen, L.: Skin lesion analysis towards melanoma detection using deep learning network. Sensors **18**(2), 556 (2018)
24. Litjens, G., et al.: A survey on deep learning in medical image analysis. Med. Image Anal. **42**, 60–88 (2017)
25. Makropoulos, A., Counsell, S.J., Rueckert, D.: A review on automatic fetal and neonatal brain MRI segmentation. NeuroImage **170**, 231–248 (2018)
26. Menze, B.H., et al.: The multimodal brain tumor image segmentation benchmark (brats). IEEE Trans. Med. Imaging **34**(10), 1993–2024 (2014)

27. Montavon, G., Binder, A., Lapuschkin, S., Samek, W., Müller, K.-R.: Layer-wise relevance propagation: an overview. In: Samek, W., Montavon, G., Vedaldi, A., Hansen, L.K., Müller, K.-R. (eds.) Explainable AI: Interpreting, Explaining and Visualizing Deep Learning. LNCS (LNAI), vol. 11700, pp. 193–209. Springer, Cham (2019). https://doi.org/10.1007/978-3-030-28954-6_10
28. Narin, A., Kaya, C., Pamuk, Z.: Automatic detection of coronavirus disease (COVID-19) using X-ray images and deep convolutional neural networks. arXiv preprint arXiv:2003.10849 (2020)
29. Oktay, O., et al.: Attention u-net: learning where to look for the pancreas. arXiv preprint arXiv:1804.03999 (2018)
30. Pereira, S., Pinto, A., Alves, V., Silva, C.A.: Deep convolutional neural networks for the segmentation of gliomas in multi-sequence MRI. In: Crimi, A., Menze, B., Maier, O., Reyes, M., Handels, H. (eds.) BrainLes 2015. LNCS, vol. 9556, pp. 131–143. Springer, Cham (2016). https://doi.org/10.1007/978-3-319-30858-6_12
31. Ronneberger, O., Fischer, P., Brox, T.: U-Net: convolutional networks for biomedical image segmentation. In: Navab, N., Hornegger, J., Wells, W.M., Frangi, A.F. (eds.) MICCAI 2015. LNCS, vol. 9351, pp. 234–241. Springer, Cham (2015). https://doi.org/10.1007/978-3-319-24574-4_28
32. Roser, M., Ritchie, H.: Cancer. Our World in Data (2020). https://ourworldindata.org/cancer
33. Russakovsky, O., et al.: ImageNet large scale visual recognition challenge. Int. J. Comput. Vision (IJCV) **115**(3), 211–252 (2015). https://doi.org/10.1007/s11263-015-0816-y
34. Schmidhuber, J.: Deep learning in neural networks: an overview. Neural Netw. **61**, 85–117 (2015)
35. Simonyan, K., Zisserman, A.: Very deep convolutional networks for large-scale image recognition. arXiv preprint arXiv:1409.1556 (2014)
36. Sun, J., Li, H., Xu, Z., et al.: Deep ADMM-net for compressive sensing MRI. In: Advances in Neural Information Processing Systems, pp. 10–18 (2016)
37. Szegedy, C., Vanhoucke, V., Ioffe, S., Shlens, J., Wojna, Z.: Rethinking the inception architecture for computer vision. In: Proceedings of the IEEE Conference on Computer Vision and Pattern Recognition, pp. 2818–2826 (2016)
38. Tajbakhsh, N., et al.: On the necessity of fine-tuned convolutional neural networks for medical imaging. In: Lu, L., Zheng, Y., Carneiro, G., Yang, L. (eds.) Deep Learning and Convolutional Neural Networks for Medical Image Computing. ACVPR, pp. 181–193. Springer, Cham (2017). https://doi.org/10.1007/978-3-319-42999-1_11
39. Wang, S., et al.: Accelerating magnetic resonance imaging via deep learning. In: 2016 IEEE 13th International Symposium on Biomedical Imaging (ISBI), pp. 514–517. IEEE (2016)
40. Zhou, X., et al.: Detection of pathological brain in MRI scanning based on wavelet-entropy and Naive Bayes classifier. In: Ortuño, F., Rojas, I. (eds.) IWBBIO 2015. LNCS, vol. 9043, pp. 201–209. Springer, Cham (2015). https://doi.org/10.1007/978-3-319-16483-0_20

# Analysis of Texture Representation in Convolution Neural Network Using Wavelet Based Joint Statistics

Yusuke Hamano and Hayaru Shouno(✉)

The University of Electro-Communications, Chofu, Tokyo, Japan
h1930100@edu.cc.uec.ac.jp, shouno@uec.ac.jp

**Abstract.** We analyze the texture representation ability in a deep convolution neural network called VGG. For analysis, we introduce a kind of wavelet-based joint statistics called minPS that applied to the visual neuron analysis. The minPS consists of 30 dimension features, which come from several types of statistics and correlations. We apply LASSO regression to the VGG representation in order to explain the minPS features. We find that the different scale type cross-correlation does not appear in the VGG representation from the regression weight analysis. Moreover, we synthesize the texture image from the VGG in the context of the style-transfer; we confirm the lack of different scale correlations influences the periodic texture to synthesize.

**Keywords:** Texture representation · Convolution neural network · Texture synthesis · Style transfer · Portilla-Simoncelli Statistics (PSS)

## 1 Introduction

The visual texture representation is essential for the interpretation of visual processing mechanisms in our brain as well as image processing applications. In image processing such as segmentation tasks, the texture plays a roll in the clue for the boundary shape of the object.

In the field of computer vision, the deep convolutional neural network (DCNN) becomes a *de facto* standard tool for image processing. The DCNN trained with massive natural scenes would acquire several texture representations in the intermediate layers [4,6]. Gatys *et al.* showed the intermediate representation of the DCNN makes a significant impact on texture image synthesis [3]. From the perceptual evaluation for synthesized texture images, we consider the DCNN would have a variety of texture representations; however, the property of the representation has not revealed yet. Zhang *et al.* suggested that the similarity of images using the DCNN internal representation is more coincident to the human perceptual evaluation rather than that of the peak-signal-noise-to-ratio (PSNR) and the structural-similarity (SSIM) [14].

On the contrary, in the field of neuroscience, such texture visual stimulus causes neuron firing, and the neuron activation raises the perception.


H. Yang et al. (Eds.): ICONIP 2020, LNCS 12532, pp. 126–136, 2020.
https://doi.org/10.1007/978-3-030-63830-6_11

The responses of the neurons in the visual pathway were investigated in this decade [1,8]. In order to evaluate the property of the neuron activity, Okazawa *et al.* introduced a psycho-physical statistics called Portilla-Simoncelli statistics, hereafter we call it as PSS [9]. The PSS is a texture feature that takes perceptual aspects and can use as a basis for texture image reconstruction. Okazawa *et al.* evaluated the texture image representation of neurons in visual area 4 (V4) [8] with the regression to the subset of the PSS. As a result, the behaviors of the V4 neurons, which can respond to the texture image, would be well explained with the small number of the combination of the PSS features. Moreover, Okazawa *et al.* also suggested that the selectivity of the neuron for the specific texture image would correspond to the higher-order statistics of the PSS.

The origin of the DCNN structure comes from "Neocognitron" proposed by Fukushima [2]. The basic structure of Neocognitron consists of "S-cell" and "C-cell" layers corresponding to the "simple cell" and "complex cell" in the primary visual cortex respectively [5]. Nowadays, the S-cell and C-cell layers are called "convolution layer" and "pooling layer" respectively. To realize a higher function of the visual cortex, such as visual classification ability, Fukushima extended the basic structure, which is a model of the primary visual cortex, hierarchically.

Here, we consider the texture image representation in the DCNN from the comparing with the visual neuron responses. In this paper, we investigate the texture representation ability of the DCNN with joint wavelet-based coefficients statistics called Portilla Simoncelli Statistics (PSS), which was aplied to analyze the visual neurons' responses. Introducing the PSS, we can discuss the property of the texture image representation in the DCNN as the feature map level. Moreover, from the viewpoint of the texture synthesis application, the feature map interpretation would bring us the synthesis's control ability.

This paper is organized as follows: In Sect. 2, we explain applying the DCNN model, which is called VGG [11], and the overview of the PSS analysis. Section 3 shows the experimental condition and the results. In Sect. 4, we summarize the conclusion and discuss the results.

## 2 Methods and Materials

Here we introduce a DCNN model called VGG16, which is one of the famous reference model [11]. Here, we summarize the model at first. Then we also explain a subset of PSS, which is for input texture image description.

### 2.1 Model Description of VGG16

Nowadays, the VGG16 is a popular reference model of the DCNN. The model is originally proposed to solve a visual classification task for large scale image data-set called ImageNet [10]. Figure 1 shows the schematic diagram of the VGG16. The model consists of two parts, that is, the feature extraction part and the classification part. The feature extraction part has a cascade structure of building blocks called "VGG block", which has several convolution layers and

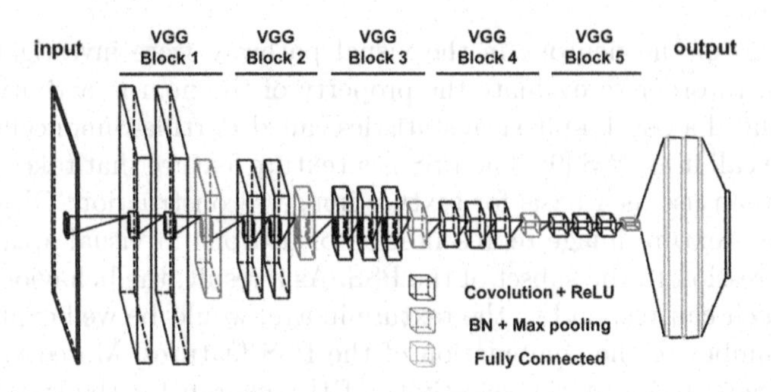


**Fig. 1.** Schematic diagram of VGG16. VGG blocks shows the feature extraction part, and the full connection part shows the classification part.

a max-pooling layer. The classification part can be considered as a multi-layer perceptron.

The original contribution of the VGG model is improving the performance for visual classification task by introducing a decomposed structure of convolution layers of small kernel size. Teramoto & Shouno analyzed the cascaded convolution structure in the VGG model for visual classification task could reproduce a kind of continuity such as in the primary visual cortex [13]. Of course, there exist a lot of improved DCNN models for the classification task. However, we would like to investigate the relationship between the DCNN model with neuron response representations in this study, so that we adopt the relatively simple VGG model as the roughly approximated of the visual pathway.

In the detail of the VGG block, the convolution layer can be considered as the cascade connection of linear convolution operation and non-linear modulation called ReLU (Rectified Linear Unit). In the following experiment, we use the ReLU output as the inner representation of the VGG, so that, we denote the $l$th ReLU layer output in $b$th VGG block as $\boldsymbol{r}_{b,l}$. The $\boldsymbol{r}_{b,l}$ has $C_b \times W_b \times H_b$ elements where $C_b$ means the number of channels, $W_b$ and $H_b$ means the size of channel in $b$th VGG block. Thus, we denote response on the location of $\boldsymbol{\xi}$ in the channel $c$ of the $\boldsymbol{r}_{b,l}$ as $r_{b,l}(c, \boldsymbol{\xi})$.

## 2.2 Overview of the PSS

The PSS is a kind of visual statistics for input image to describe the visual perception of the input in the meaning of the psychophysics. In other words, the PSS is designed to bring the same visual perception called "perceptually equivalence" when textures with the same PSS are given. Moreover, the PSS is a marginal statistics, that is, we can generate perceptually equivalent texture images from a random noise image when we obtain PSS from a texture image,

The PSS of a texture image is derived from a kind of wavelet decomposition of the image called "steerable filter pyramid", hereafter we denote it as SFP. The SFP is a multi-scale and directional image decomposition method that partially mimics the direction selectivity of the primate visual cortex. The SFP could be

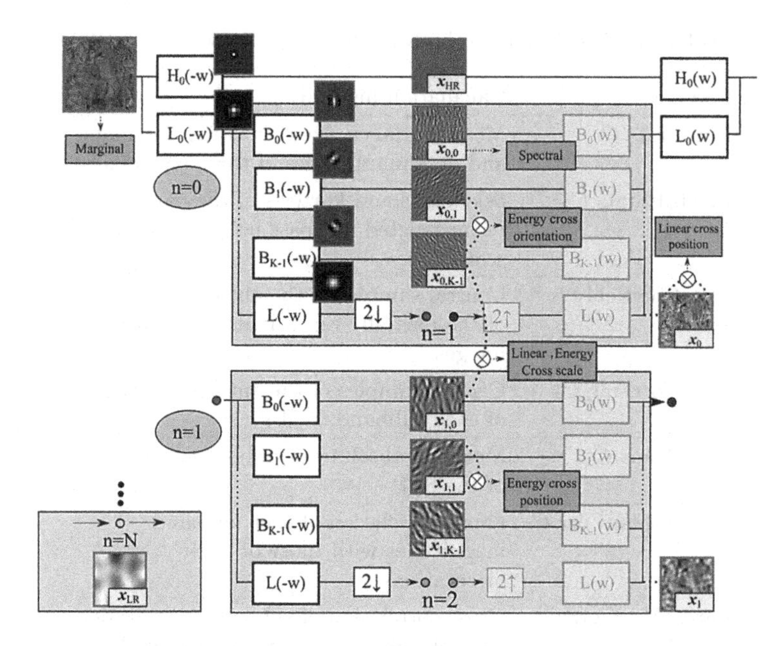


**Fig. 2.** Schemtic diagram of SFP in Fourier domain. The most left shows the input, and the large rectangle blocks show SPF levels.

"*steered*" its decomposition traits by two parameters $N$ and $K$. $N$ means the number of decomposition scales, and the $K$ means the number of decomposition in orientations/directions. These properties originated in complex orthogonal Wavelet transform and Gabor filters bank.

Figure 2 shows a Fourier domain block diagram of a steerable filter pyramid interpreted as a linear system. As shown in Fig. 2, the SFP consists of recursion of subsystems indexed by $n = 0 \ldots N$, which are drawn as large rectangles. The input image is divided with high and low spatial frequency components denoted as $H_0(-\omega)$ and $L_0(-\omega)$, where $\omega$ describes the frequency. The input for $n$th subsystem is defined as the previous scale low frequency component $L_{n-1}(-\omega)$. This recursion describes the multi-scale property of the SFP. In each scale, the input is decomposed in the $K$ orientation components, which are drawn in parallelism. In the Fig. 2, $B_k(-\omega)$ shows the orientation depending component and we draw the spatial filter in each block. Thus, we can obtain a filtered image denoted as $\boldsymbol{x}_{n,k}$ where $n$ and $k$ means scale level and orientation, respectively.

The PSS considers the following 10-types of statistics denoted as $C_1$–$C_{10}$ through the SFP. Table 1 shows a summary of the calculation.

The number of dimensions for the original PSS is controlled with the parameter $N$ and $K$. In general, the number of dimensions becomes large, even if we choose the small number of $N$ and $K$. Thus, we should consider reducing the dimensions to improve the tractability. Suzuki & Shouno showed naive dimension reduction with PCA caused missing of the texture reproducibility and proposed

**Table 1.** Calculation methods of PSS in 10 classes from SPF

| PSS class | Calculation method |
|---|---|
| PixelStats $C_1$ | Mean, variance, skew, kurtosis, minimum and maximum values of the image pixels |
| PixelLPStats $C_2$ | Sskewness and kurtosis of the partially reconstructed low-pass images at each scale |
| AutoCorrReal $C_3$ | Central samples of the auto-correlation of the partially reconstructed low-pass images |
| AutoCorrMag $C_4$ | Central samples of the auto-correlation of each subband of SFP |
| MagMeans $C_5$ | Means of absolute values of each subband of the SFP |
| CousinsMagCorr $C_6$ | Cross-correlation of each subband magnitudes with those of other orientations |
| ParentMagCorr $C_7$ | Cross-correlation of subband magnitudes with all orientations at a coarser scale |
| CousinRealCorr $C_8$ | Cross-correlation of the real part with both the real and imaginary parts at a coarser scale |
| ParentRealCorr $C_9$ | Cross-correlation of subband real parts with all orientations at a coarser scale |
| VarianceHPR $C_{10}$ | Variance of the high-pass band |

a hierarchical PCA model [12]. Okazawa *et al.* proposed a small subset of the PSS that was focus on the visual equivalence measurement of V4 neurons in the macaque monkey. They pruned PSS features based on the V4 neuron responses. In this study, we introduce a subset PSS, called "minPS", in the manner of Okazawas' method.

- **Spectral Statistics**, which corresponds to $C_5$, means the average absolute magnitude of each subband component abs $\boldsymbol{x}_{n,k}$ where abs($\boldsymbol{u}$) denotes absolute operation for each element of vector $\boldsymbol{u}$. We introduce 4 components for the minPS.
- **Marginal**, which corresponds to $C_1$ and $C_2$, means skew and kurtosis of the original input image.
- **Linear Cross Position**, which corresponds to $C_3$, are auto-correlation based statistics. We calculate auto-correlation in the center location in $M \times M$ size. In this study, we adopt $M = 3$ as the window size. Calculating the center location of auto-correlations for the training dataset, we can obtain the principal components (PC) in $M \times M$ dimensions. So, we introduce 4 PC projection values for the input image into the minPS components.

- **Linear Cross Scale**, which corresponds to $C_8$, are statistics based on the cross-correlation between different scale correlation. In the same manner, as the Linear Cross Position, we introduce four components into the minPS.
- **Energy Cross Position**, which corresponds to $C_8$, are auto-correlation based statistics. The difference from the Linear Cross Position is non-linear modulation for the correlation signal. Here, we introduce absolute modulation for each component $\boldsymbol{x}_{n,k}$, that is $\text{abs}(\boldsymbol{x}_{n,k})$ at first. Then we calculate the autocorrelation for the modulated signal in the same manner with Linear Cross Position. We introduce four components to the minPS.
- **Energy Cross Orientation**, which corresponds to $C_6$, are statistics based on the cross-correlation between different orientations with absolute modulated signal $\text{abs}(\boldsymbol{x}_{n,k})$. We obtain 6 PCs in the same manner with Linear Cross Position. Then, we introduce 6 projection values for each PC as the minPS elements.
- **Energy Cross Scale**, which corresponds to $C_7$, are statistics base on the cross-correlation between different scale with absolute modulated signal. We introduce 4 elements calculated in the same manner with Linear Cross Position.

In the Fig. 2, we draw each feature source as hatched rectangle. As the result, we obtain the minPS representation, whose number of dimension is 30, from the PSS. Hereafter, we denote the minPS representation for the image $\boldsymbol{x}$ as $\boldsymbol{m}_{\text{minPS}}(\boldsymbol{x})$

### 2.3 Image Dataset

Here we introduce a texture image dataset called Okazaki Synthetic Texture Image (OSTI) [7]. The whole images in the OSTI dataset are synthetic texture images controlled with PSS. This dataset is for measuring the response of the V4 neurons in the macaque. The dataset contains 10355 nature like textures with 128×128 [px]. In the dataset, 4170 images use textural parameters extracted from photographs of eight material categories, and the remaining 6185 are generated from the PSS interpolated from the previous 4170 images.

## 3 Experiments and Results

### 3.1 Experiment with LASSO Regression

To evaluate the DCNN representation with the PSS, we consider a regression problem from the expression of minPS to that of the VGG representation. We use a pre-trained VGG model, which is optimized to solve natural scene classification task with ImageNet. Note that the VGG does not optimize for texture images.

In the analysis, we assume the small number of the components in the DCNN representation corresponds to the PSS expression. Thus, we apply a regression analysis with a sparse selection called LASSO (Least Absolute Shrinkage and

Selection Operator) regression. We adopt the combination of the channel averages of each VGG layer as the layer representation for the translational symmetry of the convolution operation. Moreover, in the context of the style transfer method for texture patterns, Gatys *et al.* applied the 1st ReLU representation in each block for the style information [3]. Thus, we use the 1st ReLU layer output in $b$th block denoted as $\boldsymbol{r}_{b,1}$ where $b = 1 \dots 5$ for inner representation of the VGG. We denote the $c$th channel average of 1st ReLU layer in $b$ VGG block as $\bar{r}_{b,1,c} = \frac{1}{W_b H_b} \sum_{\boldsymbol{\xi}} r_{b,1}(c, \boldsymbol{\xi})$. Thus, the layer representation as a vector $\overline{\boldsymbol{r}_{b,1}} = (\bar{r}_{b,1,1}, \dots, \bar{r}_{b,1,c}, \dots \bar{r}_{b,1,C_b})$.

The loss function of the LASSO regression can describe as follows:

$$L(\boldsymbol{W}_b) = \frac{1}{P} \sum_{p=1}^{P} \| \overline{\boldsymbol{r}_{b,1}} - \boldsymbol{W}_b \boldsymbol{m}_{\mathrm{minPS}}(\boldsymbol{x}_p) \|^2 + \lambda |\boldsymbol{W}_b|_1, \quad (1)$$

where the oerator $|\cdot|_1$ means the $L_1$ norm for sparse solution. The parameter $\boldsymbol{W}_b$ means the weight matrix for optimization, $p$ shows the training pattern index, and $\lambda$ shows the hyperparameter for the sparse prior. To optimize the loss function, we carry out the 10 fold cross-validation (CV). In each fold, we determine the hyperparameter $\lambda$ with 5 fold CV at first. Then, we optimize $\boldsymbol{W}_b$ after the hyperparameter determination. In the test phase of each fold, we evaluate the averaged direction cosines between estimated vector of $\boldsymbol{W}_b \boldsymbol{m}_{\mathrm{minPS}}(\boldsymbol{x})$ and the VGG representation $\overline{\boldsymbol{r}_{b,1}}$ for test patterns.

### 3.2 Analysis of the Synthesized Image with VGG

To confirm how much the VGG internal representation has texture information, we introduce a texture generation method in the manner of the style transfer method. Here, we consider the followint loss function as the texture generation from the VGG:

$$L(\boldsymbol{x}_{\mathrm{opt}}, \boldsymbol{x}_{\mathrm{ref}}) = \sum_{b \in \mathtt{source}} \sum_{c} (\bar{r}_{b,1,c}(\boldsymbol{x}_{\mathrm{opt}}) - \bar{r}_{b,1,c}(\boldsymbol{x}_{\mathrm{ref}})^2), \quad (2)$$

where $\bar{r}_{b,1,c}(\boldsymbol{x})$ means the average output of the channel $c$ in the 1st ReLU layer of the $b$th VGG block for the VGG input $\boldsymbol{x}$. The $\boldsymbol{x}_{\mathrm{ref}}$ means the reference texture, and the $\boldsymbol{x}_{\mathrm{opt}}$ is for the optimization object. Using a gradient method for $\boldsymbol{x}_{\mathrm{opt}}$, we reduce the loss function for the texture generation. Thus, this loss function demands average channel similarity between $\boldsymbol{x}_{\mathrm{ref}}$ and $\boldsymbol{x}_{\mathrm{opt}}$.

In the numerical experiment, we introduce the LBFGS method to optimize the loss function. We consider the input VGG source as followings 9 cases. At first, we consider only a single VGG block effect denoted as the block number $b = 1 \dots 5$. Then, we also consider cumulative information up to the $b$th VGG block, that is, the source comes from $\{(1,2), (1,2,3), (1,2,3,4), (1,2,3,4,5)\}$ blocks.

### 3.3 Results

Figure 3 shows the fitness evaluation of the regression results in the meaning of the direction cosine for each block. The horizontal axis shows the VGG block

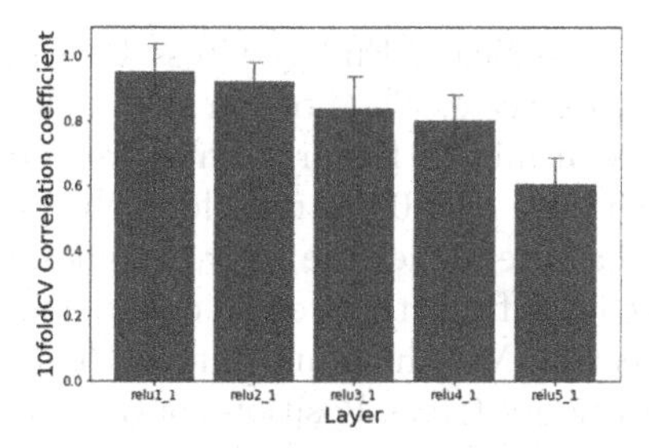


**Fig. 3.** Results of fitness evaluation of LASSO regression. The horizontal shows the block number of the VGG, and the vertical shows the direction cosine similarity between the minPS and the VGG representation.

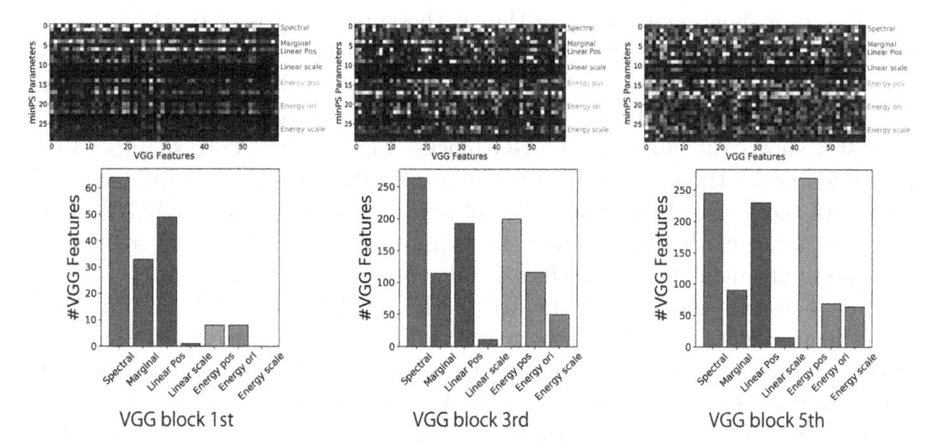


**Fig. 4.** Results of weight matrix $\boldsymbol{W}_b$. From left to right, each figure shows block $b = 1, 3, 5$, respectively. In each sub-figure, the top shows the visualization results of the weight, and the bottom shows the histogram for minPS class.

number, and the vertical shows the direction cosine values. We show the average fitness values with error bars. From the figure, we can see the fitness reduces gradually through the information process from the lower to the higher, and rapidly drop in the 5th VGG block. Thus, we confirm the texture representation ability remains in the intermediate layers except for the highest one.

Figure 4 shows the averaged weight matrix for 1st, 3rd, and 5th VGG blocks. From the left to the right, we show the averaged regression weights $\boldsymbol{W}_b$ $(b = 1, 3, 5)$. In each figure, the top shows the visualized weight in which the horizontal axis shows the VGG feature number, which we select as 60 top reliable responding channels. The vertical axis shows the minPS features. In the visualization, we aligned the minPS features as Spectral Statistics, Marginal, Linear

Cross Position, Linear Cross Scale, Energy Cross Position, Energy Cross Orientation, and Energy Cross Scale. The bottom shows the number of regression weight contributing to each minPS feature. Thus, we count the number of the regression weights, which have over 0.5 threshold in the meaning of the absolute value. We can confirm that the higher the layer becomes, the more considerable minPS contribution becomes from the weight count information. The ratio of contribution for the Spectral, Marginal, and Linear Cross Position is significant in the early layer. The Energy Cross Position contribution becomes significant stepping up the hierarchy. The Energy Cross Orientation contribution increases up to the 3rd VGG block ($b = 3$): however, it reduces after the block. The Linear Cross Scale and the Energy Cross Scale contributions do not have a substantial contribution to the minPS.

c Figure 5 shows the results of the synthesizing analysis of the VGG representations. The left figure shows the correlation of VGG representations and the minPS components as the matrix. The column axis shows the source information of the VGG blocks, in which the first 5 columns mean single VGG blocks, and the last 4 shows the cumulative VGG blocks. The row axis corresponds to the minPS features—the bright points in the matrix show strong correlations. Each column of the figure shows the ability of texture representation of VGG blocks in the meaning of the minPS. We can confirm all cumulative representations, and single intermediate blocks show good representation ability for the minPS features; however, the 1st and the 5th block of VGG does not have. Each row of the figure shows how does the specific minPS feature appears in the VGG blocks. At first, we can see the Linear Cross Scale and the Energy Cross Scale representation do not appear in any VGG representations. Moreover, we can interpret the different scale representation might exist in the intermediate VGG blocks since the whole Spectral components in any scales, which are shown as column direction of the Spectral components, have strong correlations.

The right shows some examples of the source and the synthesized images from the PSS and the VGG single block. The top shows the reference textures in the OSTI dataset. The middle shows the synthesized textures from the PSS ($N = 4, K = 4, M = 7$) corresponding to each reference texture. The bottom shows the synthesized textures from the VGG representation with Eq. (2) optimization. For these synthetic images, we select the reference textures as the images that have significant contributions to the Energy Cross Scale component to see the effect in the VGG. We can see the PSS representation can reproduce the reference well; however, the VGG can only reproduce local textures. Notably, for the index 04047 texture, the horizontal periodic structure does not appear in the VGG reproduction. Thus, the different scale energy cross-correlation would be required for the synthesis even if it does not require in the visual classification task.

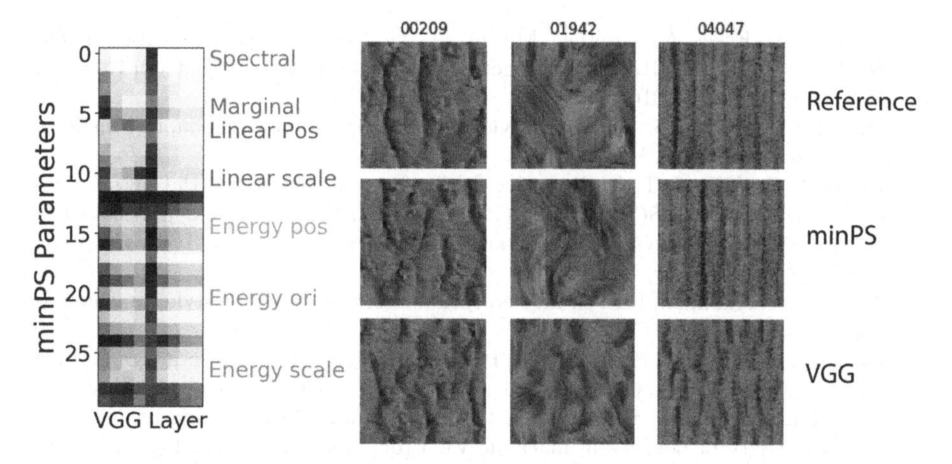


**Fig. 5.** The right shows the average correlation of the synthetic textures of the VGGs and that of the minPS for whole OSTI dataset. The right shows the synthetic examples from the minPS and the VGGs whose references have strong contribution of Energy Cross Scales.

## 4 Conclusion and Discussion

In this study, we analyze the texture representation in the VGG with the minPS, which applied for visual neurons' responses. The VGG is not trained with texture images but with a massive natural scene dataset called ImageNet. From the analysis of the VGG responses for the OSTI texture dataset, we confirm Spectral, Marginal, Linear Position, Energy Position, and Energy Orientation features appear in the representation well. On the contrary, the Linear Scale and Energy Scale cross-correlations do not appear enough. Okazawa *et al.* suggested the Energy Scale cross-correlation appears in the V4 neurons, and the Linear Cross correlation does not. Thus, we conclude the Energy cross-correlation is missing components in the VGG representations for the classification task. In other words, we should consider the Energy cross-correlation is required for texture synthesis by use of image transfer with VGGs.

**Acknowledgment.** We thank to the Mr. Satoshi Suzuki and Mr. Takahiro Kawashima for fruitful discussion about the mathematical modeling. This study was partly supported with MEXT KAKENHI, Grant-in-Aid for Scientific Research on Innovative Areas, 19H04982, Grant-in-Aid for Scientific Research (A) 18H04106.

## References

1. Freeman, J., Ziemba, C.M., Heeger, D.J., Simoncelli, E.P., Movshon, A.J.: A functional and perceptual signature of the second visual area in primates. Nat. Neurosci. **16**, 974–981 (2013)
2. Fukushima, K.: A self-organizing neural network model for a mechanism of pattern recognition unaffected by shift in position. Biol. Cybern. **36**(4), 193–202 (1980)

3. Gatys, L.A., Ecker, A.S., Bethge, M.: Image style transfer using convolutional neural networks. In: The IEEE Conference on Computer Vision and Pattern Recognition (CVPR), June 2016
4. Huang, X., Belongie, S.: Arbitrary style transfer in real-time with adaptive instance normalization. In: ICCV (2017)
5. Hubel, D.H., Wiesel, T.N.: Receptive fields of single neurones in the cat's striate cortex. J. Physiol. **148**(3), 574–591 (1959). https://doi.org/10.1113/jphysiol.1959.sp006308. https://physoc.onlinelibrary.wiley.com/doi/abs/10.1113/jphysiol.1959.sp006308
6. Johnson, J., Alahi, A., Fei-Fei, L.: Perceptual losses for real-time style transfer and super-resolution. In: ECCV (2016)
7. Okazawa, G., Tajima, S., Komatsu, H.: Okazaki synthetic texture image (OSTI) database. http://www.shitsukan-db.jp/public/texture-db/. Accessed 30 May 2020
8. Okazawa, G., Tajima, S., Komatsu, H.: Image statistics underlying natural texture selectivity of neurons in macaque v4. Proc. Natl. Acad. Sci. **112**(4), E351–E360 (2015). https://doi.org/10.1073/pnas.1415146112. https://www.pnas.org/content/112/4/E351
9. Portilla, J., Simoncelli, E.P.: A parametric texture model based on joint statistics of complex wavelet coefficients. Int. J. Comput. Vision **40**(1), 49–70 (2000)
10. Russakovsky, O., et al.: ImageNet large scale visual recognition challenge. Int. J. Comput. Vision (IJCV) **115**(3), 211–252 (2015). https://doi.org/10.1007/s11263-015-0816-y
11. Simonyan, K., Zisserman, A.: Very deep convolutional networks for large-scale image recognition (2014)
12. Suzuki, A., Shouno, H.: Generative model of textures using hierarchical probabilistic principal component analysis. In: International Conference on Parallel and Distributed Processing Techniques and Applications (PDPTA) (2017)
13. Teramoto, T., Shouno, H.: A study of inner feature continuity of the VGG model. Technical report, 470, IEICE, March 2019
14. Zhang, R., Isola, P., Efros, A.A., Shechtman, E., Wang, O.: The unreasonable effectiveness of deep features as a perceptual metric (2018)

# Auto-Classifier: A Robust Defect Detector Based on an AutoML Head

Vasco Lopes(✉) and Luís A. Alexandre

NOVA LINCS, Universidade da Beira Interior,
Rua Marquês d' Ávila e Bolama, 6201-001 Covilhã, Portugal
{vasco.lopes,luis.alexandre}@ubi.pt

**Abstract.** The dominant approach for surface defect detection is the use of hand-crafted feature-based methods. However, this falls short when conditions vary that affect extracted images. So, in this paper, we sought to determine how well several state-of-the-art Convolutional Neural Networks perform in the task of surface defect detection. Moreover, we propose two methods: CNN-Fusion, that fuses the prediction of all the networks into a final one, and Auto-Classifier, which is a novel proposal that improves a Convolutional Neural Network by modifying its classification component using AutoML. We carried out experiments to evaluate the proposed methods in the task of surface defect detection using different datasets from DAGM2007. We show that the use of Convolutional Neural Networks achieves better results than traditional methods, and also, that Auto-Classifier out-performs all other methods, by achieving 100% accuracy and 100% AUC results throughout all the datasets.

**Keywords:** Defect detection · CNNs · Deep learning · AutoML

## 1 Introduction

Visual inspection of products is crucial to ensure customer requirements and a longer product life, by removing imperfections or defects that can lead to problems like rust, sharp edges or visually deficient products. Industrial quality control and visual inspection require extreme attention to detail and are usually performed by humans, meaning that it is prone to error, requires training, and its a time-consuming task that needs to be repeated countless times in modern factories [18]. Systems capable of providing a way to automate such processes, either by completely removing the need for human labour or by complementing

This work was supported by NOVA LINCS (UIDB/04516/2020) with the financial support of FCT-Fundação para a Ciência e a Tecnologia, through national funds, and partially supported by project 026653 (POCI-01-0247-FEDER-026653) INDTECH 4.0 – New technologies for smart manufacturing, cofinanced by the Portugal 2020 Program (PT 2020), Compete 2020 Program and the European Union through the European Regional Development Fund (ERDF).


H. Yang et al. (Eds.): ICONIP 2020, LNCS 12532, pp. 137–149, 2020.
https://doi.org/10.1007/978-3-030-63830-6_12

the work conducted by humans, are essential to reduce costs and improve product quality [16]. Thus, the goal of inspection systems is to rapidly and precisely detect, classify or segment defective areas in images. However, such systems are scarce due to difficulties in acquiring real data to train Artificial Intelligence (AI) algorithms, and because industry floors require continuous changes, meaning that having controlled and unchanged environments is challenging. Traditional methods often rely on extracting hand-craft features from images, in order to represent defects and anomalies [23,24] and can be categorized into: statistical, structural or filter based. These methods are capable of detecting and or segmenting defects in images if they are acquired in a controlled environment [31]. However, these are not capable of solving the same kind of problems when applied to images with complex textures, or if the acquire data suffers slight changes or contains noise.

To mitigate the aforementioned problems, some approaches evaluate the use of deep learning to tackle the problem of detecting defects [29], more specifically, Convolutional Neural Networks (CNNs), due to the excellent results that they achieve in a multitude of tasks related to image analysis [13,21]. The use of CNNs is particularly good in the task of defect detection, because they can learn to be robust to the presence of noise and different conditions, such as light and rotation [2,26], meaning that a robust CNN that can correctly classify or detect defects, and can be invariant to the problems that undermine traditional approaches.

In this paper, we evaluate how several state-of-the-art CNNs perform in the task of surface defect detection, and propose two different approaches: 1) a CNN-fusion method, that averages the classification of all individual CNNs into a final classification, and 2) a Auto-Classifier detector, which is a novel method that integrates the use of an AutoML head to complement a CNN, by using the feature extractor of the CNN and creating a new classifier upon it. We validate our proposals in the task of surface defect detection using DAGM2007 datasets [30], and show that the propose methods can improve upon the state-of-the-art. The code for all proposed methods, as well as the used data set partitions is freely available allowing for free use and fair comparisons[1].

The rest of this paper is organized as follows: the next section discusses the state-of-the-art and related work regarding defect detection and AutoML; Sect. 3 presents the proposed methods; Sect. 4 contains the experiments and discussion, while the final sections contain the conclusions.

## 2 Related Work

**AutoML:** The field of AutoML is a domain of expertize whose aim is to develop methods and tools to provide efficient mechanisms that can be used by virtually anyone to design tailor-made machine learning algorithms to their problems [10]. Designing ML algorithms for specific problems can be a difficult task, as it can have many design choices that are both dependent and independent from

[1] www.github.com/VascoLopes/AutoClassifier.

one another. Designing efficient ML algorithms takes years of expertize and trial and error, as many optimization problems usually rely on the user, which makes it very difficult for non-expert users to do. So, AutoML is an important approach to bring machine learning algorithms closer to non-experts, but also to integrate it with other technologies, to create new and more efficient approaches for several problems. The difference between AutoML and traditional machine learning workflows, is that AutoML intends to remove all the steps between the data acquisition and getting the final model, which usually involve data processing, feature extraction and model selection, from the user.

Over the years, many AutoML methods have been proposed [3,7,12,17]. However, our proposals are not related to AutoML algorithms, but with the combination of the power that AutoML provides to create optimal, tailor-made algorithms to solve the tasks at hand. Closer to our Auto-Classifier, [11] uses AutoML to fully design a method to detect railway track defects, while our Auto-Classifier, uses AutoML to complement the feature extraction power of a CNN, by coupling a new classifier with a modified CNN.

Our work relates with Neural Architecture Search (NAS), which is a subset of AutoML that focus on automatically design deep neural architectures. NAS was initially formulated as a reinforcement learning problem, where a controller is trained over-time to sample more efficient architectures [35], requiring over 60 years of computation. A follow-up work, improved upon the base work by performing a cell-based search, where cells, which conduct some operations, are replicated to form a complete CNN [36]. In [1], the authors use Q-learning to train the sampler agent. Using a similar approach, the authors of [34], perform NAS by sampling blocks of operations instead of cells/architectures, which can then be replicated to form networks. More recently, ENAS [19], used a controller to discover architectures by searching for an optimal subgraph within a large computational graph, requiring only a few computational days to build a final architecture. DARTS, a gradient-based method, showed that by performing a continuous relaxation of the parameters, they could be optimized using a bi-level gradient optimization [15]. DARTS was then improved using regularization mechanisms [32].

The main difference between NAS and our work is that while NAS focus on designing an entire network, our focus with Auto-Classifier, is to improve upon a CNN that yields good results by leveraging the power of AutoML to design a new classifier component. This search extends the use of CNNs with other types of classifiers, and is faster than designing entire machine learning algorithms from scratch using NAS. This can be seen in the experiments section, where the search for a new classifier was extremely fast, by having AutoML to search for only two hours.

**Surface Defects:** Different types of surface defects include cracks, which can happen in a panoply of surfaces. In [22], the authors focus on detecting cracks in power plants in a private dataset, using a CNN for semantic segmentation. Similarly, in [14], a bridge cracks detection algorithm is proposed. This method uses active contours and Canny Edge detector to find the defects, and an SVM for classification. In [33], the authors trained a CNN to solve the problem of detecting road cracks, using a dataset of 500 images acquired using a smartphone.

In the task of producing gravure cylinders, it is common to have defects like holes, so, the authors of [28] proposed a method that uses a CNN to classify images acquired by a high-resolution camera. The method achieved an accuracy rate of 98.4% on a private dataset. Differently, the authors of [4], proposed a method to conduct surface quality control, by using cutting force, vibration, and acoustic emission signals information of a CNC machine. By decomposing the signals into time series, a predictor was capable of predicting the surface finish.

Using DAGM2007 set of problems, the author of [29] propose to evaluate the performance of 3 CNNs, with different network specifications, which achieve results between 96% and 99% accuracy. Traditional methods that rely on extensive feature extraction were also studied. However, their results fall short when compared to deep learning techniques. In [23], a method based on LBP achieved 95.9% accuracy, and [24], that achieved a 98.2% accuracy by using EANT2, a neuroevolution method to develop artificial neural networks for classification purposes.

Contributions of this work include the evaluation of different state-of-the-art CNNs and the proposal of two novel methods of performing surface defect detection. Moreover, the major contribution of this work, Auto-Classifier, proposes a new way of classifying images by improving the CNNs classification mechanism by using automated search to generate a new classifier that is then coupled to the CNN. To the best of our knowledge, the Auto-Classifier method proposed in this paper is the first method that combines and improves CNNs using AutoML.

## 3 Proposed Method

In this section, we present our proposed methods, which aim at classifying the presence of defects in 2D images by CNNs. We evaluate the performance of multiple state-of-the-art architectures, that yield the best results for multiple problems in the task of image and objects classification, in the task of detecting defects, which are: VGG11, VGG16 and VGG19 [25], Resnet18, Resnet34, Resnet50 and Resnet101 [8], and Densenet121 [9] (Sect. 3.1). Based on the evaluation of these CNN architectures, we propose two methods: *i)* CNN-fusion, which combines all the CNN architectures and outputs final one prediction (Sect. 3.1); *ii)* Auto-Classifier, which uses as basis, the architecture that yields the best results in the performance evaluation, and then uses Automated Machine Learning (AutoML) to automatically search a new classifier (Sect. 3.2), that is then stacked with the CNN feature extraction mechanism.

### 3.1 Convolutional Neural Architectures

CNNs are one of the most popular and prominent deep learning architectures for a variety of image processing [6,13,21]. CNNs are a variant of Multilayer

Perceptron Networks and are biologically inspired models, created to emulate how the human visual cortex processes visual information, making these networks particularly suitable for image processing. Usually, CNNs perform a series of operations, such as convolutions and pooling, and are followed by a number of fully connected layers. The idea is: CNNs start by extracting representations of the input as features maps, which gradually increase in complexity at deeper layers of the network, then, these feature maps are fed into fully connected layers that provide the output of the network (activation patterns), normally in the form of a classification map.

Provided that CNNs can extract meaningful feature maps and thus, activation patterns, the results in a multitude of tasks are usually excellent, outperforming methods based on hand-crafted features [21]. Thus, CNNs need to be trained in a set of data so that they can extract meaningful information from the input, hence, correctly solving a given problem. The most common method to train CNNs is by using gradient-descent with Back-propagation [20], where an input image is propagated forward throughout the network, and upon reaching the final layer, the loss is calculated, and retro-propagated through the network, thus adjusting the weights of the network to more accurately solve the input problem.

In this paper, the first approach sought to evaluate the performance of several state-of-the-art CNN architectures that are known to do well in a variety of classification problems, VGG11, VGG16 and VGG19 [25], Resnet18, Resnet34, Resnet50 and Resnet101 [8], and Densenet121 [9] and combine their approach.

By evaluating a set of CNNs (Sect. 4), we created a pool of networks that were specifically trained to solve the problem of defect detection, yielding excellent results individually. To harness the classification correctness of all the individual networks, we propose CNN-Fusion. This approach intends to perform a combination of all the predictions of the individually trained CNNs, into a final, unique classification.

Denoting that all the individual CNNs were trained using a training set and validated using a unique, validation set, all networks can be categorized by their Area Under The Roc Curve (AUC) obtained in the validation set, as metric for their correctness in solving the given problem. So, the proposed CNN-Fusion works by fusing all the individual predictions into a final, weighted, prediction by making a weighted sum of each class, where each network votes using a normalized weight based on the AUC obtained in the validation set. The weights are obtained with the following expression:

$$w_i = \frac{V_i}{\sum_{j=1}^{n} V_j}, i \in 1, ..., n \tag{1}$$

where $n$ is the number of CNNs, and $V$ the array that contains the AUC values (between 0 and 1) for all the CNNs.

Hence, to perform classification, using the normalized contribution of the individual for a given input, we use the following expression:

$$argmax_i(P_{ij} \cdot w_j), \begin{array}{l} i \in 1, ..., c, \\ j \in 1, ..., n \end{array} \tag{2}$$

where $c$ is the number of classes, $n$ the number of CNNs, $P_{ij}$ represents the output classification of network $j$ for class $i$.

The idea behind CNN-Fusion is that, by balancing the importance of each network through the process of normalization, where networks that have higher AUC scores in the validation set have higher confidence, we can perform a weighted voting that will improve upon the result of classifying the existence of defects using individual networks.

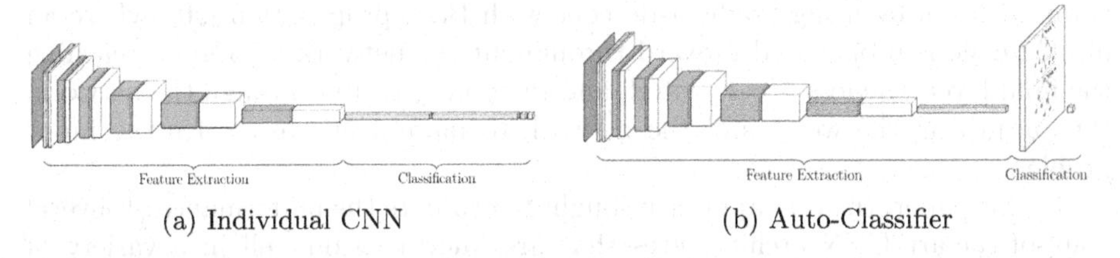


(a) Individual CNN (b) Auto-Classifier

**Fig. 1.** Visual representation of the difference between a CNN and the Auto-Classifier method. CNN is composed by two components: Feature Extraction and Classification. In the Auto-Classifier, the classification component has been replaced by another one, represented by a XGBM.

### 3.2 Auto-Classifier

The second proposed method in this paper, Auto-Classifier, focuses on improving the best individual CNN, by replacing its classification component by a new one. As mentioned before, CNNs are usually comprised of two parts: *i)* feature extraction component, which is the initial part of the CNN, and is comprised of a set of layers, normally convolutions followed by batch-normalization, pooling layers or others, and *ii)* classification component, which is the last part of a CNN and consists of a Multi-Layer Perceptron, with possible addition of regularization layers.

We hypothesize that by training a CNN from scratch and then removing its classification component partially or entirely and replacing it by other types of classification methods or even by other methods that will perform both feature extraction and classification, it might be able to outperform the initial individual CNN. This is due to the fact that other types of classifiers, such as random forests, have shown to be very good in different classification problems. The problem with many classifiers is that, to be able to process and classify images correctly, they require extensive processing power, translating into huge models.

By processing the image with the feature extraction of a CNN, complex feature maps are created, which can then be used by a classifier, without need to perform any more feature extraction or feature processing. So, the proposed method, Auto-Classifier, works by initially using an individual CNN, from which the classification component is partially removed. Then, we use AutoML to generate a new classifier, based on the output features of the modified CNN. With this, we use the feature extraction capabilities of a CNN, and also allow the first layer of the CNN's classification component to stay, ensuring that the input to the new classifier has a smaller dimensionality, which ultimately enables more types of classifiers to work with that data.

The Auto-Classifier is composed of two parts: 1) the best individual CNN without the classification component, leaving a trained CNN that outputs a representation map of the input; 2) conduct an automated search, for a new method to perform classification based on the representations generated in the previous step. Then, the final model is composed by the partial CNN, sequentially followed by the new classifier found by AutoML. In Fig. 1, we visually present the difference between a CNN and the Auto-Classifier method. In this, the Auto-Classifier method, based on the CNN presented in (a), has a new classification component, represented by a XGBM.

To perform the AutoML search, we used H2O AutoML [7], which is intended to automate the machine learning workflow, by initially performing a Random Search of different models, and then performing a post-processing step by stacking the best solutions found [5]. The models in which the AutoML performs a hyperparameter search and tuning are: 3 pre-specified XGBoost Gradient Boosting Machine (XGBM) models, a fixed grid of Generalized Linear Models, a default Random Forest, five pre-specified H2O GBMs, a near-default Deep Neural Net, an Extremely Randomized Forest, a random grid of XGBoost GBMs, a random grid of H2O GBMs, and a random grid of Deep Neural Networks.

## 4 Experiments

To evaluate our proposed methods, the performance of different CNNs, and compare them with baselines and competitive approaches, we conduct experiments on a popular set of datasets that contain surface defects, the DAGM2007 datasets. All the experiments were conducted using a computer with an NVidia GeForceGTX 1080 Ti, 16 Gb of ram, 500 GB SSD disk and an AMD Ryzen 7 2700 processor.

### 4.1 DAGM2007

The DAGM2007 consists of 6 different datasets, each with 1150 images, from which, 1000 images are of background textures without defects, and 150 images of one labelled defect each on the background texture. On each of these problems, we performed a stratified split into 3 sets: 70% for the train set, 15% for the validation set and 15% for the test set. The train and validation set were used to train the algorithms, and the test set to evaluate the final performance of the methods. The test set was never used for training purposes, and for the proposed methods, the best individual CNN was selected based on its validation AUC.

### 4.2 Results and Discussion

To validate the proposed methods, we conducted 3 experiments: *1)* evaluate the performance of multiple state-of-the-art CNN architectures, *2)* evaluate the performance of the CNN-fusion method, by fusing all individual CNNs, and *3)* evaluate the performance of the Auto-Classifier method.

To evaluate the performance of state-of-the-art CNNs, we used the same settings for each one: Stochastic Gradient Descent algorithm, batch size of 10, learning rate of $1e-3$, momentum of 0.9, and 100 epochs of training. To adjust the network's weights, we used back-propagation with gradient descent and Cross-Entropy loss. Subsequently to the training step, the model used for testing purposes is the one that yields the highest validation AUC while training. Denote that this testing step is only used to compare the different individual CNNs under the same conditions and is not used in any situation nor to select the best CNNs to be used in the proposed methods. The results, shown in the first 9 columns of Table 1, determined that amongst all the individual CNNs tested, VGG16 was the one with the best results - 99.9% mean accuracy and 99.7% mean AUC. The use of AUC as a metric for evaluating performance is extremely important because the accuracy metric is not, by its own, a good representative of a good classifier, since the dataset is not balanced and the accuracy shown does not consider that, whilst AUC-ROC is sensitive to class imbalance.

**Table 1.** Results of different state-of-the-art CNNs architectures and the two proposed methods in the task of defect detection, using the DAGM2007 dataset with test splits. Accuracy and AUC values are shown in percentages.

| Problem | VGG11 | | VGG16 | | VGG19 | | Resnet18 | | Resnet34 | | Resnet50 | | Resnet101 | | Densenet121 | | **CNN-Fusion** | | **Auto-Classifier** | |
|---|---|---|---|---|---|---|---|---|---|---|---|---|---|---|---|---|---|---|---|---|
| | Acc. | AUC | Acc. | AUC | Acc. | AUC | Acc. | AUC | Acc. | AUC | Acc. | AUC | Acc. | AUC | Acc. | AUC | Acc. | AUC | Acc. | AUC |
| 1 | 100 | 100 | 100 | 100 | 85.0 | 50.0 | 100 | 100 | 100 | 100 | 100 | 100 | 100 | 100 | 100 | 100 | 100 | 100 | 100 | 100 |
| 2 | 85.0 | 50.00 | 100 | 100 | 100 | 100 | 100 | 100 | 100 | 100 | 100 | 100 | 99.4 | 98.1 | 100 | 100 | 100 | 100 | 100 | 100 |
| 3 | 100 | 100 | 100 | 100 | 100 | 100 | 86.1 | 53.9 | 97.1 | 90.4 | 99.4 | 98.1 | 100 | 100 | 99.4 | 98.1 | 100 | 100 | 100 | 100 |
| 4 | 100 | 100 | 99.4 | 98.1 | 99.4 | 98.1 | 98.8 | 97.7 | 100 | 100 | 99.4 | 98.1 | 99.4 | 98.1 | 100 | 100 | 100 | 100 | 100 | 100 |
| 5 | 98.8 | 96.2 | 100 | 100 | 100 | 100 | 98.8 | 96.2 | 98.8 | 96.2 | 99.4 | 98.1 | 99.4 | 98.1 | 99.4 | 98.1 | 99.4 | 98.1 | 100 | 100 |
| 6 | 100 | 100 | 100 | 100 | 100 | 100 | 100 | 100 | 100 | 100 | 98.8 | 96.2 | 99.4 | 100 | 98.3 | 94.2 | 100 | 100 | 100 | 100 |
| $\mu$ | 97.3 | 91.0 | **99.9** | **99.7** | 97.4 | 91.4 | 97.3 | 91.3 | 99.3 | 97.8 | 99.5 | 98.4 | 99.6 | 99.1 | 99.5 | 98.4 | **99.9** | **99.7** | **100** | **100** |

We believe that the reason behind VGG16 having better results in almost all the six datasets and the best overall mean performances is due to the fact that, even if Resnet networks and Densenet are more powerful, their larger number of layers is a drawback when using small datasets, which is our case, since we are dealing with only 1150 images per dataset. Even though residual connects and short circuits in the mentioned networks can mitigate problems such as the vanishing gradient, their bigger complexity is a factor that will undermine their performance in problems were datasets are small and costly to acquire, e.g., defects in car painting.

By having the validation AUC values for each individual, we can complete the process of CNN-Fusion by combining the individual classifications into a final one, by first normalizing the validation AUC values, and then performing a voting using the CNN's classifications. The results for the CNN-Fusion are shown in 10th column of Table 1, where the mean accuracy was 99.9% and mean AUC was 99.7%. CNN-Fusion achieved the same mean values as VGG16. However, the difference is that CNN-Fusion was capable of perfectly identify defects in problems 1 to 4 and 6, and had miss-classifications in problem 5, while VGG16 had some miss-classifications in problem 4 and perfectly solved all the other problems. Moreover, CNN-Fusion was capable of achieving an overall high performance, but in problem 5, as many individual CNNs had classification errors, the CNN-Fusion was not capable of having 100% AUC. This can be justified because even though normalizing AUC values across all CNNs is a mean to balance individual contributions, all the CNNs had very good performances, meaning that the normalized values will not have large differences. A possible improvement would be to perform a non-linear normalization, where better models have a more normalized difference from its neighbours when compared to un-normalized values. The problem with the use of CNN-Fusion in a fast-paced environment, e.g., quality control in industrial production lines, is that it takes more time to have a final classification, as it requires all CNNs to perform their classification, which if done in parallel, will be the maximum time, $t$, from the set of times, $T$, that contain the time taken for each CNN to perform a classification for a given input, plus the time taken, $t_{fusion}$, to perform the final classification using all individual classifications: $f_{time} = max\{T(x) : x = 1, .., n\} + t_{fusion}$, where $n$ represents the number of individual CNNs. The problem is that in the vast majority of the systems, conducting a forward pass in all the individual CNNs in a parallel manner is impossible. When done sequentially, the $f_{time}$ will be increasingly higher: $f_{time} = (\sum_{x=1}^{n} T(x)) + t_{fusion}$.

The Auto-Classifier solves the problem of having an inference time that is dependant on all individual CNNs, by using the feature extraction capabilities of only the overall best CNN on the validation set (VGG16 in our experiments) and improving its classification component. We partially removed the classification component of VGG16, by removing the last two fully connected layers, leaving only one, which was used to reduce the feature maps dimensionality to speed up the search for a new classifier and also to allow a larger pool of classifier candidates. We run AutoML for 2 h for each problem, and at the end, we selected the best candidate on the validation set to be coupled to the modified VGG16 to create a final model - Auto-Classifier. The best classifiers from the AutoML step were: for problems 2 to 6, the classifier was a XGB model, and for problem 1 was a GBM. The results of Auto-Classifier on the test sets are shown in the last column of Table 1, and it is possible to see that it not only improved upon

**Table 2.** Comparison of different methods in the task of defect detection in DAGM2007 problems, using as metics the True Positive Rate (TPR), True Negative Rate (TNR), and Average Accuracy. Table adapted from [29].

| Problem | Auto-classifier (Ours) | CNN-fusion (Ours) | VGG16 (Ours) | Deep CNN [29] | Statistical features [23] | SIFT and ANN [24] | Weibull [27] |
|---|---|---|---|---|---|---|---|
| *TPR (%)* | | | | | | | |
| 1 | 100 | 100 | 100 | 100 | 99.4 | 98.9 | 87.0 |
| 2 | 100 | 100 | 100 | 100 | 94.3 | 95.7 | – |
| 3 | 100 | 100 | 100 | 95.5 | 99.5 | 98.5 | 99.8 |
| 4 | 100 | 100 | 99.3 | 100 | 92.5 | – | – |
| 5 | 100 | 99.3 | 100 | 98.8 | 96.9 | 98.2 | 97.2 |
| 6 | 100 | 100 | 100 | 100 | 100 | 99.8 | 94.9 |
| *TNR (%)* | | | | | | | |
| 1 | 100 | 100 | 100 | 100 | 99.7 | 100 | 98.0 |
| 2 | 100 | 100 | 100 | 97.3 | 80.0 | 91.3 | – |
| 3 | 100 | 100 | 100 | 100 | 100 | 100 | 100 |
| 4 | 100 | 100 | 100 | 98.7 | 96.1 | – | – |
| 5 | 100 | 100 | 100 | 100 | 96.1 | 100 | 100 |
| 6 | 100 | 100 | 100 | 99.5 | 96.1 | 100 | 100 |
| Average accuracy (%) | | | | | | | |
| | **100.0** | 99.9 | 99.9 | 99.2 | 95.9 | 98.2 | 97.1 |

the individual CNN, VGG16, but it also correctly classified all data points in each one of the 6 datasets from DAGM2007.

An important aspect relevant for industrial systems is the time required to train the models, allowing quick changes, and the inference time, as real-time inference is of utmost importance. The overall mean time and standard deviation to train each CNN, was: $50.7 \pm 0.18$ min for VGG11, $95.2 \pm 0.33$ min for VGG16, $115.6 \pm 0.71$ min for VGG19, $17.0 \pm 0.02$ min for Resnet18, $28.8 \pm 0.02$ min for Resnet34, $44.8 \pm 0.04$ min for Resnet50, $71.1 \pm 0.09$ min for Resnet101, and $47.63 \pm 0.02$ min for Densenet121. From this, we can infer that the time required to train any of the CNNs is feasible in an environment with rapid changes, since the CNN that took more time to train in the experimented datasets, was VGG19, requiring less than 2 hours. Regarding CNN-Fusion, it required no further training, as it is the combination of training all individual CNNs, and for the Auto-Classifier, the AutoML component was allowed to search for a time limited to 2 hours. By adding it to the time required by VGG16 to train, it required 235.6 min in average to create the complete model. For inference times, all CNNs were capable of running in real-time, with the fastest one being Resnet18 with an inference time of $0.057 \pm 0.016$ s, and the slowest one being VGG19, with an inference time of $0.269 \pm 0.002$ s. As for VGG16, the one selected to be the feature extraction component of Auto-Classifier, it had an inference time of $0.225 \pm 0.002$, which is enough for detecting surface defects in real-time. CNN-fusion, in a serial manner, has an $f_{time}$ of 1.152 s, which is not suitable for real-time. As for the

Auto-Classifier, which consists of the partial VGG16 and the new classification component, we found that it is extremely fast, requiring only $0.001 \pm 0.004$ s to classify an image, which is justified by having an efficient new classifier that uses less, and faster operations than the removed layers.

We also compare our two proposed methods and the best individual CNN with other methods that achieve the best results in the DAGM2007 defect classification in Table 2. In this table, it is possible to see that each one of the 3 methods studied here had the highest average accuracy, which is calculated by summing the true positive rate and true negative rate means, and divide it by two: $(TPR + TNR)/2$. It is also worth noting that our proposed method, Auto-Classifier, not only achieved a perfect classification on all DAGM2007 problems, but outperformed all other methods in this set of datasets.

## 5 Conclusions

This paper studies how different CNNs perform in the task of detecting surface defects and proposes two methods to solve the problem: 1) CNN-Fusion, which fuses the different CNN classifications into a final one, and 2) Auto-Classifier, which is a novel method that leverages the feature extraction power of a state-of-the-art CNN and complements it by performing an automated search for a new classifier component.

We initially conduct a study of how CNNs perform in a task that is usually solved by extracting hand-crafted features and then applying classifiers such as SVMs. This study showed how CNNs perform in detecting defects with low amounts of data points. Moreover, we propose a novel method, Auto-Classifier, that is capable of improving the performance of CNNs, and outperforming the current state-of-the-art in the task of detecting surface defects in DAGM2007 set of problems. With this, we can conclude that even though CNNs have exceptional results in a variety of image problems, they can be improved by partially replacing its classification component by other types of classifiers. Also, our experiments show that deep learning approaches for detecting surface defects, outperform traditional ones, and require no hand-crafted feature extraction, which removes problems that arise from environmental changes.

In short, the results of Auto-Classifier not only improve the state-of-the-art performance of surface defect detection in all problems of DAGM2007, but also show us that CNNs can be improved by replacing its inner classification component. As future work, the mechanism behind Auto-Classifier can be extended to new problems, and also study different changes in the classification component of CNNs to achieve the best possible performances in different problems.

## References

1. Baker, B., Gupta, O., Naik, N., Raskar, R.: Designing neural network architectures using reinforcement learning. In: ICLR 2017 (2017)

2. Cheng, G., Zhou, P., Han, J.: Learning rotation-invariant convolutional neural networks for object detection in VHR optical remote sensing images. IEEE Trans. Geosci. Remote Sens. **54**(12), 7405–7415 (2016)
3. Feurer, M., Klein, A., Eggensperger, K., Springenberg, J.T., Blum, M., Hutter, F.: Auto-sklearn: efficient and robust automated machine learning. In: Hutter, F., Kotthoff, L., Vanschoren, J. (eds.) Automated Machine Learning. TSSCML, pp. 113–134. Springer, Cham (2019). https://doi.org/10.1007/978-3-030-05318-5_6
4. Garcia Plaza, E., Lopez, P., Gonzalez, E.: Multi-sensor data fusion for real-time surface quality control in automated machining systems. Sensors **18**(12), 4381 (2018)
5. Gijsbers, P., LeDell, E., Thomas, J., Poirier, S., Bischl, B., Vanschoren, J.: An open source automl benchmark. In: ICMLW on Automated Machine Learning (2019)
6. Goodfellow, I., Bengio, Y., Courville, A.: Deep Learning. MIT Press, Cambridge (2016)
7. H2O.ai: H2O AutoML, June 2017. http://docs.h2o.ai/h2o/latest-stable/h2o-docs/automl.html. h2O version 3.30.0.1
8. He, K., Zhang, X., Ren, S., Sun, J.: Deep residual learning for image recognition. In: CVPR (2016)
9. Huang, G., Liu, Z., Van Der Maaten, L., Weinberger, K.Q.: Densely connected convolutional networks. In: CVPR, pp. 2261–2269 (2017)
10. Hutter, F., Kotthoff, L., Vanschoren, J.: Automated Machine Learning. Springer, Cham (2019)
11. Kocbek, S., Gabrys, B.: Automated machine learning techniques in prognostics of railway track defects. In: ICDMW. IEEE (2019)
12. Kotthoff, L., Thornton, C., Hoos, H.H., Hutter, F., Leyton-Brown, K.: Auto-weka 2.0: automatic model selection and hyperparameter optimization in weka. J. Mach. Learn. Res. **18**(1), 826–830 (2017)
13. LeCun, Y., Bengio, Y., Hinton, G.: Deep learning. Nature **521**, 7553 (2015)
14. Li, G., Zhao, X., Du, K., Ru, F., Zhang, Y.: Recognition and evaluation of bridge cracks with modified active contour model and greedy search-based support vector machine. Autom. Constr. **78**, 51–61 (2017)
15. Liu, H., Simonyan, K., Yang, Y.: DARTS: differentiable architecture search. In: ICLR (2019)
16. Malamas, E.N., Petrakis, E.G., Zervakis, M., Petit, L., Legat, J.D.: A survey on industrial vision systems, applications and tools. Image Vision Comput. **21**(2), 171–188 (2003)
17. Mendoza, H., et al.: Towards automatically-tuned deep neural networks. In: AutoML: Methods, Sytems, Challenges, December 2018
18. Mital, A., Govindaraju, M., Subramani, B.: A comparison between manual and hybrid methods in parts inspection. Integr. Manuf. Syst. **9**(6), 344–349 (1998)
19. Pham, H., Guan, M., Zoph, B., Le, Q., Dean, J.: Efficient neural architecture search via parameters sharing. In: ICML (2018)
20. Rumelhart, D.E., Hinton, G.E., Williams, R.J.: Learning representations by back-propagating errors. Nature **323**(6088), 533–536 (1986)
21. Schmidhuber, J.: Deep learning in neural networks: an overview. Neural Netw. **61**, 85–117 (2015)
22. Schmugge, S.J., Rice, L., Lindberg, J., Grizziy, R., Joffey, C., Shin, M.C.: Crack segmentation by leveraging multiple frames of varying illumination. In: WACV, March 2017
23. Scholz-Reiter, B., Weimer, D., Thamer, H.: Automated surface inspection of cold-formed micro-parts. CIRP Ann. **61**(1), 531–534 (2012)

24. Siebel, N.T., Sommer, G.: Learning defect classifiers for visual inspection images by neuro-evolution using weakly labelled training data. In: IEEE CEC (2008)
25. Simonyan, K., Zisserman, A.: Very deep convolutional networks for large-scale image recognition. In: ICLR (2015)
26. Sohn, K., Lee, H.: Learning invariant representations with local transformations. In: ICML (2012)
27. Timm, F., Barth, E.: Non-parametric texture defect detection using weibull features. In: Image Processing: Machine Vision Applications IV (2011)
28. Villalba-Diez, J., Schmidt, D., Gevers, R., Ordieres-Meré, J., Buchwitz, M., Wellbrock, W.: Deep learning for industrial computer vision quality control in the printing industry 4.0. Sensors **19**(18), 3987 (2019)
29. Weimer, D., Scholz-Reiter, B., Shpitalni, M.: Design of deep convolutional neural network architectures for automated feature extraction in industrial inspection. CIRP Ann. **65**(1), 417–420 (2016)
30. Wieler, M., Hahn, T.: Weakly supervised learning for industrial optical inspection. In: DAGM Symposium in 2007 (2007)
31. Xie, X.: A review of recent advances in surface defect detection using texture analysis techniques. ELCVIA **7**(3), 1–22 (2008)
32. Zela, A., Elsken, T., Saikia, T., Marrakchi, Y., Brox, T., Hutter, F.: Understanding and robustifying differentiable architecture search. In: ICLR (2020)
33. Zhang, L., Yang, F., Zhang, Y.D., Zhu, Y.J.: Road crack detection using deep convolutional neural network. In: ICIP (2016)
34. Zhong, Z., Yan, J., Wu, W., Shao, J., Liu, C.L.: Practical block-wise neural network architecture generation. In: CVPR (2018)
35. Zoph, B., Le, Q.V.: Neural architecture search with reinforcement learning. In: ICLR (2017)
36. Zoph, B., Vasudevan, V., Shlens, J., Le, Q.V.: Learning transferable architectures for scalable image recognition. In: CVPR (2018)

# Automating Inspection of Moveable Lane Barrier for Auckland Harbour Bridge Traffic Safety

Boris Bačić(✉), Munish Rathee, and Russel Pears

Engineering, Computing and Mathematical Sciences,
AUT University, Auckland, New Zealand
boris.bacic@aut.ac.nz

**Abstract.** A moveable lane barrier along the Auckland Harbour Bridge (AHB) enables two-way traffic flow optimisation and control. However, the AHB barrier transfer machines are not equipped with an automated solution for screening of the pins that link the barrier segments. To improve traffic safety, the aim of this paper is to combine traditional machine with deep learning approaches to aid visual pin inspection. For model training with imbalanced dataset, we have included additional synthetic frames depicting unsafe pin positions produced from collected videos. Preliminary experiments on produced models indicate that we are able to identify unsafe pin positions with precision and recall up to 0.995. To improve traffic safety beyond the AHB case study, future developments will include extended datasets to produce near-real time IoT alerting solutions using mobile and other video sources.

**Keywords:** Deep learning · Machine learning · Video surveillance · Object detection and classification · TensorFlow · Transfer learning

## 1 Introduction

Auckland Harbour Bridge is an eight-lane transport link connecting Auckland city and the North Shore. In 2019, it was estimated that more than 170,000 cars, 11,000 heavy vehicles and 1,000 busses are crossing during work days [11]. To balance the traffic load peaks, and minimise commuter's fuel consumption, the middle four lanes are divided by a moveable concrete safety barrier [6,10]. The existing $1.4 million barrier transfer machines (BTMs) are moving $16 \times 750$ kg concrete blocks at a time, 4 times during weekdays [7]. However, BTMs (Fig. 1) are not equipped with an automated solution for safety screening of pins that holds the adjacent concrete blocks together. Barrier blocks are connected with a removable pin in between the two blocks. Hence, a problem may arise if a pin is not safely connecting the two adjacent blocks. The moving barrier is regularly inspected by a staff member walking behind the BTM, but this creates safety concerns where there can be "close encounters with truck mirrors" [7].


H. Yang et al. (Eds.): ICONIP 2020, LNCS 12532, pp. 150–161, 2020.
https://doi.org/10.1007/978-3-030-63830-6_13

Inspections involve strenuous, labour-intensive observations in a static and in general anatomically awkward body position (spine laterally bent and neck tilted carrying the extra load of a hard hat) which may be additionally hindered by poor lighting conditions or glare and fast-moving traffic (approx. 80 km/h). The aim of the project is to improve the safety of Harbour Bridge traffic by providing a Computer Vision (CV) and AI-based solution to automate screening of pins connecting moving barrier segments. The intended automated solution would enable more frequent safety inspections in all weather conditions and minimise potential risks of human error as well as alleviate time-consuming, laborious and potentially hazardous work conditions.

**Fig. 1.** Auckland Harbour Bridge barrier transfer machine.

Research questions:

- Can a machine identify potentially unsafe pin position needing manual attention whether its safety ring is not securely positioned or if a pin is in elevated or any other unsafe position?
- Can we design a system that would work in lighting conditions that are less than ideal?
- Can we reduce reliance on expert's knowledge involved in machine learning feature engineering?
- What are practical implications regarding data collection and related issues?
- Can we produce a system that would not require large training and testing datasets?
- How do we provide such system with sufficient training and testing samples for minority class depicting potentially unsafe position?

For a broader audience, the follow-up section provides a background on AI, deep learning and convolutional neural networks focusing on automated feature generation from pretrained neural networks used in experimental work, reported in the Methodology section.

## 2 Background

Today, artificial intelligence (AI) is a thriving field, applicable to many context and research topics [2]. However, in the early days of AI, computers were solving tasks that were difficult for humans and could be described via (hard-coded) formal rules, while the remaining challenge was to solve tasks that people solve intuitively [2], such as maintaining balance or playing sport.

In the last decade, the AI research community has been solving tasks such as natural language processing, translations and human computer interfaces; autonomous driving, face recognition; human activity interpretation for wellbeing or body function monitoring and personalised healthcare.

### 2.1 Deep Learning and Object Detection

The quintessential example linking traditional artificial neural networks (ANN) with deep learning systems is multilayer perceptron (MLP) [2]. Considered as a universal mathematical function approximator, pretrained MLP maps input values to expected output values. Inspired by nature, its' internal operation is performed by many simpler functions (i.e. artificial neurons) that are interconnected in layered structures. Such structures as groups of artificial neurons pass their weighted output as input information to the next layer.

Common to ANN, each interconnected neuron has the same activation function, which is relatively simple and preset during the initialisation. The idea behind ANN learning algorithm is to adjust potentially large number of interconnected weights (or parameters) in a multistep fashion. Another idea that is connected to deep learning approaches is parallel execution involved in neural information processing.

A central problem to deep learning is representations learning. As a functional composition, representation learning consists of other simpler subsequent representations. Similarly, to what occurs in nature and human vision, deep learning architecture has specialised subsystems that can present the context of the processed image pixels in specialised layers generating contours, edges and other features, all of which can be seen as simpler processing tasks associated with representations learning concept.

A group of deep learning ANN commonly used in video and image processing are convolutional neural networks (CNN), which have specialised layers for convolution and pooling tasks. The convolution layer as a subsystem combines elements (such as areas of input pixels) into a smaller area, while pooling selects the elements of highest value. Collectively such information processing through (hidden) layers constitutes machine-generated feature maps, which in the end of the processing sequence provides input for classification.

Within the scope of this research, our intention was: (1) to obtain machine-generated feature maps from a pretrained CNN on a large and universal content dataset such as ImageNet [8], with over 14 million images labelled/organised according to the large lexical database hierarchy WordNet (https://wordnet.princeton.edu); (2) apply transfer learning to update more specialised features maps to improve classification; and (3) replace classification layer processing task with separate, traditional neural networks.

Historically complex ANN models, with a large number of parameters were difficult to compute. Today with multi-core CPUs and GPU parallel processing capabilities the trend is to develop more complex CNNs to improve their performance.

As a design decision within the scope of this research, we intended to use less-computationally demanding pretrained CNN, which would still provide the answer to whether we can classify video frames into PIN_OK or PIN_OUT categories. Table 1, shows comparisons of some of the recent commonly used pretrained CNN models.

**Table 1.** Comparisons of deep neural networks. Adopted from [5]

| Pretrained deep neural networks | | | | |
|---|---|---|---|---|
| Model | Input image resolution | Parameters (1,000,000) | Depth | Size [MB] |
| AlexNet | 227 × 227 | 61 | 8 | 227 |
| SqueezeNet | 227 × 227 | 1.24 | 18 | 4.6 |
| GoogleNet | 224 × 224 | 7 | 22 | 27 |
| Inception v3 | 299 × 299 | 23.9 | 23.9 | 48 |
| MobileNet v2 | 224 × 224 | 3.5 | 3.5 | 53 |

## 2.2 SqueezeNet Evaluation and Architecture

Relatively small in size, SqeezeNet (https://github.com/forresti/SqueezeNet) is able to perform AlexNet-level accuracy with approximately 50 times fewer parameters and less than 0.5 MB model size [3]. SqueezeNet architecture (Table 2) is a comparatively smaller CNN architecture which requires fewer parameters and with lower memory and resource requirements is suitable for real-time applications including low-cost computing platforms [4].

Using activations from pre-softmax 'avgpool10' layer (Table 2) is important for the intended research approach, while output size of 1000 is used as a fixed-size array of generated features.

**Table 2.** SqueezeNet layered architecture and parameters. Adopted from [3].

| Layer name/type | Output size | Filter size/stride (if not a fire layer) | Depth |
|---|---|---|---|
| input image | $224 \times 224 \times 3$ | | |
| conv1 | $111 \times 111 \times 96$ | $7 \times 7/2$ $(\times 96)$ | 1 |
| maxpool1 | $55 \times 55 \times 96$ | $3 \times 3/2$ $(\times 96)$ | 0 |
| fire2 | $55 \times 55 \times 128$ | | 2 |
| fire3 | $55 \times 55 \times 128$ | | 2 |
| fire4 | $55 \times 55 \times 256$ | | 2 |
| maxpool4 | $27 \times 27 \times 256$ | $3 \times 3/2$ | 0 |
| fire5 | $27 \times 27 \times 256$ | | 2 |
| fire6 | $27 \times 27 \times 384$ | | 2 |
| fire7 | $27 \times 27 \times 384$ | | 2 |
| fire8 | $27 \times 27 \times 512$ | | 2 |
| maxpool8 | $13 \times 13 \times 512$ | $3 \times 3/2$ | 0 |
| fire9 | $13 \times 13 \times 512$ | | 2 |
| conv10 | $1 \times 1/1(\times 1000)$ | | 1 |
| avgpool10 | $1 \times 1 \times 1000$ | | 0 |

# 3 Methodology

The key components involved in modelling of pin detection proof of concept are: (1) Augmenting dataset from collected video frames, (2) transfer learning, and (3) comparing traditional ANNs performance on extracted features from a CNN.

## 3.1 Design Decisions and Rationale of the Study

To minimise reliance of domain expertise in feature engineering and ultimately produce a solution that is robust in non-ideal lightning conditions (such as time of the day, reflected or low-in-the-horizon sun glare, rain and cloudy overcast weather) or colour-based identification, we employed deep learning approaches relying on pretrained neural networks.

The main idea behind transfer learning approach applied on pin detection was to input extracted and resized video frames into a pretrained convolutional neural network (CNN) model to obtain a vector representing image features (Fig. 2). A concept of using AI to generate features is based on a pretrained CNN on a large number of images with added new labelled images to improve discriminative power of information separation in generated feature space by a CNN. Instead of using pretrained CNN for classification of detected pin position, we considered generated features as reduced image dimensionality (into feature vectors) and used the traditional neural network approaches for further modelling and analysis. Such two-step framework and architecture allows adaptation for new training images extracted under various lighting conditions and camera views and analysis of newly generated feature space.

When eliciting requirements from the client (NZTA), pin position safety is the primary objective of this study, while other inspection points can be implemented at a later stage, leading to modifications of camera view and implementation of adaptable architecture able to work with multi-class modelling challenge by replacing traditional ML classifier (Fig. 2). To ensure incremental system adaptation the second step involves multiple classifier modelling and benchmarking on generated feature space, which would also allow adaptation from binary class ('PIN_OK', 'PIN_OUT') to multi-class classification tasks in the future.

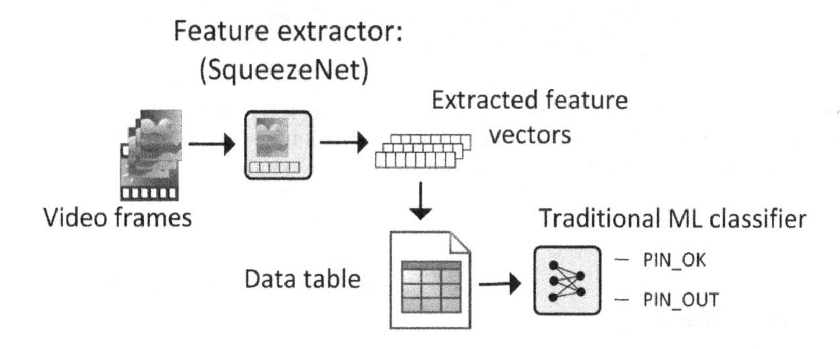


**Fig. 2.** Processing framework depicting automated feature generation, classification pipeline for pin detection from exported video frames and extraction of feature space for further analysis and visualisation.

### 3.2 Extending Minority Class and Data Pre-processing

Lack of large training data and minority class samples (Fig. 3), requiring human attention, resulted in and imbalanced dataset classification problem with insufficient data for model training. Committing to capturing and labelling video frames depicting various unsafe pin positions associated with manual attention (e.g. such as videotaping or photographing 'hammering' the pin down and ensuring that a safety ring is securely placed) was not feasible within the scope of producing a proof of concept.

As a solution, we created training data derived from the originally captured dataset. A technique for obtaining a synthetic frame (Fig. 4) allows us to produce various degrees of unsafe pin positions derived from the original frames.

Acceptable quality of a synthetic images derived from the original video frames (such as shown in Fig. 5) was ensured by visual inspection. Figure 5 also shows video frames depicting safe pin and unsafe pin positions using the cloning technique based on sampling vector from surrounding neighbourhood pixels.

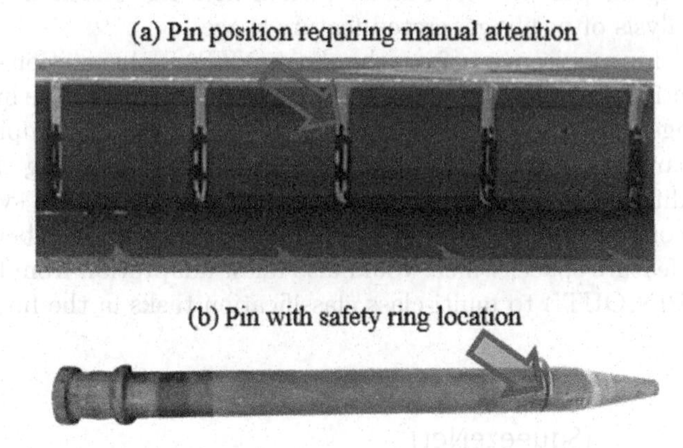

(a) Pin position requiring manual attention
(b) Pin with safety ring location


**Fig. 3.** Video frame indicating the need for visual inspection of pin position within joining interlocking system of moveable barrier segments: (a) potentially unsafe pin position located in the centre and (b) a pin with safety ring at its bottom.

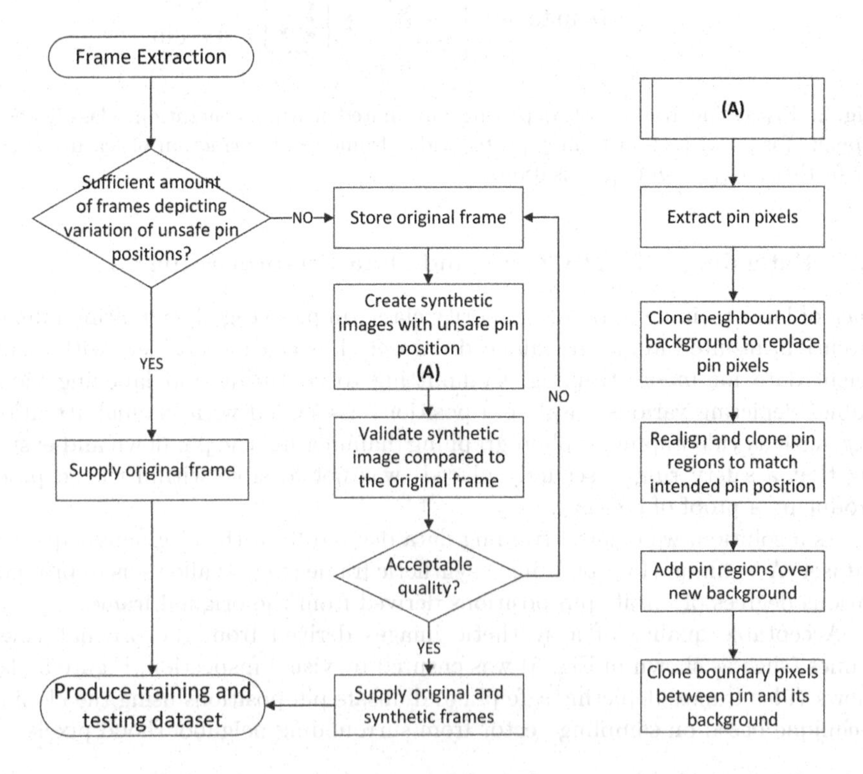


**Fig. 4.** Flowchart for creating synthetic frames depicting unsafe pin position.

**Fig. 5.** Visualization of intermediate results of data pre-processing showing an example of: (a) the original ('PIN OK') and (b) synthetic frame depicting unsafe pin position labelled as 'PIN OUT' for modelling purpose.

### 3.3 Dataset Distribution

A critical modelling aspect was developing capability to produce and augment the collected dataset with synthetic images depicting a controlled degree of unsafe position (Fig. 5 and 6). Complementing the existing dataset with synthetic images was additionally motivated by: (1) the need for achieving best possible classification results on small imbalanced and manually labelled dataset; and (2) by the circumstances where daily data collection was not possible at the time.

Another benefit of producing synthetic data around the system decision boundaries is important for Precision and Recall in model evaluation (Table 3 and Table 4). Furthermore, from a safety perspective and model classification error perspectives, detecting false positives (or a 'false alarm' for actual safe pin position) are preferred over false negatives (unsafe or 'PIN_OUT'), where positions would be undetected (Table 3).

For the task of retraining a pretrained SqueezeNet CNN and for converting input images into feature vectors, we have used a consumer-grade computer (CPU i7 2.5 GHz, 16 GB DDR3 RAM and GPU NVIDIA GeForce GT 750M, made in 2014, running Linux Mint 18.3). For pretrained SqueezeNet convolutional neural network model (https://github.com/DeepScale/SqueezeNet) documented and explained by its inventors [3], we used activations from before softmax 'avgpool10' layer (Table 2) for embedding as recommended in [5,9]. The number of extracted features or a generated feature vector size is 1000.

Within the scope of using traditional ANN for classification and producing a proof of concept for image classification into two output classes (Fig. 6), we have trained three traditional classifiers: (1) Logistic regression, (2) Multi-Layer Perceptron and Support Vector Machine.

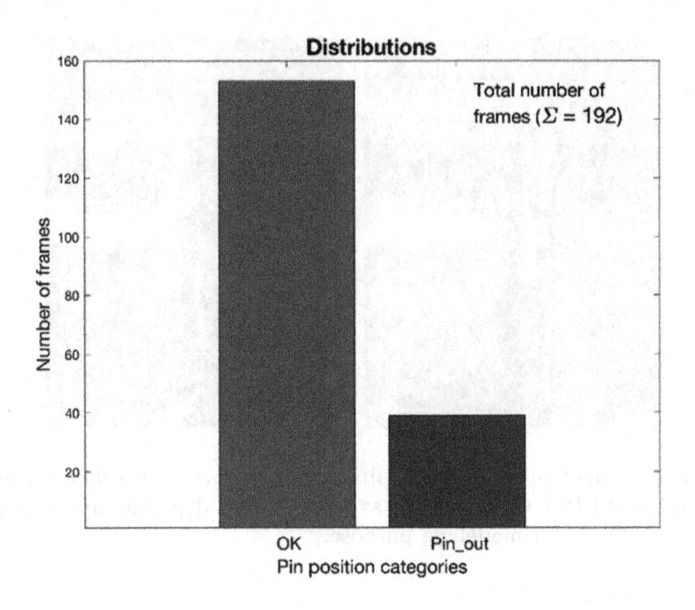


**Fig. 6.** Data distribution used for modelling and analysis.

## 4 Results

The key benefit of using AI (deep learning) in this case study is the notion of converting raw pixel values into feature vector to minimise efforts involved in feature engineering or development of feature extraction technique, which can be both time consuming and relying on multidisciplinary expertise. The second benefit was the possibility to use low-cost computing resources on relatively small and imbalanced dataset. Unlike modelling of a fully trained CNN requiring time- and extensive computational resource-consuming processing relying on large image datasets, transfer learning and feature extraction from CNN models allows modelling on small datasets achievable in short times.

A systematic approach to creating synthetic images with various degrees of unsafe pin position (Fig. 5) and system processing framework (Fig. 2) allow fine tuning of ratio between False Positives (FP, as misclassified PIN_OK positions) and False Negatives (FN, as undetected PIN_OUT positions). Achieving FP > FN ratio is possible via a labelled dataset in which the pin position is a borderline unsafe position and as such, it would still require visual assessment.

To ensure minority class distribution across each fold, in our experiments we have applied three-fold stratified cross validation (Table 4).

**Table 3.** Traditional machine learning classification results.

| Model | | | Confusion Table *Actual* PIN_OK | PIN_OUT |
|---|---|---|---|---|
| Logistic regression | *Predicted* | PIN_OK | 152 | 1 |
| | | PIN_OUT | 0 | 39 |
| MLP | | PIN_OK | 152 | 1 |
| | | PIN_OUT | 0 | 39 |
| SVM | | PIN_OK | 150 | 3 |
| | | PIN_OUT | 0 | 39 |

**Table 4.** Test and score results.

| 3-fold stratified cross validation | | |
|---|---|---|
| Model | Precision | Recall |
| Logistic regression parameters:<br>Regularization: Ridge (L2), C = 1 | 0.995 | 0.995 |
| Multilayer Perceptron (MLP) parameters:<br>Hidden layers: 2, Neurons: [180, 20]<br>Activation function: ReLu<br>Solver: Adam, Alpha: 0.02<br>Max. iterations: 200 | 0.995 | 0.995 |
| Support Vector Machine (SVM) parameters:<br>C = 1.0, $\epsilon = 0.1$, Kernel: Linear<br>Numerical tolerance: 0.001<br>Max. iteration: 100 | 0.985 | 0.984 |

## 5 Discussion

The data collection was a more demanding task than anticipated originally. Given that safety on the Auckland Harbour Bridge (along with traffic flow) is paramount, the initial data set was recorded under safety supervision of New Zealand Transport Agency (NZTA) that provided access to the BTM site with a prior safety briefing. The follow-up high speed recording attempts from public transportation and personal cars were also challenging. Examples include the traffic flow prediction as per expected times of the day it would not necessarily come to below average speed of 20 km/hr along the bridge, which allows us to collect frames depicting ideal pin positions at 240 fps. For data collection requiring a clear picture of the narrow gap between the concrete segments, a high-frame rate camera will soon be mounted on MTP, which is moving at speeds of around 10–15 km/hr, and can record video for up to four times a day. Nonetheless, the midday operation is expected to have minimal glare into the camera lenses.

For our client (NZTA), we have produced a report showing hierarchical clustering and clear visual separation of obtained feature vectors associated with minority output class by using cosine and Person correlation-based distance measures. Rather than choosing computationally inexpensive distance measure such as Euclidean, correlation-based distance measures were considered as more appropriate given relatively high feature dimensionality compared to a number of minority class samples and for computer vision problem areas where extracting relevant information should be independent from light conditions, rain, or a background colour value from a car passing behind a safety pin.

Within the scope of this project, we did not attempt to work with three or more categories of pin positions. For the presented system and future project specifications (e.g. extending binary classification into multi-class problem and incremental learning function), it would be possible to include other traditional ANN approaches as well as to inspect the barrier for other types of damages that would require attention and further maintenance operation. For such cases, we expect to collect additional dataset showing additional unsafe pin positions and to produce more synthetic data using presented flowchart (Fig. 4). Prior work on imbalanced datasets, hypergeometric distribution approach to cross-validation requirements and multiclass classification has been published in [1].

## 6 Conclusion

To optimise unevenly balanced motorway traffic during workdays as well as to minimise carbon emissions, moveable motorway barriers are an adequate solution for short distance traffic bottlenecks. Compared to motorways on land, Auckland Harbour Bridge with its moveable concrete barrier separating motorway lanes is additionally exposed to various vibrations due to long bridge movements (e.g. vertical and torque), which are most noticeable around its most elevated central segment and would raise concerns for more frequent pin inspections.

Preliminary experiments confirm that a machine can detect unsafe pin position that connects concrete barrier segments from video frames. The initial dataset was collected under non-ideal lightning conditions (low stratiform clouds and rain drizzle). The problem of lack of video frames showing a pin out of its safe position was solved by creating synthetic frames showing variations of the unsafe pin positions (with different pin heights) needed for modelling purposes. The obtained proof of concept shows high precision and accuracy ($\leq$9.995) without reliance on expert driven feature extraction technique or feature engineering. Furthermore, the experimental evidence from relatively small labelled dataset also suggests that the presented system as proof of concept does not require intensive manual labelling. Combining deep learning (SqueezeNet as pretrained CNN) to convert each video frame into a vector and traditional ANNs for classification allows for analysis of intermediate processing and flexibility for modelling:

(1) on future data and (2) with minimal labelling efforts should we modify the system to classify more than two pin position categories.

Currently with NZTA we are implementing more robust video data collection system (with supplementary photos taken by the AHB maintenance team) than used in preliminary data collection. We expect that the presented system as a foundation is universal and can be applied globally in similar traffic safety contexts with minimal modifications.

**Acknowledgement.** The authors wish to thank Gary Bonser, Angela Potae and Martin Olive from NZ Transport Agency and Auckland System Management for their assistance including safety briefings, transport to the site, video recording under their supervision, various problem insights project requirements, availability and ongoing enthusiasm to assist with this project. We also wish to express our gratitude for the comprehensive documentations provided by the contributors to ImageNet, TensorFlow, Orange Data Mining, Google Cloud, MathWorks, OpenCV and SqueezeNet and for sharing their tools and libraries.

## References

1. Bačić, B.: Predicting golf ball trajectories from swing plane: an artificial neural networks approach. Expert Syst. Appl. **65**, 423–438 (2016). https://doi.org/10.1016/j.eswa.2016.07.014
2. Goodfellow, I., Bengio, Y., Courville, A.: Deep Learning. MIT Press, Cambridge (2016). https://www.deeplearningbook.org
3. Iandola, F.N., Han, S., Moskewicz, M.W., Ashraf, K., Dally, W.J., Keutzer, K.: SqueezeNet: Alexnet-level accuracy with 50x fewer parameters and <0.5 MB model size (2016). arXiv:1602.07360
4. Lee, H.J., Ullah, I., Wan, W., Gao, Y., Fang, Z.: Real-time vehicle make and model recognition with the residual SqueezeNet architecture. Sensors **19**(5), 982 (2019)
5. MathWorks: Pretrained deep neural networks (2019). https://au.mathworks.com/help/deeplearning/ug/pretrained-convolutional-neural-networks.html
6. New Zealand Ministry for Culture and Heritage: Auckland harbour bridge opens 30 May 1959 (1959). https://nzhistory.govt.nz/the-auckland-harbour-bridge-is-officially-opened
7. NZ Transport Agency: How to move a concrete motorway barrier (2014). https://www.nzta.govt.nz/media-releases/how-to-move-a-concrete-motorway-barrier/
8. Stanford vision lab: ImageNet (2016). www.image-net.org
9. University of Ljubljana: Image embedding. https://orange.biolab.si/widget-catalog/image-analytics/imageembedding/
10. Wikipedia: Auckland harbour bridge from Wikipedia (2020). https://en.wikipedia.org/wiki/Auckland_Harbour_Bridge
11. Wilson, S.: The next harbour crossing: Road and rail, or just rail? (2019). https://www.nzherald.co.nz/nz/news/article.cfm?c_id=1&objectid=12210993

# Bionic Vision Descriptor for Image Retrieval

Guangzhe Li, Shenglan Liu(✉), Feilong Wang, and Lin Feng

Faculty of Electronic Information and Electrical Engineering,
Dalian University of Technology, Dalian, Liaoning, China
liusl@mail.dlut.edu.cn

**Abstract.** Human visual system gets remarkable performance by processing low-level features. In the last decade, many descriptors have been proposed for feature extraction. However, fewer of them get satisfying performance with low-level features. Compared to high-level ones, low-level features make use of natural underlying elements like texture and they are extracted directly, which makes low-level features more efficient in image retrieval domains. In this paper, a new descriptor named Bionic Vision Descriptor (BVD), which is based on the principle of human visual system, is proposed. The descriptor fuses uniform low-level features extracted from color, texture and gradient elements. Moreover, matrix calculation and feature selection are utilized to accelerate the calculation of BVD. Experimental results show that our method outperforms other state-of-the-art traditional descriptors with less runtime and fewer initial dimensions on benchmark datasets.

**Keywords:** Human visual system · Low-level feature · Image retrieval · Bionic vision descriptor

## 1 Introduction

We focus on the problem of feature extraction in Content-based Image Recognition (CRIB) [21], the performance of which depends mainly on the discrimination and effectiveness of features.

The features extracted from images can be falled into three levels according to the complexity of extraction process. Low-level features are usually histogram features extracted directly from global or local regions by unsupervised measures. Representative approaches with low-level features are Color Histogram (CH) [5], Hue Saturation Value (HSV) [31] and Local Binary Patterns (LBP) [25]. Other approaches process low-level information further according to the regulation of human visual system. Liu et al. propose Micro-structure Descriptor

This study was funded by National Natural Science Foundation of Peoples Republic of China(61672130, 61972064), The Fundamental Research Funds for the Central Universities(DUT19RC(3)012, DUT20RC(5)010) and LiaoNing Revitalization Talents Program(XLYC1806006).


H. Yang et al. (Eds.): ICONIP 2020, LNCS 12532, pp. 162–173, 2020.
https://doi.org/10.1007/978-3-030-63830-6_14

(MSD) [16] which obtains micro-structure according to the similarity of color pixels, and [20] introduces Perceptual Uniform Descriptor (PUD) which fuses basic frequency feature and color difference feature. Middle-level features are often local features with high dimensions and usually requires encoding process and bag-of-words (BOW) framework [28] or its variants such as bag-of-feature (BOF) [9]. Histogram of Gradient (HOG) [3] produces the histogram of sifted direction gradients in local regions. [23] introduces a local descriptor called Scale-invariant Feature Transform (SIFT), which detects key points in scale space and extracts their position, scale and rotation invariants. Many variants based on SIFT have been proposed to improve performance. Dense-SIFT [15] is a densely sampled version of SIFT, which contains more information about image content. Surf [2] is another variation of SIFT which aims to improve the speed of detecting feature points. Inspired by BOW, [30] provides bag-of-colors (BOC) that produces a color descriptor, which works by the most frequent color in different selected regions. High-level features are semantically sensitive features, which are extracted from upper layers of Convolutional Neural Network (CNN) [13] with a little data processing. [1] shows that neural codes are competitive with the state-of-the-art methods and almost achieve the best performance on benchmark datasets.

Middle-level and high-level features are most widely used in image retrieval domains, because they contain much richer information so that they can get better performance in most cases. However, they are very closely related to training datasets and require more computing resources compared to low-level features. If training samples in the datasets are scarce and not readily available or the task is simple, using middle-level and high-level features will not have a significant advantage. So they are not suitable for the tasks such as color discrimination of patches on rubik's cube [19], fruit Recognition [14] or other simple image tasks with small-scale datasets. Low-level features are extracted from images directly without complex calculation so that they are suitable for dealing with these simple problems. But in recent years, low-level features develop very slowly, and representative methods of them such as MSD and PUD do not make full use of low-level information, which results in low performance.

In a word, the middle-level and high-level methods mentioned above do not achieve reasonable tradeoffs on both performance and effciency, especially when it comes to the tasks with simple images or scanty training samples in the datasets. And existing low-level features focus too much on single information (usually color), which does not conform to human visual system well.

In order to solve above problems, we propose a low-level descriptor based on human vision system to get a balance between accuracy and effciency. The advantages of our approach are summarized as follows. First, we focus on the uniform structure of three elements color, texture and gradient, which makes our descriptor comprehensive so that we can enhance the performance with a small quantity of dimensions. Second, the essence of our approach is *seeing to learning* against *training to learning*, i.e., the approach can extract features directly without training process. Third, the features of different elements can be calculated with multithreading. Last, feature selection or extraction are used to reduce the dimensionality of our features with little accuracy loss.

## 2 Bionic Vision Descriptor

### 2.1 Motivation

Human visual system can distinguish objects in all kinds of images rapidly and accurately. Psychophysical studies have shown that at the level of minute recognizable images, a minimal change in the image can have a drastic effect on recognition [29], i.e., minimal structural differences contribute greatly to human visual system. According to Gestalt psychology [12], human visual system inclines to cluster similar bases into a whole structure, which will perform much better than the sum of these uniparted bases. These studies illustrate that in the whole process of recognition, uniform regions of the underlying elements catch most of the attention. Existing methods based on uniform regions such as MSD and PUD ignore texture information, and they have some weakness in judging similarity. For instance, if two different structures have the same uniform value as shown in Fig. 1, MSD can not discriminate them while PUD could solve this problem by calculating extra color difference bins. Motivated by the studies and problems mentioned above, this paper provides a new descriptor called Bionic Vision Descriptor (BVD) which imports texture information and optimizes the ratio of different low-level information to improve the performance of features. In addition, discrimination coefficient is introduced when calculating the values of uniform structures. The coefficient enhances the ability of features to distinguish similar local structures as shown in Fig. 1.

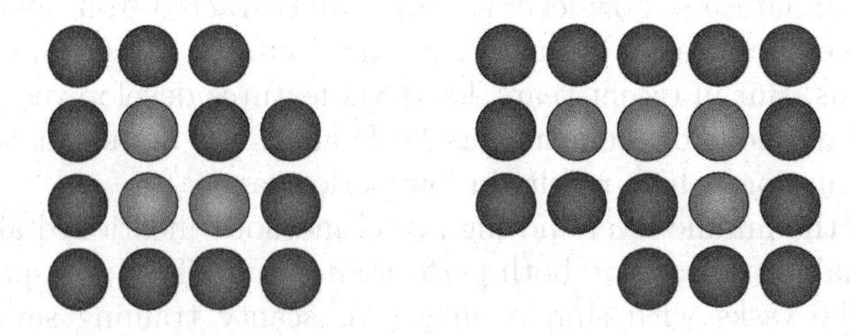

**Fig. 1.** MSD will get the same uniform value on these two structures, and PUD needs to calculate the color difference of different bins.

### 2.2 Original Bionic Vision Descriptor

Original Bionic Vision Descriptor fuses uniform structures of color, texture and gradient information to imitate human visual system. The low-level features of BVD are quantified to different bins which are encoded to uniform structures according to the difference of pixels in one neighborhood. In this paper, HSV color space is used to quantify color features. H, S and V channels are normalized to a range of 0–1, and then they are quantified to 8, 4 and 4 bins respectively, which generates $8 \times 4 \times 4 = 128$ quantified color bins. The underlying texture features are extracted by LBP. LBP quantifies the texture features to 59 bins. A method is presented in [4] to extract gradient information directly from color space instead of gray space, which reduces the loss of information. By this means, the gradient feature is quantified to 12 gradient bins in this step. After the quantization operation, three low-level feature graphs are obtained. Then the quantified pixels in these graphs are compared with neighbourhood using uniform filter structure.

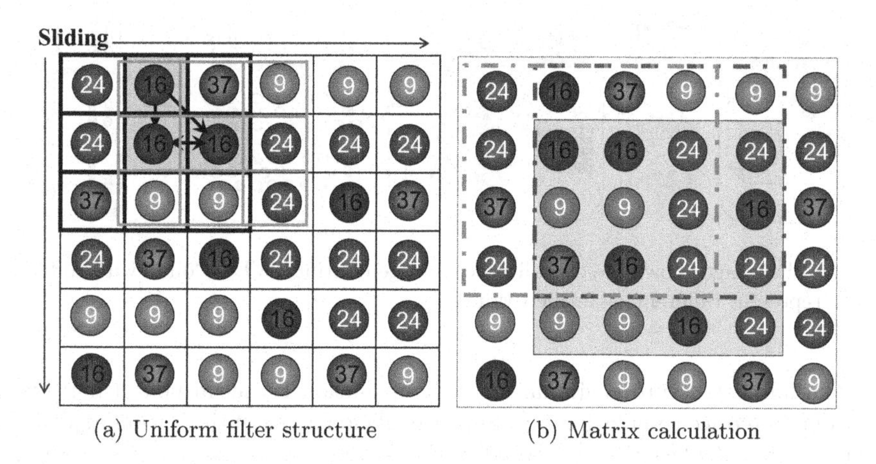


(a) Uniform filter structure (b) Matrix calculation

**Fig. 2.** (a) The image shows how the uniform filter works in a quantified feature graph. For the red quantified value 16 in the image, there are three pixels with two identical values around them in deepened boxes. One of them is at the edge of this image, so according to the formula, the uniform value of 16 is $S_{(16)} = \frac{2^\alpha + 2^\alpha}{8^\alpha + 8^\alpha}$. (b) The process of matrix calculation is shown in Fig. 2 (b). The shaded part is the middle subgraph of the whole image, subgraphs in red and blue dashed boxes will be compared with the shaded part. Around the shaded part, there are eight subgraphs like the red and blue dashed boxes in total, uniform features can be extracted after eight matrix comparison operation. (Color figure online)

The uniform filter structure is a block with a grid of size $3 \times 3$ (it can also be other size), the central digit will be compared with the surrounding points around it. The quantity of equal points may indicate the uniform degree of this structure. The filter is slided in the whole image as shown in Fig. 2 (a). And if the central digit is at the edge of image, it will be ignored. We describe this process using the following equation

$$S_{m_0}^{j} = \frac{\sum_{k=1}^{\widehat{N}\{p_0 = m_0\}} N_k \left\{ \sum_{i=1}^{n_e} (p_0 = m_0 \Lambda p_i = p_0) \right\}^{\alpha}}{{n_e}^{\alpha} \times \widehat{N}\{p_0 = m_0\}}, \tag{1}$$

where $S^j_{m_0}$ represents the uniform feature value of the quantified value $m_0$ in low-level feature graph $j$, $p_i$ is the value around the central digit $p_0$, $N$ and $\widehat{N}$ are the number of equivalent values, $n_e$ is the quantity of neighbors around the central digit and $\alpha \in [0.5, 1]$ is a discrimination coefficient. For our method, the dimensionality of $S = [S^1, n_t S^2, S^3]$ is 199, where $S^1 \in \mathbb{R}^{128}$, $S^2 \in \mathbb{R}^{59}$, $S^3 \in \mathbb{R}^{12}$, $n_t$ is the weight of texture feature. An explicit example is shown in Fig. 2 (a).

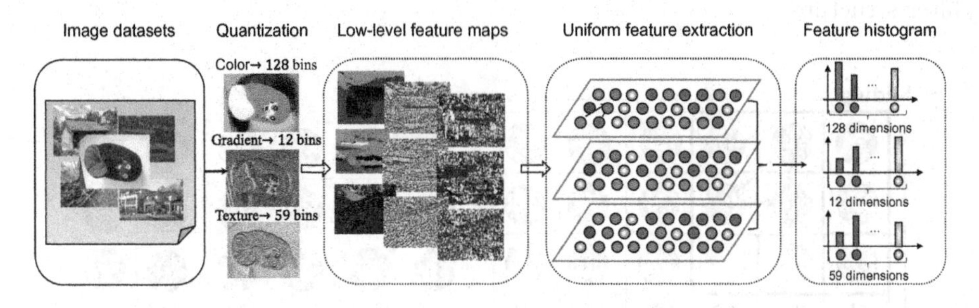


**Fig. 3.** The image shows the overall process of generating BVD. (Points with different colors represent different quantified values.)

The filter structure is similar to the convolution kernel in neural networks. In order to reduce calculation and improve efficiency, an optimization algorithm with matrix calculation is presented in this section. Experiments show that the speed of matrix calculation is 5 times faster than that of the traversal calculation. We can complete the filtering operation of one sample image by $n_e$ matrix calculations and judgements. Details of the algorithm is shown in the flow chart Algorithm 1 and Fig. 2 (b).

In conclusion, as illustrated in Fig. 3, the original generation process of BVD can be described as the following three steps:

(1) Convert original RGB to HSV color space, and then generate three quantified low-level feature graphs mentioned above;
(2) Use matrix calculation to perform uniform filter sliding on three different quantified graphs, and then get three uniform feature histograms with different dimensions. This process can be executed simultaneously by multi-threading strategy;
(3) Splice these histograms together to obtain the final 199-dimensional feature histogram BVD.

### 2.3 Extensional Bionic Vision Descriptor

The original bionic vision descriptor produces a histogram with only 199 dimensions. The features with low initial dimensions are difficult to contain sufficient information, especially when it comes to large size datasets. Experimental results

**Algorithm 1.** Uniform Filter Algorithm

**Input:** Quantization value matrix $Y \in M^{m \times n}$, discrimination coefficient $\alpha \in [0.5, 1]$, quantity of neighbors $n_e$.

1: Generate zero vectors $X^{(k)}, C^{(k)} = 0$, mark-matrix $S : \{i\%2 - 1, i/3 - 1\}$,where $i = 0, \cdots, 7$
2: $SavY = Y_{(1:m-1,1:n-1)}$
3: **for** $i \in [0, 7]$ **do**
4: $SubY = Y_{(S_{(i,1)}:S_{(i,1)}+m-2,S_{(i,2)}:S_{(i,2)}+n-2)}$
5: $Sel = SubY == Y$
6: $Sum = Sum + Sel$
7: **end for**
8: **for** $i \in [0, m-2]$ **do**
9: **for** $j \in [0, m-2]$ **do**
10: $Index = SavY_{(i,j)}$
11: $X_{(Index)} = X_{(Index)} + Sum^{\alpha}_{(i,j)}$
12: $C_{(Index)} = C_{(Index)} + {n_e}^{\alpha}$
13: **end for**
14: **end for**
15: **for** $i \in [0, k]$ **do**
16: $Res_{(i)} = X_{(i)}/C_{(i)}$
17: **end for**

**Output:** The output vector $Res$

show that integrating HSV features into the original descriptor will improve the performance [31], which is also consistent with the rules of color dominance in hunman visual system. So HSV features are added to enhance the original descriptor. Similar to the previous discussion, HSV color space is divided into $20 \times 10 \times 5 = 1000$ bins, i.e. all the pixes are quantified to 1000 bins, and then the HSV histogram is generated according to the quantity of different bins. After splicing the newly generated 1000-dimensional HSV histogram with the 199-dimensional original BVD, the 1199-dimensional extensional BVD is obtained.

### 2.4 Feature Selection and Extraction

In this section, the selection strategy of dimensionality reduction methods for the features mentioned above are introduced. In general, feature extraction and feature selection are two common approaches. According to the ratio of dimension reduction, these two approaches are selected respectively in two cases:

(1) Feature selection will be used if the dimensionality reduction ratio is small. For instance, when the dimensionality changes from 1199 to 1024, laplacian score [8] as the selection method is imported to determine which features are selected in this paper. Moreover, filtering out vectors the sum of which is small before feature selection can reduce information loss.
(2) Feature extraction will be used if the dimensionality reduction ratio is large. Because when the ratio is large, lost information by using feature selection

will be more than that by using feature extraction. For example, when the dimensionality is reduced from 199 to 64, Principal Component Analysis (PCA) [11] which is a feature extraction approach will be used.

## 3 Experimental Results

Four standard datasets (Holidays [6], UKB [6], Coil-100 [20] and ALOI [26]) are used to evaluate the performance of BVD. The basic information of the four datasets is listed in Table 1. In ranking step, L1-norm and a manifold ranking based on L1-norm (MR1) [20] are involved in our method. In all experiments, $\alpha = 0.95$.

UKB dataset is a popular dataset for image retrieval. As shown in Fig. 4, the images are non-semantically sensitive. The judgment of their category can be decided obviously by different features such as color and texture, which is fit for our method. In Table 2, we show the results obtained from various methods on the UKB dataset. It is obvious that BVD achieves better performance than other low-level features and extensional BVD even performs much better. Compared to middle-level and high-level features, BVD outperforms most methods with lower dimensions. Although our N-S value is lower than Sparse-coded features, our dimensions are much lower. The dimensions of extensional BVD is only a tenth of Sparse-coded features. The CNN involved is pre-trained on ImageNet and then retrained on related datasets using AlexNet [13], however it does not work so well on the UKB dataset. These results demonstrate the high-precision and high-effciency of our approach.

**Fig. 4.** The non-semantically sensitive images of UKB dataset can be discriminated by low-level features easily.

Holidays dataset is a semantically sensitive image dataset. The quantity of pictures with the same label is small and the judgment of different labels depends more on semantic information than low-level information. As shown in Fig. 5,

**Table 1.** The attributes of experimental datasets

| Dataset | Holidays | UKB | Coil-100 | ALOI |
|---|---|---|---|---|
| Image size | Vary | 640 × 480 | 128 × 128 | 192 × 144 |
| Class | 500 | 2550 | 100 | 1000 |
| Total image | 1491 | 10200 | 7200 | 72000 |

**Table 2.** The N-S value on UKB dataset

| Method | Category | N-S | Dimension |
|---|---|---|---|
| MSD [16] | Low-level | 3.23 | 72 |
| CDH [17] | Low-level | 2.49 | 90 |
| PUD [20] | Low-level | 3.36 | 280 |
| LBOC [30] | Middle-level | 3.5 | – |
| VLAD [10] | Middle-level | 3.15 | 6656 |
| HVLAD [22] | Middle-level | 3.56 | 32768 |
| Fisher+color [7] | Middle-level | 3.19 | 4096 |
| Sparse-coded features [6] | Middle-level | 3.76 | 11024 |
| CNN(AlexNet) [13] | High-level | 3.56 | 4096 |
| BVD | Low-level | 3.37 | 199 |
| Ext-BVD | Low-level | 3.51 | 1199 |
| Ext-BVD(MR1) | Low-level | 3.66 | 1199 |

**Fig. 5.** Color difference is large while semantical information is same.

**Fig. 6.** Low-level information is similar while semantical information is different

the two images with large color difference have the same semantic information, so they fall into the same label. And in Fig. 6, the low-level information of two images is very similar, but one of them has a Sphinx, which leads to different semantic information. So they are divided into different labels. Due to the characteristics of Holidays, it is difficult to determine by low-level features which are directly extracted by the common method. However, our scheme can also get competitive performance (shown in Table 3). Original BVD outperforms PUD by 1.8% of mAP and reduces 280 dimensions to 199 dimensions, which demonstrates the superiority of BVD in low-level features. The mAP score of extensional BVD is 3.0% higher than original BVD, because Holidays dataset is a large size dataset, the original low-dimensional descriptor is difficult to contain all information which is complemented by additional color features. Compared to middle-level features, the mAP score of extensional BVD outperforms BOC-based and VLAD-based local descriptors, which indicates that our method is effective in semantically sensitive datasets although we use low-level features.

Experiments show texture features are more important than color ones on the Holidays dataset. As shown in Table 4, mAP rises up with the increasing of $n_t$ in all dimensions. In this paper, laplacian score is applied to reduce dimensions from 1199 to 1024 and 199 to 128, PCA is used to reduce dimensions from 199 to 64. It is obvious that when $n_t$ is 1.5, the dimensionality decreases with almost no loss of precision.

Amsterdam Library of Object Images (ALOI) is a color image collection of small objects. As shown in Table 5, BVD gets a precision of 0.97 on the ALOI dataset by only 199-dimensional features, which is obviously superior to other methods. The color information of ALOI is outstanding and BVD contains a set

**Table 3.** The mAP value on Holidays dataset

| Method | mAP | Dimension |
|---|---|---|
| PUD | 0.652 | 280 |
| BOC | 0.638 | 256 |
| VLAD | 0.526 | 8192 |
| WF-VLAD [18] | 0.688 | 8192 |
| BVD | 0.670 | 199 |
| Ext-BVD | 0.702 | 1199 |

**Table 4.** Precision of different weights of texture and dimensions (%)

| $n_t$ | Dimension | | | | |
|---|---|---|---|---|---|
| | 64 | 128 | 199 | 1024 | 1199 |
| 0.5 | 57.7 | 60.6 | 63.3 | 68.0 | 69.0 |
| 1 | 58.6 | 63.6 | 65.7 | 69.2 | 69.8 |
| 1.5 | 60.4 | 66.1 | 67.0 | 70.0 | 70.2 |

**Table 5.** The precision on ALOI and Coil-100 datasets (%)

| Dataset\Method | 1k-dense | HSV | CDH | MSD | VGG [27] | BVD |
|---|---|---|---|---|---|---|
| ALOI | 83.4 | 94.1 | – | – | 88.6 | 97.5 |
| Coil-100 | 82.2 | 96.7 | 92.5 | 96.2 | 97.6 | 97.6 |

of uniform color features, so our method shows superior performance. ALOI is a simple dataset with small objects, so although our approach extracts features of very low dimensions, it has no impact on the final performance. The result also indicates that our method works well on the simple datasets.

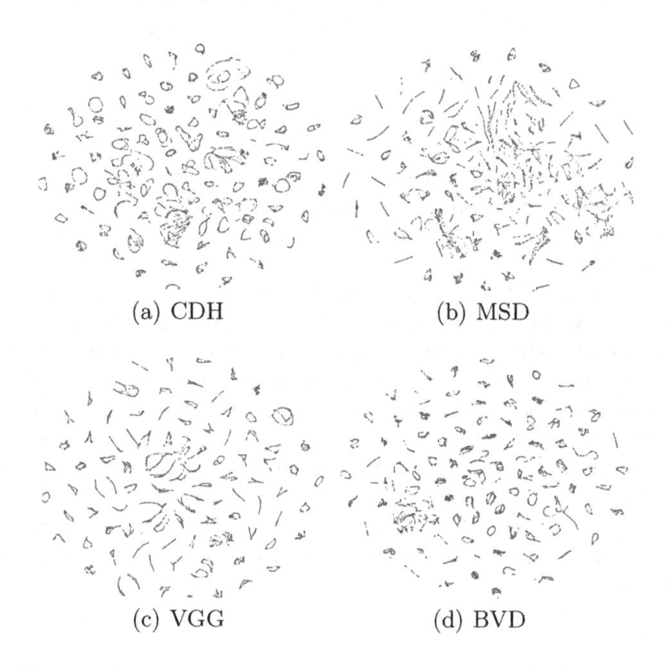


**Fig. 7.** The visualization results of four different descriptors.

The experimental results on the Coil-100 dataset also demonstrate the excellent performance of BVD as shown in Table 5. In order to get an intuitive sense of the results, the dimensionality of features extracted from Coil-100 is reduced to 2 by t-Distributed Stochastic Neighbor Embedding (t-SNE) [24]. The visualized result is shown in Fig. 7, the decision boundaries between different labels of BVD are very clear, which is comparable to VGG (4096 dimensions) and is better than CDH and MSD.

## 4 Conclusion

In this paper, we propose a low-level descriptor inspired by the human visual system and name it Bionic Vision Descriptor. As existing methods based on low-level features rely too much on color information and will have trouble in discriminating some underlying uniform structures, the extracted features are highly flawed, leading to low accuracy and efficiency. To improve this situation, we add texture information and structural discrimination coefficient in our descriptor to increase representation capability. Experiments on benchmark datasets show our descriptor exhibits very competitive performance compared to other methods based on low-level features. And our approach captures the balance between performance and efficiency better than other state-of-the-art methods. Especially when the task is not complex, our approach is more superior.

## References

1. Babenko, A., Slesarev, A., Chigorin, A., Lempitsky, V.: Neural codes for image retrieval. In: Fleet, D., Pajdla, T., Schiele, B., Tuytelaars, T. (eds.) ECCV 2014. LNCS, vol. 8689, pp. 584–599. Springer, Cham (2014). https://doi.org/10.1007/978-3-319-10590-1_38
2. Bay, H., Tuytelaars, T., Van Gool, L.: SURF: speeded up robust features. In: Leonardis, A., Bischof, H., Pinz, A. (eds.) ECCV 2006. LNCS, vol. 3951, pp. 404–417. Springer, Heidelberg (2006). https://doi.org/10.1007/11744023_32
3. Dalal, N., Triggs, B.: Histograms of oriented gradients for human detection. In: IEEE Computer Society Conference on Computer Vision and Pattern Recognition, pp. 886–893. IEEE (2005)
4. Di Zenzo, S.: A note on the gradient of a multi-image. Comput. Vis. Graph. Image Process. **33**(1), 116–125 (1986)
5. Ferman, A.M., Tekalp, A.M., Mehrotra, R.: Robust color histogram descriptors for video segment retrieval and identification. IEEE Trans. Image Process. **11**(5), 497–508 (2002)
6. Ge, T., Ke, Q., Sun, J.: Sparse-coded features for image retrieval. In: BMVC, pp. 132.1–132.11 (2013)
7. Gordoa, A., Rodríguez-Serrano, J.A., Perronnin, F., Valveny, E.: Leveraging category-level labels for instance-level image retrieval. In: 2012 IEEE Conference on Computer Vision and Pattern Recognition, pp. 3045–3052. IEEE (2012)
8. He, X., Cai, D., Niyogi, P.: Laplacian score for feature selection. In: Advances in Neural Information Processing Systems, pp. 507–514 (2006)
9. Jégou, H., Douze, M., Schmid, C.: Improving bag-of-features for large scale image search. Int. J. Comput. Vis. **87**(3), 316–336 (2010)
10. Jégou, H., Douze, M., Schmid, C., Pérez, P.: Aggregating local descriptors into a compact image representation. In: CVPR 2010–23rd IEEE Conference on Computer Vision and Pattern Recognition, pp. 3304–3311. IEEE Computer Society (2010)
11. Jolliffe, I.: Principal component analysis. Springer (2011). https://doi.org/10.1007/b98835
12. Koffka, K.: Principles of Gestalt Psychology. Routledge, London (2013)

13. Krizhevsky, A., Sutskever, I., Hinton, G.E.: Imagenet classification with deep convolutional neural networks. In: Advances in Neural Information Processing Systems, pp. 1097–1105 (2012)
14. Lewis, D.E., Pearson, J., Khuu, S.K.: The color "fruit": object memories defined by color. PloS One **8**(5), e64960 (2013)
15. Liu, C., Yuen, J., Torralba, A.: Sift flow: Dense correspondence across scenes and its applications. IEEE Trans. Pattern Anal. Mach. Intell. **33**(5), 978–994 (2010)
16. Liu, G.H., Li, Z.Y., Zhang, L., Xu, Y.: Image retrieval based on micro-structure descriptor. Pattern Recogn. **44**(9), 2123–2133 (2011)
17. Liu, G.H., Yang, J.Y.: Content-based image retrieval using color difference histogram. Pattern Recogn. **46**(1), 188–198 (2013)
18. Liu, H., Zhao, Q., Zhang, C., Mbelwa, J.T., Tang, S., Zhang, J.: Boosting vlad with weighted fusion of local descriptors for image retrieval. Multimedia Tools Appl. **78**(9), 11835–11855 (2019)
19. Liu, S., et al.: Color recognition for rubik's cube robot. arXiv preprint arXiv:1901.03470 (2019)
20. Liu, S., et al.: Perceptual uniform descriptor and ranking on manifold for image retrieval. Inf. Sci. **424**, 235–249 (2018)
21. Liu, Y., Zhang, D., Lu, G., Ma, W.Y.: A survey of content-based image retrieval with high-level semantics. Pattern Recogn. **40**(1), 262–282 (2007)
22. Liu, Z., Li, H., Zhou, W., Rui, T., Tian, Q.: Making residual vector distribution uniform for distinctive image representation. IEEE Trans. Circuits Syst. Video Technol. **26**(2), 375–384 (2015)
23. Lowe, D.G.: Distinctive image features from scale-invariant keypoints. Int. J. Comput. Vis. **60**(2), 91–110 (2004)
24. Maaten, L.V.D., Hinton, G.: Visualizing data using t-sne. J. Mach. Learn. Res. **9**, 2579–2605 (2008)
25. Ojala, T., Pietikäinen, M., Mäenpää, T.: Gray scale and rotation invariant texture classification with local binary patterns. In: Vernon, D. (ed.) ECCV 2000. LNCS, vol. 1842, pp. 404–420. Springer, Heidelberg (2000). https://doi.org/10.1007/3-540-45054-8_27
26. Rocha, A., Goldenstein, S.K.: Multiclass from binary: expanding one-versus-all, one-versus-one and ecoc-based approaches. IEEE Trans. Neural Networks Learn. Syst. **25**(2), 289–302 (2013)
27. Simonyan, K., Zisserman, A.: Very deep convolutional networks for large-scale image recognition. arXiv preprint arXiv:1409.1556 (2014)
28. Sivic, J., Zisserman, A.: Video google: A text retrieval approach to object matching in videos. In: IEEE International Conference on Computer Vision, pp. 1470–1477. IEEE (2003)
29. Ullman, S., Assif, L., Fetaya, E., Harari, D.: Atoms of recognition in human and computer vision. Proc. Nat. Acad. Sci. **113**(10), 2744–2749 (2016)
30. Wengert, C., Douze, M., Jégou, H.: Bag-of-colors for improved image search. In: Proceedings of the 19th ACM international conference on Multimedia, pp. 1437–1440. ACM (2011)
31. Zheng, L., Wang, S., Liu, Z., Tian, Q.: Packing and padding: Coupled multi-index for accurate image retrieval. In: Proceedings of the IEEE Conference on Computer Vision and Pattern Recognition, pp. 1939–1946 (2014)

# Brain Tumor Segmentation from Multi-spectral MR Image Data Using Random Forest Classifier

Szabolcs Csaholczi[1], David Iclănzan[1], Levente Kovács[2], and László Szilágyi[1,2(✉)]

[1] Computational Intelligence Research Group, Sapientia - Hungarian University of Transylvania, Tîrgu Mureş, Romania
szabolcscsaholczi55@gmail.com,
{iclanzan,lalo}@ms.sapientia.ro

[2] University Research, Innovation and Service Center (EKIK), Óbuda University, Budapest, Hungary
{kovacs.levente,szilagyi.laszlo}@nik.uni-obuda.hu

**Abstract.** The development of brain tumor segmentation techniques based on multi-spectral MR image data has relevant impact on the clinical practice via better diagnosis, radiotherapy planning and follow-up studies. This task is also very challenging due to the great variety of tumor appearances, the presence of several noise effects, and the differences in scanner sensitivity. This paper proposes an automatic procedure trained to distinguish gliomas from normal brain tissues in multi-spectral MRI data. The procedure is based on a random forest (RF) classifier, which uses 80 computed features beside the four observed ones, including morphological ones, gradients, and Gabor wavelet features. The intermediary segmentation outcome provided by the RF is fed to a twofold post-processing, which regularizes the shape of detected tumors and enhances the segmentation accuracy. The performance of the procedure was evaluated using the 274 records of the BraTS 2015 train data set. The achieved overall Dice scores between 85–86% represent highly accurate segmentation.

**Keywords:** Magnetic resonance imaging · Brain tumor detection · Tumor segmentation · Random forest

## 1 Introduction

Gliomas represent a common malignant brain tumor with low survival rate and short life expectancy. Patients with so-called high-grade (HG) gliomas live fifteen

This project was supported by the Sapientia Foundation – Institute for Scientific Research. The work of L. Kovács was supported by the European Research Council (ERC) under the European Union's Horizon 2020 research and innovation programme (grant agreement No 679681). The work of L. Szilágyi was supported by the Hungarian Academy of Sciences through the János Bolyai Fellowship program, and by the ÚNKP-19-4 New National Excellence Program of the Ministry for Innovation and Technology.


H. Yang et al. (Eds.): ICONIP 2020, LNCS 12532, pp. 174–184, 2020.
https://doi.org/10.1007/978-3-030-63830-6_15

months in average after the diagnosis, while those with low-grade (LG) gliomas can live for several years. In the current clinical practice, most of the times brain tumors are segmented manually, which is time consuming and error prone [1]. With the quickly increasing number of MRI devices deployed in hospitals and the high costs of training human experts, a strong need arising for automatic and reliable tumor detection and segmentation methods. Such algorithms could process the huge amount of acquired MRI data and select those patients which are suspected of having focal lesions in the brain, and consequently assist the medical experts in focusing on serious cases.

Multi-spectral MRI is the most frequently used and preferred medical imaging modality in brain tumor detection and segmentation, due to its fine contrast and the multiple data channels that offer complementary information. The Brain Tumor Segmentation (BraTS) Challenges organized jointly with the MICCAI conference, have provided the research community a continuously growing multi-spectral MRI data set of high quality and great challenges [2,3].

Earlier solutions to the challenge called brain tumor segmentation based on MRI data were summarized by Gordillo et al. [4]. Recent solutions usually combine advanced (mostly unsupervised) image segmentation algorithms with semi-supervised supervised classification algorithms that cover the whole arsenal of machine learning techniques, namely: graph cut segmentation algorithm [5], superpixels combined with non-parametric classifiers [6], feature fusion combined with joint label fusion [7], texture feature and kernel sparse coding [8], Gaussian mixture models [9], fuzzy $c$-means clustering in semi-supervised context [10], fuzzy $c$-means clustering combined with region growing [11], AdaBoost classifier [12], extremely random trees [13] combined with superpixel level features [14], random forests [15,16] and ensemble of random forests [17], support vector machines [18], expert systems [19], convolutional neural network [20], deep neural networks [21], generative adversarial networks [22], and tumor growth model [23].

This paper proposes a random forest based brain tumor segmentation procedure, including adequate preprocessing and post-processing tasks designed to enhance the accuracy of segmentation. Preprocessing eliminates noises and handles the spectral differences between various MRI records. Post-processing regularizes the shape of detected tumors and handles the cases with multiple detected lesions. The MICCAI BraTS 2015 train data set [2,3] is employed both for training and evaluation purposes.

The rest of the paper is structured as follows: Section 2 presents the details of the proposed automatic segmentation procedure, dedicating a subsection to every processing phase. Section 3 evaluates the accuracy of the proposed method, while Sect. 4 concludes the study.

## 2 Materials and Methods

The proposed procedure has the structure shown in Fig. 1. It starts with a multiple purpose preprocessing, which is equally applied to all MRI data records. This is followed by splitting the MRI records to train and test data records from

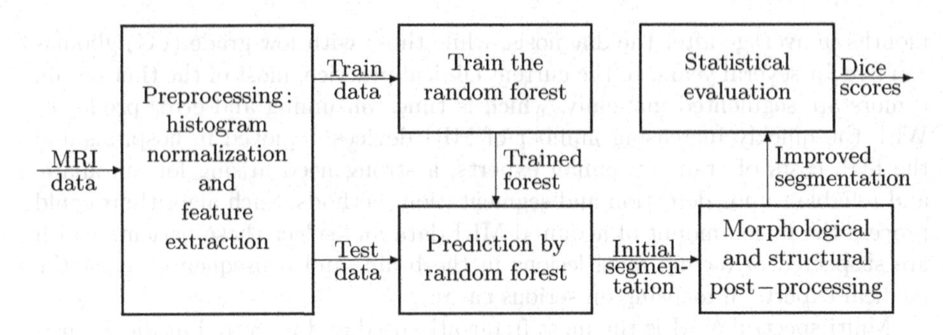


**Fig. 1.** The proposed segmentation algorithm.

test data records. Train records are fed to the random forest training process. Trained forests are employed to produce an intermediary label for all test pixels, which is then reevaluated by a two-step post-processing. Finally, the segmentation accuracy is measured using statistical indicators.

### 2.1 Data

The BraTS 2015 train data set [2,3], upon which this study relies, contains $N_{\mathrm{LG}} = 54$ low-grade and $N_{\mathrm{HG}} = 220$ high-grade glioma volumes, each consisting of 4 observed MRI channels (T1, T2, T1C, FLAIR), and human expert made labeling that can be used as ground truth. An automatic registration algorithm was used to align all data channels with the T1 data. Each volumes has a $155 \times 240 \times 240$ pixel resolution, where pixels represent the tissues from a one cubic millimeter region. According to the average size of the adult human brain, these records contain around 1.5 million brain pixels. All other tissues were eliminated. The size of the glioma within these records ranges between 10 and 330 thousand pixels.

### 2.2 Pre-processing

Pre-processing has the role to provide the input data a scenario where a successful segmentation is possible. It has to handle three problems:

- Intensity non-uniformity (INU) is a low-frequency noise with possibly high magnitude [24–26], which we suppress with the method of Tustison et al. [27].
- Absolute pixel intensities values in MRI data have no meaning by themselves, they have to be interpreted together with their context. The pre-processing uses a context dependent linear transform to treat MRI volumes and each of their data channels separately. Intensity values are mapped onto the $[0, 1]$ interval such a way, that the 25-percentile becomes 0.4 and the 75-percentile becomes 0.6, and all transformed values that fall out of the target interval are attached to the closest boundary values.

- MRI records hold much more information than the individual pixel intensities found in the 4 observed data channels. The correlation between neighbor pixels and the imperfection of data channel alignment method motivated us to generate 100 computed features for each pixels, 25 from each data channel: minima, maxima and average values from spatial neighborhood, average and median values from planar neighborhoods of various sizes, together with gradient and Gabor wavelet features. A complete description of the computed features can be found in our previous work [28].

### 2.3 Decision Making

The $N_\rho \in \{N_{\text{LG}}, N_{\text{HG}}\}$ records are randomly divided to two equal subsets, which interchangeably serve as train and test data for the deployed RF classifier. Train records are used to establish a decision making ensemble that can separate lesion pixels from normal ones. The optimal number of trees was determined empirically and finally set to $n_T = 150$. The RF was trained using a set of feature vectors that contained the same number ($n_F$) of randomly selected feature vectors from each train record. For LG glioma volumes, $n_F$ was ranging between 25k and 800k. Memory limitations determined the upper limit of $n_F$ in case of HG records at 200k. The maximum depth for the RF trees was set to $n_D = 18$, which is also a value tuned empirically during the evaluation of the procedure.

The trained RF is employed to give a prediction for each pixel of the test volumes. The decision of the RF is crisp, each pixel is fully assigned either to negatives (normal tissues) or positives (lesions), but this label is intermediary. These labels represent the input for the post-processing.

### 2.4 Post-processing

Post-processing (PP) intends to regularize the shape of detected lesions, which is likely to cause enhancement to the segmentation accuracy of the segmentation. PP has two components: a morphological phase, which is followed by a structural phase. The morphological criterion works in a cubic $11 \times 11 \times 11$ neighborhood: it extracts the number of valid brain pixels ($n_\tau$) and the number of pixels with positive intermediary label ($n_\pi$), and sets the current pixel's label to positive if and only if $n_\pi / n_\tau > 1/3$.

The structural phase may discard some of the positive labels but never makes changes in the other direction. First it identifies all contiguous regions formed by lesion pixels and then it decides whether to keep the labels for whole regions or discard them. As a first criterion, any contiguous lesion formed by less than 100 pixels is reverted, because they are too small to be reliably called tumor. Further on, principal component analysis (PCA) is applied to the coordinates of pixels belonging to a contiguous region, to establish the size of the lesion along its main spatial axis. Whenever the shortest axis indicates a radius shorter than two pixels, the lesion is discarded. All contiguous regions which remain with positive label are finally declared gliomas, and are labelled accordingly.

### 2.5 Evaluation Criteria

Let us denote by $\varGamma_i^{(\pi)}$ the set of positive pixels and by $\varGamma_i^{(\nu)}$ the set of negative pixels of volume $i$, according to the ground truth, for any $i = 1 \dots N_\rho$. Further on, let $\varLambda_i^{(\pi)}$ and $\varLambda_i^{(\nu)}$ stand for the set of pixels of volume $i$ that were labeled positive and negative, respectively. If we denote by $|X|$ the cardinality of set $X$, the main accuracy indicators extracted from volume $i$ are defined as:

1. Sensitivity and specificity, also known as true positive rate (TPR) and true negative rate (TNR), respectively, are defined as:

$$\mathrm{TPR}_i = \frac{|\varGamma_i^{(\pi)} \cap \varLambda_i^{(\pi)}|}{|\varGamma_i^{(\pi)}|} \quad \text{and} \quad \mathrm{TNR}_i = \frac{|\varGamma_i^{(\nu)} \cap \varLambda_i^{(\nu)}|}{|\varGamma_i^{(\nu)}|} . \tag{1}$$

2. Dice score (DS) is defined as:

$$\mathrm{DS}_i = \frac{2 \times |\varGamma_i^{(\pi)} \cap \varLambda_i^{(\pi)}|}{|\varGamma_i^{(\pi)}| + |\varLambda_i^{(\pi)}|} . \tag{2}$$

3. The accuracy can also be defined as the rate of correct decisions, which is given by the formula:

$$\mathrm{ACC}_i = \frac{|\varGamma_i^{(\pi)} \cap \varLambda_i^{(\pi)}| + |\varGamma_i^{(\nu)} \cap \varLambda_i^{(\nu)}|}{|\varGamma_i^{(\pi)}| + |\varGamma_i^{(\nu)}|} . \tag{3}$$

All the above accuracy indicators are defined in the $[0, 1]$ interval. In case of perfect segmentation, all indicators have the maximum value of 1. For any accuracy indicator $\mathcal{X} \in \{\mathrm{TPR}, \mathrm{TNR}, \mathrm{DS}, \mathrm{ACC}\}$, we will call its average and we will denote by $\overline{\mathcal{X}}$ the value given by the formula

$$\overline{\mathcal{X}} = \frac{1}{n_V} \sum_{i=1}^{n_V} \mathcal{X}_i . \tag{4}$$

The overall Dice score is denoted by $\widetilde{\mathrm{DS}}$ and is extracted with the formula:

$$\widetilde{\mathrm{DS}} = \frac{2 \times \left| \bigcup_{i=1}^{n_V} \varGamma_i^{(\pi)} \cap \bigcup_{i=1}^{n_V} \varLambda_i^{(\pi)} \right|}{\left| \bigcup_{i=1}^{n_V} \varGamma_i^{(\pi)} \right| + \left| \bigcup_{i=1}^{n_V} \varLambda_i^{(\pi)} \right|} . \tag{5}$$

The overall values of the other accuracy indicators, denoted by $\widetilde{\mathrm{TPR}}$, $\widetilde{\mathrm{TNR}}$, and $\widetilde{\mathrm{ACC}}$, whose formulas can be expressed analogously to Eq. (5).

## 3 Results and Discussion

The proposed algorithm underwent two separate evaluation processes, involving the 54 LG glioma records and the 220 HG glioma records of the BraTS 2015 data set, respectively. Several different values of the train data size parameter $n_F$ were deployed, and other parameters of the RF were set as presented in Sect. 2.3. Detailed results are presented in the following figures and tables.

Table 1 presents the global performance indicators of the proposed segmentation procedure, average and overall values. The train data size does not have large impact on the segmentation accuracy: rising parameter $n_F$ above 100k does not bring any benefit, neither for LG nor for HG glioma records. Average Dice scores obtained for LG data are slightly above 85%, while those for HG data are in the proximity of 82.5%. This difference can be justified by the quality of image data in the two data sets. However, the overall Dice scores are much closer to each other. The difference between average and overall sensitivity is only relevant in case of HG data. The average and overall values of specificity and correct decision rate (accuracy) hardly show any differences.

Figure 2 exhibits the Dice scores achieved for individual HG glioma volumes. Figure 2(a) plots Dice scores against the size of the tumor (according to the ground truth). Each cross ($\times$) represents the outcome of one MRI record. The dashed line indicates the linear trend identified by linear regression. The linear trend indicates that accuracy if better for larger gliomas, but even for small ones the Dice score is almost 80%. Figure 2(b) plots the individual value of all four accuracy indicators obtained for HG data, in increasing order. These graphs clarify the distribution of these values and reveals that their median value is higher than the average.

Similarly, Fig. 3 gives the same representations for the segmentation outcome of LG glioma records. The shape of the graphs are quite similar to the ones obtained for HG data. The biggest difference is in the linear trend against tumor size: Dice scores hardly change with the glioma size, but even small gliomas are segmented with high accuracy characterized by a Dice score over 85%. The accuracy indicator values presented in Figs. 2 and 3 were achieved with train data size $n_F = 100k$.

Specificity values around 99% are very important, as the number of negative pixels is very high. Lower values in specificity would mean the presence of a lot of false positives in the segmentation outcome. Correct decision rates are in the proximity of 98%, meaning that only one pixel out of fifty is misclassified.

Figure 4 shows some examples of segmented brains. One selected slice from six different HG volumes are presented, the four observed data channels, namely T1, T2, T1C and FLAIR, followed by the segmentation outcome drawn in colors. Colors represent the following: true positives are drawn in green, false positives in blue, false negatives in red, and true negatives in gray. Mistaken pixels are either red or blue.

The proposed segmentation procedure was implemented in Python 3 programming language. The segmentation of a single MRI record, including all steps according to Fig. 1, requires approximately two minutes without using any parallel computation.

**Table 1.** Main global accuracy indicators

| BraTS volumes | Train data size | Dice score | | Sensitivity | | Specificity | | Accuracy | |
|---|---|---|---|---|---|---|---|---|---|
| | | $\overline{DS}$ | $\widetilde{DS}$ | $\overline{TPR}$ | $\widetilde{TPR}$ | $\overline{TNR}$ | $\widetilde{TNR}$ | $\overline{ACC}$ | $\widetilde{ACC}$ |
| LG | 25k | 0.8484 | 0.8591 | 0.8248 | 0.8257 | 0.9923 | 0.9926 | 0.9806 | 0.9808 |
| | 50k | 0.8486 | 0.8582 | 0.8296 | 0.8287 | 0.9918 | 0.9922 | 0.9804 | 0.9806 |
| | 100k | 0.8502 | 0.8598 | 0.8349 | 0.8346 | 0.9915 | 0.9918 | 0.9805 | 0.9807 |
| | 200k | 0.8500 | 0.8593 | 0.8364 | 0.8353 | 0.9914 | 0.9917 | 0.9804 | 0.9806 |
| | 400k | 0.8502 | 0.8589 | 0.8389 | 0.8366 | 0.9914 | 0.9915 | 0.9803 | 0.9805 |
| | 800k | 0.8500 | 0.8586 | 0.8392 | 0.8367 | 0.9911 | 0.9914 | 0.9802 | 0.9804 |
| HG | 25k | 0.8248 | 0.8543 | 0.8111 | 0.8477 | 0.9891 | 0.9892 | 0.9789 | 0.9789 |
| | 50k | 0.8248 | 0.8539 | 0.8118 | 0.8477 | 0.9890 | 0.9891 | 0.9789 | 0.9789 |
| | 100k | 0.8254 | 0.8536 | 0.8146 | 0.8500 | 0.9887 | 0.9888 | 0.9788 | 0.9787 |
| | 200k | 0.8253 | 0.8537 | 0.8131 | 0.8486 | 0.9889 | 0.9890 | 0.9788 | 0.9788 |

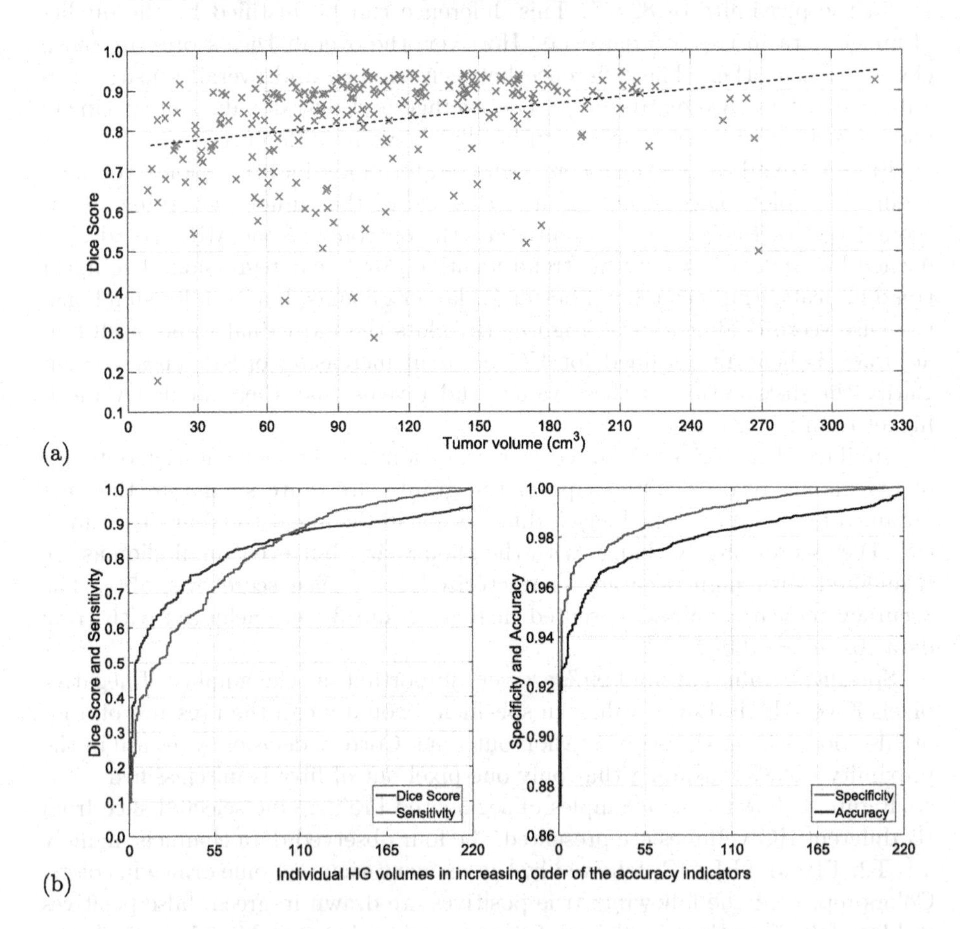


**Fig. 2.** Accuracy indicators obtained for individual HG records: (a) Dice scores plotted against the true size of the glioma; (b) Individual DS, TPR, TNR and ACC values plotted in increasing order.

**Table 2.** Comparison with state-of-the-art methods

| Method | Year | Classifier | Data | Dice scores |
|---|---|---|---|---|
| Tustison et al. [17] | 2015 | RF, MRF | BraTS 2013 | $\overline{DS} = 0.87$ |
| Pereira et al. [20] | 2016 | CNN | BraTS 2013 | $\overline{DS} = 0.88$ |
| Pinto et al. [13] | 2018 | ERT | BraTS 2013 | $\overline{DS} = 0.85$ |
| Pereira et al. [20] | 2016 | CNN | BraTS 2015 | $\overline{DS} = 0.78$ |
| Zhao et al. [21] | 2018 | CNN, CRF | BraTS 2015 | $\overline{DS} = 0.84$ |
| Pei et al. [7] | 2020 | RF, boosting | BraTS 2015 | $\overline{DS} = 0.850$ |
| Proposed method | | RF | BraTS 2015 | $\overline{DS} = 0.85$, $\widetilde{DS} = 0.86$ (LG) |
| | | | | $\overline{DS} = 0.826$, $\widetilde{DS} = 0.854$ (HG) |

MRF - Markov random field, CRF - conditional random field
ERT - extremely randomized trees

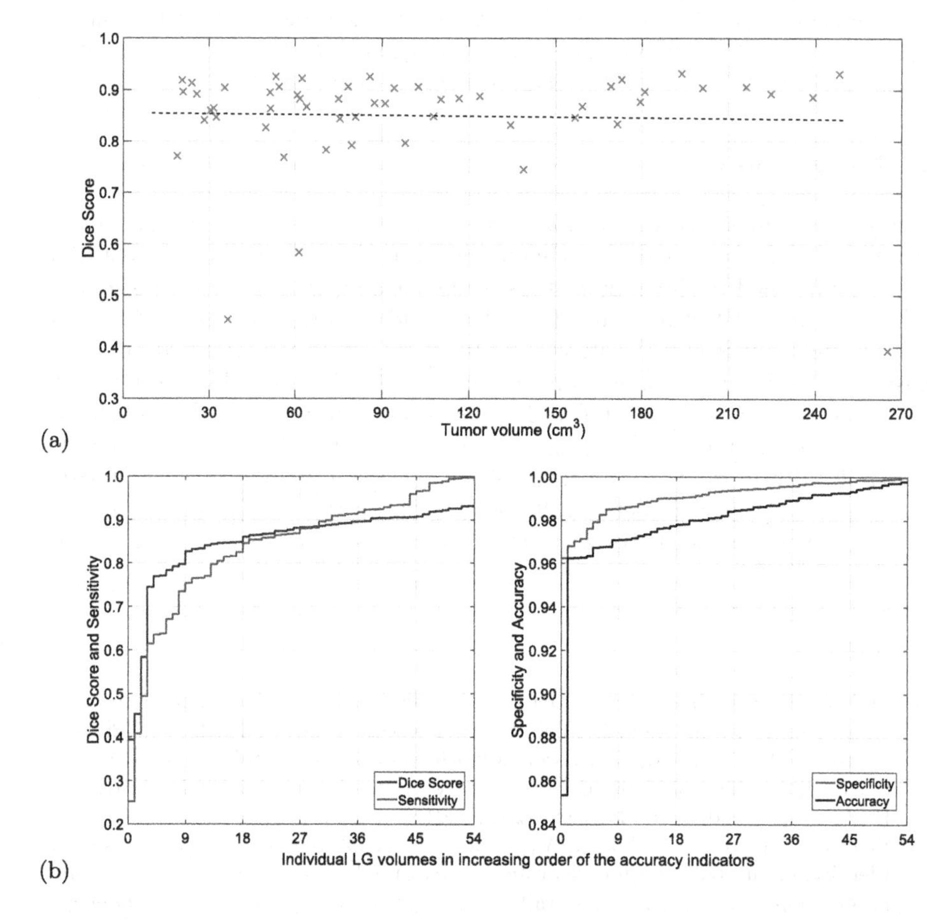


**Fig. 3.** Accuracy indicators obtained for individual LG records: (a) Dice scores plotted against the true size of the glioma; (b) Individual DS, TPR, TNR and ACC values plotted in increasing order.

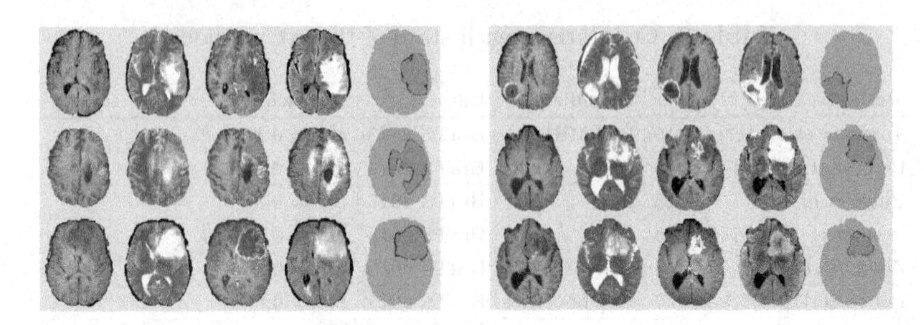

**Fig. 4.** One slice from six different HG tumor volumes, the four observed data channels and the segmentation result. The first four columns present the T1, T2, T1C and FLAIR channel data of the chosen slices. The last column shows the segmented slice, representing true positives ($|\Gamma_i^{(\pi)} \cap \Lambda_i^{(\pi)}|$) in green, false negatives ($|\Gamma_i^{(\pi)} \cap \Lambda_i^{(\nu)}|$) in red, false positives ($|\Gamma_i^{(\nu)} \cap \Lambda_i^{(\pi)}|$) in blue, and true negatives ($|\Gamma_i^{(\nu)} \cap \Lambda_i^{(\nu)}|$) in gray, where $i$ is the index of the current MRI record. (Color figure online)

## 4 Conclusions

This paper proposed a random forest based procedure for fully automatic segmentation of brain tumors from multi-spectral MRI data. The segmentation was accomplished in three main steps. Preprocessing was aimed at image data enhancement and feature generation. The random forest was trained to separate normal pixels from positive ones, based on which it performed an initial classification of test pixels, providing them intermediary labels. Finally, the post-processing reevaluated the intermediary labels and produced regularized shapes to the detected tumors. The procedure was trained and evaluated using the BraTS 2015 train records. The achieved segmentation accuracy is characterized by an overall Dice score between 85–86%, both in case of LG and HG glioma records, which is competitive with respect to state-of-the-art methods, as indicated in Table 2.

## References


1. Mohan, G., Subashini, M.M.: MRI based medical image analysis: survey on brain tumor grade classification. Biomed. Signal Process. Control **39**, 139–161 (2018)
2. Menze, B.H., Jakab, A., Bauer, S., Kalpathy-Cramer, J., Farahani, K., Kirby, J., et al.: The multimodal brain tumor image segmentation benchmark (BRATS). IEEE Trans. Med. Imag. **34**, 1993–2024 (2015)
3. Bakas, S., Reyes, M., Jakab, A., Bauer, S., Rempfler, M., Crimi, A., et al.: Identifying the best machine learning algorithms for brain tumor segmentation, progression assessment, and overall survival prediction in the BRATS challenge. arXiv: 1181.02629v3, 23 Apr 2019
4. Gordillo, N., Montseny, E., Sobrevilla, P.: State of the art survey on MRI brain tumor segmentation. Magn. Res. Imaging **31**, 1426–1438 (2013)
5. Njeh, I., et al.: 3D multimodal MRI brain glioma tumor and edema segmentation: a graph cut distribution matching approach. Comput. Med. Imaging Graph. **40**, 108–119 (2015)

6. Rehman, Z.U., Naqvi, S.S., Khan, T.M., Khan, M.A., Bashir, T.: Fully automated multi-parametric brain tumour segmentation using superpixel based classification. Expert Syst. Appl. **118**, 598–613 (2019)
7. Pei, L.M., Bakas, S., Vossough, A., Reza, S.M.S., Murala, C., Iftekharuddin, K.M.: Longitudinal brain tumor segmentation prediction in MRI using feature and label fusion. Biomed. Signal Process. Control **55**, 101648 (2020)
8. Tong, J.J., Zhao, Y.J., Zhang, P., Chen, L.Y., Jiang, L.R.: MRI brain tumor segmentation based on texture features and kernel sparse coding. Biomed. Signal Process. Control **47**, 387–392 (2019)
9. Menze, B.H., van Leemput, K., Lashkari, D., Riklin-Raviv, T., Geremia, E., Alberts, E., et al.: A generative probabilistic model and discriminative extensions for brain lesion segmentation - with application to tumor and stroke. IEEE Trans. Med. Imaging **35**, 933–946 (2016)
10. Szilágyi, L., Lefkovits, L., Benyó, B.: Automatic brain tumor segmentation in multispectral MRI volumes using a fuzzy $c$-means cascade algorithm. In: Proceedings 12th International Conference on Fuzzy Systems and Knowledge Discovery, pp. 285–291. IEEE (2015)
11. Li, Q.N., et al.: Glioma segmentation with a unified algorithm in multimodal MRI images. IEEE Access **6**, 9543–9553 (2018)
12. Islam, A., Reza, S.M.S., Iftekharuddin, K.M.: Multifractal texture estimation for detection and segmentation of brain tumors. IEEE Trans. Biomed. Eng. **60**, 3204–3215 (2013)
13. Pinto, A., Pereira, S., Rasteiro, D., Silva, C.A.: Hierarchical brain tumour segmentation using extremely randomized trees. Pattern Recogn. **82**, 105–117 (2018)
14. Imtiaz, T., Rifat, S., Fattah, S.A., Wahid, K.A.: Automated brain tumor segmentation based on multi-planar superpixel level features extracted from 3D MR images. IEEE Access **8**, 25335–25349 (2020)
15. Lefkovits, L., Lefkovits, S., Szilágyi, L.: Brain tumor segmentation with optimized random forest. In: Crimi, A., Menze, B., Maier, O., Reyes, M., Winzeck, S., Handels, H. (eds.) Brainlesion: Glioma, Multiple Sclerosis, Stroke and Traumatic Brain Injuries. LNCS, vol. 10154, pp. 88–99. Springer, Cham (2017). https://doi.org/10.1007/978-3-319-55524-9_9
16. Lefkovits, S., Szilágyi, L., Lefkovits, L.: Brain tumor segmentation and survival prediction using a cascade of random forests. In: Crimi, A., Bakas, S., Kuijf, H., Keyvan, F., Reyes, M., van Walsum, T. (eds.) BrainLes 2018. LNCS, vol. 11384, pp. 334–345. Springer, Cham (2019). https://doi.org/10.1007/978-3-030-11726-9_30
17. Tustison, N.J., et al.: Optimal symmetric multimodal templates and concatenated random forests for supervised brain tumor segmentation (simplified) with ANTsR. Neuroinformatics **13**, 209–225 (2015). https://doi.org/10.1007/s12021-014-9245-2
18. Zhang, N., Ruan, S., Lebonvallet, S., Liao, Q., Zhou, Y.: Kernel feature selection to fuse multi-spectral MRI images for brain tumor segmentation. Comput. Vis. Image Understand. **115**, 256–269 (2011)
19. Sert, E., Avci, D.: Brain tumor segmentation using neutrosophic expert maximum fuzzy-sure entropy and other approaches. Biomed. Signal Process. Control **47**, 276–287 (2019)
20. Pereira, S., Pinto, A., Alves, V., Silva, C.A.: Brain tumor segmentation using convolutional neural networks in MRI images. IEEE Trans. Med. Imaging **35**, 1240–1251 (2016)
21. Zhao, X.M., Wu, Y.H., Song, G.D., Li, Z.Y., Zhang, Y.Z., Fan, Y.: A deep learning model integrating FCNNs and CRFs for brain tumor segmentation. Med. Image Anal. **43**, 98–111 (2018)

22. Nema, S., Dudhane, A., Murala, S., Naidu, S.: RescueNet: an unpaired GAN for brain tumor segmentation. Biomed. Signal Process. Control **55**, 101641 (2020)
23. Lê, M., et al.: Personalized radiotherapy planning based on a computational tumor growth model. IEEE Trans. Med. Imaging **36**, 815–825 (2017)
24. Vovk, U., Pernuš, F., Likar, B.: A review of methods for correction of intensity inhomogeneity in MRI. IEEE Trans. Med. Imaging **26**, 405–421 (2007)
25. Szilágyi, S.M., Szilágyi, L., Iclănzan, D., Dávid, L., Frigy, A.: Benyó, Z: Intensity inhomogeneity correction and segmentation of magnetic resonance images using a multi-stage fuzzy clustering approach. Neural Netw. World **09**(5), 513–528 (2009)
26. Szilágyi, L., Szilágyi, S.M., Benyó, B.: Efficient inhomogeneity compensation using fuzzy $c$-means clustering models. Comput. Methods Progr. Biomed. **108**, 80–89 (2012)
27. Tustison, N.J., et al.: N4ITK: improved N3 bias correction. IEEE Trans. Med. Imaging **29**, 1310–1320 (2010)
28. Szilágyi, L., Iclănzan, D., Kapás, Z., Szabó, Z., Győrfi, Á., Lefkovits, L.: Low and high grade glioma segmentation in multispectral brain MRI data. Acta Univ. Sapientia, Informatica **10**(1), 110–132 (2018)

# CAU-net: A Novel Convolutional Neural Network for Coronary Artery Segmentation in Digital Substraction Angiography

Rui-Qi Li[1,2], Gui-Bin Bian[1,2], Xiao-Hu Zhou[1,2], Xiaoliang Xie[1,2], Zhen-Liang Ni[1,2], and Zengguang Hou[1,2,3,4](✉)

[1] State Key Laboratory of Management and Control for Complex Systems, Institute of Automation, Chinese Academy of Sciences, Beijing 100190, China
zengguang.hou@ia.ac.cn

[2] University of Chinese Academy of Sciences, Beijing 100049, China

[3] CAS Center for Excellence in Brain Science and Intelligence Technology, Beijing 100190, China

[4] Joint Laboratory of Intelligence Science and Technology, Institute of Systems Engineering, Macau University of Science and Technology, Macau, China

**Abstract.** Coronary artery analysis plays an important role in the diagnosis and treatment of coronary heart disease. Coronary artery segmentation, as an important part of quantitative researc h on coronary heart disease, has become the main topic in coronary artery analysis. In this paper, a deep convolutional neural network (CNN) based method called Coronary Artery U-net (CAU-net) is proposed for the automatic segmentation of coronary arteries in digital subtraction angiography (DSA) images. CAU-net is a variant of U-net. Based on the observation that coronary arteries are composed of many vessels with the same appearance but different thicknesses, a novel multi-scale feature fusion method is proposed in CAU-net. Besides, a new dataset is proposed to solve the problem of no available public dataset on coronary arteries segmentation, which is also one of our contributions. Our dataset contains 538 image samples, which is relatively large compared with the public datasets of other vessel segmentation tasks. In our dataset, a new labeling method is applied to ensure the purity of the labeling samples. From the experimental results, we prove that CAU-net can make significant improvements compared with the vanilla U-net, and achieve the state-of-the-art performance compared with other traditional segmentation methods and deep learning methods.

**Keywords:** Machine learning · Coronary artery · Segmentation

## 1 Introduction

Coronary heart disease (CHD) has been one of the leading causes of death worldwide. As the standard modality to diagnose CHD, Digital Substraction


H. Yang et al. (Eds.): ICONIP 2020, LNCS 12532, pp. 185–196, 2020.
https://doi.org/10.1007/978-3-030-63830-6_16

Angiography (DSA) examination has been widely used and considered as the "gold standard". By observing the coronary arteries in the DSA image, the doctor can determine whether there is stenosis in the coronary arteries and get some useful information about the stenosis. Besides, DSA examination has been implemented at all stages of percutaneous coronary intervention (PCI), a kind of minimally invasive treatment and also the primary treatment for CHD. During PCI, by observing and analyzing the coronary arteries in DSA images, surgeons can: 1. determine the surgical plan and select the appropriate surgical instruments before the intervention; 2. get navigation information for delivering surgical instruments during the intervention, 3. check the surgical results after the intervention. However, all of the above useful information can be obtained from the segmentation results of coronary arteries. In conclusion, as a key component of the quantitative study of CHD, coronary artery segmentation in DSA images plays a crucial role in clinical diagnosis and surgical planning.

Recent years have witnessed the rapid development of deep convolutional neural networks (CNN) for medical segmentation. Especially after the appearance of U-net [1], a large number of its variants were proposed for various segmentation tasks and achieved the state-of-the-art results on different medical datasets. As a supervised method, deep learning algorithms need sufficient training samples with annotations to obtain good performance. For the segmentation task, the training image needs to be manually labeled by category at pixel-level.

However, the pixel-wise annotation of blood vessels is time-consuming, because blood vessels have complex shapes and various edge details. This is the reason why there is still no public dataset for coronary artery segmentation, and also the reason why there are so few deep learning methods for coronary artery segmentation. The same difficulty also occurred in other vessel segmentation tasks: the main public datasets of the retinal vessel segmentation task basically contain fewer than 100 samples. In order to solve the lack of dataset, a dataset containing 538 samples is established in this paper. More details on the dataset will be introduced in the Sect. 4.

Except for the dataset, there are also some difficulties that need to be solved in the task of coronary artery segmentation in DSA images. First, the signal-to-noise ratio of DSA images is low, which is a great challenge for accurate segmentation. Because noise blurs the boundaries of coronary arteries. Second, DSA images belong to grayscale images, so the low diversity of color makes many things in DSA images have similar appearances with vessels and are easily misclassified into coronary arteries, such as the outline of the spine and ribs. Third, there are also many motion artifacts caused by heart beating and respiration. These problems make accurate segmentation of the coronary artery a challenging problem. To solve these problems, the algorithm should have more denoising and recognition capabilities, rather than the ability to extract small vessels.

Our contributions are as follows: (1) We set up a new dataset for the usage of coronary artery segmentation. The dataset will be introduced in Sect. 4. (2) CAU-net, a newly designed CNN modified on U-net, is proposed. In CAU-net, a novel multi-scale feature fusion method is proposed. (3) We apply both traditional segmentation methods and deep learning methods on our dataset for comparison. From the comparison results, it can be seen that our method achieves the state-of-the-art performance.

## 2 Related Work

Up to now, the existing methods for coronary artery segmentation in DSA images are mainly the traditional segmentation methods. These methods have similar workflows. First vessel enhancement filters [2,3] based on Hessian measures are used to analyze the local second-order profile. These morphological filters can provide vesselness features. Then thresholding or other post-processing methods (level set [4], region growing [5], active contour [6]) are utilized to obtain the final segmentation result. These traditional segmentation approaches have two common problems. First, the generalization performance of these methods is poor. This is because these methods have many hyperparameters that need to be selected, and the selection of these parameters is entirely empirical. Besides, for these methods, performance varies greatly under different hyperparameters. Second, these methods only extract appearance features and cannot extract the semantic features, so any structure with a similar appearance with blood vessels is easy to be classified as coronary arteries. Because of the lack of available public datasets for training, few deep learning methods have been used for coronary artery segmentation. Fan *et al.* [7] and Yang *et al.* [8] both propose a deep learning method to achieve the segmentation of coronary arteries. But different from our task, their inputs not only include the image to be segmented but also include a background image without the contrast agent (they call it 'mask image').

Although there are few supervised learning methods for coronary artery segmentation, many supervised learning and even deep learning methods have emerged in other blood vessel segmentation tasks with public datasets. The most common task is retinal vessel segmentation in the fundus images. For retinal vessel segmentation, researchers have established some public datasets, including DRIVE [9] and STARE [10]. So a number of supervised methods have emerged, including traditional supervised learning methods (k-nearest neighbors [9], support vector machine (SVM) [11], conditional random fields (CRFs) [12]), and many deep learning methods [13,14].

## 3 Method

U-net is commonly used for medical image segmentation tasks because of their good performance and efficient use of GPU memory. Compared with other segmentation methods, U-net has a small number of parameters (the vanilla U-net only has 31M parameters), which make U-net is very suitable for the training of medical image datasets which only contain a small number of samples. In our network, we further reduce the channel numbers of all convolution layers to a

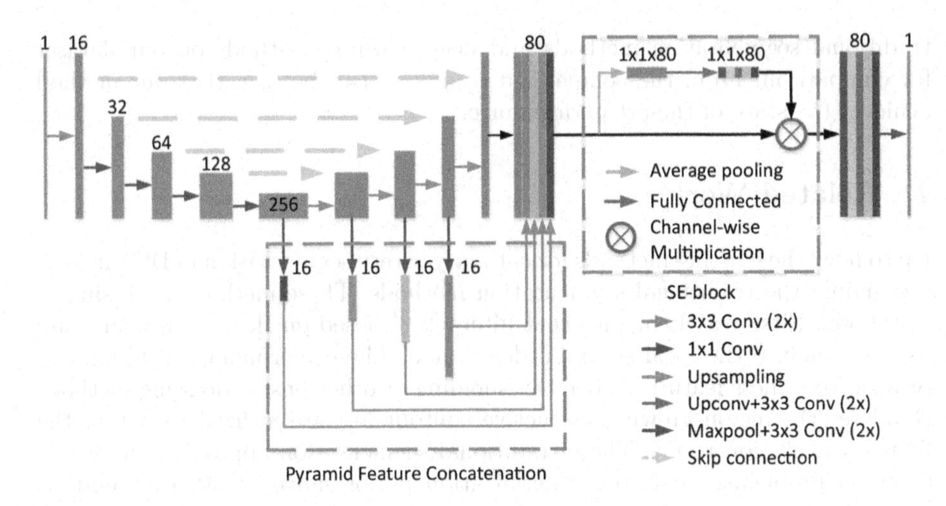


**Fig. 1.** Architecture of CAU-net

quarter of the former number to reduce the inference time and the risk of overfitting, and the parameter number of U-net is reduced to about 2M. In order to make U-net more suitable for the task of coronary artery segmentation, we propose a novel feature fusion module on it and named it Coronary Artery U-net (CAU-net), as shown in Fig. 1.

### 3.1 Feature Fusion Module

Coronary arteries are composed of many vessels with the same appearance but different thicknesses. These vessels with different thickness will have different activations in the feature maps at different scales. Based on this observation, we believe that the idea of multi-scale feature fusion is helpful to improve the coronary artery segmentation results.

The most famous multi-scale feature fusion method is FPN (Feature Pyramid Networks) [15], as shown in Fig. 2(a), which has been widely used in many detection models. In fact, from the perspective of structure, U-net's decoder structure itself is very similar to the structure of the FPN, as shown in Fig. 2(b). The main difference between them is that the feature maps of different scales in FPN are used to generate the detection results, while U-net only uses the last feature maps to predict the final segmentation results. As shown in Fig. 2(c), we also hope to propose a network that can use the feature maps of each scale to predict a segmentation result, and then fuse all segmentation results into one, just like what FPN did. But the segmentation results at different scales can not be fused like the way of detection task. Therefore, in implementation, we choose to fuse all feature maps at different scales first, and then generate the final segmentation results, as shown in Fig. 2(d).

As shown in Fig. 1, we first unify all the feature maps of different scales to the same size and concatenate them, which we called PFC (Pyramid Feature

Concatenation). And then an SE-block is applied to assign the fusion weights to the feature maps of different scales. This fusion weights act as the weights for the fusion of the final segmentation results of different scales. The final segmentation result will be predicted by the fusion feature maps. We treat this result as the fusion of multiple segmentation results generated by feature maps of different scales. In summary, our feature fusion module consists of two parts: PFC and SE-block.

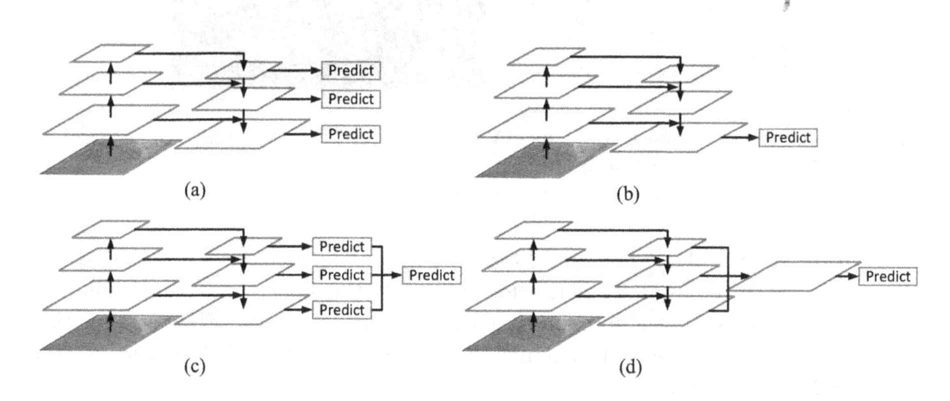


**Fig. 2.** Some schematic structures of networks: (a) Schematic diagram of FPN. (b) Schematic diagram of U-net. (c) The structure we want to propose. (d) The structure we actually propose

### 3.2 Pyramid Feature Concatenation

PFC is used to fuse all feature maps of each scale. The feature maps are first reduced to the same channels as the output feature maps by a $1 \times 1$ convolution and then are upsampled to the same size as the segmentation result, as shown in Fig. 1. The reason for using concatenation instead of addition in PFC is to ensure that the features of each scale can be completely retained to generate the final segmentation results.

In addition to drawing on the idea of FPN, we explain the role of PFC from another perspective. Because high-level semantic information is the key to removing noise in DSA images, PFC can help the segmentation results obtain more semantic features to reduce the noise, which is the main reason why PFC works.

### 3.3 SE-Block

SE-block was first introduced in [16], the features are first squeezed by global average pooling to generate channel-wise statistics, then two fully connected (FC) layers are applied to fully capture channel-wise dependencies. At last, multiply the channel-wise dependencies by the original feature maps to get the excitation feature maps, as shown in Fig. 1.

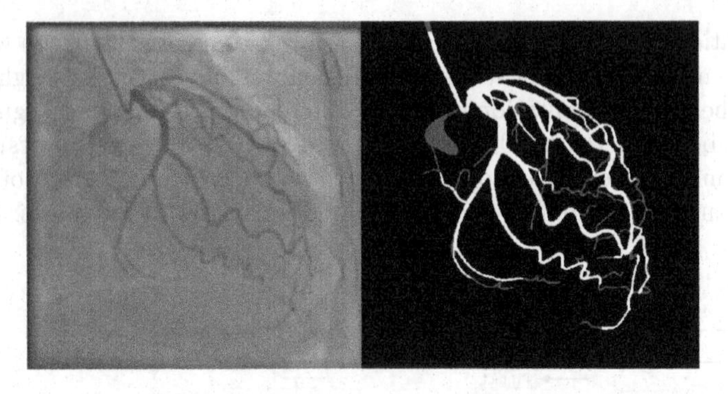

**Fig. 3.** An example of the sample in our dataset. (left) The original image. (right) Corresponding label (Color figure online)

By using an SE-block, we can obtain the importance of the feature maps at different levels to the final segmentation result. At the same time, it can balance the importance of semantic information and appearance information, so that the final result can get a good performance in removing the noise and preserving the details of blood vessels.

### 3.4 Loss Function

Following the SE-block, the final feature maps are generated. A $1 \times 1$ convolution layer is applied to the final feature maps to calculate the segmentation result. We use sigmoid cross-entropy as the loss function of the segmentation result:

$$Loss = -(ylogy^* + (1 - y)log(1 - y^*)) \tag{1}$$

In this equation, $y$ is the segmentation result. $y^*$ is the ground truths of segmentation.

## 4 Dataset

We established a dataset to verify the superiority of our model in coronary artery segmentation. Our dataset contains 538 samples from 36 DSA sequences in total, including all 412 samples from 7 DSA sequences and 126 samples randomly selected from 29 DSA sequences. All samples were labeled at the pixel-level.

Because of the special image-forming principle of DSA, we used a new way to label our dataset. Instead of labeling all pixels into background and foreground, all pixels are labeled into three categories: coronary artery (White), background (Black), and uncertainty (Red), as you can see from Fig. 3. Uncertainty means that we cannot determine whether some pixels should be marked as background or foreground when labeling. There are several reasons why a pixel may be labeled as uncertainty:

1. The contrast agent, which makes the coronary artery visible, is liquid. The liquid doesn't have a definite boundary. It's just because of the constraints of the vessel wall that make the boundary of the vessel appears in DSA images. But the axial direction of the vessel, also the direction of blood flow, is unconstrained, so along this direction, there are always some pixels where it's hard to say whether it should be labeled as the vessel or not.
2. Again, since the contrast agent is liquid, there will be a phenomenon that the contrast agent will partially backflow at the catheter orifice and produce smoke, makes it difficult to label the category of some pixels covered by smoke.
3. There are often shadows on the border of DSA images. When the blood vessels go out of the image range, the pixels at the boundaries will be covered by shadows. We also think that these pixels are difficult to mark.
4. Because not enough contrast agent flowing in, some small vessels can barely show, which makes some pixels hard to label.

These pixels are annotated as a new category because they are regarded as bad samples. In other deep learning tasks bad samples should be removed from datasets. Due to the special image-forming principle of DSA, these pixels are inevitable in DSA images, so we annotate them as a new category. Loss generated by these pixels is not calculated in the training, and the performance of these pixels is not calculated in the test.

In our experiments, we use 337 samples from 24 DSA sequences as training set, and 201 samples from 12 DSA sequences as testing set.

# 5 Experiments

## 5.1 Implementation Details

The network is implemented using Tensorflow, and for optimization, SGD optimizer with a momentum of 0.9 is applied and batch size is 8. We use a initial learning rate of 0.1, and the initial learning rate is multiplied by 0.8 every 4000 steps to avoid overfitting. For data augmentation, random flip, random crop, random grayscale adjustment $[-20, 20]$ and random contrast ratio $[0.8, 1.2]$ are adopted. Training takes about 9 h on an NVIDIA Titan XP for 2000 epochs.

## 5.2 Evaluation Metrics

Similar to the retinal vessel segmentation task, we also use the metrics including Specificity (Sp), Sensitivity (Se) and Accuracy (Acc) to evaluate the segmentation result. These metrics are defined below:

$$Sp = \frac{TN}{TN + FP}, Se = \frac{TP}{TP + FN}, Acc = \frac{TP + TN}{TP + FP + TN + FN} \tag{2}$$

TP, FN, TN, FP denote true positive, false negative, true negative and false positive, respectively.

**Table 1.** Results of ablation experiment

| Methods | Acc | Se | Sp | Dice |
|---|---|---|---|---|
| U-net(baseline) | 0.9910 | **0.8905** | 0.9956 | 0.8817 |
| U-net+PFC | 0.9913 | 0.8836 | 0.9963 | 0.8968 |
| U-net+SE | 0.9907 | 0.8873 | 0.9955 | 0.8670 |
| U-net+PFC+SE(Ours) | **0.9919** | 0.8863 | **0.9969** | **0.9035** |

*PFC stands for Pyramid Feature Concatenation, SE stands for SE-Block

Since the coronary artery segmentation task contains most of the easily segmented pixels, the improvements of Sp, Se and Acc are not significant. In order to more intuitively reflect the performance differences between different methods, Dice score is used to measure the segmentation performance.

$$Dice = \frac{2 \times TP}{2 \times TP + FP + FN} \quad (3)$$

### 5.3 Ablation Experiments

We run a number of ablations to analyze the performance gains of our feature fusion module. Results are shown in Table 1. Because DSA images contain many pixels that can be easily segmented correctly, the performance of U-net on Acc, Se and Sp is already good. But both of PFC and SE-block can significantly improve the segmentation performance, indicating that our method performs better on the pixels that are hard to classify correctly. Good detail performance is important for coronary artery segmentation.

It should be pointed out that we also try to add only SE-block on U-net, but we find that the result gets even worse. That is to say, the SE-block can only work if it is used together with the PFC. This result also indicates that the function of the SE-block is to assign importance to the feature maps at different scales, so it is meaningless to use the SE-block on feature maps at a single scale.

### 5.4 Comparing with Other Methods

We do not find any specific CNN design for coronary vessel segmentation, so we compared our method with some classic CNN methods for medical image segmentation to prove that our network is superior to other methods for coronary vessel segmentation. We also compare our network with some traditional methods, including an unsupervised method and a supervised method. These competing methods are briefly introduced as follows:

1. **Frangi** [2]: Frangi filter is one of the most popular Hessian-based multiscale filter used for vessel enhancement. The output processed by Frangi filter gives

**Table 2.** Performance of all segmentation methods on our dataset

| Methods | Acc | Se | Sp | Dice |
|---|---|---|---|---|
| Frangi [2] | 0.9591 | 0.7141 | 0.9706 | 0.5532 |
| CRF [12] | 0.9676 | 0.7236 | 0.9790 | 0.6213 |
| Attention U-net [18] | 0.9908 | 0.8809 | 0.9959 | 0.8811 |
| U-net [1] | 0.9910 | **0.8905** | 0.9956 | 0.8817 |
| Resnet50 U-net | 0.9912 | 0.8846 | 0.9962 | 0.8850 |
| Ours | **0.9919** | 0.8863 | **0.9969** | **0.9035** |

a possibility that each pixel belongs to a blood vessel. To get the segmentation results, a threshold is used to form an unsupervised method. The threshold and other hyperparameters of Frangi filter were optimized using random search.

2. **CRF** [12]: Conditional random field (CRF) is a widely used image segmentation method. In this method, each pixel represents a node, and each node is connected with an edge to its neighbors. Then, energy minimization helps to achieve segmentation based on the graphs. Although the CRF-based method proposed in [12] is for the retinal vessel segmentation, the method is essentially used for the segmentation of blood vessels, so we still regard it as a comparison method.
3. **Resnet50 U-net**: In many computer vision fields, the use of pre-trained models to speed up training and improve training results has become a common method, especially when training data is insufficient. To verify if this method works on our task, we also tried to replace the encoder part of U-net with a pre-trained model and fine-tune with our data. The pre-trained model we use here is Resnet50 [17], so we name the model Resnet50 U-net.
4. **Attention U-net** [18]: Attention U-net is one of the state-of-the-art algorithms for medical segmentation. A novel attention gate (AG) was proposed in Attention U-net. AGs can implicitly learn to suppress irrelevant regions in an input image while highlighting salient features useful for the segmentation task. This network has been shown to be more effective than original U-net in many organ segmentation tasks. As a variant of U-net, we also apply this model for comparison with our method.

The final results are shown in Table 2. We visualized some results in Fig. 4. According to the results, we can see that whether supervised or unsupervised, the traditional segmentation method does not work well, and many background pixels are segmented into coronary arteries. This is because the features used in the traditional methods are handcraft features, which can only extract the appearance features but not the semantic features, so that all the structures that look like vessels are classified into coronary arteries, and such similar structures happen to be very common in DSA images.

Similarly, U-net also misclassifies a small number of background pixels into coronary arteries, especially when there are only a few positive pixels in the image. We believe this is because U-net only use the feature maps of high resolution to predict segmentation result, and the high-level semantic information are not fully utilized.

From the results, Attention U-net and Resnet50 U-net, as two commonly used methods to improve the accuracy of feature extraction, do not significantly improve the segmentation results compared with U-net. We believe that this is mainly due to the characteristics of DSA images and the characteristics of the coronary arteries. DSA images are grayscale images and relatively simple, so there is little knowledge that can be transferred from the model pre-trained on ImageNet. At the same time, as a single-class segmentation task, it is not significant to calculate the attention of coronary arteries.

According to the results in Table 2, our CAU-net is more suitable for coronary vessel segmentation than other methods because of our novel feature fusion module. Compared to the best performing network, CAU-net improved by 1.85% in the Dice score. As shown in Fig. 4, we could see that our method performed better in the details of coronary vessel segmentation and removal of impurities.

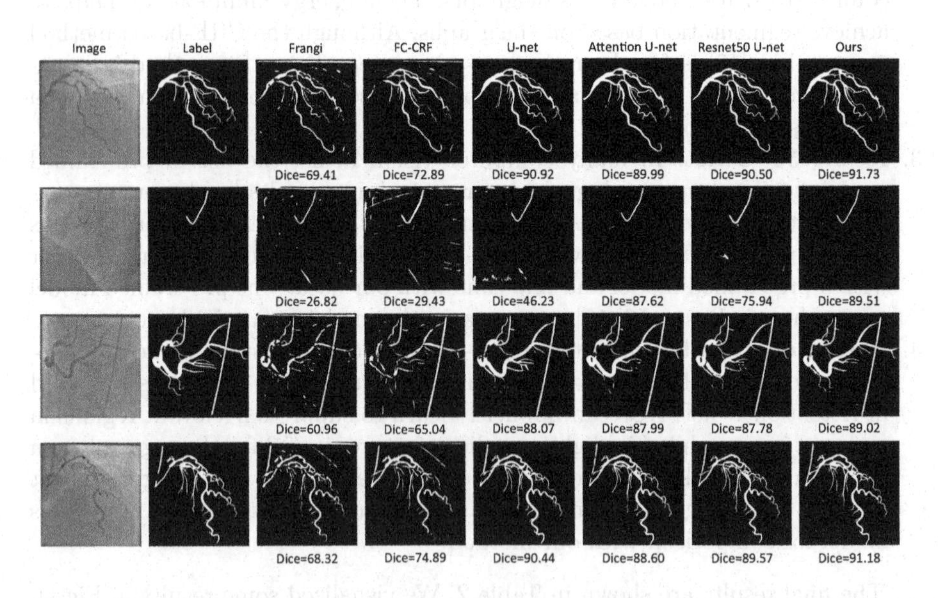


**Fig. 4.** Illustration of the segmentation results achieved by Frangi, CRF, U-net, Attention U-net, Resnet50 U-net and our method; the four rows indicate four sets of results

## 6 Conclusion

We propose a new CAU-net for automatic segmentation of coronary arteries in DSA images. In this network a new feature fusion module is proposed to fuse multi-scale features, which can significantly improve the performance in coronary artery segmentation. We also establish a coronary artery segmentation dataset with a special labeling method and compared many methods with our CAU-net using this dataset. The experimental results show that our method achieves the state-of-the-art segmentation results in coronary artery segmentation.

**Acknowledgments.** This work was supported in part by the National Key Research and Development Program of China under Grant 2019YFB1311700, the National Natural Science Foundation of China under Grants 61533016, U1913601, and 61421004, the Youth Innovation Promotion Association of CAS under Grant 2020140 and the Strategic Priority Research Program of CAS under Grant XDBS01040100.

## References

1. Ronneberger, O., Fischer, P., Brox, T.: U-net: convolutional networks for biomedical image segmentation. In: Navab, N., Hornegger, J., Wells, W.M., Frangi, A.F. (eds.) MICCAI 2015. LNCS, vol. 9351, pp. 234–241. Springer, Cham (2015). https://doi.org/10.1007/978-3-319-24574-4_28
2. Frangi, A.F., Niessen, W.J., Vincken, K.L., Viergever, M.A.: Multiscale vessel enhancement filtering. In: Wells, W.M., Colchester, A., Delp, S. (eds.) MICCAI 1998. LNCS, vol. 1496, pp. 130–137. Springer, Heidelberg (1998). https://doi.org/10.1007/BFb0056195
3. Manniesing, R., Viergever, M.A., Niessen, W.J.: Vessel enhancing diffusion: a scale space representation of vessel structures. Med. Image Anal. **10**(6), 815–825 (2006)
4. Brieva, J., Gonzalez, E., Gonzalez, F., Bousse, A., Bellanger, J.J.: A level set method for vessel segmentation in coronary angiography. In: IEEE Engineering in Medicine and Biology 27th Annual Conference, Shanghai, pp. 6348–6351 (2005)
5. Kerkeni, A., Benabdallah, A., Manzanera, A., Bedoui, M.H.: A coronary artery segmentation method based on multiscale analysis and region growing. Comput. Med. Imag. Graph. **48**, 49–61 (2016)
6. Dehkordi, M.T., Mohamad, A., Hoseini, D., Sadri, S., Soltanianzadeh, H.: Local feature fitting active contour for segmenting vessels in angiograms. IET Comput. Vis. **8**(3), 161–170 (2014)
7. Fan, J., et al.: Multichannel fully convolutional network for coronary artery segmentation in x-ray angiograms. IEEE Access **6**, 44635–44643 (2018)
8. Yang, S., et al.: Automatic coronary artery segmentation in X-ray angiograms by multiple convolutional neural networks. In: Proceedings of the 3rd International Conference on Multimedia and Image Processing, pp. 31–35 (2018)
9. Staal, J., Abramoff, M.D., Niemeijer, M., Viergever, M.A., van Ginneken, B.: Ridge-based vessel segmentation in color images of the retina. IEEE Trans. Med. Imaging **23**(4), 501–509 (2004)
10. Hoover, A., Kouznetsova, V., Goldbaum, M.: Locating blood vessels in retinal images by piecewise threshold probing of a matched filter response. IEEE Trans. Med. Imaging **19**(3), 203–210 (2000)
11. You, X., Peng, Q., Yuan, Y., Cheung, Y., Lei, J.: Segmentation of retinal blood vessels using the radial projection and semi-supervised approach. Pattern Recogn. **44**(10–11), 2314–2324 (2011)
12. Orlando, I., Prokofyeva, E., Blaschko, M.B.: A discriminatively trained fully connected conditional random field model for blood vessel segmentation in fundus images. IEEE Trans. Biomed. Eng. **64**(1), 16–27 (2016)

13. Li, Q., et al.: A cross-modality learning approach for vessel segmentation in retinal images. IEEE Trans. Med. Imaging **35**(1), 109–118 (2016)
14. Wu, Y., Xia, Y., Song, Y., Zhang, Y., Cai, W.: Multiscale network followed network model for retinal vessel segmentation. In: Frangi, A.F., Schnabel, J.A., Davatzikos, C., Alberola-López, C., Fichtinger, G. (eds.) MICCAI 2018. LNCS, vol. 11071, pp. 119–126. Springer, Cham (2018). https://doi.org/10.1007/978-3-030-00934-2_14
15. Lin, T., Dollár, P., Girshick, R., He, K., Hariharan, B., Belongie, S.: Feature pyramid networks for object detection. In: 2017 IEEE Conference on Computer Vision and Pattern Recognition, pp. 936–944. IEEE (2017)
16. Hu, J., Shen, L., Sun, G.: Squeeze-and-excitation networks. In: 2018 IEEE/CVF Conference on Computer Vision and Pattern Recognition, pp. 7132–7141. IEEE (2018)
17. He, K., Zhang, X., Ren, S., Sun, J.: Identity mappings in deep residual networks. In: Leibe, B., Matas, J., Sebe, N., Welling, M. (eds.) ECCV 2016. LNCS, vol. 9908, pp. 630–645. Springer, Cham (2016). https://doi.org/10.1007/978-3-319-46493-0_38
18. Oktay, O., et al.: Attention U-net: learning where to look for the pancreas. arXiv preprint arXiv:1804.03999 (2018)

# Combining Filter Bank and KSH for Image Retrieval in Bone Scintigraphy

Hang Xu[1,2], Yu Qiao[1,2(✉)], Yueyang Gu[1,2], and Jie Yang[1,2]

[1] Institute of Image Processing and Pattern Recognition, Department of Automation, Shanghai Jiao Tong University, Shanghai, China
{sjtu.xuhang,qiaoyu,guyueyang,jieyang}@sjtu.edu.cn
[2] Key Laboratory of System Control and Information Processing, Ministry of Education, Shanghai, China

**Abstract.** Bone scintigraphy is widely used to diagnose bone tumor and metastasis. Accurate bone scan image retrieval is of great importance for tumor metastasis diagnosis. In this paper, we propose a framework to retrieve images by integrating the techniques of texture feature extraction and supervised hashing with kernels (KSH). We first use a filter bank to extract the texture features. Then KSH is used to train a set of hashing functions with constructed features, which can convert images to hashing codes. We can obtain the most similar retrieval images by comparing Hamming distance of these hashing codes. We evaluate the proposed framework quantitatively on the testing dataset and compare it with other methods.

**Keywords:** Bone scintigraphy · Image retrieval · Filter bank · KSH

## 1 Introduction

Bone scintigraphy makes it easier for doctors to diagnose cancer and tumor metastases [17]. In the process of diagnosis, doctors often encounter uncertain cases. In these cases, if we can provide with similar images in the past cases as reference, such as Fig. 1, doctors can effectively improve the rate of successful diagnosis. The hospital has accumulated a large number of cases in the process of clinical diagnosis and treatment. But these valuable cases have not been well utilized. Therefore, we can establish a medical image retrieval system to help doctors quickly find the information about similar cases they need from the database. This can provide valuable reference information for doctors' diagnosis.

With the increasing attention on medical image processing, there have been researches about classification [2] and segmentation [4,15] of bone scan images. But the area of medical image retrieval is lack of sufficient attention, in which the

This research is partly supported by NSFC, China (No: 61375048), National Key R&D Program of China (No. 2019YFB1311503), Committee of Science and Technology, Shanghai, China (No. 19510711200).


H. Yang et al. (Eds.): ICONIP 2020, LNCS 12532, pp. 197–206, 2020.
https://doi.org/10.1007/978-3-030-63830-6_17

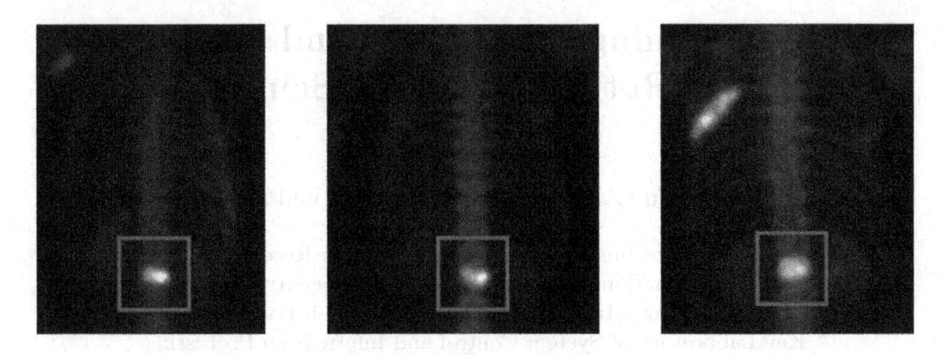

**Fig. 1.** An example of similar cases.

most $k$ similar images compared to the query image are found from the dataset, is lack of sufficient attention. It belongs to content-based image retrieval (CBIR).

Unsupervised hashing methods have been attempted for this task [16]. Hashing methods try to represent images with binary hashing codes that occupy little storage and can rapidly search images. Consequently, hashing methods is effective for practical use.

Based on how the data is used, hashing methods are mainly divided into unsupervised and supervised. One of the common unsupervised methods is Local Sensitive Hashing (LSH) [5,7]. Iterative Quantization (ITQ) [6] seeks an optimal rotating matrix to minimize the loss function by iteration. PCA Hashing [14] uses PCA to reduce the dimension of the data, and computes the projective hyperplane by top $k$ principle orientations. Supervised methods combine data's label information in the training process, thus achieve better performance. Supervised hashing with kernels (KSH) [10] proposed a kernel-based supervised algorithm that employs the label information. Latent factor hashing (LFH) [18] use a latent factor model to learn similarity-preserving codes.

Feature extraction is an important part of many image processing tasks. The features not only represent characteristics of images in some cases, but also can reduce computational complexity and occupy less storage. Apart from widely-used operators such as SIFT [12], texture feature is a promising choice for bone scan images. Common texture feature operators include Histograms of Oriented Gradient (HoG) [3] and Local Binary Patterns (LBP) [1].

In this paper, we utilize a filter bank to extract texture features. Specifically, we divide the local bone scan images into small patches and compute convolution with the filters. Feature vectors of all the patches are spliced to form the texture features of the local image. In addition, we employ supervised hashing with kernels to train hashing functions, which can transform the feature vectors to hash codes. The final retrieval is performed by comparing the Hamming distance between the query image and others.

## 2 Methodology

This paper proposed a two-stage framework for bone scan image retrieval. We utilized a filter bank [9,15] to extract the texture features of bone scan images. Combined with KSH [10], these features are effective for retrieval task. Figure 2 gives an illustration of our framework. We select a hot spot and tailor it from an image. The tailored local image is the input. Each input image will be divided into small patches, whose features will be extracted independently through the filter bank. KSH also train hashing functions with such features of the dataset. The hashing functions can transform the feature into a hash code. Finally, we find the images most similar to the input image by comparing the Hamming distance between the images.

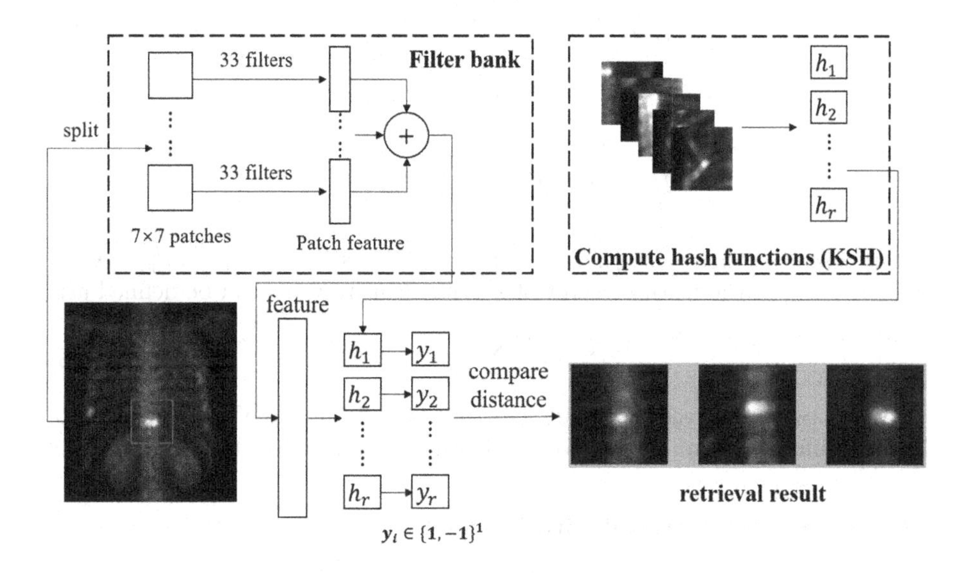


**Fig. 2.** Framework of our algorithm.

### 2.1 Texture Feature Extraction with Filter Bank

The filter bank consists of 24 directional filters, 6 Gaussian Laplacian filters and 3 Gaussian filters. We combine such filters to extract texture features of local bone scan images. These filters can be used to calculate the features of small images. However, it is not suitable for extracting features of large images directly with filter bank, such as the local bone scan images. Therefore, we divide the local images into several $m \times m$ patches, $m \ll n$ (size of local images), the size of which is suitable for filtering directly.

Sobel operator is a well known directional filter and will be used for feature extraction in our approach. Sobel operator is a classical linear filter for edge detection, which can be regarded as the measurement of image changes in vertical and horizontal directions. Suppose the small patch is $I^k, k \in \{1, ..., n^2/m^2\}$. Directional filter $f_{Direction}$ combines two components $G_x$ and $G_y$:

$$G_x = \begin{vmatrix} -1, 0, 1 \\ -2, 0, 2 \\ -1, 0, 1 \end{vmatrix} * I^k, \quad G_y = \begin{vmatrix} 1, & 2, & 1 \\ 0, & 0, & 0 \\ -1, & -2, & 1 \end{vmatrix} * I^k \tag{1}$$

$$f_{Direction} * I^k = \sqrt{G_x^2 + G_y^2} \tag{2}$$

In addition, Gaussian Laplacian filter $f_{LoG}$ and Gaussian filter $f_{Gaussian}$ have similar usage:

$$f_{LoG} * I^k = \begin{vmatrix} 0, 0, & 1, & 0, 0 \\ 0, 1, & 2, & 1, 0 \\ 1, 2, & -16, & 2, 1 \\ 0, 1, & 2, & 1, 0 \\ 0, 0, & 1, & 0, 0 \end{vmatrix} * I^k \tag{3}$$

$$f_{Gaussian} * I^k = \frac{1}{16} \begin{vmatrix} 1, 2, 1 \\ 2, 4, 2 \\ 1, 2, 1 \end{vmatrix} * I^k \tag{4}$$

Suppose the filter is $f_i, i \in \{1, 2, ..., 33\}$, $\mathbf{x^k} = [x_1^k, ..., x_{33}^k]^T, k \in \{1, ..., n^2/m^2\}$ is the feature vector of $I^k$. The feature value can be defined as:

$$x_i^k = \frac{1}{m^2} \sum (f_i * I^k) \tag{5}$$

We splice feature vectors of all the patches to form the effective feature of the input image. It can be used for KSH training and case testing.

### 2.2 Supervised Retrieval with Kernels

The algorithm of supervised retrieval with kernels can generate hashing function quickly and achieve enough accuracy with short hashing code. Therefore, this algorithm is applied in many medical image retrieval systems. Bone scan image retrieval system will also utilize KSH to complete the retrieval task.

Given feature set of images in the database $X = \{\mathbf{x_1}, ..., \mathbf{x_n}\} \in \mathbb{R}^d$. We denote the feature vector of an unlabeled image $\mathbf{x} \in \mathbb{R}^d$, a set of hash functions $H = \{h_1, ..., h_r\}, h : \mathbb{R}^d \mapsto \{1, 0\}^1$, which can compute $r$-bit hashing code $\mathbf{y} = \{y_1, ..., y_r\}$. Each hashing code can be expressed as follows:

$$y_k = sgn(h_k(\mathbf{x})) \tag{6}$$

There is a problem that samples may be linearly indivisible. This can be solved by introducing Gaussian kernel function. Another advantage of introducing a kernel function is to reduce the number of parameters, which should be learned, by selecting fewer anchors. KSH algorithm extracts $m$ anchors

$\{\mathbf{x}_{(1)}, ..., \mathbf{x}_{(m)}\}$ $(m \ll n)$ from the labeled samples to compute the kernel function. We select the kernel function $\kappa : \mathbb{R}^d \times \mathbb{R}^d \mapsto \mathbb{R}$ [8] . The hashing function with kernel function is defined as follows:

$$h_k(\mathbf{x}) = \sum_{j=1}^{m} \kappa(\mathbf{x}_{(j)}, \mathbf{x}) a_j - b \tag{7}$$

where $a_j$ and $b$ are both undetermined coefficients. In order to ensure that the generated hashing code has enough information, each bit of the hashing code $y_k$ should satisfy $\sum_{i=1}^{n} y_k(\mathbf{x_i}) = 0$ [11,13], so $b$ can be defined as follows:

$$b = \frac{1}{n} \sum_{i=1}^{n} \sum_{j=1}^{m} \kappa(\mathbf{x}_{(j)}, \mathbf{x}) a_j \tag{8}$$

In this way, we can express the hash function as follows:

$$h_k(\mathbf{x}) = \sum_{j=1}^{m} (\kappa(\mathbf{x}_{(j)}, \mathbf{x}) - \frac{1}{n} \sum_{i=1}^{m} \kappa(\mathbf{x}_{(j)}, \mathbf{x})) a_j = \mathbf{a}^T \overline{k}(x) \tag{9}$$

where $\overline{k}(x) = [\kappa(\mathbf{x_1}, \mathbf{x}) - \mu_1, ..., \kappa(\mathbf{x_m}, \mathbf{x}) - \mu_m]^T : \mathbb{R}^d \mapsto \mathbb{R}^m$, in which $\mu_j = \sum_{i=1}^{m} \kappa(\mathbf{x}_{(j)}, \mathbf{x_i}, \mathbf{x_i})/n$. $\mathbf{a} = [a_1, ..., a_m]^T$ is the coefficient of the kernel function and decides the final hashing functions. Therefore, we need to calculate the optimal $\mathbf{a}$ depending on the labeled data. In order to establish the supervised information matrix $S \in \mathbb{R}^{l \times l}$, we select $l$ samples from the sample set. $S$ can be expressed as follows:

$$S(i,j) = \begin{cases} 1, & i, j \, are \, similar \\ -1, & otherwise \end{cases} \tag{10}$$

It is difficult to minimize the Hamming distance $D_y(\mathbf{x_i}, \mathbf{x_j})$, so the problem is transformed into the problem of optimizing the inner product of codes. The inner product of two codes $code_r(\mathbf{x_i})$ and $code_r(\mathbf{x_j})$ is:

$$\begin{aligned} code_r(\mathbf{x_i}) \cdot code_r(\mathbf{x_j}) &= \sum (y_k(\mathbf{x_i}) = y_k(\mathbf{x_j})) - \sum (y_k(\mathbf{x_i}) \neq y_k(\mathbf{x_j})) \\ &= (r - D_y(\mathbf{x_i}, \mathbf{x_j})) - D_y(\mathbf{x_i}, \mathbf{x_j}) = r - 2D_y(\mathbf{x_i}, \mathbf{x_j}) \end{aligned} \tag{11}$$

Therefore, the problem is equivalent to solving the maximum value of the codes' inner product. The objective function of optimization is defined as the following least squares form:

$$\min_{H_l \in \{1,-1\}^{l \times r}} Q = \left| \frac{1}{r} H_l H_l^T - S \right|_F^2 \tag{12}$$

where $H_l = sgn(\overline{K}_l A) \in \mathbb{R}^{l \times r}$, representing $r$-bit Hamming code of $l$ supervised samples. $A = [\mathbf{a_1}, ..., \mathbf{a_r}] \in \mathbb{R}^{m \times r}$, is the parameter matrix of the model.

$\overline{K}_l = [\overline{k}(x_1), ..., \overline{k}(x_l)]^T \in \mathbb{R}^{l \times m}$, is the matrix of kernel functions. In this way, the target formula can be transformed into:

$$\min_{A \in \mathbb{R}^{m \times r}} Q(A) = \left| \frac{1}{r} sgn(\overline{K}_l A) sgn(\overline{K}_l A)^T - S \right|_F^2 \tag{13}$$

Finally, the greedy algorithm is used to optimize the kernel function coefficients of each hash function: $\mathbf{a_i} = [a_{i1}, ..., a_{im}]$.

## 3 Experiments

In this section, we compare our proposed framework with other algorithms to test the effectiveness of our method. We start by presenting the details of experimental settings. Comparison results will then be demonstrated.

### 3.1 Dataset and Settings

The dataset used in this paper is from the Department of nuclear medicine, Renji Hospital, Shanghai Jiao Tong University. In order to build a bone scan image retrieval system, we need to establish a local image database of original images. Besides, we need to label the relevant information about the local image, such as diagnosis results and location in the original image, etc.

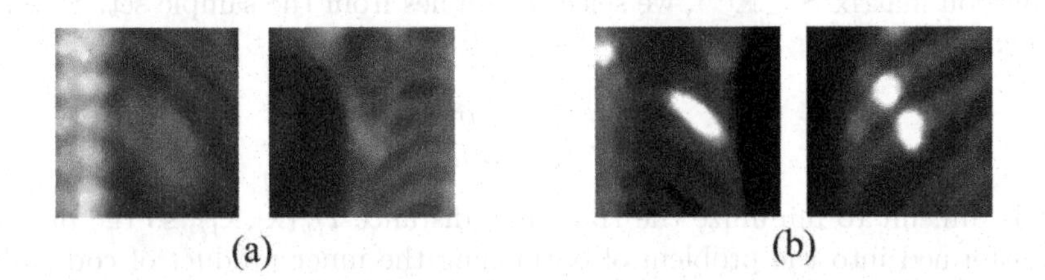


**Fig. 3.** Examples of local normal and sick images: (a) normal; (b) sick.

We select 628 abnormal thoracic bone scan images and 786 normal ones. Then we randomly cut 49 × 49 local images from thoracic images. A total of 8091 local images are extracted and labeled. Among them, those cropped from normal images are automatically labeled as no hot spots. Those local images from sick images are labelled according to doctor verification. Finally, a total of 3768 local images were labeled as containing hot spots, while 4323 local images without hot spots were labeled. Figure 3 illustrates some examples of local normal and sick images.

### 3.2 Quantitative Comparison

Each local image is divided into small blocks for using filter bank. Firstly, texture features are extracted from each small block. Then feature vectors are spliced to form the feature vector of the final image. Among them, each block and the matrix after convolution of the filter is averaged as a feature.

All samples are divided into training set and test set. Finally, we train 48 hashing functions by using supervised hashing functions with kernels. We use these functions to extract hashing code from image in database in order to build a set of bone scanning image retrieval system. When using the query image for retrieval in the database, the image retrieval system first extracts the features of the query image. The system then maps the extracted features into hashing codes using hashing functions. Finally it searches for $K$ retrieved images with the minimum Hamming distance from the database.

**Table 1.** Performance comparison on bone scan image retrieval.

| Method | K = 3 | K = 5 | K = 11 |
|---|---|---|---|
| SIFT [12] + KSH | 0.5599 | 0.5832 | 0.5877 |
| LBP [1] + KSH | 0.5821 | 0.5957 | 0.5872 |
| HoG [3] + KSH | 0.6087 | 0.6014 | 0.6167 |
| Sparse coding + KSH | 0.6102 | 0.6370 | 0.6394 |
| Our method | 0.6258 | 0.6354 | **0.6587** |

After the establishment of local bone scan image database, several retrieval methods apart from our method are also tested in this paper, such as SIFT [12] + KSH, LBP [1] + KSH, HoG [3] + KSH and Sparse coding + KSH. 428 images are randomly selected from the local image database to form a test set. $K$ images with the most similar features are extracted and retrieved from the database. If we calculate the Hamming distance between the hash code of the query image and the result images, we can judge whether two images are similar by evaluating the distance.

The accuracy of retrieval results is shown in Table 1. It demonstrates that our method implements more accurate bone scan image retrieval and outperforms the other methods. Among them, when $K = 11$, the accuracy is the highest.

**Table 2.** Performance comparison on different bits of hashing codes.

| Code length | 24-bit | 32-bit | 48-bit | 64-bit |
|---|---|---|---|---|
| Accuracy (K = 11) | 0.644 | 0.6457 | 0.6587 | 0.6695 |

Table 2 presents the results about the influence of the hashing code's length. The increasing length of hashing codes actually improves the retrieval accuracy slightly. But the impact of code length on the retrieval results is little.

### 3.3 Subjective Comparison

In order to test visual evaluation of our method, we select a query image and return the top 5 similar images using our method and other algorithms. The results are shown in Fig. 4. This figure demonstrates that our method perform better than other methods. LBP + KSH and HoG + KSH have poor retrieval results. Besides, only one of five results is desired with SIFT + KSH method. Sparse coding method may find the local sick images, which stay in the same region compared with the query image. Our method achieves much better results in this case. Most of the retrieval results are quite similar to the query image.

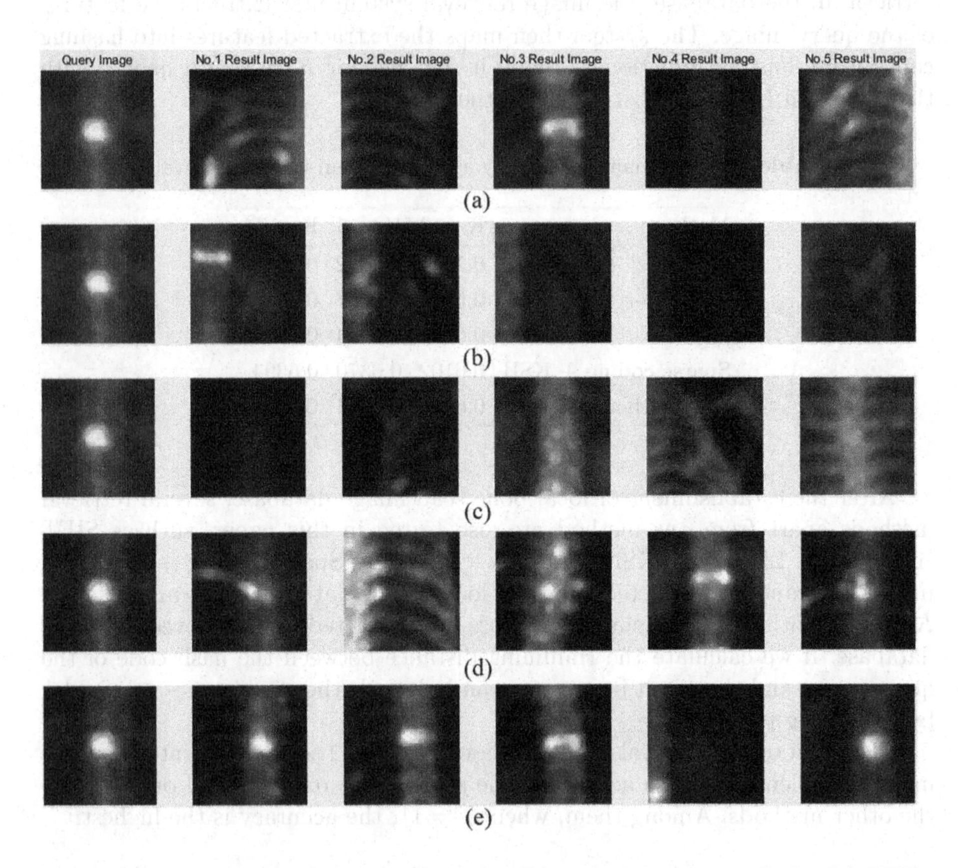


**Fig. 4.** Comparison of retrieval results with 5 feature extracting methods: (a) SIFT; (b) LBP; (c) HoG; (d) Sparse coding; (e) Our method.

## 4 Conclusion

In this paper, we present a framework combining texture feature extraction with a supervised hashing retrieval method for accurate bone scan image retrieval. In order to obtain more effective texture features, we take a filter bank to compute the features. The images need to be divided into small patches, in which feature vectors will be extracted with the filter bank. These feature vectors will be used to construct the final effective feature. The approach of supervised hashing with kernels will compute the hashing function that will output the hashing codes. Experimental results show that our framework outperform other methods and the filter bank is helpful for effective feature extraction.

## References


1. Ahonen, T., Hadid, A., Pietikainen, M.: Face description with local binary patterns: application to face recognition. IEEE Trans. Pattern Anal. Mach. Intell. **28**(12), 2037–2041 (2006)
2. Chang, Q., et al.: Adaptive detection of hotspots in thoracic spine from bone scintigraphy. In: Lu, B.-L., Zhang, L., Kwok, J. (eds.) ICONIP 2011. LNCS, vol. 7062, pp. 257–264. Springer, Heidelberg (2011). https://doi.org/10.1007/978-3-642-24955-6_31
3. Dalal, N., Triggs, B.: Histograms of oriented gradients for human detection. In: CVPR 2005–2005 IEEE Computer Society Conference on Computer Vision and Pattern Recognition (CVPR), San Diego, CA, USA, pp. 886–893. IEEE (2005)
4. Geng, S., Ma, J., Niu, X., Jia, S., Qiao, Y., Yang, J.: A mil-based interactive approach for hotspot segmentation from bone scintigraphy. 2016 IEEE International Conference on Acoustics. Speech and Signal Processing (ICASSP), Shanghai, pp. 942–946. IEEE (2016)
5. Gionis, A., Indyk, P., Motwani, R.: Similarity search in high dimensions via hashing. In: Proceedings of the 25th International Conference on Very Large Data Bases (VLDB), Edinburgh, pp. 518–529. ACM (1999)
6. Gong, Y., Lazebnik, S., Gordo, A., Perronnin, F.: Iterative quantization: a procrustean approach to learning binary codes for large-scale image retrieval. IEEE Trans. Pattern Anal. Mach. Intell. **35**(12), 2916–2929 (2013)
7. Indyk, P., Motwani, R.: Approximate nearest neighbors: towards removing the curse of dimensionality. Theor. Comput. **1**, 604–613 (2000)
8. Kulis, B., Grauman, K.: Kernelized locality-sensitive hashing. IEEE Trans. Pattern Anal. Mach. Intell. **34**(6), 1092–1104 (2012)
9. Leung, T., Malik, J.: Representing and recognizing the visual appearance of materials using three-dimensional textons. Int. J. Comput. Vis. **43**(1), 29–44 (2001)
10. Liu, W., Wang, J., Ji, R., Jiang, Y., Chang, S.: Supervised hashing with kernels. In: 2012 IEEE Conference on Computer Vision and Pattern Recognition (CVPR), Providence, RI, USA, pp. 2074–2081. IEEE (2012)
11. Liu, W., Wang, J., Kumar, S., Chang, S.: Hashing with graphs. In: Proceedings of the 28th International Conference on Machine Learning (ICML), Bellevue, Washington, USA, pp. 1–8. ACM (2011)
12. Lowe, D.G.: Distinctive image features from scale-invariant keypoints. Int. J. Comput. Vis. **60**(2), 91–110 (2004)
13. Wang, J., Kumar, S., Chang, S.: Semi-supervised hashing for large-scale search. IEEE Trans. Pattern Anal. Mach. Intell. **34**(12), 2393–2406 (2012)
14. Wang, X., Zhang, L., Jing, F., Ma, W.: Annosearch: image auto-annotation by search. In: 2006 IEEE Computer Society Conference on Computer Vision and Pattern Recognition (CVPR), New York, USA, pp. 1483–1490. IEEE (2006)

15. Xu, H., Geng, S., Qiao, Y., Xu, K., Gu, Y.: Combining CGAN and mil for hotspot segmentation in bone scintigraphy. In: ICASSP 2020–2020 IEEE International Conference on Acoustics. Speech and Signal Processing (ICASSP), Barcelona, pp. 1404–1408. IEEE (2020)
16. Xu, K., Qiao, Y., Niu, X., Fang, X., Han, Y., Yang, J.: Bone scintigraphy retrieval using sift-based fly local sensitive hashing. In: 2018 IEEE 27th International Symposium on Industrial Electronics (ISIE), Cairns, pp. 735–740. IEEE (2018)
17. Yin, T.K., Chiu, N.T.: A computer-aided diagnosis for locating abnormalities in bone scintigraphy by a fuzzy system with a three-step minimization approach. IEEE Trans. Med. Imaging **23**(5), 639–654 (2004)
18. Zhang, P., Zhang, W., Li, W., Guo, M.: Supervised hashing with latent factor models. In: Proceedings of the 37th international ACM SIGIR conference on Research & development in information retrieval (SIGIR), Gold Coast, Australia, pp. 173–182. ACM (2014)

# Contrastive Learning with Hallucinating Data for Long-Tailed Face Recognition

Zhao Liu[1(✉)], Zeyu Zou[1], Yong Li[1], Jie Song[2], Juan Xu[1], Rong Zhang[1], and Jianping Shen[1]

[1] Ping An Life Insurance Of China, Ltd., Shenzhen, Guangdong, China
{liuzhao556,zouzeyu313,liyong457,xujuan635,zhangrong272, shenjianping324}@pingan.com.cn
[2] Zhejiang University, Hangzhou, Zhejiang, China
sjie@zju.edu.cn

**Abstract.** Face recognition has been well studied over the past decades. Most existing methods focus on optimizing the loss functions or improving the feature embedding networks. However, the long-tailed distribution problem, i.e, most of the samples belong to a few identities, while the remaining identities only have limited samples, is less explored, where these datasets are not fully utilized. In this paper, we propose a learning framework to balance the long-tailed distribution problem in public face datasets. The proposed framework learns the diversity from head identity samples to generate more samples for identifying persons' identities in the tail. The generated samples are used to finetune face recognition models through a contrastive learning process. The proposed framework can be adapted to any feature embedding networks or combined with different loss functions. Experiments on both constrained and unconstrained datasets have proved the efficiency of the proposed framework.

**Keywords:** Face recognition · Long-tailed distribution · Contrastive learning

## 1 Introduction

Face recognition (FR) has been an extensively studied subject in computer vision. With the success of deep learning based methods and large scale face datasets [1] available for training, FR has obtained significant improvement in recent years. However, some problems still remain unsolved, for instance, FR in unconstrained environments such as surveillance scenes. Since most of the existed FR models are trained on high-quality web-crawled faces, when applied to unconstrained FR, accuracy usually drops obviously.

Besides the unconstrained FR in real applications, the long-tailed distribution has been a non-negligible problem in existed FR datasets. As shown in Fig. 1, most of the collected face images belong to a few head identities, while other identities only have limited images. This problem is especially apparent in web-crawled datasets. In general FR training frameworks, using such data for feature


H. Yang et al. (Eds.): ICONIP 2020, LNCS 12532, pp. 207–217, 2020.
https://doi.org/10.1007/978-3-030-63830-6_18

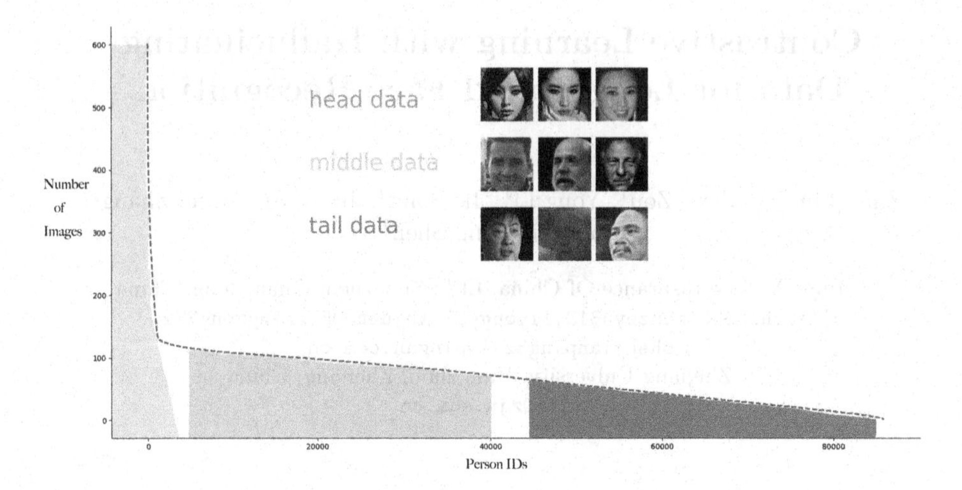


**Fig. 1.** Visualization of the long-tailed distribution problem for the triaining set of MS1M [1] database, the identity numbers are sorted in descending order according to number of samples. Most of the samples belong to a few identities, while the remaining identities only have limited samples for each.

extracting will introduce bias, that is, the samples of the tail identities are so sparse that they are easily to be confused with huge number of samples from head identities.

A common way to overcome the long-tailed data distribution is to increase samples of tail identities, either by data re-sampling or augmentation. Also, Generative Adversarial Networks (GAN) can be used to generate many good cases, but due to the unstable performance for large amount of long tail face identities. However, such methods usually introduce noisy samples, making the distribution more artificial.

Different from previous methods, in this paper, we propose a framework to balance the long-tailed distribution in existed FR datasets at the feature embedding level. It consists of the data hallucination module and the contrastive learning module. The data hallucination module learns the diversity (including variations of orientation, illumination, occlusion and blurry) from sampls in head identities, then the learned diversity is transferred to the tail identities to generate more training samples. The contrastive learning module works with a modified loss function and uses the augmented samples to finetune the FR models. The main contributions of our proposed method include:

- We propose a novel framework to solve the long-tailed distribution problem in public FR datasets. The proposed framework is able to transfer the data diversity in head identities to desired tail identities to incease both number and diversity thus the training dataset is balanced to some extent.
- We design a contrastive learning process to utilize the augmented samples for fine-tuning FR models, along with a modified form of the loss function in training the feature embedding networks.

- We perform several experiments on public face recognition datasets, including constrained and unconstrained scenes, where our proposed method is able to outperform most state-of-the-art face recognition methods on public FR datasets.

## 2 Related Work

The proposed framework in this paper mainly focuses on Face Recognition and Contrastive Learning. In this section, we will review previous related work briefly.

### 2.1 Face Recognition

Early face recognition methods, such as [2,3], relied on hand-crafted features and classifiers, which were restricted in tiny database and lacked generalization ability. Most of the CNN based methods focused on the optimization of loss functions for better discriminative features, including triplet loss [14], cosine loss [4], center loss [5] and arc-loss [6]. The novel Probabilistic Face Embeddings in [7] worked on the uncertainty of face features and proposed a module to estimate distribution of features in the latent space. Besides, pose-invariant FR problem has also attracted many researchers, such as work in [8,9].

In real-world applications, face recognition algorithms are always deployed in surveillance systems, which bring new challenges such as identities in distance with low-resolution face images. Some methods [10,11] used high resolution and low resolution together to learn a unified feature space, while others [12,13] tried to recover higher resolution face images. However, the recovery process usually inevitably bring distortions, making it hard to increase recognition accuracy.

### 2.2 Contrastive Learning

The goal of contrastive learning is to learn a feature representation which is able to accumulate data with the same classes tight and push apart data with different classes. Earlier idea includes metric learning and triplet losses [14,15]. In [16], the authors proposed an efficient contrastive loss and was followed by some famous framework such as SimCLR [17] and MoCo [18]. Most of the existing contrastive learning based methods [19,20] learned the data difference through a self-supervised process. These methods ignored the variations existed in the same class, thus would result in mis-classifcation. Recently, Khosla et al. proposed the supervised contrastive learning [21], which focused on optimizating contrastive loss on the class level rather than image level. However, how to accelerate the process of contrastive learning and keep the stability still remains an unsolved problem.

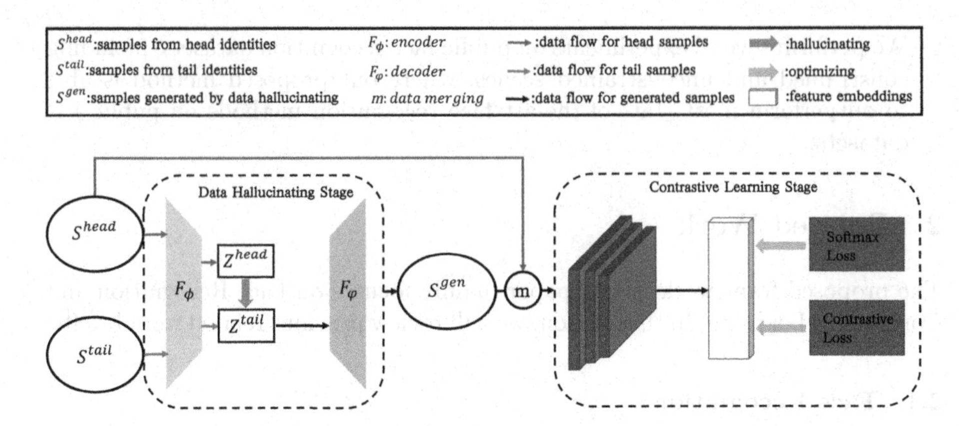


**Fig. 2.** Framework of the proposed method. The data hallucinating module learns and transfers the data diversity from samples of head identities, and the contrastive learning module is used to finetune the feature embedding network.

# 3 Method

## 3.1 Framework

Figure 2 shows the whole process of our proposed method for surveillance face recognition. The proposed method consists of the data hallucinating module and the contrastive learning module. Given the input face image datasets, we first divide them to the head samples $S^{head} = \{I_1^{head}, ..., I_M^{head}\}$ and the tail samples $S^{tail} = \{I_1^{tail}, ..., I_N^{tail}\}$. The data hallucinating module learns the attributes hiding in samples of head identities, and then uses them to generate more diverse samples $S^{gen}$ from the tail samples. Additionally, a contrastive learning module is used to manipulate the few confused samples generated from hallucinating module, and guide the fine-tuning process.

## 3.2 Data Hallucinating

We use an encoder-decoder structure network to perform the representation disentangling. The encoder $F_\phi$ is the pre-trained feature embedding network. The input samples from head identities can be encoded as a distribution of representations:

$$F_\phi(Z^{head}) = \frac{1}{M}\sum_{m=1}^{M} \mathcal{E}(Z^{head}|I_k^{head}), \tag{1}$$

where M is the total number of samples in $S^{head}$. $Z^{head} = \{Z_1^{head}, ..., Z_r^{head}\}$ are the real-valued code vectors of individual representations. The decoder $F_\varphi$ decodes the combination of code vectors to reconstruct original samples with different attributes. We follow the strategies proposed in [22] to disentangle the representations. Based on the disentangled repesentations, we can generate new samples for tail identities as:

$$S^{gen} = F_{\varphi}[\sum_{r=1}^{R}(\alpha_r Z_r^{head} + Z_{img}^{tail})], \tag{2}$$

where $Z_{img}^{tail}$ is the original face feature embedding of tail identities. $\alpha_r$ is the parameter for controlling the degree of disentangled attribute code vectors.

### 3.3 Contrastive Learning

The goal of fine-tuning FR models is to learn a general feature representation space, in which two individual images are far away if they belong to different identities, while they are nearby if belong to the same identity.

After the data hallucination, we are able to obtain mass of samples with labeled identities. Inspired by [21], we use these samples to perform a supervised contrastive learning process. For each individual training image, we add the class level supervised loss:

$$L^{contrastive} = \frac{-1}{N_{c_j}-1}\sum_{j=1}^{N}\mathbb{1}_{n\neq j}\mathbb{1}_{c_n=c_j} log\frac{e^{F_n\cdot F_j/\tau}}{\sum_{k=1}^{N}\mathbb{1}_{n\neq k}e^{F_n\cdot F_k/\tau}}, \tag{3}$$

where $N_{c_j}$ is the number of images with the same identity as $I_n$, $\mathbb{1}_B \in \{0,1\}$ is a binary conditional judgement function which is 1 if B satisfy the judgement, $\tau$ is a scalar temperature parameter, $F_n \cdot F_j$ represents the inner computation between the extracted feature vectors $F_n$ and $F_j$. Finally, we use a combination of the contrastive learning loss and the original softmax loss as the optimization loss L for the training process:

$$L = \lambda_1 L^{softmax} + \lambda_2 L^{contrastive}, \tag{4}$$

where $\lambda_1$ and $\lambda_2$ represent the weights for controlling the two losses.

### 3.4 Training and Inference

In order to obtain a stable feature extractor, we adopt a multi stage pre-train policy. In the training process, we first process the supervised contrastive learning on the samples belonging to head identities. The sufficient variations of these samples make it easy to get a stable feature extractor. We then finetune the model on the orginal images of tail identities to get a relatively stable model. Finally, we use the hallucinated images to further optimize the model. In the inference process, we remove the final classfication layer of the trained FR model and use the embedding network to extract features for test samples and reference samples, and we use Euclidean distance to evaluate the similarity of them.

**Table 1.** Comparison with the SOTA methods on constrained datasets.

| Method | Training data | LFW | CFP-FP | AgeDB-30 |
|---|---|---|---|---|
| FaceNet [14] | 200 M | 99.65 | – | – |
| SphereFace [23] | 0.5 M | 99.11 | 94.38 | 91.70 |
| CenterFace [5] | 0.7 M | 99.28 | 93.84 | – |
| CosFace [4] | 5.0 M | 99.51 | 95.44 | 94.56 |
| PFE [7] | 4.4 M | 99.82 | 93.34 | – |
| Baseline | 5.8 M | 99.83 | 92.24 | 97.82 |
| Proposed | 5.8 M | **99.83** | **97.46** | **97.85** |

**Table 2.** Comparison with the SOTA methods on IJB-C dataset.

| Method | Training data | IJB-C(TPR@FPR) | | | | |
|---|---|---|---|---|---|---|
| | | 0.0001% | 0.001% | 0.01% | 0.1% | 1% |
| VggFace2 [24] | 3.3 M | – | 74.7 | 84.0 | 91.0 | 96.0 |
| Multicolumn [26] | 3.3 M | – | 77.1 | 86.2 | 92.7 | 96.8 |
| DCN [25] | 3.3 M | – | – | 88.5 | 94.7 | 98.3 |
| PFE [7] | 4.4 M | – | 89.6 | 93.3 | 95.5 | 97.2 |
| Baseline | 5.8 M | 77.67 | 87.65 | 91.89 | 94.83 | 96.8 |
| Proposed | 5.8 M | **85.02** | **91.06** | **94.30** | **96.31** | **97.8** |

# 4 Experiments

## 4.1 Experiment Settings

**Datasets and Implementation Details.** To obtain the best recognition results, we used the cleaned and augmented MS-Celeb-1M datasets proposed by [6] as the training set. For the test sets, we use LFW and CFP to evaluate the performance on constrained scenes, and IJB-C as the unconstrained scene. We use the standard TAR@FAR as the evaluation metric. We sort the identities according to the number of samples, and use 2000 identities as the head set, each of which has 300 images correspondingly. We choose Arcface [6] as our baseline for comparison.

## 4.2 Performance on Constrained Datasets

Table 1 shows the performance of different SOTA methods and our proposed method on contrained datasets. From the results we can see that the proposed method is able to outperform all the methods in comparison. In fact, most of the SOTA face recognition methods are able to achieve very high accuracy on LFW dataset, where no exception exists for our proposed method. On the other hand, our proposed method shows obvious improvements on the relatively challenging CFP-FP and AgeDB_30 datasets. Moreover, compared with the baseline Arcface, our proposed method shows significant improvement on all evalutaions, which cetifies the improvements by adding our proposed framework.

**Table 3.** Comparison of proposed framework with different backbones.

| Model | Backbone | IJB-C(TPR@FPR) | | | | |
|---|---|---|---|---|---|---|
| | | 0.0001% | 0.001% | 0.01% | 0.1% | 1% |
| Base | Resnet100 | 87.61 | 93.09 | 95.42 | 96.90 | 97.91 |
| Proposed | | 88.46 | 94.14 | 96.13 | 97.38 | 98.33 |
| Base | Mobilenet-V1 | 54.01 | 70.45 | 80.37 | 87.33 | 92.87 |
| Proposed | | 65.10 | 79.38 | 87.93 | 92.76 | 95.92 |

**Table 4.** Comparison of proposed framework with different loss functions. (Backbone model is ResNet50.)

| Model | Loss | IJB-C(TPR@FPR) | | | | |
|---|---|---|---|---|---|---|
| | | 0.0001% | 0.001% | 0.01% | 0.1% | 1% |
| Base | Softmax | 35.48 | 56.61 | 76.45 | 89.11 | 95.90 |
| Proposed | | 47.89 | 67.15 | 81.28 | 90.94 | 96.42 |
| Base | CenterFace [5] | 42.02 | 62.60 | 78.40 | 89.73 | 96.11 |
| Proposed | | 52.07 | 69.24 | 82.05 | 91.45 | 96.47 |
| Base | CosFace [4] | 82.47 | 88.86 | 92.90 | 95.56 | 97.30 |
| Proposed | | 84.08 | 90.63 | 93.63 | 95.79 | 97.32 |

### 4.3 Performance on Unconstrained Datasets

IJB-C is an challenging unconstrained dataset with verification protocal of more heterogeneous identity pairs, making it possible to evaluate TAR@FAR at different levels. Table 2 shows the comparison of our proposed method with different methods on IJB-C dataset. Compared with the results on constrained datasets, our method shows more significant improvement over all the levels. To be noticed, the proposed method is able to get TAR improvement of 7.35% at FAR 0.0001%. The results certify our proposed method is able to achieve better results on more challenging occasions.

### 4.4 Ablation Studies

**Performance on Different Backbones.** Table 3 shows the performance on IJB-C dataset with ResNet100 and MobileNet-V1. From the results we can see by adding the proposed framework, we can obtain comprehensively improvements for both backbones compared with the base models.

**Performance with Different Losses.** We add the proposed method with several general loss functions, including the original Softmax loss, Center loss, and Cosine loss [4]. Results of the base models and the modified methods on IJB-C dataset are shown in Table 4. By adding the proposed framework, we are able to obtain significant improvements on the selected base models.

### 4.5 Visualization Results

**Impact on Data Distributions.** Figure 3 shows the t-SNE visualization results of distributions of the feature emmbeddings on selected tail identities. From the results we can see that the data hallucinating module can generate more samples with abundant diversity, but it might create some confusions among nearby samples. By using the contrastive learning module, this situation is improved. We can pull the feature embeddings belonging to the same identities closer while pushing the others beloning to different identities further away.

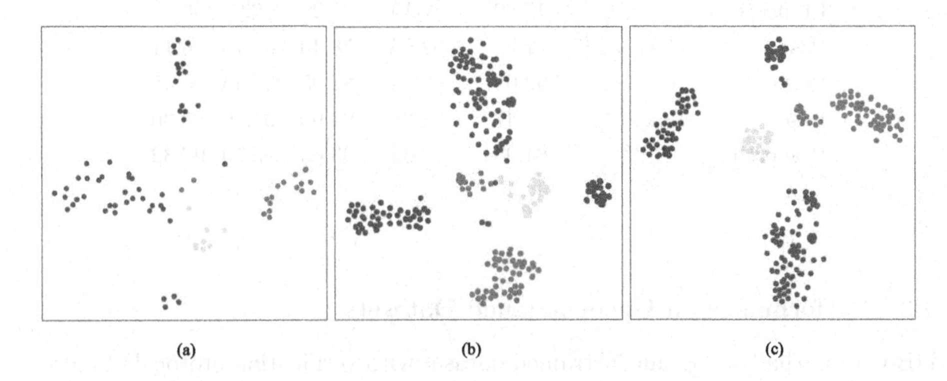


**Fig. 3.** t-SNE visualization results of the feature embeddings in 2D space: different colors indicate different identities. (a) Original samples from the tail identities; (b) Generated samples by using the data hallucinating module; (c) Re-clusterred samples after the contrastive learning module.

**Visualization of Generated Faces.** Figure 4 shows some generated sample faces of the tail identities. By using the data hallucinating module and controlling the degree of attributes correspondingly, we are able to generate various samples from the disentangled representations of the head identities. Some attributes are hard to be disentangled individually, for example the mouth opening in the second row and the face orientation in the last row. Nevertheless, these generated face images play an important role in improving FR accuracy since they enrich the variation of the training datasets and therefore the feature embedding space.

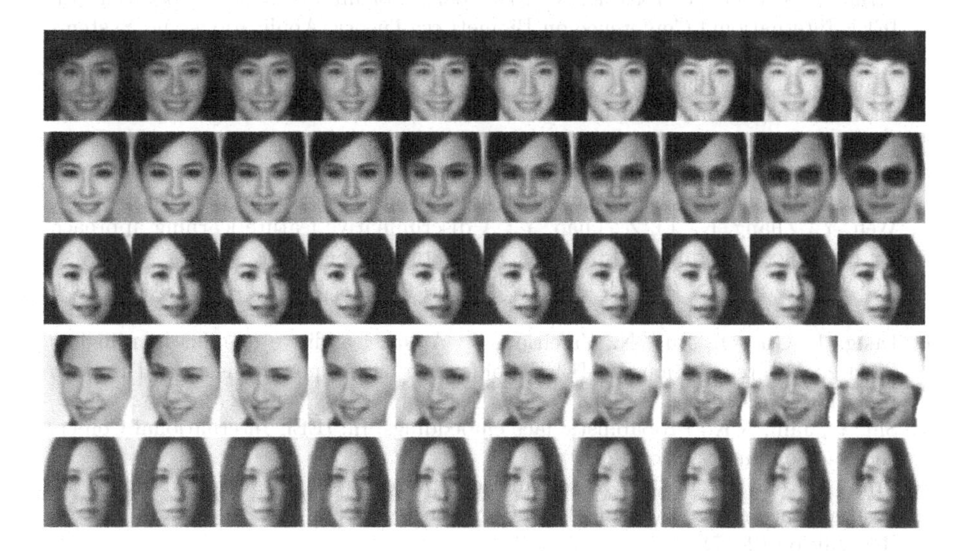

**Fig. 4.** Visualization results of the data hallucinating module: from left to right are original image and successively generated images with different degrees of attributes. From the top to bottom are generated face images corresponding to different lighting conditions, with or without glasses, different orientations, with or without hat and different color of hair.

## 5 Conclusion

In this work, we propose a framework to handle the long-tailed distribution problem existed in public face recognition datasets. This framework utilizes an encoder-decoder structure to transfer the data diversity from head identities to tail identities. It then uses a contrastive learning process to finetune the FR models. The proposed framework can be used as an independent module adapted to any feature embedding networks and loss functions. We have performed several experiments on constrained and unconstrained datasets. By adding the proposed framework, the existing face recognition methods have shown significant improvements.

## References


1. Guo, Y., Zhang, L., Hu, Y., He, X., Gao, J.: MS-Celeb-1M: a dataset and benchmark for large-scale face recognition. In: Leibe, B., Matas, J., Sebe, N., Welling, M. (eds.) ECCV 2016. LNCS, vol. 9907, pp. 87–102. Springer, Cham (2016). https://doi.org/10.1007/978-3-319-46487-9_6
2. Struc, V., Gajsek, R., Pavesic, N.: Principal gabor filters for face recognition. In: IEEE International Conference on Biometrics: Theory, Applications and Systems (2009)
3. Manuele, B., Castellani, U., Murino, V.: Using hidden Markov models and wavelets for face recognition. In: IEEE International Conference on Image Analysis and Processing (2003)
4. Wang, H., et al: Cosface: large margin cosine loss for deep face recognition. In: IEEE Conference on Computer Vision and Pattern Recognition (2018)
5. Wen, Y., Zhang, K., Li, Z., Qiao, Y.: A discriminative feature learning approach for deep face recognition. In: Leibe, B., Matas, J., Sebe, N., Welling, M. (eds.) ECCV 2016. LNCS, vol. 9911, pp. 499–515. Springer, Cham (2016). https://doi.org/10.1007/978-3-319-46478-7_31
6. Deng, J., Guo, J., Xue, N., Zafeiriou, S.: Arcface: additive angular margin loss for deep face recognition. In: IEEE Conference on Computer Vision and Pattern Recognition (2019)
7. Shi, Y., Jain, A.K.: Probabilistic face embeddings. In: IEEE International Conference on Computer Vision (2019)
8. Tran, L., Yin, X., Liu, X.: Disentangled representation learning GAN for pose-invariant face recognition. In: IEEE Conference on Computer Vision and Pattern Recognition (2017)
9. Yin, X., Liu, X.: Multi-task convolutional neural network for pose-invariant face recognition. IEEE Trans. Image Process **27**(2), 964–975 (2018)
10. Li, P., Brogan, J., Flynn, P.J.: Toward facial re-identification: Experiment with data from an operational surveillance camera plant. In: IEEE International Conference on Biometrics: Theory, Applications and Systems (2016)
11. Shekhar, S., Patel, V.M., Chellappa, R.: Synthesis-based recognition of low resolution faces. In: IEEE International Joint Conference on Biometrics (2011)
12. Zhang, K., et al.: Super-identity convolutional neural network for face hallucination. In: European Conference on Computer Vision (2018)
13. Wu, J., Ding, S., Xu, W., Chao, H.: Deep joint face hallucination and recognition. arXiv:1611.08091v1 (2016)
14. Schroff, F., Kalenichenko, D., Philbin, J.: Facenet: a unified embedding for face recognition and clustering. In: IEEE conference on computer vision and pattern recognition (2015)
15. Weinberger, K.Q., Saul, L.K.: Distance metric learning for large margin nearest neighbor classification. J. Mach. Learn. Res. **10**, 207–244 (2009)
16. Oord, A.V.D., Li, Y., Vinyals, O.: Representation learning with contrastive predictive coding. arXiv:1807.03748 (2018)
17. Chen, T., Kornblith, S., Norouzi, M., Hinton, G.: A simple framework for contrastive learning of visual representations. arXiv:2002.05709 (2020)
18. He, K., Fan, H., Wu, Y., Xie, S., Girshick, R.: Momentum contrast for unsupervised visual representation learning. arXiv:1911.05722 (2019)
19. Henaff, O.J., Razavi, A., Doersch, C., Eslami, S.M., Oord, A.V.D: Data-efficient image recognition with contrastive predictive coding. arXiv preprint arXiv:1905.09272 (2019)

20. Tian, Y., Krishnan, D., Isola, P.: Contrastive multiview coding. arXiv preprint arXiv:1906.05849 (2019)
21. Khosla, P., et al.: Supervised Contrastive Learning. arXiv preprint arXiv:2004.11362 (2020)
22. Kim, H., Mnih, A.: Disentangling by factorising. In: Proceedings of the 35th International Conference on Machine Learning (2018)
23. Liu, W., Wen, Y., Yu, Z., Li, M., Raj, B., Song, L.: Sphereface: deep hypersphere embedding for face recognition. IEEE conference on computer vision and pattern recognition (2017)
24. Cao, Q., Shen, L., Xie, W., Parkhi, O.M., Zisserman, A.: Vggface2: a dataset for recognising faces across pose and age. In: International Conference on Automatic Face and Gesture Recognition (2018)
25. Xie, W., Shen, L., Zisserman, A.: Comparator networks. In: European Conference on Computer Vision (2018)
26. Xie, W., Zisserman, A.: Multicolumn networks for face recognition. In: The British Machine Vision Conference (2018)

# Deep Cascade Wavelet Network for Compressed Sensing-MRI

Zhao Li[1,2], Qinjia Bao[3,4], and Chaoyang Liu[1(✉)]

[1] State Key Laboratory of Magnetic Resonance and Atomic and Molecular Physics, Wuhan Center for Magnetic Resonance, Wuhan Institute of Physics and Mathematics, Innovation Academy for Precision Measurement Science and Technology, Chinese Academy of Sciences, Wuhan, China
chyliu@wipm.ac.cn

[2] University of Chinese Academy of Sciences, Beijing, China

[3] Wuhan United Imaging Healthcare Co., Ltd., Wuhan, China

[4] Weizmann Institute of Science, 76001 Tel Aviv-Yafo, Israel

**Abstract.** Compressed sensing (CS) theory can accelerate magnetic resonance imaging (MRI) by sampling partial k-space measurements. Recently, deep learning models have been introduced to solve CS-MRI problem. It is noticed that the wavelet transform can obtain the coarse and detail information of the image, so we designed a deep cascade wavelet network (DCWN) to solve the CS-MRI problem. Our network consists of several sub-networks and each sub-network is delivered to the next one by dense connection. The input of each sub-network comprises 4 sub-bands of the former predictions in wavelet coefficients and outputs are residuals of 4 sub-bands of reconstructed MR images in wavelet coefficients. Wavelet transform enhances the sparsity of feature maps, which may greatly reduce the training burden for reconstructs high-frequency information, and provide more structural information. The experimental results show that DCWN can achieve better performance than previous methods, with fewer parameters and shorter running time.

**Keywords:** Deep learning · Wavelet transform · Compressed sensing · Magnetic resonance imaging

## 1 Introduction

Magnetic resonance imaging (MRI) is one of the important diagnostic methods in modern medicine. Unlike other imaging methods like computed tomography (CT), it can provide anatomical images with various contrast and function information. However, MRI requires multiple phase encodings in sampling, and each phase encoding requires waiting time TR, which make MRI relatively slow, and long scan time can also cause discomfort and motion artifacts. An effective approach to accelerate the acquisition process is to under sample k-space data at a frequency encoding which is lower than the Nyquist rate by compressed sensing (CS).


H. Yang et al. (Eds.): ICONIP 2020, LNCS 12532, pp. 218–228, 2020.
https://doi.org/10.1007/978-3-030-63830-6_19

Conventional CS-MRI builds models by using different sparsity constraints in fix transform, such as SparseMRI [1] and RecPF [2]. In addition, this has been extended to a more flexible sparse representation learned directly from data using dictionary learning, such as PBDW [3], GBRWT [4] and PANO [5]. However, conventional CS algorithm is often limited by a low acceleration factors and suffers from the need for multiple iterations, which are time-consuming, and make it hard to be used in scenarios requiring real-time performance, such as in surgical navigation.

Recently, convolutional neural networks (CNN) models have been used to solve CS-MRI problem. CNN framework can address these limitations of CS-based algorithms as it can obtain deeper image information through the highly nonlinear activation function, and can make full use of the sparsity of image information. Some excellent network structures have been proposed. Wang et al. proposed the first structure MRI based on deep learning (MRI-DL) [6], and have established a non-linear mapping from zero-filled images to full-sampling results directly. Schlemper et al. proposed a multi-level cascade structure called DC-CNN [7], which connects a data consistency layer (DC layer) after each CNN blocks to enforces consistency between the reconstruction and the k-space measurements. Huang et al. proposed a cascade structure U-CA [8], which utilized Channel-wise attention layer into cascade Unet [9] to obtain better performance.

Although some recent works have taken deepening the network as the direction to improve performance, stacking more convolutional layers when the network is already deep has a very limited improvement or even cause overfitting. In this paper, we explore the advantages of exploiting sparsity transform to solve CS-MRI problem. It is worth noting that as an efficient sparsity transform, wavelet transform plays an important role in conventional CS-MRI, especially it enhances the sparsity of MRI data. In CNNs, that will cause more activation sparsity in middle and output layer. What's more, wavelet transform decompose MRI data into sub-bands which provide more detail structural information. We propose the Deep Cascade Wavelet Network (DCWN)to combine the advantages of wavelet transform and residual network and make sure obtain a better CS-MRI reconstruction performance. To the best of our knowledge, DCWN is the first wavelet CNN-based method for CS-MRI. Results show that the proposed DCWN model achieved the state-of-the-art PSNR/SSIM on Calgary-Campinas dataset [10].

The main contributions of this paper are listed as follows:

1. We propose a cascade structure with dense connections to generate low-level and high-level features flow into deeper layers to get more accurate detail for MRI reconstruction.
2. We propose a deep wavelet block (DWB). Wavelet transform is used to enhance the sparsity of feature maps, so that the coarse contents and sharp details are separated explicitly in training.
3. We use a channel-wise attention layer (CA layer) to adaptive features refinement in channel dimensions, which squeeze information by average-pooling, and extract channel features by 1D convolution layers.

# 2 Methods

## 2.1 Problem Formulation

From [7], let $x \in \mathbb{C}^{kx \times ky}$represent a sequence of 2D complex-valued MR images, $k_u \in \mathbb{C}^{kx \times ky}$ represent acquired measurements from k-space, the problem is to reconstruct from $k_u$ to $x$. The formulation is force $x$ to be well-approximated by the CNN reconstruction as follows:

$$\arg\min_{x,\theta} \|x - f_{CNN}(x_u, \theta)\|_2^2 + \lambda \|F_u(x) - k_u\|_2^2 \tag{1}$$

Where $F_u \in \mathbb{C}^{kx \times ky}$ denoted undersampled Fourier encoding matrix, $f_{CNN}$ is the forward mapping of the CNN parameterized by $\theta$, containing amount of adjustable network weights, which takes in the zero-filled reconstruction $x_u = H(F_u(k_u))$ as input and directly produces a reconstruction as an output ($H$ denoted Hermitian matrix). The second term is regularization term for data consistency of measured data and $\lambda$ denotes regularization parameter.

## 2.2 Overall Structure

The proposed DCWN for CS-MRI mainly consists of several sub-networks, each sub-network consists of cascade deep wavelet block (DWB) and DC layer. For all sub-networks after the second one, former sub-networks predictions are transmitted by dense connection (red arrow in Fig. 1.a). A concatenate operation is used to summarize these predictions as the inputs of DWB. Follow with DWB, data consistency item was defined as a layer (DC layer) of network. The overview of DCWN framework is shown in Fig. 1.a.

## 2.3 Deep Wavelet Block

As a traditional image processing technique, wavelet transform is widely used for image analysis. For Haar wavelet kernel, the input image X is first passed through a low-pass filter GL and a high-pass filter GH, then half downsampled along columns. After that, two paths are both go through by low-pass and high-pass filters respectively, and further half downsampled along rows [11,12]. Finally, the output are four sub-band coefficients, denoted as LL, HL, LH, LL. Figure 2 shows an example of the 2D discrete wavelet transform, the size of original MR image A is $h \times w$, after wavelet transform, A is decomposed to four sub-bands: LL donated the result of 2 GL, and comprise average information of A. Similarly, HL, LH, and HH comprise vertical, horizontal, and diagonal information details from the original image. In addition, each sub-band's size is $\frac{h}{2} \times \frac{w}{2}$. It is worth noting that wavelet transform and its inverse transform are both invertible, which lead no information loss in training.

In the DWB, wavelet transform wad embedded into network. For the 2nd to n th sub-network, the size of input is $h \times w \times 2(n-1)$ where n donated n th sub-network and 2 donated the separated real and imaginary channels. After

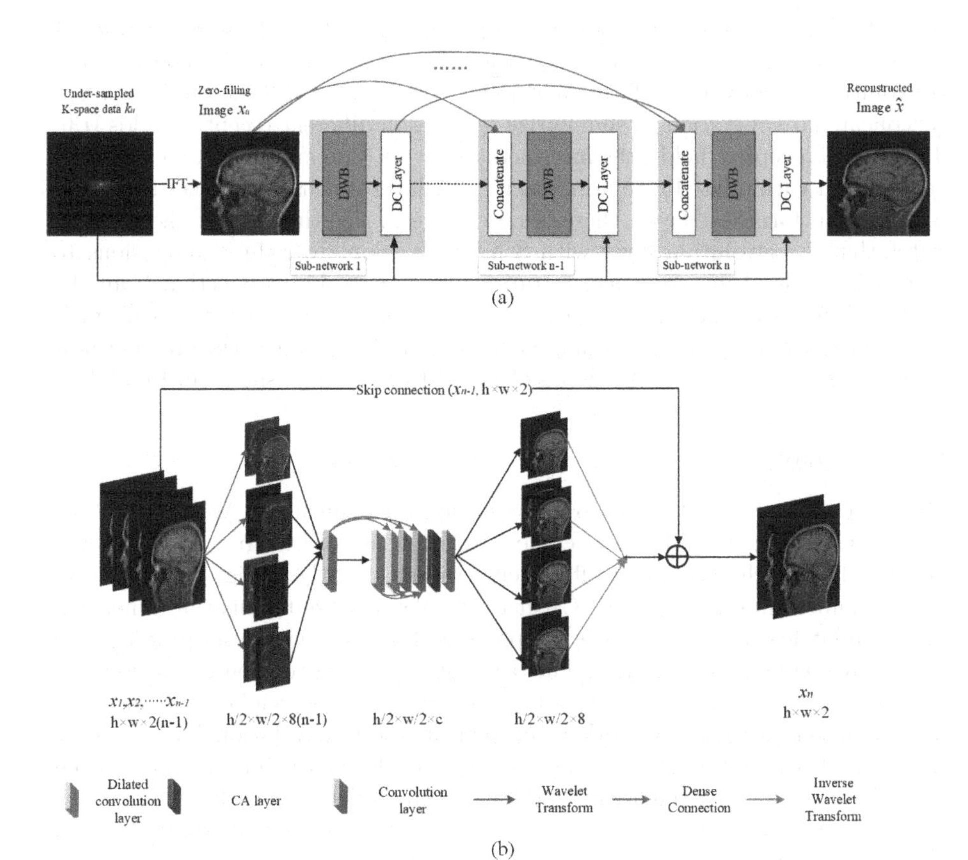


(a)

(b)

**Fig. 1.** (a) The proposed DCWN framework. (b) The detail of DWB framework, $x_1 \cdots x_{n-1}$ donated the predictions of the first to n−1 th sub-network.

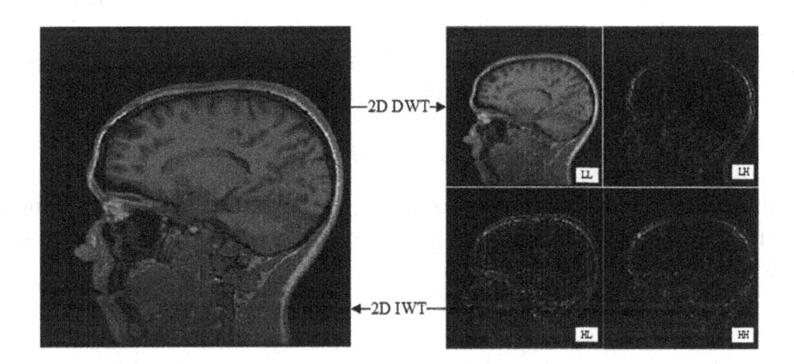


**Fig. 2.** Sub-band images by 2D discrete wavelet transform.

wavelet transform, the size of feature map is $\frac{h}{2} \times \frac{w}{2} \times 8(n-1)$, and a dense-net like network was employed to extract these features. The network consists of 5 convolution layers with 32 filters and kernel size is 3, each layer was connected by concatenate operation to summarize features in different receptive fields [13]. To enhance receptive fields, the intermediate layers (2nd and 3rd layer) is dilated convolution with dilated rate 3. A specific CA layer was used to select necessary feature map form last layer, there is no change in size throughout the process. After that a convolution layer with 8 filters is adopt to reduce dimensions to meet the needs of inverse wavelet transform. A long skip connection from the n−1 th DWB's prediction ($x_{n-1}$) to the result of inverse wavelet transform is utilized to learn the global residual information and to stabilize the gradient flow in deep residual network. The detail of DWB framework is shown in Fig. 1.b.

### 2.4 CA Layer

CA layer is used as gate unit to adaptive feature refinement. We produce the CA layer by exploiting the inter-channel relationship of features. To adjust the weight of each channel more effectively, we refer to the design idea of local cross-channel interaction from ECA-net [14]. We squeeze the spatial dimension of the input feature map by average-pooling [15], then uses descriptor $F_{avg}^{c}$ to denote average-pooling features. The descriptor is reshaped then forwarded to a weight-adjustment network to produce our final channel weight $M_c \in \mathbb{R}^{1\times1\times c}$. The weight-adjustment network composed of 2 1D convolution layers with c filters and kernel size is 3, which means only considering the interaction between each channel and its 3 neighbors, and followed by different activation layers: the first activation layer is ReLU [8] and the second is sigmoid. In short, the CA layer is computed as:

$$\begin{aligned} M_c &= sig(CNN(Avgpool(F_{in}^{c})) \\ &= sig(w_1(\sigma(w_0(F_{avg}^{c})))) \end{aligned} \tag{2}$$

where $\sigma$ denotes ReLU function, $sig$ donated sigmoid function, $w_0, w_1$ and are weights of convolution layers which are shared for both inputs. Finally, the descriptor $M_c$ is applied to the input of this module by element-wise product; i.e. each weight multiplies one input feature map, written as:

$$F_{out}^{c} = M_c \otimes F_{in}^{c} \tag{3}$$

where $\otimes$ donated element-wise product, $F_{in}^{c}$, $F_{out}^{c}$ donated the input and output feature which have the same size $\frac{h}{2} \times \frac{w}{2} \times c$ . The detail of CA layer is shown in Fig. 3.

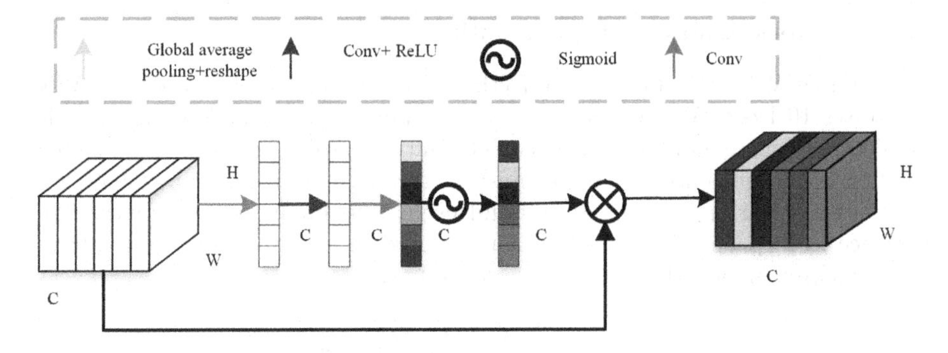


**Fig. 3.** The structure of proposed CA layer. H, W, C stands for height, width and channel for input feature maps.

### 2.5 DC Layer

In addition, the DC layer is used in every sub-network of DCWN to keep raw data consistency. The DC layer can be written as:

$$\begin{aligned} k_{rec}(c_x, c_y) &= \hat{k}(c_x, c_y) && \text{if } c_x, c_y \notin M \\ k_{rec}(c_x, c_y) &= \frac{\hat{k}(c_x, c_y) + \lambda k_u(c_x, c_y)}{1+\lambda} && \text{if } c_x, c_y \in M \end{aligned} \tag{4}$$

where $M$ denotes the binary undersampled mask, $c_x, c_y$ represents the index of k-space, $\hat{k}(c_x, c_y)$ donated $FT(f_{CNN}(x_u, \theta))$, as the value of k-space which obtained by the network, $k_{rec}$ denotes the reconstructed k-space by data consistency, and $\lambda$ is a positive constant which is related to imaging noise level. i.e. smaller value for stronger noise, and it was determined by training. Through these operations, the value of the row corresponding to the originally acquired k-space can be maintained after training, and the remaining values are filled by the training result, so that the ground-truth can be further approached.

## 3 Experiments

### 3.1 Dataset

The data set we used was the Calgary Brain MRI dataset (Calgary-Campinas dataset), which was acquired by a clinical MR scanner (Discovery MR750; General Electric (GE) Healthcare, Waukesha, WI) and the acquisition matrix of size 256 × 256, T1 weighted images from 45 volumetric. We use 35 cases from this data set and randomly select 23 cases as the training set, 10 cases as validation set and 2 cases as test set. In order to enhance the training performance, we used the data generation, which including left and right, up and down reversal, 20° rotation and 0.075 width and height shifting.

### 3.2 Network Training and Evaluation

We build DCWN on Keras[1] for the Python 3.6 environment. The whole DCWN comprises 10 Deep Wavelet blocks. Adam optimizer with default parameters is applied in training and batch size is 4. The initial learning rate was set to 0.001, the first-order momentum to be 0.9 and the second momentum to be 0.999. We set the total learning epochs as 100 and use early-stopping strategy when the patience is 5 epochs. The loss function we used in this work were normalized mean squared error (NRMSE), which defined as:

$$\text{NMRSE}(x, \hat{x}) = \sqrt{\frac{1}{B}\sum_{t=1}^{B}[\frac{x(t)-\hat{x}(t)}{x(t)}]^2} \tag{5}$$

$x(t)$,$\hat{x}(t)$ are the t-th fully-sampled image in the training set and the corresponding reconstructed image by our method, and $B$ is the number of batch size. The metrics we used are peak SNR (PSNR) and Structural Similarity (SSIM [16]). They are the most commonly used indicators for evaluating image similarity.

## 4 Results

### 4.1 Comparison of Different Methods

To verify the robustness of our method, we conduct experiments on 2 kinds of undersampled trajectory (Cartesian and radial). Since we concerned more about very aggressive undersampled rates, the reduction factor(R) we set were 5 and 8. To evaluate the proposed DCWN framework, we compare DCWN with 3 other methods: CS based on patch-based nonlocal operator (PANO) [5] is one of state-of-the-art and the most time-saving conventional CS-MRI algorithm, DC-CNN [7] and U-CA [8] are the most advanced deep learning methods for CS-MRI. A quantitative evaluation of the results obtained on our experiments is presented in Table 1. The experimental result shows that the proposed DCWN achieves the highest PSNR and SSIM in every under sampled trajectories(for Cartesian at R = 5, the central region is 24 lines, and for Cartesian at R = 8 is 18 lines). Our method increases PSNR by 1.19 dB than DC-CNN and 0.97 dB than UCA in Cartesian trajectory at R = 5. Especially in aggressive undersampled rates (R = 8), our method increases PSNR by 2.37 dB than DC-CNN and 0.42 dB than UCA in radial trajectory. One reconstruction example of Cartesian and radial trajectory at R = 5 are presented in Fig. 4 and Fig. 5, our method achieves the best performance of detail information restoring and image clarity.

We compare the network architecture used in experiment, and average run time, the result is presented in Table 2. Our method has the least parameters and the average run time is much less then U-CA. That means our proposed network has low computational complexity.

[1] https://github.com/fchollet/keras.

**Table 1.** The reconstruction performance of different undersampled trajectory by different methods, PSNR(dB)/SSIM.

| Method | Cartesian R = 5 | Cartesian R = 8 | Radial R = 5 | Radial R = 8 |
|---|---|---|---|---|
| PANO | 31.61/0.8834 | 28.32/0.8173 | 33.06/0,9134 | 30.04/0.7910 |
| DC-CNN | 35.18/0.9349 | 30.87/0.8679 | 37.47/0.9529 | 32.94/0.8953 |
| U-CA | 35.40/0.9386 | 31.53/0.8851 | 37.86/0.9562 | 34.89/0.9311 |
| DCWN | **36.37/0.9478** | **32.25/0.8972** | **37.96/0.9575** | **35.31/0.9349** |

**Table 2.** The comparison of network architecture and run time used in experiment.

| Method | Depth | Channels | Average run time | Parameters (kilobyte) |
|---|---|---|---|---|
| DC-CNN | 25 | 2/64 | 0.0207 s | 745k |
| U-CA | 75 | 2/32/64/128 | 0.0294 s | 3303k |
| DCWN | 50 | 2/8/32 | 0.0210 s | 620k |

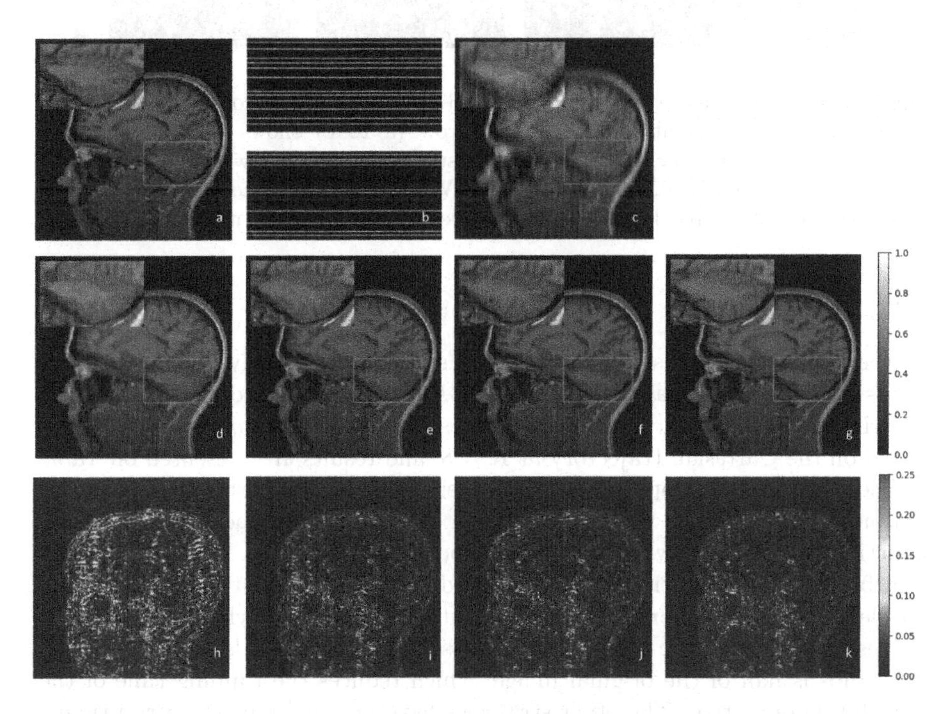


**Fig. 4.** Reconstructions results of different CS-MRI methods on complex-valued data using Cartesian trajectory at R = 5. The first row (a to c) shows the fully-sampled image, sampled trajectory and zero-filling result. The second row shows the results of PANO(d), DC-CNN(e), U-CA(f) and DCWN(g) with local zooming views. The third row shows their results of absolute difference maps of above methods.

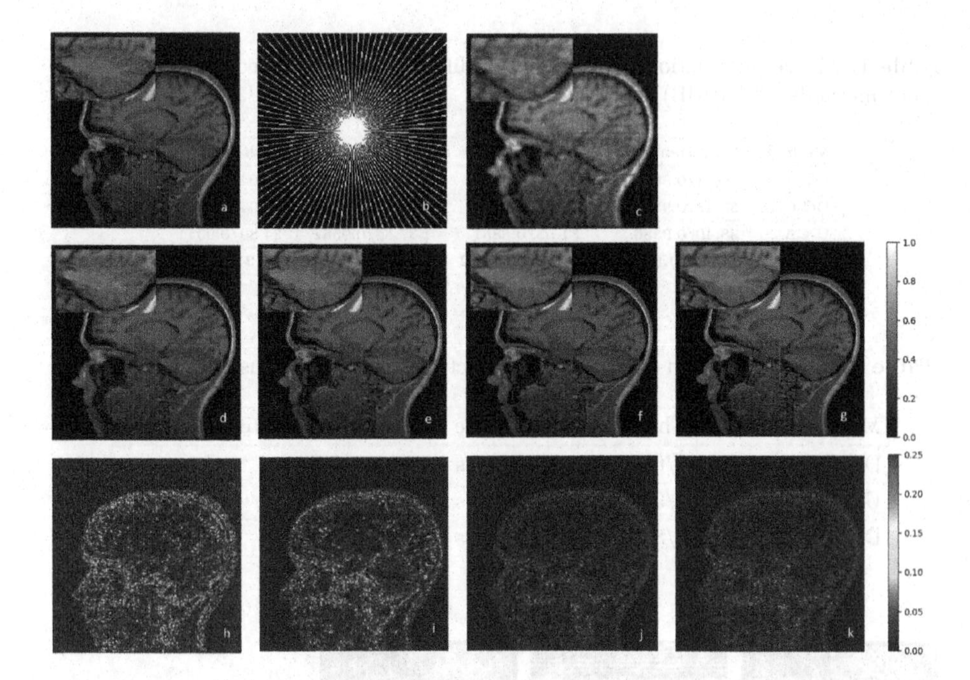


**Fig. 5.** Reconstructions results of different CS-MRI methods on complex-valued data using radial trajectory at R=5. The first row (a to c) shows the fully-sampled image, sampled trajectory and zero-filling result. The second row shows the results of PANO(d), DC-CNN(e), U-CA(f) and DCWN(g) with local zooming views. The third row shows their results of absolute difference maps of above methods.

### 4.2 Ablation Study

To verify the effectiveness of our proposed strategies, we analyzed the enhancements of 3 strategies (wavelet transform, CA layer and dense connection) to the experimental results, these three structures are removed from the DCWN and tested on the Cartesian trajectory at R = 8, and results are presented on Table 3. The structure we proposed can effectively improve the performance in most instances, of which the wavelet transform contributes the most to the reconstruction performance when it works on both trajectories at R = 8. It can be seen from the running time of various modified methods that the model using wavelet transform is more than twice shorter than the one which not using it. This is because after the image is decomposed into 4 sub-bands, the size of each sub-bands is half of the original image, which reduces the running time of the convolution operation. The other strategies have minimal impact on run time.

We also analyzed the impact of different DWB numbers on the reconstruction performance, and found that the performance dropped when using 11 DWBs, At this point, we believe the model has been over-fitted (Table 4).

**Table 3.** The reconstruction performance and run time of modified network, PSNR/SSIM/s.

| Modified model | Cartesian R = 8 | Radial R = 8 | Cartesian R = 5 | Radial R = 5 | Run time |
|---|---|---|---|---|---|
| DCWN w/o wavelet transform | 31.84/0.8875 | 35.07/0.9312 | 36.01/0.9439 | **38.00/0.9576** | 0.0513 |
| DCWN w/o dense connection | 31.92/0.8909 | 35.04/0.9314 | 35.80/0.9425 | 37.49/0.9542 | 0.0195 |
| DCWN w/o CA layer | 32.10/0.8940 | 35.12/0.9325 | 36.02/0.9445 | 37.79/0.9562 | 0.0208 |
| DCWN complete | **32.25/0.8972** | **35.31/0.9349** | **36.37/0.9478** | 37.96/0.9575 | 0.0210 |

**Table 4.** The reconstruction performance of different number of DWB, PSNR/SSIM.

| Modified model | Cartesian R = 8 | Radial R = 8 | Cartesian R = 5 | Radial R = 5 |
|---|---|---|---|---|
| DCWN with 9 DWBs | 31.88/0.8875 | 35.03/0.9307 | 36.03/0.9439 | 37.82/0.9561 |
| DCWN with 11 DWBs | 31.98/0.8913 | 35.15/0.9330 | 35.91/0.9430 | 37.92/0.9572 |
| DCWN complete (10 DWBs) | **32.25/0.8972** | **35.31/0.9349** | **36.37/0.9478** | **37.96/0.9575** |

## 5 Conclusion

We propose DCWN in this paper to improve the performance of CS-MRI. DCWN contains several sub-networks, which uses Wavelet transform to enhance the sparsity of feature maps and alleviate the learning difficulty in training. Besides, the CA layer and dense connections are applied to select necessary features for performance and guide image restoration. All the designs improve the interpretability of network. Our results indicate that the proposed method has significantly outperformed the for conventional and other deep learning-based CS-MRI methods on complex-valued dataset.

**Acknowledgments.** This work was supported in part by The National Major Scientific Research Equipment Development Project of China (81627901), the National key of R&D Program of China (Grant 2018YFC0115000, 2016YFC1304702), National Natural Science Foundation of China (11575287, 11705274), and Youth Innovation Promotion Association CAS (YZ201677).

The authors would like to express gratitude to the medical workers who fight against COVID-19 all over the world.

## References

1. Lustig, M., Donoho, D., Pauly, J.M.: Sparse MRI: the application of compressed sensing for rapid MR imaging. Magn. Resonan. Med. **58**(6), 1182–1195 (2007)
2. Huang, J., Zhang, S., Metaxas, D.: Efficient MR image reconstruction for compressed MR imaging. Med. Image Anal. **15**(5), 670–679 (2011). https://doi.org/10.1016/j.media.2011.06.001
3. Qu, X., et al.: Undersampled MRI reconstruction with patch-based directional wavelets. Magn. Reson. Imaging **30**(7), 964–977 (2012). https://doi.org/10.1016/j.mri.2012.02.019
4. Lai, Z., et al.: Image reconstruction of compressed sensing MRI using graph-based redundant wavelet transform. Med. Image Anal. **27**, 93C104 (2016). https://doi.org/10.1016/j.media.2015.05.012

5. Qu, X., Hou, Y., Lam, F., Guo, D., Zhong, J., Chen, Z.: Magnetic resonance image reconstruction from undersampled measurements using a patch-based nonlocal operator. Med. Image Anal. **18**(6), 843–856 (2014). https://doi.org/10.1016/j.media.2013.09.007
6. Wang, S., et al.: Accelerating magnetic resonance imaging via deep learning, in 2016 IEEE 13th International Symposium on Biomedical Imaging (ISBI), vol. 36, no. 4, pp. 514–517 (2016) https://doi.org/10.1109/ISBI.2016.7493320
7. Schlemper, J., Caballero, J., Hajnal, J.V., Price, A.N., Rueckert, D.: A deep cascade of convolutional neural networks for dynamic MR image reconstruction. IEEE Trans. Med. Imaging, 37(2), 491–503 (2018). https://doi.org/10.1109/TMI.2017.2760978
8. Huang, Q., Yang, D., Wu, P., Qu, H., Yi, J., Metaxas, D.: MRI reconstruction via cascaded channel-wise attention network. In: Proceedings - International Symposium Biomedical Imaging, vol. 2019-April, pp. 1622–1626 (2019). https://doi.org/10.1109/ISBI.2019.8759423
9. Ronneberger, O., Fischer, P., Brox, T.: U-Net: convolutional networks for biomedical image segmentation. In: Navab, N., Hornegger, J., Wells, W.M., Frangi, A.F. (eds.) MICCAI 2015. LNCS, vol. 9351, pp. 234–241. Springer, Cham (2015). https://doi.org/10.1007/978-3-319-24574-4_28
10. Souza, R., et al.: An open, multi-vendor, multi-field-strength brain MR dataset and analysis of publicly available skull stripping methods agreement. Neuroimage **170**, 482–494 (2018). https://doi.org/10.1016/j.neuroimage.2017.08.021
11. Xue, S., Qiu, W., Liu, F., Jin, X.: Wavelet-based residual attention network for image super-resolution. Neurocomputing, 382, 116–126 (2020). https://doi.org/10.1016/j.neucom.2019.11.044
12. Guo, T., Mousavi, H.S., Vu, T.H., Monga, V.: Deep wavelet prediction for image super-resolution. In: Proceedings of the IEEE Conference on Computer Vision and Pattern Recognition Workshops, vol. 2017-July, pp. 1100–1109 (2017). https://doi.org/10.1109/CVPRW.2017.148
13. Huang, G., Liu, Z., Maaten, L.V.D., et al.: Densely Connected Convolutional Networks. Computer Era (2017)
14. Wang, Q., Wu, B., Zhu, P., et al.: ECA-Net: efficient channel attention for deep convolu-tional neural networks (2019)
15. Woo, S., Park, J., Lee, J.Y., Kweon, I.S.: CBAM: convolutional block attention module, Lecture Notes Computer Science (including Subseries Lecture Notes Artificial Intelligence Lecture Notes Bioinformatics), vol. 11211 LNCS, pp. 3–19 (2018). https://doi.org/10.1007/978-3-030-01234-2
16. Wang, Z., Bovik, A.C., Sheikh, H.R., Simoncelli, E.P.: Image quality assessment: from error visibility to structural similarity. IEEE Trans. Image Process. **13**(4), 600–612 (2004). https://doi.org/10.1109/TIP.2003.819861

# Deep Patch-Based Human Segmentation

Dongbo Zhang[1], Zheng Fang[2], Xuequan Lu[3](✉), Hong Qin[4], Antonio Robles-Kelly[3], Chao Zhang[5], and Ying He[2]

[1] Beihang University, Beijing, China
zhangdongbo9212@163.com
[2] Nanyang Technological University, Singapore, Singapore
fz0420@hotmail.com, YHe@ntu.edu.sg
[3] Deakin University, Geelong, Australia
{xuequan.lu,antonio.robles-kelly}@deakin.edu.au
[4] Stony Brook University, Stony Brook, USA
qin@cs.stonybrook.edu
[5] University of Fukui, Fukui, Japan
zhang@u-fukui.ac.jp

**Abstract.** 3D human segmentation has seen noticeable progress in recent years. It, however, still remains a challenge to date. In this paper, we introduce a deep patch-based method for 3D human segmentation. We first extract a local surface patch for each vertex and then parameterize it into a 2D grid (or image). We then embed identified shape descriptors into the 2D grids which are further fed into the powerful 2D Convolutional Neural Network for regressing corresponding semantic labels (e.g., head, torso). Experiments demonstrate that our method is effective in human segmentation, and achieves state-of-the-art accuracy.

**Keywords:** Human segmentation · Deep learning · Parameterization · Shape descriptors

## 1 Introduction

3D human segmentation is a fundamental problem in human-centered computing. It can serve many other applications such as skeleton extraction, editing, interaction etc.,. Given that traditional optimization methods have limited segmentation outcomes, deep learning techniques have been put forwarded to achieve better results.

Recently, a variety of human segmentation methods based upon deep learning have emerged [1,2,4,5]. The main challenges are twofold. Firstly the "parameterization" scheme and, secondly, the feature information as input. Regarding the parametrization scheme, some methods convert 3D geometry data to 2D image style with brute force [1]. Methods such as [4] convert the whole human

D. Zhang and Z. Fang—Joint first author.


H. Yang et al. (Eds.): ICONIP 2020, LNCS 12532, pp. 229–240, 2020.
https://doi.org/10.1007/978-3-030-63830-6_20

model into an image-style 2D domain using geometric parameterization. However, it usually requires certain prior knowledge like the selection of different groups of triplet points. Some methods like [2] simply perform a geodesic polar map. Nevertheless, such methods often need augmentation to mitigate origin ambiguity and sometimes generate poor patches for non-rigid humans. Regarding the input feature information, one simple solution is using 3D coordinates for learning which highly relies on data augmentation [5]. Other methods [1,4] employ shape descriptors like WKS [12] as their input.

In this paper, we propose a novel deep learning approach for 3D human segmentation. In particular, we first cast the 3D-2D mapping as a geometric parameterization problem. We then convert each local patch into a 2D grid. We do this so as to embed both global features and local features into the channels of the 2D grids which are taken as input for powerful image-based deep convolutional neural networks like VGG [34]. In the testing phase, we first parameterize a new 3D human shape in the same way as training, and then feed the generated 2D grids into the trained model to output the labels.

We conduct experiments to validate our method and compare it with state-of-the-art human segmentation methods. Experimental results demonstrate that it achieves highly competitive accuracy for 3D human segmentation. We also conduct further ablation studies on different features and different neural networks.

## 2 Related Work

### 2.1 Surface Mapping

Surface mapping approaches solve the mapping or parameterization, ranging from local patch-like surfaces to global shapes. The Exponential Map is often used to parameterize a local region around a central point. It defines a bijection in the local region and preserves the distance with low distortion. Geodesic Polar Map (GPM) describes the Exponential Map using polar coordinates. [14–17] implemented GPM on triangular meshes based on approximate geodesics. Exact discrete geodesic algorithms such as [18,35] are featured with relatively accurate tracing of geodesic paths and hence polar angles. The common problem with GPM is that it easily fails to generate a one-to-one map due to the poor approximation of geodesic distances and the miscalculation of polar angles. To overcome the problem one needs to find the inward ray of geodesics mentioned in [19]. However, sometimes the local region does not form a topological disk and the tracing of the isocurve among the triangles is very difficult. To guarantee a one-to-one mapping in a local patch, one intuitive way is to adapt the harmonic maps or the angle-preserving conformal maps. A survey [20] reviewed the properties of these mappings. The harmonic maps minimize deformation and the algorithm is easy to implement on complex surfaces. However, as shown in [21,22], in the discrete context (i.e. a triangle mesh) if there are many obtuse triangles, the mapping could be flipped over. [23–27] solved the harmonic maps on closed surfaces with zero genus, which is further extended to arbitrary-genus by [23,28]. These global shapes are mapped to simple surfaces with the same genus. If the domains are not homeomorphous, one needs to cut or merge pieces into another topology [29,30]. These methods are globally injective and maintain the harmonicity while producing greater distortion around the cutting points.

### 2.2 Deep Learning on Human Segmentation

Inspired by current deep learning techniques, there have been a number of approaches attempting to extend these methods to handle the 3D human segmentation task. Limited by irregular domain of 3D surfaces, successful network architecture can not be applied straightforwardly. By leveraging Convolutional Neural Networks (CNNs), Guo et al. [1] initially handled 3D mesh labeling/segmentation in a learning way. To use CNNs on 3D meshes, they reshape per-triangle hand-crafted features (e.g. Curvatures, PCA, spin image) into a regular gird where CNNs are well defined. This approach is simple and flexible for applying CNNs on 3D meshes. However, as the method only considers per-triangle information, it fails to aggregate information among nearby triangles which is crucial for human segmentation. At the same time, Masci et al. [2] designed the network architecture, named GCNN (Geodesic Convolutional Neural Networks), so as to deal with non-Euclidean manifolds. The convolution is based on a local system of geodesic polar coordinates to parameterize a local surface patch. This convolution requires to be insensitive to the origin of angular coordinates, which means it disregards patch orientation. Following [2], anisotropic heat kernels were introduced in [3] to learn local descriptor with incorporating patch orientation. To use CNNs on surface setting, Maron et al. [4] introduced a deep learning method on 3D mesh models via parameterizating a surface to a canonical domain (2D domain) where the successful CNNs can be applied directly. However, their parameterization rely on the choice of three points on surfaces, which would involve significant angle and scale distortion. Later, an improved version of parameterization was employed in [5] to produce a low distortion coverage in the image domain. Recently, Rana et al. [6] designed a specific method for triangle meshes by modifying traditional CNNs to operate on mesh edges.

## 3 Method

### 3.1 Overview

In this work, we address 3D human segmentation by assigning a semantic label to each vertex with the aid of its local structure (patch). Due to intrinsic irregularity of surfaces, traditional 2D CNNs can not be applied to this task immediately. To this end, we map a surface patch into a 2D grid (or image), in which we are able to leverage successful network architectures (e.g. ResNet [33], VGG [34]).

As shown in Fig. 1, for each vertex on a 3D human model, a local patch is built under geodesic measurement. We then convert each local patch into a 2D grid (or image) via a 3D-2D mapping step, to suit the powerful 2D CNNs.

To preserve geometric information both locally and globally, we embed local and global shape descriptors into the 2D grid as input features. Finally, we establish the relation between per-vertex (or per-patch) feature tensor and its corresponding semantic label in a supervised learning manner. We first introduce the surface mapping step for converting a local patch into 2D grid in Sect. 3.2, and then explain the neural network and implementation details in Sect. 3.3.

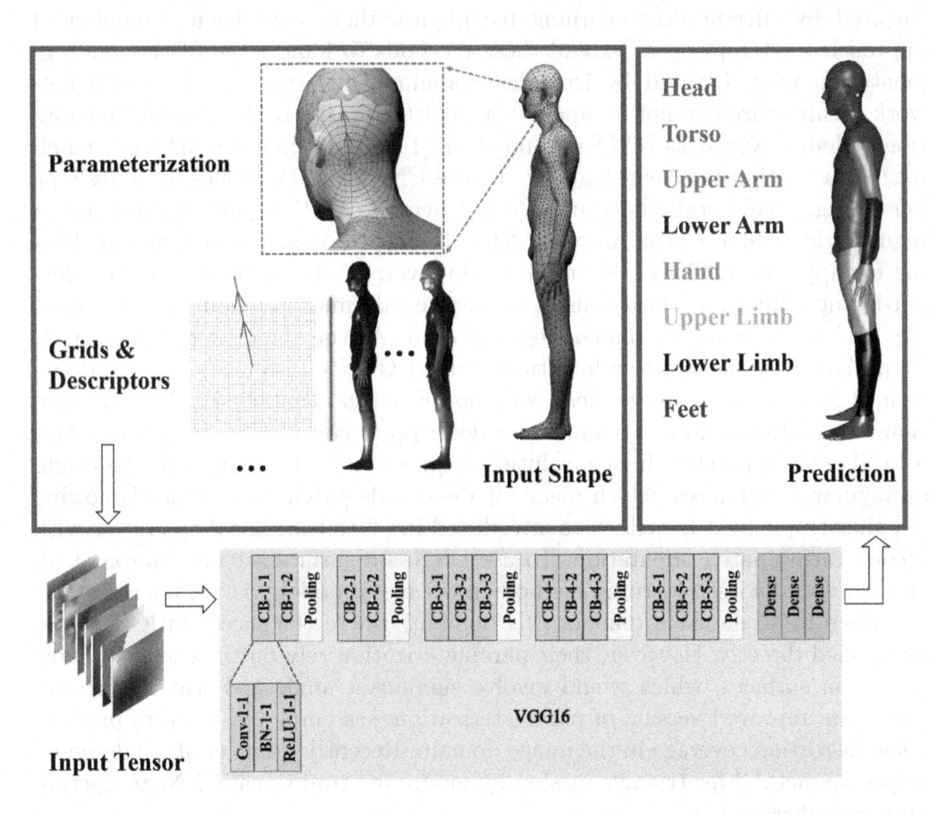


**Fig. 1.** Overview of our method. For each vertex, we first build a local patch on surface and then parameterize it into a 2D grid (or image). We embed the global and local features (WKS, Curvatures, AGD) into the 2D grid which is finally fed into VGG16 [34] to regress its corresponding semantic label.

### 3.2 Surface Mapping

**Patch Extraction.** Given a triangular mesh $M$, we compute the local patch $P$ for each vertex $v \in M$ based on the discrete geodesic distance $d$ by satisfying $d(v_p) < r_p$ for all $v_p \in P$. $r_p$ is an empirically fixed radius for all patches. Assume the area of $M$ is $\alpha$, $r_p = \sqrt{(\alpha/m)}$, where $m$ is set to 1000 in this work. The geodesic distance $d$ is computed locally using the ICH algorithm [35] due to its efficiency and effectiveness.

**Parameterization.** There are two cases for parameterization in our context, whereby $P$ is a topological disk and otherwise. For the former case, we denote the 2D planar unit disk by $D$, we compute the harmonic maps $\mu : P \to D$ by solving the Laplace equations

$$\sum_{(v_j,v_i)\in M} c_{ij}(\mu(v_j) - \mu(v_i)) = 0, \tag{1}$$

with Dirichlet boundary condition

$$\mu(v'_k) = (\cos\theta_k, \sin\theta_k), \theta_k = 2\pi \frac{\sum_{l=1}^{k} |v'_l - v'_{l-1}|}{\sum_{o=1}^{m} |v'_o - v'_{o-1}|}, \tag{2}$$

where $v_i$ is an interior vertex of $P$ (Eq. (1)) and $c_{ij}$ is the cotangent weight on edge $(v_i, v_j)$. In Eq. (2), $v'_k$ ($k \in [1, m]$) belongs to the boundary vertex set of $P$. The boundary vertex set contains $m$ vertices, which are sorted in a clockwise order according to the position on the boundary of $P$. Suppose $(v_i, v_j)$ is an interior edge. $(v_i, v_j, v_k)$ and $(v_i, v_j, v_l)$ are two adjacent triangles, $c_{ij}$ is calculated as

$$c_{ij} = \frac{1}{2}(\cot\beta_k + \cot\beta_l), \tag{3}$$

where $\beta_k$ and $\beta_l$ is the angle between $(v_i, v_k)$ and $(v_j, v_k)$, and between $(v_i, v_l)$ and $(v_j, v_l)$, respectively.

There are cases where the local patch $P$ is not a topological disk and the harmonic maps can not be computed. In this case, we trace the geodesic paths for each $v_p \in P$, by reusing the routing information stored by the ICH algorithm when computing $d$. See Fig. 2 for illustration of parameterization. Similar to [2], we then obtain a surface charting represented by polar coordinates on $D$. We next perform an alignment and a $32 \times 32$ grid discretization on $D$.

**Alignment and Grid Discretization.** The orientation of $D$ is ambiguous in the context of the local vertex indexing. We remove the ambiguity by aligning each patch with a flow vector field $\phi$ on $M$. For each vertex $v \in M$ and its associated patch $P_v$, the flow vector $\phi(v)$ serves as the reference direction of $P_v$ when mapping to $\mathbb{R}^2$. Figure 2 illustrates the reference direction as an example. $\phi$ is defined as a vector field flowing from a set of pre-determined sources $v_s \in M$ to the sinks $v_t \in M$. We initially solve a scalar function $u$ on $M$ using the following Laplace equation.

$$\Delta u(v) = 0,$$
$$u(v_s) = 0,\ u(v_t) = 1,$$

and the flow vector field is $\phi = \nabla u$. We further calibrate the polar angles of $D$ with $\phi$. From the mapping $\mu : P \to D$, we have the polar angle of the local polar axis $\theta_{axis} = 0$. The goal is to assign zero to the reference direction on $D$ such that $\theta_{ref} = 0$ and adjust the polar angle $\theta_v$ of the other vertices $v \in D$ accordingly. For easier calculation, we introduce an auxiliary edge $e_{base}$

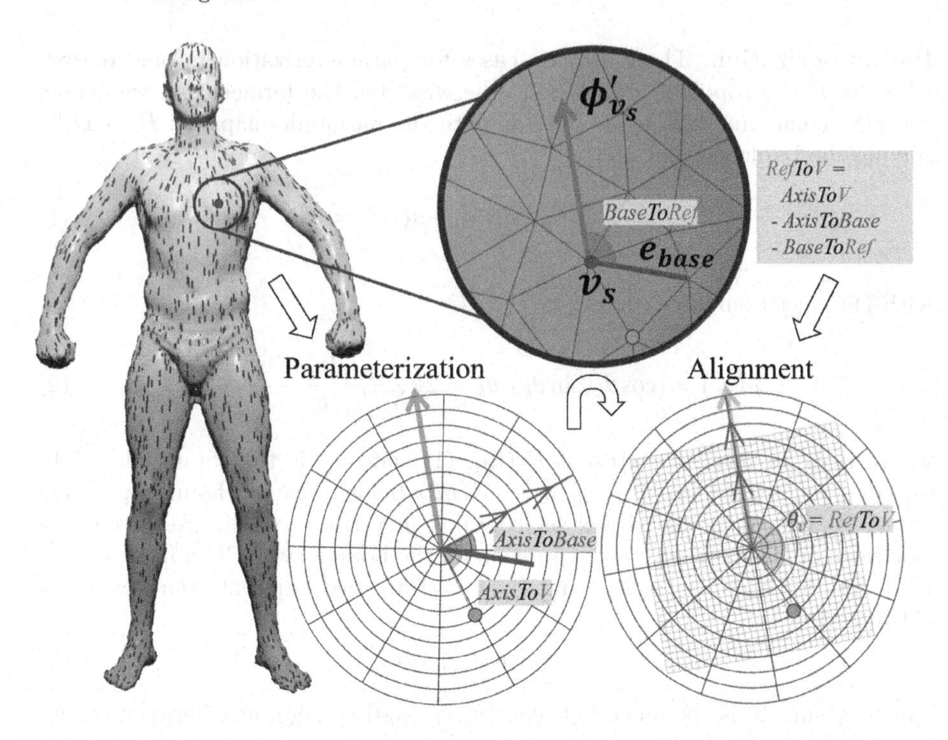


**Fig. 2.** Left: flow vector field $\phi$ rendered in red color on a human model. Top right: the close-up view of a local patch $P$ around $v_s$. We show the projected flow vector $\phi'(v_s)$ emanating from $v_s$ in orange and the base edge $e_{base}$ in blue. Bottom left: surface mapping. The red double-arrow line indicates the polar axis of the local polar coordinate system. Green dots represent the parameterized vertices from $P$. Bottom right: alignment of polar angles and grid discretization. The polar axis is rotated to overlap $\phi'(v_s)$. The angle of rotation is indicated in the yellow box. $\theta_v$ is the new polar angle for the vertex. After the alignment, the grid with $32 \times 32$ cells are embedded to the unit disk $D \in \mathbb{R}^2$. (Color figure online)

such that the angles between $e_{base}$ and the local axis, $e_{base}$ and the reference direction are known. Considering an arbitrary edge $e_{base}$ around a source vertex $v_s$ as a base edge, *BaseToRef* is the angle between the projected flow vector $\phi'(v_s)$ and $e_{base}$. $\phi'$ is the projected $\phi$ onto a random adjacent face of the base edge. From the harmonic maps $\mu$, we easily obtain the polar angles of the local, randomly-oriented polar coordinate system. The polar angles are represented by *AxisToV* for all $v \in D$ and *AxisToBase* for $e_{base}$. To align the local polar axis to the reference direction, the calibrated polar angle for all $v \in D$ is calculated as $\theta_v = AxisToV - AxisToBase - BaseToRef$. As the relevant angles to $e_{base}$ are eliminated in the equation, the selection of $e_{base}$ is robust to triangulation and any transformation.

The grid with $32 \times 32$ cells is embedded inside the calibrated $D$ such that $D$ is the circumcircle of the grid. We build a Cartesian coordinate system in $D$, and the origin is the pole in the polar system. The x-axis and the y-axis overlap the polar axis and $\theta = \pi/2$, respectively. The vertices and triangles on $D$ are converted to this Cartesian coordinate system. Some cells belong to a triangle if the cell centers are on the triangle. We compute the barycentric coordinates of each involved cell (center) with respect to the three vertices of that triangle. The barycentric coordinates will be used for calculating cell features based on vertex features later.

**Shape Descriptors.** After generating $32 \times 32$ grids (or images), we embed shape descriptors as features into them. The features of each cell are calculated with linear interpolation using the barycentric coordinates computed above. The descriptors include Wave Kernel Signature (WKS) [12], curvatures (minimal, maximal, mean, Gaussian) and average geodesic distance (AGD) [13]. We normalize each kind of descriptors on a global basis, that is, the maximum and minimum values are selected from the descriptor matrix, rather than simply from a single row or column of the matrix.

### 3.3 Neural Network and Implementation

**Neural network.** As a powerful and successful network, we adopt VGG network architecture with 16 layers (see Fig. 1) as our backbone in this work. The cross-entropy loss is employed as our loss function for the VGG16 net. It is worth noting, however, that the surface parameterization presented in this work is quite general in nature, being applicable to many other CNNs elsewhere in the literature.

**Implementation Details.** We implement the VGG16 network in PyTorch on a desktop PC with an Intel Core i7-9800X CPU (3.80 GHz, 24GB memory). We set a training epoch number of 200 and a mini-batch size of 64. SGD is set as our optimizer and the learning rate is decreased from $1.0 \times 10^{-3}$ to $1.0 \times 10^{-9}$ with increasing epochs. To balance the distribution of each label, in the training stage we randomly sample $5,000$ samples per label in each epoch. Training takes about 3.5 hours on a GeForce GTX 2080Ti GPU (11GB memory, CUDA 9.0).

Once the model is trained, we can infer semantic labels of a human shape in a vertex-wise way. Given a human shape, we first compute the involved shape descriptors for each vertex. For each vertex, we build a local surface patch and parameterize it into a 2D grid (or image) as described in Sect. 3.2. We embed all the shape descriptors into a 2D grid and feed it into our trained model for prediction.

## 4 Experimental Results

In this section, we first introduce the dataset used in our experiments, and then explain the evaluation metric. We then show the visual and the quantitative results. We also perform ablation studies for the input features and different neural networks.

### 4.1 Dataset Configuration

In this work, we use dataset from [4] which consists of 373 train human models from SCAPE [7], FAUST [8], MIT [9] and Adobe Fuse [10], and 18 test human models from SHREC07 [11]. Some examples of our training dataset are shown in Fig. 3. For each human model, there are 8 semantic labels (e.g., Head, Arm, Torso, Limb, Feet), as shown in Fig. 1. To represent geometric information of a human model both globally and locally, we concatenate a set of shape descriptors as input features: 26 WKS features [12], 4 curvature features ($C_{min}$, $C_{max}$, $C_{mean}$, $C_{gauss}$) and AGD [13].

**Fig. 3.** Examples from the training set.

### 4.2 Evaluation Metric

To provide a fair comparison, we also evaluate our segmentation results in an area-aware manner [4]. For each segmentation result, the accuracy is computed as a weighted ratio of correctly labeled triangles over the sum of all triangle area. Therefore, the overall accuracy on all involved human shapes is defined as

$$ACC = \frac{1}{N} \sum_{i=1}^{N} \frac{1}{A_i} \sum_{j \in J_i} a_{ij}, \tag{4}$$

where $N$ denotes the number of test human models and $A_i$ is the sum of triangle area of the $i$-th human model. $J_i$ is the set including the indices of correctly labeled triangles of the $i$-th human model and $a_{ij}$ represents the $j$-th triangle area of the $i$-th human model. Since we address the human segmentation task in a vertex-wise manner, the per-vertex labels need to be transferred into per-face labels for the quantitative evaluation. The face label is simply estimated by using a voting strategy among its three vertex labels. We immediately set the label with two or three vertices as the label on the face. We randomly select a vertex label as the face label, if three vertex labels are totally different.

**Fig. 4.** Some visual results of our method on the test set. The top row and the bottom row respectively show the results of our method and the corresponding ground truth models.

### 4.3 Visual and Quantitative Results

In this section, we show the visual and quantitative results. As shown in Fig. 4, the top row lists several of our results in the test set, and the bottom row displays the corresponding ground-truth models. To further evaluate our method for 3D human segmentation, a quantitative comparison with recent human segmentation techniques are summarized in Table 1. As we can see from Table 1, our method achieves an accuracy of 89.89%, ranking the second place among all methods. Our approach is a bit inferior to the best method [5] which certainly benefits from its data augmentation strategy.

**Table 1.** Comparisons with recent methods for 3D human segmentation. "#Features" denotes the number of channels of the input local patch.

| Method | #Features | ACC |
|---|---|---|
| DynGCNN [32] | 64 | 86.40% |
| Toric CNN [4] | 26 | 88.00% |
| MDGCNN [31] | 64 | 89.47% |
| SNGC [5] | 3 | 91.03% |
| GCNN [2] | 64 | 86.40% |
| Our Method | 31 | 89.89% |

### 4.4 Ablation Study

Besides the above results, we also evaluate different selection choices for input features. Table 2 shows that the input features including WKS, curvatures and AGD obtain the best performance, in terms of accuracy. Moreover, we evaluate the performance of two different neural networks in 3D human segmentation, as shown in Table 3. It is obvious that the VGG16 obtains a better accuracy than the ReseNet50, and we thus employ VGG16 as the backbone in this work.

**Table 2.** Comparisons for different input features. For simplicity, S, W, C and A are respectively short for SI-HKS (Scale-Invariant Heat Kernel Signature), WKS, Curvatures (Cmin, Cmax, Cmean, Cgauss) and AGD.

| Features Used | #Features | ACC |
|---|---|---|
| SWCA | 50 | 89.25% |
| SWA | 46 | 89.81% |
| WCA (Our) | 31 | 89.89% |

**Table 3.** Comparisons for two different network architectures.

| Network | Features | ACC |
|---|---|---|
| ResNet50 | WKS, Curvatures, AGD | 87.60% |
| VGG16 | WKS, Curvatures, AGD | 89.89% |

## 5 Conclusion

We have presented a deep learning method for 3D human segmentation. Given a 3D human mesh as input, we first parameterize each local patch in the shape into 2D image style, and feed it into the trained model for automatically predicting the label of each patch (i.e., vertex). Experiments demonstrate the effectiveness of our approach, and show that it can achieve state-of-the-art accuracy in 3D human segmentation. In the future, we would like to explore and design more powerful features for learning the complex relationship between the non-rigid 3D shapes and the semantic labels.

**Acknowledgements.** This research is supported in part by the National Key R&D Program of China under Grant No. 2017YFF0106407, Deakin University (Australia) internal grant (CY01-251301-F003-PJ03906-PG00447) and research grant (PJ06625), National Natural Science Foundation of China under Grant No. 61532002, and National Science Foundation of USA under Grant IIS-1715985 and IIS-1812606.

## References

1. Guo, K., Zou, D., Chen, X.: 3D mesh labeling via deep convolutional neural networks. ACM Trans. Graph. **35**(1), 1–12 (2015)
2. Masci, J., Boscaini, D., Bronstein, M., Vandergheynst, P.: Geodesic convolutional neural networks on Riemannian manifolds. In: Proceedings of the IEEE International Conference on Computer Vision Workshops, pp. 37–45 (2015)
3. Boscaini, D., Masci, J., Rodolà, E., Bronstein, M.: Learning shape correspondence with anisotropic convolutional neural networks. In: Advances in Neural Information Processing Systems, pp. 3189–3197 (2016)
4. Maron, H., et al.: Convolutional neural networks on surfaces via seamless toric covers. ACM Trans. Graph. **36**(4), 71–1 (2017)
5. Haim, N., Segol, N., Ben-Hamu, H., Maron, H., Lipman, Y.: Surface networks via general covers. In: Proceedings of the IEEE International Conference on Computer Vision, pp. 632–641 (2019)
6. Hanocka, R., Hertz, A., Fish, N., Giryes, R., Fleishman, S., Cohen-Or, D.: MeshCNN: a network with an edge. ACM Trans. Graph. **38**(4), 1–12 (2019)
7. Anguelov, D., Srinivasan, P., Koller, D., Thrun, S., Rodgers, J., Davis, J.: Scape: shape completion and animation of people. In: ACM SIGGRAPH 2005 Papers, pp. 408–416 (2005)
8. Bogo, F., Romero, J., Loper, M., Black, M.J.: Faust: Dataset and evaluation for 3D mesh registration. In: Proceedings of the IEEE Conference on Computer Vision and Pattern Recognition, pp. 3794–3801 (2014)
9. Vlasic, D., Baran, I., Matusik, W., Popovic, J.: Articulated mesh animation from multi-view silhouettes. In: ACM SIGGRAPH 2008 Papers, pp. 1–9 (2008)
10. Adobe fuse 3D characters. https://www.mixamo.com
11. Giorgi, D., Biasotti, S., Paraboschi, L.: Shape retrieval contest 2007: watertight models track. SHREC competition **8**(7) (2007)
12. Aubry, M., Schlickewei, U., Cremers, D.: The wave kernel signature: a quantum mechanical approach to shape analysis. In: 2011 IEEE International Conference on Computer Vision Workshops, pp. 1626–1633. IEEE (2011)
13. Hilaga, M., Shinagawa, Y., Kohmura, T., Kunii, T.L.: Topology matching for fully automatic similarity estimation of 3D shapes. In: Proceedings of the 28th Annual Conference on Computer Graphics and Interactive Techniques, pp. 203–212 (2001)
14. Melvær, E.L., Reimers, M.: Geodesic polar coordinates on polygonal meshes. In: Computer Graphics Forum, vol. 31, pp. 2423–2435. Wiley Online Library (2012)
15. Schmidt, R., Grimm, C., Wyvill, B.: Interactive decal compositing with discrete exponential maps. In: ACM SIGGRAPH 2006 Papers, pp. 605–613 (2006)
16. Floater, M.S.: Mean value coordinates. Comput. Aided Geometr. Des. **20**(1), 19–27 (2003)
17. Ju, T., Schaefer, S., Warren, J.: Mean value coordinates for closed triangular meshes. In: ACM SIGGRAPH 2005 Papers, pp. 561–566 (2005)
18. Surazhsky, V., Surazhsky, T., Kirsanov, D., Gortler, S.J., Hoppe, H.: Fast exact and approximate geodesics on meshes. ACM Trans. Graph. **24**(3), 553–560 (2005)
19. Kokkinos, I., Bronstein, M.M., Litman, R., Bronstein, A.M.: Intrinsic shape context descriptors for deformable shapes. In: 2012 IEEE Conference on Computer Vision and Pattern Recognition, pp. 159–166. IEEE (2012)
20. Floater, M.S., Hormann, K.: Surface parameterization: a tutorial and survey. In: Dodgson, N.A., Floater, M.S., Sabin, M.A. (eds.) Advances in Multiresolution for Geometric Modelling, pp. 157–186. Springer, Heidelberg (2005). https://doi.org/10.1007/3-540-26808-1_9

21. Duchamp, T., Certain, A., DeRose, A., Stuetzle, W.: Hierarchical computation of pl harmonic embeddings. preprint (1997)
22. Floater, M.S.: Parametric tilings and scattered data approximation. Int. J. Shape Model. **4**(03n04), 165–182 (1998)
23. Gu, X., Yau, S.T.: Global conformal surface parameterization. In: Proceedings of the 2003 Eurographics/ACM SIGGRAPH Symposium on Geometry Processing, pp. 127–137 (2003)
24. Haker, S., Angenent, S., Tannenbaum, A., Kikinis, R., Sapiro, G., Halle, M.: Conformal surface parameterization for texture mapping. IEEE Trans. Vis. Comput. Graph. **6**(2), 181–189 (2000)
25. Praun, E., Hoppe, H.: Spherical parametrization and remeshing. ACM Trans. Graph. **22**(3), 340–349 (2003)
26. Sheffer, A., de Sturler, E.: Parameterization of faceted surfaces for meshing using angle-based flattening. Eng. Comput. **17**(3), 326–337 (2001)
27. Sheffer, A., Gotsman, C., Dyn, N.: Robust spherical parameterization of triangular meshes. Computing **72**(1–2), 185–193 (2004)
28. Khodakovsky, A., Litke, N., Schroder, P.: Globally smooth parameterizations with low distortion. ACM Trans. Graph. **22**(3), 350–357 (2003)
29. Tutte, W.T.: How to draw a graph. Proc. London Math. Soc. **3**(1), 743–767 (1963)
30. Floater, M.: One-to-one piecewise linear mappings over triangulations. Math. Comput. **72**(242), 685–696 (2003)
31. Poulenard, A., Ovsjanikov, M.: Multi-directional geodesic neural networks via equivariant convolution. ACM Trans. Graph. **37**(6), 1–14 (2018)
32. Wang, Y., Sun, Y., Liu, Z., Sarma, S.E., Bronstein, M.M., Solomon, J.M.: Dynamic graph CNN for learning on point clouds. ACM Transa. Graph. **38**(5), 1–12 (2019)
33. He, K., Zhang, X., Ren, S., Sun, J.: Deep residual learning for image recognition. In: Proceedings of the IEEE conference on Computer Vision and Pattern Recognition, pp. 770–778 (2016)
34. Simonyan, K., Zisserman, A.: Very deep convolutional networks for large-scale image recognition. arXiv preprint arXiv:1409.1556 (2014)
35. Xin, S.Q., Wang, G.J.: Improving Chen and Han's algorithm on the discrete geodesic problem. ACM Trans. Graph. **28**(4), 1–8 (2009)

# Deep Residual Local Feature Learning for Speech Emotion Recognition

Sattaya Singkul[1], Thakorn Chatchaisathaporn[2], Boontawee Suntisrivaraporn[2], and Kuntpong Woraratpanya[1(✉)]

[1] Faculty of Information Technology, King Mongkut's Institute of Technology Ladkrabang, Bangkok, Thailand
{59070173,kuntpong}@it.kmitl.ac.th

[2] Data Analytics, Siam Commercial Bank, Bangkok, Thailand
thakorn.chatchaisathaporn@scb.co.th, meng234@gmail.com

**Abstract.** Speech Emotion Recognition (SER) is becoming a key role in global business today to improve service efficiency, like call center services. Recent SERs were based on a deep learning approach. However, the efficiency of deep learning depends on the number of layers, i.e., the deeper layers, the higher efficiency. On the other hand, the deeper layers are causes of a vanishing gradient problem, a low learning rate, and high time-consuming. Therefore, this paper proposed a redesign of existing local feature learning block (LFLB). The new design is called a deep residual local feature learning block (DeepResLFLB). DeepResLFLB consists of three cascade blocks: LFLB, residual local feature learning block (ResLFLB), and multilayer perceptron (MLP). LFLB is built for learning local correlations along with extracting hierarchical correlations; DeepResLFLB can take advantage of repeatedly learning to explain more detail in deeper layers using residual learning for solving vanishing gradient and reducing overfitting; and MLP is adopted to find the relationship of learning and discover probability for predicted speech emotions and gender types. Based on two available published datasets: EMODB and RAVDESS, the proposed DeepResLFLB can significantly improve performance when evaluated by standard metrics: accuracy, precision, recall, and F1-score.

**Keywords:** Speech Emotion Recognition · Residual feature learning · CNN network · Log-Mel spectrogram · Chromagram

## 1 Introduction

Emotional analysis has been an active research area for a few decades, especially in recognition domains of text and speech emotions. Even if text and speech emotions are closely relevant, both kinds of emotions have different challenges. One of the challenges in text emotion recognition is ambiguous words, resulting

**Electronic supplementary material** The online version of this chapter (https://doi.org/10.1007/978-3-030-63830-6_21) contains supplementary material, which is available to authorized users.


H. Yang et al. (Eds.): ICONIP 2020, LNCS 12532, pp. 241–252, 2020.
https://doi.org/10.1007/978-3-030-63830-6_21

from omitted words [1,2]. On the other hand, one of the challenges in speech emotion recognition is creating an efficient model. However, this paper focuses on only the recognition of speech emotions. In this area, two types of information, linguistic and paralinguistic, were mainly considered in speech emotion recognition. The linguistic information refers to the meaning or context of speech. The paralinguistic information implies the implicit message meaning, like the emotion in speech [3–6]. Speech characteristics can interpret the meaning of speech; therefore, behavioral expression was investigated in most of the speech emotion recognition works [7–9].

In recent works, local feature learning block (LFLB) [10], one of the efficient methods, has been used in integrating local and global speech emotion features, which provide better results in recognition. Inside LFLB, convolution neural network (CNN) was used for extracting local features, and then long short-term memory (LSTM) was applied for extracting contextual dependencies from those local features to learn in a time-related relationship. However, vanishing gradient problems may occur with CNN [11]. Therefore, residual deep learning was applied to the CNN by using skip-connection to reduce unnecessary learning and add feature details that may be lost in between layers.

Furthermore, the accuracy of speech recognition does not only rely on the efficiency of a model, but also of a speech feature selection [12]. In terms of speech characteristics, there are many distinctive acoustic features that usually used in recognizing the speech emotion, such as continuous features, qualitative features, and spectral features [12–16]. Many of them have been investigated to recognize speech emotions. Some researchers compared the pros and cons of each feature, but no one can identify which feature was the best one until now [3,4,17,18].

As previously mentioned, we proposed a method to improve the efficiency of LFLB [11] for deeper learning. The proposed method, deep residual local feature learning block (DeepResLFLB), was inspired by the concept of human brain learning; that is, 'repeated reading makes learning more effective,' as the same way that Sari [19] and Shanahan [20] were used. Responding to our inspired concept, we implemented a learning method for speech emotion recognition with three parts: Part 1 is for general learning, like human reading for the first time, Part 2 is for further learning, like additional readings, and the last part is for associating parts learned to decide types of emotions. Besides, the feature selection is compared with two types of distinctive features to find the most effective feature in our work: the normal and specific distinctive features are log-mel spectrogram (LMS), which is fully filtered sound elements, and MFCC deltas, delta-deltas, and chromagram (LMSDDC) are more clearly identify speech characteristics extracted based on the human mood.

Our main contributions of this paper are as follows: (i) Deep residual local feature learning block (DeepResLFLB) was proposed. DeepResLFLB was arranged its internal network as LFLB, batch normalization (BN), activation function, normalization-activation-CNN (NAC), and deep layers. (ii) Learning sequences of DeepResLFLB were imitated from human re-reads. (iii) Speech emotion features, based on human mood determination factors such as LMS and LMSDDC, were applied and compared their performances.

## 2 Literature Reviews

Model efficiency is one of the important factors in SER. Many papers focused on learning methods of machine learning or deep learning. Demircan [17] introduced fuzzy c-mean as a preprocessing step to group and add characteristics before using machine learning. Venkataramanan [21] studied of using deep learning in SER. The findings of the study revealed that CNN outperformed the traditional machine learning. Also, Huang [15] showed that semi-CNN in SER can increase accuracy. Zhao [10] presented the use of CNN in conjunction with LSTM to extract and learn features. Zhao's method used a sequence of CNN in a block style, consisting of CNN, BN, activation function, and pooling, for local feature learning, and then used LSTM for extracting contextual dependencies in a time-related relationship. In this way, both local and global features are extracted and learned.

It is undeniable that the effectiveness of deep learning mainly depends on the data size for training [22]. Recently, Google brain research [23] proposed data augmentation, one of the efficient techniques that can increase the amount of data, by adding spectrogram characteristics, also known as "Spectrogram Augmentation." This augmentation consists of time warping to see more time shift patterns, time masking to reduce the overfitting rate of the model and improve the sound tolerance that may have characteristics of silence, frequency masking to reduce the overfitting rate and increase sound resistance from concealing characteristics of a specific wavelength. The spectrogram is a basic feature of sound that can lead to various specific features. Therefore, by using above methods, the model can learn more perspectives of the data.

Also, different features lead to different performances in speech emotion recognition. Among the features of speech, mel-frequency cepstral coefficient (MFCC) [21], which can be characterized by the frequency filter in the range 20 Hz to 20 kHz, similar to human hearing, is widely used to obtain coefficients from the filtered sound. Recent research papers [17,21] used the difference of MFCC to get more specific details, but, in the aspect of MFCC, it has no time relationship. Therefore, many papers [10,15,21] used mel spectrogram (MS) instead. MS can respond to the time relationship, thus providing better results than just using MFCC. Besides, music can be looked different from speech; therefore, chromagram is widely used instead of MFCC, since it can provide better features than normal MFCC and MS.

Our work is different from the previously mentioned works in that the deep residual local feature learning block (DeepResLFLB) was redesigned from LFLB. This method helps reduce the chance of feature and updated losses caused by CNN model in the LFLB, especially in deeper layers. DeepResLFLB uses a repeated learning style that local features extracted from a bias frame with silent voice (see Subsect. 3.3) can be learned through a residual deep learning approach. Moreover, we extracted distinctive features based on a concept of determining human emotions, consisting of prosodic [24], filter bank [21,24], and glottal flow [25]. These three features in conjunction with our ResLFLB can improve learning efficiency.

## 3 The Proposed Model

To enable SER as efficiently as possible, the following factors: raw datasets, environments, and features are included in our system design. Based on such factors, a new designed framework, called DeepResLFLB, was proposed as shown in Fig. 1. This framework consists of five parts: (i) raw data preparation, (ii) voice activity detection, (iii) bias frame cleaning, (iv) feature extraction, and (v) deep learning.

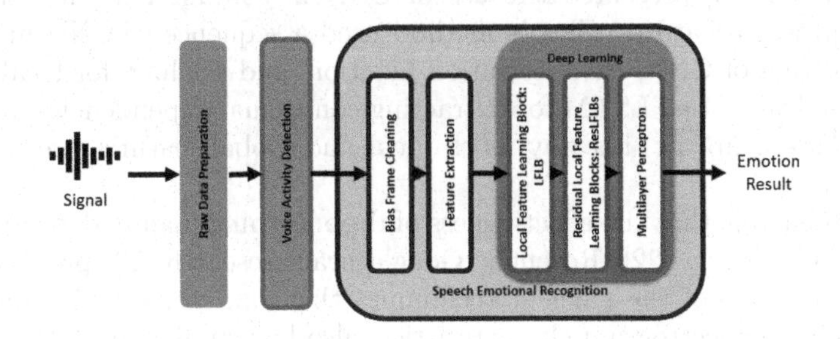


**Fig. 1.** A deep residual local feature learning framework.

### 3.1 Raw Data Preparation

Due to the complex nature of datasets and the difference of languages like EMODB in Berlin German and RAVDESS in English, the representation of the original datasets may not be enough for training a model based on deep learning. Therefore, increasing a variety of data to see more new dimensions or characteristics is essential. Responding to this, various data augmentation techniques, including noise adding, pitch tuning, and spectrogram, were used in this work to make the model more robust to noise and unseen voice patterns.

### 3.2 Voice Activity Detection

Although both datasets, EMODB and RAVDESS, were produced in a closed environment, quite a bit of noise, they were found that noise remains at the starting and stopping points of sound records. Indeed, noise is not related to speakers' voice, so it could be removed. Here, voice activity detection [26] was used to detect only voice locations, i.e., excluding noise locations. As a result of the voice activity detection, selected frames can be efficiently analyzed and classified male and female voices by energy-base features.

### 3.3 Bias Frame Cleaning

Bias frame cleaning is used as a postprocessing of voice activity detection; that is, each frame segmented by the voice activity detection is identified its loudness through Fourier transform (FT). If FT coefficients of a segmented frame are zero, that frame is identified as no significant information for emotional analysis, so it is rejected.

### 3.4 Feature Extraction

Model performance of deep learning mainly depends on features. The good features usually gain more model performance. Thus, this paper focuses on efficient extraction of human emotion features. Naturally, speech signals always contain human emotions. In other words, we can extract human emotions from speech signals. Here, we briefly describe three important components of speech signals: glottal flow, prosody, and human hearing. Glottal flow can be viewed as a source of speech signals [25]. It mainly produces fundamental frequencies [27] or latent sounds within the speech. Prosody is vocal frequencies, which are produced from the air pushed by the lung [25]. It contains important characteristics, such as intonation, tone, stress, and rhythm. On the other hand, for human hearing, MFCC is one of the analytical tools that can mimic the behavior of human ears by applying cepstral analysis [28]. Based on our assumption of extracting better emotion features, two important factors are included for feature extraction design: (i) the wide band frequencies of speech signals are regarded as much as possible to cover important features of speech emotions, and (ii) time-frequency processing is used for extracting speech emotions. Here, log-mel spectrogram (LMS) was used as time-frequency representation for emotion features. Two additional features extracted based on MFCC were delta and delta-delta. Furthermore, chromagram feature [29,30] was extracted as one of the emotion features. Figure 2 shows our emotion feature extraction. As a result, four features, LMS, delta, delta-delta, and chromagram were used as emotional representation.

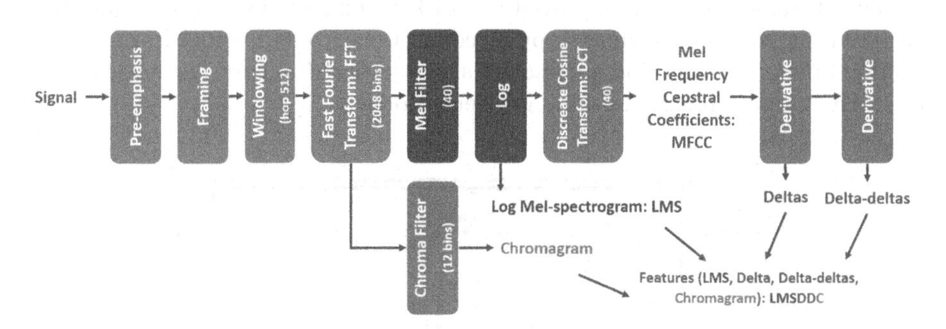


**Fig. 2.** Feature extraction of LMS and LMSDDC

### 3.5 Deep Learning

Inspired by learning characteristics of human brain activity, i.e., the more repeated reading, the more comprehension. It is similar with Shanahan's definition, called "repeatedly reads" or "re-reads" [20]. Responding to the use of re-reading theory for improving the accuracy of SER, we designed a feature learning method as shown in Fig. 3, consisting of three sections: (i) main feature learning (MFL), (ii) sub-feature learning (SFL), and (iii) extracted relation of feature distribution (ERFD).

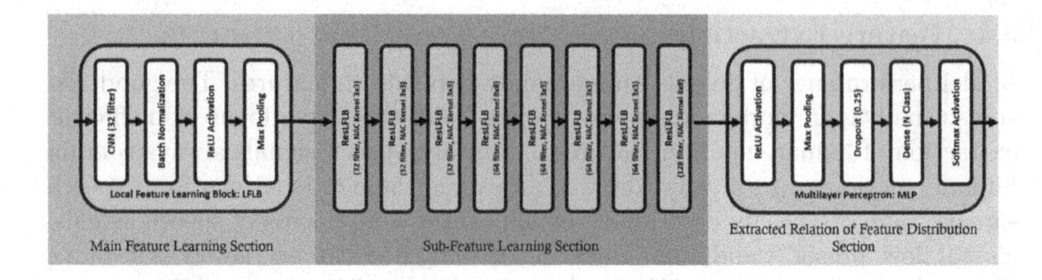


**Fig. 3.** A feature learning structure based on the re-reading theory.

**Main Feature Learning (MFL) Section.** MFL was designed based on behavior of human brain, like first reading that a human brain starts to learn things. We designed MFLS similar with the LFLB procedure to learn locally basic information as the following steps: (i) 2D-CNN was used to extract necessary features; (ii) BN was applied to enhance learning efficiency of a model; (iii) activation functions converted data to suit for the learning model; and (iv) pooling was for reducing feature size and increased learning speed.

**Sub-Feature Learning (SFL) Section.** SFL was a further learning process that plays a role in assembling repeated reading for deeper learning. In general, LFLB may be at risk of a vanishing gradient problem that affects learning efficiency. Therefore, we have improved the LFLB's efficient by means of residual deep learning, or also known as skipping connections, to skip deeper learning layers that are unnecessary and add more feature details after passing each learning layer; this can avoid the vanishing gradient problem.

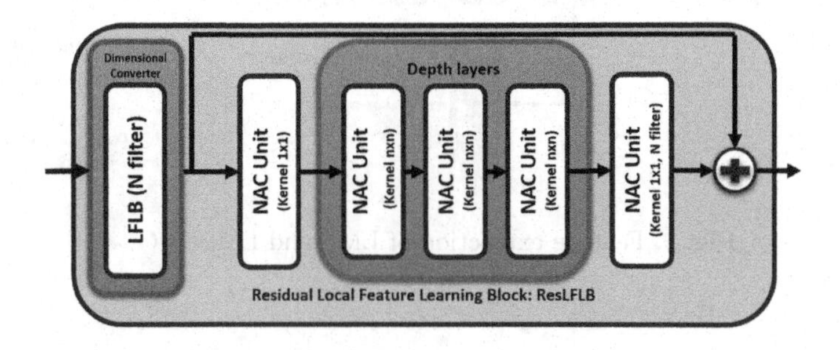


**Fig. 4.** A residual local feature learning block structure.

For sub-feature learning, there are two important sections: (i) A normal LFLB was used as a preprocessing phase of block to transform input data dimensions into a suitable form for the next section. (ii) Depth layers learned features in more depth by using the network sequence as normalization-activation-CNN (NAC) [31], which is a sequence of sub-layers in residual deep learning. With this structure, it can reduce test error. In addition, more deeper learning layers may be at risk of a vanishing gradient problem; therefore, the skipping connection was added at an input of this section to compensate features lost in deeper layers. This arrangement of two sections is called residual local feature block (ResLFLB) as shown in Fig. 4. The LFLB also known as a dimensional converter and the NAC final layer had the same output filters to determine the number of output filters of ResLFLB. The kernel size in the first and last layer in deep layers were one. Furthermore, in between the first and last learning layers, the bottleneck design, the compression/decompression strategy, was applied to achieve higher learning performance.

**Extracted Relation of Feature Distribution (ERFD) Section.** ERFD was implemented with multilayer perceptron (MLP) to extract the relationship of learning results. A sequence of processes are as follows: (i) Activation ReLU transformed the data to the suitable relationship; (ii) Max pooling reduced data size and extracted important data based on maximum values; (iii) Dropout reduced the overfitting of the model; (iv) Dense layers obtained relationships in which neurons equals the number of classes needed to predict; and (v) Activation softmax determined the probability of predicting the emotions.

## 4 Experiments and Discussion

The proposed DeepResLFLB and LMSDDC were evaluated with two main objectives: (i) classification performance and (ii) model performance. In classification performance, four metrics, accuracy, precision, recall, and F1-score as defined by (1), (2), (3), and (4), respectively, were used for evaluation. In model performance, validation loss was used as monitoring vanishing gradient problems and the number of CNN layer parameters was used as indicating resource-consuming. The experiments were conducted in comparison with three models: the normal ML with fuzzy c-mean [17], traditional LFLB [10], and DeepResLFLB, and two different features: LMS and LMSDDC. Note that, in Tables 2 and 4, Dermircan's method was excluded from those experiments due to a mismatch of feature dimensions.

$$Accuracy = \frac{true\,positive + true\,negative}{number\,of\,data} \tag{1}$$

$$Precision = \frac{true\,positive}{total\,predicted\,positive} \tag{2}$$

$$Recall = \frac{true\,positive}{total\,actual\,positive} \tag{3}$$

$$F1 = 2 \times \frac{precision \times recall}{precision + recall} \tag{4}$$

**Dataset Preparation.** Two available published datasets: Berlin emotional data-base (EMODB) [32] and Ryerson audio-visual database (RAVDESS) [33] were used to evaluate speech emotion performance of our and baseline methods. Two key factors of both selected datasets are the difference in data size and language vocalization that can prove the performance of test methods. EMODB is a German speech in Berlin with 535 utterances and RAVDESS is English speech with 1440 utterances. EMODB dataset was recorded by male and female voices and contained seven different emotions: happiness, sadness, angry, neutral, fear, boredom, and disgust while RAVDESS dataset has one more emotion than EMODB, that is the calm emotion. Here, each dataset was divided into three subsets: 80% for training set, 10% for validation set, and 10% for test set.

**Parameter Settings for Learning Model.** All learning models were set up with the following parameter settings. Learning rate (LR) is very important in deep learning, when compared to step rate to find the minimum gradient. Generally, a high LR may make it over the minimum point. On the other hand, a low LR may take a long time to reach the goal. Here, we choose Adam optimizer for our experiments. Its learning rate and maximum epoch were set to 0.001 and 150, respectively. In addition, plateau strategy was used for reducing the LR and for avoiding overstepping the minimum point. In this case, we set the minimum LR to 0.00001. Batch size of models was set to 10. Finally, if an error value tends to increase, the early stopping criteria is active and then take the model weight with the previous minimum error.

**Table 1.** A performance comparison of DeepResLFLB and baseline methods with LMS feature, tested on Berlin EMODB dataset.

| Method | Accuracy | Precision | Recall | F1-score |
|---|---|---|---|---|
| Demircan [17] | 0.6755±0.0351 | 0.7549±0.0375 | 0.6295±0.0346 | 0.6295±0.0346 |
| 1D-LFLB [10] | 0.7577±0.0241 | 0.7609±0.0224 | 0.7574±0.0318 | 0.7514±0.0275 |
| 2D-LFLB [10] | 0.8269±0.0214 | 0.831±0.0228 | 0.824±0.0215 | 0.8233±0.0.0233 |
| DeepResLFLB | **0.8404±0.0225** | **0.8481±0.0225** | **0.8298±0.0236** | **0.8328±0.0244** |

**Table 2.** A performance comparison of DeepResLFLB and baseline methods with LMSDDC feature, tested on Berlin EMODB dataset.

| Method | Accuracy | Precision | Recall | F1-score |
|---|---|---|---|---|
| Demircan [17] | – | – | – | – |
| 1D-LFLB [10] | 0.8355±0.0186 | 0.8385±0.0170 | 0.8313±0.0205 | 0.8322±0.0198 |
| 2D-LFLB [10] | 0.8754±0.0232 | 0.8802±0.0237 | 0.8733±0.0237 | 0.8745±0.0226 |
| DeepResLFLB | **0.8922±0.0251** | **0.8961±0.0212** | **0.8856±0.0322** | **0.8875±0.0293** |

**Result Discussion.** Based on dataset preparation and parameter setup for learning models, all experiments were used 5-fold validation. Tables 1 and 2

**Table 3.** A performance comparison of DeepResLFLB and baseline methods with LMS feature, tested on RAVDESS dataset.

| Method | Accuracy | Precision | Recall | F1-score |
|---|---|---|---|---|
| Demircan [17] | 0.7528±0.0126 | 0.7809±0.0109 | 0.7422±0.0076 | 0.7479±0.0114 |
| 1D-LFLB [10] | 0.9487±0.0138 | 0.9491±0.0134 | 0.948±0.0123 | 0.9478±0.0133 |
| 2D-LFLB [10] | 0.9456±0.0128 | 0.9438±0.0136 | 0.946±0.0129 | 0.9442±0.0135 |
| DeepResLFLB | **0.9602±0.0075** | **0.9593±0.0072** | **0.9583±0.0066** | **0.9584±0.0071** |

**Table 4.** A performance comparison of DeepResLFLB and baseline methods with LMSDDC feature, tested on RAVDESS dataset.

| Method | Accuracy | Precision | Recall | F1-score |
|---|---|---|---|---|
| Demircan [17] | – | – | – | – |
| 1D-LFLB [10] | 0.9367±0.0225 | 0.9363±0.0196 | 0.9352±0.0218 | 0.9347±0.0217 |
| 2D-LFLB [10] | 0.9466±0.0159 | 0.9475±0.0171 | 0.9441±0.0162 | 0.9449±0.0168 |
| DeepResLFLB | **0.949±0.0142** | **0.9492±0.0143** | **0.9486±0.016** | **0.9484±0.0154** |

**Table 5.** A comparison of a number of parameters in DeepResLFLB and LFLB models, tested on EMODB and RAVDESS datasets.

| Method | EMODB | | RAVDESS | |
|---|---|---|---|---|
| | LMS | LMSDDC | LMS | LMSDDC |
| 2D-LFLB [10] | 260544 | 262272 | 260544 | 262272 |
| DeepResLFLB | **156068** | **163268** | **156074** | **164608** |

shows performance comparison between LMS and LMSDDC features, respectively, tested on EMODB dataset. It can be seen that LMSDDC feature (Table 2) provided the improvement of accuracy, precision, recall, and F1-score, when compared with LMS feature (Table 1). In the same way, when the same learning models with different features, LMS and LMSDDC, were tested on RAVDESS dataset, as shown in Tables 3 and 4, the evaluation results were comparable, not much of an improvement. One of the main reasons is that RAVDESS has less speech variation than EMODB, as reported by Breitenstein research [34]. The less variation of speech leads to the lower quality of features. This made no difference in quality of LMS and LMSDDC features. These results have proved that the LMSDDC feature extracted with three components of human emotions: glottal flow, prosody, and human hearing, usually provided wider speech band frequencies, can improve the speech emotion recognition, especially with high speech variation datasets.

When considering the efficiency of the learning model, Tables 1, 2, 3, and 4 show that DeepResLFLB outperforms the baselines with the highest accuracy, precision, recall, and F1-score. This achievement proved that a learning sequence of DeepResLFLB, imitated from "repeatedly reading" concept of human, is efficient. In addition, DeepResLFLB can avoid a vanishing gradient problem and reduce resource-consuming. Figure 5 shows that DeepResLFLB had better validation loss and generalization; it can be seen from the graph that has less fluctuation than 2D-LFLB, and Table 5 reports that DeepResLFLB still used fewer parameters than the baseline model around 40%. These results have proved that DeepResLFLB used residual deep learning by arranging its internal network as LFLB, BN, activation function, NAC, and deep layers can solve vanishing gradient and resource-consuming. Besides, when regarding resource-consuming between LMS and LMSDDC, LMSDDC parameters were slightly more than a baseline.

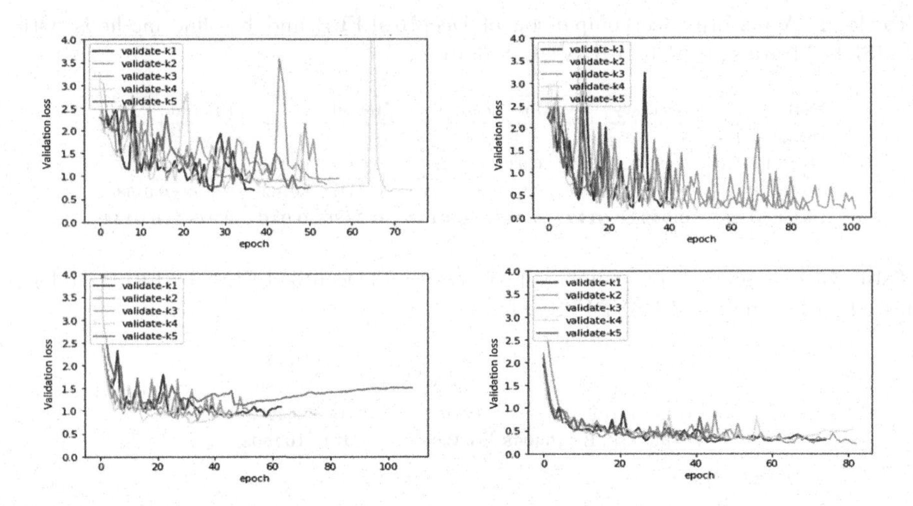


**Fig. 5.** Validation loss of learning models: conventional LFLB tested on EMODB (top-left), DeepResLFLB tested on EMODB (bottom-left), conventional LFLB tested on RAVDESS (top-right), and DeepResLFLB tested on RAVDESS (bottom-right). Note that only LMS feature was used to test in this experiment.

## 5 Conclusion

This paper has described a DeepResLFLB model and LMSDDC feature for speech emotion recognition. The DeepResLFLB was redesigned from LFLB based on the 'repeatedly reads' concept while the LMSDDC was emotional feature extracted from speech signals based on human glottal flow and human hearing. Performance of our model and emotional feature was tested on two well-known databases. The results show that the DeepResLFLB can perform better than baselines and use fewer resources in learning layers. In addition, the proposed LMSDDC can outperform conventional LMS.

Although DeepResLFLB presented in this paper have provided better performance in speech emotion recognition, many aspects still can be improved, especially activation function. In future work, we will apply different kinds of activation function in each section of neural network; this will improve the performance of DeepResLFLB.

**Acknowledgments.** We would like to thank Science Research Foundation, Siam Commercial Bank, for partial financial support to this work.

## References

1. Singkul, S., Khampingyot, B., Maharattamalai, N., Taerungruang, S., Chalothorn, T.: Parsing thai social data: a new challenge for thai NLP. In: 2019 14th International Joint Symposium on Artificial Intelligence and Natural Language Processing (iSAI-NLP), pp. 1–7 (2019)
2. Singkul, S., Woraratpanya, K.: Thai dependency parsing with character embedding. In: 2019 11th International Conference on Information Technology and Electrical Engineering (ICITEE), pp. 1–5 (2019)
3. El Ayadi, M., Kamel, M.S., Karray, F.: Survey on speech emotion recognition: features, classification schemes, and databases. Pattern Recogn. **44**(3), 572–587 (2011)
4. Anagnostopoulos, C.-N., Iliou, T., Giannoukos, I.: Features and classifiers for emotion recognition from speech: a survey from 2000 to 2011. Artif. Intell. Rev. **43**(2), 155–177 (2012). https://doi.org/10.1007/s10462-012-9368-5
5. Zhang, Z., Coutinho, E., Deng, J., Schuller, B.: Cooperative learning and its application to emotion recognition from speech. IEEE/ACM Trans. Audio, Speech Lang. Proces. **23**(1), 115–126 (2014)
6. Guidi, A., Vanello, N., Bertschy, G., Gentili, C., Landini, L., Scilingo, E.P.: Automatic analysis of speech f0 contour for the characterization of mood changes in bipolar patients. Biomed. Signal Process. Control **17**, 29–37 (2015)
7. Gunes, H., Piccardi, M.: Bi-modal emotion recognition from expressive face and body gestures. J. Netw. Comput. Appl. **30**(4), 1334–1345 (2007)
8. Bong, S.Z., Wan, K., Murugappan, M., Ibrahim, N.M., Rajamanickam, Y., Mohamad, K.: Implementation of wavelet packet transform and non linear analysis for emotion classification in stroke patient using brain signals. Biomed. Signal Process. Control **36**, 102–112 (2017)
9. Yuvaraj, R., Murugappan, M., Ibrahim, N.M., Sundaraj, K., Omar, M.I., Mohamad, K., Palaniappan, R.: Detection of emotions in parkinson's disease using higher order spectral features from brain's electrical activity. Biomed. Signal Process. Control **14**, 108–116 (2014)
10. Zhao, J., Mao, X., Chen, L.: Speech emotion recognition using deep 1d & 2d CNN LSTM networks. Biomed. Signal Process. Control **47**, 312–323 (2019)
11. He, K., Zhang, X., Ren, S., Sun, J.: Deep residual learning for image recognition. In: Proceedings of the IEEE conference on computer vision and pattern recognition, pp. 770–778 (2016)
12. Wu, S., Falk, T.H., Chan, W.Y.: Automatic speech emotion recognition using modulation spectral features. Speech Commun. **53**(5), 768–785 (2011)
13. He, L., Lech, M., Maddage, N.C., Allen, N.B.: Study of empirical mode decomposition and spectral analysis for stress and emotion classification in natural speech. Biomed. Signal Process. Control **6**(2), 139–146 (2011)
14. Pérez-Espinosa, H., Reyes-Garcia, C.A., Villaseñor-Pineda, L.: Acoustic feature selection and classification of emotions in speech using a 3d continuous emotion model. Biomed. Signal Process. Control **7**(1), 79–87 (2012)
15. Huang, Z., Dong, M., Mao, Q., Zhan, Y.: Speech emotion recognition using CNN. In: Proceedings of the 22nd ACM international conference on Multimedia. pp. 801–804 (2014)

16. Huang, Y., Wu, A., Zhang, G., Li, Y.: Extraction of adaptive wavelet packet filter-bank-based acoustic feature for speech emotion recognition. IET Signal Proc. **9**(4), 341–348 (2015)
17. Demircan, S., Kahramanli, H.: Application of fuzzy c-means clustering algorithm to spectral features for emotion classification from speech. Neural Comput. Appl. **29**(8), 59–66 (2018)
18. Sun, Y., Wen, G., Wang, J.: Weighted spectral features based on local hu moments for speech emotion recognition. Biomed. Signal Process. Control **18**, 80–90 (2015)
19. Sari, S.W.W.: The influence of using repeated reading strategy towards student's reading comprehension. In: Proceeding 1st Annual International Conference on Islamic Education and Language: The Education and 4.0 Industrial Era in Islamic Perspective, p. 71 (2019)
20. Shanahan, T.: Everything you wanted to know about repeated reading. Reading Rockets. https://www.readingrockets.org/blogs/shanahan-literacy/everything-you-wanted-know-about-repeated-reading (2017)
21. Venkataramanan, K., Rajamohan, H.R.: Emotion recognition from speech (2019)
22. Soekhoe, D., Putten, P., Plaat, A.: On the impact of data set size in transfer learning using deep neural networks, pp. 50–60 (2016)
23. Park, D.S., et al: Specaugment: a simple data augmentation method for automatic speech recognition. arXiv preprint arXiv:1904.08779 (2019)
24. Jagini, N.P., Rao, R.R.: Exploring emotion specific features for emotion recognition system using pca approach. In: 2017 International Conference on Intelligent Computing and Control Systems (ICICCS), pp. 58–62 (2017)
25. Degottex, G.: Glottal source and vocal-tract separation. Ph.D. thesis (2010)
26. Doukhan, D., Carrive, J., Vallet, F., Larcher, A., Meignier, S.: An open-source speaker gender detection framework for monitoring gender equality. In: 2018 IEEE International Conference on Acoustics, Speech and Signal Processing (ICASSP). pp. 5214–5218. IEEE (2018)
27. Doval, B., d'Alessandro, C., Henrich, N.: The spectrum of glottal flow models. Acta acustica united with acustica **92**(6), 1026–1046 (2006)
28. Wang, Y., Guan, L.: Recognizing human emotional state from audiovisual signals. IEEE Trans. Multimedia **10**(5), 936–946 (2008)
29. Robinson, K., Patterson, R.D.: The stimulus duration required to identify vowels, their octave, and their pitch chroma. J. Acoust. Soc. Am. **98**(4), 1858–1865 (1995)
30. Wakefield, G.H.: Chromagram visualization of the singing voice. In: International Workshop on Models and Analysis of Vocal Emissions for Biomedical Applications (1999)
31. He, K., Zhang, X., Ren, S., Sun, J.: Identity mappings in deep residual networks. In: Leibe, B., Matas, J., Sebe, N., Welling, M. (eds.) ECCV 2016. LNCS, vol. 9908, pp. 630–645. Springer, Cham (2016). https://doi.org/10.1007/978-3-319-46493-0_38
32. Burkhardt, F., Paeschke, A., Rolfes, M., Sendlmeier, W.F., Weiss, B.: A database of german emotional speech. In: Ninth European Conference on Speech Communication and Technology (2005)
33. Livingstone, S.R., Russo, F.A.: The ryerson audio-visual database of emotional speech and song (ravdess): a dynamic, multimodal set of facial and vocal expressions in north american english. PloS one **13**(5) (2018)
34. Breitenstein, C., Lancker, D.V., Daum, I.: The contribution of speech rate and pitch variation to the perception of vocal emotions in a German and an American sample. Cogn. Emotion **15**(1), 57–79 (2001)

# Densely Multi-path Network for Single Image Super-Resolution

Shi Xu[1] and Li Zhang[1,2](✉)

[1] School of Computer Science and Technology and Joint International Research Laboratory of Machine Learning and Neuromorphic Computing, Soochow University, Suzhou 215006, Jiangsu, China
20185227038@stu.suda.edu.cn

[2] Provincial Key Laboratory for Computer Information Processing Technology, Soochow University, Suzhou 215006, Jiangsu, China
zhangliml@suda.edu.cn

**Abstract.** Recently, deep convolutional neural networks (CNNs) make many breakthroughs in accuracy and speed for single image super-resolution (SISR). However, we observe that the fusion of information on different receptive fields have not been fully exploited in current SR methods. In this paper, we propose a novel deep densely multi-path network (DMPN) for SISR that introduces densely multi-path blocks (DMPBs). A DMPB contains several multi-path subnets (MPSs) with dense skip connections, and concatenates the outputs of MPSs that are fed into the next DMPB. A MPS uses convolution kernels of different sizes in each path, and exchanges information through cross-path skip connections. Such a multi-path fusion strategy allows the network to make full use of different levels of information and better adapt for extracting high-frequency features. Quantitative and qualitative experimental results indicate the effectiveness of the proposed DMPN, which achieves better restoration performance and visual effects than state-of-the-art algorithms.

**Keywords:** Deep convolutional neural networks · Computer vision · Super-resolution · Multi-path

## 1 Introduction

Single image super-resolution (SISR) is designed to recover a satisfactory high-resolution (HR) image from its associated low-resolution (LR) image, which is an important kind of application in computer vision and image processing. Super-resolution (SR) technology is of great value because it can be effectively used for

Supported by the Natural Science Foundation of the Jiangsu Higher Education Institutions of China under Grant No. 19KJA550002, the Six Talent Peak Project of Jiangsu Province of China under Grant No. XYDXX-054, and the Priority Academic Program Development of Jiangsu Higher Education Institutions.


H. Yang et al. (Eds.): ICONIP 2020, LNCS 12532, pp. 253–265, 2020.
https://doi.org/10.1007/978-3-030-63830-6_22

a variety of applications, such as surveillance [23,33], medical imaging [6,11], and ultra-high-definition content generation [21,25]. Even after decades of extensive researches, SISR is still a very challenging and open research problem owing to its inherently ill-posed inverse problem. Given an LR image, there exist numerous HR images that can be down sampled to the same LR image.

Inspired by the achievement of other computer vision tasks tackled in deep learning, convolutional neural network (CNN) has been introduced into SISR in these years. Dong et al. [4] first proposed a 3-layer super-resolution convolutional neural network (SRCNN) to optimize nonlinear mapping and feature extraction, which has attracted great attention for its promising results. Much recent work on SISR shows that a deeper model would lead to the better performance. Increasing the depth or width of a network can enrich receptive fields that provide more features information to help to reconstruct SR images.

Although the above methods for SISR have achieved prominent results, there are still some drawbacks. First, a deeper or wider network achieves the better performance by adding more new layers, which requires higher computational costs and larger storage space. The huge model may be unacceptable in practice. Second, the traditional CNN structure tends to adopt a simple stacked way. Thus, a deeply-recursive convolutional network (DRCN) [15] was proposed to remedy the two issues by constructing recursive units and sharing weights parameters. However, it is still not easy for DRCN to converge for the issue of vanishing or exploding gradient during the training procedure. On the basis of DRCN, a deep recursive residual network (DRRN) [26] was presented by applying the global and local residual learning to help with both feature extraction and gradient flow. However, these methods need to resize the inputs (LR images) to the desired size before training, which results in higher computational complexity, and over-smooths or blurs LR images. Tong et al. [29] took the architecture of densely connected convolutional network (DenseNet) [9] as the building block for SISR instead of residual neural network (ResNet) [7] and presented super-resolution DenseNet (SRDenseNet). Additionally, SRDenseNet uses the deconvolutional layers for upscaling in the end of network, which helps to infer the high frequency details. However, DenseNet was specially designed for image classification instead of SISR. Thus, it may be inappropriate to directly employ DenseNet to SISR that is one kine of the low-level vision problem. Lai et al. [17] proposed a Laplacian Pyramid Super-Resolution Network (LapSRN) that consists of a feature extraction path and an image reconstruction path. LapSRN progressively reconstructs the sub-band residuals of HR images in one feed-forward network. However, LapSRN neglects to use hierarchical features for reconstruction. Li et al. [18] proposed a multi-scale residual network (MSRN) that uses a two-bypass network to explore the image features in different aspects, which provides better modeling capabilities. But it is hard to directly extract the output of each convolutional layer in the LR space.

To address the aforementioned issues, this paper proposes a densely multi-path network (DMPN). In the training procedure, DMPN directly takes LR images as inputs without upscaling them to the desired spatial resolution using the pre-defined upsampling operator. In DMPN, a module of densely multi-path blocks (DMPBs) is the key component and aggregates different types of features to obtain more abundant and efficient information, which helps to solve the image reconstruction tasks. A reconstruction module generates SR images from DMPBs and trains the network with a combined $L_1$ loss function. The main contributions of our work are summarized as follows:

(1) We design a new module that consists of multiple DMPBs, sharing parameters among them to avoid the number of parameters growing fast. A multi-path subnet (MPS) provides high-level feature representations in cross-path flows through skip connections. In each DMPB, several MPSs are connected through dense skip connections for fusing different levels of feature information, which provides richer information to reconstruct high-quality details.
(2) On the basis of the DMPB module, we propose a novel DMPN with 59 convolutional layers for image super-resolution. We utilize all generated SR images from DMPBs to train DMPN, which is conducive to obtaining powerful early reconstruction ability. Our model achieves the high performance and competitive results in common benchmarks.

## 2 Related Work

For decades, numerous super-resolution methods have been proposed, such as prediction-based methods [13], edge-based methods [12], statistical methods [30], patch-based methods [5], sparse representation methods [31], and CNN-based methods. Here, we focus on the CNN-based methods.

With the rapid development of deep learning techniques in recent years, SR models based on data-driven deep learning have been actively explored to restore the required reconstruction details, which often reach the most advanced performance on various super-resolution benchmark datasets. In contrast to modeling the LR-HR mapping in the patch space, Dong et al. [4] proposed a super-resolution convolutional neural network (SRCNN) that learns the nonlinear LR-HR mapping in an end-to-end manner and outperforms classical shallow methods. Kim et al. [14] proposed a very deep CNN model known as very deep super-resolution (VDSR), which increases the depth from 3 layers in SRCNN to 20 layers. VDSR utilizes the gradient clipping and residual learning to accelerate network convergence during training. Tai et al. [27] proposed a very deep end-to-end new type of persistent memory network (MemNet), which can be divided into three parts similar to SRCNN. The stacked memory blocks play the most crucial role in the network, which consists of a recursive unit and a gate unit where the outputs of each recursion are concatenated to go through an extra $1 \times 1$ convolution for memorizing and forgetting. The densely connected structure helps to compensate for mid-to-high frequency signals and ensures that information flows adequately between memory blocks. Shi et al. [24] proposed an efficient sub-pixel convolutional network (ESPCN), which performs feature extraction in the LR space. The sub-pixel convolution operation is used to increase the spatial resolution of input feature maps and results in lower computational requirements.

Although a deeper or wider network would achieve better performance by adding more layers, it requires higher computational costs and larger storage space. To remedy it, Kim et al. [15] presented a deeply recursive convolutional network (DRCN) that consists of three smaller sub-networks and utilizes the same convolutional layer multiple times. In fact, DRCN adopts a very deep recursive layer. The advantage of this technique is that the number of parameters remains the same for more recursions. In addition, the recursive supervision and skip connections can be used to further improve performance. Tai et al. [26] designed a deep recursive residual network (DRRN) by introducing a deeper architecture that contains up to 52 convolutional layers.

DRRN effectively alleviates the difficulty of training deep networks by using global and local residual learning, and passes rich image details to subsequent layers. Tong et al. [29] proposed a method called SRDenseNet in which the current layer can directly operate on the outputs from all previous layers via dense connections. SRDenseNet improves information flows through the network, alleviating the gradient vanishing problem. Motivated by the inception module, Li et al. [18] invented a multi-scale residual network (MSRN) that builds a new block for local multi-scale feature extraction. MSRN, the first multiple scale method based on the residual learning, can adaptively extract the features mappings and achieve the fusion of features at different scales.

# 3 Proposed Method

This section presents our method DMPN. We first describe the framework of DMPN and then explain its modules that are the core of the proposed method.

## 3.1 Network Framework

As shown in Fig. 1, the proposed DMPN mainly consists of three parts: a feature extraction (FE) module, a DMPB module, and a reconstruction module. Let $I_{LR}$ and $I_{SR}$ be the input and the output of DMPN, respectively.

In the feature extraction module, we utilize one $3 \times 3$ convolutional layer to extract the shallow feature maps from the original LR images. This procedure can be expressed as

$$F_0 = f_s(I_{LR}) \tag{1}$$

where $f_s(\cdot)$ represents the feature extraction function, and $F_0$ denotes the shallow coarse features and servers as the input to the next stage.

In the DMPB module, multiple DMPBs gradually generates complex and abstract features. Namely,

$$F_1 = R_1(F_0) \tag{2}$$

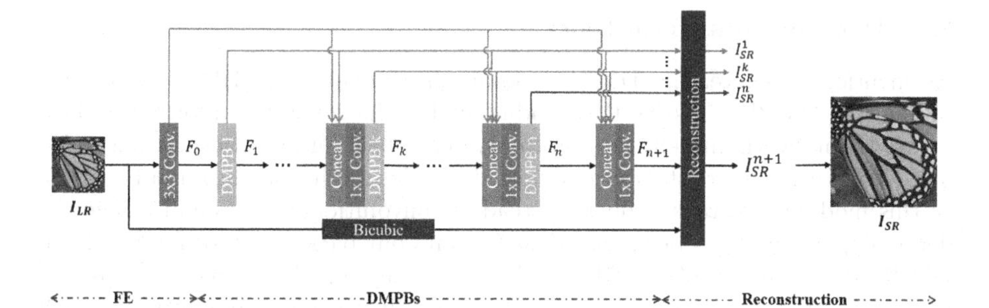


**Fig. 1.** Architecture of the densely multi-path network (DMPN).

and

$$F_k = R_k(f_1([F_0, F_1, \cdots, F_{k-1}])),\ k = 2, 3, \cdots, n \tag{3}$$

where $f_1(\cdot)$ denotes the convolution function with $1 \times 1$ kernel, $[\cdot]$ denotes the concatenation in the channels, $R_k(\cdot)$ denotes the $k$-th DMPB function, $F_{k-1}$ and $F_k$ indicates the input and the output of the $k$-th DMPB, respectively. We unite the outputs from each DMPB and the coarse features $F_0$ to construct $F_{n+1}$ that is the key result of the DMPB module. Namely,

$$F_{n+1} = f_1([F_0, F_1, \cdots, F_n]) \tag{4}$$

In the reconstruction module, we restore the residual images $F_k, k = 1, \cdots, n$ generated by each DMPB and $F_{n+1}$ instead of learning the direct maps from the LR images to the HR images. Hence, the SR images can be formulated as

$$I_{SR}^k = f_{res}(F_k) + f_{up}(I_{LR}),\ k = 1, 2, \cdots, n+1 \tag{5}$$

where $f_{res}(\cdot)$ denotes the reconstruction function and $f_{up}(\cdot)$ denotes an specific interpolation operation. Here, we choose a bicubic interpolation method for $f_{up}(\cdot)$. Note that we take the SR image $I_{SR}^{n+1}$ as the final SR image.

In the training procedure, we need a loss function to guide the optimization of DMPN. Presently, both $L_1$ and $L_2$ loss functions have been widely applied to CNN. Because Lim et al. [19] experimentally demonstrated that the $L_2$ loss is not a good choice, we adopt the $L_1$ loss to train DMPN. Given a batch training set $\{(I_{LR}^i, I_{HR}^i)\}_{i=1}^N$, the loss function in the network can be formulated as:

$$L(\Theta) = \frac{1}{N}\sum_{i=1}^{N}\sum_{k=1}^{n+1} W^k \|I_{SR}^{i,k} - I_{HR}^i\|_1 \tag{6}$$

where $\|\cdot\|_1$ denotes the $L_1$ loss, $I_{SR}^{i,k}$ is the output of the $k$-th stage in the DMPB module for $I_{LR}^i$, $\Theta$ denotes parameters of our network, $N$ is the size of each batch, $n$ is the number of DMPBs, and $W^k$ denotes the weight factor for the $k$-th SR image.

### 3.2 Densely Multi-path Block

As mentioned before, the DMPBs module is the key of DMPN and consists of multiple DMPBs. The DMPB module makes the generated information flow across the network and encourages the gradient back propagation during training. Now, we give more structure details about this module as shown in Fig. 2(a). In this module, we stack subnets instead of convolutional layers and utilize the dense skip connections between MPSs to build our backbone flow for SR. Each DMPB that consists of $m$ MPSs takes advantage of all the hierarchical features through dense skip connections. Additionally, the input and output of each DMPB are consistent in dimensions due to sharing parameters among DMPBs. Hence, the number of the parameters in this method could be obviously reduced while a satisfactory reconstruction performance could be maintained on the corresponding information.

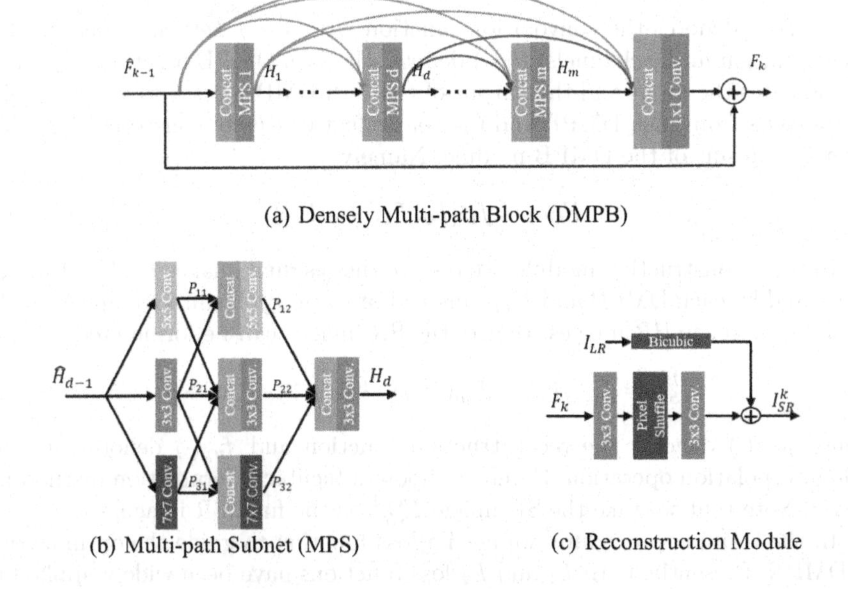


**Fig. 2.** Specific modules of DMPN.

Each convolution in a DMPB is followed by the activation function, parametric rectified linear unit (PReLU) [8], which is omitted here for simplification. The operations can be expressed as:

$$H_d = M_d([\hat{F}_{k-1}, H_1, \cdots, H_{d-1}]),\ d = 1, 2, \cdots, m \tag{7}$$

and

$$F_k = \hat{F}_{k-1} + f_1([\hat{F}_{k-1}, H_1, \cdots, H_m]),\ k = 1, \cdots, n \tag{8}$$

where $M_d(\cdot)$ denotes the function of $d$-th MPS, $\hat{F}_{k-1}$ and $H_d$ denote the input of $k$-th DMPB and the output of the $k$-th MPS, respectively.

The role of MPS is to mix together features in each different paths and distill more efficacious information from receptive fields with different levels. The design of an MPS is illustrated in Fig. 2(b). An MPS adopts three paths with different sizes of convolution kernels, which can be described as follows:

$$P_{12} = f_5([P_{11}, P_{21}]) = f_5([f_5(\hat{H}_{d-1}), f_3(\hat{H}_{d-1})]) \quad (9)$$

$$P_{22} = f_3([P_{11}, P_{21}, P_{31}]) = f_3([f_5(\hat{H}_{d-1}), f_3(\hat{H}_{d-1}), f_7(\hat{H}_{d-1})]) \quad (10)$$

$$P_{32} = f_7([P_{21}, P_{31}]) = f_7([f_3(\hat{H}_{d-1}), f_7(\hat{H}_{d-1})]) \quad (11)$$

$$H_d = f_1([P_{12}, P_{22}, P_{32}]) \quad (12)$$

where $P_{rj}$, $r = 1, 2, 3$, $j = 1, 2$, is the intermediate result in each path, $\hat{H}_{d-1}$, $H_d$ are the input and output of the MPS, respectively; $f_3$, $f_5$, and $f_7$ are the convolution functions with the sizes of 3, 5, and 7.

### 3.3 Reconstruction Block

In order to reduce the computational burden, DMPN performs in the low-dimensional space without the predefined upsampling operations. As shown in Fig. 2(c), the final reconstruction block utilizes a sub-pixel layer [24] and two $3 \times 3$ convolutional layers, which can be formulated as

$$I_{SR}^k = f_3(PS(f_3(F_k))) + f_{up}(I_{LR}), \ k = 1, \cdots, n+1 \quad (13)$$

where $PS(\cdot)$ is an operation of sub-pixel layer and $I_{SR}^k$ is the output of the $k$-th stage in the DMPB module.

## 4 Experiments

In this section, we evaluate the performance of the proposed method on four benchmark datasets. A description of these datasets is first provided, followed by the introduction of the experimental implementation details. The benefit of using different levels of features is then introduced. After that, comparisons with state-of-the-art results are presented.

### 4.1 Datasets and Metrics

In the training progress, we combine the open high-quality (2K resolution) datasets DIV2K (800 images) [1] and Flickr2K (2650 images) [28] as a new training dataset for the SISR task. During testing, we evaluate the proposed method on four publicly available benchmark datasets: Set5 [3], Set14 [32], BSD100 [22], and Urban100 [10]. As commonly done in SISR, we evaluated DMPN results with two standard metrics: the peak signal-to-noise ratio (PSNR) and the structural similarity index (SSIM) [34]. The training and testing of DMPN are performed on the RGB channels.

**Table 1.** Average PSNR/SSIM evaluation of state-of-the-art SR algorithms

| Algorithm | Scale | Set5 | Set14 | BSD100 | Urban100 |
|---|---|---|---|---|---|
| Bicubic | 2 | 33.66/0.9299 | 30.24/0.8688 | 29.56/0.8431 | 26.88/0.8403 |
| VDSR[14] | 2 | 37.53/0.9587 | 33.03/0.9124 | 31.90/0.8960 | 30.76/0.9140 |
| DRCN[15] | 2 | 37.63/0.9588 | 33.04/0.9118 | 31.85/0.8942 | 30.75/0.9133 |
| DRRN[26] | 2 | 37.74/0.9591 | 33.23/0.9136 | 32.05/0.8973 | 31.23/0.9188 |
| LapSRN [17] | 2 | 37.52/0.9581 | 33.08/0.9109 | 31.80/0.8949 | 30.41/0.9112 |
| MemNet [27] | 2 | 37.78/0.9597 | 33.28/0.9142 | 32.08/0.8978 | 31.31/0.9195 |
| CARN [2] | 2 | 37.76/0.9590 | 33.52/0.9166 | 32.09/0.8978 | 31.51/0.9312 |
| NLRN [20] | 2 | 38.00/0.9603 | 33.46/0.9159 | 32.19/0.8992 | 31.81/0.9249 |
| MSRN [18] | 2 | 38.08/0.9605 | 33.74/0.9170 | 32.23/0.9013 | 32.22/0.9326 |
| DMPN (ours) | 2 | 38.12/0.9608 | 33.70/0.9183 | 32.27/0.9009 | 32.51/0.9320 |
| Bicubic | 3 | 30.39/0.8682 | 27.55/0.7742 | 27.21/0.7385 | 24.46/0.7349 |
| VDSR [14] | 3 | 33.66/0.9213 | 29.77/0.8314 | 28.82/0.7976 | 27.14/0.8279 |
| DRCN [15] | 3 | 33.82/0.9226 | 29.76/0.8311 | 28.80/0.7963 | 27.15/0.8276 |
| DRRN [26] | 3 | 34.03/0.9244 | 29.96/0.8349 | 28.95/0.8004 | 27.53/0.8378 |
| LapSRN [17] | 3 | 33.82/0.9207 | 29.89/0.8304 | 28.82/0.7950 | 27.07/0.8298 |
| MemNet [27] | 3 | 34.09/0.9248 | 30.00/0.8350 | 28.96/0.8001 | 27.56/0.8376 |
| CARN [2] | 3 | 34.29/0.9255 | 30.29/0.8407 | 29.06/0.8034 | 27.38/0.8404 |
| NLRN [20] | 3 | 34.27/0.9266 | 30.16/0.8374 | 29.06/0.8026 | 27.93/0.8453 |
| MSRN [18] | 3 | 34.38/0.9262 | 30.34/0.8395 | 29.08/0.8041 | 28.08/0.8554 |
| DMPN (ours) | 3 | 34.53/0.9283 | 30.43/0.8445 | 29.19/0.8077 | 28.48/0.8601 |
| Bicubic | 4 | 28.42/0.8104 | 26.00/0.7027 | 25.96/0.6675 | 23.14/0.6577 |
| VDSR [14] | 4 | 31.35/0.8838 | 28.01/0.7674 | 27.29/0.7251 | 25.18/0.7524 |
| DRCN [15] | 4 | 31.53/0.8854 | 28.02/0.7670 | 27.23/0.7233 | 25.14/0.7510 |
| DRRN [26] | 4 | 31.68/0.8888 | 28.21/0.7721 | 27.38/0.7284 | 25.44/0.7638 |
| LapSRN [17] | 4 | 31.54/0.8811 | 28.19/0.7635 | 27.32/0.7162 | 25.21/0.7564 |
| MemNet [27] | 4 | 31.74/0.8893 | 28.26/0.7723 | 27.40/0.7281 | 25.50/0.7630 |
| SRDenseNet [29] | 4 | 32.02/0.8934 | 28.50/0.7782 | 27.53/0.7337 | 26.05/0.7819 |
| CARN [2] | 4 | 32.13/0.8937 | 28.60/0.7806 | 27.58/0.7349 | 26.07/0.7837 |
| NLRN [20] | 4 | 31.92/0.8916 | 28.36/0.7745 | 27.48/0.7306 | 25.79/0.7729 |
| MSRN [18] | 4 | 32.07/0.8903 | 28.60/0.7751 | 27.52/0.7273 | 26.04/0.7896 |
| DMPN (ours) | 4 | 32.33/0.8966 | 28.72/0.7844 | 27.65/0.7387 | 26.33/0.7942 |

* Red indicates the best performance, and blue the second best.

### 4.2 Training Details

During training, we randomly crop 16 LR RGB images with a size of $48 \times 48$ as inputs for each batch. The corresponding HR patches are sub-images with $48s \times 48s$ size, where $s$ is a upscaling factor and $s = 2, 3, 4$. Additionally, we adopt the data augmentation technique to diversify the training images by randomly rotation and flipping. We build DMPN with 59 layers. There are 4 DMPBs in the DMPB module and 4 MPSs in one DMPB (i.e., $n = 4$ and $m = 4$). Each convolutional layer produces 64-channel feature maps except for the layers generating $64s^2$-channel feature maps or a 3-channel RGB image in the reconstruction module.

We subtract the mean image of all training ones from all inputs and add the mean image to all outputs in the final. The proposed DMPN is trained by applying the Adam algorithm [16] with $\beta_1 = 0.9$, $\beta_2 = 0.999$ and $\epsilon = 10^{-8}$. We initialize the weights using the scheme described in [8] and the biases with zero. The learning rate is initially set to $10^{-4}$ and then is decreased to half in every 200 epochs of back-propagation before 1000 epochs. We implement our DMPN with the PyTorch framework in a TITAN V GPU.

### 4.3 Comparisons with State-of-the-arts

We compare the proposed networks with other SR algorithms, including Bicubic, VDSR [14], DRCN [15], DRRN [26], LapSRN [17], MemNet [27], SRDenseNet [29], CARN [2], NLRN [20] and MSRN [18].

We calculate the average PSNR and SSIM values on four test datasets mentioned above. The results are shown in Table 1. Observation on Table 1 indicates that our DMPN exhibits the improvement compared to the other methods. DMPN achieves the best PSNR on all scaling factors and all datasets except for $\times 2$ on Set14, where DMPN is 0.04 dB worse in PSNR than MSRN. On the Urban100 dataset that consists of regular patterns from buildings, DMPN is 0.29 dB higher in PSNR and $\times 2$ than the second best model. Moreover, DMPN is superior in terms of SSIM values, which implies that the model can recover the structural information more effectively.

Apart from the quantitative comparison, Fig. 3 shows qualitative visually comparison results in the $\times 4$ enlargement, which clearly demonstrates the advantage of DMPN over other CNN models. For the images "barbara" and "8023", most of previous methods generate distorted blurry artifacts at the fringes, especially for those in the cropped images. Our DMPN can distinguish the mixture of lines more effectively to restore the structural information, which explains why the model can achieve a promising SSIM result. For the image "img_004", DMPN can still produce the result much closer to the HR image, while the results of other methods have an obvious loss in the structure and the texture blur.

**Fig. 3.** Qualitative comparison of our model with other works on ×4 upscaling.

## 5 Conclusions and Future Works

In this paper, we present a novel densely multi-path network for single image super-resolution. To obtain high-quality performance, our model focuses on using different levels of information to enhance the ability of feature representations. For this, we design the DMPB that can effectively improve the flow of information across paths and provides more powerful representation ability. Meanwhile, the loss function urges each DMPB to reconstruct fine structures as soon as possible. The proposed DMPN achieves competitive results with different scale factors, in terms of both efficiency and accuracy.

It is well known that the human visual system does not capture information from the entire scene, but captures information from significant parts to improve efficiency. In the future work, we will introduce attention mechanism into the multi-path structure. Our DMPN has been shown to be effective for bicubic (BI) degradation scenarios. Next, we will improve the model so that it is also suitable for blur-downscale (BD) degradation and downscale-noise (DN) degradation scenarios.

## References


1. Agustsson, E., Timofte, R.: Ntire 2017 challenge on single image super-resolution: dataset and study. In: Computer Vision and Pattern Recognition Workshops, pp. 1122–1131. IEEE (2017). https://doi.org/10.1109/CVPRW.2017.150
2. Ahn, N., Kang, B., Sohn, K.-A.: Fast, accurate, and lightweight super-resolution with cascading residual network. In: Ferrari, V., Hebert, M., Sminchisescu, C., Weiss, Y. (eds.) ECCV 2018. LNCS, vol. 11214, pp. 256–272. Springer, Cham (2018). https://doi.org/10.1007/978-3-030-01249-6_16
3. Bevilacqua, M., Roumy, A., Guillemot, C., Alberi Morel, M.L.: Low-complexity single-image super-resolution based on nonnegative neighbor embedding. In: British Machine Vision Conference, pp. 135.1-135.10. BMVA (2012). https://doi.org/10.5244/C.26.135
4. Dong, C., Loy, C.C., He, K., Tang, X.: Learning a deep convolutional network for image super-resolution. In: Fleet, D., Pajdla, T., Schiele, B., Tuytelaars, T. (eds.) ECCV 2014. LNCS, vol. 8692, pp. 184–199. Springer, Cham (2014). https://doi.org/10.1007/978-3-319-10593-2_13
5. Freeman, W.T., Jones, T.R., Pasztor, E.C.: Example-based super-resolution. IEEE Comput. Graphics Appl. **22**(2), 56–65 (2002)
6. Greenspan, H.: Super-resolution in medical imaging. Comput. J. **52**(1), 43–63 (2008)
7. He, K., Zhang, X., Ren, S., Sun, J.: Deep residual learning for image recognition. In: Computer Vision and Pattern Recognition, pp. 770–778. IEEE (2016). https://doi.org/10.1109/CVPR.2016.90
8. He, K., Zhang, X., Ren, S., Sun, J.: Delving deep into rectifiers: surpassing human-level performance on imagenet classification. In: International Conference on Computer Vision, pp. 1026–1034. IEEE (2015). https://doi.org/10.1109/ICCV.2015.123

9. Huang, G., Liu, Z., Van Der Maaten, L., Weinberger, K.Q.: Densely connected convolutional networks. In: Computer Vision and Pattern Recognition, pp. 2261–2269. IEEE (2017). https://doi.org/10.1109/CVPR.2017.243
10. Huang, J., Singh, A., Ahuja, N.: Single image super-resolution from transformed self-exemplars. In: Computer Vision and Pattern Recognition, pp. 5197–5206. IEEE (2015). https://doi.org/10.1109/CVPR.2015.7299156
11. Huang, Y., Shao, L., Frangi, A.F.: Simultaneous super-resolution and cross-modality synthesis of 3d medical images using weakly-supervised joint convolutional sparse coding. In: Computer Vision and Pattern Recognition, pp. 5787–5796. IEEE (2017). https://doi.org/10.1109/CVPR.2017.613
12. Sun, J., Xu, Z, Shum, H.Y.: Image super-resolution using gradient profile prior. In: Computer Vision and Pattern Recognition, pp. 1–8. IEEE (2008). https://doi.org/10.1109/CVPR.2008.4587659
13. Keys, R.: Cubic convolution interpolation for digital image processing. IEEE Trans. Acoust. Speech Signal Process. **29**(6), 1153–1160 (1981)
14. Kim, J., Lee, J.K., Lee, K.M.: Accurate image super-resolution using very deep convolutional networks. In: Computer Vision and Pattern Recognition, pp. 1646–1654. IEEE (2016). https://doi.org/10.1109/CVPR.2016.182
15. Kim, J., Lee, J.K., Lee, K.M.: Deeply-recursive convolutional network for image super-resolution. In: Computer Vision and Pattern Recognition, pp. 1637–1645. IEEE (2016). https://doi.org/10.1109/CVPR.2016.181
16. Kingma, D.P., Ba, J.: Adam: A method for stochastic optimization. In: International Conference on Learning Representations. arXiv:1412.6980 (2015)
17. Lai, W.S., Huang, J.B., Ahuja, N., Yang, M.H.: Deep laplacian pyramid networks for fast and accurate super-resolution. In: Computer Vision and Pattern Recognition, pp. 5835–5843. IEEE (2017). https://doi.org/10.1109/CVPR.2017.618
18. Li, J., Fang, F., Mei, K., Zhang, G.: Multi-scale residual network for image super-resolution. In: Ferrari, V., Hebert, M., Sminchisescu, C., Weiss, Y. (eds.) ECCV 2018. LNCS, vol. 11212, pp. 527–542. Springer, Cham (2018). https://doi.org/10.1007/978-3-030-01237-3_32
19. Lim, B., Son, S., Kim, H., Nah, S., Lee, K.M.: Enhanced deep residual networks for single image super-resolution. In: Computer Vision and Pattern Recognition Workshops, pp. 1132–1140. IEEE (2017). https://doi.org/10.1109/CVPRW.2017.151
20. Liu, D., Wen, B., Fan, Y., Loy, C., Huang, T.: Non-local recurrent network for image restoration. In: Neural Information Processing Systems, pp. 1673–1682. Curran (2018)
21. Lobanov, A.P.: Resolution limits in astronomical images. arXiv:astro-ph/0503225 (2005)
22. Martin, D., Fowlkes, C., Tal, D., Malik, J.: A database of human segmented natural images and its application to evaluating segmentation algorithms and measuring ecological statistics. In: International Conference on Computer Vision. ICCV 2001, vol. 2, pp. 416–423. IEEE (2001). https://doi.org/10.1109/ICCV.2001.937655
23. Rasti, P., Uiboupin, T., Escalera, S., Anbarjafari, G.: Convolutional neural network super resolution for face recognition in surveillance monitoring. In: Perales, F.J.J., Kittler, J. (eds.) AMDO 2016. LNCS, vol. 9756, pp. 175–184. Springer, Cham (2016). https://doi.org/10.1007/978-3-319-41778-3_18
24. Shi, W., et al.: Real-time single image and video super-resolution using an efficient sub-pixel convolutional neural network. In: Computer Vision and Pattern Recognition, pp. 1874–1883. IEEE (2016). https://doi.org/10.1007/978-3-319-41778-3_18

25. Swaminathan, A., Wu, M., Liu, K.J.R.: Digital image forensics via intrinsic fingerprints. IEEE Trans. Inf. Forensics Secur. **3**(1), 101–117 (2008)
26. Tai, Y., Yang, J., Liu, X.: Image super-resolution via deep recursive residual network. In: Computer Vision and Pattern Recognition, pp. 2790–2798. IEEE (2017). https://doi.org/10.1109/CVPR.2017.298
27. Tai, Y., Yang, J., Liu, X., Xu, C.: Memnet: a persistent memory network for image restoration. In: International Conference on Computer Vision, pp. 4549–4557. IEEE (2017). https://doi.org/10.1109/ICCV.2017.486
28. Timofte, R., Agustsson, E., Gool, L.V.: Ntire 2017 challenge on single image super-resolution: Methods and results. In: Computer Vision and Pattern Recognition Workshops, pp. 1110–1121. IEEE (2017). https://doi.org/10.1109/CVPRW.2017.149
29. Tong, T., Li, G., Liu, X., Gao, Q.: Image super-resolution using dense skip connections. In: International Conference on Computer Vision, pp. 4809–4817. IEEE (2017). https://doi.org/10.1109/ICCV.2017.514
30. Xiong, Z., Sun, X., Wu, F.: Robust web image/video super-resolution. IEEE Trans. Image Process. **19**(8), 2017–2028 (2010)
31. Yang, J., Wright, J., Huang, T.S., Ma, Y.: Image super-resolution via sparse representation. IEEE Trans. Image Process. **19**(11), 2861–2873 (2010)
32. Zeyde, R., Elad, M., Protter, M.: On single image scale-up using sparse-representations. In: Boissonnat, J.-D., Chenin, P., Cohen, A., Gout, C., Lyche, T., Mazure, M.-L., Schumaker, L. (eds.) Curves and Surfaces 2010. LNCS, vol. 6920, pp. 711–730. Springer, Heidelberg (2012). https://doi.org/10.1007/978-3-642-27413-8_47
33. Zhang, L., Zhang, H., Shen, H., Li, P.: A super-resolution reconstruction algorithm for surveillance images. Sig. Process. **90**(3), 848–859 (2010)
34. Wang, Z., Bovik, A.C., Sheikh, H.R., Simoncelli, E.P.: Image quality assessment: from error visibility to structural similarity. IEEE Trans. Image Process. **13**(4), 600–612 (2004)

# Denstity Level Aware Network for Crowd Counting

Wencai Zhong[1,2], Wei Wang[1], and Hongtao Lu[1(✉)]

[1] Department of Computer Science and Engineering, Shanghai Jiao Tong University, Shanghai, China
{emperorwen,ieee-wangwei,htlu}@sjtu.edu.cn
[2] Alibaba Group, Hangzhou, China

**Abstract.** Crowd counting has wide applications in video surveillance and public safety, while it remains an extremely challenging task due to large scale variation and diverse crowd distributions. In this paper, we present a novel method called Density Level Aware Network (DLA-Net) to improve the density map estimation in varying density scenes. Specifically, we divide the input into multiple regions according to their density levels and handle the regions independently. Dense regions (with small scale heads) require higher resolution features from shallow layers, while sparse regions (with large heads) need deep features with broader receptive filed. Based on this requirement, we propose to predict multiple density maps focusing on regions of varying density levels correspondingly. Inspired by the U-Net architecture, our density map estimators borrow features of shallow layers to improve the estimation of dense regions. Moreover, we design a Density Level Aware Loss (DLA-Loss) to better supervise those density maps in different regions. We conduct extensive experiments on three crowd counting datasets (ShanghaiTech, UCF-CC-50 and UCF-QNRF) to validate the effectiveness of the proposed method. The results demonstrate that our DLA-Net achieves the best performance compared with other state-of-the-art approaches.

**Keywords:** Crowd counting · Denstity map estimation · Multi-level supervision

## 1 Introduction

Crowd counting has attracted considerable attention due to its wide range of applications, such as video surveillance, congestion monitoring and crowd control. Due to multiple issues like background clutters, scale variation, non-uniform crowd distributions and perspective distortions, crowd counting remains a difficult task in computer vision, especially in highly congested scenes. Earlier works considered crowd counting as an object detection problem [1,15] or regression problem [7,11]. While these methods are based on handcrafted features that

W. Zhong—Work down as an intern at Alibaba Group.


H. Yang et al. (Eds.): ICONIP 2020, LNCS 12532, pp. 266–277, 2020.
https://doi.org/10.1007/978-3-030-63830-6_23

only contain low-level information. In recent years, with tremendous progress in deep learning, much more Convolutional Neural Networks (CNN) based crowd counting approaches [3] are proposed to learn representative features for density map estimation.

One of the most critical issues of crowd counting is scale variation (as shown in Fig. 1), which includes both variations among different images and within a single image. Many works [12,14,21] have been proposed to tackle this problem. MCNN [21] proposed a multi-column architecture with each column using different kernel sizes to capture multi-scale features. Switch-CNN [12] adopted the same architecture of MCNN but further trained a classifier to decide which column to estimate density maps. Albeit great progress achieved by multi-column networks, they suffer from the disadvantages of information redundancy and inefficiency. Therefore, more recent works [6,20] adopted single-column architectures to alleviate scale variation by extracting multi-scale features. For instance, SaCNN [20] proposed to combine the feature maps of multiple layers to adapt to the changes in scale and perspective. TEDNet [6] presented a trellis encoder-decoder network architecture with multi-scale encoder modules and multiple decoder paths to capture multi-scale features.

**Fig. 1.** The scale variation problem lies in two aspects: 1) Within an image. The sizes of heads in the left image vary widely from near to far. 2) Among images. The overall sizes of heads in the left image are obviously larger than that in the right image.

In this paper, we aim to address the scale variation by predicting density maps by regions. The crowd counts of local regions in images vary hugely. Therefore, we divide the density of local regions into 4 levels (from sparse to dense) and then estimate those regions with the idea of divide and conquer. To better estimate the density of those regions, our DLA-Net adopts an encoder-decoder architecture with multiple outputs (see Fig. 2). With the network goes deeper, the features contain more high-level semantic information while less spatial localization information due to the down-sampling operations. Therefore, the outputs of our decoder are predicted from features of different levels. For example, we use small resolution features with broader receptive filed to predict low-density regions where the head sizes are larger. As for dense regions, we borrow features from earlier layers which contains more spatial details to make the prediction. In addition to the basic loss for each output, we also propose a DLA-Loss according to the region partitions, which can make each output focus on regions of corresponding density levels. At the testing stage, we calculate probability maps for density levels from outputs and then use it to estimate the final density map.

In summary, our contributions are three-fold: 1) We propose to employ a multiple outputs encoder-decoder architecture to generate accurate density maps, which is robust to head scale variation across different scenes and within a single scene. 2) We propose a DLA-Loss to supervise regions belongs to different density levels separately, which makes the outputs more focusing on specific regions (sparse regions or dense regions). 3) We conduct extensive experiments on three crowd counting datasets: ShanghaiTech [21], UCF-CC-50 [4] and UCF-QNRF [5]. The results clearly demonstrate the effectiveness of our proposed method. Besides, our model (DLA-Net) achieves state-of-the-art results on all datasets.

## 2 Related Works

Earlier crowd counting approaches rely on slide window to detect human body or body parts by with handcrafted descriptors like HOG [1] and Haar wavelets [15]. Generally, those methods do not work well in congested scenes. Therefore, some prior works deploy regression models to map the low-level image features to total count [11] or density map [7].

Recently, CNN is widely used in crowd counting and other computer vision algorithms [16,17]. Cross-Scene [19] proposed to train a deep convolutional neural network for crowd counting with two objectives: density maps and counts. Fully convolutional networks are explored in [10] to generate density maps. CP-CNN [14] tried to combine global and local contextual information to improve the quality of density maps. CSRNet [8] adopted dilation convolution layers to enlarge the receptive field of features, which is effective in crowd counting. The attention mechanism has recently been incorporated into crowd counting models to enhance performance. ADCrowdNet [9] first utilized negative samples to train an attention map generator to suppress background clutters and then predicted density maps with a multi-scale deformable decoder. ANF [18] proposed attentional neural field which incorporates CRF and self-attention mechanism for crowd counting.

Besides, a large group of density estimation models is designed to address the scale variation problem. MCNN [21] proposed a multi-column architecture with varying kernel sizes for each column to extract multi-scale features. Switch-CNN [12] trained a switch classifier to select the best density regressor for input image patches from several independent density regressors. ACSCP [13] introduced a scale-consistency regularizer to guarantee the collaboration of different columns. Our DLA-Net is more efficient in comparison by predicting multiple outputs for image regions of varying density levels with a single column architecture. TED-Net [6] designed a trellis encoder-decoder network with a multi-scale encoder and a multi-path decoder. The encoder blocks are implemented with kernels of different sizes, and the decoder produces intermediate estimation maps to provide better supervision.

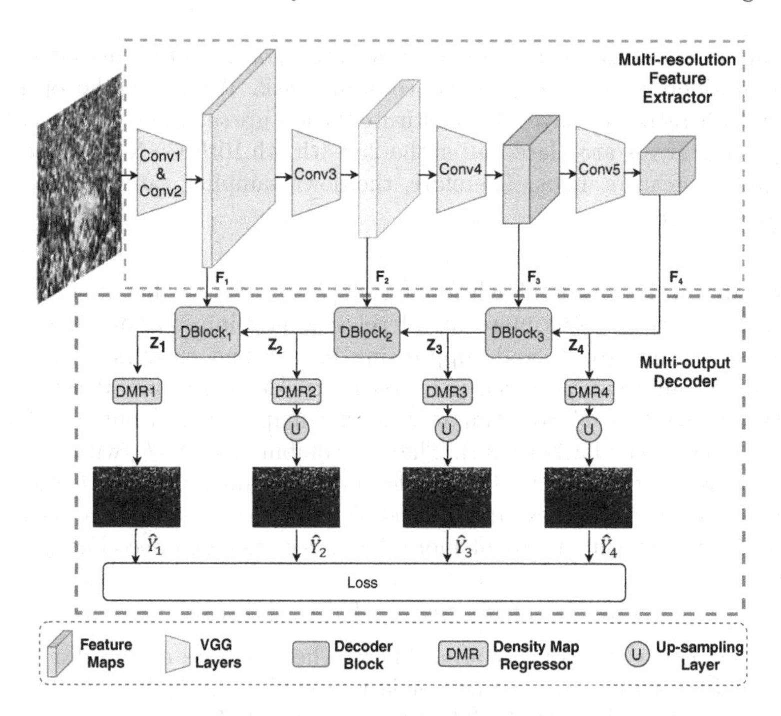


**Fig. 2.** The architecture of the proposed DLA-Net which is composed of a multi-resolution encoder and a multi-output decoder.

## 3 Proposed Method

In this section, we describe our proposed method in detail. We will first elaborate on the architecture of DLA-Net, which contains a multi-resolution feature extractor and a multi-output decoder as shown in Fig. 2. Then we introduce our density level estimator to estimate the probability map of density levels, which is used to fuse the multiple outputs to get the final prediction. Finally, we present our loss function.

### 3.1 Architecture

The architecture contains a multi-resolution feature extractor and a multi-output decoder. We choose VGG-16-bn as our backbone for the feature extractor since its popularity in crowd counting. The decoder includes three Decoding Blocks (DBlock) to fuse features of different levels and four Density Map Regressors (DMR) to generate final outputs.

**Multi-resolution Feature Extractor.** Instead of extracting multi-scale features by a multi-column network, we employ a single column backbone to generate features of different semantic levels. As shown in Fig. 2, we use the first 13

CNN layers of pre-trained VGG-16 with batch normalization as our encoder, which is widely used in the crowd counting task. We adopt the outputs of 4th,7th,10th,13th CNN layers to constitute the feature sets: $\{F_1, F_2, F_3, F_4\}$. The max-pooling layers are placed after the 2nd,4th,7th,10th CNN layers to down-sampling the feature maps. Therefore, the down-sampling rates of the outputs are $\{\frac{1}{2}, \frac{1}{4}, \frac{1}{8}, \frac{1}{16}\}$.

**Multi-output Decoder.** Each feature of $\{F_1, F_2, F_3, F_4\}$ has its own merits. For instance, $F_4$ contains more high-level semantic information which is crucial for detecting crowds and suppressing background clutters, while $F_1$ has more spatial information which is beneficial for dense regions. Based on that, we directly use $Z_4 = F_4$ to estimate a density map $\hat{Y}_4$ which puts emphasis on sparse regions (more background). Then we combine $Z_4$ and $F_3$ with a decoding block (DBlock3) to generate decoded features $Z_3$, which is applied to generate the density map of the next dense level. $F_2$ and $F_1$ are also borrowed to better predict the density maps of more dense regions. The decoding process is formulated as:

$$Z_i = DBlock_i(Z_{i+1}, F_i), i \in \{1, 2, 3\}, \quad (1)$$

where $DBlock_i$ is the $i$-th decoding block. The structure of decoding block is shown in Fig. 3 (a). The features of last layer $Z_{i+1}$ is up-sampled to match the high-resolution features $F_i$. Then the concatenated features are feed to two convolutional layers with batch normalization and ReLU activation function, resulting in fused feature $Z_i$ containing both high-level semantic information and spatial details. The output channels $C_i$ of the blocks are $64, 128, 256$ respectively.

For each output of decoding blocks, we use an independent Density Map Regressor (DMR) to generate the density map $\hat{Y}_i$:

$$\hat{Y}_i = DMRi(Z_i). \quad (2)$$

Each DMR consists of two convolutional layers (see Fig. 3 (b)). The first one will reduce the channels and is employed with batch normalization and activation function. Then the second layer generates final density prediction. All outputs are resized by bilinear interpolation to the same size of the ground truth density map.

### 3.2 Density Level Estimator

DLE aims to categorize each pixel into four density levels for region partition. Given a density map $Y$, DLE first calculate the density of a local neighborhood for all positions by local sum operations, which is formulated as:

$$D_{ij} = \sum_{k \in N(i,j,r)} Y_k, \quad (3)$$

where $i, j$ is the coordinates of a spatial location and $N(i, j, r)$ means the local neighborhood of size $(r+1)*(r+1)$. We set $r = 15$ in this paper. Then we map $D$

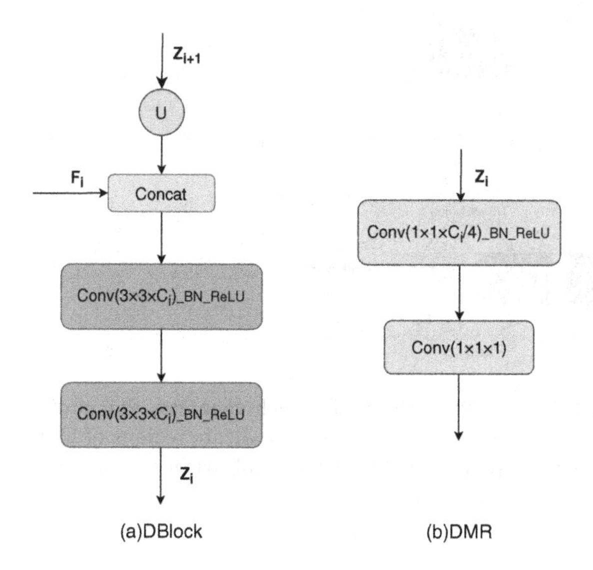


**Fig. 3.** An illustration of the decoding block (a) and the density map regressor (b). A convolutional layer is denoted as Conv(kernel size × kernel size × output channels). BN and ReLU represent batch normalization layer and relu function respectively.

into four manually defined intervals $(0, t_1), (t_1, t_2), (t_2, t_3), (t_3, +\infty)$, which represent the 1st,2nd,3rd,4th density levels respectively. Instead of directly generating density level mask by comparing between $D$ and thresholds $t_1, t_2, t_3$, we estimate a density level distribution $S_1, S_2, S_3, S_4$ by a soft thresholding process (see Fig. 4 (a)), which is more robust when fusing four outputs. The values of $S_i$ indicates the probability of pixels belong to the $i$-th density level, which is generated with the help of Sigmoid function:

$$Sigmoid(x) = \frac{1}{1 + exp(-\delta x)}, \tag{4}$$

where $\delta$ is set to 1 in this paper. Then $S_i$ is calculated as:

$$P_i = Sigmoid(D - t_i), i \in \{1, 2, 3\}, \tag{5}$$
$$S_1 = P_3, \tag{6}$$
$$S_2 = (1 - P_3) \times P_2, \tag{7}$$
$$S_3 = (1 - P_3) \times (1 - P_2) \times P_1, \tag{8}$$
$$S_4 = (1 - P_3) \times (1 - P_2) \times (1 - P_1). \tag{9}$$

From $S_1$ to $S_4$, the corresponding region's density level becomes lower. Figure 4 (b) shows a mapping example from $D$ to $S_i$ when the thresholds are set as $2, 5, 8$. In this way, $S_4$ has higher response on low density regions while $S_1$ activates on dense regions.

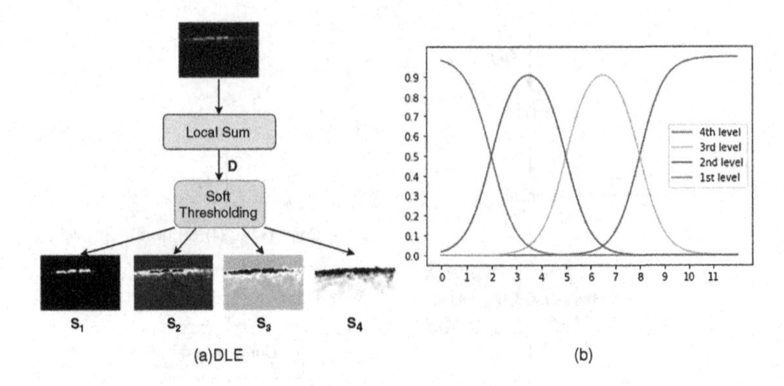


**Fig. 4.** (a): An illustration of the DLE which generates four probability maps corresponding to four density levels, white pixels indicate regions of higher probability. (b) An example of the mapping function from local density to the probability of density levels when the thresholds are 2,5,8.

### 3.3 Loss Function

**Ground Truth Density Maps.** Most datasets of crowd counting only have head points annotation. Therefore, it's essential to generate density maps from head locations to provide better supervision. We follow [21] to generate ground truth density maps by convolving head annotations with Gaussian kernels, which is formulated as:

$$Y = \sum_{i=1}^{C} \delta(x - x_i) * G_\sigma(x), \tag{10}$$

where $C$ is the number of heads, $\delta(x - x_i)$ is the indicator function and $G_\sigma(x)$ is the Gaussian kernel with standard deviation $\sigma$. In this paper, we set $\sigma = 5$ for ShanghaiTech Part-B and $\sigma = 15$ for UCF-CC-50 and UCF-QNRF. For ShanghaiTech Part-A dataset, we use the geometry-adaptive kernels as in [21].

**Euclidean Loss.** We choose euclidean distance to measure the estimation error of pixels between $\hat{Y}_i$ and generated ground truth $Y$, which is defined as:

$$L_{euc} = \sum_{i=1}^{4} \sum_{j=1}^{N} ||\hat{Y}_{ij} - Y_j||^2, \tag{11}$$

where $N$ is the number of pixels and $\hat{Y}_{ij}$ means the j-th pixel of $i$-th prediction. The euclidean loss provide supervision for all outputs of the decoder, which ensures that the outputs are valid density maps. Another advantage of such multiple supervision scheme is that it can help optimize the network by alleviating gradients vanishing of earlier layers because the outputs targeting at dense regions are connected to shallow layers.

**Density Level Aware Loss.** Although the outputs are based on features of different layers, the supervision is distributed to all pixels equally which will homogenize the four predictions. We hence use a DLA-Loss to explicitly help the $i$-th output $\hat{Y}_i$ focus on regions with high probability of $S_i$. We use euclidean distance weighted by $S_i$ for the $i$-th output $Y_i$, which is formulated as:

$$L_d^i = \sum_{j=1}^{N} S_{ij} ||\hat{Y}_{ij} - Y_j||^2, \tag{12}$$

where $S_{ij}$ is value of j-th pixel of $S_i$. Take $S_1$ as an example, the regions with higher probability of the 4-th density level will get larger loss weights.

**Final Loss.** The final loss of our DLA-Net is:

$$L_f = L_{euc} + \lambda \sum_{i=1}^{4} L_d^i, \tag{13}$$

where $\lambda$ is the parameter to balance the contribution of DLA-Loss, which is set to 1 in this paper.

# 4 Experiments

## 4.1 Datasets

In the experiments, we evaluate our method on three crowd counting benchmark datasets which are elaborated below.

**ShanghaiTech** [21] is divided into two parts: Part-A and Part-B. Part-A includes 482 images of highly congested scenes with 241677 head annotations. Part-B includes 716 images with 88488 annotated people, which is about sparse scenes taken from streets in Shanghai.

**UCF-CC-50** [4] contains 50 images of different resolutions. The number of heads per image ranges from 94 to 4543 with an average number of 1280.

**UCF-QNRF** [5] is a large crowd counting dataset which consists of 1535 images of dense scenes with a total number of 1251642 person annotations.

## 4.2 Experimental Settings

**Implementation Details.** We use the first 13 layers of VGG16-bn pretrained in Imagenet as our encoder. We initialize decoder layers with Gaussian distributions with mean 0 and standard deviation 0.01. The thresholds of density levels are set as 2,5,8 for all datasets except ShanghaiTech-Part-A which is 0.01,1,3 since it mainly contains sparse scenes. We use Adam optimizer with an initial learning rate 0.0001 to train our model and the batch size is set to 16.

At the training stage, we use random scale augmentation with ratio (0.8, 1.2) and then crop input images into $400 \times 100$ patches at random locations. Random horizontal flipping and random gamma contrast transformation are then used for augmentation. For ShanghaiTech Part-A which contains grayscale images, we randomly convert the input to gray with ratio 0.1.

We use $\hat{Y}_1$ to estimate the density level distributions with DLE when inference. Figure 5 shows three examples of estimated test results. The final estimated density map is obtained as:

$$\hat{Y} = \sum_{i=1}^{4} S_i \hat{Y}_i. \tag{14}$$

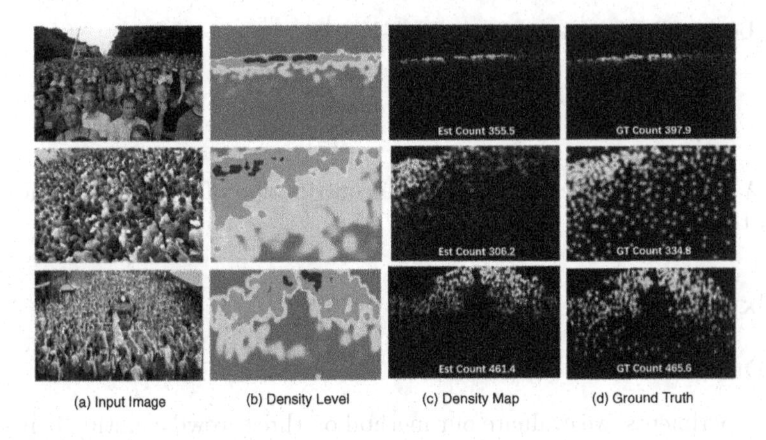


**Fig. 5.** Three visual examples of test images on ShanghaiTech Part-A.

**Evaluation Metrics.** Similar to other works, we calculate mean absolute error (MAE) and mean squared error (MSE) for evaluation:

$$MAE = \frac{1}{N}\sum_{i=1}^{N}|C_i - \hat{C}_i|,\ MSE = \sqrt{\frac{1}{N}\sum_{i=1}^{N}(C_i - \hat{C}_i)^2}, \tag{15}$$

where $N$ is the number of images, $C_i$ and $\hat{C}_i$ represent ground truth count and estimated count of the $i$-th image respectively. $\hat{C}_i$ is obtained by the summation operation on density maps.

### 4.3 Performance Comparison

To validate the effectiveness of our proposed DLA-Net, we compare it with other state-of-the-art methods including: MCNN [21], Switch-CNN [12], CSRNet [8],

ADCrowdNet [9], TEDnet [6], ANF [18], etc. The comparison results of MAE and MSE on all three datasets are summarized in Table 1. Our method outperforming all current methods by a clear margin in terms of MAE and MSE on datasets with large varying scales. For instance, our DLA-Net achieves 59.6 and 103 in MAE on ShanghaiTech Part-A and UCF-QNRF respectively, which indicates that our method performs well under congested scenes. For the dataset ShanghaiTech Part-B which is sparse and suffer less from scale variation, our method is still competitive compared with recent models. Similar improvement can be observed on the very small dataset UCF-CC-50.

**Table 1.** Comparison Results on ShanghaiTech, UCF-CC-50 and UCF-QNRF datasets. The best results are marked in bold. Our method achieves the best performance by clear margins.

| Method | Part-A | | Part-B | | UCF-CC-50 | | UCF-QNRF | |
|---|---|---|---|---|---|---|---|---|
| | MAE | MSE | MAE | MSE | MAE | MSE | MAE | MSE |
| Cross-Scene [19] | 181.8 | 277.7 | 32.0 | 49.8 | 467.0 | 498.5 | – | – |
| MCNN [21] | 110.2 | 173.2 | 26.4 | 41.3 | 377.6 | 509.1 | 277 | – |
| FCN [10] | 126.5 | 173.5 | 23.8 | 33.1 | 338.6 | 424.5 | – | – |
| Switch-CNN [12] | 90.4 | 135.0 | 21.6 | 33.4 | 318.6 | 439.2 | 228 | 445 |
| CP-CNN [14] | 73.6 | 106.4 | 20.1 | 30.1 | 295.8 | 320.9 | – | – |
| SaCNN [20] | 86.8 | 139.2 | 16.2 | 25.8 | 314.9 | 424.8 | – | – |
| DR-ResNet [2] | 86.3 | 124.2 | 14.5 | 21.0 | 307.4 | 421.6 | – | – |
| ACSCP [13] | 75.7 | 102.7 | 17.2 | 27.4 | 291.0 | 404.6 | – | – |
| PADFF [3] | 70.6 | 117.5 | **7.4** | 11.4 | 268.2 | 384.4 | 130 | 208 |
| CSRNet [8] | 68.2 | 115.0 | 10.6 | 16.0 | 266.1 | 397.5 | 120 | 208 |
| ADCrowdNet [9] | 63.2 | 98.9 | 8.2 | 15.7 | 266.4 | 358.0 | – | – |
| TEDnet [6] | 64.2 | 109.1 | 8.2 | 12.8 | 249.4 | 354.5 | 113 | 188 |
| ANF [18] | 63.9 | 99.4 | 8.3 | 13.2 | 250.2 | 340.0 | 110 | **174** |
| Our DLA-Net | **59.6** | **95.2** | **7.4** | **11.7** | **194.2** | **272.8** | **103** | 187 |

## 4.4 Ablation Study

To gain a better understanding of our proposed method, we conduct ablation experiments on ShanghaiTech Part-A. The results of four model variants in terms of MAE and MSE are reported in Table 2. We use the output $F_4$ from encoder with $DMR_4$ to generate a density map (that is $\hat{Y}_4$) as our baseline model. We then report the result of adding our decoder to make the prediction ($\hat{Y}_1$), which shows that combining features from earlier layers is beneficial to our model. We also test the model that supervises all outputs but without DLA-Loss. The results compared to our full model indicate that the DLA-Loss is crucial and the performance gain isn't brought by multiple supervisions.

**Table 2.** Ablation study results on ShanghaiTech Part-A.

| Method | MAE | MSE |
|---|---|---|
| Baseline | 63.2 | 105.9 |
| Baseline+Decoder | 61.4 | 97.8 |
| Baseline+Multi-output Decoder | 61.7 | 105.2 |
| Our DLA-Net | **59.6** | **95.2** |

## 5 Conclusion

In this paper, we present a novel Density Level Aware Network for the crowd counting task. We design a multi-output decoder to generate multiple predictions with each one focusing on regions of a specific density level. The decoding blocks of DLA-Net borrow different high-resolution features from earlier layers for different outputs. Besides, DLA-Loss is proposed to explicitly supervise the outputs in their responsible regions. Our experiments on three popular datasets demonstrate the effectiveness and robustness of our method when varying crowd scenes from sparse to dense.

**Acknowledgments.** This work is supported by Alibaba Group (Grant No. SccA50202002101), NSFC (No. 61772330, 61533012, 61876109), and Major Scientific Research Project of Zhejiang Lab (No. 2019DB0ZX01).

## References

1. Dalal, N., Triggs, B.: Histograms of oriented gradients for human detection. In: 2005 IEEE computer society conference on computer vision and pattern recognition (CVPR 2005), vol. 1, pp. 886–893. IEEE (2005)
2. Ding, X., Lin, Z., He, F., Wang, Y., Huang, Y.: A deeply-recursive convolutional network for crowd counting. In: 2018 IEEE International Conference on Acoustics, Speech and Signal Processing (ICASSP), pp. 1942–1946. IEEE (2018)
3. Guo, H., He, F., Cheng, X., Ding, X., Huang, Y.: Pay attention to deep feature fusion in crowd density estimation. In: Gedeon, T., Wong, K.W., Lee, M. (eds.) ICONIP 2019. CCIS, vol. 1142, pp. 363–370. Springer, Cham (2019). https://doi.org/10.1007/978-3-030-36808-1_39
4. Idrees, H., Saleemi, I., Seibert, C., Shah, M.: Multi-source multi-scale counting in extremely dense crowd images. In: Proceedings of the IEEE conference on computer vision and pattern recognition, pp. 2547–2554 (2013)
5. Idrees, H., et al.: Composition loss for counting, density map estimation and localization in dense crowds. In: Proceedings of the European Conference on Computer Vision (ECCV), pp. 532–546 (2018)
6. Jiang, X., Xiao, Z., Zhang, B., Zhen, X., Cao, X., Doermann, D., Shao, L.: Crowd counting and density estimation by trellis encoder-decoder networks. In: Proceedings of the IEEE Conference on Computer Vision and Pattern Recognition. pp. 6133–6142 (2019)

7. Lempitsky, V., Zisserman, A.: Learning to count objects in images. In: Advances in neural information processing systems, pp. 1324–1332 (2010)
8. Li, Y., Zhang, X., Chen, D.: Csrnet: Dilated convolutional neural networks for understanding the highly congested scenes. In: Proceedings of the IEEE conference on computer vision and pattern recognition, pp. 1091–1100 (2018)
9. Liu, N., Long, Y., Zou, C., Niu, Q., Pan, L., Wu, H.: Adcrowdnet: an attention-injective deformable convolutional network for crowd understanding. In: Proceedings of the IEEE Conference on Computer Vision and Pattern Recognition, pp. 3225–3234 (2019)
10. Marsden, M., McGuinness, K., Little, S., O'Connor, N.E.: Fully convolutional crowd counting on highly congested scenes. arXiv preprint arXiv:1612.00220 (2016)
11. Ryan, D., Denman, S., Fookes, C., Sridharan, S.: Crowd counting using multiple local features. In: 2009 Digital Image Computing: Techniques and Applications, pp. 81–88. IEEE (2009)
12. Sam, D.B., Surya, S., Babu, R.V.: Switching convolutional neural network for crowd counting. In: 2017 IEEE Conference on Computer Vision and Pattern Recognition (CVPR), pp. 4031–4039. IEEE (2017)
13. Shen, Z., Xu, Y., Ni, B., Wang, M., Hu, J., Yang, X.: Crowd counting via adversarial cross-scale consistency pursuit. In: Proceedings of the IEEE conference on computer vision and pattern recognition, pp. 5245–5254 (2018)
14. Sindagi, V.A., Patel, V.M.: Generating high-quality crowd density maps using contextual pyramid CNNs. In: Proceedings of the IEEE International Conference on Computer Vision, pp. 1861–1870 (2017)
15. Viola, P., Jones, M.J.: Robust real-time face detection. Int. J. Comput. Vision **57**(2), 137–154 (2004)
16. Yan, Y., et al.:Unsupervised image saliency detection with gestalt-laws guided optimization and visual attention based refinement. Pattern Recogn. **79**, 65– 78 (2018). https://doi.org/10.1016/j.patcog.2018.02.004,http://www.sciencedirect.com/science/article/pii/S0031320318300517
17. Zabalza, J., et al.: Novel segmented stacked autoencoder for effectivedimensionality reduction and feature extraction in hyperspectral imaging. Neurocomputing **185**, 1–10 (2016).https://doi.org/10.1016/j.neucom.2015.11.044,http://www.sciencedirect.com/science/article/pii/S0925231215017798
18. Zhang, A., et al.: Attentional neural fields for crowd counting. In: Proceedings of the IEEE International Conference on Computer Vision, pp. 5714–5723 (2019)
19. Zhang, C., Li, H., Wang, X., Yang, X.: Cross-scene crowd counting via deep convolutional neural networks. In: Proceedings of the IEEE Conference on Computer Vision and Pattern Recognition, pp. 833–841 (2015)
20. Zhang, L., Shi, M., Chen, Q.: Crowd counting via scale-adaptive convolutional neural network. In: 2018 IEEE Winter Conference on Applications of Computer Vision (WACV), pp. 1113–1121. IEEE (2018)
21. Zhang, Y., Zhou, D., Chen, S., Gao, S., Ma, Y.: Single-image crowd counting via multi-column convolutional neural network. In: Proceedings of the IEEE conference on computer vision and pattern recognition, pp. 589–597 (2016)

# Difficulty Within Deep Learning Object-Recognition Due to Object Variance

Qianhui Althea Liang(✉) and Tai Fong Wan

SCSE, Nanyang Technological University, 50 Nanyang Avenue, Singapore 639798, Singapore
{qhliang,twan002}@ntu.edu.sg

**Abstract.** It is one of the areas where deep learning models demonstrate possible performance bottleneck to learn objects with variations rapidly and precisely. We research on the variances of visual objects in terms of the difficulty levels of learning performed by deep learning models. Multiple dimensions and levels of variances are defined and formalized in the form of categories. We design "variance categories" and "quantitative difficulty levels" and translate variance categories into difficulty levels. We experiment how multiple learning models are affected by the categorization of variance separately and in combination (of several categories). Our experimental analysis on learning of the dataset demonstrates not only the expected way of utilization of variance of the data differs from models, the amount of learning or information gained for each data fed into models also varies significantly. Our results suggest it matters to search for a possible key explicit representation of the "invariance" part of objects (or of the respective cognitive mechanism) and for the pertinent elements and capabilities in the deep learning architectures. It can be used to make the learning models a match to humans on complex object recognition tasks.

**Keywords:** Deep learning · Variance · Difficulty in deep learning · Object recognition

## 1 Introduction

According to the studies from neuroscience and behavioural science, with regards to the existing implementation of Convolutional Neural Network (CNN), it lacks a certain mechanism to cater for an invariant representation [13] of a subject that has to be learnt. This, as a result, explains that CNN is unable to compete against the proficiency of human beings in tasks such as classifying or learning objects with greater variations. In the case of visual objects, it would seem that

---

Research work is sponsored by Academic Research Funds of MOE Singapore. Authors thank the sponsorship.


H. Yang et al. (Eds.): ICONIP 2020, LNCS 12532, pp. 278–289, 2020.
https://doi.org/10.1007/978-3-030-63830-6_24

CNN has difficulties in learning the 3D variations, where the variations include the rotation of the 3D objects and changes to the surface of the objects.

Towards finding a remedy to overcoming such a difficulty, we have conducted research on the different variances of objects in terms of the difficulty levels of learning performed by learning models. We study multiple dimensions and levels of variances and formalize them in the form of categories. Using a specially designed and carefully controlled data set, the impacts of the variance in the data set on learning in general deep learning are examined. We study how multiple learning models including DNN, CNN and RNN are affected by the categorization of variance separately and in combination (of several categories). The aim of our study is to come up with different levels of variations for the object recognition tasks or subjects, with respect to the variations of the neural network performance. Using several neural network stacks, our study will quantitatively evaluate the models with the measures that include complexity of neural network stack, duration of training time, accuracy and model output pattern. The study has been conducted in the domain of spatial image data. The novelty of our work lies on a definitive investigation on the difficulty levels of learning of various deep learning models in support of our notion of categories of variance in the objects and its impacts on learning. We claim that not only the expected way of utilization of variance of the data differs from models, the amount of learning or information gained for each data fed into models also varies significantly. Presenting limited amount of data but with maximized amount of information on variance, and/or of its categories, can help the learning models to gain progress notably. Our results suggest it matters to search for a possible key explicit representation of the "invariance" part of objects (or of the respective cognitive mechanism) and for the pertinent elements and capabilities in the deep learning architectures in order to make them a match to humans on complex object recognition tasks.

## 2 Related Work

The major operations that have been studied in the image processing community and that result in the variances that we are interested in mostly include position, scale, pose variability [1]. Illumination, clutter and intra-class variability [2] are also considered. One study [3] had images that are generated from a variety of objects with an irrelevant background with randomized parameters following a defined range. The parameters controlled the position of the object (on the x and y axis), the rotation (on the x, y and z axis) and the size of object. With the amount or the degree of the operations of low, medium and high, the images were tested from volunteers. The higher the degree, the worse the performance of the volunteers. The tasks performed by the volunteers included object recognition (including facial recognition), object differentiation and facial differentiation, ordered from the best performed task. This shows to a strong extent that human proficiency at object recognition can be affected by the amount of variance introduced onto the object in the image.

Most of the related work focuses on designing datasets for tuning and understanding trained models' performance on the specific datasets and on their reliance to the amount of the image data. For example, in ObjectNet [4], the objects are placed in more varied backgrounds, and some objects have gone through operations of rotations and are taken from different viewpoints. They show that trained models do generally better on datasets than on real-world applications, to the extent of a 40–45% drop in performance. They demonstrate that models may possibly previously take advantage of bias in the available datasets and thus to perform well (e.g. objects are typically found with specific type of background [5]). As also shown [6], deep neural networks can easily fit the random labels. This may support the opinion that trained models are very likely to show good results, misleading its capability on true generalization.

However, it also notable that most existing work has a strong reliance on huge datasets to reduce the correlation of features from the training set (to achieve generalization) [4], such that the invariant factors are learnt implicitly [1]. This might stem from when the well-known dataset ImageNet [8] was regularly used for its vast labelled dataset to train neural networks, where most models would be pre-trained and further fine-tuned to the task required. However, a large varied collection of data that may not relate to the actual task of the model, may not always bring the supposed benefits through pre-training.

When testing multiple models with different portions of the data [9], it is found that the reduction in classes or data provided to the models did not lead to a performance degradation in the CNN architecture. The study then suggests CNN could possibly be not as "data-hungry", which might imply that CNN could extract the needed generalization information for the given task with minimal data. Yet, with a more recent study into ImageNet pre-training [7,10], it shows that ImageNet pre-training may not be required as models trained from scratch do not necessarily per-form worse off. It is able to help models reach convergence faster but does not show performance improvement unless the target dataset to be trained on directly is small. This might then suggest that given a neural network, it can possibly perform averagely well given any form of data but would require task-relevant data to achieve better results.

Through the large amount of data provided in ImageNet, deep CNN was shown to be able to perform well solely through the use of supervised learning [11]. Deep CNN's performance degradation would be observed, had any of the convolutional layers were to be removed. As pointed out by them, "any given object can cast an infinite number of different images onto the retina" [12], thus the number of images required to represent this amount of variations present in the real world is far larger than the number of images our systems can store. Therefore, it is suggested [12] that synthetic images are better able to present this variation found in the transformation process. By having a greater control on the information present in our data, we can then rigorously test our models of different architecture to see how each internalises the information and thereby affecting the results.

## 3 Method

We have defined variances via variance categorization with its associated dimensions and degrees. We regulate the variances that are to be introduced in the sub-task of "figure-ground" segregation in a numerical range of numbers in order to further study the magnitude of variance impact. The smallest value zero (0) represents invariant or no variance, while the largest value five (5) represents maximal variance, each indicating a variance category. In Sect. 3.1. we have included a brief introduction about such categorization. We then define the impacts of variance in terms of difficulty in learning. In particular, the difficulty in learning is quantified into difficulty levels, defined as "learning improvement" rate, i.e. $I_l$. $I_l$ is defined as the average improvement upon every epoch divided by the most recent loss value as the following. $I_l$ uses the values from the dataset, and can be used to see how well the model explains the variance given such information.

$$I_l = \frac{Average\ (LossValueImprovementOfEveryEpoch)}{MostRecentEpoch'sLossValue} \tag{1}$$

We have derived the difficulty level treatment in deep learning corresponding to the variances presented to the neural network stacks in their learning process based on the following: There are obvious performance gaps among the varieties of variances [1] in learning. In addition, the gaps also demonstrate different magnitudes which may suggest a form of certain kind of hierarchy of performance differences. It is also observed that humans progressively perform worse at the task of object recognition when the variance was increased [1,2]. Therefore, our treatment of difficulty is to transfer it to different levels of categories.

On the other hand, among the three deep learning models (stacks) we have selected, CNN may experience greater difficulties in handling variant sub-tasks that is demonstrated by using very deep structure and longer time and bigger training set to complete the sub-tasks. We will use the output of neural network stacks or sub-structures to provide indications as to the magnitude of task difficulty from the perspective of the given neural network stack. The sub-structure can be as simple as a single layer of dense neurons or as complex as models with feedback mechanisms, where feedback is employed both in the training and executing phase. Recurrent Processing is such an example of the neural network stack with feedback mechanism that should perform reasonably well to handle variant sub-tasks. We will study what kind of indications are these, how these indications can be quantified (in ways such as loss pattern, consistency of accuracy pattern). With respect to the neural network architecture, the structure of the CNN model uses a bottom-up or feed-forward neural network stacks that do not form cycles. Conversely, the RNN model would be viewed as recurrent neural network stack that involves formation of cycles. Such that the complexities as viewed by one architecture such as CNN can be regarded as additional capabilities from the opposite perspective such as the RNN architecture where it processes the task from a different perspective. There could therefore exist multiple valid interpretation in the decomposition of the same task, so the goal is to then optimize and minimize the difficulty levels. Therefore, our treatment on difficulty also includes to find the relationship between the levels of categories and the learning capabilities of individual learning models on the sub tasks.

### 3.1 Datasets of Variances for Various NN Architectures

Here we present a brief summary of the development of our dataset first. The dataset is composed of 3430 images, each made up of a number of different objects, that are rendered with different backgrounds. They have been generated with the help of computer graphics software, in our case Pov-Ray [14]. The image creation utilizes ray-tracing to render 3D objects and place them within a 3D space. Therefore, the creation is done strictly in a XYZ coordinate system as well as light sources in the same system. Our dataset of pictures is designed to not only be "real-life" like, but also allow us to focus on the task of studying the variance of objects and its impacts on learning difficulty.

Given the variation level as defined previously, we are able to break down a given task into its components and give them a variant rating with the variant scale. Given a task of known variant components, we will then evaluate how the CNN and RNN stacks performs by observing the several indicators, where we define a numerical range of numbers (1–5) for the magnitude of difficulty. The indicators we will measure from the neural network model includes the training time of the model, accuracy of the model, output pattern of the model and complexity of the model (depth of model and number of units).

There are several operations that we have used to manipulate the variances of the objects during the creation of our data set. We will first try to map these operations to a corresponding variance category by following a theme of degrees of the visible change of the target object. These five categories and their mapped image processing operations are listed below and they will be analyzed individually: Category 5, the maximum variance category (Background Clutter), Category 4 (Surface Changing), Category 3 (Rotation), Category 2 (Scale), Category 1, minimum variance category (Movement). Based on the above categories, we study how difficult the different architectures of neural network are able to adapt and generalize the learning of specific tasks of a given domain. With this difficulty scale, we study how well the neural network models handle variant components of given tasks. For example, we may expect that both the CNN and RNN model will receive a difficulty rating of one (1) if tasked to differentiate geometric objects apart (such as a triangle from a circle). Where given the task of "figure-ground" segregation may be rated as a difficulty of five (5) for the CNN and a difficulty of three (3) for the RNN.

Given each variance category, we have also extended our dataset with another test dataset in order to test the impact of multiple variance categories. The focus of the work described here is to see if the effects of concurrent variance categories, whether they might be of additive or multiplicative in nature or more. Using Category 1 (Movement) as an example, the object translated without regards to the camera's position will result in movement with minor rotational effect.

To test the datasets, we will be focusing on the two types of neural network (NN) architectures. Top-down RNN model and bottom-up CNN model, while using a basic Dense (DNN) model as control.

The models will be tested with low complexity to prevent immediate overfitting, as the dataset is considerably small. We hope that the simpler the model, the more likely it is to capture interesting behavior during the testing phase.

For each NN architecture, we will record the loss value based on binary cross-entropy for both training and validation over 3 epochs. We will avoid using accuracy as a measure of the model because the dataset only contains positive samples (all pictures contain the target object) and models are likely able to blindly guess all positive and get a good result as a result eventually. The implementation for the models will be considerably different as well, and so using raw loss values will not be a good enough measure across the models.

### 3.2 Distribution of Data with Sensitivity Analysis

We have designed three different distributions with different amounts of the data in our data sets that are to be fed into the neural networks. We want to examine if there is any notable impact of feeding models with more data that are progressive more modified in terms of variance operations.

**Model B.** Deep NN architectures trained on a single non-modified image. (The original image of the object).

**Model B+[2-6].** The dataset is split 5 parts, where for Model B+[2] NN architectures will receive the first 2 parts of the dataset and so on. For Model B+[6], Deep NN architectures trained will receive almost all of the data; a very small portion will be left out for testing.

**Model B*.** Deep NN architectures are trained on a small amount of, heavily modified images. (Such as shifted to the corner of the picture in Movement, or rotated away from the camera in Rotation).

This shows how much information is required to achieve the desired performance and which architecture is able to perform better with the least amount of information provided. We theorize that if given a dataset of an insignificant variance category, the confidence of the model on the testing set should be explainable. For example, in Category 2 (Scaling), we would expect to see that the confidence of the model would be affected by some fixed explainable (linear or exponential) factor with respect to how to chair is scaled. Where in a dataset of a significant variance category, the confidence of the model on the increasing variance magnitude data will be difficult to explain (Table 1).

**Table 1.** Experimental verifications

| Variance category | Data distribution model | Architecture | Confidence graph |
|---|---|---|---|
| Move | B, B+[2-6], B* | CNN, RNN, DNN | ✓ |
| Scale | B, B+[2-6], B* | CNN, RNN, DNN | ✓ |
| Rotation | B, B+[2-6], B* | CNN, RNN, DNN | ✓ |
| Colour | B, B+[2-6], B* | CNN, RNN, DNN | ✓ |
| Background (BG) | B, B+[2-6], B* | CNN, RNN, DNN | ✓ |

# 4 Experiments

In our experiments, we have chosen to use CNN and RNN as the neural network architectures to be studied in our setting. The Dense (DNN) architecture is also studied and used as a comparison. We have performed extensively experiments to study the difficulty of learning in terms of their $I_l$ with the change of variance categories. Below we will only present some of our experiment results and use such results to support our findings.

In our previous study, we have proven that with the base data distribution model (Model B), our targeted network architectures are able to achieve an equivalently good training $I_l$ over all variance categories. We, in this experiment, increase the amount of information on variance in the training set by presenting more pictures with possible variance, i.e. applying the distribution models from B+[2] all the way to B+[6] . We then examine how the training $I_l$ changes as it trains on more such pictures. A selection of results of these experiments are shown in Figs. 1, 2, 3, and 4.

There is a trend for each targeted architecture that for a particular variance category it adapts in a way as seen in Fig. 1. For example, in the Dense architecture, Rotation learns at the rate of 11.57% in Model B+[3] and increases to 25.11% in Model B+[6], as more information of variance are presented. Background that learns at with $I_l$ of 0.44% in Model B+[3] and increases to 1.14% in Model B+[5] in Fig. 3. We see that CNN architecture, when handling Background variance, learns with $I_l$ of 3327.45% in Model B+[3] and increases to 244057.20% in Model B+[6]. While some of the validation $I_l$ is 0% in CNN, we assume that the model may have perfectly fitted (or overfitted) the data, as we still see 38106.51% validation learning in Model B+[5] in Fig. 3.

From our results, we can see that all the neural network models will fully utilize the information given them as expected. This can be observed by the notable difference in loss value from a model that did not receive it. For example, the loss rate of Rotation decreases from 0.0599 of Model B+[3] in Fig. 1 to 0.0231 of Model B+[6] in Fig. 4. From the results in Fig. 1 to Fig. 4, we see evidences that further support our claim that certain architectures are more adept at handling certain kinds of variance than others. This is particularly important when considering for commercial use, as we tend to push large amounts of data to models and assume that the invariant factors can be learnt implicitly. The different type of architectures might have different learning curves

| Model B+[3] | Dense | | | | | |
|---|---|---|---|---|---|---|
| | loss | Average | $l_l$ | val_loss | Average | $l_l$ |
| Move | 0.6568 | 0.01465 | 2.23% | 0.6495 | 0.0144 | 2.22% |
| MoveFlip | 0.6222 | 0.0284 | 4.56% | 0.6092 | 0.02765 | 4.54% |
| MoveRotate | 0.5581 | 0.053 | 9.50% | 0.5341 | 0.0505 | 9.46% |
| Scale | 0.4983 | 0.07545 | 15.14% | 0.4669 | 0.07005 | 15.00% |
| Rotation | 0.5348 | 0.06185 | 11.57% | 0.5076 | 0.0583 | 11.49% |
| Colour | 0.5385 | 0.0604 | 11.22% | 0.5119 | 0.05705 | 11.14% |
| BG | 0.6859 | 0.00305 | 0.44% | 0.6912 | 0.0031 | 0.45% |
| **Model B+[3]** | **CNN** | | | | | |
| | loss | Average | $l_l$ | val_loss | Average | $l_l$ |
| Move | 0.00002047 | 0.052539765 | 256667.15% | 0.000013522 | 3.254E-05 | 240.65% |
| MoveFlip | 0.00001605 | 0.031141975 | 194031.00% | 5.5995E-06 | 1.218E-06 | 21.76% |
| MoveRotate | 0.000012947 | 0.015393527 | 118896.47% | 6.4264E-06 | 2.49E-06 | 38.74% |
| Scale | 9.0086E-06 | 0.010295496 | 114285.19% | 0.000030959 | 1.646E-05 | 53.18% |
| Rotation | 0.000013067 | 0.020043467 | 153389.96% | 8.2962E-06 | 3.6E-06 | 43.39% |
| Colour | 0.000016864 | 0.018641568 | 110540.61% | 0.000006357 | 3.087E-06 | 48.56% |
| BG | 0.0051 | 0.1697 | 3327.45% | 0.00056577 | 0.0255671 | 4518.99% |
| **Model B+[3]** | **RNN** | | | | | |
| | loss | Average | $l_l$ | val_loss | Average | $l_l$ |
| Move | 0.6854 | 0.00385 | 0.56% | 0.6814 | 0.00395 | 0.58% |
| MoveFlip | 0.6854 | 0.00385 | 0.56% | 0.6814 | 0.00395 | 0.58% |
| MoveRotate | 0.6759 | 0.0079 | 1.17% | 0.6694 | 0.008 | 1.20% |
| Scale | 0.0599 | 0.1621 | 270.62% | 0.0502 | 0.0438 | 87.25% |
| Rotation | 0.6755 | 0.0079 | 1.17% | 0.6694 | 0.008 | 1.20% |
| Colour | 0.6756 | 0.00785 | 1.16% | 0.6694 | 0.008 | 1.20% |
| BG | 0.6854 | 0.00385 | 0.56% | 0.7405 | 0.0026 | 0.35% |

**Fig. 1.** Difficulty results from model B+[3] by each architecture.

| Model B+[4] | Dense | | | | | |
|---|---|---|---|---|---|---|
| | loss | Average | $l_l$ | val_loss | Average | $l_l$ |
| Move | 0.6393 | 0.02155 | 3.37% | 0.6287 | 0.02115 | 3.36% |
| MoveFlip | 0.5897 | 0.0409 | 6.94% | 0.5716 | 0.03945 | 6.90% |
| MoveRotate | 0.502 | 0.07405 | 14.75% | 0.4705 | 0.0689 | 14.64% |
| Scale | 0.4253 | 0.10165 | 23.90% | 0.3861 | 0.09085 | 23.53% |
| Rotation | 0.4712 | 0.0854 | 18.12% | 0.4364 | 0.0781 | 17.90% |
| Colour | 0.4758 | 0.0837 | 17.60% | 0.4411 | 0.0769 | 17.43% |
| BG | 0.6822 | 0.00455 | 0.67% | 0.689 | 0.0046 | 0.67% |
| **Model B+[4]** | **CNN** | | | | | |
| | loss | Average | $l_l$ | val_loss | Average | $l_l$ |
| Move | 0.000017813 | 0.036891094 | 207102.08% | 0.000012827 | 5.212E-06 | 40.63% |
| MoveFlip | 0.000023868 | 0.019538066 | 81858.83% | 0.000013031 | 3.957E-06 | 30.36% |
| MoveRotate | 9.1399E-06 | 0.01059543 | 115925.01% | 3.8584E-06 | 2.808E-06 | 72.77% |
| Scale | 5.4265E-06 | 0.006647287 | 122496.76% | 0.000018831 | 1.74E-05 | 92.41% |
| Rotation | 9.4586E-06 | 0.012895271 | 136333.82% | 8.8207E-06 | 9.048E-06 | 102.58% |
| Colour | 5.3807E-06 | 0.01079731 | 200686.03% | 4.9411E-07 | 1.262E-07 | 25.53% |
| BG | 0.00053327 | 0.124733365 | 23390.28% | 0.000060159 | 0.0060199 | 10006.68% |
| **Model B+[4]** | **RNN** | | | | | |
| | loss | Average | $l_l$ | val_loss | Average | $l_l$ |
| Move | 0.6869 | 0.0031 | 0.45% | 0.6833 | 0.0035 | 0.51% |
| MoveFlip | 0.6768 | 0.00785 | 1.16% | 0.6694 | 0.008 | 1.20% |
| MoveRotate | 0.6688 | 0.01105 | 1.65% | 0.6597 | 0.0118 | 1.79% |
| Scale | 0.0408 | 0.12555 | 307.72% | 0.0331 | 0.0285 | 86.10% |
| Rotation | 0.6654 | 0.012 | 1.80% | 0.657 | 0.0122 | 1.86% |
| Colour | 0.6681 | 0.0111 | 1.66% | 0.6597 | 0.0118 | 1.79% |
| BG | 0.6854 | 0.00385 | 0.56% | 0.7602 | 0.0022 | 0.29% |

**Fig. 2.** Difficulty results from model B+[4] by each architecture.

| Model B+[5] | Dense | | | | | |
|---|---|---|---|---|---|---|
| | loss | Average | $l_l$ | val_loss | Average | $l_l$ |
| Move | 0.6223 | 0.02825 | 4.54% | 0.6087 | 0.0275 | 4.52% |
| MoveFlip | 0.5588 | 0.0528 | 9.45% | 0.535 | 0.05025 | 9.39% |
| MoveRotate | 0.4523 | 0.09215 | 20.37% | 0.4155 | 0.08355 | 20.11% |
| Scale | 0.3644 | 0.1223 | 33.56% | 0.3216 | 0.1051 | 32.68% |
| Rotation | 0.4167 | 0.10465 | 25.11% | 0.3769 | 0.09305 | 24.69% |
| Colour | 0.4216 | 0.1029 | 24.41% | 0.3823 | 0.09175 | 24.00% |
| BG | 0.6785 | 0.00595 | 0.88% | 0.6891 | 0.0061 | 0.89% |
| **Model B+[5]** | **CNN** | | | | | |
| | loss | Average | $l_l$ | val_loss | Average | $l_l$ |
| Move | 0.00001605 | 0.031141975 | 194031.00% | 0.000012926 | 2.464E-06 | 19.06% |
| MoveFlip | 0.000015899 | 0.016392051 | 103101.14% | 0.000015105 | 7.271E-06 | 48.14% |
| MoveRotate | 6.9979E-06 | 0 | 0.00% | 4.3402E-06 | 0 | 0.00% |
| Scale | 4.3855E-06 | 0.005797807 | 132204.02% | 0.000012428 | 2.22E-05 | 178.64% |
| Rotation | 8.0722E-06 | 0.008195984 | 101533.21% | 4.8348E-06 | 6.864E-06 | 141.97% |
| Colour | 7.9338E-06 | 0.008496033 | 107086.55% | 5.2217E-07 | 5.767E-07 | 110.45% |
| BG | 0.000075333 | 0.103612334 | 137539.10% | 2.2848E-07 | 8.707E-05 | 38106.51% |
| **Model B+[5]** | **RNN** | | | | | |
| | loss | Average | $l_l$ | val_loss | Average | $l_l$ |
| Move | 0.6869 | 0.0031 | 0.45% | 0.6833 | 0.0035 | 0.51% |
| MoveFlip | 0.6759 | 0.0079 | 1.17% | 0.6694 | 0.008 | 1.20% |
| MoveRotate | 0.6592 | 0.01525 | 2.31% | 0.6471 | 0.01625 | 2.51% |
| Scale | 0.0298 | 0.1047 | 351.34% | 0.0236 | 0.0241 | 102.12% |
| Rotation | 0.6551 | 0.0162 | 2.47% | 0.6443 | 0.01655 | 2.57% |
| Colour | 0.658 | 0.01535 | 2.33% | 0.6471 | 0.01625 | 2.51% |
| BG | 0.6869 | 0.0031 | 0.45% | 0.8005 | 0.00115 | 0.14% |

**Fig. 3.** Difficulty results from model B+[5] by each architecture.

| Model B+[6] | Dense | | | | | |
|---|---|---|---|---|---|---|
| | loss | Average | $l_l$ | val_loss | Average | $l_l$ |
| Move | 0.6111 | 0.03265 | 5.34% | 0.5956 | 0.0317 | 5.32% |
| MoveFlip | 0.5345 | 0.062 | 11.60% | 0.5073 | 0.0584 | 11.51% |
| MoveRotate | 0.4115 | 0.1065 | 25.88% | 0.3714 | 0.0943 | 25.39% |
| Scale | 0.3163 | 0.1375 | 43.47% | 0.2722 | 0.11395 | 41.86% |
| Rotation | 0.3736 | 0.1193 | 31.93% | 0.3312 | 0.10315 | 31.14% |
| Colour | 0.3773 | 0.11805 | 31.29% | 0.3351 | 0.10235 | 30.54% |
| BG | 0.6749 | 0.0077 | 1.14% | 0.671 | 0.00735 | 1.10% |
| **Model B+[6]** | **CNN** | | | | | |
| | loss | Average | $l_l$ | val_loss | Average | $l_l$ |
| Move | 0.000015709 | 0.024692146 | 157184.71% | 0.000014472 | 2.265E-06 | 15.65% |
| MoveFlip | 8.8676E-06 | 0.015795566 | 178126.73% | 8.5434E-06 | 4.139E-06 | 48.45% |
| MoveRotate | 6.5323E-06 | 0.006846734 | 104813.52% | 5.8651E-06 | 6.89E-06 | 117.48% |
| Scale | 3.3664E-06 | 0.005398317 | 160358.75% | 8.3447E-06 | 1.452E-05 | 174.00% |
| Rotation | 4.4065E-06 | 0.007247797 | 164479.67% | 0.000001514 | 2.557E-06 | 168.89% |
| Colour | 4.5473E-06 | 0.006747726 | 148389.73% | 1.1921E-07 | 0 | 0.00% |
| BG | 0.00003097 | 0.075584515 | 244057.20% | 1.1921E-07 | 0 | 0.00% |
| **Model B+[6]** | **RNN** | | | | | |
| | loss | Average | $l_l$ | val_loss | Average | $l_l$ |
| Move | 0.6854 | 0.00385 | 0.56% | 0.6814 | 0.00395 | 0.58% |
| MoveFlip | 0.6755 | 0.0079 | 1.17% | 0.6694 | 0.008 | 1.20% |
| MoveRotate | 0.655 | 0.0162 | 2.47% | 0.6443 | 0.01655 | 2.57% |
| Scale | 0.0231 | 0.0915 | 396.10% | 0.0179 | 0.02105 | 117.60% |
| Rotation | 0.6448 | 0.0205 | 3.18% | 0.6312 | 0.0211 | 3.34% |
| Colour | 0.6449 | 0.02055 | 3.19% | 0.6312 | 0.0211 | 3.34% |
| BG | 0.7005 | 0.0039 | 0.56% | 0.8265 | 0.0045 | 0.54% |

**Fig. 4.** Difficulty results from model B+[6] by each architecture.

based on the type of information it receives, where we would hope to use the architecture that can exhibit an exponential learning curve. In Fig. 5, models are trained on a small amount of selected images that have received the heaviest modification. The training learning values for Dense has increased all across, from the largest increase to 11.99% in Background and lowest increase to 0.14% in Scaling. Validation learning values have increased by very small values in comparison, where Background validation learning value did not improve at all. Whereas in CNN and RNN the training learning values have both increased by a larger comparison. For example, it is 0.48% in RNN. Such results from Model B* are encouraging as it shows that given very few pictures but with the information provided maximized, notable progress can be made. This is hopefully a notable point to raise as the existing trend in neural network training leans towards providing large chunks of data to facilitate implicit generalization learning.

| Model B* | Dense | | | | | |
|---|---|---|---|---|---|---|
| | loss | Average | $l_r$ | val_loss | Average | $l_r$ |
| Move | 0.6921 | 0.0033 | 0.48% | 0.6941 | 0.0011 | 0.16% |
| MoveFlip | 0.6921 | 0.0033 | 0.48% | 0.6935 | 0.00085 | 0.12% |
| MoveRotate | 0.6921 | 0.0087 | 1.26% | 0.6931 | 0.0006 | 0.09% |
| Scale | 0.6921 | 0.00095 | 0.14% | 0.6922 | 0.00075 | 0.11% |
| Rotation | 0.6921 | 0.0108 | 1.56% | 0.6927 | 0.0007 | 0.10% |
| Colour | 0.691 | 0.0029 | 0.42% | 0.6902 | 0.00095 | 0.14% |
| BG | 0.6922 | 0.083 | 11.99% | 0.6917 | 0.00045 | 0.07% |
| Model B* | RNN | | | | | |
| | loss | Average | $l_r$ | val_loss | Average | $l_r$ |
| Move | 0.6854 | 0.00385 | 0.56% | 0.6814 | 0.00395 | 0.58% |
| MoveFlip | 0.6854 | 0.00385 | 0.56% | 0.6814 | 0.00395 | 0.58% |
| MoveRotate | 0.6869 | 0.0031 | 0.45% | 0.6833 | 0.0035 | 0.51% |
| Scale | 0.6854 | 0.00385 | 0.56% | 0.6814 | 0.00395 | 0.58% |
| Rotation | 0.6869 | 0.0031 | 0.45% | 0.6833 | 0.0035 | 0.51% |
| Colour | 0.6869 | 0.0031 | 0.45% | 0.6833 | 0.0035 | 0.51% |
| BG | 0.8482 | 0.00405 | 0.48% | 0.7146 | 0.0029 | 0.41% |

**Fig. 5.** Difficulty results from model B* by each architecture.

| Model B+[6] | Dense | | |
|---|---|---|---|
| | Training $l_r$ | Validation $l_r$ | Difficulty |
| Move | 5.34% | 5.32% | 4 |
| MoveFlip | 11.60% | 11.51% | 4 |
| MoveRotate | 25.88% | 25.39% | 3 |
| Scale | 43.47% | 41.86% | 1 |
| Rotation | 31.93% | 31.14% | 2 |
| Colour | 31.29% | 30.54% | 2 |
| BG | 1.14% | 1.10% | 5 |
| Model B+[6] | CNN | | |
| | Training $l_r$ | Validation $l_r$ | Difficulty |
| Move | 157184.71% | 15.65% | 4 |
| MoveFlip | 178126.73% | 48.45% | 2 |
| MoveRotate | 104813.52% | 117.48% | 5 |
| Scale | 160358.75% | 174.00% | 3 |
| Rotation | 164479.67% | 168.89% | 3 |
| Colour | 148389.73% | 0.00% | 4 |
| BG | 244057.20% | 0.00% | 1 |
| Model B+[6] | RNN | | |
| | Training $l_r$ | Validation $l_r$ | Difficulty |
| Move | 0.56% | 0.58% | 5 |
| MoveFlip | 1.17% | 1.20% | 4 |
| MoveRotate | 2.47% | 2.57% | 3 |
| Scale | 396.10% | 117.60% | 1 |
| Rotation | 3.18% | 3.34% | 2 |
| Colour | 3.19% | 3.34% | 2 |
| BG | 0.56% | 0.54% | 5 |

**Fig. 6.** Learning improvement rates and difficulty from Model B+[6] ranked by the training learning rates.

Our study might suggest that it might be possible to achieve a sufficient enough standard with specially tailored dataset, albeit not adequate standards given today's expectations of neural network.

Through our results in Fig. 6, it might seem that inter-variance categories may not always necessarily be more difficult. That the Dense and RNN architecture would deal with the Movement with Rotation better than the Pure 2D movement dataset. However, it is also true that our results show the opposite for CNN. One possible reason might be that there might potentially by newer information when 2 operations are applied on a dataset, then whether the model is able to make use of the information to explain the data depends heavily on the architecture and how it views and encode the information. Our results might have been due to the dataset being notably small that the Dense and RNN are capable of memorizing certain aspects of the data instead of true learning. Unfortunately, our results also do not prove or disprove any relationship of inter-variance category to their parent categories. As we observe the values from the Dense and RNN Architecture as seen in Fig. 6, the tasks can be grouped into different categories of similar impact. Scaling has the highest learning rate, followed by Rotational and Color in one group, the group of movement, and lastly background. Then comparing to the CNN architecture, it is clear that CNN performs significantly better at the background category, but the order of the other categories is debatable. Another notable point from Fig. 5 is the inter-variance category Movement with Rotation, it would seem easier to the Dense and RNN architecture in terms of training learning value but not for the CNN architecture. CNN in this case then has better validation learning rate as it likely did not overfit the training data.

Among the outcome of our experiments, the confidence graphs of the CNN architectures with Model B are shown in Figs. 7, 8, 9, 10 and 11 have the most reasonable pattern. Our results also shown that Dense architecture has some unexplainable fluctuations and RNN's being a bit flat for almost all graphs. Movement without Flip has the chair shifted from the center to the top left corner, yet the graphs have notable peaks and troughs that need further explanation. The Movement with Flip shows clearly in the center a clear drop, but the same pattern persists. Background has various local peaks and trough depending on the additional object accompanying the target object, the subsequent spike is due to later images where the target object is accompanied by a background scene.

The confidence graphs see higher variance impacts with some tasks. Background has various local peaks and trough depending on the additional object accompanying the target object, the subsequent spike is due to later images where the target object is accompanied by a background scene. Scaling is done by alternating between larger and smaller. Therefore, we can see a slight trend upwards and downwards while observing overlapping. Interestingly, the confidence goes up when the target object is scaled upwards and goes down when scaled downwards. This might suggest that scaling down objects causes some loss of information that the model uses. Movement with Rotation is steadily moving downwards. The rotating angle in the transition has helped; however, this graph can be studied further to be more explainable. Rotational has the target object rotated such that it returns to original position at the center and at the end. We can see that the confidence trends towards these two points and trending downwards when the rotation is further away from the original positions. Color increases somewhat in trend with brightness, the more visible the target object is, the more confident the model is. As the target object starts as black color (same as the background), and we see that when two of the RGB values are zero, the model drops back to the lowest point.

We have also presented the selected results of our study of the learning gain per data from all our distribution data models in Fig. 12. To perform this study, we took all the learning values and divided by the number of images there were in each dataset. Throughout Model B to Model B+[6] for Dense, this might be the expected pattern most would have in mind if asked how Dense treats its information, all information given is equally important. Or possibly it just reflects the trajectory of the loss function towards a local minima. A similar story is seen in RNN except that it has a higher learning gain per data for the Scaling dataset. Figure 12 shows a spike in learning at Model B+[2] followed with a rather consistent decrease all the way till Model B+[6] with the exception of the Background dataset. The vital information was sufficient for CNN at Model B+[2] and it did not have much else to learn. Unlike the background dataset, especially in Model B+[5] and B+[6], where the images with background scene are added are likely the causes of the spike in values. Our study thus manages to suggest that there seems to be a mechanism in the architectures that models take advantage of to better analyses and explain the data in the domain of images.

To summarize, the described outcome of our experiments in this section will help in critical decision-making processes. Understanding the nature of variances and their impact on learning as variance categorization in general and in various deep learning network architectures results in explainable learning output, and

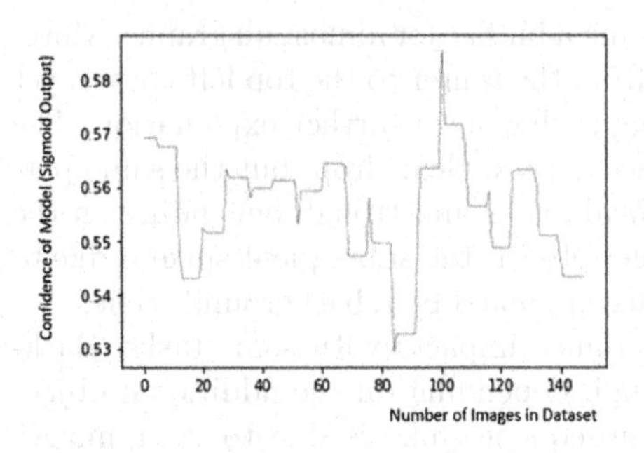


Fig. 7. Confidence for movement.

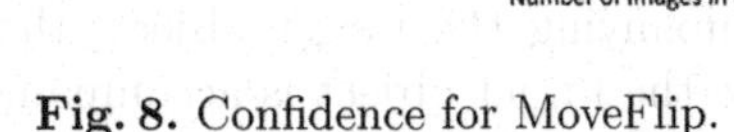

Fig. 8. Confidence for MoveFlip.

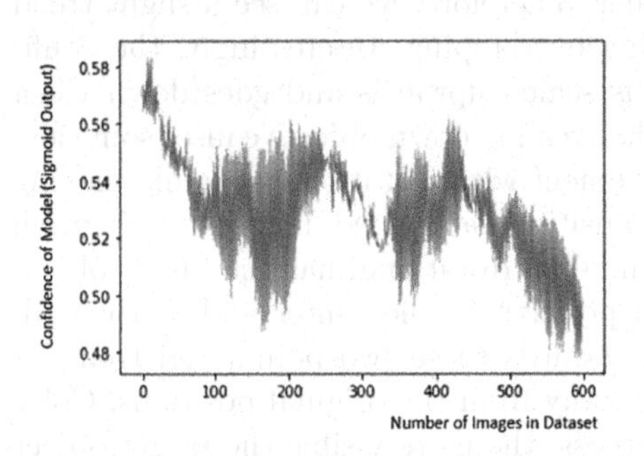


Fig. 9. Confidence for MoveRotation.

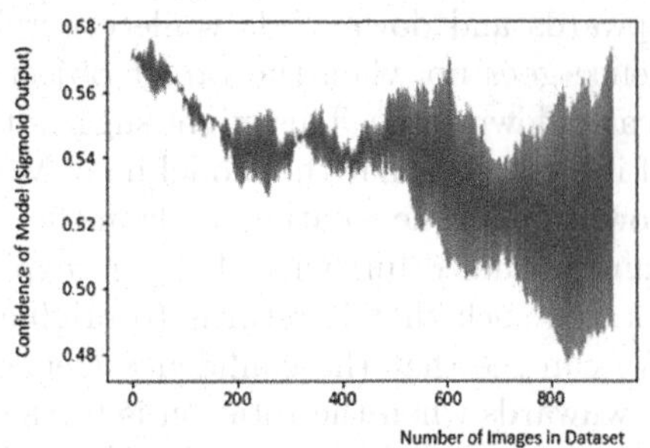


Fig. 10. Confidence for scaling.

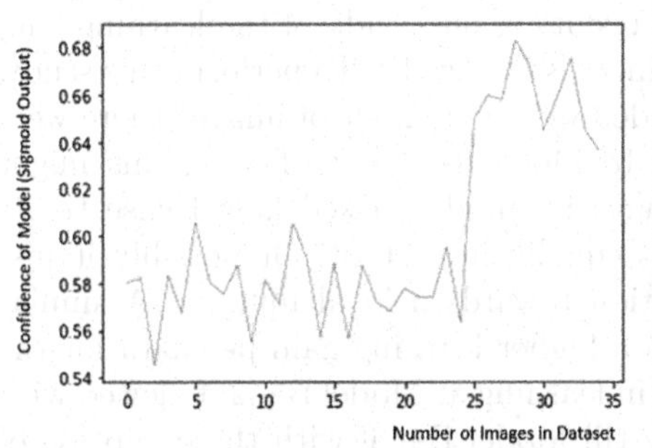


Fig. 11. Confidence for background.

| Model B+[2] | Dense | | | Model B+[6] | Dense | | |
|---|---|---|---|---|---|---|---|
| | Dataset # | Training $I_r$ | Validation $I_r$ | | Dataset # | Training $I_r$ | Validation $I_r$ |
| Move | 30 | 0.039% | 0.037% | Move | 140 | 0.038% | 0.038% |
| MoveFlip | 60 | 0.037% | 0.037% | MoveFlip | 290 | 0.040% | 0.040% |
| MoveRotate | 120 | 0.039% | 0.038% | MoveRotate | 590 | 0.044% | 0.043% |
| Scale | 185 | 0.039% | 0.039% | Scale | 911 | 0.048% | 0.046% |
| Rotation | 144 | 0.039% | 0.039% | Rotation | 707 | 0.045% | 0.044% |
| Colour | 141 | 0.038% | 0.038% | Colour | 696 | 0.045% | 0.044% |
| BG | 6 | 0.034% | 0.034% | BG | 30 | 0.038% | 0.037% |
| Model B+[2] | CNN | | | Model B+[6] | CNN | | |
| | Dataset # | Training $I_r$ | Validation | | Dataset # | Training $I_r$ | Validation $I_r$ |
| Move | 30 | 9303.62% | 1222.67% | Move | 140 | 1122.75% | 0.11% |
| MoveFlip | 60 | 4277.79% | 4.47% | MoveFlip | 290 | 614.23% | 0.17% |
| MoveRotate | 120 | 2226.22% | 0.15% | MoveRotate | 590 | 177.65% | 0.20% |
| Scale | 185 | 867.33% | 0.11% | Scale | 911 | 176.02% | 0.19% |
| Rotation | 144 | 1198.18% | 0.23% | Rotation | 707 | 232.84% | 0.24% |
| Colour | 141 | 1026.64% | 0.16% | Colour | 696 | 213.20% | 0.00% |
| BG | 6 | 38.59% | 83.42% | BG | 30 | 8135.24% | 0.00% |
| Model B+[2] | RNN | | | Model B+[6] | RNN | | |
| | Dataset # | Training $I_r$ | Validation | | Dataset # | Training $I_r$ | Validation $I_r$ |
| Move | 30 | 0.0187% | 0.0193% | Move | 140 | 0.0040% | 0.0041% |
| MoveFlip | 60 | 0.0075% | 0.0085% | MoveFlip | 290 | 0.0040% | 0.0041% |
| MoveRotate | 120 | 0.0047% | 0.0048% | MoveRotate | 590 | 0.0042% | 0.0044% |
| Scale | 185 | 0.0063% | 0.0065% | Scale | 911 | 0.4348% | 0.1291% |
| Rotation | 144 | 0.0039% | 0.0040% | Rotation | 707 | 0.0045% | 0.0047% |
| Colour | 141 | 0.0032% | 0.0036% | Colour | 696 | 0.0046% | 0.0048% |
| BG | 6 | 0.0752% | 0.0582% | BG | 30 | 0.0186% | 0.0181% |

Fig. 12. Learning gain per data.

will be highly desired. Further, we also expect that our results will be used as a guidance for researchers to design a mechanism to cater for better learning efforts. This shall be achieved by having learning models to acquire more intrinsic and higher responsiveness to the data but also their variance in their respective learning processes. Our experiment results have also suggested that it is likely such required mechanisms for every architecture may differ.

## 5 Conclusion

In this paper, we study how variance of datasets are translated into difficulties in deep learning. We experiment on how different deep learning models and their architectures respond to such variance in their respective learning activities in terms of the learning performance of object recognition. We claim that the expected way of utilization of variance information of datasets in our specific learning task and learning environment differs from models. The amount of learning or information gained for each data fed into models also varies significantly in our setting. The investigation of tasks other than object recognition is being conducted and we hope to extract common elements to shed light on a possible invariance representation that can help machine to match the performance of human beings in similar learning tasks.

## References

1. Amit, Y., Felzenszwalb, P.: Object Detection, pp. 537–542 (2014)
2. DiCarlo, J.J., Zoccolan, D., Rust, N.C.: How does the brain solve visual object recognition? Neuron **73**(3), 415–434 (2012)
3. Majaj, N.J., Hong, H., Solomon, E., DiCarlo, J.: Simple learned weighted sums of inferior temporal neuronal firing rates accurately predict human core object recognition performance. J. Neurosci. **35**(39), 13402 (2015)
4. Barbu, A., et al.: ObjectNet: a large-scale bias-controlled dataset for pushing the limits of object recognition models. In: Neural Information Processing Systems (2019)
5. Zhu, Z., Xie, L., Yuille, A.L.: Object recognition with and without objects. arXiv e-prints (2016)
6. Zhang, C., Bengio, S., Hardt, M., Recht, B., Vinyals, O.: Understanding deep learning requires rethinking generalization. In: International Conference on Learning Representations. OpenReview.net, Toulon, France (2017)
7. Geirhos, R., et al.: ImageNet-trained CNNs are biased towards texture; increasing shape bias improves accuracy and robustness. In: International Conference on Learning Representations. OpenReview.net (2019)
8. Deng, J., Dong, W., Socher, R., Li, L., Li, K., Li, F.: ImageNet: a large-scale hierarchical image database. In: 2009 IEEE Conference on Computer Vision and Pattern Recognition, pp. 248–255 (2009)
9. Huh, M., Agrawal, P., Efros, A.: What makes ImageNet good for transfer learning? arXiv e-prints (2016)
10. He, K., Girshick, R., Dollar, P.: Rethinking ImageNet Pre-Training. In: 2019 IEEE International Conference on Computer Vision, pp. 4917–4926 (2019)
11. Krizhevsky, A., Sutskever, I., Hinton, G.: ImageNet classification with deep convolutional neural networks. Association for computing machinery. Commun. ACM, **60**(6), 84–90 (2017)
12. Pinto, N., et al.: Why is real-world visual object recognition hard? (real-world visual object recognition). PLoS Comput. Biol. **4**(1), e27 (2008)
13. Kheradpisheh, S., et al.: Deep networks can resemble human feed-forward vision in invariant object recognition. Sci. Rep. **6**, 32672 (2016)
14. Pov-Ray Website. http://www.povray.org/download/

# Drawing Dreams

Jingxian Wu[1], Zhaogong Zhang[1(✉)], and Xuexia Wang[2]

[1] School of Computer Science and Technology,
Heilongjiang University, Harbin, China
18860455718@163.com, 2013010@hlju.edu.cn

[2] Department of Mathematics, University of North Texas, Denton, USA
xuexia.wang@unt.edu

**Abstract.** Dreams have been responsible for some major creative and scientific discoveries in the course of human history. Recording dreams in the form of images is an interesting and meaningful thing. Our task is to generate images based on the description of dreams. Recently, there has been exciting progress in generating images from descriptions of birds and flowers, but the dream scene is more fantasy than the real scene. The challenge to reproduce complex sentences with many objects and relationships remain. To truthfully reappear the dream scene, we process sentences into scene graphs that are a powerful structured representation for both images and language; then using a graph convolution network to obtain layout information, combining the layout information and a single feedforward network to generate the image. Subsequently, we apply Cycle-Consistent Adversarial Net (CycleGAN) to change the image into different styles according to the mood of users when dreaming. According to the experimental results, our method can generate complex and diverse dreams.

**Keywords:** Image synthesis · Scene graphs · GAN

## 1 Introduction

Inspiration can strike in the most unexpected places. Oftentimes, the best creative ideas occur while we are sleeping. Dreams can be a rich source of inner wisdom, and they can be useful in a variety of contexts, from problem-solving to reducing stress. Our dreams can function on many different levels, from telling us which parts of our psyche are out of balance to anticipating our future needs. Some directors used their dreams as inspiration for some of their greatest films. Drawing dreams is an interesting and meaningful thing. However, it is difficult for ordinary people to paint their dreams, and memories of dreams are easy to disappear. For these reasons, we were inspired to record dreams in the form of pictures in time. Dreams are different from real-life scenes. The scenes that appear in dreams may be imaginary, and the mood when dreaming will affect the scenes of dreams.


H. Yang et al. (Eds.): ICONIP 2020, LNCS 12532, pp. 290–300, 2020.
https://doi.org/10.1007/978-3-030-63830-6_25

Dream generation from textual descriptions could be regarded as an image synthesis problem. Recently, numerous methods of text to image synthesis have emerged. For instance, one of the state-of-the-art image synthesis has given stunning results on fine-grained descriptions of birds [20], and the real scenes trained on the coco dataset [8] have also achieved good results. For our task of dream generation, we need to solve the following questions: (1) The previous task of generating images from the text was trained on the coco dataset, which contains a large number of pictures in real scenes. However, the dream scene is different from the real scene. The relationship of objects in the dream scene may be illogical. For example, in the dream, people can fly in the sky, but there are no such images in the real scene. The first problem we need to solve is how to reproduce the dream scene without the large dream dataset. For this problem, we use the scene graph [4], the expression of dreams is reflected in the relative positional relationship between the objects, and scene graphs can reason explicitly about objects and relationships, and generate complex images with many recognizable objects. We synthesize the scene graph by extracting the objects and the relationship of objects from the text and use a graph convolution network to predict bounding boxes and masks for all objects by inputting the scene graph, so we use this method to get the layout information of the image. Then combining the layout information and a cascaded refinement network (CRN) [1] for obtaining the final image. (2) The second problem is the mood when dreaming. The generation of images includes not only the scenes of dreams but also the emotions during dreaming. When we wake up from a dream, the scene of the dream may be vague but the feelings while dreaming are still clear. For this problem, We apply the style transfer. We use a cycle-consistent adversarial network (CycleGAN) [19] to change the style of the image. We chose the five most common emotions when dreaming: peaceful, anxious, sad, grotesque, and pleasant. We change the style of the image according to different emotions.

To summarize, as shown in Fig. 1, we use the description of the dream and the emotions during dreaming as the input of the model, we extract the objects and relationships from the text to generate the scene graph, and use the scene graph as the input of the graph convolution network to obtain the layout information, the cascaded refinement network (CRN) [1] generates the final image from the layout information. We transfer the style of the image according to different emotions.

The major contributions of this paper are summarized as follows: (1) We propose a method combined with scene graphs to generate more complex images from sentences. (2) We propose the idea of generating dream images to record and study the effects of dreams on people. (3) The generated image can be recognized by more users.

## 2 Related Work

With the recent successes of generative adversarial networks (GANs) [2,18], a large number of methods have been proposed to handle image generation.

Recently, generating images based on the text descriptions gained interest in the research community [7,16,20]. Synthesizing images from a given text description utilizing only the content usually produce the inferior result, thus there's a need for incorporating a dual adversarial inference procedure to improve text-to-image synthesis [6] in order to learn disentangled representations of content and style in an unsupervised way.

Image generation based on scene graphs was recently presented in [4]. Scene graphs represent scenes as directed graphs, where nodes are objects and edges represent relationships between objects. Scene graphs enable explicit reasoning about objects and their relationships. Methods have been developed for converting sentences to scene graphs [13]. These methods are more closely related to our work.

Neural style transfer synthesizes a novel image by combining the content of one image with the style of another image by matching the Gram matrix statistics of pre-trained deep features. The cycle-consistent adversarial network [19] achieves unpaired image-to-image translation and builds on the pix2pix framework [3], which uses a conditional generative adversarial network [10] to learn a mapping from input to output images. The Graph Neural Networks (GNNs) [12] can operate on arbitrary graphs, and we utilize the network to process the scene graph.

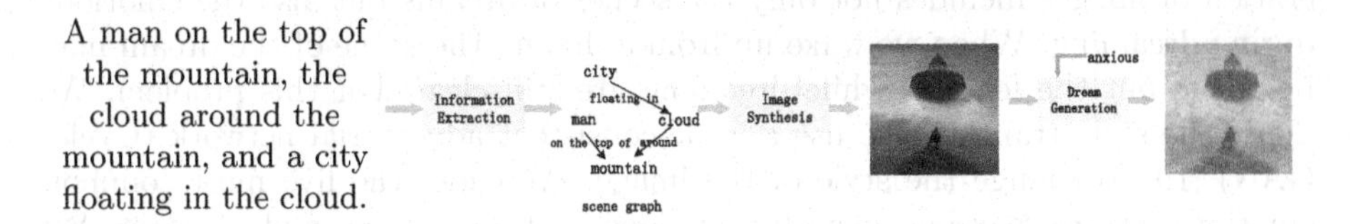


**Fig. 1.** The modules and the workflow of our method. Our model extracts the relationship information of objects from text to generate a scene graph, and predicts the layout information of the image from the scene graph, then generates the image. Finally, our model changes the style of the image according to the mood during dreaming.

## 3 Method

Our goal is to develop a model which takes text as input and generates an image corresponding to the description of our dream. Our model is divided into three sub-modules: Information Extraction, Image Synthesis, and Dream Generation.

In Sect. 3.1, we discuss the method of information extraction, which outputs scene graphs from text. In Sect. 3.2, we describe the network structure of image synthesis that generates images by processing the scene graph. In Sect. 3.3, we introduce the sub-module of dream generation. We show the final image according to the mood of users. The Fig.1 shows the workflow of our method.

### 3.1 Information Extraction

The first step is to extract information from text. For this issue, we use the Stanford CoreNLP [9] which provides a set of human language technology tools for word segmentation. Given a set of object categories $C$ and a set of relationship categories $R$, $O = \{o_1, ..., o_n\}$ is a set of objects with each $o_i \in C$. We obtain relationship $E$ where $E \subseteq O \times R \times O$ is set of directed edges of the form $(o_i, r, o_j)$ where $o_i, o_j \in O$ and $r \in R$ by using Stanford CoreNLP [9]. The result lists all relationships. In some cases, an object has multiple relationships. For example, a man standing on a mountain, the cloud around a mountain, and one mountain above another mountain, after the word segmentation, the relationships include (man, standing on, mountain), (cloud, around, mountain), and (mountain, above, mountain). In this case, there is semantic ambiguity if we do not make constraints; it is not clear whether it is the same mountain or different mountains. So, we make a restraint that $o_i$ is the object that first appeared and $o_j$ is the object that appears for the first time or has recently appeared. We define a list *obj* to record objects that appeared in text. We traverse every relationship. In the loop, directly put $o_i$ into *obj* and return the index. Before processing $o_j$, we will check whether $o_j$ exists in the *obj* from back to front. If it exists, it will return the index of the existing object. Otherwise, it will put $o_j$ in the *obj* and then return the index.

In the end, we output a dictionary that includes two lists, $obj = [o_i, o_j, ...]$ and $rela = [[x_i, r, x_j], ...]$, where $x_i$ is the index of $o_i$, $x_j$ is the index of $o_j$ in the *obj*, and $r \in R$. The specific process is shown in the Algorithm 1.

---

**Algorithm 1:** Information extraction

---

**Input**: text
**Output**: A dictionary includes information of objects and relationships.

```
1  obtain E from CoreNLP.
2  Let dic = {}, objects = [], relationships = [].
3  for E do
4      obtain (o_i, r, o_j)
5      objects.append(o_i)
6      x_i = objects.index(o_i)
7      if o_j in objects then
8          x_j = max(objects.index(o_j))
9      else
10         objects.append(o_j)
11         x_j = objects.index(o_j)
12     end
13     relationships.append([x_i, r, x_j])
14 end
15 dic['objects'] = object
16 dic['relationships] = relationships
17 return dic
```

### 3.2 Image Synthesis

In this part, our goal is to generate an image corresponding to the scene graph. In the previous sub-module, we obtained the information of objects and relationships; this information composes the scene graph. We use a graph convolution network [4,12] composed of several graph convolution layers to process the scene graph.

The scene graph is passed to the graph convolution network $G$ to create the object embedding vector $u_i$ for each object, $u_i = G(obj, rela)$, where $obj$ and $rela$ record the information of objects and relationships from the sub-module of information extraction. The object layout network $L$ [4] receives an embedding vector $u_i$ for object $o_i$ and passes it to a box regression network $B$ to predict a bounding box $\widehat{b_i} = (x_0, y_0, x_1, y_1)$, the box regression network is a multilayer perceptron. The mask regression network $M$ predicts a binary mask per object. The mask regression network is used to predict the shape of the object, and the box regression network is used to determine the position of the object in the image. The scene layout is the sum of all object layouts. As shown in Fig. 2. We synthesize an image given in the layout. Using the cascaded refinement network [1] $E$ that outputs an image $\widehat{p} \in R^{H \times W \times 3}$ based on scene layout.

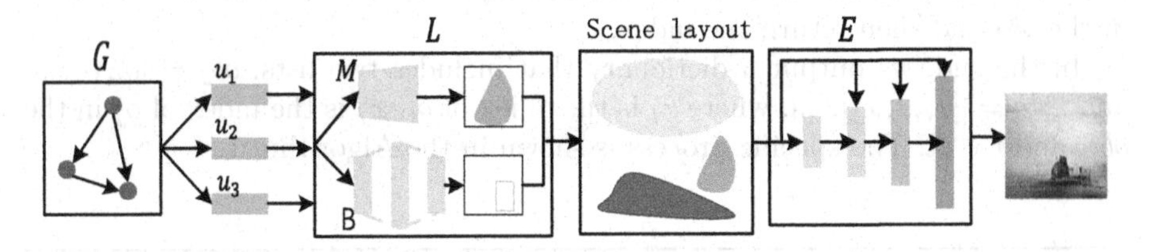


**Fig. 2.** The scene graph is passed to the network $G$ to create the embedding vector $u_i$. The object layout network L predicts a layout for the object from this embedding, summing all objects layouts gives the scene layout. The autoencoder $E$ generates the final image from the scene layout.

**Loss and Training.** The loss is used to optimize the networks and generate images corresponding to the information. The box loss $L_{box}$ is the L1 norm difference between the computed $\widehat{b_i}$ and the ground truth bounding box $b_i$.

$$L_{box} = \sum_{i=1}^{n} \left\| b_i - \widehat{b_i} \right\|_1 \tag{1}$$

where $n$ is the number of all objects.

The reconstruction loss $L_{rec}$ compares the reconstructed image $\widehat{p}$ and the ground truth training image $p$. Let $\Phi$ be a trained VGG-19 network [14], $\{\Phi_l\}$ is a collection of layer $l$ in the network $\Phi$.

$$L_{rec} = \sum_{l} \frac{1}{l} \left\| \Phi_l(p) - \Phi_l(\widehat{p}) \right\|_1 \tag{2}$$

We employ two discriminators $D_{image}$ and $D_{object}$. A discriminator $D$ attempts to classify its input $p$ as real or fake by maximizing the objective. $D_{image}$ ensures the image is recognizable.

$$L_{image} = \sum_{i=1} log\left(D_{image}\left(p_i\right)\right) + log\left(1 - D_{image}\left(\widehat{p}_i\right)\right) \tag{3}$$

where $p_i$ is the ground truth image and $\widehat{p}_i$ is the fake image that is generated using the network. $D_{image}$ maximizes this loss during training, and the network attempts to generate outputs $\widehat{p}_i$ that will fool the discriminator.

The discriminator $D_{object}$ guarantees that generated objects are identifiable. We create object image $\widehat{I}_i$ by cropping $\widehat{p}_i$ using the bounding boxes $\widehat{b}_i$, $I_i$ is ground truth image that obtained from the ground truth image $p_i$, using the ground truth bounding boxes $b_i$.

$$L_{object} = \sum_{i=1} log\left(D_{object}\left(I_i\right)\right) + log\left(1 - D_{object}\left(\widehat{I}_i\right)\right) \tag{4}$$

We minimize all of the weighted losses given the need to train the network and two discriminators.

### 3.3 Dream Generation

We use the cycle-consistent adversarial network (CycleGAN) [19] to change the visual sense of the image through the mood entered by the user. The network learns to translate an image from a source domain $X$ to a target domain $Y$.

We have generated images from the module of image synthesis. We have two domains $X$ and $Y$, the generated images $\{x_i\}_{i=1}^{N} \in X$ and the target images $\{y_j\}_{j=1}^{M} \in Y$. The model includes two generators, $G(x)$ and $F(y)$, and two adversarial discriminators, $D_X$ and $D_Y$, where $D_X$ discriminates between images $x$ and translated images $\widehat{x} = F(y)$ and $D_Y$ discriminates between images $y$ and translated images $\widehat{y} = G(x)$. We apply a cycle consistency loss [19]:

$$L_{cyc} = E_{x \sim p_{data}(x)} \left\|F\left(G\left(x\right)\right) - x\right\|_1 + E_{y \sim p_{data}(y)} \left\|G\left(F\left(y\right) - y\right)\right\|_1 \tag{5}$$

This learned mapping functions should be cycle-consistent, for each image x from domain X, the image translation cycle $F\left(G\left(x\right)\right)$ should be able to bring x back to the original image.

We combine adversarial losses [2] to train $G\left(x\right)$ and $D_Y$.

$$L_{gan} = E_{y \sim p_{data}(y)} log\left(D_Y\left(y\right)\right) + E_{x \sim p_{data}(x)} log\left(1 - D_Y\left(G\left(x\right)\right)\right) \tag{6}$$

$G\left(x\right)$ and $D_Y$ are both trained simultaneously, adjusting parameters for $G\left(x\right)$ to minimize $log\left(1 - D_Y\left(G\left(x\right)\right)\right)$ and to minimize $log\left(D_Y\left(y\right)\right)$ for $D_Y$. Through training, $G\left(x\right)$ generates images that are more like images from domain Y. $F_Y$ and $G\left(x\right)$ have a similar adversarial loss.

# 4 Experiment

In this section, we show our results with comparisons to recent related works. All experiments were executed with Pytorch 0.4, CUDA v10, Cudnn v7 and Python 3. We train our model to generate 128 × 128 images on Visual Genome [5] datasets.

## 4.1 Dataset

The second module is conducted on the Visual Genome [5] dataset, which comprises of 108,077 images annotated with scene graphs. Each image has an average of 21 objects, 18 attributes, and 18 pairwise relationships between objects. It allows for a multi-perspective study of an image, from pixel-level information like objects to relationships that require further inference, and to even deeper cognitive tasks like question answering. In our experiment, we use the information about objects and relationships to express the scene of the dream. The third module is trained on the photographs downloaded from some photo sharing websites, such as Flickr, Pexels, and Baidu. We chose different styles of images that match the mood of the dream from the website, approximately 1,000 pictures per style.

## 4.2 Quantitative Evaluation

We use the Inception score as the quantitative evaluation metrics. The IS [11] uses a pre-trained Inception v3 network [15] to compute the KL-divergence between the conditional class distribution and the marginal class distribution. A large IS means that the generated model outputs a high diversity of images for all classes and each image clearly belongs to a specific class.

We split our result images into 10 groups, each group calculates the inception score and then we calculate the mean and standard deviation of these inception scores as the final measurement criteria. Our model achieves 5.6 $\pm.2$ inception score on the Visual Genome [5] dataset. We use the AttnGAN [17] model as a reference. AttnGAN boosts the best-reported inception score on COCO [8] dataset to 25.89 $\pm.47$. We are unaware of any previous methods for generating images from texts to apply the VG dataset as the dataset. The VG dataset has a certain ability to analyze more complex visual scenes. For our task, it is more suitable than the coco dataset.

## 4.3 Qualitative Comparison and User Studies

We compare our method with the previous state-of-the-art GAN model for text-to-image generation. Our method has the same function as AttnGAN [17] that can generate images from text. AttnGAN applies multi-stage refinement for fine-grained text-to-image generation. It considers important fine-grained information at the word level. We display the qualitative result in Fig.3. For more complex images, such as mountains floating on the sky, our method can also predict the bounding box of the target, and generate an image that matches the predicted location. The advantage of our method is that the images we generate can reflect the bizarreness of our dreams.

A cat sitting in the boat on the ocean. | A cow eating grass below the sky. | The sky above mountain and sand below the mountain and a boy standing on sand. | A mountain on top of the sky. | A cabinet has a hole in a room.

**Fig. 3.** Qualitative comparisons with AttnGAN [17]. We use the pre-trained model of the AttnGAN and select the corresponding scene from the example sentences published by AttnGAN at GitHub, we generate our images by using the same text.

It is important to point out that none of the quantitative metrics are perfect. The generated picture should be understandable and recognizable by humans. Based on this idea, we take personal choice as one of the metrics for generating images.

We use questionnaires to collect the user's choices. In the process of sample collection, we set 1000 as the number of samples, there is no limit to the target population, and each ID can only submit the questionnaire once. We show some results in Fig. 4.

### 4.4 Results Display

In the development of dreams, the emotions in dreaming are much more important than the scenes in the dream. Sometimes we may not remember the main events in the dream, but we still remember how we felt during the dream. We provide inputs of emotion to express the mood while dreaming.

Dreams have been responsible for some major creative and scientific discoveries in the course of human history. In Fig. 5. We show several famous dreams. The second row of images shows the dream of Einstein. As a young man, Einstein

1.A cat sitting in the boat on the ocean.

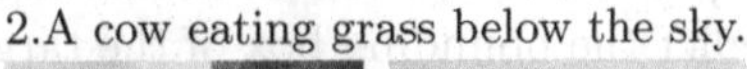

Please choose the image that matches the caption better.

| image | AttnGAN | our |
|---|---|---|
| user | 375/1000 | 625/1000 |

| image | AttnGAN | our |
|---|---|---|
| user | 169/1000 | 831/1000 |

**Fig. 4.** We conducted a user survey to compare the intelligibility of the pictures of our method against AttnGAN [17], the bottom tables show the user's choice of two pictures.

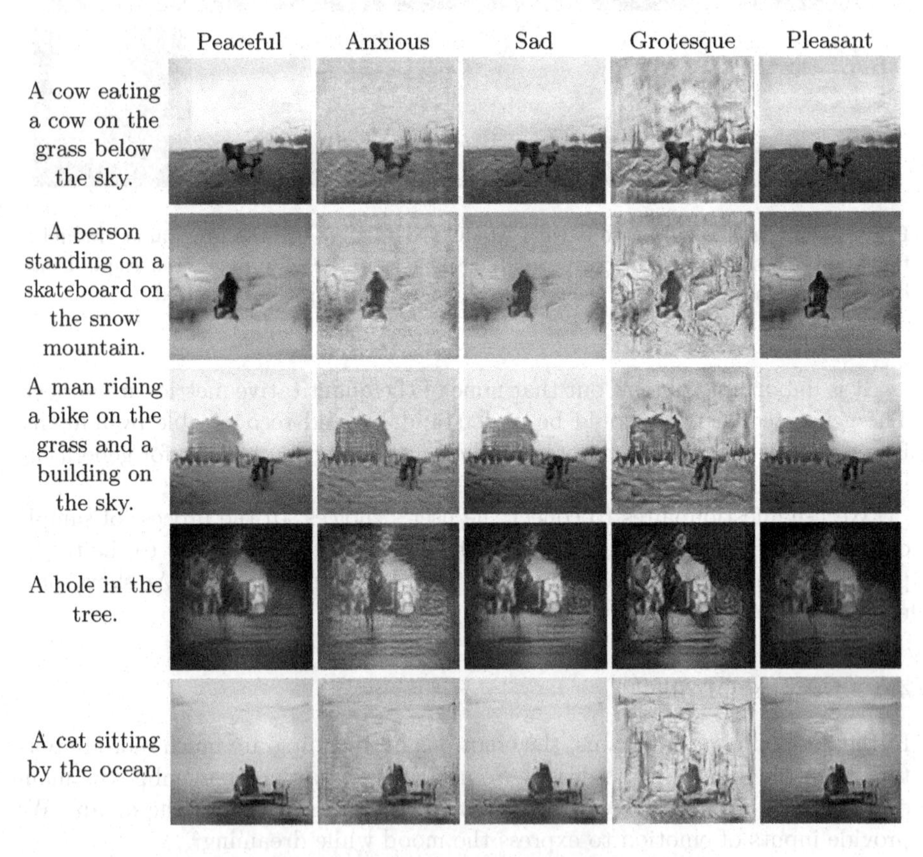


**Fig. 5.** Dream generated by adding mood options. We use the 5 most common emotions in dreams to change the generation of dream images.

dreamed he was sledding down a steep mountainside, he approached the speed of light. He awoke and meditated on this idea, soon formulating what would become one of the most famous scientific theories in the history of mankind. We provide five kinds of moods when dreaming: peaceful, anxious, sad, grotesque, and pleasant. In this part, we use the models on the generator to transform the original image into different styles. We put this generator into our network, and the user only needs to enter a description and mood of the dream to generate the final image.

## 5 Conclusion and Future Work

We propose a method that can generate dream images through text description. Experiments show that our method can generate complex images and different styles of images according to the dreamer's mood. In the future, we will add user feedback to modify the image, and further train the network so that the generated image can more accurately restore the dream scene. We will use the generated images to further study the impact of dreams on people.

## References


1. Chen, Q., Koltun, V.: Photographic image synthesis with cascaded refinement networks. In: Proceedings of the IEEE International Conference on Computer Vision, pp. 1511–1520 (2017)
2. Goodfellow, I., et al.: Generative adversarial nets. In: Advances in neural information processing systems, pp. 2672–2680 (2014)
3. Isola, P., Zhu, J.Y., Zhou, T., Efros, A.A.: Image-to-image translation with conditional adversarial networks. In: Proceedings of the IEEE Conference on Computer Vision and Pattern Recognition, pp. 1125–1134 (2017)
4. Johnson, J., Gupta, A., Fei-Fei, L.: Image generation from scene graphs. In: Proceedings of the IEEE Conference on Computer Vision and Pattern Recognition, pp. 1219–1228 (2018)
5. Krishna, R., et al.: Visual genome: connecting language and vision using crowdsourced dense image annotations. Int. J. Comput. Vision **123**(1), 32–73 (2017)
6. Lao, Q., Havaei, M., Pesaranghader, A., Dutil, F., Jorio, L.D., Fevens, T.: Dual adversarial inference for text-to-image synthesis. In: Proceedings of the IEEE International Conference on Computer Vision, pp. 7567–7576 (2019)
7. Li, B., Qi, X., Lukasiewicz, T., Torr, P.: Controllable text-to-image generation. In: Advances in Neural Information Processing Systems, pp. 2063–2073 (2019)
8. Lin, T.-Y., et al.: Microsoft COCO: common objects in context. In: Fleet, D., Pajdla, T., Schiele, B., Tuytelaars, T. (eds.) ECCV 2014. LNCS, vol. 8693, pp. 740–755. Springer, Cham (2014). https://doi.org/10.1007/978-3-319-10602-1_48
9. Manning, C.D., Surdeanu, M., Bauer, J., Finkel, J., Bethard, S.J., McClosky, D.: The stanford corenlp natural language processing toolkit. In: Association for Computational Linguistics (ACL) System Demonstrations, pp. 55–60 (2014). http://www.aclweb.org/anthology/P/P14/P14-5010
10. Mirza, M., Osindero, S.: Conditional generative adversarial nets. arXiv preprint arXiv:1411.1784 (2014)

11. Salimans, T., Goodfellow, I., Zaremba, W., Cheung, V., Radford, A., Chen, X.: Improved techniques for training GANs. In: Advances in Neural Information Processing Systems, pp. 2234–2242 (2016)
12. Scarselli, F., Gori, M., Tsoi, A.C., Hagenbuchner, M., Monfardini, G.: The graph neural network model. IEEE Trans. Neural Networks **20**(1), 61–80 (2008)
13. Schuster, S., Krishna, R., Chang, A., Fei-Fei, L., Manning, C.D.: Generating semantically precise scene graphs from textual descriptions for improved image retrieval. In: Proceedings of the fourth workshop on vision and language, pp. 70–80 (2015)
14. Simonyan, K., Zisserman, A.: Very deep convolutional networks for large-scale image recognition. Computer Science (2014)
15. Szegedy, C., Vanhoucke, V., Ioffe, S., Shlens, J., Wojna, Z.: Rethinking the inception architecture for computer vision. In: Proceedings of the IEEE Conference on Computer Vision and Pattern Recognition, pp. 2818–2826 (2016)
16. Tan, H., Liu, X., Li, X., Zhang, Y., Yin, B.: Semantics-enhanced adversarial nets for text-to-image synthesis. In: Proceedings of the IEEE International Conference on Computer Vision, pp. 10501–10510 (2019)
17. Xu, T., et al.: Attngan: Fine-grained text to image generation with attentional generative adversarial networks. In: Proceedings of the IEEE Conference on Computer Vision and Pattern Recognition, pp. 1316–1324 (2018)
18. Zhao, J., Mathieu, M., LeCun, Y.: Energy-based generative adversarial network. arXiv preprint arXiv:1609.03126 (2016)
19. Zhu, J.Y., Park, T., Isola, P., Efros, A.A.: Unpaired image-to-image translation using cycle-consistent adversarial networks. In: Proceedings of the IEEE International Conference on Computer Vision, pp. 2223–2232 (2017)
20. Zhu, M., Pan, P., Chen, W., Yang, Y.: DM-GAN: dynamic memory generative adversarial networks for text-to-image synthesis. In: Proceedings of the IEEE Conference on Computer Vision and Pattern Recognition, pp. 5802–5810 (2019)

# Encoder-Decoder Based CNN Structure for Microscopic Image Identification

Dawid Połap[1(✉)], Marcin Wozniak[1], Marcin Korytkowski[2], and Rafał Scherer[2]

[1] Faculty of Applied Mathematics, Silesian University of Technology, Kaszubska 23, 44-100 Gliwice, Poland
{dawid.polap,marcin.wozniak}@polsl.pl

[2] Department of Intelligent Computer Systems, Częstochowa University of Technology, Częstochowa, Poland
{marcin.korytkowski,rafal.scherer}@pcz.pl

**Abstract.** The significant development of classifiers has made object detection and classification by using neural networks more effective and more straightforward. Unfortunately, there are images where these operations are still difficult due to the overlap of objects or very blurred contours. An example is images obtained from various microscopes, where bacteria or other biological structures can merge, or even have different shapes. To this end, we propose a novel solution based on convolutional auto-encoders and additional two-dimensional image processing techniques to achieve better efficiency in the detection and classification of small objects in such images. In our research, we have included elements such as very weak contours of shapes that may result from the merging of biological objects. The presented method was compared with others, such as a faster recurrent convolutional neural network to indicate the advantages of the proposed solution.

**Keywords:** Convolutional neural networks · Encoder-decoder · Bacteria shape detection

## 1 Introduction

The use of image processing is found in almost every industrial field. Particularly, object detection and possible recognition are essential from the perspective of quality estimation or objects counting. Objects on the production line can be counted and evaluated using image processing. Hence, image processing and analysis are important elements of today's computing. Unfortunately, this type of image may have noise or shifts, which cause difficulties in the analysis. An example is images containing many similar objects where they merge or many different shapes or types of the same object class.

Recent years have brought various image analysis techniques. Attempts were made to bring out the features of individual objects. An example is the use of


H. Yang et al. (Eds.): ICONIP 2020, LNCS 12532, pp. 301–312, 2020.
https://doi.org/10.1007/978-3-030-63830-6_26

SURF or BRISK algorithms to extract points that indicate important areas of the image [7,12]. In [1], the idea of combining the reduction of feature vectors with the Bag of Words search method was presented. The authors compared the technique with existing feature search algorithms and pointed to possible competitiveness of the presented idea. Acquiring points is only the first stage because based on the indicated areas, it is necessary to decide on the size of the object for detection, or the class for classification problems. In [9], authors obtained features using SURF and then used them in image classification by Support Vector Machine for the problem of sorting garbage. Acquiring features through image processing techniques were also used to recognize activity in [4].

Acquiring features involves using a classifier to evaluate found areas or features. The most commonly used are neural networks [11] or fuzzy controllers [2,14]. Another approach is to use derivative structures called Convolution Neural Networks (CNN), whose architecture not only performs classification but also indirectly acquires maps of features at a later stage. An example is an idea of using CNN for multiclass classification [15], where the classifier was trained for making a decision about breast cancer. The application of CNN boils down to using the structure itself in the practical problem [18] or designing solutions artificially expanding the training database to increase the effectiveness of the network [17]. Moreover, many researchers focus on image reconstruction using such solutions [5,8] or spatial regularization [19]. Another issue is analyzing autoencoders, which reduce the incoming image to a vector with numerical values and then reproduce the image from it what was presented in [3]. Additionally, in [10], the authors proposed a fuzzy approach to autoencoders solutions.

In this paper, we focus on processing images of small biological objects obtained through special microscopes. These types of images are used to count the number of objects in different phases, and above all, the number of individual objects which are being combined with others. These types of images are difficult to analyze through obstructing shapes, where one biological type can have different shapes by its arrangement. Another problem is the actual analysis of the objects by determining whether they are on top of each other or already merged. Additionally, we propose a fusion of image processing techniques with an auto-encoder based on convolution networks for verification of individual objects and their counting. The described solution was tested and compared with other existing methods for showing the advantages of the presented technique. The contribution of this paper is:

- a hybrid model of processing microscopy images,
- the idea of extending autoencoders to the classification layer.

## 2 Neural Identification Technique

The main problem with analyzing photos of microorganisms is too many different positions of objects for evaluation what can be seen in Fig. 1. For selection, we propose a fusion of image processing techniques and convolutional encoders.

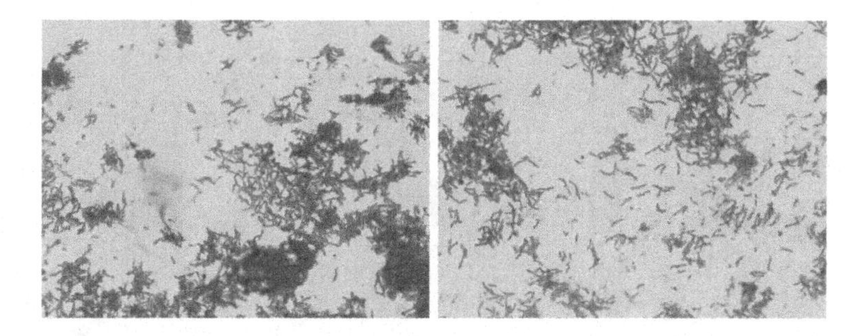

**Fig. 1.** Sample images of biological organisms obtained through microscope.

### 2.1 Image Processing

At the beginning resulting microscope images contain organisms of different sizes and colors. In order not to miss any of them during the evaluation, the skewed image is subjected to simple binarization according to the following equation

$$P'_{ij} = \begin{cases} 0 & \text{if } \delta(P_{ij}) \geq \alpha \\ 255, & \text{if } \delta(P_{ij}) < \alpha \end{cases}, \tag{1}$$

where $P'_{ij}$ is new value of pixel on each component of RGB (*Red Green Blue*) model, and funcion $\delta(\cdot)$ changes the pixel value in position $(i, j)$ to grayscale and normalizes the obtained value (using *min-max*) to the range $\leq 0, 1 \geq$ as

$$\delta(P_{ij}) = \frac{(0.21 \cdot R(P_{ij}) + 0.72 \cdot G(P_{ij}) + 0.07 \cdot B(P_{ij}))}{255}, \tag{2}$$

where function $R(\cdot)$, $G(\cdot)$ and $B(\cdot)$ are the value components in RGB model of input pixel and the coefficients are standard values in the luminosity conversion method [13]. Sample result is presented in Fig. 2.

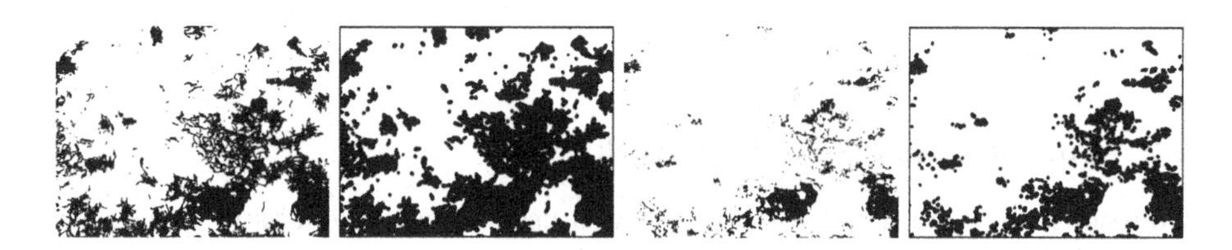

**Fig. 2.** Sample images of the initial part of binarization where the input image is filtered and normalized with two different *min-max* values: 0.6 in the first two images, and 0.4 in the third and fourth images are used.

Then, depending on the value $\alpha$ we extract each object by marking it in the image through a red box. We assume that $\alpha \in \{0.1, 0.2, \ldots, 0.9\}$ because the limit values mean the complete removal of each object from the image. As a result

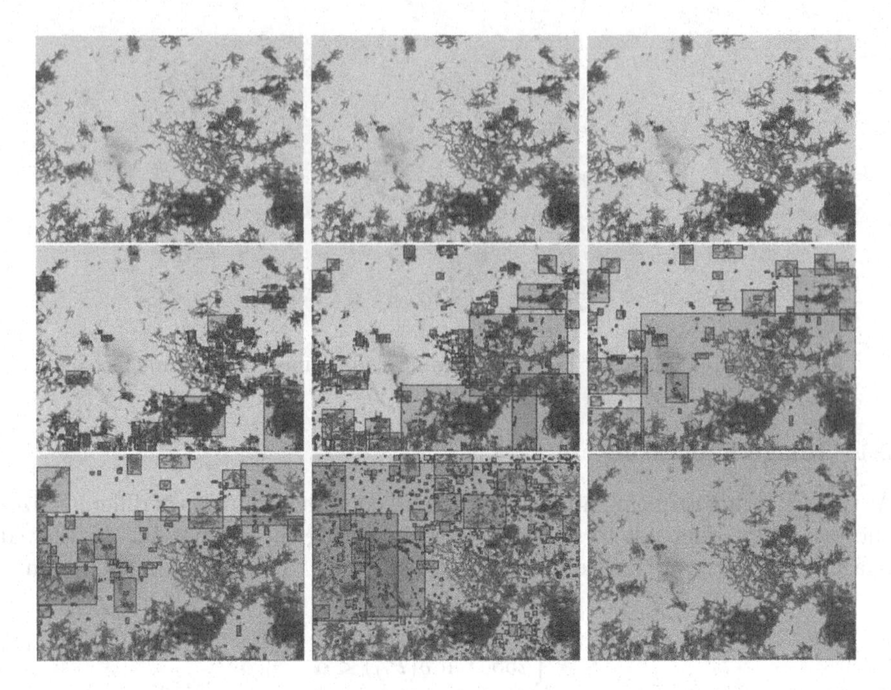

**Fig. 3.** Sample resulting images of binarization where parameter $\alpha \in \{0.1, 0.2, \ldots, 0.9\}$ as presented from top left corner image to bottom right corner image various number of frames is marked on the image.

of binarization, 9 different images are created, where the smaller the value of the parameter, the less objects marked (due to binarization), see Fig. 3. This also allows noticing the fact that reducing the alpha value causes that smaller and smaller fragments remain in the image. Such object degradation removes not only smaller, neighboring objects, but also border objects. Consequently, this allows for further analysis of the objects inside larger ones, which is important for biological organisms, especially in a situation where the whole body can be in a large, merged group.

### 2.2 Convolutional Neural Network

Convolutional Neural Network (CNN) was used in our model as it works with spatial information sourced in the microscope input image. At each par of the encoder-decoder structure, images are processed similarly, however in the opposite way. In the encoder, the image is simplified by the use of standard CNN operations like pooling and convolution, after which on the output we receive digital information about the input image. This information is processed on two levels. First is the classifier composed as a traditional fully connected layer. Second is inverted CNN, which is composed to reproduce the input image working in an inverted CNN way, so first we have a fully connected layer to process digital

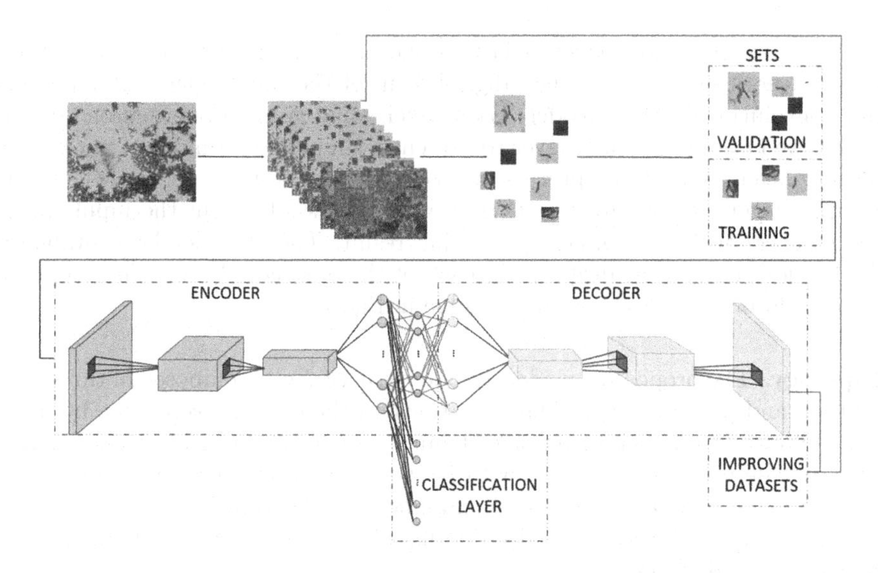


**Fig. 4.** The proposed model of encoder-decoder CNN structure used for bacteria evaluation from input microscopic images. At the beginning, the input image is binarized with 9 different values of a parameter. For this collection resulting segments covering whole bacteria shapes are selected for validation and training of decoder CNN part. When the system of two cooperating CNN is trained, the classification layer is used to perform the task of bacteria evaluation from the input microscopic image.

information for following operations of pooling and convolution, which restore the image.

In each segment, convolution layers extract important features by image filtering. This operation maintains image size but changes the content. Applied filtering modifies pixels values by the Gaussian and edge detection filters. In result, the output resolution $s_{output}$ of the image is related to image and filter size as

$$s_{output} = \frac{N - m}{S} + 1, \tag{3}$$

where the image size is $N \times M$ and filter $\omega$ size is $m \times m$, $S$ is $\omega$ step size over the image. Convolution layer $l$ returns Feature Maps, where unit $x_{ij}^{l}$ ($i$ and $j$ are neurons in the layer) forwards a summation of products from previous layer $l-1$

$$x_{ij}^{l} = \sum_{a=0}^{m-1} \sum_{b=0}^{m-1} \omega_{ab} y_{(i+a)(j+b)}^{l-1}, \tag{4}$$

$$y_{ij}^{l} = \sigma(x_{ij}^{l}). \tag{5}$$

The pooling layer reduces information that resizes the image. In our encoder-decoder model, we have used max-pooling of size $a \times a$. It goes over the Feature Maps with step $a$ and returns the maximum value in each step which is forwarded as replacement of these maximum filter windows to the next layer.

The fully connected layer is last in the encoder and first in the decoder. These layers receive a simplified digital form of the input image after pooling and convolution in the encoder. Each pixel value goes to one neuron. In our model, we have one encoder layer to present a simplified form of the image, one decoder layer to start the process of reconstruction from these values and one classifier layer to compare results and evaluate the object in the input image due to encoder-decoder image processing result. The classifier layer evaluates similarities and returns decisions in each of the abstract classes. The proposed encoder-decoder CNN structure is visible in Fig. 4.

**Topology.** The proposed encoder-decoder structure was composed of three convolution and pooling layers. The operations on the encoder were done by using $5 \times 5$ filter run on each of 3 color channels. First, we used blur filtering and after edge detector, each of them with 1-pixel step. Pooling layer was doing $4 \times 4$ filtering, which reduced image into simplification. Feature Map was forwarded to fully connected layer for the decision based on decoding results from another inverse neural network.

**Training.** CNN structure was trained using the Back Propagation Algorithm. Error function $f(\cdot)$ value is minimized for neuron outputs $\sigma(x_{ij}^l)$ in last layer in accordance to gradient value $\frac{\partial f}{\partial y_{ij}^l}$. We have used cross entropy

$$f(x_{ij}^l) = -\sum_{j=0}^{m-N} [E_j^r \ln \sigma(x_{ij}^L) + (1 - E_j^r) \ln(1 - \sigma(x_{ij}^l))], \tag{6}$$

which has gradient value

$$\nabla f(x_{ij}^l) = \frac{\sigma(x_{ij}^l) - E_j^r}{(1 - \sigma(x_{ij}^l))\sigma(x_{ij}^l)} \tag{7}$$

calculated for neuron $j$ output $E_j^r$ for $r$ training image.

The chain rule is used to calculate gradients for inputs sums $x_{ij}^l$ using $\omega_{ab}$ filter as

$$\frac{\partial f}{\partial \omega_{ab}} = \sum_{i=0}^{N-m} \sum_{j=0}^{N-m} \frac{\partial f}{\partial x_{ij}^l} \frac{\partial x_{ij}^l}{\partial \omega_{ab}} = \sum_{i=0}^{N-m} \sum_{j=0}^{N-m} \frac{\partial f}{\partial x_{ij}^l} y_{(i+1)(j+b)}^{l-1}. \tag{8}$$

Therefore input signal change $\frac{\partial f}{\partial x_{ij}^l}$ is

$$\frac{\partial f}{\partial x_{ij}^l} = \frac{\partial f}{\partial y_{ij}^l} \frac{\partial y_{ij}^l}{\partial x_{ij}^l} = \frac{\partial f}{\partial y_{ij}^l} \frac{\partial \left(\sigma(x_{ij}^l)\right)}{\partial x_{ij}^l} = \frac{\partial f}{\partial y_{ij}^l} \sigma'(x_{ij}^l), \tag{9}$$

where $\sigma(x_{ij}^l)$ is neuron output (sigmoid activation function over the input signal)

$$\sigma(x_{ij}^l) = \frac{1}{1 + e^{-x_{ij}^l}}. \tag{10}$$

this function approximates large positive inputs by 1 and large negative inputs by 0. Signal change is calculated from back propagated errors

$$\frac{\partial f}{\partial y_{ij}^{l-1}} = \sum_{a=0}^{m-1}\sum_{b=0}^{m-1} \frac{\partial f}{\partial x_{(i-a)(j-b)}^{l}} \frac{\partial x_{(i-a)(j-b)}^{l}}{\partial y_{ij}^{l-1}} = \sum_{a=0}^{m-1}\sum_{b=0}^{m-1} \frac{\partial f}{\partial x_{(i-a)(j-b)}^{l}} \omega_{ab}, \tag{11}$$

error on the previous layer equals filtering

$$\frac{\partial x_{(i-a)(j-b)}^{l}}{\partial y_{ij}^{l-1}} = \omega_{ab}. \tag{12}$$

On the pooling layer change of the error does not happen so it only forwards the error values back.

### 2.3 CNN Encoder-Decoder Structure

For each image from the set of binarized ones, all objects are extracted as small images. The value in each of them is transferred from the original image (colored one). Then, each extracted image is resized to $k \times k$. All of these images are used for training convolutional auto-encoders. Auto-encoders can be defined as two, main layers as encoder $f_W(\cdot)$ and decoder $g_V(\cdot)$ and embedded layer between them. The encoder is composed of convolutional layers, which means there is a matrix understood as image filter and after using the filter, image is resized (called pooling), and then all units are flattened to a vector. After the encoder, there is a single layer of neurons which combines encoder and decoder. The decoder design is an inversion of the encoder. This architecture is trained by minimizing the reconstruction error defined as follows

$$L_{reconstruction} = \frac{1}{n}\sum_{i=1}^{n} ||G_{\omega'}(F_{\omega}(x_i)) - x_i||_2^2, \tag{13}$$

where $x_i$ is $i$-th image in the dataset composed of $n$ images, and parameters of encoder/decoder are marked as $F_{\omega}(x)$ and $G_{\omega'}(h)$ where for the fully-connected layer there is

$$\begin{aligned} f_W(x) &= \Delta(Wx) \equiv h \\ g_V(h) &= \Delta(Vh), \end{aligned} \tag{14}$$

where $x$ and $h$ are vectors, and $\Delta(\cdot)$ is activation function (Fig. 5).

This auto-encoder is trained in obtaining some value of the reconstruction error. This kind of structure is needed for analyzing obtained form with original and binary version – comparison is made by checking the structure of these images. All pixels are compared - if in one of the images is a pixel different from

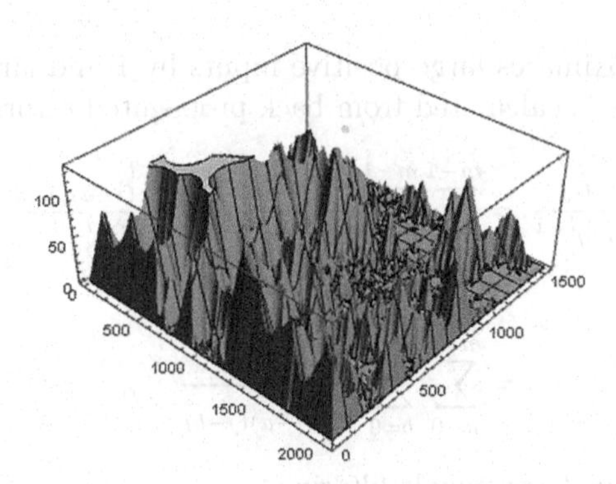


**Fig. 5.** The value of comparison function between sample input microscopic image shown in Fig. 1 and the results of binarization in the proposed method.

Split dataset in random way into two sets – training and validation;
**for** *each image in both sets* **do**
  **for** *each* $\alpha \in \{0.1, 0.2, \ldots, 0.9\}$ **do**
    Binarize image with $\alpha$ parameter;
    Extract all object in binarized image;
  **end**
**end**
Train autocoder using training set;
**for** *each object's image in both sets* **do**
  Calculate reconstraction image using autocoder;
  **if** *on reconstraction are some pixel diffrent from backgroud* **then**
    Change or improve image in a set;
  **end**
**end**
**for** *each object's image in validation set* **do**
  Calculate result using coder part and classification layer;
  **if** *object is classified as merged one* **then**
    Delete all images that define the same area;
  **end**
**end**
Count all results;

**Algorithm 1:** Proposed technique for counting biological organism in microscopy images.

the background, this pixel is put into a binary version as a white pixel. It this way, the reconstruction image can be used to improve incoming images to the classifier. Of course, this comparison occurs when the auto-encoder is trained.

Having nine sets full of smaller fragments of organisms (after binarization), each image is reconstructed. Then, each of the images is rated and classified by the encoder itself, and an additional layer, which is called the classification layer, is used. This is presented in the architecture visualization in Fig. 4.

The reclassifier assigns the image after reconstruction to one of the defined classes - a single organism, many organisms, merged organisms, no organism. However, adding layer minimizes the function defined in Eq. (13) loses sense. For this purpose, another loss function is defined at the level of the classification layer as

$$L_{classification} = -\frac{1}{n}\sum_{i=0}^{N} \log \frac{e^{W_{y_i}^T x_i + b_{y_i}}}{\sum_{j=1}^{n} e^{W_j^T x_i + b_j}}, \tag{15}$$

where $W$ are the weight matrix, $b$ is bias and the global error is defined as a sum of these two loss functions, so it can be described as

$$L_{complete} = L_{reconstruction} + L_{classification}. \tag{16}$$

Artificial increase of the set by additional images after binarization allows to increase the probability of analyzing individual elements of the image, and in particular to focus on the exterior of larger, probably merged organisms. In particular, this architecture allows counting organisms. Having nine collections, we check whether the objects are not on themselves. If a smaller image is marked as merged, it will belong to an image with a lower binarization value. The proposed encoder-decoder CNN structure used in our experiments is presented in Fig. 4 (Fig. 6).

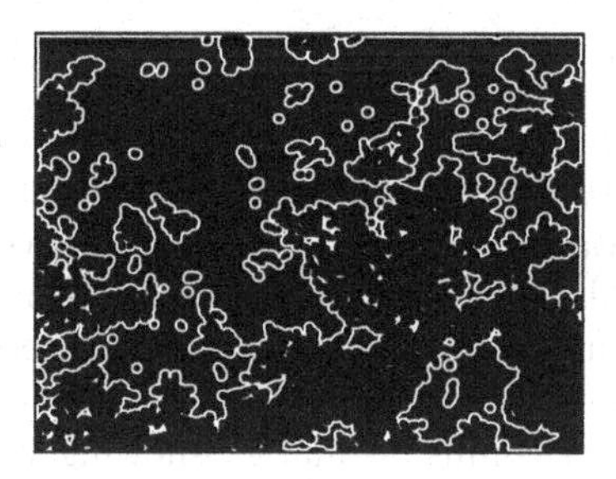

**Fig. 6.** Differences between the original input microscopic image and performed binarizations visible in shapes of bacteria.

In the training process, we binarize the input image using various $\alpha$ parameter so that we can extract bacterias in various segments. This training set is used for the encoder to train it for the recognition of bacteria. The decoder, at the same time, is trained to reconstruct images. After this operation, both images (original and reconstructed) are compared to find differences in pixels. If in both

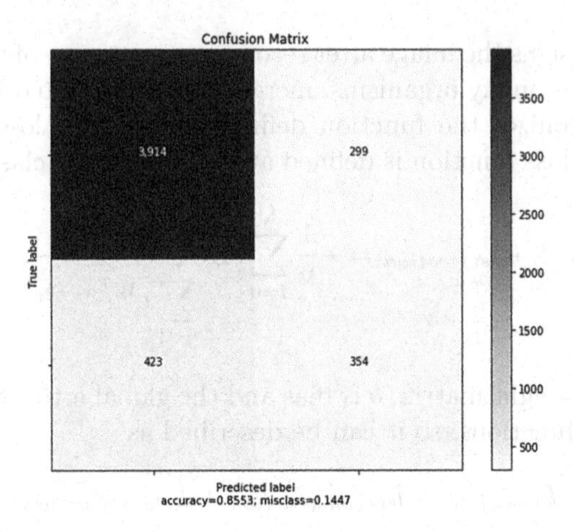


**Fig. 7.** Confusion matrix of research results for evaluation set of 23 microscopic images of bacteria strain, containing about 4213 individuals each.

images compared pixels match we know this is the same bacteria, therefore the classification layer learns to recognize this shape. As a result, we can calculate shapes visible in the microscopy image. The algorithm used for training our system is defined in Algorithm 1.

## 3 Experiments

In our experiments, we have used the original data set of images showing mixed bacteria strains. Presented bacteria images have dimensions of $807 \times 768$ pixels and 96 dpi resolution. Each one of the images in our collection presents about 4213 bacteria individuals. The experiments were done to train the proposed encoder-decoder structure to be able to classify bacteria in the image even if shapes overlap or merge in the image.

In Fig. 7 we can see a confusion matrix showing how the proposed model works for 23 images used as an evaluation set. Presented results are achieved for encoder-decoder trained at $\alpha = 0.01$ level. We can see that most bacteria were correctly classified, which gives above 92% of correct evaluations of bacteria shapes in sample images.

In Table 1 we can see a comparison of our proposed method to other approaches. We can see that encoder-decoder architecture gives very good results, achieving at least 2% to 20% higher number of recognized bacterias in input images. Beside proposed architecture has less complex processing, since we use only CNN architecture and it's reversed version. One of them is used to perform the main task while the other one is trained to inverse simplification. Results are compared by a fully connected layer to give recognition.

**Table 1.** Comparison of the number of recognized bacteria from input microscopic images.

| Technique | Number of found organisms | Effectiveness |
|---|---|---|
| SIFT+SVM | 2142 | 51% |
| SURF+SVM | 2120 | 50% |
| CNN for extracted images | 2892 | 69% |
| Faster rCNN | 3782 | 90% |
| Proposed technique | 3914 | 92% |

## 4 Conclusions

In this article, we have proposed a CNN architecture modified to perform recognition tasks over microscopic images. The proposed solution was developed to overcome the problem of overlapping or merging objects – like bacteria in microscopic images. Because of using the proposed encoder-decoder model, the system was able to compare shapes of an object between original and reconstructed images, therefore final evaluation was much less influenced by typical problems with the recognition of similar, small objects. Conducted experiments show that the fusion of image processing with auto-encoders can be very promising. We reached a better accuracy level than faster rCNN due to additional image processing techniques and the extraction of individual areas for further analysis.

Future works will be oriented on different strains, which may have different shapes and sizes. In current work, we have solved the problem of recognition when the object is similar in shape and size, but more complex compositions need more experiments. Another important task will be the further development of the proposed model to fit also other objects from microscopic images like blood cells, cancer cells, tissues, and capillaries. These objects are often recorded on various medical screenings and effective methods that may help in their recognition will be a valuable help in medical systems.

**Acknowledgements.** This work is supported by the Silesian University of Technology under grant BKM-504/RMS2/2020.

## References

1. Ahmed, K.T., Ummesafi, S., Iqbal, A.: Content based image retrieval using image features information fusion. Inf. Fusion **51**, 76–99 (2019)
2. Aminikhanghahi, S., Shin, S., Wang, W., Jeon, S.I., Son, S.H.: A new fuzzy Gaussian mixture model (FGMM) based algorithm for mammography tumor image classification. Multimed. Tools Appl. **76**(7), 10191–10205 (2017)
3. Chen, L.C., Zhu, Y., Papandreou, G., Schroff, F., Adam, H.: Encoder-decoder with atrous separable convolution for semantic image segmentation. In: Proceedings of the European Conference on Computer Vision (ECCV), pp. 801–818 (2018)

4. Costa, M., Oliveira, D., Pinto, S., Tavares, A.: Detecting driver's fatigue, distraction and activity using a non-intrusive ai-based monitoring system. J. Artif. Intell. Soft Comput. Res. **9**(4), 247–266 (2019). https://doi.org/10.2478/jaiscr-2019-0007
5. Dirvanauskas, D., Maskeliūnas, R., Raudonis, V., Damaševičius, R., Scherer, R.: Hemigen: Human embryo image generator based on generative adversarial networks. Sensors **19**(16), 3578 (2019). https://doi.org/10.3390/s19163578. http://dx.doi.org/10.3390/s19163578
6. Gao, Q., Wang, J., Ma, X., Feng, X., Wang, H.: CSI-based device-free wireless localization and activity recognition using radio image features. IEEE Trans. Veh. Technol. **66**(11), 10346–10356 (2017)
7. Grycuk, R., Scherer, R.: Software framework for fast image retrieval. In: 2019 24th International Conference on Methods and Models in Automation and Robotics (MMAR), pp. 588–593, August 2019. https://doi.org/10.1109/MMAR.2019.8864722
8. Li, Y., Xie, W., Li, H.: Hyperspectral image reconstruction by deep convolutional neural network for classification. Pattern Recogn. **63**, 371–383 (2017)
9. Liu, Y., Fung, K.C., Ding, W., Guo, H., Qu, T., Xiao, C.: Novel smart waste sorting system based on image processing algorithms: surf-bow and multi-class SVM. Comput. Inf. Sci. **11**(3), 35–49 (2018)
10. Lu, H., Liu, S., Wei, H., Tu, J.: Multi-kernel fuzzy clustering based on auto-encoder for fMRI functional network. Expert Syst. Appl. 113513 (2020)
11. Mou, L., Ghamisi, P., Zhu, X.X.: Deep recurrent neural networks for hyperspectral image classification. IEEE Trans. Geosci. Remote Sens. **55**(7), 3639–3655 (2017)
12. Najgebauer, P., Grycuk, R., Scherer, R.: Fast two-level image indexing based on local interest points. In: 2018 23rd International Conference on Methods Models in Automation Robotics (MMAR), pp. 613–617, August 2018. https://doi.org/10.1109/MMAR.2018.8485831
13. Rygał, J., Najgebauer, P., Romanowski, J., Scherer, R.: Extraction of objects from images using density of edges as basis for GrabCut algorithm. In: Rutkowski, L., Korytkowski, M., Scherer, R., Tadeusiewicz, R., Zadeh, L.A., Zurada, J.M. (eds.) ICAISC 2013. LNCS (LNAI), vol. 7894, pp. 613–623. Springer, Heidelberg (2013). https://doi.org/10.1007/978-3-642-38658-9_56
14. Scherer, R.: Multiple Fuzzy Classification Systems. Springer, Heidelberg (2014). https://doi.org/10.1007/978-3-642-30604-4
15. Spanhol, F.A., Oliveira, L.S., Cavalin, P.R., Petitjean, C., Heutte, L.: Deep features for breast cancer histopathological image classification. In: 2017 IEEE International Conference on Systems, Man, and Cybernetics (SMC), pp. 1868–1873. IEEE (2017)
16. Tareen, S.A.K., Saleem, Z.: A comparative analysis of SIFT, SURF, KAZE, AKAZE, ORB, and BRISK. In: 2018 International Conference on Computing, Mathematics and Engineering Technologies (iCoMET), pp. 1–10. IEEE (2018)
17. Wang, J., Perez, L.: The effectiveness of data augmentation in image classification using deep learning. In: Convolutional Neural Networks and Vision Recognition (2017)
18. Yu, S., Jia, S., Xu, C.: Convolutional neural networks for hyperspectral image classification. Neurocomputing **219**, 88–98 (2017)
19. Zhu, F., Li, H., Ouyang, W., Yu, N., Wang, X.: Learning spatial regularization with image-level supervisions for multi-label image classification. In: Proceedings of the IEEE Conference on Computer Vision and Pattern Recognition, pp. 5513–5522 (2017)

# Face Manipulation Detection via Auxiliary Supervision

Xinyao Wang[1,2], Taiping Yao[2], Shouhong Ding[2], and Lizhuang Ma[1(✉)]

[1] Shanghai Jiao Tong University, Shanghai, China
{WXYjj789,ma-lz}@sjtu.edu.cn
[2] Youtu Lab, Tencent, Shanghai, China
{taipingyao,ericshding}@tencent.com

**Abstract.** The rapid progress of face manipulation technology has attracted people's attention. At present, a reliable edit detection algorithm is urgently needed to identify real and fake faces to ensure social credibility. Previous deep learning approaches formulate face manipulation detection as a binary classification problem. Many works struggle to focus on specific artifacts and generalize poorly. In this paper, we design reasonable auxiliary supervision to guide the network to learn discriminative and generalizable cues. A multi-scale framework is proposed to estimate the manipulation probability with texture map and blending boundary as auxiliary supervisions. These supervisions will guide the network to focus on the underlying texture information and blending boundary, making the learned features more generalized. Experiments on FaceForensics and FaceForensics++ datasets have demonstrated the effectiveness and generalization of our method.

**Keywords:** Forgery detection · Deepfake · Artifacts

## 1 Introduction

Recently, face manipulation technology has made tremendous progress, which enables people to arbitrarily modify face content in images and videos. Advanced deepfake methods can produce synthetic images even infeasible for humans to discriminate, but in the meanwhile they give change to malicious attackers, e.g. faces of public characters may be blended onto some indecent videos, leading to serious security and trust issues. Therefore, it's significantly important to develop effective methods for face manipulation detection.

Many countermeasures have been proposed to detect fake images and videos. Some works [3,15,17] solve the problem with a simple binary classification network and can achieve high performance on one specific manipulation dataset. However, such solutions haven't digged into the essence of the face forgery procedure and may easily overfit to a certain dataset. When applied to an unseen manipulation type, they usually suffer from significant performance drop.


H. Yang et al. (Eds.): ICONIP 2020, LNCS 12532, pp. 313–324, 2020.
https://doi.org/10.1007/978-3-030-63830-6_27

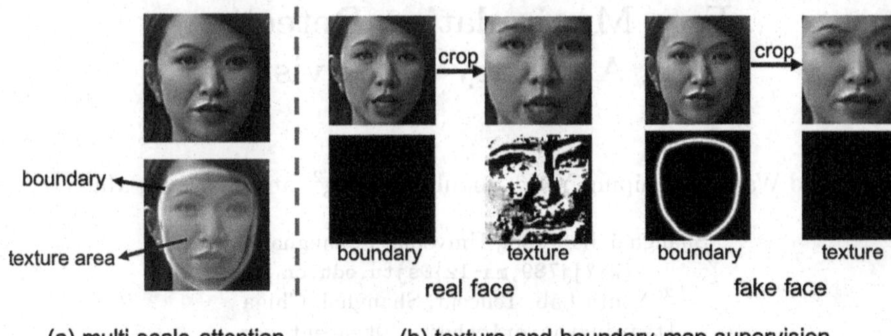


**Fig. 1.** The texture and boundary supervision. Texture artifacts always exist at the center area of the face. A fake face has an abnormal texture map(represented by a zero map) and a blending map. A real face has a normal texture map(denoted by an LBP map) without blending map

Some works have noticed the generalization problem and try to focus on the intrinsic artifacts from fake face generation process, such as head pose inconsistency [24], face blending artifacts [12] and abnormal spectrum from up-sampling layers [25]. However, these proposed methods usually focus on a certain artifact, which may fail when facing low resolutions images [12] and increasing developed forgery approaches.

In this paper, we propose to use texture map and blending boundary as auxiliary supervisions, as well as a multi-scale strategy to guide the model to learn more discriminative and generative features. The key observation behind the thought is that face manipulation operation mainly contains two steps: 1) synthesizing a fake face, 2) blending the altered face onto the original background. In the first step, the mimetic generation process will result in abnormal texture, like skin inconsistency, light direction disorder or specific low-level texture [25]. Secondly, discrepancies at the blending boundary always exist, which may come from the texture mismatch or malposition. Thus we pay more attention on center face and boundary area for better learning, as shown in Fig. 1(a).

Based on the above observations, we propose a multi-task framework to detect manipulated facial images, as shown in Fig. 2. For an input face and its cropped center face, two encoders are used to simultaneously capture the blending boundary and texture information. As shown in Fig. 1(b), Texture decoder estimates Local Binary Patterns (LBP) map for the real image and zero map for fake one, which reinforces the model to learn the essential texture difference. Blending boundary of the fake face is predicted by another decoder and then performs as an attention map onto the classification feature. The information gained from these auxiliary tasks covers two comprehensive artifacts and finally contributes to the main classification performance improvement. Our experiments on FaceForensic++ [17] and Faceforensic [16] dataset demonstrate that the proposed method can not only effectively detect forgery on certain face manipulation type, but also show robust transferablity on unseen forgery methods and unknown dataset.

The main contributions of the paper are:

1. We propose a multi-task framework with auxiliary supervisions for face forgery detection.
2. We utilize texture map as auxiliary supervision to guide the model learning the comprehensive artifacts during the forgery generation.
3. We also reconstruct the blending boundary map, which leads the model paying attention to the blending discrepancies.
4. Experiments shows the detection ability and generalization of the proposed method.

## 2 Related Work

### 2.1 Face Manipulation Methods

With the help of 3D model reconstruction, Dale et al. [8] present one of the first attempts about automatic face swapping. Since then, digital face manipulation has made huge improvement. Thies et al. propose Face2Face [21], which can real-timely alter facial movements. Deep video portraits [11] can fully alter the head position, rotation, face expression and even eye blinking of the source face to the target one.

Deep learning also plays an important role in face manipulated field. One of the landmarks is Deepfake [2] with autoencoder-decocer pairs structure. In this method, the source and target face share a common encoder to learn the latent feature representing similarities of the face, then exchange decoders to reconstruct a syncretic face. Generative adversarial networks (GANs) has also show its powerful ability on synthesising realistic content, for example, StyleGAN [10] can alter high level facial attributes(identity, pose, as well as freckles and hair) based on a face generator.

### 2.2 Forgery Detection Methods

Aimed at face manipulation detection, many countermeasures have been introduced. A kind of typical approaches focuses on the artifacts during fake content generation. Ciftci's work [4] shows basic biological signals, such as PPG, can neither be spatially nor temporally preserved in fake faces. Li and Lyu [13] denote existing DeepFake algorithms can only generate images with limited resolutions. Considering the blending procedure, face X-ray [12] generates an novel dataset with blending artifacts based on only real images, but its performance drops rapidly with low resolutions.

Methods based on CNN also take effect. Some binary classification networks are specifically designed for this task, such as Meseso-Net [3] and capsule network [15]. Some auxiliary supervisions, like manipulation mask [14] and facial landmarks [18] are also employed for multi-task learning. To improve transferability,

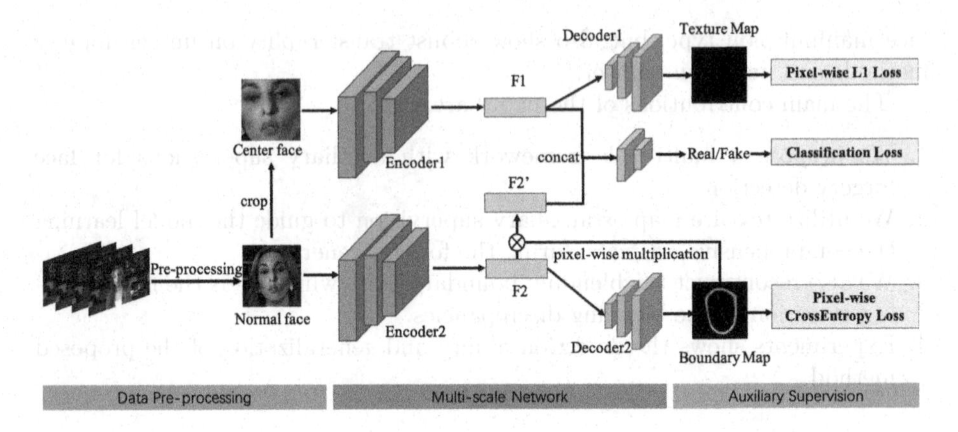


**Fig. 2.** The overall structure of the proposed method (take a fake face for example). The network structure contains two encoder-decoder branches: 1) the above texture branch estimates the texture map. 2) the bottom blending boundary branch estimates the boundary map and then acts as an attention map by multiplying with the latent feature. Two latent features are concatenated for final classification task

[22] emphasizes on data pre-processing and augmentation. Forensictransfer [6] uses a weakly-supervised encoder-decoder framework and activates specific half part of the latent feature for real and fake images respectively. However, such methods haven't take the comprehensive artifacts into consideration, resulting in limited performance.

## 3 Methods

The overview of our proposed face manipulation detection network with comprehensively auxiliary supervisions is shown in Fig. 2. Given a fake video, first we extract each frame, then detect the face and crop its center face as two inputs. The network mainly consists of two encoder-decoder branches: the texture branch and the blending boundary branch. The texture branch, designed for learning the artifacts during the tampering process, takes a center face as input and predicts corresponding texture map. The blending boundary branch is trained on a normal face and outputs the blending boundary, which is latter utilized as an attention map to highlight the blending traces of the feature. Finally the latent features from two branches are fused to do classification task, predicting the spoofing probability of the input image.

### 3.1 Texture Map Supervision

Forgery images generated by computer graphics or deep learning algorithms commonly share some distortion traces. For example, abnormal spectra [25] like checkerboard patterns may occur in synthetic content. Instead of focusing on one specific artifact, we take the comprehensive representation, texture, to distinguish real and fake images. Texture artifacts only appear in the fake content, and thus we only use center face for network input, avoiding the disturbance from background.

We utilize Local Binary Patterns (LBP) to represent the texture of real images. LBP [9] thresholds the neighboring pixels based on the current value, which usually acts as an effective texture descriptor for images. For fake images, the texture may be disordered or blurry. In order to take an effectively distinguishable measure, we forcibly set zero maps for fake contents. This method can not only guide the network to learn the semantic information, but also learn the actual physical texture, which enhances the generalization of the network.

Formally, the texture map $T$ is an LBP map if the input is real, a zero map if the input is fake:

$$T(I) = \begin{cases} LBP \quad map, \ I = real \\ zero \quad map, \ I = fake \end{cases} \tag{1}$$

where I denotes the input image.

Let $\hat{T} = ED_T(I)$ be the predicted texture map of input image $I$ where $ED_T$ is an encoder-decoder network. We use pixel-wise L1 loss function in the training process:

$$L_t = \frac{1}{N} \sum_{i,j} \left| T_{i,j} - \hat{T}_{i,j} \right| \tag{2}$$

where N denotes the pixel number in texture map.

### 3.2 Blending Boundary Supervision

Except for texture artifacts in the center face, we also leverage discrepancies hidden at the blending border. Recent research [7] shows each photographed image has typical fingerprint for its camera equipment and post-processing softwares. When two different-source images are blended, the boundary is inevitably discontinuous in some way, which can be detected as the evidence for face forgery identification.

Inspired by face X-ray [12], the boundary map should be an image $B$ such that for a fake face it shows the blending border, while for a real face, it's a zero map. Formally, the boundary map $B$ is defined as $B = 4 \cdot M \cdot (1 - M)$, where $M$ is the mask denoting the manipulated region and its value of each pixel is between 0 and 1. For a real face, $M$ is a zero map. For the fake one, we first binarize the original face mask (if exists in dataset, otherwise use convex hull of landmarks instead), then turn it into a soft mask with a $23 * 23$ Gaussian kernel.

For an input Image $I$, let $\hat{B} = ED_B(I)$ denote the predicted blending boundary map where $ED_B$ is an encoder-decoder network with the same structure as Sect. 3.1. We use a weighted pixel-wise cross-entropy loss to guide the training:

$$L_b = -\frac{1}{N} \sum_{i,j} w((B_{i,j} log \hat{B}_{i,j}) + (1 - B_{i,j}) log(1 - \hat{B}_{i,j})) \tag{3}$$

where $N$ denotes the total number of pixels in map $B$. As most pixels are 0, to avoid outputting all zero maps, we apply weight $w$ for each pixel: $w = 5$ if the pixel value is greater than 0 and $w = 1$ otherwise.

**Attention Mechanism.** The predicted blending boundary map can highlight the blending artifacts in the image, and thereby it's used as the spatial attention map on the latent feature. For the high-dimensional latent feature $F \in \mathbf{R}^{C \times H \times W}$ and the blending boundary map $B \in \mathbf{R}^{1 \times H' \times W'}$, we first resize the map $B$ to the same size of $H \times W$ and repeat $C$ times to get the attention map $B_{att} \in \mathbf{R}^{C \times H \times W}$, then apply pixel-wise multiplication with feature $F$ and finally output $F'$ as part of the input to the subsequent classification branch.

### 3.3 Loss Function

Finally, the latent features from two encoders are concatenated and fed into the classification branch. The binary cross entropy loss is used for classification:

$$L_c = -(clog(\hat{c}) + (1 - c)log(1 - \hat{c})) \tag{4}$$

where $c$ and $\hat{c}$ denotes the ground truth predicted probability respectively.

The overall loss function is:

$$L = \lambda_t L_t + \lambda_b L_b + \lambda_c L_c \tag{5}$$

where $\lambda_t$, $\lambda_b$ and $\lambda_c$ are loss weights for balance. During training we increase $\lambda_c$ from 0 to 1, and decrease $\lambda_t$ and $\lambda_b$ from 1 to 0 in first 5 epochs, which makes the network focus on learning texture and boundary artifacts at beginning and then try to improve classification performance at later period.

## 4 Experimental Results

In this section, we provide the datasets used and network implementation details, then show the forgery detection and generalization performance of our method.

### 4.1 Experimental Setting

**Dataset.** In the experiment we use Faceforensics++[17] for basic forgery detection and unseen manipulation methods evaluation, and Faceforensic [16] for cross-dataset test. Faceforensic++ is a large video dataset with 1000 original videos and corresponding manipulated videos from DeepFakes [2], Face2Face [21] and FaceSwap [1]. Faceforensic [16] contains 1004 real videos and 1004 fake videos manipulated by Face2Face. Each video in the two dataset has three compression levels: no compression(raw), light compression(c23), and hard compression(c40).

**Implementation Details.** For the network, we adopt EfficientNet-b4 [20] with noisy-student [23] pretrained weights as encoders and classification branch. The input image size is resized to $299 \times 299$. The outputs from block 10, 22 and last layer are skip-connected as the latent feature. Our decoder comprises of two bilinear up-sampling operations followed by two $1 * 1$ convolutional layers. Empirically, the normal face is enlarged by a factor of 1.3 and the center face is cropped by 0.7 for images in Faceforensics++, which can keep the center face and avoid the background in most cases. Empirically, the size of boundary and texture map are both set to $72 \times 72$, which can keep the main information and avoid complex computations as well. In the training process, the batch size is set to be 64 and the learning rate is initialized as $2e - 4$ with Adam, then it decreases by half every 3 epochs.

### 4.2 Ablation Study

We take experiments on c40 data of Faceforensic++ for ablation study and strictly follow the experiment setting of [17], i.e. 720 videos and 270 frames per video for training, 140 videos and 100 frames per video for validation and testing. EfficientNet-b4 with normal face input is set as the baseline and Table 1 summarizes the results.

**Texture Map Supervision.** We first add texture map decoder onto EfficientNet-b4 with center faces as input (b4+TM Pred) and this model outperforms the baseline in both ACC (2.43%) and AUC (1.62%). Notice that a center face contains less information than a normal one, but the result can still keep improvement, which shows the benefit of texture map supervision.

Instead of direct classification output, an alternative is to use the average of the estimated texture map for decision (b4+TM Pred - map). Table 1 shows the accuracy keeps relative stable (90.89% vs 90.48%). Here the AUC metric decreases a lot because the average of the map is either zero or near zero, and the threshold is 0.09 instead of normal 0.5.

**Blending Boundary Map Supervision.** Similarly, We evaluate the effect of the boundary map decoder: the accuracies of classier (b4+BM Pred) and map average (b4+BM Pred - map) improve to 90.43% and 91.26% respectively, which denotes the map estimation works better than classification in some sense.

To take the advantage of the decoder, attention mechanism is tested (b4+BM Pred+Attention), which further improves the result (92.91%). These experiments demonstrate boundary map estimation with attention on the latent feature can lead the network paying more attention on the blending artifacts and improve the performance.

**Overall Framework.** When taking both normal and center faces as input, Table 1 shows the performance of a basic EfficientNet-b4 classifier, b4 with

**Table 1.** Ablation Study Evaluation. 'TM Pred' means texture map prediction and 'BM Pred' means boundary map prediction

| Input face | Model | ACC (%) | AUC (%) |
|---|---|---|---|
| Normal face | b4 | 88.46 | 93.50 |
| Center face | b4+TM Pred | 90.89 | 95.12 |
| Center face | b4+TM Pred - map | 90.48 | 88.10 |
| Normal face | b4+BM Pred | 90.43 | 95.20 |
| Normal face | b4+BM Pred - map | 91.26 | 88.02 |
| Normal face | b4+BM Pred+Attention | 92.91 | 96.85 |
| Normal & center face | b4 | 91.12 | 96.17 |
| Normal & center face | b4+ TM & BM Pred | 93.46 | 97.57 |
| Normal & center face | b4+ TM & BM Pred+Attention | **94.38** | **98.17** |

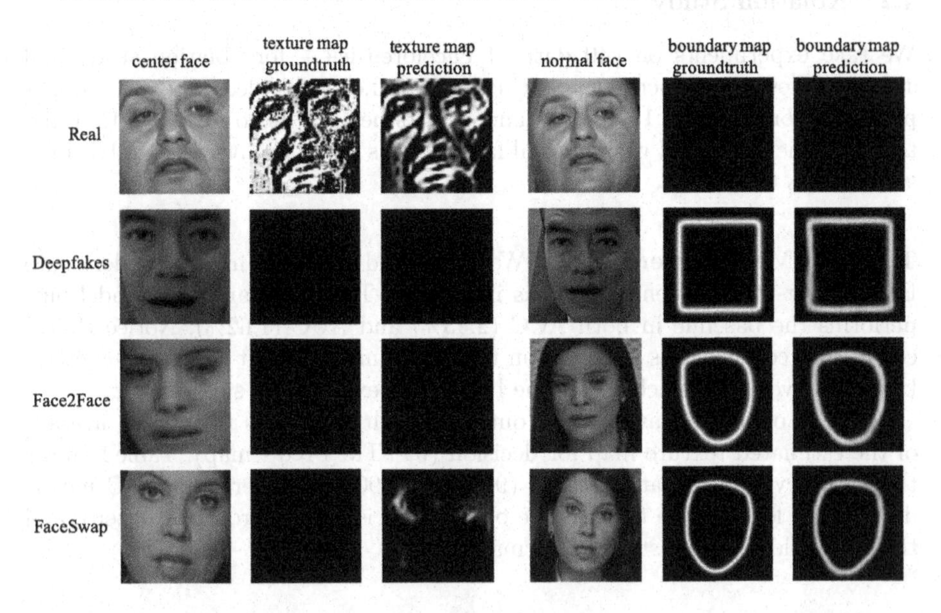


**Fig. 3.** Visualization of map prediction. Most maps can be correctly predicted and the last line also shows one failure sample in FaceSwap

two decoders (b4+ BM&TM Pred) and the overall framework (b4+ BM&TM Pred+Attention). The overall framework finally achieves accuracy 94.38% and AUC 98.17%, making significantly improvement compared to baseline (accuray 88.46% and AUC 93.5%), which shows the importance of auxiliary information and attention feedback mechanism.

Figure 3 shows the visualization of estimated texture and boundary maps. We can find that in most cases the model can not only exactly distinguish real and fake images, but also predict maps accurately. The last row gives an failure example in FaceSwap, but we can see that the network has almost predict the correct map.

### 4.3 Evaluation on Faceforensic++

Some recent related works [3,5,17,18] also present solutions for forgery detection problem. Cozzolino et al. [5] feed the hand-crafted Steganalysis features into a CNN-based network. MesoInception-4 [3] is a modified version from InceptionNet with two inception modules and two classic convolution layers interlaced with max-pooling layers designed for face tampering detection. Songsri-in et al. [18] employ spatial facial landmarks as auxiliary information for classification and localization predictions. We present the comparison experiment with these recent works in Table 2.

**Table 2.** Evaluation results on FaceForensic++ dataset of different compression levels. Each category is trained on all manipulation types together. The metric is accuracy (%)

| Models | No Compression (raw) | light compression (c23) | hard compression (c40) |
|---|---|---|---|
| Cozzolino et al. [5] | 98.56 | 79.56 | 56.38 |
| MesoInception-4 [3] | 96.51 | 85.51 | 75.65 |
| Xception [17] | **99.41** | 97.53 | 85.49 |
| Songsri-in et al. [18] | 96.58 | 95.85 | 89.33 |
| **Ours** | 99.35 | **98.72** | **94.38** |

The experiment is performed on the data of different compression levels (raw, c23, c40) in Faceforensics++ [17], and each category is trained on all manipulation types together (Deepfakes, Face2Face, FaceSwap). From Table 2, we observe that the performance of other work usually decrease rapidly on higher compression level, while the proposed method can keep a relative high accuracy (94.38%) even in low image quality.

**Table 3.** Generalization ability evaluation on unseen manipulated methods in Faceforensic++. 'DF'- Deepfakes,'F2F'- Face2Face,'FS'- FaceSwap. The metric is AUC (%)

| Train | DF | | | F2F | | | FS | | |
|---|---|---|---|---|---|---|---|---|---|
| Test | DF | F2F | FS | DF | F2F | FS | DF | F2F | FS |
| Xception [17] | 99.38 | 75.05 | 49.13 | 87.56 | 99.53 | 65.23 | 70.12 | 61.7 | 99.36 |
| HRNet [12] | 99.26 | 68.25 | 39.15 | 83.64 | 99.5 | 56.6 | 63.59 | 64.12 | 99.24 |
| FaceXray [12] | 99.17 | 94.14 | 75.34 | 98.52 | 99.06 | 72.69 | 93.77 | 92.29 | 99.09 |
| **Ours** | **99.8** | **98.05** | **97.13** | **98.53** | **99.99** | **78.63** | **99.98** | **99.64** | **1.00** |

### 4.4 Generalizability Evaluation

Except for basic experiment on certain dataset, we also do generalization evaluation on both unseen manipulated methods and dataset.

**Evaluation on Unseen Manipulated Methods.** To evaluate the generalization ability on unseen forgery methods, we take experiments on three manipulation algorithms on Faceforensic++. Following [12], we train on one specific method and test on others with raw data. Xception [17], HRNet [19], FaceXray [12] are set as baselines. The best result of FaceXray is trained on an extra dataset and thus we choose the version trained on original Faceforensic++ for fair comparison. Table 3 shows the comparison result.

For binary classification networks Xception and HRNet, they can get high accuracy (99%) on known specific dataset, but they suffer from rapid performance drop when meeting unseen manipulated methods due to overfitting. The result of FaceXray and our method can keep a relative stable result even on unknown forgery datasets, which may credit to the boundary map supervision. Furthermore, our method can achieve higher results than FaceXray, for example, when training on Deepfakes and testing on FaceSwap, the performance in terms of AUC of proposed method is 21.79% higher than faceXray. This improved generalization ability may come from two factors: 1) we use attention mechanism to further lead the network paying attention on boundary discrepancies; 2) The auxiliary texture map supervision can simultaneously guide model to learn generation artifacts, improving the adaptation ability.

**Evalution on Unseen Dataset.** To evaluate the adaptation ability on unknown dataset, we train the model on Face2Face dataset in FaceForensics and test on Deepfakes and FaceSwap dataset in FaceForensics++. The setting strictly follows [14] and we only use c23 data for fair comparison. Table 4 gives the result and we observe that our method can significantly improve the performance compared with other existing works on generalization issue.

**Table 4.** Evaluation on unseen dataset and manipulated methods. The model is trained on Faceforensics (FF) and test on Faceforensics++ (FF++). The metric is accuracy (%)

| Models | Face2Face (FF) | Deepfakes (FF++) | FaceSwap (FF++) |
|---|---|---|---|
| FT [6] | 88.43 | 62.61 | 52.29 |
| FT-res [6] | 82.30 | 64.75 | 53.50 |
| MTDS [14] | 92.77 | 52.32 | 54.070 |
| **Ours** | **95.30** | **73.11** | **65.39** |

## 5 Conclusion

This paper introduces a new perspective for face manipulation detection that adopts auxiliary supervisions with a multi-scale structure. A novel multi-task framework is proposed with texture map and blending boundary supervision, which facilitates the extraction of discriminative and generalizable features. Experiments on FaceForensics and FaceForensics++ datasets demonstrate our method makes significant improvement.

**Acknowledgments.** We thank for the support from National Natural Science Foundation of China (61972157, 61902129), Shanghai Pujiang Talent Program (19PJ1403100), Economy and Information Commission of Shanghai (XX-RGZN-01-19-6348), National Key Research and Development Program of China (No. 2019YFC1521104).

## References

1. Faceswap. https://github.com/MarekKowalski/FaceSwap/. Accessed 26 Apr 2020
2. Deepfakes. https://www.deepfakes.club/openfaceswap-deepfakessoftware/. Accessed 26 Apr 2020
3. Afchar, D., Nozick, V., Yamagishi, J., Echizen, I.: MesoNet: a compact facial video forgery detection network. In: 2018 IEEE International Workshop on Information Forensics and Security (WIFS), p. 17. IEEE (2018)
4. Ciftci, U.A., Demir, I.: FakeCatcher: detection of synthetic portrait videos using biological signals. arXiv preprint arXiv:1901.02212 (2019)
5. Cozzolino, D., Poggi, G., Verdoliva, L.: Recasting residual-based local descriptors as convolutional neural networks: an application to image forgery detection. In: Proceedings of the 5th ACM Workshop on Information Hiding and Multimedia Security, pp. 159–164 (2017)
6. Cozzolino, D., Thies, J., Röossler, A., Riess, C., Nießner, M., Verdoliva, L.: ForensicTransfer: weakly-supervised domain adaptation for forgery detection. arXiv preprint arXiv:1812.02510 (2018)
7. Cozzolino, D., Verdoliva, L.: Noiseprint: a CNN-based camera model fingerprint. IEEE Trans. Inf. Forensics Secur. **15**, 144–159 (2019)
8. Dale, K., Sunkavalli, K., Johnson, M.K., Vlasic, D., Matusik, W., Pfister, H.: Video face replacement. In: Proceedings of the 2011 SIGGRAPH Asia Conference, pp. 1–10 (2011)
9. He, D.C., Wang, L.: Texture unit, texture spectrum, and texture analysis. IEEE Trans. Geosci. Remote Sens. **28**(4), 509–512 (1990)
10. Karras, T., Laine, S., Aila, T.: A style-based generator architecture for generative adversarial networks. In: Proceedings of the IEEE Conference on Computer Vision and Pattern Recognition, pp. 4401–4410 (2019)
11. Kim, H., et al.: Deep video portraits. ACM Trans. Graph. (TOG) **37**(4), 1–14 (2018)
12. Li, L., et al.: Face x-ray for more general face forgery detection. arXiv preprint arXiv:1912.13458 (2019)
13. Li, Y., Lyu, S.: Exposing deepfake videos by detecting face warping artifacts. arXiv preprint arXiv:1811.00656 (2018)
14. Nguyen, H.H., Fang, F., Yamagishi, J., Echizen, I.: Multi-task learning for detecting and segmenting manipulated facial images and videos. arXiv preprint arXiv:1906.06876 (2019)
15. Nguyen, H.H., Yamagishi, J., Echizen, I.: Use of a capsule network to detect fake images and videos. arXiv preprint arXiv:1910.12467 (2019)

16. Röossler, A., Cozzolino, D., Verdoliva, L., Riess, C., Thies, J., Nießner, M.: FaceForensics: a large-scale video dataset for forgery detection in human faces. arXiv preprint arXiv:1803.09179 (2018)
17. Rossler, A., Cozzolino, D., Verdoliva, L., Riess, C., Thies, J., Nießner, M.: FaceForensics++: learning to detect manipulated facial images. In: Proceedings of the IEEE International Conference on Computer Vision, pp. 1–11 (2019)
18. Songsri-in, K., Zafeiriou, S.: Complement face forensic detection and localization with faciallandmarks. arXiv preprint arXiv:1910.05455 (2019)
19. Sun, K., et al.: High-resolution representations for labeling pixels and regions. arXiv preprint arXiv:1904.04514 (2019)
20. Tan, M., Le, Q.V.: EfficientNet: rethinking model scaling for convolutional neural networks. arXiv preprint arXiv:1905.11946 (2019)
21. Thies, J., Zollhofer, M., Stamminger, M., Theobalt, C., Nießner, M.: Face2Face: real-time face capture and reenactment of RGB videos. In: Proceedings of the IEEE Conference on Computer Vision and Pattern Recognition, pp. 2387–2395 (2016)
22. Wang, S.Y., Wang, O., Zhang, R., Owens, A., Efros, A.A.: CNN-generated images are surprisingly easy to spot... for now. arXiv preprint arXiv:1912.11035 (2019)
23. Xie, Q., Hovy, E., Luong, M.T., Le, Q.V.: Self-training with noisy student improves imagenet classification. arXiv preprint arXiv:1911.04252 (2019)
24. Yang, X., Li, Y., Lyu, S.: Exposing deep fakes using inconsistent head poses. In: ICASSP 2019–2019 IEEE International Conference on Acoustics, Speech and Signal Processing (ICASSP), pp. 8261–8265. IEEE (2019)
25. Zhang, X., Karaman, S., Chang, S.F.: Detecting and simulating artifacts in GAN fake images. arXiv preprint arXiv:1907.06515 (2019)

# Fine-Grained Scene-Graph-to-Image Model Based on SAGAN

Yuxin Ding(✉), Fuxing Xue, Weiyi Li, and Cai Nie

School of Computer Science and Technology,
Harbin Institute of Technology (Shenzhen), Shenzhen, China
yxding@hit.edu.cn, {xuefuxing,19s151099,19s051001}@stu.hit.edu.cn

**Abstract.** Text-to-image has become one of the most active research fields in recent years. Although current technology can generate complex images with multiple objects, how to generate detailed images with multiple objects still faces enormous challenges. In this paper, we propose a model driven by the attention mechanism to generate multi-object images from scene graphs. In the proposed model we introduce self-attention network into the mask regression network to overcome the limitation of the local receptive field of convolution Generative Adversarial Networks. Self-attention network can extract the long-range dependencies in an image, so the proposed model can generate image with more details. In addition, our model improves the stability of image generation and accelerates the process of image generation by gradually increasing the resolution of the cascading refinement network. Experiments on the Visual Genome and COCO-Stuff datasets show that the proposed model can generate more detailed images.

**Keywords:** Text-to-image · Self-attention · Progressive grown · Deep learning

## 1 Introduction

Text-to-image technology has wide applications in reality. Once the technology is ready for commercial applications, content creators can use natural language to guide machines to create images people want. For example, instead of spending a lot of time searching for the design they need, people can get their home design drawings through text descriptions. People can also choose a creative title and then get a creative picture right away.

Current most of text-to-image synthesis methods are based on Generative Adversarial Networks [1]. In the early stage most text-to-image researches [3,7,12] focused on generating images with a single object. Johnson J et al. [10] proposed a scene graph based model for generating an image with multiple objects from a scene graph. In this paper we study how to improve the scene graph based model to generate more detailed images and how to improve the stability of the image generation process of the model.


H. Yang et al. (Eds.): ICONIP 2020, LNCS 12532, pp. 325–337, 2020.
https://doi.org/10.1007/978-3-030-63830-6_28

In order to enhance the details of objects in the generated image, we introduce self-attention network [15] into the scene graph based model [10]. Due to the limitations of the local receptive field of the convolutional network, convolutional networks cannot generate accurate feature representation for objects having a wide range in an image. The self-attention module calculates the weight of each position feature using the weighted sum of all position features. Therefore, the self-attention module can find global dependencies of objects in an image, which is helpful to generate more detailed images.

In order to improve the stability of the image generation process, we gradually increase the number of cascaded refinement modules (CRM) to enhance the resolution of an image. We separately train CRM modules for generating different resolution images, and at last train all CRM modules together, which can accelerate the convergence of model training and improve the stability of the network.

We used two datasets Visual Genome [14], which provides a hand-annotated scene graph, and COCO Stuff [17], which contains composite scene graphs for describing the positional relationships between objects, to evaluate the performance of the proposed model. The experimental results on both data sets show that our methods can generate more detailed images.

## 2 Related Work

In general the text-to-image method mainly includes the generation method based on the Variational Auto-encoder (VAE) model, and the method based on the GAN model. In the early research of text-to-image models, the VAE models were the most commonly used methods, currently more and more test-to-image models are based on GAN. Compared with the VAE based model, the GAN based model can generate high quality images.

Reed et al. [3] developed a novel text-to-image architecture with GAN model to effectively represent the conversion from characters to pixels. Then, Reed et al. [4] proposed a new model, the Generative Adversarial What-Where Network (GAWWN), which can synthesize images and give instructions describing where to draw content. The authors presented a high-quality $128 \times 128$ generated image on the Caltech-UCSD Birds dataset given the informal text description and object location. Nguyen et al. [5] proposed a model that similarity to GAWWN, called the generic class model "Plug and Play Generation Network" (PPGN). The approach improves the visualization of multifaceted features, which generates a set of synthetic inputs that allow activated neurons to understand the operation of the deep neural network. Xu et al. [7] proposed an Attentional Generative Adversarial Networks (AttnGAN) that allows attention-driven multi-stage refinement to generate fine-grained images.

Most GAN based models [3,7,12,13] generate lower resolution images from text description. In order to improve the quality of the generated image, Wang et al. [8] decomposed the image generation process and proposed the style and structure Generative Adversarial Networks (S2-GAN) model. The S2-GAN model is interpretable, which can produce more realistic images, and can be used to learn unsupervised RGBD representations. Similarly, in order to improve the quality of image, Karras et al. [9] proposed to gradually increase the hidden layer of the generator and discriminator during the training process. As the training goes on, the image becomes more and more refined. Zhang et al. [11] (2017) proposed Stacked Generative Adversarial Networks (StackGANs), which aims to produce high-resolution photo-realistic images. First, the authors proposed a two-stage generating model StackGAN-v1 for text-to-image synthesis. In stage I, GAN plots the original shape and color of the scene based on a given text description, resulting in a lower resolution image. In stage II, GAN takes Stage-I results and text descriptions as input and produces high-resolution images with photo-realistic detail. An advanced multi-stage GAN StackGANv2 [11] is proposed for conditional and unconditional generation tasks. StackGAN-v2 shows a more stable training process than StackGAN-v1 [12] by jointly approximating multiple distributions.

One problem of text-to-image is how to improve the details of generated images. Most papers [3,4] use upsampling interpolation to generate more detailed images. At present, the best result is generated by the stackGAN-v2 model [11]. In [11] the image detail is improved by adding the condition GAN to process the text information again. However, the model can only be used to generate image with a single object and cannot generate multiple objects. Johnson J et al. [10] used scene graphs to better handle complex textual information. Because the scene graph contains the relation-ship between objects, the model [10] can not only generate multiple objects, but also can ensure the positional relationship between the objects. In this paper the proposed model is based on the model [10], and our goal is improving the detail of the generated image with multi-object.

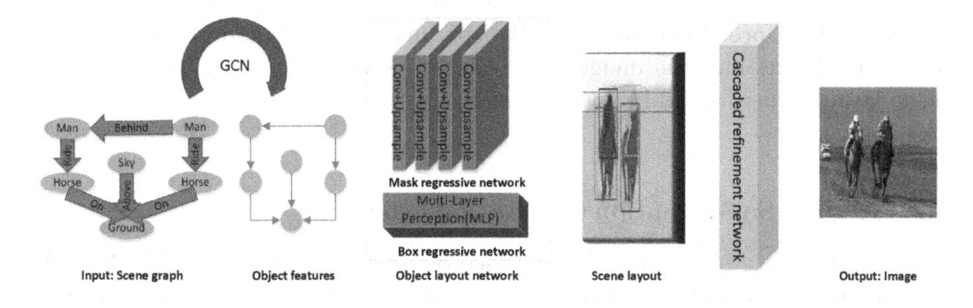


**Fig. 1.** Overview of the network structure of scene graph to image model

## 3 Methodology

Figure 1 show the whole process of generating image from scene graph. We take the object and relationship as input, transfer it to the graph convolution network and get the object's embedding vector. We enter the object's embedding vector into the mask regressive network and box regressive network to get the position and mask of the object. The mask and position of all objects in an image form the scene layout. Finally, we get the generated image by inputting the scene layout to Cascaded refinement network.

### 3.1 The Overall Architecture of Text-to-Image

To generate a detailed image with multiple objects, we intend to use the scene graph as input. The scene graph consists of a set of nodes and edges. The nodes represent objects and the relationship between objects is represented as edges. Scene graphs are used in many applications, such as image retrieval, image captions, and extracting scene graphs from images.

As shown in Fig. 1, we use a graphical convolutional network (GCN) to process the scene graph. The GCN processes graphical information using deep learning technologies. The goal of GCN is to learn the feature representation of nodes and edges in the graph $G = (V, E)$. Each node represents an object and the features of all objects can be represented as an $N \times D$ feature matrix, where $N$ represents the number of objects and $D$ represents the number of object features. The relationship between objects (edges) can be represented by an adjacent matrix. Finally, a node-level output $Z$ is generated by the graph convolutional neural network, which is an $N \times F$ feature matrix, where $F$ represents the dimension of output features for each node. We input object's embedding vector into the box regression network and the mask regression network respectively, and get the bounding box $\widehat{b_i} = (x_0, y_0, x_1, y_1)$ and soft binary mask $k\hat{m}_i$ of shape $M \times M$ where $x_0$, $x_1$ are the left and right coordinates of the box and $y_0$, $y_1$ are the top and bottom coordinates of the box. The box regression network is a Multi-Layer Perception and the mask regression network consists of a series of non-linear transposed convolutions. We combine the mask and bounding boxes of all the objects into the image to get the scene layout. Finally, the generated scene layout is used as an input to cascade the refinement network and generate an image.

### 3.2 Adding Self-attention in Object Layout

Since most text-to-image models are based on deep convolutional GAN [3,7,12], GAN is good at generating image classes with few structural constraints, such as ocean, sky, landscape, etc. However, it's hard for GAN to capture the geometry features that appears in objects, such as birds that usually have well-defined feet and realistic feather textures. The previous models [3,11] relied heavily on convolution to model the features of different regions of the image. Since the convolution operator operates on a local receptive field, long range dependencies can only be handled after passing through multiple convolutional layers.

Learning for long range dependencies can be prevented for a variety of reasons, such as smaller models may not be able to represent long-term dependencies or optimization algorithms may not be able to coordinate multiple layers to

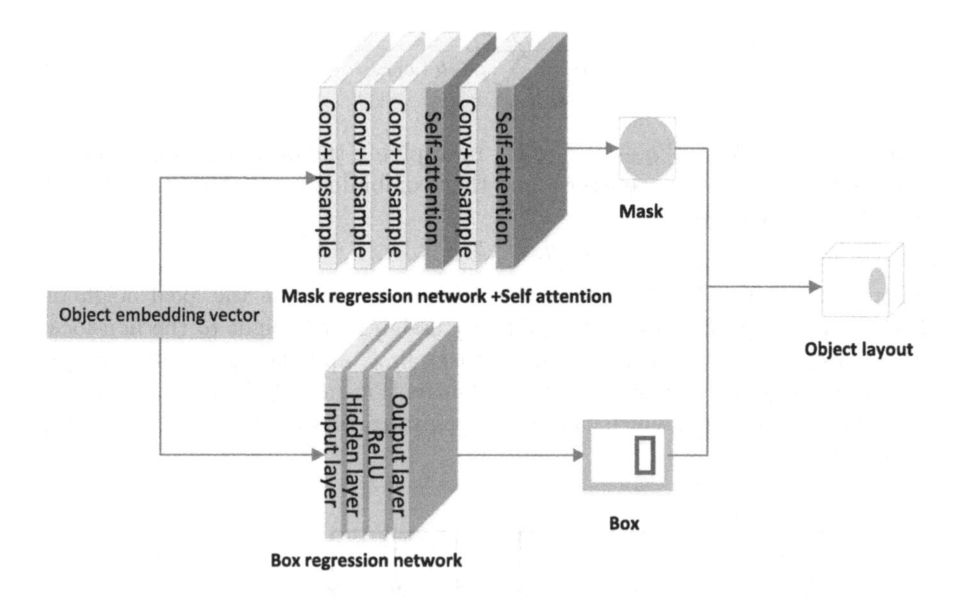


**Fig. 2.** Overview of the network structure of object layout

capture parameter values for these dependencies. Although increasing the size of the convolution kernel can reduce the local receptive field limitations, it also loses computational and statistical efficiency obtained by using local convolution structures. In addition, the calculation efficiency is low when only the convolutional layers are used to calculate the long-range dependence in the image. To address the ability of long-range dependencies, we borrow the idea of Self-Attention Generative Adversarial Networks (SAGAN) [15] and add self-attention to the mask regression network to improve the details of the object mask.

As shown in Fig. 2, the object embedding vector is input to the Masking regression network with self-attention and the Box regression network to produce the object mask and object box, respectively. Finally, by combining the object mask and the object box, the object layout is obtained.

Self-Attention Generative Adversarial Networks (SAGAN) combines the self-attention model in NLP with GAN. The formula is as follows:

$$\beta_{j,i} = \frac{\exp\left(s_{ij}\right)}{\sum_{i=1}^{N} \exp\left(s_{ij}\right)}, \tag{1}$$

where $s_{ij} = f\left(x_i\right)^T g\left(x_j\right)$. In Eq. (1) $x \in \mathbb{R}^{C \times N}$ represents the output of the last convolutional layer, and the image feature $x$ is input into two different $1 \times 1$ convolutional layers to obtain two different feature spaces $f(x) = W_f x$, $g(x) = W_g x$ where $W_f \in \mathbb{R}^{C \times C}$ and $W_f \in \mathbb{R}^{C \times C}$ are the learned weight matrix. $\beta_{j,i}$ represents the extent to which the $i^{th}$ region participates in the $j^{th}$ region when synthesizing the $j^{th}$ region. The output of the attention layer is $o = (o_1 \dots o_j \dots o_N) \in \mathbb{R}^{C \times N}$ where

$$o_j = \sum_{i=1}^{N} \beta_{j,i} h(x_i), \tag{2}$$

where $h(x_i) = W_h x$, $W_h \in \mathbb{R}^{C \times C}$. By multiplying the output $o$ of the layer of attention by the scale parameter $\gamma$ and then combining it with the image feature $x$, the final output is:

$$y_i = \gamma o_i + x_i, \tag{3}$$

where $\gamma$ from 0 to 1. This allows the network to focus on the local neighborhood at the beginning and then gradually assign more weight to the non-local. The reason for this is to learn simple tasks first, and then gradually increase the complexity of the task [15]. The self-attention module is a complement to convolution operation, which can model long range and multi-level dependencies in an image.

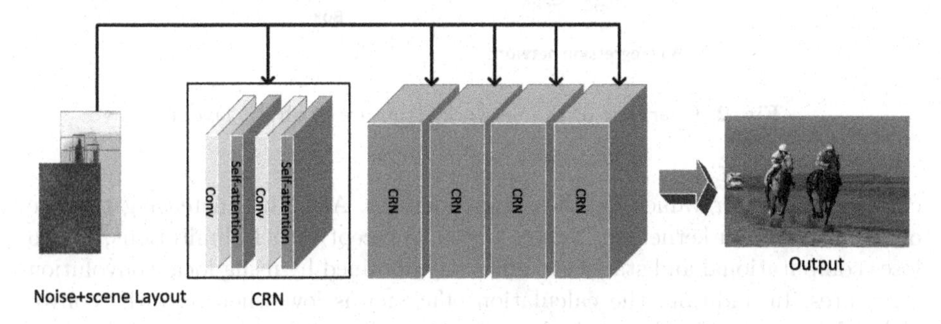


**Fig. 3.** An overview of the network structure of the Cascaded Refinement Network

### 3.3 Improving the Stability of Cascaded Refinement Network

As shown in Fig. 3, the scene Layout and Noise are input into each CRM module, after passing through multiple CRMs, an image is generated. From Fig. 3 we can see generating detailed images is difficult because it is easier for the discriminator to distinguish the generated image with more detail information from the real image, which greatly magnifies the gradient problem [6]. Due to memory limitation, training models generating detailed images require the use of smaller batches, which further affect the stability of model training. To solve this issue, we gradually increase the image resolution to improve the learning stability of the generator and discriminator. At first, the lower resolution image is generated, then we gradually increasing the CRM layer in the CRN to generate an image with higher resolution. This incremental nature allows the training to discover the whole structure of an image and then turn their attention to finer details. In our model, firstly the scene layout is input to the first CRM module to generate a $4 \times 4$ image. After the first CRM module has been trained for a certain number of epochs, an $8 \times 8$ image is generated by adding the second CRM module. Repeat the above steps until a $128 \times 128$ image is generated. While adding each CRM module to the generator, the discriminator also needs to gradually add hidden layers to identify images having the corresponding resolutions. In theory, it is possible to continuously increase the resolution by adding CRM module during the training process. The network structure is shown in Fig. 4.

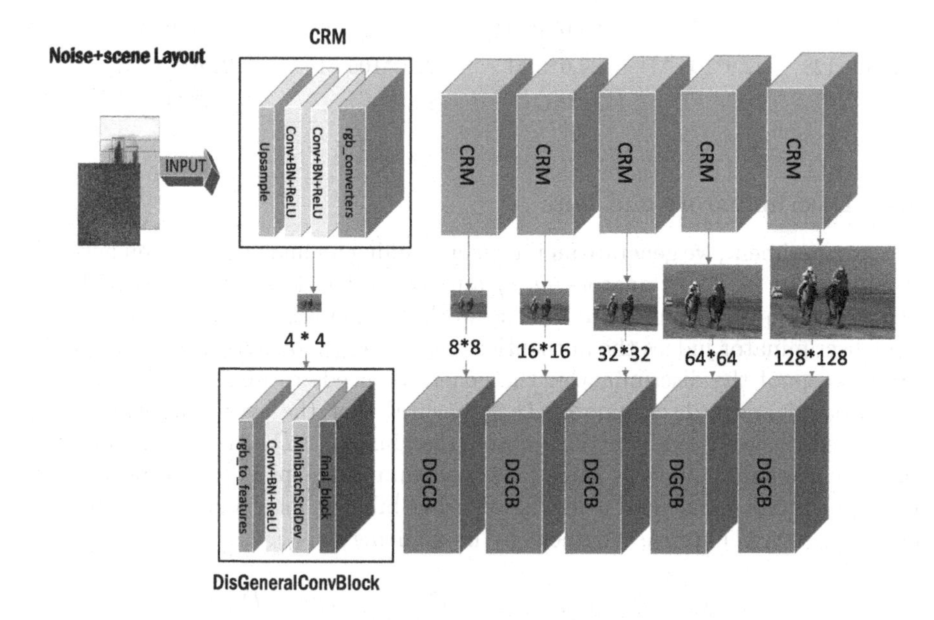


**Fig. 4.** Overview of the network structure of the improved cascading refinement network

By continuously adding hidden layers to the generator and discriminator during training, we can accelerate the speed of network training and reduce the computation for training. In paper [10], all CRM modules are trained simultaneously. Compared with the training method in [10], the proposed training method of gradually increasing the resolution can significantly reduce the numbers of parameters needed to be learned. Table 1 shows the parameter number needed to be adjusted in both training methods. In addition, the training method of gradually increasing the resolution can also improve the stability of the cascaded refinement network (see Sect. 4). As shown in Table 1, our method reduces about 1.2G learning parameters compared to the method [10] for the same epoch. In our model the number of training parameters in the generator and the discriminator is the same.

**Table 1.** Number of training parameters in both training methods

| Parameters | Generator [10] | Our generator | Discriminator [10] | Our discriminator |
|---|---|---|---|---|
| $4 \times 4$ | – | 10.4*30 M | – | 10.4*30 M |
| $8 \times 8$ | – | 17.8*25 M | – | 17.8*25 M |
| $16 \times 16$ | – | 19.8*20 M | – | 19.8*20 M |
| $32 \times 32$ | – | 20.4*15 M | – | 20.4*15 M |
| $64 \times 64$ | – | 22.0*10 M | – | 22.0*10 M |
| $128 \times 128$ | 22.6*105 M | 22.6*5 M | 22.6*105 M | 22.6*105 M |
| total | 2.3G | **1.75G** | 2.3G | **1.75G** |

### 3.4 Discriminators and Loss

In the experiment, we generate an image by training an image generation network $f$ and a pair of discriminators $D_{img}$ and $D_{obj}$. The loss function of GAN is defined as Eq. (4). The generation network $f$ generates the image $x \sim p_{fake}$. The discriminator judges the authenticity of $x \sim p_{fake}$ by maximizing $L_{GAN}$. In order to spoof the discriminator, the generator needs to minimize $L_{GAN}$. The patch-based image discriminator $D_{img}$ ensures that the overall appearance of generated images is realistic. The object discriminator $D_{obj}$ is added to ensure that the generated object is realistic, whose input is the pixel of an object which is cropped and rescaled to a fixed size using bilinear interpolation. Both $D_{obj}$ and $D_{img}$ try to improve judgment by maximizing $L_{GAN}$.

$$L_{GAN} = E_{x \sim preal} log(x) + E_{x \sim p_{fake}} log(1 - D(x)) \tag{4}$$

## 4 Experiments

### 4.1 Datasets

In the proposed model, we add self-attention to the mask regressive network and train the model by gradually increasing the CRM layer in the cascaded refinement network (see Sect. 3.3). We train our model to generate $128 \times 128$ images. Same as [10], we used the COCO dataset [3] and the Visual Genome dataset [14] to evaluate the performance of text-to-image models. The data partitions are shown in Table 2.

**Table 2.** Detailed parameters of the data set we used in the experiment

| Datasets | Train | Validation | Test | No. object | Relationship |
|---|---|---|---|---|---|
| COCO | 62565 | 5506 | 5088 | 171 | 6 |
| VG | 24972 | 1024 | 2048 | 178 | 45 |

### 4.2 Evaluation

We use the criteria 'inception score' [2] to evaluate the quality of the generated image, which is the widely used evaluation criterion for image generation. The formula is as follows:

$$IS(G) = \exp\left(\mathbb{E}_{x \sim p_g} D_{KL}(p(y \mid x) \| p(y))\right) \tag{5}$$

where $x \sim p_g$ is the generated image, and $p(y|x)$ is the predicted label distribution. KL divergence is used to measure the distance between two probability distributions.

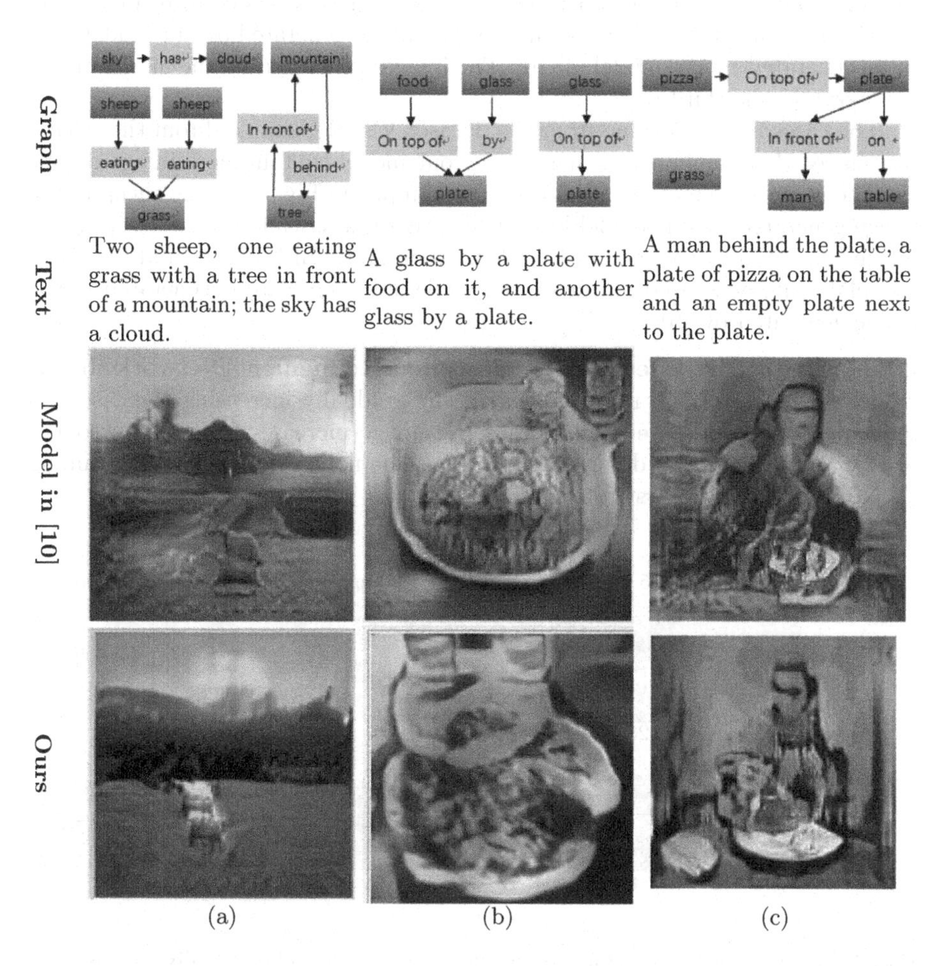


**Fig. 5.** Examples of 128 × 128 images generated by our model and the model in [10]

### 4.3 Experimental Results

We evaluate the performance of the proposed model from two aspects. First, evaluate the effect of self-attention network on the performance of model. Second, evaluate the effect of Cascaded Refinement Network on the performance of model.

**Evaluation of the Effect of Self-attention Network:** To learn the long-range dependencies, we add the self-attention network in the mask regression network. Figure 5 shows four cases in the experiments. In Fig. 5 the second row gives the text descriptions, the first row gives the scene graphs generated from the corresponding text, the fourth row shows the images generated by our model (mask with self-attention). We compare the images generated by our model with these generated by the model in [10]. The images generated by the model in [10] are shown in the third row.

The cases in Fig. 5 come from the dataset Visual Genome. From the results we can see that the images generated by our model contain more image details and the images are more continuous. For example, in Fig. 5(a), one sheep in the image generated by our model looks like eating grass, while the two seep in the image generated by the model [10] all stand on grass. In Fig. 5(b) and Fig. 5(c), two plates are generated in our image, while only one plate is generated in the image generated by [10].

**Evaluation of the Effect of the Cascaded Refinement Network.** In our model we generate high resolution image by gradually increasing the cascaded refinement modules. The resolution of the image increases from 4 × 4 to 128 × 128. For images with different resolutions, the number of epochs for training CRN is shown on the first row in Fig. 6.

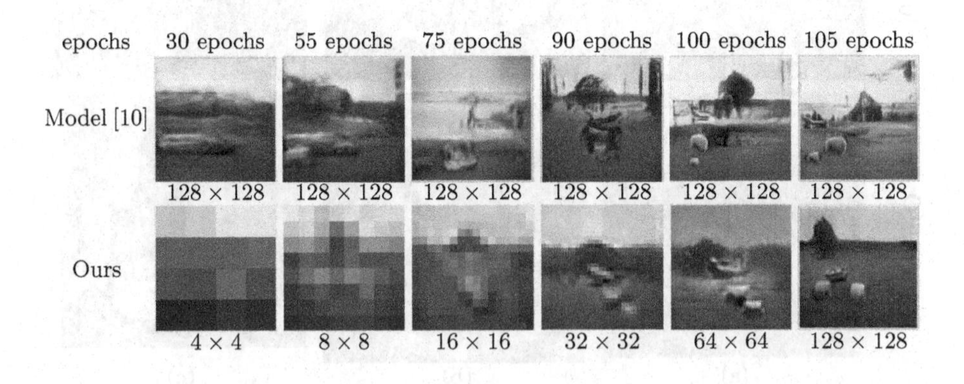


**Fig. 6.** Examples of 128 × 128 images generated by our model (gradually increasing in CRN) and the model in [10]. (The text of the image is described as "sky above grass; sheep standing on grass; tree behind sheep; sheep' by sheep; ocean by tree; boat in ocea")

As shown in Fig. 6, the image quality generated by the model [10] is not stable during the generation of 128 × 128 images. The objects (two sheep) have been generated at 75 epochs, but at 90 epochs the objects overlap again and cannot be distinguished from each other. In our model, the process of generating images is more stable, and the objects become clearer by gradually increasing the CRM layer.

**Table 3.** Evaluation on the COCO dataset and the VG dataset using the Inception score

| | Model | Inception score | |
|---|---|---|---|
| | | COCO | VG |
| 64 × 64 | Real Images | $16.3 \pm 0.4$ | $13.9 \pm 0.5$ |
| | Sg2im [10] | $6.7 \pm 0.1$ | $5.5 \pm 0.1$ |
| | StackGAN++ [11] | $8.4 \pm 0.2$ | - |
| | Ours (mask with attention) | $6.7 \pm 0.2$ | $5.5 \pm 0.2$ |
| | Ours (gradually increasing in CRN) | $6.8 \pm 0.1$ | $5.6 \pm 0.1$ |
| 128 × 128 | Real Images | $25.3 \pm 0.4$ | $22.1 \pm 0.5$ |
| | Sg2im [10] | $6.3 \pm 0.1$ | $5.1 \pm 0.3$ |
| | Ours (mask with attention) | $6.6 \pm 0.2$ | $6.2 \pm 0.2$ |
| | Ours (gradually increasing in CRN) | $\mathbf{7.7 \pm 0.3}$ | $\mathbf{6.7 \pm 0.3}$ |

We also evaluate the performance of different models using inception score. We divided the test data set into 10 parts and then calculated the mean and standard deviation of inception score. As shown in Table 3, our model can achieve higher values in inception scores than that generated by the model [10], especially for image with higher resolution.

## 5 Conclusions

In this paper, we propose a model driven by the attention mechanism to generate multi-object images from scene graphs. In the proposed model we introduce self attention network into the mask regression network to overcome the limitation of the local receptive field of convolution Generative Adversarial Networks. Self-attention network can extract the long-range dependencies in an image, so the proposed model can generate image with more details. In addition, we improve the stability of image generation and accelerate the process of image generation by gradually increasing the resolution of the cascading refinement network. The experiments on the Visual Genome and COCO-Stuff datasets show that the proposed model can generate more detailed images.

**Acknowledgements.** This work was partially supported by the National Key R&D Program of China under Grant no. 2018YFB1003800, 2018YFB1003805, Scientific Research Foundation in Shenzhen (Grant No. JCYJ20180306172156841, JCYJ20180507183608379), the National Natural Science Foundation of China (Grant No. 61872107).

## References

1. Goodfellow, I.J., Pouget-Abadie, J., Mirza, M., et al.: Generative adversarial networks. In: Advances in Neural Information Processing Systems, pp. 2672–2680 (2014)
2. Salimans, T., Goodfellow, I., Zaremba, W., et al.: Improved techniques for training GANs. In: Advances in Neural Information Processing Systems, pp. 2234–2242 (2016)
3. Reed, S., Akata, Z., Yan, X., et al.: Generative adversarial text to image synthesis. arXiv preprint arXiv:1605.05396 (2016)
4. Reed, S.E., Akata, Z., Mohan, S., et al.: Learning what and where to draw. In: Advances in Neural Information Processing Systems, pp. 217–225 (2016)
5. Nguyen, A., Clune, J., Bengio, Y., et al.: Plug & play generative networks: conditional iterative generation of images in latent space. In: Proceedings of the IEEE Conference on Computer Vision and Pattern Recognition, pp. 4467–4477 (2017)
6. Mirza, M., Osindero, S.: Conditional generative adversarial nets. Computer Science, pp. 2672–2680 (2014)
7. Xu, T., Zhang, P., Huang, Q., et al.: Attngan: fine-grained text to image generation with attentional generative adversarial networks. In: Proceedings of the IEEE Conference on Computer Vision and Pattern Recognition, pp. 1316–1324 (2018)
8. Wang, X., Gupta, A.: Generative image modeling using style and structure adversarial networks. In: Leibe, B., Matas, J., Sebe, N., Welling, M. (eds.) ECCV 2016. LNCS, vol. 9908, pp. 318–335. Springer, Cham (2016). https://doi.org/10.1007/978-3-319-46493-0_20
9. Karras, T., Aila, T., Laine, S., et al.: Progressive growing of GANs for improved quality, stability, and variation. arXiv preprint arXiv:1710.10196 (2017)
10. Johnson, J., Gupta, A., Fei-Fei, L.: Image generation from scene graphs. In: Proceedings of the IEEE Conference on Computer Vision and Pattern Recognition, pp. 1219–1228 (2018)
11. Zhang, H., Xu, T., Li, H., et al.: Stackgan++: realistic image synthesis with stacked generative adversarial networks. arXiv preprint arXiv:1710.10916 (2017)
12. Zhang, H., Xu, T., Li, H., et al.: Stackgan: text to photo-realistic image synthesis with stacked generative adversarial networks. In: Proceedings of the IEEE International Conference on Computer Vision, pp. 5907–5915 (2017)
13. Kingma, D.P., Welling, M.: Auto-encoding variational bayes. arXiv preprint arXiv:1312.6114 (2013)
14. Krishna, R., Zhu, Y., Groth, O., et al.: Visual genome: connecting language and vision using crowd sourced dense image annotations. Int. J. Comput. Vision **123**(1), 32–73 (2017)

15. Zhang, H., Goodfellow, I., Metaxas, D., et al.: Self-attention generative adversarial networks. arXiv preprint arXiv:1805.08318 (2018)
16. Isola, P., Zhu, J.Y., Zhou, T., et al.: Image-to-image translation with conditional adversarial networks. In: Proceedings of the IEEE Conference on Computer Vision and Pattern Recognition, pp. 1125–1134 (2017)
17. Caesar, H., Uijlings, J., Ferrari, V.: Coco-stuff: thing and stuff classes in context. In: Proceedings of the IEEE Conference on Computer Vision and Pattern Recognition, pp. 1209–1218 (2018)

# Generative Adversarial Network Using Multi-modal Guidance for Ultrasound Images Inpainting

Ruiguo Yu[1,2,3], Jiachen Hu[1,2,3], Xi Wei[4], Mei Yu[1,2,3], Jialin Zhu[4], Jie Gao[1,2,3], Zhiqiang Liu[1,2,3], and Xuewei Li[1,2,3](✉)

[1] College of Intelligence and Computing, Tianjin University, Tianjin 300354, China
{rgyu,hujc,tjubeisong,yumei,gaojie,lixuewei}@tju.edu.cn
[2] Tianjin Key Laboratory of Cognitive Computing and Application, Tianjin, China
[3] Tianjin Key Laboratory of Advanced Networking (TANK Lab), Tianjin, China
[4] Tianjin Medical University Cancer Hospital, Tianjin 300060, China
weixi198204@126.com, sally2010zhu@126.com

**Abstract.** Medical image inpainting not only helps computer-aided diagnosis systems to eliminate the interference of irrelevant information in medical images, but also helps doctors to prognosis and evaluate the operation by blocking and inpainting the lesion area. However, the existing diffusion-based or patch-based methods have poor performance on complex images with non-repeating structures, and the generate-based methods lack sufficient priori knowledge, which leads to the inability to generate repair content with reasonable structure and visual reality. This paper proposes a generative adversarial network via multi-modal guidance (MMG-GAN), which is composed of the multi-modal guided network and the fine inpainting network. The multi-modal guided network obtains the low-frequency structure, high-frequency texture and high-order semantic of original image through the structure reconstruction generator, texture refinement generator and semantic guidance generator. Utilizing the potential attention mechanism of convolution operation, the fine inpainting network adaptively fuses features to achieve realistic inpainting. By adding the multi-modal guided network, MMG-GAN realizes the inpainting content with reasonable structure, reliable texture and consistent semantic. Experimental results on Thyroid Ultrasound Image (TUI) dataset and TN-SCUI2020 dataset show that our method outperforms other state-of-the-art methods in terms of PSNR, SSIM, and relative l1 measures. Code and TUI dataset will be made publicly available.

**Keywords:** Image inpainting · Ultrasound images · Generative adversarial network

## 1 Introduction

Image inpainting or image completion is one of the important problems in computer vision systems. Its goal is to generate content with reasonable


H. Yang et al. (Eds.): ICONIP 2020, LNCS 12532, pp. 338–349, 2020.
https://doi.org/10.1007/978-3-030-63830-6_29

structure, reliable texture and consistent semantic for the missing areas of damaged image [10,14]. High-quality image inpainting has a wide range of applications, such as image editing [3], object removal, image-based rendering and computational photography [16]. In medical image analysis community, medical images with interference information, such as patient information or labels made by doctor, are usually used by computer-aided diagnosis systems. However, the generalization of neural network is limited. Experiments by [2,8] proved that the performance of neural network degrades rapidly after adding some small perturbations, which are not recognized by naked eyes. And the difference between pixels in medical images is extremely small, any tiny disturbance will have a fatal effect on the computer-aided diagnosis model. Therefore, it is very important to remove and inpaint the areas that interfere with the computer-aided diagnosis model in medical images. In addition, removing and inpainting the lesion area of medical images can also help doctors to prognosis and evaluate the operation.

Image inpainting methods are mainly divided into two categories: diffusion-based [5] and patch-based [20] traditional methods, and generative-based deep learning methods. Diffusion-based and patch-based methods perform well when common textures or missing regions are small, but perform poorly on complex images with non-repetitive structures. With the rapid development of generative adversarial networks [7] and encoder-decoders [17,24], generative-based deep learning methods have made great progress in image inpainting systems. However, these methods lack information in different dimensions, resulting in locally blurred or distorted images. Nazeri et al. [15] first separate the high-frequency information from low-frequency information in missing area and then gradually recover. Ren et al. [18] first restore the approximate structure of missing area and then gradually recover the texture details. Song et al. [21] prove that high-dimensional information in semantic space provide structural guidance for pixel-level prediction. However, there is no way to guide the image inpainting by using the priori information of different dimensions, so as to generate inpainting content with reasonable structure, reliable texture and consistent semantics.

To solve the above problems, we propose a novel two-stage generative adversarial network for medical image inpainting, which is composed of the multi-modal guided network and the fine inpainting network. The multi-modal guided network consists of the structure reconstruction generator, texture refinement generator and semantic guidance generator to mine the priori knowledge of missing regions. The information obtained by using the multi-modal guided network is incorporated into the fine inpaninting network to guide the prediction of missing areas. The structure reconstruction generator trained using smooth images to complete the low-frequency structure of missing area. The texture refinement generator trained using edge images to recover high-frequency textures of missing areas. The semantic guidance generator trained using semantic segmentation mask to achieve high-order consistency of inpainting content.

We conducted experiments using TN-SCUI2020 dataset [1] and TUI dataset provided by cooperative hospital. Our approach is verified by comparison with state-of-the-art methods and comprehensive ablation experiments. Experimental results show that MMG-GAN yields more convincing inpainting content compared with other image inpainting methods.

The main contributions of this paper are as follows:

- The MMG-GAN proposed in this paper uses the different dimensional information extracted by the multi-modal guided network, effectively fuses and gradually guides the fine inpaninting network to predict the missing areas.
- The multi-modal guided network proposed in this paper fully mines the low-frequency structure, high-frequency texture and high-order semantic of missing regions, and provides a priori knowledge for image inpainting tasks.

## 2 Relate Work

Existing image inpainting methods can be roughly divided into two categories: traditional methods using pixel diffusion or patch matching techniques, deep learning methods using generative-based techniques.

The diffusion-based methods [5,12] inpaint the damaged image by gradually propagating the neighborhood information to the missing area. The above methods lack the long-distance pixel information and overall information, and cannot generate a meaningful structure, so it only deal simple images with small holes. However, the patch-based methods [4,9] fill the lost regions by searching and copying similar image patches from the undamaged regions of original image. PatchMatch [3] designs a fast nearest neighbor searching algorithm that uses natural coherence in image as priori information. Without using any high-order information to guide the search, the above method is only suitable for pure background inpainting tasks [22].

The generative-based deep learning methods [6,11,19,22] model the image inpainting task as a conditional generation problem. These methods extract the meaningful information from undamaged image and generate new content for the damaged area. Context Encoder [17] using encoder-decoder structure first extracts low-dimensional features and then reconstructs the outputs. To avoid the existence of visual artifacts in the generated images, Iizuka et al. [7] introduce global and local discriminators to maintain the consistency of generated images and generate real alternative content for the missing areas respectively. To address the shortcomings of convolutional neural networks in building long-term correlations, Yu et al. [23] propose a contextual attention module to borrow features from long distances. To obtain a inpainting image with fine details, Nazeri et al. [15] suggest generating edge structures to help fill the missing areas. Ren et al. [18] consider that the edge image representation ability is limited and propose StructureFlow method, which sequentially generates structural information and edge details for the missing area. However, sequentially recovering low-frequency structures and high-frequency textures are not suitable for all missing areas. We hope that the models will fuse the multi-modal priori information and adaptively guide image inpainting tasks. In addition, the above methods fail to make use of the semantic information of original image and achieve high-order consistency of the inpainted image.

# 3 Multi-modal Guided Generative Adversarial Network

The framework of our proposed MMG-GAN is shown in Fig. 1. MMG-GAN consists of a two-stage network: the multi-modal guided network $I_g$ and the fine inpainting network $I_f$. The multi-modal guided network $I_g$ includes the structure reconstruction generator $G_s$, the texture refinement generator $G_t$ and the semantic guidance generator $G_e$. The structure reconstruction generator $G_s$ is used to reconstruct the low-frequency structure of missing region. The texture refinement generator $G_t$ is used to recover the high-frequency texture of missing area. The semantic guidance generator $G_e$ is used to achieve high-order consistency of inpainted content. The fine inpainting network $I_f$ uses the information extracted by the multi-modal guided network $I_g$ to finely inpaint the missing areas of damaged image.

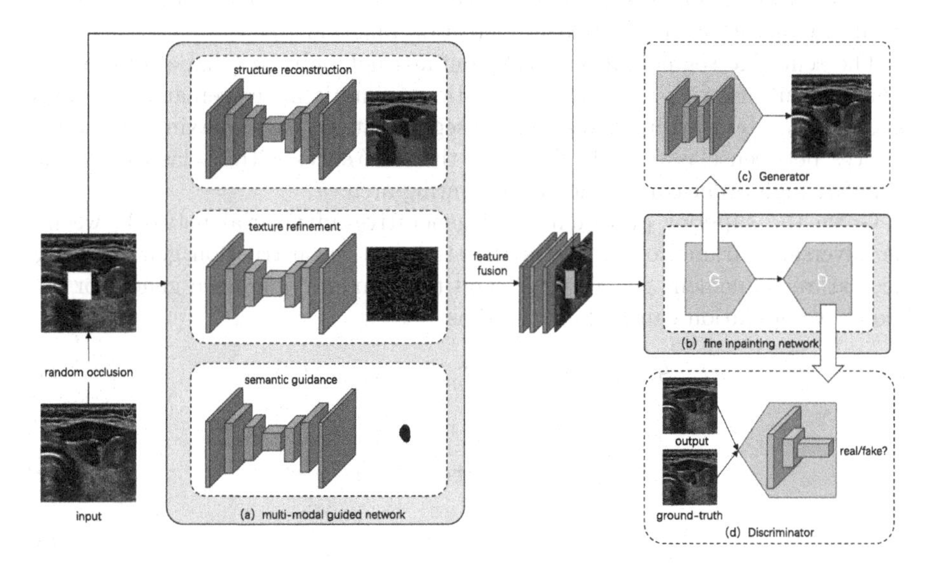


**Fig. 1.** Overall framework of MMG-GAN. (a) multi-modal guided network $I_g$, including the structure reconstruction generator $G_s$, the texture refinement generator $G_t$ and the semantic guidance generator $G_e$. (b) fine inpainting network $I_f$. (c) Generator G for fine inpainting network. (d) Discriminator D for fine inpainting network.

## 3.1 Multi-modal Guided Network

One of the challenges of image inpainting tasks is that the models lack sufficient priori knowledge of the missing areas of damaged image. Through experiments, we find that low-frequency structures, high-frequency textures and high-order semantics are significantly helpful for image inpainting tasks. Therefore, we design the structure reconstruction generator $G_s$, the texture refinement generator $G_t$ and the semantic guidance generator $G_e$ to obtain priori knowledge, guiding the fine inpainting network to predict the missing areas of damaged image.

The goal of the structure reconstruction generator $G_s$ is to reconstruct the global structure information of missing area according to the undamaged image. The edge-preserved smooth method can remove the high-frequency texture while retaining the low-frequency structure, so we use the edge-preserved smooth method as the structure extraction function $f_{se}(\cdot)$.

The goal of the texture refinement generator $G_t$ is to recover the local texture information of missing area according to the undamaged image. Gaussian smoothing first is used to reduce the error rate of input image. Then, the gradient magnitude and direction are calculated to estimate the edge strength and direction at each point in image. According to the gradient direction, we perform non-maximum suppression on the amplitude of gradient. Finally, the output is obtained through double threshold processing and edge connection. We define the above operations as the texture detection function $f_{td}(\cdot)$.

The semantic segmentation mask contains not only the shape information of object, but also rich high-order semantic information. The semantic guidance generator $G_e$, by learning the semantic segmentation mask, maximizes the differences between classes and minimizes the differences within classes, thereby achieving high-order consistency of inpainting area.

Given the superior performance of generative adversarial network, we use the adversarial framework to train the structure reconstruction generator $G_s$, the texture refinement generator $G_t$ and the semantic guidance generator $G_e$. The above operation can be formalized as:

$$\hat{S} = G_s\left(I_{in}, (1 - I_m)^{\circ} f_{se}\left(I_{gt}\right), I_m\right) \tag{1}$$

$$\hat{T} = G_t\left(I_{in}, (1 - I_m)^{\circ} f_{td}\left(I_{gt}\right), I_m\right) \tag{2}$$

$$\hat{E} = G_e\left(I_{in}, (1 - I_m)^{\circ} I_e, I_m\right) \tag{3}$$

Here, $I_{in}$ indicates the input image. $I_{gt}$ indicates the ground-truth image. $I_e$ indicates the semantic segmentation mask corresponding to the input image $I_{in}$. $I_m$ indicates the mask of the input image$I_{in}$, which is a binary matrix, where 1 represents the missing area and 0 represents the background. $^{\circ}$ means Hadamard product. $f_{se}(\cdot)$ represents the structure extraction function. $(1 - I_m)^{\circ} f_{se}\left(I_{gt}\right)$ represents the structure information of the input image $I_{in}$, $\hat{S}$ represents the predicted structure of the missing region. $f_{td}(\cdot)$ represents the texture detection function. $(1 - I_m)^{\circ} f_{td}\left(I_{gt}\right)$ represents the texture information of the input image $I_{in}$, $\hat{T}$ represents the predicted texture of missing area. $(1 - I_m)^{\circ} I_e$ represents the semantic information of the input image $I_{in}$, $\hat{E}$ represents the predicted high-order semantic of missing area.

$L_s^{l1}$ represents the $L1$ distance between the prediction structure $\hat{S}$ and the ground-truth structure $S_{gt}$, $L_t^{l2}$ represents the $L2$ distance between texture $\hat{T}$

and the ground-truth texture $T_{gt}$, $L_e^{l1}$ represents the $L1$ distance between the predicted semantic $\hat{E}$ and the semantic segmentation mask $E_{gt}$:

$$L_s^{l1} = \left\|\hat{S} - S_{gt}\right\|_1 \tag{4}$$

$$L_t^{l2} = \left\|\hat{T} - T_{gt}\right\|_2 \tag{5}$$

$$L_e^{l1} = \left\|\hat{E} - E_{gt}\right\|_1 \tag{6}$$

To better learn the low-frequency structure, high-frequency texture and high-order semantic of missing region, we use the form of adversarial loss to measure the performance of the multi-modal guided network $I_g$. $L_s^{adv}$ represents the adversarial loss of the structure reconstruction generator $G_s$, $L_t^{adv}$ represents the adversarial loss of the texture refinement generator $G_t$, $L_e^{adv}$ represents the adversarial loss of the semantic guidance generator $G_e$:

$$L_s^{adv} = E\left[\log\left(1 - D_s(\hat{S})\right)\right] + E\left[\log D_s\left(S_{gt}\right)\right] \tag{7}$$

$$L_t^{adv} = E\left[\log\left(1 - D_t(\hat{T})\right)\right] + E\left[\log D_t\left(T_{gt}\right)\right] \tag{8}$$

$$L_e^{adv} = E\left[\log\left(1 - D_e(\hat{E})\right)\right] + E\left[\log D_e\left(E_{gt}\right)\right] \tag{9}$$

Here, $D_i$ is the discriminator of the generator $G_i$. We use the following optimization strategy to jointly train the generator $G_i$ and the discriminator $D_i$:

$$\min_{G_i} \max_{D_i} L_i\left(G_i, D_i\right) = \omega_i^l L_i^l + \omega_i^{adv} L_i^{adv} \tag{10}$$

Here, $\omega_i^l$ and $\omega_i^{adv}$ are regularization parameters. Unless otherwise noted, we set $\omega_i^l = 1$ and $\omega_i^{adv} = 0.25$ by default.

### 3.2 Fine Inpainting Network

After obtaining the priori information of the reconstructed structures $\hat{S}$, texture details $\hat{T}$ and high-order semantic information $\hat{E}$, the fine inpainting network $I_f$ is used to predict the content of missing area. The fine inpainting network $I_f$ can be formalized as:

$$\hat{O} = I_f\left(I_{in}, \hat{S}, \hat{T}, \hat{E}, I_m\right) \tag{11}$$

Here, $I_{in}$ indicates the input image. $I_m$ indicates the mask of the input image $I_{in}$. $\hat{S}$, $\hat{T}$, $\hat{E}$ represent the low-frequency structure, high-frequency texture and high-order semantic obtained by the multi-modal guided network $I_g$. $\hat{O}$ represents the final output.

We use the $L1$ distance between the predicted output $\hat{O}$ and ground-truth image $I_{gt}$ to measure the performance of the fine inpainting network $I_f$:

$$L_f^{l1} = \left\| \hat{O} - I_{gt} \right\|_1 \tag{12}$$

To obtain inpainting content with reasonable structure and visual reality, we use the form of adversarial loss to train the fine inpainting network $I_f$:

$$L_f^{adv} = E\left[\log\left(1 - G_f(\hat{O})\right)\right] + E\left[\log D_f\left(I_{gt}\right)\right] \tag{13}$$

Here, $D_f$ is the discriminator of the generator $G_f$. We use the following optimization strategy to jointly train the fine inpainting network$I_f$:

$$\min_{G_f}\max_{D_f} L_f\left(G_f, D_f\right) = \omega_f^{l1} L_f^{l1} + \omega_f^{adv} L_f^{adv} \tag{14}$$

Here, $\omega_f^{l1}$ and $\omega_f^{adv}$ are regularization parameters. Unless otherwise noted, we set $\omega_f^{l1} = 1$ and $\omega_f^{adv} = 0.2$ by default.

### 3.3 Training Detail

Our proposed model is implemented in Pytorch. The network is trained using $256 \times 256$ images with a batch size of eight. We use the ADAM optimizer to optimize the model, which is a stochastic optimization method with adaptive moment estimation. Adam optimizer parameters $\beta_1$ and $\beta_2$ are set to 0 and 0.999 respectively. Train the generator with a learning rate of $10^{-4}$ until the loss is stable, then we reduce the learning rate and continue training the generator until convergence.

## 4 Experiment

### 4.1 Dataset and Metrics

TUI dataset uses thyroid ultrasound images of cooperative hospital from 2013 to 2018 as dataset. Each ultrasound image contains a nodule. All thyroid ultrasound images are scanned, labeled and pathologically sliced by doctors. Finally, with the help of ultrasound diagnostician and radiologist who has more than ten years of clinical experience, we accurately label lesion areas as high-order semantic labels. The TUI dataset contains a total of 3627 thyroid ultrasound images.

TN-SCUI2020 dataset [1] will provide the biggest public dataset of thyroid nodule with over 4500 patient cases from different ages, genders, and were collected using different ultrasound machines. Each ultrasound image is provided with its ground truth class (benign or malignant) and a detailed delineation of the nodule. The dataset comes from the Chinese Artificial Intelligence Alliance for Thyroid and Breast Ultrasound (CAAU) which was initiated by Dr. Jiaqiao Zhou, Department of Ultrasound, Ruijin Hospital, School of Medicine, Shanghai Jiaotong University.

We randomly select 80% of the data as train set, and the remaining 20% as test set. Our model is trained on the irregular mask dataset [13].

We evaluate the performance of the proposed method MMG-GAN and conventional methods by measuring the peak signal-to-noise ratio (PSNR), structural similarity index (SSIM) and relative $l1$ coefficients of the inpainting content. PSNR is the most widely used objective measurement method for evaluating inpainting images. SSIM evaluates the image quality by measuring the brightness, contrast and structure of the inpainting content, and overcomes the situation where the evaluation results produced by PSNR are inconsistent with human subjective perception. The relative $l1$ coefficient reflects the ability of model to reconstruct the original image content.

## 4.2 Comparison with State-of-the-Arts Methods

In this section, we compare MMG-GAN with other state-of-the-art image inpainting models on TUI test set and TN-SCUI2020 test set, including EdgeConnect [15], StructureFlow [18]. The performance comparison of each method is shown in Table 1.

When using randomly shaped occlusions on TUI test set, MMG-GAN achieves 33.435, 0.979 and 0.012 on PSNR, SSIM and relative $l1$ respectively, which outperforms all other state-of-the-art image inpainting methods. For example, compared to StructureFlow and EdgeConnect, MMG-GAN has improved 11.5% and 17.5% on PSNR respectively. A significant increase indicated that compared with the method of only using structure information or texture information, our model MMG-GAN, which obtain simultaneously priori information (low-frequency structure, high-frequency texture and high-order semantic) and make use of the advantage of adaptive selection of feature fusion, achieves better inpainting results. We show some qualitative results of MMG-GAN in Fig. 2.

**Table 1.** Performance comparison with current state-of-the-art image inpainting methods on TUI and TN-SCUI2020 dataset. We evaluate the performance of image inpainting under different occlusions, including random shaped masks, squares of size [32, 32] and squares of size [96, 96].

| Mask type | Method | TUI datasets | | | TUSCI datasets | | |
|---|---|---|---|---|---|---|---|
| | | psnr | ssim | $l1$ | psnr | ssim | $l1$ |
| Random | EdgeConnect | 28.465 | – | 0.023 | 28.260 | – | 0.027 |
| | StructureFlow | 29.978 | 0.963 | 0.020 | 28.210 | 0.942 | 0.025 |
| | MMG-GAN | **33.435** | **0.979** | **0.012** | **33.798** | **0.979** | **0.012** |
| [32, 32] | EdgeConnect | 29.630 | – | 0.021 | 29.468 | – | 0.022 |
| | StructureFlow | 31.191 | 0.976 | 0.018 | 28.857 | 0.955 | 0.024 |
| | MMG-GAN | **35.760** | **0.990** | **0.010** | **35.741** | **0.988** | **0.010** |
| [96, 96] | EdgeConnect | 24.491 | – | 0.031 | 25.121 | – | 0.032 |
| | StructureFlow | 25.960 | 0.885 | 0.028 | 25.569 | 0.867 | 0.031 |
| | MMG-GAN | **27.011** | **0.898** | **0.020** | **28.251** | **0.907** | **0.018** |

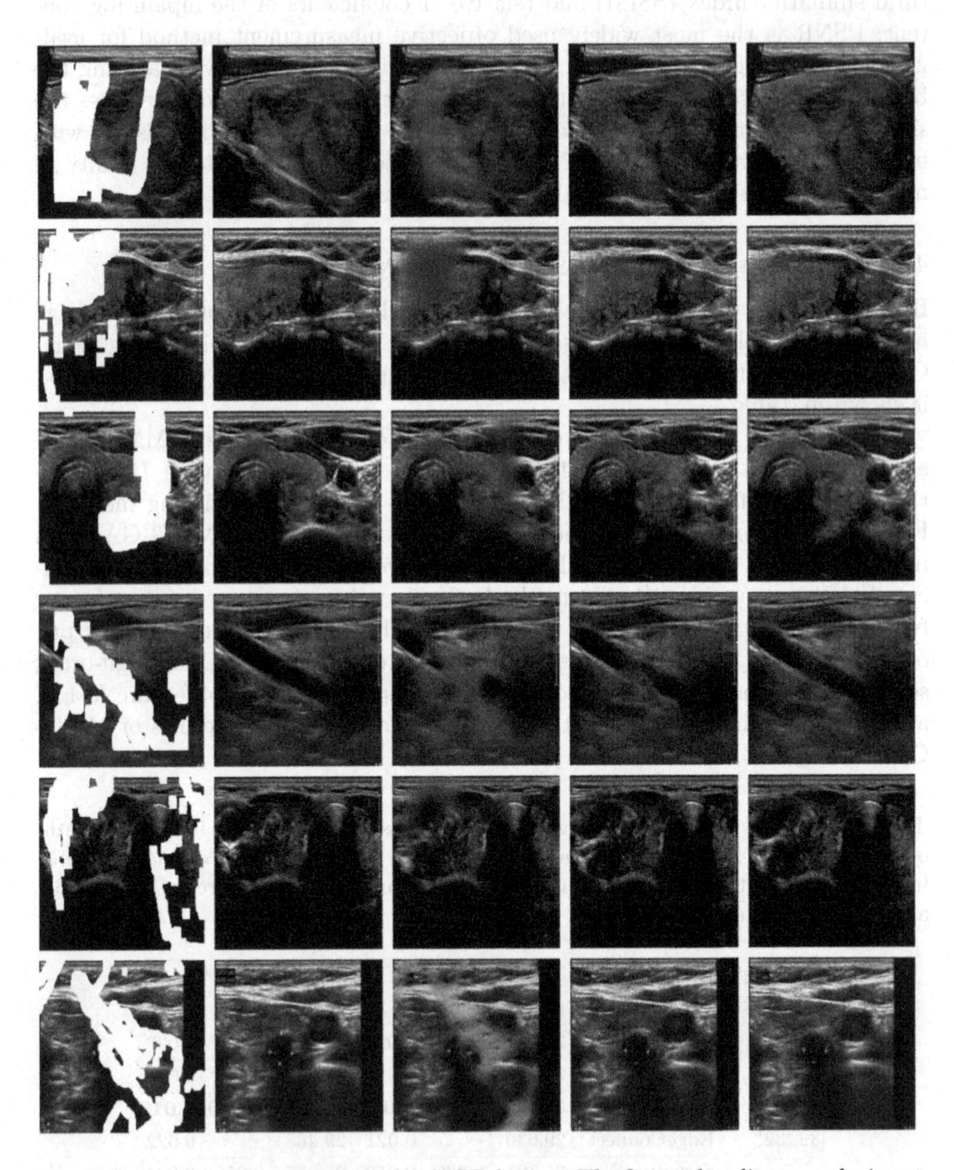

**Fig. 2.** Examples of inpainting result of TUI dataset. The first to last lines are the input image, ground-truth, the inpainting results of EdgeConnect [15], StructureFlow [18] and MMG-GAN (ours).

### 4.3 Ablation Experiment

To verify whether different priori information has an impact on medical image inpainting tasks, we add the structure reconstruction generator, texture refinement generator and semantic guidance generator to provide low-frequency structures, high-frequency textures and high-order semantics before the fine inpainting network. Through the performance comparison of inpainting results, we judge whether the priori information of different dimensions is helpful for medical image inpainting task. The experimental results are shown in Table 2.

**Table 2.** On TUI test set, the influence of different modal priori information for inpainting network.

| Priori information | PSNR | SSIM | $l1$ |
|---|---|---|---|
| – | 28.785 | 0.954 | 0.025 |
| +semantic | 29.887 | 0.962 | 0.020 |
| +texture | 30.543 | 0.964 | 0.019 |
| +structure | 29.978 | 0.963 | 0.020 |

To make full use of the guiding role of multi-modal priori knowledge, we use the information obtained through the structure reconstruction generator, texture refinement generator and semantic guidance generator to find a way that is most suitable for medical image inpainting tasks. The experimental results are shown in Table 3.

Notably, as shown in row 3 of Table 2 and row 4 of Table 3, the model performance of adding the structure reconstruction generator and texture refinement generator is lower than that of adding the texture refinement generator. We think that is because the information generated between the structure reconstruction generator and the texture refinement generator interferes with each other. If no other information is used to help, the performance of model will be degraded. As shown in the last row in Table 3, with the addition of information from the structural reconstruction generator, texture refinement generator and semantic guidance generator as priori knowledge, the model achieves an optimal 33.435 PSNR, 0.979 SSIM and 0.012 relative $l1$.

**Table 3.** On TUI test set, performance comparison of image inpainting network when using different modal priori information

| $G_e$ | $G_t$ | $G_s$ | PSNR | SSIM | $l1$ |
|---|---|---|---|---|---|
| √ | √ |  | 29.888 | 0.963 | 0.021 |
| √ |  | √ | 30.580 | 0.964 | 0.018 |
|  | √ | √ | 30.365 | 0.963 | 0.018 |
| √ | √ | √ | **33.435** | **0.979** | **0.012** |

## 5 Conclusion

In this paper, we propose a novel end-to-end generative adversarial network MMG-GAN for medical image inpainting tasks. This model can generate inpainting content with reasonable structure, reliable texture and consistent semantic. The proposed MMG-GAN is a two-stage model consisting of the multi-modal guided network and fine inpainting network. The multi-modal guided network, which is composed of structure reconstruction generator, texture refinement generator and semantic guidance generator, fully mines the implicit representation in damaged image. The fine inpainting network uses the above implicit representation as a priori knowledge to guide the inpaint of missing areas in an adaptive fusion way. Experimental results on TUI and TN-SCUI2020 dataset show that our model achieves reliable inpainting results. In the future, we hope that the model acquires more high-dimensional information as priori knowledge to guide inpainting tasks. And we hope that MMG-GAN has better robustness and generalization. Namely, it only uses minimal data to achieve explicit inpainting.

**Acknowledgment.** This work is supported by National Natural Science Foundation of China (Grant No. 61976155), and Key Project for Science and Technology Support from Key R&D Program of Tianjin (Grant No. 18YFZCGX0 0960).

## References

1. TN-SCUI2020 homepage (2020). https://tn-scui2020.grand-challenge.org/Home
2. Akhtar, N., Mian, A.S.: Threat of adversarial attacks on deep learning in computer vision: a survey. IEEE Access **6**, 14410–14430 (2018). https://doi.org/10.1109/ACCESS.2018.2807385
3. Barnes, C., Shechtman, E., Finkelstein, A., Goldman, D.B.: Patchmatch: a randomized correspondence algorithm for structural image editing. ACM Trans. Graph. **28**(3), 24 (2009). https://doi.org/10.1145/1531326.1531330
4. Darabi, S., Shechtman, E., Barnes, C., Goldman, D.B., Sen, P.: Image melding: combining inconsistent images using patch-based synthesis. ACM Trans. Graph. **31**(4), 82:1–82:10 (2012). https://doi.org/10.1145/2185520.2185578
5. Efros, A.A., Freeman, W.T.: Image quilting for texture synthesis and transfer. In: SIGGRAPH 2001, pp. 341–346 (2001). https://dl.acm.org/citation.cfm?id=383296
6. Grigorev, A., Sevastopolsky, A., Vakhitov, A., Lempitsky, V.S.: Coordinate-based texture inpainting for pose-guided human image generation. In: CVPR 2019, pp. 12135–12144 (2019). http://openaccess.thecvf.com/CVPR2019.py
7. Iizuka, S., Simo-Serra, E., Ishikawa, H.: Globally and locally consistent image completion. ACM Trans. Graph. **36**(4), 107:1–107:14 (2017). https://doi.org/10.1145/3072959.3073659
8. Jia, X., Wei, X., Cao, X., Foroosh, H.: Comdefend: an efficient image compression model to defend adversarial examples. In: CVPR 2019, pp. 6084–6092 (2019). http://openaccess.thecvf.com/CVPR2019.py
9. Jin, D., Bai, X.: Patch-sparsity-based image inpainting through a facet deduced directional derivative. IEEE Trans. Circuits Syst. Video Tech. **29**(5), 1310–1324 (2019). https://doi.org/10.1109/TCSVT.2018.2839351

10. Ke, J., Deng, J., Lu, Y.: Noise reduction with image inpainting: an application in clinical data diagnosis. In: SIGGRAPH 2019, Posters, pp. 88:1–88:2 (2019). https://doi.org/10.1145/3306214.3338593
11. Li, A., Qi, J., Zhang, R., Ma, X., Ramamohanarao, K.: Generative image inpainting with submanifold alignment. In: IJCAI 2019, pp. 811–817 (2019). https://doi.org/10.24963/ijcai.2019/114
12. Li, H., Luo, W., Huang, J.: Localization of diffusion-based inpainting in digital images. IEEE Trans. Inf. Forensics Secur. **12**(12), 3050–3064 (2017). https://doi.org/10.1109/TIFS.2017.2730822
13. Liu, G., Reda, F.A., Shih, K.J., Wang, T., Tao, A., Catanzaro, B.: Image inpainting for irregular holes using partial convolutions. In: ECCV 2018, Proceedings, Part XI, pp. 89–105 (2018). https://doi.org/10.1007/978-3-030-01252-6_6
14. Ma, Y., Liu, X., Bai, S., Wang, L., He, D., Liu, A.: Coarse-to-fine image inpainting via region-wise convolutions and non-local correlation. In: IJCAI 2019, pp. 3123–3129 (2019). https://doi.org/10.24963/ijcai.2019/433
15. Nazeri, K., Ng, E., Joseph, T., Qureshi, F.Z., Ebrahimi, M.: Edgeconnect: generative image inpainting with adversarial edge learning. CoRR abs/1901.00212 (2019). http://arxiv.org/abs/1901.00212
16. Park, E., Yang, J., Yumer, E., Ceylan, D., Berg, A.C.: Transformation-grounded image generation network for novel 3D view synthesis. In: CVPR 2017, pp. 702–711 (2017). https://doi.org/10.1109/CVPR.2017.82
17. Pathak, D., Krähenbühl, P., Donahue, J., Darrell, T., Efros, A.A.: Context encoders: feature learning by inpainting. In: CVPR 2016, pp. 2536–2544 (2016). https://doi.org/10.1109/CVPR.2016.278
18. Ren, Y., Yu, X., Zhang, R., Li, T.H., Liu, S., Li, G.: Structureflow: image inpainting via structure-aware appearance flow. In: ICCV 2019, pp. 181–190 (2019). https://doi.org/10.1109/ICCV.2019.00027
19. Sagong, M., Shin, Y., Kim, S., Park, S., Ko, S.: PEPSI: fast image inpainting with parallel decoding network. In: CVPR 2019, pp. 11360–11368 (2019). http://openaccess.thecvf.com/CVPR2019.py
20. Simakov, D., Caspi, Y., Shechtman, E., Irani, M.: Summarizing visual data using bidirectional similarity. In: CVPR 2008 (2008). https://doi.org/10.1109/CVPR.2008.4587842
21. Song, Y., Yang, C., Shen, Y., Wang, P., Huang, Q., Kuo, C.J.: SPG-Net: segmentation prediction and guidance network for image inpainting. In: BMVC 2018, p. 97 (2018). http://bmvc2018.org/contents/papers/0317.pdf
22. Xiong, W., Yu, J., Lin, Z., Yang, J., Lu, X., Barnes, C., Luo, J.: Foreground-aware image inpainting. In: CVPR 2019, pp. 5840–5848 (2019). http://openaccess.thecvf.com/CVPR2019.py
23. Yu, J., Lin, Z., Yang, J., Shen, X., Lu, X., Huang, T.S.: Generative image inpainting with contextual attention. In: CVPR 2018, pp. 5505–5514 (2018). https://doi.org/10.1109/CVPR.2018.00577
24. Zeng, Y., Fu, J., Chao, H., Guo, B.: Learning pyramid-context encoder network for high-quality image inpainting. In: CVPR 2019, pp. 1486–1494 (2019). http://openaccess.thecvf.com/CVPR2019.py

# High-Level Task-Driven Single Image Deraining: Segmentation in Rainy Days

Mengxi Guo[1], Mingtao Chen[1], Cong Ma[2], Yuan Li[2], Xianfeng Li[1], and Xiaodong Xie[2(✉)]

[1] School of Electronic and Computer Engineering, Peking University, Shenzhen 518055, China
{nicolasguo,mingtaochen,lixianfeng.sz}@pku.edu.cn

[2] School of Electronics Engineering and Computer Science, Peking University, Beijing 100871, China
{Cong-Reeshard.Ma,yuanli,donxie}@pku.edu.cn

**Abstract.** Deraining driven by semantic segmentation task is very important for autonomous driving because rain streaks and raindrops on the car window will seriously degrade the segmentation accuracy. As a pre-processing step of semantic segmentation network, a deraining network should be capable of not only removing rain in images but also preserving semantic-aware details of derained images. However, most of the state-of-the-art deraining approaches are only optimized for high PSNR and SSIM metrics without considering objective effect for high-level vision tasks. Not only that, there is no suitable dataset for such tasks. In this paper, we first design a new deraining network that contains a semantic refinement residual network (SRRN) and a novel two-stage segmentation aware joint training method. Precisely, our training method is composed of the traditional deraining training and the semantic refinement joint training. Hence, we synthesize a new segmentation-annotated rain dataset called Raindrop-Cityscapes with rain streaks and raindrops which makes it possible to test deraining and segmentation results jointly. Our experiments on our synthetic dataset and real-world dataset show the effectiveness of our approach, which outperforms state-of-the-art methods and achieves visually better reconstruction results and sufficiently good performance on semantic segmentation task.

**Keywords:** Single image deraining · Semantic segmentation · High-level task driven application · Deep learning

## 1 Introduction

Semantic segmentation is a key task in automatic driving. However, complicated rain weather can dramatically affect the robustness of the segmentation. In recent years, some efforts concentrate on heightening accuracy of the segmentation system in rainy days [24,29]. All their methods are training a deraining


H. Yang et al. (Eds.): ICONIP 2020, LNCS 12532, pp. 350–362, 2020.
https://doi.org/10.1007/978-3-030-63830-6_30

network independently and use it as a pre-processing module before the semantic segmentation network to improve the performance of segmentation systems affected by rain.

However, this kind of approach is not a perfect solution, because the separate deraining models are inclined to pursue higher image-quality metrics [11], but seldom systematically considered whether this is helpful for advanced tasks. Some works [8,18,22] demonstrated that only consider image-quality metrics con not help with the high-level tasks. That is, there is a gap between the low-level and high-level tasks. Although some efforts [5,8,17,20,30] have been made to close the gap through joint training, we observe that there are still some deficiencies: (1) A dataset with rainy-clean pair images and the annotations of high-level task is not enough. (2) The traditional way to close the gap is the one-stage joint training for the low-level and high-level vision tasks networks. However, it often leads to the PSNR/SSIM degradation when getting good performance for high-level tasks [8,17,20,30].

To address above limitations, firstly, we focus on the single image-based deraining of the urban road scenes and synthesize a high resolution rainy dataset based on Cityscapes [2] to simulate the real-world autonomous driving in rainy day. In particular, different from the datasets of [12,21,24] which only contain rain streaks or raindrop, our dataset simulates a complete process of rainy day with rain streaks of different densities and raindrops of different sizes. This makes our dataset more real and more challenging. And this dataset has semantic segmentation labels, which is convenient for us to explore the combination of deraining task and segmentation task simultaneously.

Secondly, in order to better integrate the deraining with the high-level task, we propose a novel high-level task-driven framework consisting of the low-level deraining network, the semantic refinement network and the high-level segmentation network, which follows a two-stage training process. In particular, at the Stage I, the low-level deraining network is trained to remove rain roughly. Subsequently, at the Stage II, the parameters of the previous deraining network are fixed, but the semantic refinement network trains with the high-level vision task network jointly to gets rid of residual rain while retaining high-level semantic information within the images. To achieve this goal, we present a semantic refinement residual network (SRRN) which repairs the semantic information of the image while keeping its vision quality. Our experiments show that compared to traditional one-stage joint training, our proposed model with the two-stage training method, as a bridge of deraining and segmentation task, can balance the two to enhance the performance of segmentation while maintaining relatively high image-quality metrics and achieve better results on both two tasks.

**The contributions of this paper are as follows:**

1. We create a large-scale synthetic rainy dataset of road scenes for training and evaluation, which contains both rain streaks and raindrops with well segmentation-annotation. It is worth mentioning that this dataset simulates continuous changes in a rainy process and covers more complex rainy conditions.

2. We propose a novel semantic segmentation task-driven deraining framework with two-stage segmentation aware joint training, in which the semantic refinement residual network (**SRRN**) is the bridge to balance the performance of low-level deraining and high-level segmentation. Experimental results show our deraining framework can contribute to segmentation dramatically while achieving better image quality than the one-stage joint training.
3. We conduct a benchmark test and analysis by evaluating our proposed deraining framework and state-of-the-art methods on our synthetic rainy dataset and the other open real-world rainy dataset. The experimental results show that our method quantitatively and qualitatively outperforms existing works on both visual performance and segmentation accuracy.

## 2 Related Work

*Rain Streak and Raindrop Removal.* Recently, many researches focus on single image rain streak removal by adopting image priors [15,19] or deep learning approaches [6,27]. Unfortunately, rain streaks have quite different characteristics from raindrops, the state-of-the-art rain streak removal methods cannot be used for raindrop removal directly [18]. Most traditional de-raindrop methods exploit physical characteristics and context information of multiple frames [16,32]. As for a deep learning based method, Eigen et al. [4] first separate raindrops via a simple CNN. Popular image-to-image translation methods like Pix2pix [13] are used for reference in learning raindrop-free images [24,25]. Other mainstream methods are inclined to combine their models with attention mechanism (Attentive-GAN [25], shape and channel attention [26], depth-guided attention [12]). Furthermore, Liu et al. [21] proposes a versatile Dual Residual Network, which can be applied to both rain-streak and raindrop removal tasks.

*Bridging Low-Level and High-Level Vision Task.* Lately, Some researches begin to combine low-level task with high-level vision task. [5,9,31] fuse high-level semantic information into the low-level network to help it to obtain better visual effects. [17,18,22,24] utilize the low-level network as an independent preprocessor before the high-level network. However, [18,22] show that, in this independent pipeline, the output of the low-level network will degrade the performance of subsequent high-level vision system. Therefore, some efforts have been made to close the gap between them using an end-to-end joint training method. Liu et al. [20] train a denoising model by cascading a fixed pre-trained segmentation network and tune with a joint loss function of segmentation awareness and MSE loss. Wang et al. [30] propose an unsupervised segmentation-aware denoising network using joint training without needing segmentation labels. Similar high-level application-driven method [8,17] have been proposed in the fields of super-resolution and dehazing. However, to the best of our knowledge, there is no related deraining method using our proposed two-stage semantic refinement joint training so far.

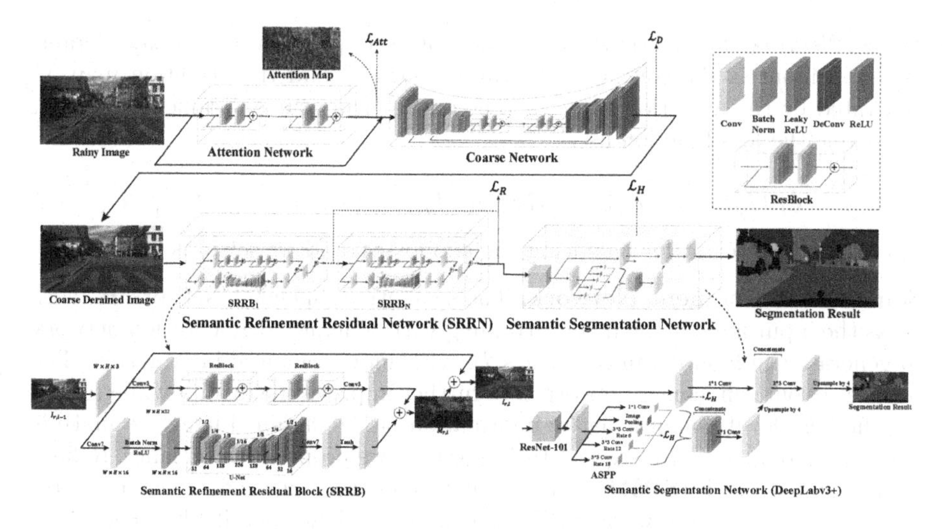


**Fig. 1.** The detail architectures of our networks.

# 3 Method

## 3.1 Model Architecture

As shown in Fig. 1, the architecture of our cascaded network consists of three parts: the deraining network, semantic refinement network and the semantic segmentation network for high-level vision task. The deraining network is first applied to generate the coarse derained image. And then the derained image is fed into the semantic refinement network to further repair the semantic information. Finally, the network for semantic segmentation gets the refined derained image as input and generates the segmentation result.

**Deraining Network:** The detailed structure of the whole deraining network can be seen in Fig. 1. It consists of two networks: (1) the attention Network **A** takes a rainy image $I_r$ as input and generates an attentive map $M$ which could localize the regions with rain streaks and raindrops in an image; (2) the coarse Network **C** is responsible for the preliminary rain removal of the rainy image. It takes $M$ and $I_r$ as input and generates a coarse derained image $I_C$.

*Attention Network.* We use a full-resolution ResNet [23] as the attention network to generate a binary attention map $M$. As the supervisory information, the binary attention map, $M_{gt}$, is simply generated by subtracting the gray-scale rainy image $I_{r,g}$ with the gray-scale clean image$I_{G,g}$. We set the threshold as 5 to determine whether a pixel is rainy region. The loss function is defined as the mean squared error (MSE) between $M$ and $M_{gt}$:

$$\mathcal{L}_{Att} = MSE(M, M_{gt}) = MSE(A(I_r), M_{gt}) \tag{1}$$

*Coarse Network.* The role of the coarse network is to perform single image deraining. The input is the concat of the rainy image $I_r$ and the attention map $M$ generated by the attention network. The output is a coarse derained image $I_C$. We define the loss function as:

$$\mathcal{L}_D(I_C, I_{gt}) = \alpha(1 - SSIM(I_C, I_{gt})) + (1 - \alpha) \left\| I_C - I_{gt} \right\|_1 \quad (2)$$

where $\alpha$ is the weight for balancing the two losses, and we set $\alpha$ as 0.85.

**Semantic Refinement Network.** The semantic refinement Network **R** takes $I_C$ as the input and conducts joint training with semantic segmentation network to generate the final derained images $I_R$ with more semantic information. The purpose of our refinement network is to further improve the quality while restoring the high-level semantic-aware information of the derained image to make it better adapted to the subsequent task. For this, we propose the semantic Refinement residual block (**SRRB**) and use it to construct our refinement network. As shown in Fig. 1, the whole SRRB is composed of two parallel branches: One U-Net structure branch and one full-resolution convolutional branch. It allows the network to acquire semantic information from a large receptive field. In order to make the output of SRRB not only as much as possible keep the pixel-level detail, but also further improve the semantic-aware information of the local pixels, we use three tips: (1) We use skip connections in the U-Net structure, which has been proved to be helpful in keeping most of the details of the output images; (2) We add a parallel branch, which is composed of two full-resolution convolutional layers and two Res-blocks. Such a branch can maintain the image details and obtain the local semantic information; (3) The outputs of the two branches of SRRB are added to generate a 3-channel residual semantic map $M_r$ with rich semantic and pixel-level restoring information.

We connect multiple SRRBs to make up our semantic refinement residual network (**SRRN**). SRRB can accomplish local and global information restoring. However, we find that one SRRB can not be enough for restoring well. There are two purposes we connect multiple SRRBs: first, in the refinement stage, the superposition of the blocks can maintain a certain depth of the network, which ensure the quality of the images. Second, what we find is that if only one SRRB is used for the refinement network, it is still difficult to achieve better consistency between the quality of images and the performance of high-level tasks. Therefore, we use $N$ blocks to form the SRRN and use a step process to generate the final output. In the SRRN, the $i$-th SRRB takes the output of $i-1$-th SRRB $I_{R,i-1}$ as the input and generates the more refined image $I_{R,i}$. When $i = 0$, the input of SRRB is the output of coarse network $I_C$. The reconstruction step loss of the SRRN has two parts: One is the sum of mean squared error between the outputs of the $N-1$ SRRBs and the clean image. The other is a $L_D$ loss between the output of the $N$-th SRRB and the clean image. It can be expressed as:

$$\mathcal{L}_R = 0.5 \sum_{i=0}^{N-1} (I_{R,i-1} - I_{gt})^2 + \mathcal{L}_D(I_{R,N}, I_{gt}) \quad (3)$$

Finally, we set N = 2 because what we find in our experiment is that when N > 2, the performance doesn't increase significantly.

**Semantic Segmentation Network:** The detailed structure of the semantic segmentation network can be seen in Fig. 1. We use DeepLabv3+ [1] as our semantic segmentation network and use Resnet-101 [10] as a backbone. The network is well trained in the rainless settings and the weights of the network are fixed while training the deraining network. The Astrous spatial pyramid pooling (ASPP) which contains several parallel atrous convolution kernels with different ratios is used to extract semantic information at different scales in DeepLabv3+. To some extent, the feature maps from ASPP have multi-scale high-level semantic information and the feature maps generated by the last layer of the backbone contain the low-level semantic information. We regard these feature maps as important semantic supervised information and use them in our training process. The loss of the semantic segmentation network can be defined as:

$$\mathcal{L}_H = \frac{1}{N_{AF}} \sum_{i=1}^{N_{AF}} \left\| SS(I_{gt})_{AFi} - SS(I_{R,N})_{AFi} \right\|_1 + \lambda_{LF} \left\| SS(I_{gt})_{LF} - SS(I_{R,N})_{LF} \right\|_1 \tag{4}$$

where SS is the semantic segmentation network. $SS(I)_{AF}$ means the feature maps generated from ASPP. $N_{AF}$ is the number of parallel channels of ASPP and it is 5 in this model. $SS(I)_{LF}$ represents the feature maps from the last layer of the backbone. $\lambda_{LF}$ is the weight for balancing low-level and high-level semantic losses and we set it as 0.1 in our experiments.

It is worth mentioning that, unlike the previous methods [5,20], $L_H$ can be calculated without the actual need of any segmentation ground truth during the training process which is similar to minimizing the perceptual loss [14] for image super-resolution [3,7].

### 3.2 Training Method

For training our cascaded network, we design a brand new corresponding two-stage segmentation aware joint training method.

*Stage I: Traditional Deraining Training.* At this training stage, we use a normal deraining training method to train our attentive network and coarse network. We use the end-to-end training method and the pixel-level image processing loss functions to adjust the parameters of our network. The loss function in this stage is expressed as:

$$\mathcal{L}_1 = \mathcal{L}_{Att} + \mathcal{L}_D \tag{5}$$

*Stage II: Semantic Refinement Joint Training.* After the stage I training process, we leave the parameters of the attention network and the coarse network fixed. On the basis of stage I, stage II cascades SRRN for refining the coarse derained image and the semantic segmentation network for the high-level vision task, aiming to simultaneously further reconstruct visually pleasing results of deraining and attain sufficiently good performance in semantic segmentation task. The training at this stage is achieved by minimizing the joint loss function and updating the parameters of SRRN:

$$\mathcal{L}_2 = \mathcal{L}_R + \mathcal{L}_H \tag{6}$$

# 4 A New Raindrop-Cityscapes Dataset

It is almost impossible to capture a pair of photos with and without rain in real life. Many researches [12,24,25,33] try to construct rainy datasets through simulation methods. However, the existing datasets are not good enough to simulate the complex rain streaks and raindrops in real life. Moreover, except [12,18,24], there is no open rainy dataset with complete annotation information of high-level vision task which makes it difficult to test cascaded high-level network performance. We model a rain streak and raindrop degraded image as the combination of a clean image and effect of the raindrops and a rain streaks layer:

$$I_r = (1 - M) \odot I_{gt} + R + S \tag{7}$$

where $I_r$ is the synthetic rain image, $I_{gt}$ is the clean image, $M$ is the binary rain map, $R$ is the effect brought by the raindrops and $S$ is the rain streak layer.

In this paper, we aim at a task-driven rain removal problem in road scenes, therefore, based on Cityscapes [2], we use the synthetic model in [24,28] to generate raindrops and use model in [33] to generate rain streaks in images. On the basis, we add the rain streak layer which randomly changes with each timestep. It is worth mentioning that, unlike other synthesis rainy dataset, our dataset contains a complete process of rainy change from no rain to gradual occurrence of rain streaks, to accumulation of small raindrops on the car window and slow generation for large raindrops which finally flows down. This allows our dataset to cover more rainy day situations, making it more complex and challenging. Altogether, our Raindrop-Cityscapes dataset has 3,479 high-resolution (1024 × 2048) images that cover various rainy scenes.

# 5 Experiments

*Datasets.* We use our Raindrop-Cityscapes as a synthetic dataset and RobotCar(R) [24] as a real-world dataset in our experiment. Specially. The Raindrop-Cityscapes contains 3,479 images and we randomly select 479 images as the testing dataset and the remaining 3,000 images as the training dataset.

*Semantic Segmentation Test.* For semantic segmentation test, following the evaluation protocol of Cityscapes [2], 19 output of 30 semantic labels are used for evaluation. The final performance is measured in terms of pixel intersection-over-union (mIoU) averaged across these classes.

*Settings.* The proposed network is implemented by PyTorch and executed on four NVIDIA TITAN X GPUs. In our experiments, the semantic segmentation network-DeepLabv3+ is trained in their default setting. The two-stage training for the deraining network and semantic segmentation network share the same training setting: the patch size is $512 \times 512$, and the batch size is 6 for the limitations by the GPU memory. For gradient descent, Adam is applied with an initial learning rate of $1 \times 10^{-3}$, and ends after 100 epochs. When reaching 30, 50 and 80 epochs, the learning rate is decayed by multiplying 0.2.

### 5.1 Ablation Studies

All the ablation studies are conducted on our Raindrop-Cityscapes dataset. We discuss the effects of the refinement network and the two-stage joint training method. As shown in Table 1, A/C means we only use attention network and coarse network to do deraining and use traditional deraining training as Eq. 5. A/C/R means adding the semantic refinement network R to A/C and training same as A/C. In addition to discussing the role of the refinement network, we also test the effect of high-level task aware jointing training. We abbreviate the traditional one-stage joint training [20] and our two-stage joint training to OJT and TJT in the experiments section. It should be noted that because R is not introduced into A/C, the joint training of A/C can only be trained with OJT.

**Ablation for the Semantic Refinement Network.** The comparisons between A/C and A/C/R, and between A/C + OJT and A/C/R + OJT show the importance of the refinement network. As shown in Table 1, the models with R get higher PSNR and SSIM which means that the semantic refinement network indeed has a function of further optimizing the quality of the derained image. Moreover, the higher mIoU shows that the semantic refinement network strengthens the link between deraining and semantic segmentation tasks.

**Ablation for Joint Training.** Table 1 lists the quantitative evaluations of these four different tests. Unsurprisingly, A/C + OJT, A/C/R + OJT, A/C/R + TJT outperform A/C and A/C/R in terms of the performance of derained images in semantic segmentation task. However, PSNR/SSIM of OJT group are lower than these without joint training. Particularly, A/C + OJT are significantly lower than A/C. Similar phenomenon also occurs in another works [8,17,20,30]. This shows the disadvantage of OJT. We think there are two reasons for this phenomenon: (1) PSNR/SSIM depend on the reconstruction error between the derained output and the clean image. Compared with the traditional image optimized loss function which always is MSE loss or L1 loss, the introduction of joint loss from the cascaded high-level task will make it difficult for the single deraining network to reach the maximum PSNR/SSIM with a one-stage training process. (2) A single small deraining network limits the optimal performance of the network in both tasks. The comparison between A/C/R, A/C/R + OJT and A/C/R + TJT can support our first explanation to some extent, because compared to A/C/R + OJT, A/C/R + TJT shows slight PSNR decrease while mIoU increase which shows the advantage of our TJT. And from the higher

PSNR and SSIM of A/C/R + OJT than A/C + OJT, we can see that increasing the number of network parameters can make up for this shortcoming of OJT, but it is still not as good as TJT.

**Table 1.** The results of the ablation studies.

| Method | Rainy image | A/C | A/C+OJT | A/C/R | A/C/R+OJT | A/C/R+TJT |
|---|---|---|---|---|---|---|
| PSNR | 23.23 | 33.88 | 28.84 | 34.61 | 31.26 | 34.57 |
| SSIM | 0.8002 | 0.9573 | 0.9324 | 0.9581 | 0.9521 | 0.9583 |
| mIoU(%) | 34.50 | 66.60 | 67.54 | 66.79 | 67.50 | 67.60 |

### 5.2 Evaluation on Synthetic Dataset

Our proposed network is evaluated on our Raindrop-Cityscapes datasets. Four competing methods: AGAN [25], ICSC [24], PReNet [27] and DuRN [21] are considered. Our model is the whole network with attention network, coarse network and SRRN. As we can see the top table in Fig. 2, our network not only achieves significant PSNR and SSIM gains over all the competing methods but also shows the best performance in the successor semantic segmentation task. As can be seen from Fig. 2, our method has fewer residual traces on the derained image. In terms of semantic segmentation results, our network, especially trained with semantic refinement joint training method, retains more complete semantic information. We also test the semantic segmentation performance through training segmentation network directly, but the performance is not good as ours.

### 5.3 Evaluation on the Real-World Dataset

Since we focus on a rain removal problem of the urban street scene driven by segmentation task, We use RobotCar(R) dataset [24], which is a real-world raindrop dataset of road scenes constructed by using a double-lens stereo camera mounted on a vehicle. However, this dataset doesn't have strictly annotated semantic segmentation labels. Therefore, we use it in well-trained semantic segmentation network on Cityscapes [2] to generate the loosely annotated ground truth on the rainless images of RobotCar(R), and we pick 500 images manually as the dataset based on intuition. Of these, 352 images are the training dataset and 148 images are the testing dataset. The competing methods are similar to those in Sect. 5.2 and our model is the whole networks trained by our two-stage training method. The top table in Fig. 3 presents the qualitative results of these five methods. It is seen that our method achieves the best result. As shown in Fig. 3, the visual quality improvement by our method is also significant in this dataset and the semantic segmentation performance achieves the best among the five methods.

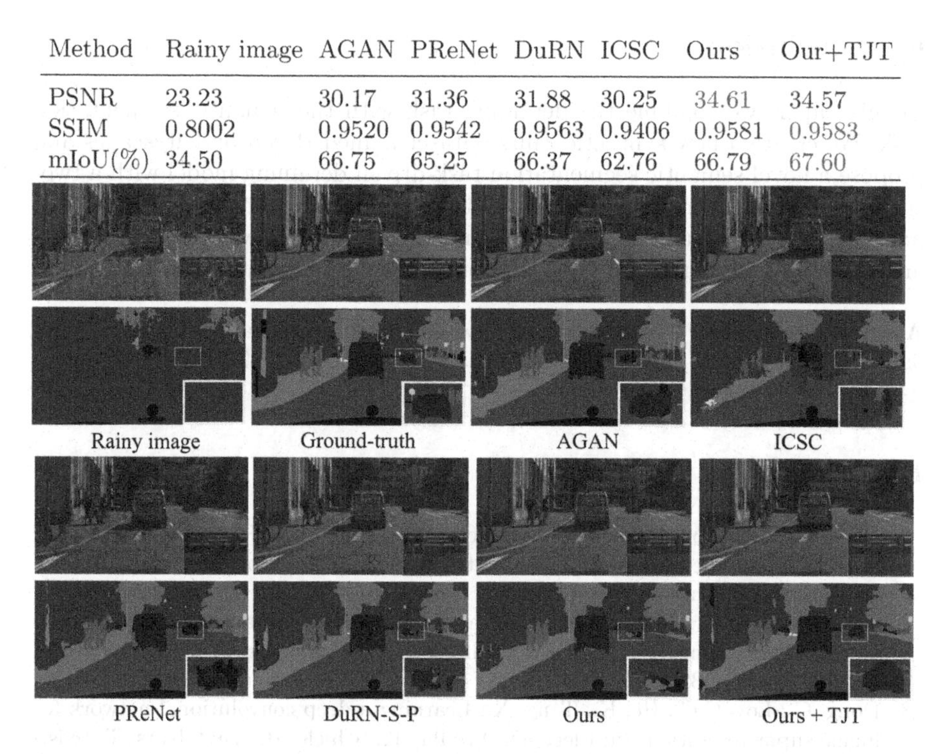

| Method | Rainy image | AGAN | PReNet | DuRN | ICSC | Ours | Our+TJT |
|---|---|---|---|---|---|---|---|
| PSNR | 23.23 | 30.17 | 31.36 | 31.88 | 30.25 | 34.61 | 34.57 |
| SSIM | 0.8002 | 0.9520 | 0.9542 | 0.9563 | 0.9406 | 0.9581 | 0.9583 |
| mIoU(%) | 34.50 | 66.75 | 65.25 | 66.37 | 62.76 | 66.79 | 67.60 |



**Fig. 2.** Examples of deraining and semantic segmentation results obtained by ours and others on our Raindrop-Cityscapes dataset. **Top:** Mean PSNR/SSIM and mIoU results of our model and other state-of-arts. **Bottom:** Deraining and semantic segmentation examples of our model and other state-of-arts.

| Method | Rainy image | AGAN | PReNet | DuRN-S-P | ICSC | Ours |
|---|---|---|---|---|---|---|
| PSNR | 14.20 | 20.87 | 19.02 | 21.69 | 20.15 | 22.01 |
| SSIM | 0.5927 | 0.7870 | 0.7252 | 0.8047 | 0.7269 | 0.8085 |
| mIoU(%) | 13.86 | 33.32 | 17.10 | 33.39 | 28.52 | 35.68 |

Rainy image Ground-truth AGAN ICSC PReNet DuRN-S-P Ours

**Fig. 3.** Examples of deraining and semantic segmentation results obtained by five methods including ours on RobotCar(R). **Top:** Mean PSNR/SSIM and mIoU of our model and other state-of-arts. **Bottom:** Deraining and semantic segmentation examples of our model and other state-of-arts.

## 6 Conclusion

In this paper, we combine the deraining task with the semantic segmentation task. We create a new synthetic rainy dataset named Raindrop-Cityscapes and propose a novel semantic segmentation task-driven deraining model with a two-stage training method. We demonstrate that our proposed method is capable of not only deraining but also semantic segmentation and achieves the state-of-the-art on both deraining visual performance and semantic segmentation accuracy.

**Acknowledgement.** This work is partially supported by the National Science Foundation of China under contract No. 61971047 and the Project 2019BD004 supported by PKU-Baidu Fund.

## References

1. Chen, L.C., Zhu, Y., Papandreou, G., Schroff, F., Adam, H.: Encoder-decoder with atrous separable convolution for semantic image segmentation. In: Proceedings of the European Conference on Computer Vision (ECCV), pp. 801–818 (2018)
2. Cordts, M., et al.: The cityscapes dataset for semantic urban scene understanding. In: Proceedings of the IEEE Conference on Computer Vision and Pattern Recognition, pp. 3213–3223 (2016)
3. Dong, C., Loy, C.C., He, K., Tang, X.: Learning a deep convolutional network for image super-resolution. In: Fleet, D., Pajdla, T., Schiele, B., Tuytelaars, T. (eds.) ECCV 2014. LNCS, vol. 8692, pp. 184–199. Springer, Cham (2014). https://doi.org/10.1007/978-3-319-10593-2_13
4. Eigen, D., Krishnan, D., Fergus, R.: Restoring an image taken through a window covered with dirt or rain. In: Proceedings of the IEEE International Conference on Computer Vision, pp. 633–640 (2013)
5. Fan, Z., Sun, L., Ding, X., Huang, Y., Cai, C., Paisley, J.: A segmentation-aware deep fusion network for compressed sensing MRI. In: Proceedings of the European Conference on Computer Vision (ECCV), pp. 55–70 (2018)
6. Fu, X., Huang, J., Zeng, D., Huang, Y., Ding, X., Paisley, J.: Removing rain from single images via a deep detail network. In: Proceedings of the IEEE Conference on Computer Vision and Pattern Recognition, pp. 3855–3863 (2017)
7. Glasner, D., Bagon, S., Irani, M.: Super-resolution from a single image. In: 2009 IEEE 12th International Conference on Computer Vision, pp. 349–356. IEEE (2009)
8. Haris, M., Shakhnarovich, G., Ukita, N.: Task-driven super resolution: object detection in low-resolution images. arXiv preprint arXiv:1803.11316 (2018)
9. Harley, A.W., Derpanis, K.G., Kokkinos, I.: Segmentation-aware convolutional networks using local attention masks. In: Proceedings of the IEEE International Conference on Computer Vision, pp. 5038–5047 (2017)
10. He, K., Zhang, X., Ren, S., Sun, J.: Deep residual learning for image recognition. In: Proceedings of the IEEE Conference on Computer Vision and Pattern Recognition, pp. 770–778 (2016)
11. Hore, A., Ziou, D.: Image quality metrics: PSNR vs. SSIM. In: 2010 20th International Conference on Pattern Recognition, pp. 2366–2369. IEEE (2010)

12. Hu, X., Fu, C.W., Zhu, L., Heng, P.A.: Depth-attentional features for single-image rain removal. In: Proceedings of the IEEE Conference on Computer Vision and Pattern Recognition, pp. 8022–8031 (2019)
13. Isola, P., Zhu, J.Y., Zhou, T., Efros, A.A.: Image-to-image translation with conditional adversarial networks. In: Proceedings of the IEEE Conference on Computer Vision and Pattern Recognition, pp. 1125–1134 (2017)
14. Johnson, J., Alahi, A., Fei-Fei, L.: Perceptual losses for real-time style transfer and super-resolution. In: Leibe, B., Matas, J., Sebe, N., Welling, M. (eds.) ECCV 2016. LNCS, vol. 9906, pp. 694–711. Springer, Cham (2016). https://doi.org/10.1007/978-3-319-46475-6_43
15. Kang, L.W., Lin, C.W., Fu, Y.H.: Automatic single-image-based rain streaks removal via image decomposition. IEEE Trans. Image Process. **21**(4), 1742–1755 (2011)
16. Kurihata, H., et al.: Rainy weather recognition from in-vehicle camera images for driver assistance. In: IEEE Proceedings of Intelligent Vehicles Symposium, pp. 205–210. IEEE (2005)
17. Li, B., Peng, X., Wang, Z., Xu, J., Feng, D.: Aod-net: all-in-one dehazing network. In: Proceedings of the IEEE International Conference on Computer Vision, pp. 4770–4778 (2017)
18. Li, S., et al.: Single image deraining: a comprehensive benchmark analysis. In: Proceedings of the IEEE Conference on Computer Vision and Pattern Recognition, pp. 3838–3847 (2019)
19. Li, Y., Tan, R.T., Guo, X., Lu, J., Brown, M.S.: Rain streak removal using layer priors. In: Proceedings of the IEEE Conference on Computer Vision and Pattern Recognition, pp. 2736–2744 (2016)
20. Liu, D., Wen, B., Liu, X., Wang, Z., Huang, T.S.: When image denoising meets high-level vision tasks: a deep learning approach. arXiv preprint arXiv:1706.04284 (2017)
21. Liu, X., Suganuma, M., Sun, Z., Okatani, T.: Dual residual networks leveraging the potential of paired operations for image restoration. In: Proceedings of the IEEE Conference on Computer Vision and Pattern Recognition, pp. 7007–7016 (2019)
22. Pei, Y., Huang, Y., Zou, Q., Lu, Y., Wang, S.: Does haze removal help CNN-based image classification? In: Proceedings of the European Conference on Computer Vision (ECCV), pp. 682–697 (2018)
23. Pohlen, T., Hermans, A., Mathias, M., Leibe, B.: Full-resolution residual networks for semantic segmentation in street scenes. In: Proceedings of the IEEE Conference on Computer Vision and Pattern Recognition, pp. 4151–4160 (2017)
24. Porav, H., Bruls, T., Newman, P.: I can see clearly now: image restoration via de-raining. arXiv preprint arXiv:1901.00893 (2019)
25. Qian, R., Tan, R.T., Yang, W., Su, J., Liu, J.: Attentive generative adversarial network for raindrop removal from a single image. In: Proceedings of the IEEE Conference on Computer Vision and Pattern Recognition, pp. 2482–2491 (2018)
26. Quan, Y., Deng, S., Chen, Y., Ji, H.: Deep learning for seeing through window with raindrops. In: Proceedings of the IEEE International Conference on Computer Vision, pp. 2463–2471 (2019)
27. Ren, D., Zuo, W., Hu, Q., Zhu, P., Meng, D.: Progressive image deraining networks: a better and simpler baseline. In: Proceedings of the IEEE Conference on Computer Vision and Pattern Recognition, pp. 3937–3946 (2019)
28. Roser, M., Kurz, J., Geiger, A.: Realistic modeling of water droplets for monocular adherent raindrop recognition using bézier curves (2010)

29. Valada, A., Vertens, J., Dhall, A., Burgard, W.: Adapnet: adaptive semantic segmentation in adverse environmental conditions. In: 2017 IEEE International Conference on Robotics and Automation (ICRA), pp. 4644–4651. IEEE (2017)
30. Wang, S., Wen, B., Wu, J., Tao, D., Wang, Z.: Segmentation-aware image denoising without knowing true segmentation. arXiv preprint arXiv:1905.08965 (2019)
31. Wu, J., Timofte, R., Huang, Z., Van Gool, L.: On the relation between color image denoising and classification. arXiv preprint arXiv:1704.01372 (2017)
32. You, S., Tan, R.T., Kawakami, R., Mukaigawa, Y., Ikeuchi, K.: Adherent raindrop modeling, detection and removal in video. IEEE Trans. Pattern Anal. Mach. Intell. **38**(9), 1721–1733 (2015)
33. Zhang, H., Patel, V.M.: Density-aware single image de-raining using a multi-stream dense network. In: Proceedings of the IEEE Conference on Computer Vision and Pattern Recognition, pp. 695–704 (2018)

# Hybrid Training of Speaker and Sentence Models for One-Shot Lip Password

Kavin Ruengprateepsang, Somkiat Wangsiripitak(✉), and Kitsuchart Pasupa

Faculty of Information Technology,
King Mongkut's Institute of Technology Ladkrabang, Bangkok, Thailand
{59070009,somkiat,kitsuchart}@it.kmitl.ac.th

**Abstract.** Lip movement can be used as an alternative approach for biometric authentication. We describe a novel method for lip password authentication, using end-to-end 3D convolution and bidirectional long-short term memory. By employing triplet loss to train deep neural networks and learn lip motions, representation of each class is more compact and isolated: less classification error is achieved on one-shot learning of new users with our baseline approach. We further introduce a hybrid model, which combines features from two different models; a lip reading model that learns what phrases uttered by the speaker and a speaker authentication model that learns the identity of the speaker. On a publicly available dataset, AV Digits, we show that our hybrid model achieved an 9.0% equal error rate, improving on 15.5% with the baseline approach.

**Keywords:** Lip password · Speaker verification · One-shot learning

## 1 Introduction

Today, authentication using features, for example, such as face and fingerprints, is a key technology for identifying people. It has been shown to be superior to classic authentication models over the last decades in terms of security and convenience. There are two types of biometrics that can be used for authentication: physical biometrics (e.g., fingerprints, face, iris, or DNA) and behavioral biometrics (e.g., gait, keystrokes, or handwriting). However, these biometric features are prone to varying degrees of attacks [8], where someone impersonates another.

A lip motion password or lip password [15] relies on a password embedded in lip movement. A user can create a password by moving his or her lips in a specific way. Later, the same user can move his or her lips in the same way for authentication. This scheme uses both behavioral characteristics from lip movement and private password information. Due to the private password information, as another layer of security, a lip password is less vulnerable to attacks than many biometric techniques [15]. Moreover, if the password has been leaked, it can be changed similarly to regular text-based passwords—this is not possible for physical biometrics. Lip password works in a noisy environment and requires


H. Yang et al. (Eds.): ICONIP 2020, LNCS 12532, pp. 363–374, 2020.
https://doi.org/10.1007/978-3-030-63830-6_31

no special hardware like fingerprint or iris scanner and can be implemented on typical cameras in smartphones, laptops, and ATMs.

Lip passwords or lip related authentications have not received the same attention as other biometrics, e.g., faces, fingerprints, and voices. The work of Liu and Cheung [15] is closest to our lip password. They extracted various visual lip features and mouth area variations. Then, they used a multi-boosted hidden Markov model to verify speakers and achieved an equal error rate (EER) of 3.37%. Their private dataset contained 20 repetitions of a single "3175" password phrase and another random ten different four-digit phrases (as wrong passwords) all spoken in English. However, they relied on distinguishable subunits of numeric lip passwords; thus, whether their method applies to continuous sentences has yet to be studied. Afterwards, Cheung and Zhou [4] discussed how lip password can be classified in any language and proposed lip password based on sparse representation obtained by Lasso algorithm [27]. They adopted linear support vector machine as a classifier and achieved 8.86% EER on a private dataset consisted of three correct passwords (two in English and one in Chinese) and five incorrect passwords in Chinese.

Other lip-based speaker verification systems have been reported [13,14,25]. Nevertheless, reported performances used only a single phrase or digits for every speaker and did not consider when a speaker saying an incorrect password. Thus, the advantage of being able to change a lip password may not be applicable to those methods. Moreover, they used private datasets, which makes performance comparison more difficult. Recently, Wright and Stewart [31] proposed one-shot learning for lip-based authentication using LipNet [1] on XM2VTS dataset [16], but, again, they only used a digit sequence from the dataset.

Therefore, we described a new approach to lip passwords, in which users can select their own password. Our method uses convolutional neural networks (CNN) and bidirectional long-short term memory (BLSTM), which so far has not been used with lip passwords. Moreover, we applied one-shot learning on our lip password model; it allowed us to use only one or a very few samples to register a new user and did not require as many as ten or more samples.

Lastly, we proposed a hybrid model which combines features from a speaker model and a sentence model, so that it effectively learned the speaker identity and the spoken password at the same time. According to the result in Sect. 4, our hybrid model performed better than the baseline method.

## 2 Related Work

### 2.1 Neural Networks

**Separable 3D Convolutional Neural Networks.** Convolutional neural networks (CNN) have been used on multiple time series problems, e.g., video, sound signal, and human action recognition (HAN) [2,9]. Instead of conventional 2D filters, 3D convolution uses 3D filters and applies them to a video input. However, 3D convolution still suffers from the complexity of the additional temporal dimension. Many works [19,26,28,32] have generated factorized versions of

3D convolution—composed of separated spatial and temporal convolutions. For example, a $k_t \times k_h \times k_w$ convolution can be factored into a $1 \times k_h \times k_w$ convolution followed by a $k_t \times 1 \times 1$ one. The factorization considerably reduced computational complexity. Moreover, performances were generally increased [19,32].

**Bidirectional Long Short-Term Memory.** A variant of a recurrent neural network (RNN), long short-term memory (LSTM), was introduced by Hochreiter and Schmidhuber [7]. A special connection called cell state was added between time steps to help information flowing through them as far as needed. In each connection, there is also a gate which controls whether to let a piece of information go through it or not—this allows LSTM to either remember or forget information in the previous time steps. While there are many variants of classic LSTM, e.g., gated recurrent unit (GRU) [5], we found that LSTM performed acceptably at learning lip motion in our work.

Bidirectional recurrent neural network (BRNN) [24] is a structural improvement over regular RNN. It includes forward time steps to solve the limitation of a typical RNN, that can only look backward in the previous time steps. Furthermore, it can be used in any RNN, in particular with LSTM in our work.

## 2.2 Loss Function

**Categorical Cross-Entropy Loss.** Categorical cross-entropy has been used in many classification tasks using neural networks. It classifies data into categories in a supervised fashion. The loss, $L(y, \hat{y})$, can be calculated as

$$L(y, \hat{y}) = -\frac{1}{n}\sum_{i=1}^{n} y_i \cdot log(\hat{y}_i) \tag{1}$$

where $y \in \{0,1\}^m$ is a one-hot encoded ground truth label, $\hat{y} \in [0,1]^m$ is a predicted value from the softmax function, $n$ is the number of samples and $m$ is a number of class.

**One-Shot Learning and Triplet Loss.** One-shot learning or few-shot learning [30] has often been used in deep learning in recent years. Usually, deep learning requires large amounts of data to achieve good performance. On many occasions, it is more convenient to learn from a small set of data.

Cross-entropy loss is one of the most common loss functions for classification tasks in deep neural networks. Unfortunately, kernel parameters in a feature extractor, that was trained by cross-entropy and then used in one-shot learning normally, had led to inferior results, especially when internal variations in a data class were larger than difference across classes [29]. Thus, triplet loss was introduced in FaceNet [23]. It solved the problem by directly penalize large distances of samples in the same classes and help to increase distances between the different classes, resulting in lower intra-variation and higher inter-difference in embedding space.

In this approach, a triplet consists of an anchor sample, $a$, and positive sample, $p$, from the same class, and negative sample, $n$, from a different class. The loss of a triplet is:

$$L(a, p, n) = max(D_{a,p} - D_{a,n} + \alpha, 0) \tag{2}$$

where $D_{i,j}$ is a Euclidean distance between two samples, $x_i$ and $x_j$, which is mapped to embedding space by $f$:

$$D_{i,j} = ||f(x_i) - f(x_j)||_2 \ . \tag{3}$$

## 3 Method

We divide our work into three phases: learning, registration and authentication phases—see Fig. 1. In the learning phase, we train the feature extractor using preprocessed lip motion videos and ground-truth labels. Later, in the registration and authentication phases, we use the feature extractor to create an embedding vector from each lip motion video. Finally, we use a distance function to compute the embedding distance between the lip motion video previously registered and the lip motion video being examined. Later, a threshold is used to test whether two lip motions are the same.

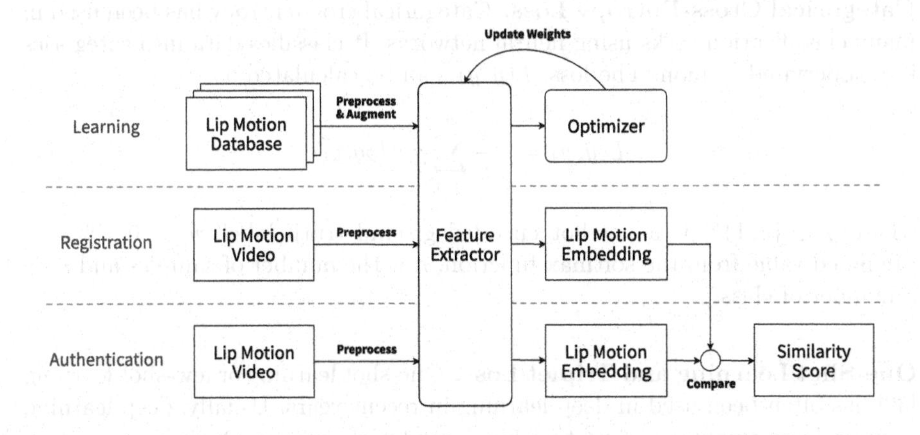


**Fig. 1.** Data flow in our system.

We can also represent the registration and authentication phase as a Siamese neural network [3] and written as follows:

$$D[F(P(V_i)), F(P(V_j))] \tag{4}$$

where $V_i$ and $V_j$ are two lip motion videos, $P$ is a lip motion preprocessor, $F$ is a deep feature extractor, $D$ is a Euclidean distance function. Each of these modules are explained in Sect. 3.1 and Sect. 3.2.

### 3.1 Data Preparation

**Data Preprocessing.** We first segment and crop only the lip region from each video frame using Dlib's facial landmark detector [10,11] with pre-trained model [20–22]. Then, we perform a bicubic interpolated downsampling on each cropped lip image into $40 \times 20$ pixels.

**Data Augmentation.** To improve training performance on a small dataset, we perform data augmentation to create more samples. The following video augmentation is performed sequentially to create eight additional videos from each original video:

(i) Random cropping into $32 \times 16$ pixels.
(ii) Horizontal flipping all frames. (randomly choose only 50% of videos)
(iii) Frame resampling with randomly chosen rate. (0.8 to 1.2)

**Batch Training.** Each video in the dataset has a variable length. To train samples in batch, we pad each video until it has 150 frames. The padding frames are inserted on both the start and the end of the video using duplicated frames from the first and the last frame, respectively. The ratio between head and tail padding is uniformly randomized.

### 3.2 Feature Extractor

**Architecture.** For feature extractor $F$, we slightly modify low-resource deep neural network proposed by Fung and Mak [6] to extract lip motion feature from lip frames. The architecture consists of 2 parts (see Fig. 2).

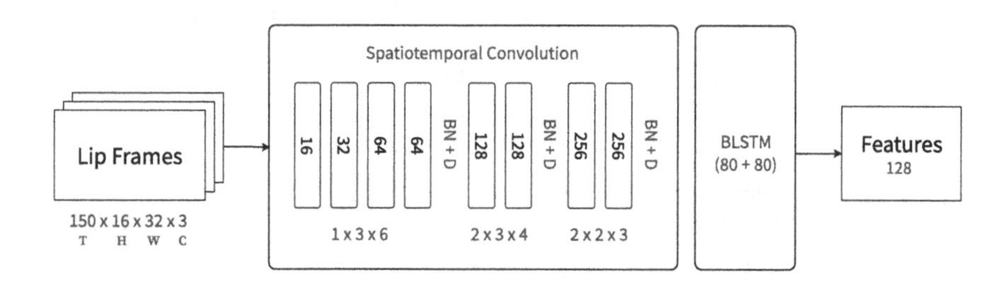


**Fig. 2.** Network architecture.

The first part contains 8 convolutional layers. We factorize 3D convolutional filter from $k_t \times k_h \times k_w$ into $1 \times k_h \times k_w$ and $k_t \times 1 \times 1$. Each convolution is followed by ReLU activation function. The dropout rate is 50%.

The second part is the BLSTM layer, which is LSTM in the BRNN structure. The outputs from bidirectional streams are merged using vector concatenation. Finally, the output is forwarded to a dense layer with 128 units output.

**Training.** In order to create a feature extractor capable of one-shot learning, we need to train the model to gain knowledge from a decent amount of data in the lip motion database. Then use it for other sets of unseen data both in the registration phase and the authentication phase.

First, we train the network using cross-entropy loss for faster convergence. Then, we continue training the network with triplet loss. For triplet mining, triplet loss tends to learn better with hard or semi-hard triplets [23]. In this paper, we select semi-hard triplets where a negative sample is farther from an anchor than a positive sample, but still within margin $\alpha$:

$$D_{a,p} < D_{a,n} < D_{a,p} + \alpha \; . \tag{5}$$

To train our feature extractor for lip password, we use data with the label of spoken phrase $y_{p_i}$ and speaker identity $y_{s_i}$. Let $y_p$ be the set of phrase labels and $y_s$ be the set of speaker identity labels, the combined labels

$$(y_{p_i}, y_{s_i}) \in y_p \times y_s \mid i \in \{1, 2, \ldots, n\} \tag{6}$$

are used for optimization with cross-entropy and triplet loss, so that the optimizer will consider this information during the optimization of embedding distances.

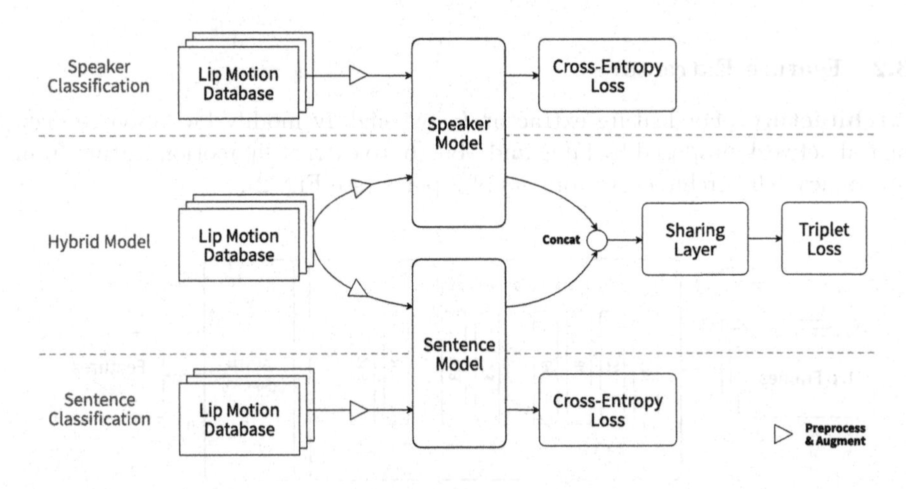


**Fig. 3.** Hybrid model.

### 3.3 Hybrid Model

In the single network model, the combination of two labels created by the Cartesian product $y_p \times y_s$ in (6) produces a very large set of classes, which can hinder triplet loss convergence—results in inferior performance. In order to reduce the complexity of vast class space of $y_p \times y_s$, we separate it into two subspaces and propose a hybrid model, which contains two dedicated deep neural networks for each task—this makes the convergence focus mainly on a spoken phrase and identity of a speaker, respectively. Each network is pre-trained using different label $y_p$ and $y_s$ (see Fig. 3). The sentence model is trained using $y_p$, and the speaker model is trained using $y_s$.

After we pre-train both networks using cross-entropy loss, we remove the last softmax layer of both networks. Then, the output vector of each pre-trained model is concatenated together and passed to the next sharing layer (Fig. 4). The sharing layer contains 3 layers of fully-connected network (FCN) and ReLU activation functions. Finally, we use labels from (6) to train the framework using triplet loss.

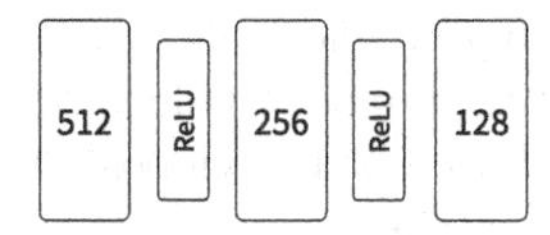


**Fig. 4.** Sharing layer.

# 4 Experiment and Results

## 4.1 Dataset

For our experiments, we selected, from the publicly available audio-visual AV Digits database [18], ten phrases: "Excuse me", "Goodbye", "Hello", "How are you", "Nice to meet you", "See you", "I'm sorry", "Thank you", "Have a good time" and "You are welcome". Each phrase was spoken by 39 participants and repeated five times in 3 different speech modes: silent, normal, and whisper. There are 5,850 utterances in total.

We ran a $k$-fold cross-validation (Fig. 5) with $k = 5$ speakers, and used one part on each fold to evaluate the model as unseen speakers.

## 4.2 Experiment Settings

For model training, we used Adam [12] as an optimizer with the default $\beta_1 = 0.9$ and $\beta_2 = 0.999$. We used learning rate $= 2 \times 10^{-4}$ for cross-entropy loss in the single model, and when the sentence and speaker models were trained separately. Later, we fine-tuned the networks using triplet loss with learning rate $2 \times 10^{-5}$ and margin $= 1$. We trained our model in an Nvidia T4 GPU with 16 GB of memory using mixed-precision [17] floating-point, which consists of half-precision floating-point (FP16) and full-precision floating-point (FP32), to make the training faster with less memory usage on the Nvidia GPU.

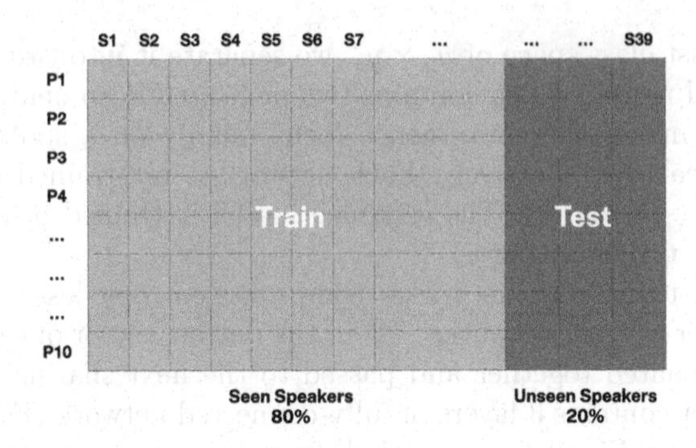


**Fig. 5.** $k$-fold cross-validation with $k = 5$. $S_i$ is the $i$th speaker and $P_j$ is the $j$th phrase.

**Training.** In each training batch selection, we randomly sampled 8 speakers and 4 phrases, then randomly sampled 4 videos corresponding to each speaker and phrase, so that we had a batch of 128 videos. Our random did not allow the same sample in two different batches.

**Evaluation.** We used every video in the test data to create embedding vectors of size 128. Then, we computed a distance matrix for every possible pair of samples using Euclidean distance. Each pair denoted samples from registration and authentication phases. We evaluated our model performance in terms of equal error rate (EER).

To simulate real-world scenarios, we used an evaluation scheme similar to that used by Liu and Cheung [15]. Our sample pairs were evaluated in 4 cases, as follows.

(i) Case A — The same speaker saying a wrong password.
(ii) Case B — Different speaker saying a correct password.
(iii) Case C — Different speaker saying a wrong password.
(iv) Case D — All cases.

### 4.3 Performance Comparison of Single and Hybrid Model

From Table 1, we saw that the end-to-end hybrid model had lower EER than the baseline in single model in all cases.

**Table 1.** Performance comparison of proposed models.

| Training methods | Equal error rate (%) | | | |
|---|---|---|---|---|
| | Case A | Case B | Case C | Case D |
| Single model | 22.4 | 20.3 | 8.4 | 15.5 |
| Hybrid | **16.8** | **17.0** | **5.6** | **9.0** |

In most of our error rate curves, e.g., Fig. 6, we noticed that the curve between false acceptance rate (FAR) and false rejection rate (FRR) was not balanced. The FRR was less sensitive to threshold variation than FAR. Thus, in many real-world use cases, we can sacrifice FRR for FAR in a sensitive application, when high security is required.

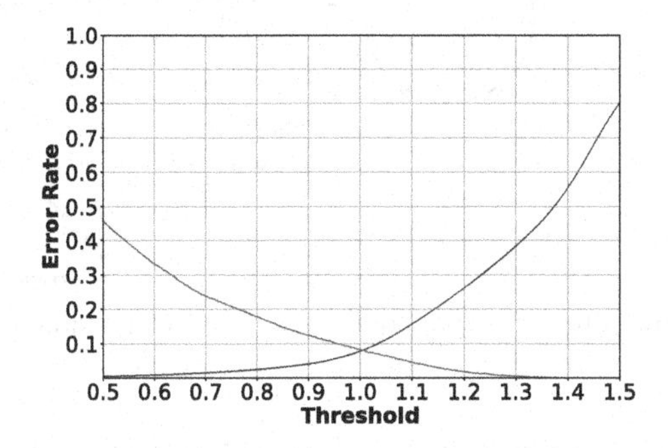


**Fig. 6.** False acceptance rate (blue) and false rejection rate (green) of hybrid model (in case D). (Color figure online)

## 4.4 Effect of Separable 3D Convolution

Using a separable 3D convolution for the dedicated sentence classification model, we reduced 20% of the parameters in the network and gained 34% faster convergence. We also achieved 3.4% more accuracy, from 83.1% to 86.5%. However, this effect was insubstantial in speaker classification model.

## 4.5 Speaker Model and Sentence Model in Hybrid Method

To develop a better approach in the future, we also investigated the performance of sentence classification and speaker classification in the hybrid model. Speaker network performed at 95% in classification accuracy. Meanwhile, sentence network performed at 86.5% in classification accuracy. We applied principal component analysis to the output of the layer before the last layer of our sentence classification model, then projected the feature vectors onto the 2D plane (Fig. 7a). We found that some sentences had a similar lip movement, producing a similar feature vector, e.g., "Thank you", "See you", and "Hello". The confusion matrix (Fig. 7b) created from the classification result also reflects this same problem. While other sentences had a wrong prediction at a single-digit percentage, the sentences "Thank you", "See you", and "Hello" were confused to each other more than 10%; for example, 20% of "Thank you" and 14% of "Hello" phrases were predicted as "See you".

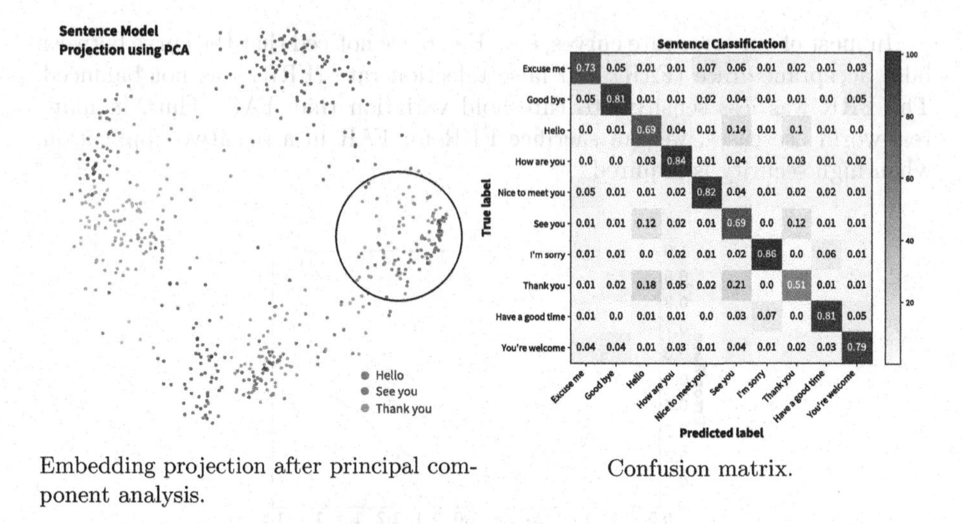


Embedding projection after principal component analysis.

Confusion matrix.

**Fig. 7.** Performance analysis of sentence classification.

# 5 Summary

We proposed a lip password framework with one-shot learning for new client registration. Our model achieved good performance, even with only one training sample for a new user. Since a user can register by speaking a chosen password only once, we see better usability. We also showed that our hybrid model, which uses dedicated networks for speaker classification and sentence classification tasks, significantly improved user authentication accuracy over a single model in all scenarios. When selecting distance thresholds, we noted that some sensitive applications, e.g., banking applications, can sacrifice FRR to reduce FAR, leading to lower impostor access rate. Our model also favored reducing the threshold without increasing FRR significantly.

We also investigated various points to create solutions to improve our method furthur, e.g., how the sentence and speaker model perform individually. As we found that the dedicated sentence classification had more room to be improved, we planned to use separate lip reading dataset to increase its performance. In this work, since AV Digits has limited unique phrases, we did not perform an experiment where a spoken password was unlisted in the training data; we shall leave this for future work.

## References


1. Assael, Y.M., Shillingford, B., Whiteson, S., de Freitas, N.: LipNet: end-to-end sentence-level lipreading. arXiv:1611.01599 [cs] (2016)
2. Baccouche, M., Mamalet, F., Wolf, C., Garcia, C., Baskurt, A.: Sequential deep learning for human action recognition. In: Salah, A.A., Lepri, B. (eds.) HBU 2011. LNCS, vol. 7065, pp. 29–39. Springer, Heidelberg (2011). https://doi.org/10.1007/978-3-642-25446-8_4
3. Bromley, J., et al.: Signature verification using a "siamese" time delay neural network. Int. J. Pattern Recogn. Artif. Intell. 669–688 (1993). https://doi.org/10.1142/S0218001493000339
4. Cheung, Y., Zhou, Y.: Lip password-based speaker verification without a priori knowledge of speech language. In: Li, K., Li, W., Chen, Z., Liu, Y. (eds.) ISICA 2017. CCIS, vol. 874, pp. 461–472. Springer, Singapore (2018). https://doi.org/10.1007/978-981-13-1651-7_41
5. Chung, J., Gulcehre, C., Cho, K., Bengio, Y.: Empirical Evaluation of Gated Recurrent Neural Networks on Sequence Modeling. arXiv:1412.3555 [cs] (2014)
6. Fung, I., Mak, B.: End-to-end low-resource lip-reading with Maxout CNN and LSTM. In: 2018 IEEE International Conference on Acoustics, Speech and Signal Processing (ICASSP), Calgary, AB, pp. 2511–2515. IEEE (2018). https://doi.org/10.1109/ICASSP.2018.8462280
7. Hochreiter, S., Schmidhuber, J.: Long short-term memory. Neural Comput. 1735–1780 (1997). https://doi.org/10.1162/neco.1997.9.8.1735
8. Jain, R., Kant, C.: Attacks on biometric systems: an overview. Int. J. Adv. Sci. Res. 283 (2015). https://doi.org/10.7439/ijasr.v1i7.1975
9. Ji, S., Xu, W., Yang, M., Yu, K.: 3D convolutional neural networks for human action recognition. IEEE Trans. Pattern Anal. Mach. Intell. 221–231 (2013). https://doi.org/10.1109/TPAMI.2012.59
10. Kazemi, V., Sullivan, J.: One millisecond face alignment with an ensemble of regression trees. In: 2014 IEEE Conference on Computer Vision and Pattern Recognition, Columbus, OH, pp. 1867–1874. IEEE (2014). https://doi.org/10.1109/CVPR.2014.241
11. King, D.E.: Dlib-ml: a machine learning toolkit. J. Mach. Learn. Res. 1755–1758 (2009). https://doi.org/10.1145/1577069.1755843
12. Kingma, D.P., Ba, J.: Adam: A Method for Stochastic Optimization. arXiv:1412.6980 [cs] (2017)
13. Lai, J.Y., Wang, S.L., Liew, A.W.C., Shi, X.J.: Visual speaker identification and authentication by joint spatiotemporal sparse coding and hierarchical pooling. Inf. Sci. 219–232 (2016). https://doi.org/10.1016/j.ins.2016.09.015
14. Liao, J., Wang, S., Zhang, X., Liu, G.: 3D convolutional neural networks based speaker identification and authentication. In: 2018 25th IEEE International Conference on Image Processing (ICIP), Athens, pp. 2042–2046. IEEE (2018). https://doi.org/10.1109/ICIP.2018.8451204
15. Liu, X., Cheung, Y.M.: Learning multi-boosted HMMs for lip-password based speaker verification. IEEE Trans. Inf. Forensics Secur. 233–246 (2014). https://doi.org/10.1109/TIFS.2013.2293025
16. Maître, G., Valais-Wallis, H.S., Luettin, J., GmbH, R.B.: XM2VTSDB: the extended M2VTS database, p. 7 (2000)
17. Micikevicius, P., et al.: Mixed Precision Training. arXiv:1710.03740 [cs, stat] (2018)

18. Petridis, S., Shen, J., Cetin, D., Pantic, M.: Visual-Only Recognition of Normal, Whispered and Silent Speech. arXiv:1802.06399 [cs] (2018)
19. Qiu, Z., Yao, T., Mei, T.: Learning Spatio-Temporal Representation with Pseudo-3D Residual Networks. arXiv:1711.10305 [cs] (2017)
20. Sagonas, C., Antonakos, E., Tzimiropoulos, G., Zafeiriou, S., Pantic, M.: 300 faces in-the-wild challenge: database and results. Image Vision Comput. 3–18 (2016). https://doi.org/10.1016/j.imavis.2016.01.002
21. Sagonas, C., Tzimiropoulos, G., Zafeiriou, S., Pantic, M.: 300 faces in-the-wild challenge: the first facial landmark localization challenge. In: 2013 IEEE International Conference on Computer Vision Workshops, Sydney, Australia, pp. 397–403. IEEE (2013). https://doi.org/10.1109/ICCVW.2013.59
22. Sagonas, C., Tzimiropoulos, G., Zafeiriou, S., Pantic, M.: A semi-automatic methodology for facial landmark annotation. In: 2013 IEEE Conference on Computer Vision and Pattern Recognition Workshops, OR, USA, pp. 896–903. IEEE (2013). https://doi.org/10.1109/CVPRW.2013.132
23. Schroff, F., Kalenichenko, D., Philbin, J.: FaceNet: a unified embedding for face recognition and clustering. In: 2015 IEEE Conference on Computer Vision and Pattern Recognition (CVPR), pp. 815–823 (2015). https://doi.org/10.1109/CVPR.2015.7298682
24. Schuster, M., Paliwal, K.: Bidirectional recurrent neural networks. IEEE Trans. Signal Process. 2673–2681 (1997). https://doi.org/10.1109/78.650093
25. Shi, X.X., Wang, S.L., Lai, J.Y.: Visual speaker authentication by ensemble learning over static and dynamic lip details. In: 2016 IEEE International Conference on Image Processing (ICIP), Phoenix, AZ, USA, pp. 3942–3946. IEEE (2016). https://doi.org/10.1109/ICIP.2016.7533099
26. Sun, L., Jia, K., Yeung, D.Y., Shi, B.E.: Human Action Recognition using Factorized Spatio-Temporal Convolutional Networks. arXiv:1510.00562 [cs] (2015)
27. Tibshirani, R., Johnstone, I., Hastie, T., Efron, B.: Least angle regression. Ann. Stat. 407–499 (2004). https://doi.org/10.1214/009053604000000067
28. Tran, D., Wang, H., Torresani, L., Ray, J., LeCun, Y., Paluri, M.: A Closer Look at Spatiotemporal Convolutions for Action Recognition. arXiv:1711.11248 [cs] (2017)
29. Wang, M., Deng, W.: Deep Face Recognition: A Survey. arXiv:1804.06655 [cs] (2018)
30. Wang, Y., Yao, Q., Kwok, J., Ni, L.M.: Generalizing from a Few Examples: A Survey on Few-Shot Learning. arXiv:1904.05046 [cs] (2019)
31. Wright, C., Stewart, D.: One-shot-learning for visual lip-based biometric authentication. In: Bebis, G., et al. (eds.) ISVC 2019. LNCS, vol. 11844, pp. 405–417. Springer, Cham (2019). https://doi.org/10.1007/978-3-030-33720-9_31
32. Xie, S., Sun, C., Huang, J., Tu, Z., Murphy, K.: Rethinking Spatiotemporal Feature Learning: Speed-Accuracy Trade-offs in Video Classification. arXiv:1712.04851 [cs] (2018)

# Identifying Real and Posed Smiles from Observers' Galvanic Skin Response and Blood Volume Pulse

Renshang Gao, Atiqul Islam, Tom Gedeon, and Md Zakir Hossain(✉)

The Australian National University (ANU), Canberra, Australia
{Renshang.Gao,Atiqul.Islam,Tom.Gedeon,Zakir.Hossain}@anu.edu.au

**Abstract.** This study addresses the question whether galvanic skin response (GSR) and blood volume pulse (BVP) of untrained and unaided observers can be used to identify real and posed smiles from different sets of smile videos or smile images. Observers were shown smile face videos/images, either singly or paired, with the intention to recognise each viewed as real or posed smiles. We created four experimental situations, namely single images (SI), single videos (SV), paired images (PI), and paired videos (PV). The GSR and BVP signals were recorded and processed. Our machine learning classifiers reached the highest accuracy of 93.3%, 87.6%, 92.0%, 91.7% for PV, PI, SV, and SI respectively. Finally, PV and SI were found to be the easiest and hardest way to identify real and posed smiles respectively. Overall, we demonstrated that observers' subconscious physiological signals (GSR and BVP) are able to identify real and posed smiles at a good accuracy.

**Keywords:** Physiological signals · Affective computing · Classification · Machine learning · Smile

## 1 Introduction

Philosophers have often referred to the smile as being the reflection of the soul (state of mind), as smiles are one of our most common and easily distinguishable expressions. But the emotions stored in a smile are not just the obvious one. That is, these smiles can be genuine or deceptive. Surprisingly we still know very little about the influence of smiles on others when the smile is deceptive, and people's acceptance or rejection of deceptive smiles as genuine. In different social situations, people may use smiles to cover up their real feelings, disappointment or tension or embarrassment and so on. Some posed smiles are not easily recognized through our eyes. So if we can more accurately identify the meaning of these smiles, it will be of great help in many social situations we (or our AIs or robots) need to face.

Analysis and classification of facial emotions are important research questions in behavioural science. With the advent of computer vision, the literature on facial expressions or emotions is broad, we refer our reader to these literature


H. Yang et al. (Eds.): ICONIP 2020, LNCS 12532, pp. 375–386, 2020.
https://doi.org/10.1007/978-3-030-63830-6_32

surveys presented in [13,26,34]. Authors in [32] attempted to develop more reliable smile detection in realistic situations with the advent of machine learning. This type of smile detector is usually embedded in commercial digital cameras. They only can detect the expression of a smile without identifying whether the smile is genuine or posed. However, classifying real and posed smile through the application of machine learning is comparatively a recent research question. Different researchers have tried different approaches. Eyelid movement has been used as a classifying factor for differentiating between spontaneous and posed smiles [7]. Authors have reached up to 92% accuracy for posed smile detection with neural networks. Authors in [2] utilized machine learning on subjects' physiological signals while watching emotionally evocative films to distinguish between different emotions that arose within the subject. Real and posed smiles have been classified from observer's physiological signals in [16]. The authors have used observers' galvanic skin response and blood volume pulse along with pupillary responses. From twenty videos of a benchmark database with twenty four subjects using neural network classification, the authors have reached around 93% accuracy.

This study focuses on the research and analysis of subjects' physiological signals. Through processing of the physiological signals generated by observing posed and real smiles, feature extraction, and training classification, create a high-precision classifier to improve the accuracy of predicting real and posed smiles, thereby helping people to more accurately judge human smiles. People are surprisingly bad at differentiating posed and real smiles. In this paper, we have analyzed the accuracy of the proposed classifier, and investigated which training method has higher efficiency in differentiating between real and posed smiles. Also, we analyse feature distributions to find the reasons behind low performances. First, by comparing the accuracy of different individual features, low-precision features are excluded, and high-precision multiple features are used to train a new classifier jointly. The results show that if only low-precision features are excluded, this does not improve the accuracy of the final classifier very well. In the second part of the study, to investigate the distribution of features, and to experiment with independent parameters in the feature list, the experiment found that the final accuracy has been greatly improved. The highest accuracy classifier performs data prediction, and finally shows that the pair data has better predictive power, with an accuracy of 93.3%, while the single data achieved up to 92%.

## 2 Background

Emotions are one of the essential factors in our lives. Positive emotions can help people maintain a healthy and high quality of life. Negative emotions often bring negative life conditions and even affect physical health. For example, long-term depression, claustrophobia, or irritability may transform into mental illness. If it is not controlled in time, serious life-threatening situations such as injury or suicide may occur! Therefore, emotion recognition has been widely researched in many Fields, such as safe driving, psychological disease monitoring, social safety and many more. The extensive state-of-the-art literature surveys in the last few decades is an indication of the wide range of research that is happening in the field of emotion recognition from speech, audiovisual, facial, smile etc. [11,22, 27,33].

However, it is difficult to judge human emotion changes based only on human body movements or facial expressions, because people often cover their true emotions in different ways, such as smiling [10]. Therefore, the discovery of physiological signals is an important help for emotion recognition. Because when the human body has different emotional changes, physiological signals will have different changes, and generally these changes are spontaneous and not controllable. Therefore, it can reflect the true human emotions without being covered up, helping us to identify human emotion changes more easily. Being one of the emotional cues, a considerable amount of research has been done on the smile. That research can be grouped into a few categories such as smiles in consumer research [31], smile detection [32] and physiological responses during a smile [9]. A significant amount of study suggests that smiles are influential social forces that positively impact interpersonal judgments in many ways. For example, researchers found that people who express genuine smiles are perceived to be more polite, carefree and pleasant [5,6,23]. The substantial amount of evidence supporting the social benefits of smiles implies that a smile always convey some form of information. Indeed, researchers believed that people sometimes intentionally strengthen emotional displays to get favourable interpersonal feedback [25]. For instance, service workers often exaggerate their positive emotional displays in order to increase consumer experiences [3]. Smiles can therefore be considered as an important facial capacity in successful social communication. Thus it is important to recognize real and posed smiles not only in human to human communication but also in human machine communication. In this paper, real and posed smiles have been classified from observers' physiological signals using machine learning algorithms.

### 2.1 Physiological Signals

Users' emotions including smile can be recognized in a myriad of ways like facial expression, body pose, audiovisual, gesture recognition and many more but physiological signals are the most useful in the field of human-computer interaction because they cannot be controlled intentionally. There are many physiological signals that can be collected from the human body to get information about internal behaviour and mental states. The main physiological signals include cardiac function, temperature, muscle electrical activity, respiration, electrical activity of the skin and brain, blood volume pulse and so on. This paper will mainly focus on the analysis of two physiological signals: skin conductance and blood volume pulse.

GSR (galvanic skin resistance), a general term for the galvanic activity or EDA (electrodermal activity) refers to changes in sweat gland activity, reflecting the intensity of our emotional state, also known as emotional arousal [4]. Our level of emotional arousal changes depending on the environment if something is frightening, threatening, happy or otherwise related to emotions, then our subsequent emotional response changes also increase endocrine sweat gland activity. Studies have shown that this is related to emotional arousal. It is worth noting that both positive ("joyful" or "happy") and negative ("threat" or "frightening") stimuli can cause an increase in arousal and lead to an increase in skin response. Therefore, the GSR signal does not represent the type of emotion, but rather its intensity. An extensive review on literature shows that GSR has been very popular in the last two decades for working with emotion or smiles [8,20,28].

Photoplethysmography (PPG) allows infrared light to pass through the tissue and measure the absorption of light by blood flowing through the blood vessels below the skin. It can be used to detect the blood flow rate of the heartbeat, also known as blood volume pulse (BVP). BVP is not only a convenient measurement of heart rate variability but also a valuable information source for emotion recognition. The BVP signal indicates the changes in blood volume and blood flow through the human body [14]. Many researchers include heart rate or BVP in their experiment or database of psycho-physiological classification or recognition [21].

## 3 Experimental Methodology

The aim of our study is to analyze the physiological signal data produced by feeling as evoked by watching real and posed smiles. After finding a suitable classification method, we have identified real and posed smiles from observers' physiological signals to demonstrate observers' emotional communication.

### 3.1 Smile Videos and Images Stimuli

The smiles stimuli were collected from the UvA-NEMO dataset [15]. A total of 60 videos were randomly selected for this experiment. We created four categories out of these 60 videos, namely PV (paired videos), PI (paired images), SV (single video), and SI (single image). When the same person was viewed by observers in both real and posed smile videos, we report this as "paired videos", otherwise we use the term "single video". A similar rule was applied for the images we used. Here, *image* means a single frame and *video* means a sequence of frames. The selected videos/images were processed using oval masks [17] to keep the face portion only, and presented to the observer in an order balanced way to avoid any order effects. We kept the length of each video the same as the source video in the dataset. The length of the videos usually spans from a minimum of 2 seconds to a maximum of 8 seconds. In case of images, we chose one of the middle frames from each video and showed the frame for five seconds. Overall, we considered 20 videos for PV, 10 videos for SV, 20 images for PI, and 10 images for SI respectively.

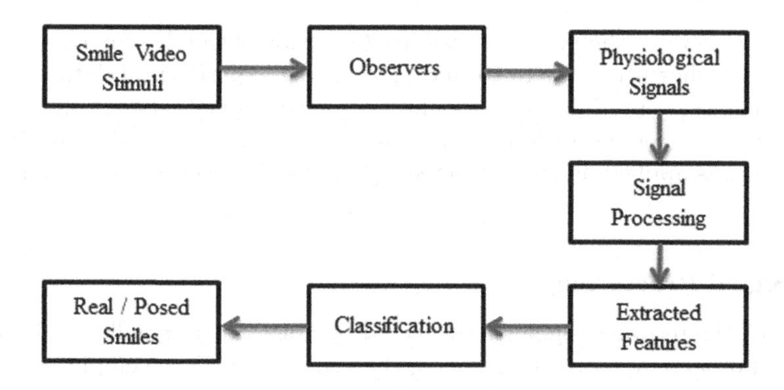


**Fig. 1.** Experimental procedure.

### 3.2 Participants/Observers

We invited 25 participants (also called *observers* as they watch or observe visual stimuli) who were the university students and volunteered to take part in the experiment. Amongst the participants, we had 13 males and 12 females with age range $24.6 \pm 2.54$ (average $\pm$ standard deviation). All the participants had normal or corrected to normal vision and provided written consent prior to their participation. Approval from our University's Human Research Ethics Committee was received before the experiment.

### 3.3 Physiological Data Acquisition

In order to detect real smiles from mixed presentations of real and posed smiles, we recorded galvanic skin response (GSR) and blood volume pulse (BVP) using an Empatica E4 [12] on the wrist of each observers' non-dominant arm. The GSR and BVP data were recorded at a sampling rate 4 Hz and 64 Hz respectively.

### 3.4 Experimental Procedure

Observers were briefed about the experiment after arrival at the laboratory, and asked to sign the consent form. They were seated on a static chair, facing a 15.6 in. ASUS laptop in a sound-attenuated, dimly lit, closed room. Sensors were attached to measure their GSR and BVP signals. Their chairs were moved forward or backwards to adjust the distance between the chair and laptop. Observers were asked to limit their body movements in order to reduce undesired artefacts in the signals. The smile videos or images were presented to the observers in a randomised fashion considering each experiment separately (i.e SI, SV, PI, and PV). At the end of each (or pair) video or image, the observer was asked to choose whether the smile was real or posed by answering the question on the laptop into a web interface. The total duration of the experiment was around 45 min. After completing the experiment, the sensors were removed, and the observers were thanked for their participation. Overall, our experimental procedure is shown in Fig. 1. We have stored the experimental data for future post processing. Filtering and normalization were applied on each signal. After that we have extracted several features from each signal. Then the classification algorithm was applied on the extracted features and smiles are identified as real or posed.

### 3.5 Signal Processing

Generally, the recorded signals are affected by noise due to small signal fluctuations and observers' movements during the experiment. So we employed median filters to reduce the effect of noise: a one-dimensional median filter, with parameter use of $20^{th}$-order. Finally to overcome the individual human differences on the data, we normalise the data into the range of [0 1]. The normalization technique used here is Linear Scaling, which is obtained by calculating the maximum and minimum values as shown in Eq. 1. Here $x_i$ is the physiological signal vector.

$$Normalization = \frac{x_i - min(x_i)}{max(x_i) - min(x_i)} \tag{1}$$

In this experiment, we used Matlab for signal processing. The signal noise has been reduced by removing all the outliers after filtering. There are four sets of physiological signals (SI, SV, PI and PV). So after normalizing the data, we can compare those different categories of data as they are normalised to the same levels by observer.

### 3.6 Feature Extraction

After the normalization and filtering, a number of features have been calculated from the filtered and normalized signals. We extracted several features from each video/image for each observer's physiological signals. The features have been selected based on the state-of-the-art literature in human emotion, physiological signal and affective computing [1,19,24]. Initially 20 features have been extracted from each physiological signal. The features set includes average, root mean square (RMS), variance, standard deviation, Hjorth mobility, sum, simple square integral (SSI), approximate Entropy (ApEn), mean absolute deviation (MAD), Kurtosis, sample Entropy, Shannon Entropy, peak to peak, root-sum-of-squares (Rssq), mean frequency, occupied bandwidth, equivalent noise bandwidth, band power, Fuzzy Entropy and detrended fluctuation analysis (DFA). DFA determines the self similarity level of physiological signals. We extracted a sample for each of the experimental condition such as SI, SV, PI, PV from each participant. A total of 20 features have been extracted from each sample of a participant for each physiological signal. So, from each participant 160 features (4 samples (SI, SV, PI and PV) $\times$ 20 Features $\times$ 2 Physiological signals (GSR, BVP)) have been extracted. For each condition the real and posed smiles are balanced; for example, condition SI got 5 real smile images and 5 posed smile images.

## 4 Experimental Results and Discussion

The data is analysed according to four groups, namely SI, SV, PI, and PV as mentioned earlier. The four sets of data were classified separately using normalized features. After initial training the prediction accuracy of all the data were quite low as shown in Table 1. In the table, we have shown the classification method that achieve the maximum accuracy amongst the methods we used. As an example, we used all four methods (Fine KNN, Kernel Naive Bayes, Coarse Tree Logistic Regression) for PV, but we achieved maximum accuracy with Coarse Tree. We can see that the highest accuracy is 62.5% for PV, which is quite low. Therefore, we focus on two factors: the type of features selected by the classifier, and the number of features for training the classifier.

**Table 1.** Initial accuracy of classification using all the normalized features

| Type of data | Accuracy | Method of classification |
|---|---|---|
| SI | 53.6% | Fine KNN |
| SV | 58.6% | Kernel Naive Bayes |
| PV | 62.5% | Coarse tree |
| PI | 57.0% | Logistic regression |

### 4.1 Analysis of the Feature Performance

As a first approach, we re-analyzed each of the features to find the features' performance. It is impossible to judge the accuracy by comparing the classification methods of different categories, rather we need to compare the accuracy of the same classification method/condition. There are four main types of classification methods with good results: Coarse Tree, Logistic regression, Fine KNN, and Kernel Naive Bayes. Therefore, we computed the accuracy for each situations (i.e. SI, PI, SV, and PV) separately considering each classification methods where we achieved the highest accuracies, i.e Fine KNN for SI, Logistic Regression for PI, and so on.

First, we trained all of the features and recorded the accuracy. After excluding some features with low accuracy and using the features with the highest accuracy, we found that removing some features can improve accuracy. We have selected 10 high accuracy features based on their individual performance. The selected features set includes average, root mean square (RMS), variance, standard deviation, Hjorth mobility, approximate Entropy (ApEn), mean absolute deviation (MAD), Kurtosis and mean frequency. Some of these results indicate that the accuracy of a single feature can be slightly higher than a classifier that used multiple features for training. The increased range is about 2.3% to 6% as shown in Table 2. So we can conclude that removing some low accuracy features is not the main reason for improving accuracy.

**Table 2.** After feature selection accuracy has increased

| Type of data | Accuracy with all features | Accuracy after feature selection | Increase range |
|---|---|---|---|
| SI | 53.6% | 59.5% | +5.9% |
| SV | 58.6% | 60.9% | +2.3% |
| PV | 62.5% | 65.5% | +3.0% |
| PI | 57.0% | 61.1% | +4.1% |

### 4.2 Feature Distributions

To examine the reason of a feature with respect to another feature, we checked the distribution of features with trained and untrained data as shown in Fig. 2 as an example considering PV. By observation, we can see that the posed smile data shows a lot of loss during training as displayed in Fig. 2(b), and the real smile data is alright during training as shown in Fig. 2(d). In other words, the classifier erroneously assigns posed smile data to real smiles data during the classification process. Therefore, in the entire training set, the data assigned to real smiles set also includes the data of some posed smiles. It is because posed smiles take longer time to generate as well as recognise compared to the real smiles [7,16].

Due to the uneven features distribution, the classification performances are also different considering the real and posed smiles separately. For instance, our classifier achieves 90% and 41% correctness for real and posed smiles separately for PV (i.e 65.5% on average as shown in Table 2). It is because the classifier chose the 30% test data randomly and unevenly, which did not match with training data. For this reason, a few posed smile features were considered as real smile features. To overcome this lagging, we applied leave-one-participant-out method (the preferred method when using human sensor data [29,30]) to improve the accuracies. The leave-one-out approach was applied to validate the performance, which used each subject's data as testing set while others as training set. Therefore the validation ran a total of 25 loops, and the result was the average of each loop. The leave-one-out validation process was repeated for 10 times and we average the results to get the final result. We used k-nearest neighbour (KNN), Naive Bayes, Decision Tree and Logistic Regression classifiers. The performance parameters were: 1 nearest neighbours, kernel function, Coarse function and logistic function respectively.

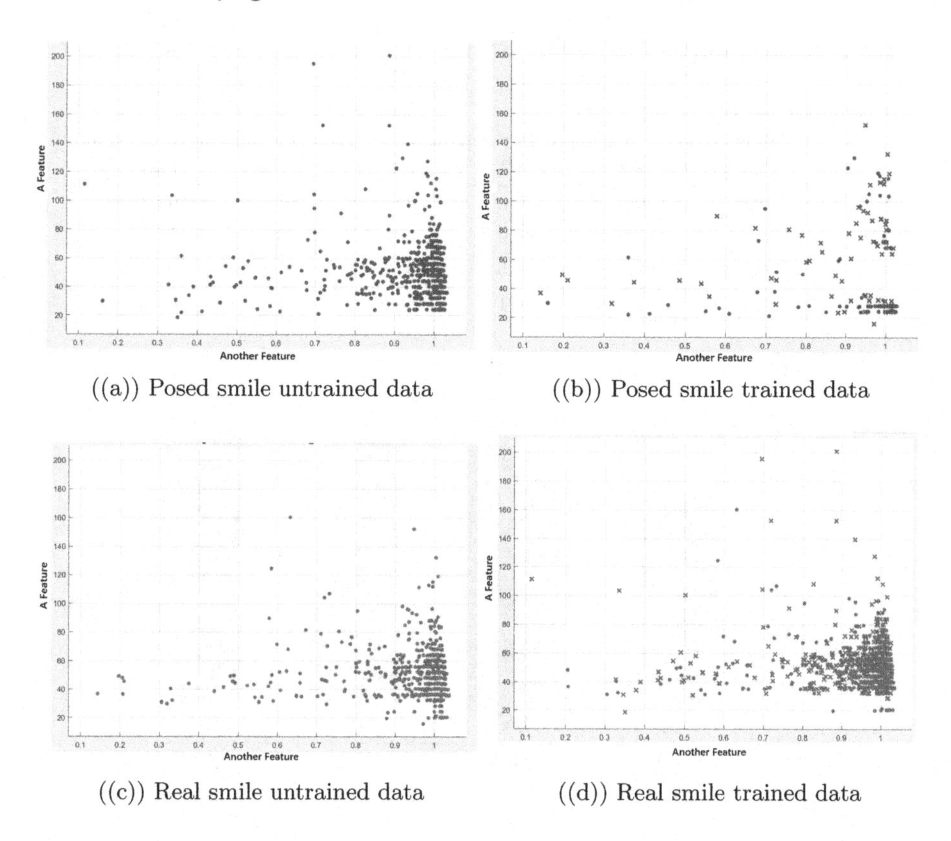


((a)) Posed smile untrained data

((b)) Posed smile trained data

((c)) Real smile untrained data

((d)) Real smile trained data

**Fig. 2.** Features distribution before and after classification for real and posed smiles

### 4.3 Predictive Data Testing

According to the above analysis of features, the number of features is not the main reason for improving accuracy, and now the classifier with the highest accuracy is selected for predictive data testing. The test method removes one observer's data for testing and remaining 24 observers' data to train the classifier, thus full-filling the requirements for the leave-one-participant-out method. The Table 3 records the accuracy of the predictions for GSR and BVP signals. It is important to note that, we illustrated the results from first three test participants (P01, P02, and P03) as an example as shown in Table 3, because the classifiers achieve similar results for all test participants. Average accuracy of all 25 participants is shown in Fig. 3. It can be seen from Fig. 3 that highest accuracies are recorded from BVP data considering all situations except PI. This is likely to be because the PI data were noisy and corrupted for most of the observers for this case. For all other cases, it can be seen that highest accuracies were found from PV followed by PI, SV, and SI respectively. BVP was found to be more reliable compared to GSR in this case.

**Table 3.** Classification Accuracies (in %) of Three Participants (as an example)

| Conditions | GSR | | | BVP | | |
|---|---|---|---|---|---|---|
| | Participant P01 | Participant P02 | Participant P03 | Participant P01 | Participant P02 | Participant P03 |
| SI | 78.6 | 79.0 | 79.0 | 92.3 | 91.1 | 92.3 |
| SV | 81.1 | 82.9 | 80.9 | 91.3 | 89.8 | 94.5 |
| PV | 90.0 | 88.1 | 89.5 | 94.8 | 91.1 | 95.2 |
| PI | 88.5 | 87.7 | 87.1 | 58.8 | 65.4 | 65.5 |

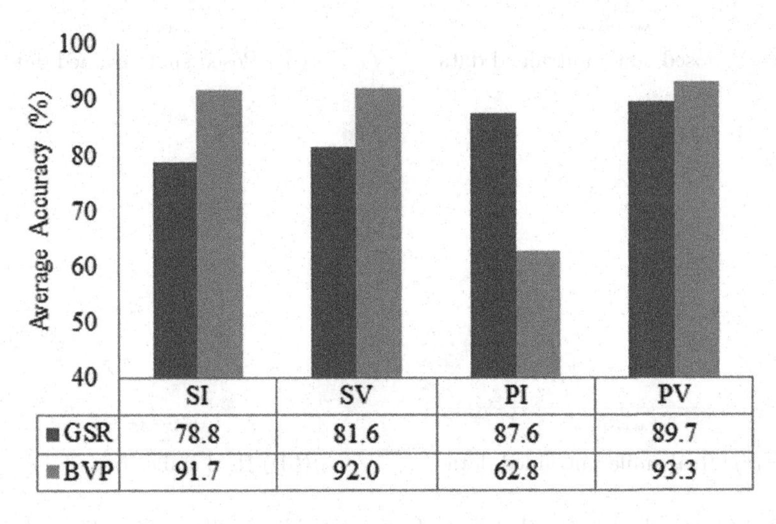


**Fig. 3.** Overall accuracy of classification for all participants for both GSR and BVP signals.

## 5 Conclusions

In this study, we have computed the accuracy of classifiers when differentiating between real and posed smile from participants' GSR and BVP data. An exclusion method was used to remove low-efficiency features, leaving high-precision features for training the classifiers. The results showed that this method does not improve the accuracy very well. In subsequent experiments, the classifier was trained using the leave-one-participant-out approach. We can see that the results have greatly improved for all four situations namely SI, SV, PI, and PV. After overall analysis, it was shown that in condition PV it is most easy to recognise genuine smiles and differentiate them from posed smiles (accuracy 93.3%) followed by PI, SV, and SI. In other words, paired videos/images are easy compared to the single videos/images, and videos are easier to recognise compared to single images. This demonstrates experimentally a general hypothesis of human intuition, which tells us about the applicability of AI techniques for human facial expression recognition from only looking at (or observing) them without connecting any devices to the person displaying the emotion. In our future research we will include more observers from diversified social groups and age ranges. We will also include other state-of-the-art databases, diverse features, and well-known feature selection methods [8,18,33,34].

## References


1. Ayata, D., Yaslan, Y., Kamaşak, M.: Emotion recognition via galvanic skin response: comparison of machine learning algorithms and feature extraction methods. Istanbul Univ.-J. Electr. Electron. Eng. **17**(1), 3147–3156 (2017)
2. Bailenson, J.N., et al.: Real-time classification of evoked emotions using facial feature tracking and physiological responses. Int. J. Hum. Comput. Stud. **66**(5), 303–317 (2008)
3. Barger, P.B., Grandey, A.A.: Service with a smile and encounter satisfaction: emotional contagion and appraisal mechanisms. Acad. Manag. J. **49**(6), 1229–1238 (2006)
4. Braithwaite, J.J., Watson, D.G., Jones, R., Rowe, M.: A guide for analysing electrodermal activity (EDA) & skin conductance responses (SCRs) for psychological experiments. Psychophysiology **49**(1), 1017–1034 (2013)
5. Bugental, D.B.: Unmasking the "polite smile" situational and personal determinants of managed affect in adult-child interaction. Pers. Soc. Psychol. Bull. **12**(1), 7–16 (1986)
6. Deutsch, F.M., LeBaron, D., Fryer, M.M.: What is in a smile? Psychol. Women Q. **11**(3), 341–352 (1987)
7. Dibeklioglu, H., Valenti, R., Salah, A.A., Gevers, T.: Eyes do not lie: spontaneous versus posed smiles. In: Proceedings of the 18th ACM International Conference on Multimedia, pp. 703–706 (2010)
8. Dzedzickis, A., Kaklauskas, A., Bucinskas, V.: Human emotion recognition: review of sensors and methods. Sensors **20**(3), 592 (2020)
9. Ekman, P., Davidson, R.J.: Voluntary smiling changes regional brain activity. Psychol. Sci. **4**(5), 342–345 (1993)
10. Ekman, P., Friesen, W.V.: Felt, false, and miserable smiles. J. Nonverbal Behav. **6**(4), 238–252 (1982)
11. El Ayadi, M., Kamel, M.S., Karray, F.: Survey on speech emotion recognition: features, classification schemes, and databases. Pattern Recogn. **44**(3), 572–587 (2011)
12. A medical-grade wearable device that offers real-time physiological data acquisition (2020). https://www.empatica.com/en-int/research/e4/
13. Fasel, B., Luettin, J.: Automatic facial expression analysis: a survey. Pattern Recogn. **36**(1), 259–275 (2003)
14. Gouizi, K., Bereksi Reguig, F., Maaoui, C.: Emotion recognition from physiological signals. J. Med. Eng. Technol. **35**(6–7), 300–307 (2011)
15. Dibeklioglu, H., Salah, A.A., Gevers, T.: Recognition of genuine smiles. IEEE Trans. Multimedia **17**, 279–294 (2015)
16. Hossain, M.Z., Gedeon, T.: Classifying posed and real smiles from observers' peripheral physiology. In: Proceedings of the 11th EAI International Conference on Pervasive Computing Technologies for Healthcare, pp. 460–463 (2017)
17. Hossain, M.Z., Gedeon, T.: Discriminating real and posed smiles: human and avatar smiles. Technical report, Brisbane, QLD, Australia, November 2017

18. Hossain, M.Z., Kabir, M.M., Shahjahan, M.: A robust feature selection system with Colin's CCA network. Neurocomputing **173**, 855–863 (2016)
19. Islam, A., Ma, J., Gedeon, T., Hossain, M.Z., Liu, Y.H.: Measuring user responses to driving simulators: a galvanic skin response based study. In: 2019 IEEE International Conference on Artificial Intelligence and Virtual Reality (AIVR), pp. 33–337. IEEE (2019)
20. Jerritta, S., Murugappan, M., Nagarajan, R., Wan, K.: Physiological signals based human emotion recognition: a review. In: 2011 IEEE 7th International Colloquium on Signal Processing and its Applications, pp. 410–415. IEEE (2011)
21. Koelstra, S., et al.: Deap: a database for emotion analysis; using physiological signals. IEEE Trans. Affect. Comput. **3**(1), 18–31 (2011)
22. Mehta, D., Siddiqui, M.F.H., Javaid, A.Y.: Facial emotion recognition: a survey and real-world user experiences in mixed reality. Sensors **18**(2), 416 (2018)
23. Mueser, K.T., Grau, B.W., Sussman, S., Rosen, A.J.: You're only as pretty as you feel: facial expression as a determinant of physical attractiveness. J. Pers. Soc. Psychol. **46**(2), 469 (1984)
24. Picard, R.W., Vyzas, E., Healey, J.: Toward machine emotional intelligence: analysis of affective physiological state. IEEE Trans. Pattern Anal. Mach. Intell. **23**(10), 1175–1191 (2001)
25. Pugh, S.D.: Service with a smile: emotional contagion in the service encounter. Acad. Manag. J. **44**(5), 1018–1027 (2001)
26. Samal, A., Iyengar, P.A.: Automatic recognition and analysis of human faces and facial expressions: a survey. Pattern Recogn. **25**(1), 65–77 (1992)
27. Sebe, N., Cohen, I., Gevers, T., Huang, T.S.: Multimodal approaches for emotion recognition: a survey. In: Internet Imaging VI, vol. 5670, pp. 56–67. International Society for Optics and Photonics (2005)
28. Shu, L., et al.: A review of emotion recognition using physiological signals. Sensors **18**(7), 2074 (2018)
29. Song, T., Lu, G., Yan, J.: Emotion recognition based on physiological signals using convolution neural networks. In: Proceedings of the 2020 12th International Conference on Machine Learning and Computing, pp. 161–165 (2020)
30. Teichmann, D., Klopp, J., Hallmann, A., Schuett, K., Wolfart, S., Teichmann, M.: Detection of acute periodontal pain from physiological signals. Physiol. Meas. **39**(9), 095007 (2018)
31. Wang, Z., Mao, H., Li, Y.J., Liu, F.: Smile big or not? Effects of smile intensity on perceptions of warmth and competence. J. Consum. Res. **43**(5), 787–805 (2017)
32. Whitehill, J., Littlewort, G., Fasel, I., Bartlett, M., Movellan, J.: Toward practical smile detection. IEEE Trans. Pattern Anal. Mach. Intell. **31**(11), 2106–2111 (2009)
33. Wu, C.H., Lin, J.C., Wei, W.L.: Survey on audiovisual emotion recognition: databases, features, and data fusion strategies. APSIPA Trans. Signal Inf. Process. **3** (2014)
34. Zeng, Z., Pantic, M., Roisman, G.I., Huang, T.S.: A survey of affect recognition methods: audio, visual, and spontaneous expressions. IEEE Trans. Pattern Anal. Mach. Intell. **31**(1), 39–58 (2008)

# Image Generation with the Enhanced Latent Code and Sub-pixel Sampling

Ting Xu, Dibo Shi, Yi Ji(✉), and Chunping Liu(✉)

College of Computer Science and Technology, Soochow University, Suzhou, China
{txu7,dbshi}@stu.suda.edu.cn, {jiyi,cpliu}@suda.edu.cn

**Abstract.** Synthesizing realistic images from coarse layout is a challenging problem in computer vision. For generating the image with multiple objects, there still exists the problem of the shortage of object details and certain artifacts in existing methods. In this paper, we propose an image generation approach based on the Modified Variational Lower Bound and Sub-pixel Sampling (MVLBSS). Firstly, we modify variational lower bound to form the enhanced latent code with the semantic of object appearance. The change of variational lower bound is implemented by introducing the object category information in the crop estimator and layout encoder to capture the details of object appearance. Secondly, to reduce artifacts, we propose a sub-pixel sampling module with up-sampling and convolution operation. Lastly, our proposed approach is trained in an end-to-end manner to the global optimal image generation. The comparison experimental results with the state-of-the-art methods on both COCO-stuff and Visual Genome datasets show that the proposed method can effectively improve the quality of generation image. Moreover, ablation experiments demonstrate that the improved strategy is effective for enhancing image details and eliminating artifacts.

**Keywords:** Image generation · Layout · CVAE-GAN · Sub-pixel sampling · Object category

## 1 Introduction

Image generation is a reconstruction task based on the deep understanding of the given thing, such as text, image etc. Similar conditional models for image generation have a wide range of real-world applications such as art assist, image in-painting and aid of photography.

Supported by National Natural Science Foundation of China (61972059, 61773272), The Natural Science Foundation of the Jiangsu Higher Education Institutions of China (19KJA230001), Key Laboratory of Symbolic Computation and Knowledge Engineering of Ministry of Education, Jilin University (93K172016K08), Suzhou Key Industry Technology Innovation-Prospective Application Research Project SYG201807), the Priority Academic Program Development of Jiangsu Higher Education Institutions.


H. Yang et al. (Eds.): ICONIP 2020, LNCS 12532, pp. 387–398, 2020.
https://doi.org/10.1007/978-3-030-63830-6_33

Nowadays most models generate natural-like images by capturing the high-level structure of the training data. For instance, text-to-image (T2I) approaches [7–13] can build image from natural language description by combining recurrent neural network and Generative Adversarial Network (GAN). Although realistic images have been successfully synthesized on simple datasets, such as birds and flowers, these approaches do not specifically model objects and their relations in images and thus have difficulties in generating realistic complex results with multi-objects on complex datasets such as COCO-stuff [14] and Visual Genome [15]. Because sentence is a linear structure, the various relationships between objects can not been well present. Furthermore, natural language description in the dataset is usually very short.

Some scene graphs based methods [2,24] try to compensate for this shortcoming. Since scene graphs can better represent objects and their relationships to provide a structured description of complex scene and the semantic relationship between objects. So Johnson *et al.* [2] propose sg2im, which introduces a graph convolutional neural network for handling scene graphs. And Tripathi *et al.* [24] introduce a scene context network to harness scene graphs context for improving image generation from scene graphs. Although the scene graphs can provide more useful information, building a complete and precise scene graph for non-professional workers is almost impossible. Besides, there are only limited relationship types in sg2im and it lacks of specification of core spatial properties, e.g. location, appearance and object size.

Compared with text description and scene graphs, the layout is a more flexible and convenient representation. Hong *et al.* [16] and Li *et al.* [13] regard the layout as an intermediate representation between original domain (text or scene graph) and target domain (image). Zhao *et al.* [3] propose a coarse layout-based approach (layout2im) to generate a set of complicated realistic images with the correct object in the desired locations. Although the proposed layout2im method has the advantages of flexibly controlling image generation, there are serious artifacts in the generated image. To address this problem, taking the Layout2im model as backbone, we propose a novel model called MVLBSS that adopts a conditional variational auto-encoder (CVAE)-GAN network based on our modified variational lower bound (MVLB). Different from the Layout2im, firstly, our object latent code posterior distributions are estimated from the cropped each object and object category. Secondly, we design a sub-pixel sampling module for substituting deconvolution operation.

Conclusively, our contributions are three-fold: 1) We propose an improved image generation model from coarse layout by constructing enhanced object latent code with object category information and sub-pixel sampling. 2) We introduce a modified variational lower bound to ensure that we can generate images with more details. 3) We successfully solve the artifacts problems in the generated images by a sub-pixel sampling module.

## 2 Related Work

**Image Synthesis with Layouts.** Image generation from layout is a relative novel task. Existing GAN-based T2I synthesis is usually conditioned only the global sentence vector, which misses important fine-grained information and lacks the ability of generating high-quality images. There are also some approaches that can be seen as models that generate images from layouts such as semantic layout (bounding boxes + object shapes) [2,13,16] and coarse spatial layout (bounding boxes + object category) [3]. To explicitly encode the semantic layout into the generator, Hong *et al.* [16] and Li *et al.* [13] first construct a semantic layout from the text and generate an image conditioned on the layout and text description. Park *et al.* [17] propose a novel spatial-adaptive normalization layer for generating images from an input semantic layout. Johnson *et al.* [2] introduce the concept of scene graphs. Similar to [13] used the caption to infer a scene layout. Zhao *et al.* [3] adopt a network which combine variational auto-encoder and Generative Adversarial Network (VAE-GAN) to generate images from coarse spatial layout.

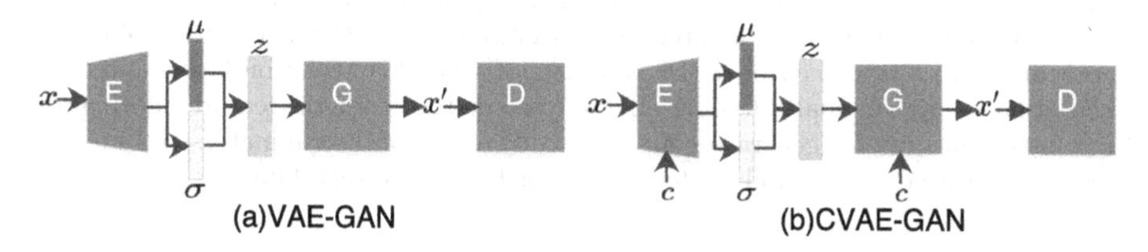


**Fig. 1.** Illustration of the structure of VAE-GAN and CVAE-GAN. Where $x$ and $x'$ are real data and generated sample, E, G, D are encoder, generator, and discriminator, respectively. $\mu$ and $\sigma$ are the mean and standard deviation vector, $c$ is the condition, for example, the category label or text.

**CVAE-GAN.** VAE-GAN [18], combined a variational auto-encoder and a generative adversarial network, as in Fig. 1(a). The loss function is formulated as,

$$\mathcal{L} = -\mathbb{E}_{q(\boldsymbol{z}|\boldsymbol{x})}\left[\log p\left(\mathrm{D}(\boldsymbol{x})|\boldsymbol{z}\right)\right] + D_{\mathrm{KL}}(q(\boldsymbol{z}|\boldsymbol{x})\|p(\boldsymbol{z})) + \log(\mathrm{D}(\boldsymbol{x})) + \log(1-\mathrm{D}(\mathrm{G}(\boldsymbol{z}))) \tag{1}$$

where $\boldsymbol{x}$ is the ground-truth, $\boldsymbol{z}$ represents the latent code by feeding $\boldsymbol{x}$ into the encoder, $\boldsymbol{D}$ is the discriminator, and $\boldsymbol{G}$ refers to generator. The former term describes the reconstruction data for the posterior $q(\boldsymbol{z}|\boldsymbol{x})$, the intermediate term makes sure that the prior does not deviate from the posterior, the latter term indicates that the GAN objective. VAE-GAN is limited to randomly generate images and failed to produce fine-grained image. To synthesize images in fine-grained categories with randomly sampled values on a latent code, Bao *et al.* [19] present a CVAE-GAN, as shown in Fig. 1(b). The difference is that the posterior changes from $q(\boldsymbol{z}|\boldsymbol{x})$ to $q(\boldsymbol{z}|\boldsymbol{x},\boldsymbol{c})$ and the generator from $p\left(\mathrm{D}(\boldsymbol{x})|\boldsymbol{z}\right)$

to $p(\mathrm{D}(\boldsymbol{x})|\boldsymbol{z}, \boldsymbol{c})$. Bao *et al.* analyse the efficacy of CVAE-GAN and find that the model is able to keep the identity information. Inspired this, we also introduce class-level information to help our model generates more accurate object appearance.

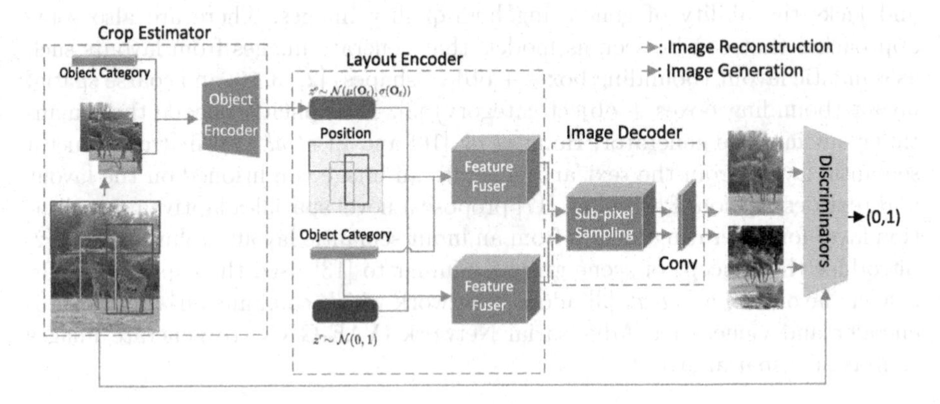


**Fig. 2.** Overview of our generation framework. Our framework includes four modules: (1) a crop estimator (CE) with category information to get the distribution of object appearance. (2) a layout encoder (LE) consists of several convolution layers and ConvLSTM layers. (3) an image decoder (IE) only needs to decode the feature information from layout encoder for generating an image. (4) a pair of discriminators (D) are utilized to determine the possibility that the input belongs to real data.

# 3 Methodology

## 3.1 Network Architecture

Our main framework (as illustrated in Fig. 2) keeps high-level architecture design of Layout2im: 1) A crop estimator with the modified variational lower bound. Different from Layout2im, our model not only takes the real image patches as input, but also feeds category labels into the crop estimator. It takes care of fitting the distribution of real image patches. 2) A layout encoder composes of ConvLSTM. It converts the layout into a hidden state. 3) Given the hidden state, an image decoder is responsible for decoding and getting the image. The composition of image decoder (Sect. 3.3) is completely different from that in Layout2im. 4) A pair of discriminators. In Fig. 2, we utilize a block to represent two discriminators $D_{img}$ and $D_{obj}$. The discriminator $D_{img}$ ensures that the generated images are high-quality. The discriminator $D_{obj}$ makes sure that the results with fine-grained objects.

### 3.2 Modified Variational Lower Bound for Reliable Object Appearance

Here for the purpose of generating images with fine-grained objects, we add a crop estimator which utilizes a modified variational lower bound to map the real image patches to latent vector. Therefore, the relationships between real data and latent code are explicitly set in our model. The modified variational lower bound is formulated as:

$$\mathcal{L} = -\mathbb{E}_{q(z^o|I^o,c)}\left[\log p\left(\mathrm{D}(\boldsymbol{I^o}, \boldsymbol{c})|\boldsymbol{z^o}\right)\right] + D_{\mathrm{KL}}(q(\boldsymbol{z^o}|\boldsymbol{I^o}, \boldsymbol{c})\|p(\boldsymbol{z^o})) \tag{2}$$

where $\boldsymbol{I^o}$ refers to the real image patches, $\boldsymbol{c}$ represents the class labels and $\boldsymbol{z^o}$ is the latent code. Thus, our model can also benefit from the given category information, while in Layout2im only depends on the real image patches. With the enhanced crop estimator, we can get an enhanced latent code for representing the appearance of the objects. In Sect. 4.3, we will show that this modification of crop estimator helps to learn more detailed textures.

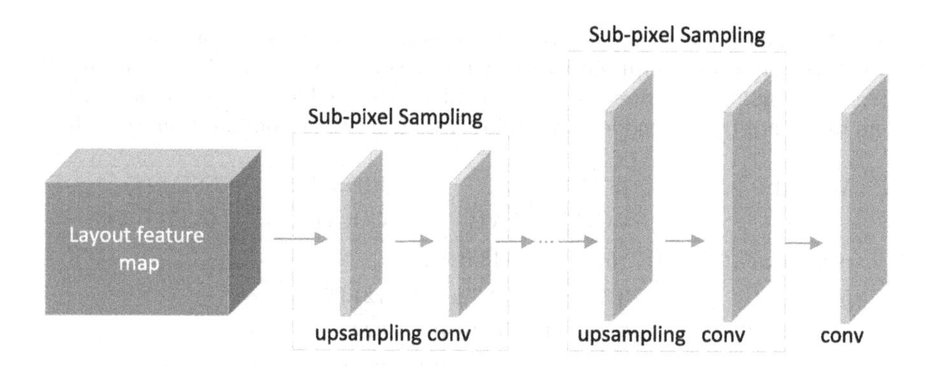


**Fig. 3.** Image decoder based on sub-pixel sampling.

### 3.3 Sub-pixel Sampling for Eliminating Artifacts

As depicted in Fig. 3, while the coarse layout $(B_{1:\mathrm{T}}, z^r, c)$ is already fed as an input, we then convert layout feature maps (hidden state) H to images with an image decoder. Our image decoder module consists of the sub-pixel sampling module which works like the sub-pixel convolutional layer in ESPCN [21]. Most methods [2,3] adopt several deconvolutional layers to build the decoder. Unfortunately, as stated in [20,21], deconvolution can easily produce "uneven overlap" and bring artifacts especially when the kernel size cannot be removed by the stride. In theory, the model can learn to handle these uneven overlaps, but in fact it is not so well handled. It may be possible to avoid artifacts by reducing the number of filters, but the results are poor because the capacity becomes lower. In one word, as long as the deconvolution is used, artifacts must "attend" punctually, which is hard to avoid. To overcome this limitation, we provide a sub-pixel sampling layer which consists of one upsampling layer and one convolution layer instead of the deconvolutional layer in the baseline [3]. Concretely, We use the interpolation method to achieve the upsampling. Then the convolution processing is performed on the up-sampling result, the values in each window are weighted, effectively suppressing artifacts.

# 4 Experiments and Analysis

To evaluate our model, extensive experiments are conducted to evaluate our model on two common datasets COCO-stuff [14] and Visual Genome [15]. In the experiments, we follow the same objective function as layout2im. In this section, we introduce the two datasets, evaluation metrics briefly, and then compare our method with previous state-of-the-art image generation methods. Finally, we display the results of the ablation study.

**Table 1.** Quantitative comparisons using Inception score (IS), Fréchet Inception Distance (FID) and structural similarity index (SSIM), based on $64 \times 64$ images. For pix2pix and sg2im, We directly use the values reported in the paper. layout2im-p refers to the pre-trained model. layout2im-o: we train the layout2im from scratch.

| Method | IS (↑) | | FID (↓) | | SSIM (↑) | |
|---|---|---|---|---|---|---|
| | COCO | VG | COCO | VG | COCO | VG |
| real image ($64 \times 64$) | $16.3 \pm 0.4$ | $13.9 \pm 0.5$ | – | – | – | – |
| pix2pix [1] | $3.5 \pm 0.1$ | $2.7 \pm 0.02$ | 121.97 | 142.86 | – | – |
| sg2im (GT) [2] | $7.3 \pm 0.1$ | $6.3 \pm 0.2$ | 67.96 | 74.61 | – | – |
| layout2im [3] | $9.1 \pm 0.1$ | $8.1 \pm 0.1$ | **38.14** | **31.25** | – | – |
| layout2im-p | $9.1 \pm 0.1$ | $8.1 \pm 0.1$ | 42.80 | 40.07 | 24.2 | 36.1 |
| layout2im-o | $9.1 \pm 0.1$ | $8.1 \pm 0.1$ | 43.80 | 39.39 | 24.1 | 36.1 |
| ours | $\mathbf{9.5 \pm 0.05}$ | $\mathbf{8.3 \pm 0.2}$ | 40.01 | 35.75 | **25.3** | **37.1** |

## 4.1 Datasets and Evaluation Metrics

**Datasets.** COCO-stuff [14] is a subset of COCO dataset. Each image in the dataset is annotated with bounding boxes and segmentation masks. For Visual Genome, there are 108k images, and each image contains an average of 35 objects, 26 attributes. We follow the data preprocessing operation in [3] and [2].

**Evaluation Metrics.** Judging whether a generative model for image generation is good or bad mainly depends on image clarity and diversity. In this paper, we choose the Inception score (IS), Fréchet Inception Distance (FID) and structural similarity index (SSIM) as the metrics. Inception score [22] is the earliest indicator of image clarity assessment. The higher the score, the more realistic the image is. Fréchet Inception Distance (FID) is introduced by Heusel *et al.* [23] for evaluating GANs. FID is more sensitive to noise and focuses on measuring the diversity of generated pictures. The lower the score, the better. SSIM [25] is an index to measure the similarity of two images, and it's an image quality evaluation metric that accords with human intuition. The higher the value, the more similar the two images are.

### 4.2 Comparing Analysis

Table 1, with quantitative results on IS, FID, SSIM over two datasets, respectively, presents a comparison between our model and three state-of-the-art methods: pix2pix [1], sg2im [2], layout2im [3]. Pix2pix is a conditional image translation model based on GAN. Sg2im takes scene graphs as input and processes it with graph convolution. Layout2im is the original model that gets the spatial layout as input directly and VAE-GAN as the basic frame to generate realistic images. To ensure fair comparison, for pix2pix [1] and sg2im [2], we list the values reported in the paper. As for layout2im, we not only include the pre-training model but also train it from scratch.

As shown in Table 1, our MVLBSS is obviously superior to the existing approaches with a substantial margin in IS and SSIM on COCO-stuff and VG datasets. On COCO-Stuff dataset, our model increases Inception score from 9.1 to 9.5, about four percent improvement achieved. For SSIM index, our model improves from 24.2 to 25.3. On Visual Genome dataset, the inception score is improved from 8.1 to 8.3 and the corresponding SSIM index is increased from 36.1 to 37.1. This comparison indicates that our model can create images with more details. At the same time, we also notice that the introduction of category label-level information may sacrifice the diversity of images generated by the network in some way, so the FID of layout2im is slightly lower than MVLBSS. So how to further improve the diversity of generated images will become one of the focus of our follow-up work. Nevertheless, in the same experimental environment, our model still significantly surpasses sg2im and layout2im.

We visualize the outputs of three different image generation models in Fig. 4. It displays the image generated from same layouts, which observe that each complex layout contains multiple objects. Even so our MVLBSS can synthesize images with high quality. It is undeniable that with the help of VAE-GAN, layout2im can produce images with multiple objects, but as depicted in Fig. 4 column (a), (d), (f) and (n), most of the shapes in the generated samples are not clear or have serious deformation. As displayed in Fig. 4 column (k), layout2im paints purple for the road. It clearly not conforms to usual understanding. In column (m), layout2im colorizes the water wave directly. The images generated by sg2im seem to be without structure on the whole, especially on VG, because the input of sg2im is only scene graphs, lacking more details. Though, of course, there are some bad cases in our generated sample. For instance, in Fig. 4 column (g) the headless horse is produced in generated sample. How to further synthesize

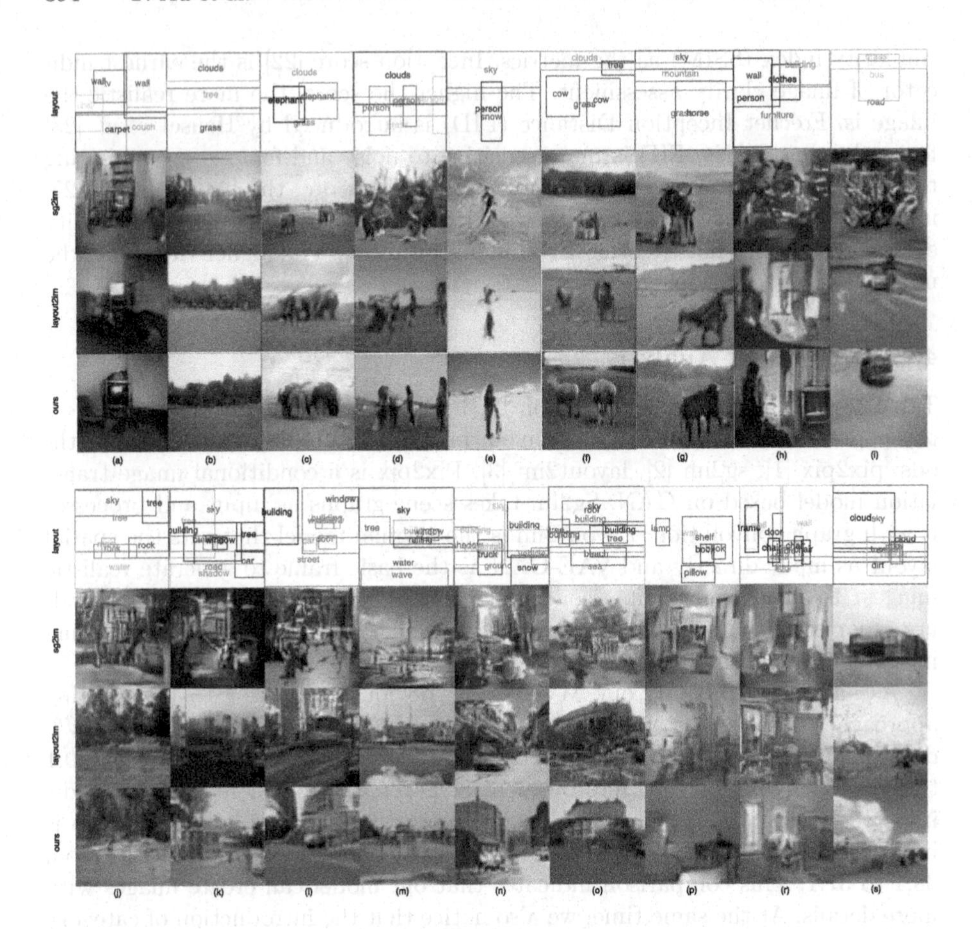


**Fig. 4.** Examples of $64 \times 64$ images generated by sg2im, layout2im and our model from same coarse layouts on COCO-Stuff (top) and Visual Genome (bottom) datasets.

**Fig. 5.** Illustration of artifacts in images generated by sg2im, Layout2im on Visual Genome (top) and COCO-Stuff datasets (bottom). Our model generates non-artifact images.

**Table 2.** Ablation Study. MVLBSS1: introducing the modified variational lower bound into the crop estimator, MVLBSS2, area: using sub-pixel sampling with area interpolation, MVLBSS2, bilinear: using the bilinear interpolation, MVLBSS2, nearest: the nearest interpolation.

| Method | IS (↑) | | FID (↓) | | SSIM (↑) | |
|---|---|---|---|---|---|---|
| | COCO | VG | COCO | VG | COCO | VG |
| baseline-o | $9.1 \pm 0.1$ | $8.1 \pm 0.1$ | 43.80 | 39.39 | 24.1 | 36.1 |
| MVLBSS1 | $9.2 \pm 0.2$ | $8.1 \pm 0.1$ | 42.56 | 38.24 | 24.9 | 36.3 |
| MVLBSS2, nearest | $9.3 \pm 0.1$ | $8.2 \pm 0.1$ | 41.53 | 36.92 | 25.7 | 36.4 |
| MVLBSS2, area | $9.2 \pm 0.2$ | $7.9 \pm 0.3$ | 42.68 | 38.51 | 24.3 | 36.9 |
| MVLBSS2, bilinear | $8.9 \pm 0.1$ | $7.7 \pm 0.1$ | 43.74 | 39.31 | 23.7 | 35.6 |
| ours | $\mathbf{9.5 \pm 0.1}$ | $\mathbf{8.3 \pm 0.2}$ | **40.01** | **35.75** | **25.3** | **37.1** |

objects with more accurate appearance is an interesting work in the future. On balance, our MVLBSS can create visually attractive images with consistent colors and shapes.

### 4.3 Ablation Study

In order to study the efficacy of each component in the proposed MVLBSS, we gradually modify the Layout2im and compare their differences. The qualitative comparison is illustrated in Fig. 5 and Fig. 6. Table 2 presents the quantitative comparison among different ablative versions.

**The Modified Variational Lower Bound.** Our intention is to demonstrate the effectiveness of the crop estimator with modified variational lower bound. For this, we implement an ablation version, named MVLBSS1. With the category label-level information, the IS and SSIM of the model are increased to 9.2 and 24.9, separately. This reveals that the model is able to generate more attractive object appearance under the guidance of these additional category label-level information. Layout2im omits label-level information, resulting in the appearance of some objects in images created by the model is not so reliable. In Fig. 5 column (a), layout2im draws sea looks like broken glass. Another case is the person in column (d), layout2im "integrates" these people and "ignores" the legs.

**Sub-pixel Sampling.** The sub-pixel sampling module aims to eliminate the artifacts in layout2im. So does it really tackle the problem? From the data stated in Table 2, the IS score is increased to 9.3 and 8.2 on two different datasets via the sub-pixel sampling module. In Fig. 6 column (a), (b) and (c), the image generated by layout2im has a strange checkerboard pattern of artifacts (the position marked by the red box), which is due to the "uneven overlap" brought by deconvolution in the generation stage. In contrast, the images generated by our model look smoother.

**Fig. 6.** Comparison of images generated by sg2im, layout2im and our model. The object in the image generated by our model has a visually more satisfactory appearance (Color figure online)

As listed in the Table 2 (row 5, 6), we have implemented two other models ablation versions based on other interpolation methods. The nearest neighbor interpolation method wins the best result. Remarkably, in the other two interpolation methods, our results are slightly lower on IS, which may simply mean that the nearest neighbor interpolation is exactly "matched" with the hyper-parameters optimized or updated for deconvolution operation. Another possible interpretation is that bilinear interpolation suppresses the high frequency image features.

In addition, we also dispose $5 \times 5$ convolution kernel in the sub-pixel sampling module, and the improvement of experimental results are not ideal.

## 5 Conclusion

In this paper, we propose a novel and end-to-end method for synthesizing realistic images from coarse layout. This method adopts the modified variational lower bound to generate objects corresponding to the input category labels and a novel component called sub-pixel sampling module for removing the artifacts in the generated image. Experimental results reveal that our model outperforms the state-of-the-art methods in image generation from layout task.

## References

1. Isola, P., Zhu, J.-Y., Zhou, T., Efros, A.A.: Image-to-image translation with conditional adversarial networks. In: Proceedings of the IEEE Conference on Computer Vision and Pattern Recognition, pp. 1125–1134 (2017)

2. Johnson, J., Gupta, A., Fei-Fei, L.: Image generation from scene graphs. In: Proceedings of the IEEE Conference on Computer Vision and Pattern Recognition, pp. 1219–1228 (2018)
3. Zhao, B., Meng, L., Yin, W., Sigal, L.: Image generation from layout (2019)
4. Gulrajani, I., Ahmed, F., Arjovsky, M., Dumoulin, V., Courville, A.C.: Improved training of wasserstein GANs. In: Advances in Neural Information Processing Systems, pp. 5767–5777 (2017)
5. Odena, A., Olah, C., Shlens, J.: Conditional image synthesis with auxiliary classifier GANs. In: Proceedings of the 34th International Conference on Machine Learning, vol. 70, JMLR. org, pp. 2642–2651 (2017)
6. Simonyan, K., Zisserman, A.: Very deep convolutional networks for large-scale image recognition, arXiv preprint arXiv:1409.1556 (2014)
7. Reed, S., Akata, Z., Yan, X., Logeswaran, L., Schiele, B., Lee, H.: Generative adversarial text to image synthesis, arXiv preprint arXiv:1605.05396 (2016)
8. Zhang, H., et al.: Stackgan: text to photo-realistic image synthesis with stacked generative adversarial networks. In: Proceedings of the IEEE International Conference on Computer Vision, pp. 5907–5915 (2017)
9. Tan, F., Feng, S., Ordonez, V.: Text2scene: generating abstract scenes from textual descriptions, arXiv preprint arXiv:1809.01110 (2018)
10. Reed, S.E., Akata, Z., Mohan, S., Tenka, S., Schiele, B., Lee, H.: Learning what and where to draw. In: Advances in Neural Information Processing Systems, pp. 217–225 (2016)
11. Xu, T., et al.: Attngan: fine-grained text to image generation with attentional generative adversarial networks. In: Proceedings of the IEEE Conference on Computer Vision and Pattern Recognition, pp. 1316–1324 (2018)
12. Zhang, Z., Xie, Y., Yang, L.: Photographic text-to-image synthesis with a hierarchically-nested adversarial network. In: Proceedings of the IEEE Conference on Computer Vision and Pattern Recognition, pp. 6199–6208 (2018)
13. Li, W., et al.: Object-driven text-to-image synthesis via adversarial training. In: Proceedings of the IEEE Conference on Computer Vision and Pattern Recognition, pp. 12174–12182 (2019)
14. Caesar, H., Uijlings, J., Ferrari, V.: Coco-stuff: thing and stuff classes in context (2016)
15. Krishna, R., et al.: Visual genome: connecting language and vision using crowdsourced dense image annotations. Int. J. Comput. Vision **123**(1), 32–73 (2017)
16. Hong, S., Yang, D., Choi, J., Lee, H.: Inferring semantic layout for hierarchical text-to-image synthesis. In: Proceedings of the IEEE Conference on Computer Vision and Pattern Recognition, pp. 7986–7994 (2018)
17. Park, T., Liu, M.-Y., Wang, T.-C., Zhu, J.-Y.: Semantic image synthesis with spatially-adaptive normalization. In: Proceedings of the IEEE Conference on Computer Vision and Pattern Recognition, pp. 2337–2346 (2019)
18. Larsen, A.B.L., Sønderby, S.K., Larochelle, H., Winther, O.: Autoencoding beyond pixels using a learned similarity metric, arXiv preprint arXiv:1512.09300 (2015)
19. Bao, J., Chen, D., Wen, F., Li, H., Hua, G.: CVAE-GAN: fine-grained image generation through asymmetric training. In: Proceedings of the IEEE International Conference on Computer Vision, pp. 2745–2754 (2017)
20. Gauthier, J.: Conditional generative adversarial nets for convolutional face generation. Class Project for Stanford CS231N: Convolutional Neural Networks for Visual Recognition. Winter Semester **2014**(5), 2 (2014)

21. Shi, W., et al.: Real-time single image and video super-resolution using an efficient sub-pixel convolutional neural network. In: The IEEE Conference on Computer Vision and Pattern Recognition (CVPR), June 2016
22. Salimans, T., Goodfellow, I., Zaremba, W., Cheung, V., Radford, A., Chen, X.: Improved techniques for training GANs. In: Advances in Neural Information Processing Systems, pp. 2234–2242 (2016)
23. Heusel, M., Ramsauer, H., Unterthiner, T., Nessler, B., Hochreiter, S.: GANs trained by a two time-scale update rule converge to a local nash equilibrium. In: Advances in Neural Information Processing Systems, pp. 6626–6637 (2017)
24. Tripathi, S., Bhiwandiwalla, A., Bastidas, A., Tang, H.: Using scene graph context to improve image generation, arXiv preprint arXiv:1901.03762 (2019)
25. Wang, Z., Bovik, A.C., Sheikh, H.R., Simoncelli, E.P., et al.: Image quality assessment: from error visibility to structural similarity. IEEE Trans. Image Process. **13**(4), 600–612 (2004)

# Joint Optic Disc and Optic Cup Segmentation Based on New Skip-Link Attention Guidance Network and Polar Transformation

Yun Jiang, Jing Gao(✉), and Falin Wang

College of Computer Science and Engineering, Northwest Normal University, Lanzhou, Gansu, China
2234254566@qq.com

**Abstract.** The challenge faced by the joint optic disc and optic cup segmentation is how to learn an efficient segmentation model with good performance. This paper proposes a method based on a new type of Skip-Link attention guidance network and polar transformation, called Skip-Link Attention Guidance Network (SLAG-CNN) model, which implements the simultaneous segmentation of the optic disc and optic cup. In SLAG-CNN, the Skip-Link Attention Gate (SLAG) module was first introduced, which was used as a sensitive extension path to transfer the semantic information and location information of the previous feature map. Each SLAG module of the SLAG-CNN model combines channel attention and spatial attention, and adds Skip-Link to form a new attention module to enhance the segmentation results. Secondly, multi-scale input images are constructed by spatial pyramid pooling. Finally, a weighted cross-entropy loss function is used at each side output layer to sum up to calculate the total model loss. On the DRISHTI-GS1 dataset, the joint optic disc and optic cup segmentation task proves the effectiveness of our proposed method.

**Keywords:** Convolutional neural network · Image segmentation · Skip-link attention gate · Extended path

## 1 Introduction

Glaucoma is a chronic eye disease that causes irreversible vision loss. The Cup-to-Disc Ratio (CDR) plays an important role in the screening and diagnosis of glaucoma [1]. Therefore, it is a basic task to accurately and automatically segment the Optic Disc (OD) and Optic Cup (OC) from the fundus image. However, manual evaluations by trained clinicians are time-consuming and expensive, and are not suitable for population screening [2]. For ophthalmologists, the automatic segmentation and analysis of OD and OC play an important role in the diagnosis and treatment of retinal diseases [3].


H. Yang et al. (Eds.): ICONIP 2020, LNCS 12532, pp. 399–410, 2020.
https://doi.org/10.1007/978-3-030-63830-6_34

A series of methods for segmenting OD and OC in retinal images have been developed: template matching-based and superpixel-based methods [4–10], and deep learning methods [11–15,17,18].

**Template Matching-Based and Superpixel-Based Methods:** These methods are relatively representative in the early days. However, the acquired images may have different colors, uneven intensity, and the presence of lesion areas, permeates, and blood vessels in the OD area, making these segmentation methods less robust.

**Deep Learning Methods:** These methods segment OD and OC by training a large number of data samples to automatically extract features. Among them, the popular U-Net [18] is a variant of FCN (Fully Convolution Network). U-Net [18] has its multi-scale skip connection and learnable deconvolution layer. It and its variants [18–21] have become a popular neural network architecture for biomedical image segmentation tasks. Although these methods have made some progress in the field of OD and OC, for too large segmented regions, the receptive field in these methods is not large enough to fully understand the global information, and some large segmented regions cannot be accurately identified. For example, the simple skip connection in U-Net [18] as the network expansion path cannot effectively use the structural information of the feature map, such as the edge structure that is important for retinal image analysis, which may hinder the segmentation performance. Therefore, it is expected to design a better extension path to save structural information.

In order to alleviate these problems, we also studied another aspect of architecture design, namely the attention mechanism. Many works [24–26,31,32] have proved that the attention mechanism can highlight the key areas in the image, tell us the focus of attention, and have excellent positioning ability. Abraham N [31] and others used the improved attention U-shaped network for skin lesion segmentation, and proposed a generalized focus loss function based on Tversky index to solve the problem of data imbalance in medical image segmentation. Chaitanya [32] and others proposed a fully convolutional network based on attention FocusNet. It uses feature maps generated by separate convolutional auto-encoders to focus attention on convolutional neural networks for medical image segmentation. Therefore, we propose to use the attention mechanism to effectively improve the feature extraction ability, focus on important structural features and suppress unnecessary features.

This paper proposes a method based on a new type of Skip-Link attention guidance network and polar transformation, called SLAG-CNN model, which realizes the simultaneous segmentation of OD and OC. Combined with the idea of polar transformation, the feature map in polar coordinate system is used to train the model, which improves the segmentation performance of the model.

In summary, there are five contributions in our paper:

1. A new method based on the new Skip-Link attention guidance network and polar transformation is proposed, called Skip-Link Attention Guidance

Network (SLAG-CNN) model, which realizes the simultaneous segmentation of OD and OC.
2. We took the attention method and designed an attention mechanism, called Skip-Link Attention Gate (SLAG) module, which was used as a sensitive expansion path to transfer the semantic information of the previous feature map And location information. We verified the effectiveness of the SLAG module through ablation experiments.
3. In SLAG-CNN, a multi-scale input image is constructed by spatial pyramid pooling, and a weighted cross-entropy loss function is used at each side output layer to calculate the total loss of the model.
4. Finally, the effectiveness and generalization of the proposed method are evaluated on the DRISHTI-GS1 dataset.

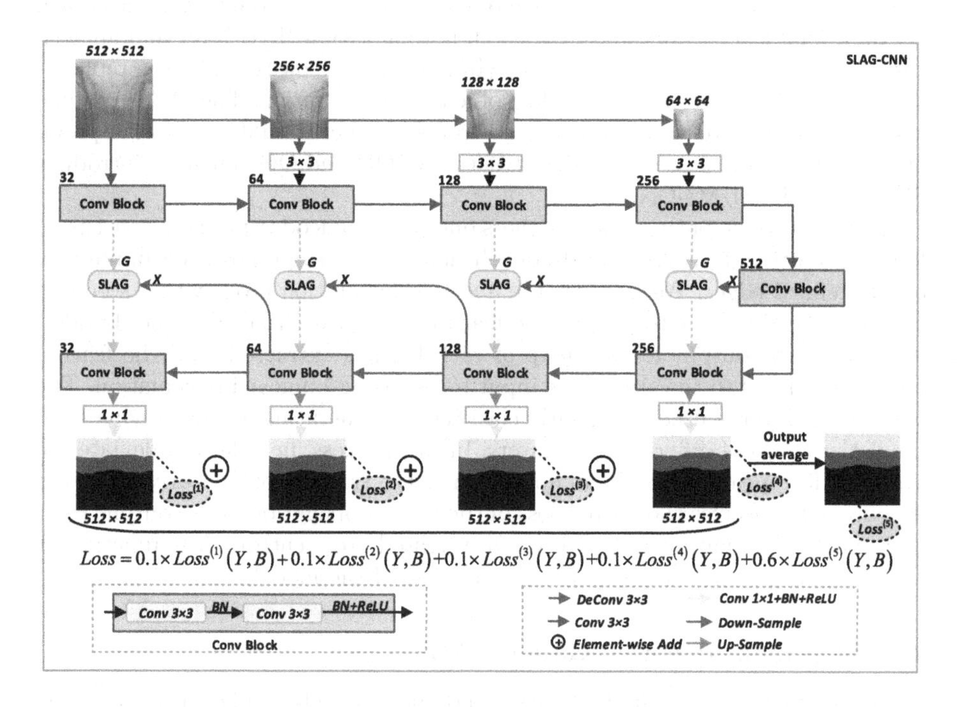


**Fig. 1.** Structure model of SLAG-CNN.

## 2 SLAG-CNN Network Structure

For the OD and OC segmentation tasks, we designed a new end-to-end Skip-Link attention guidance network model SLAG-CNN based on CNNs [11] to segment the retinal images OD and OC simultaneously. First, we use the YOLOv2 [30] method to locate the center of the optic disc, and then perform data augmentation on the located image. Then we use the polar transformation [21] to convert the fundus image into an image under polar coordinate system and input it into the network SLAG-CNN. Finally, after the side output, we will restore the segmentation image under cartesian coordinates by reverse polar transformation of the segmentation image. Figure 1 shows the proposed SLAG-CNN architecture, where U-Net [18] is used as the basic network, which contains the encoder path and the decoder path.

In the encoder path, we give the original image with an input scale of $512 \times 512$, after three downsampling processes to obtain three different scale images of $256 \times 256$, $128 \times 128$ and $64 \times 64$. Adding multi-scale input to the encoder path can ensure the feature transfer of the original image and improve the quality of segmentation. The encoder path contains four convolutional blocks. Each convolutional block is composed of two $3 \times 3$ convolutional layers, and each convolutional layer is followed by a batch normalization (BN) layer and an activation function Rectified Linear Unit (ReLU) layer. Downsampling is replaced by $3 \times 3$ convolution to reduce the loss of feature information. The SLAG module is used as a sensitive expansion path to transfer structural information in previous feature maps, such as the edge structure of OD and OC. Then we introduce SLAG into the SLAG-CNN.

In the decoder path, which is the same as the encoder path, it consists of four convolutional blocks, and deconvolution is used to perform upsampling to obtain refined edges. We first use the $3 \times 3$ deconvolution layer with a step size of 2 and the BN layer to output the feature map of each layer of the decoder. After that, the output feature map of each layer is extracted, and the feature map is expanded to the size of the input image using bilinear interpolation. The side output layer is used as an early classifier for generating accompanying local prediction maps of different scale layers. In this process, in order to alleviate the gradient disappearance problem and enhance the training of the early layer, the decoder path receives the output loss of the back propagation of the output layer and updates the parameters. We use a weighted cross-entropy loss function at each side output layer to sum up the total loss of the model.

## 2.1 Skip-Link Attention Gate (SLAG)

SLAG modified the original attention module [25], which played a great role in improving the entire model. The structure of SLAG is shown in Fig. 2. In SLAG-CNN, we use SLAG as a structure-sensitive skip connection instead of skip connection and upsampling layer to better merge information.

By filtering low-resolution feature maps and high-resolution feature maps, SLAG recovers spatial information from different resolution levels and merges structural information, which can effectively learn better attention weights for OD and OC fine boundaries. First, SLAG uses two inputs to recover feature information from feature maps with different resolutions. Second, SLAG uses Skip-Link to merge features skipped from the corresponding resolution level to solve the problem of boundary blur caused by upsampling. Third, SLAG

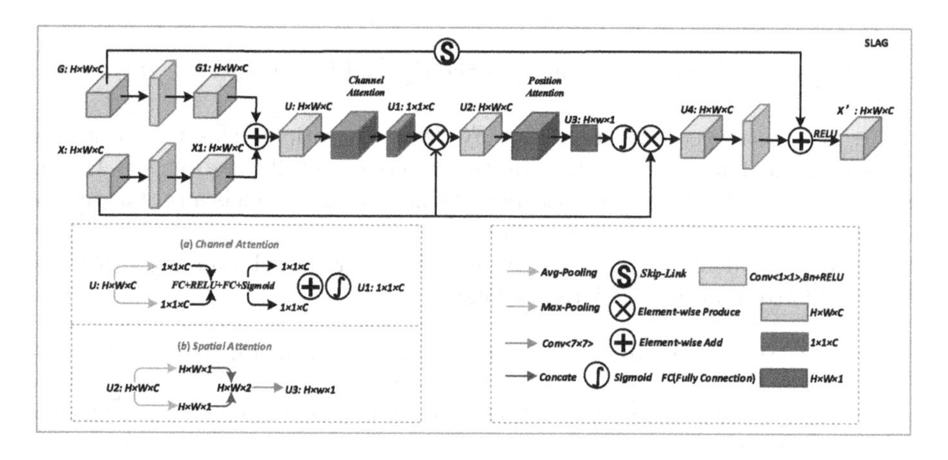


**Fig. 2.** Structure module of SLAG.

infers the attention map in sequence along two independent dimensions, channels and spaces. Fourth, SLAG is a lightweight general-purpose module that can be seamlessly integrated into any CNN architecture with negligible overhead and can be trained end-to-end with basic CNNs.

SLAG has two input features. One is a high-resolution feature map $X \in R^{H \times W \times C}$. $X$ utilizes a $1 \times 1$ convolution and undergo $BN$ and $RELU$ transformation operations to obtain a feature map $X_1 \in R^{H \times W \times C}$ with the same resolution. Another input $G \in R^{H \times W \times C}$ is the low-resolution feature map, after same operations, can obtain the feature map $G_1 \in R^{H \times W \times C}$ with the same resolution. Then add $X_1$ and $G_1$ to get the special diagnosis map $U \in R^{H \times W \times C}$. Compared with the series connection, the addition can reduce the convolution parameters, which helps to reduce the calculation cost.

Since convolution operations extract information features by mixing cross-channel and spatial information together, we use our module to emphasize meaningful features along these two main dimensions: channel and space [25]. Therefore, given an intermediate feature map $U$, our module SLAG infers the attention map in sequence along this two independent dimensions. After each dimension, the attention map is multiplied by the input feature map $X$ for adaptive feature refinement. The channel attention module aims to enhance the specific semantic response of the channel and emphasize it by establishing links between the channels and focusing more on the meaningful parts. As a supplement to channel attention, the spatial attention module aims to use the association between any two points of features to mutually enhance the expression of their respective features, so it pays more attention to the features of spatial location. The two attention sub-modules are shown in Fig. 2.

Channel attention is calculated as follows:

$$A_c(U) = \sigma(W_\beta(W_\alpha(AvgPool(U))) + W_\beta(W_\alpha(MaxPool(U)))) \quad (1)$$

In the above equation, $\sigma$ represents the sigmoid function, $W_\alpha \in R^{\frac{C}{r} \times C}$ and $W_\beta \in R^{C \times \frac{C}{r}}$ respectively represent FC layer operations, this $r$ is a scaling parameter. In this paper, drawing on the experience of [26], we set $r = 16$, the purpose of this parameter $r$ is to reduce the number of channels and thus reduce the amount of calculation. The RELU activation is required after $W_\alpha \in R^{\frac{C}{r} \times C}$.

Spatial attention is calculated as follows:

$$A_s(U_2) = \sigma(Conv^{7\times7}(Concat[AvgPool(U_2); MaxPool(U_2)])) \tag{2}$$

In the above equation, $\sigma$ represents the sigmoid function, $Conv^{7\times7}$ means $7 \times 7$ convolution operation.

Next, $U_3 \in R^{H\times W\times 1}$ undergoes a sigmoid operation and is multiplied with the original feature map $X \in R^{H\times W\times C}$ to perform adaptive feature refinement and output the feature map $U_4 \in R^{H\times W\times C}$. It should be noted that for feature images, average pooling is usually used to summarize spatial information, and maximum pooling can be used to infer a finer channel attention. Therefore, in both modules, we choose to use the average pooling and maximum pooling functions at the same time.The channel attention uses the maximum pool output and average pool output of the shared network, the spatial attention uses two similar outputs, which are gathered along the channel axis and transmitted to the convolutional layer. Finally, after $1 \times 1$ convolution, $BN$ and $RELU$ transformation operations, we add $U_4$ to the feature map $G$ to calibrate the attention feature map, and output the final attention feature map $X^{'}$ after the $RELU$ operation.

## 3 Experiments and Analysis

### 3.1 Training Parameter Settings and Dataset

We use the PyTorch deep learning framework [28] to implement SLAG-CNN. The server configuration of the training model is: Intel(R) Xeon(R) E5-2620 v3 2.40 GHz CPU, Tesla K80 GPU, and Ubuntu64 system. In the experiment, the learning rate is 0.0001, and the SLAG-CNN is trained from scratch using stochastic gradient descent (SGD). The momentum is 0.9 and the weight attenuation coefficient is 0.0005. The training iterated for a total of 400 cycles, and the size of the output segmented image is $512 \times 512 \times 3$.

The Drishti-GS1 dataset [27] contains 101 fundus images, 31 normal and 70 diseased images. The train set contains 50 images and the test set contains 51 images. We employ the ORIGA dataset [33] containing 650 fundus images with 168 glaucomatous eyes and 482 normal eyes.

### 3.2 Evaluation Method

For quantitative analysis of experimental results, we compared evaluation indicators such as $F1$, and boundary distance localization error ($BLE$) [31], which are defined as:

$$F1 = 2 \times \frac{Precision \times Recall}{Precision + Recall} \tag{3}$$

Here, Precision and Recall are defined as:

$$Precision = \frac{TP}{TP + FP} \tag{4}$$

$$Recall = \frac{TP}{TP + FN} \tag{5}$$

Among them, $TP$, $TN$, $FP$, and $FN$ represent true positive, true negative, false positive, and false negative respectively. It should be noted that the $F1$ index here is the same as the *dice* score in the field of OD and OC segmentation. Similarly, BLE is used to measure the error (in pixels) between the predicted OD and OC boundary ($C_0$) and the true label boundary ($C_g$). BLE is better able to embody the local (boundary) level of segmentation. It is defined as:

$$BLE(C_0, C_g) = -\frac{1}{T}\sum_{\theta=1}^{N} \left| d_g^\theta - d_o^\theta \right| \tag{6}$$

Here, $d_g^\theta$, $d_o^\theta$ denotes the euclidean distance between $C_g$, $C_0$ and the center point in the direction of $\theta$, respectively, with 24 equidistant points being considered ($T = 24$). The desirable value for BLE is 0.

**Table 1.** OD and OC segmentation results on Drishti-GS1 dataset.

| Method | Year | F1(OD) | BLE(OD) | F1(OC) | BLE(OC) |
|---|---|---|---|---|---|
| U-Net [18] | 2015 | 0.9600 | 7.23 | 0.8500 | 19.53 |
| BCRF [29] | 2017 | 0.9700 | 6.61 | 0.8300 | 18.61 |
| RACE-net [23] | 2018 | 0.9700 | 6.06 | 0.8700 | 16.13 |
| DenseNet FCN [20] | 2018 | 0.949 | – | 0.8283 | – |
| LARKIFCM [22] | 2019 | 0.940 | 9.882 | 0.811 | 23.335 |
| LAC [16] | 2019 | 0.947 | 8.885 | 0.826 | 21.98 |
| HANet [3] | 2019 | 0.9721 | – | 0.8513 | – |
| SLAG-CNN (Proposed) | – | **0.9791** | **4.20** | **0.8929** | **14.01** |

### 3.3 Analysis of Experimental Results

In order to verify the performance of SLAG-CNN, in this group of experiments, on the two indicators of $F1$ and $BLE$, we compared the proposed method with several state-of-the-art segmentation methods such as LARKIFCM [22], LAC [16], and HAnet [3].

The segmentation results of these 51 methods in the testset are shown in Table 1. It can be seen that the proposed method SLAG-CNN has obtained the best performance on these two indicators. Compared with the most advanced attention method HANet [3], its $F1$ has increased by 0.7% on OD. On OC, $F1$ has increased by 4.16%. This proves the effectiveness of the SLAG-CNN.

From the results, we were pleasantly surprised to find that first, the performance of SLAG-CNN is superior to the original U-Net [18], which shows that SLAG is superior to skipping connections in terms of transmitting structural information. On OD, $F1$ increased by 1.91% and $BLE$ decreased by 3.03. On OC, $F1$ increased by 4.29% and $BLE$ decreased by 5.52. This illustrates the effectiveness of SLAG as a network extension path and has superiority in transmitting structural information.

## 4 Ablation Study

### 4.1 Is the Attention Module Effective?

The SLAG module is the main part that improves the segmentation performance. This module combines channel attention and spatial attention, and adds Skip-Link to form a new attention module to enhance the segmentation result. To illustrate the advantages of the SLAG module over other attention methods, we designed a set of ablation experiments based on different attention methods. We use the SLAG-CNN network as a quantitative model, and select two typical attention modules that can be embedded in other models to compare with our SLAG module. Among them, the first is the Squeeze-and-Excitation attention module of SE-Net [26], which is often used in classification tasks. We call it the SE module. It uses the interdependence between convolutional feature channels in the network to It improves the presentation ability of the network. It belongs to the channel attention module. Another attention module is the CBAM module proposed by Woo S [25] and others, which infers the attention map in sequence along two independent dimensions (channel and space), and then multiplies the attention map by the input feature map for adaptive feature improvement. It is a combination of channel attention module and spatial attention module.

In Table 2, we compare the performance of the three attention modules SLAG, SE, and CBAM on OD and OC segmentation. Although two attention modules, SE and CBAM, improve the performance of the model to a certain extent, from the two evaluation indicators of $F1$ and $BLE$, the overall segmentation results of SLAG are higher than those of the two attention modules embedded in the SLAG-CNN network. The SLAG module modifies the traditional channel and spatial attention learning mechanism.

**Table 2.** Using SLAG-CNN as the basic architecture, $F1$ and $BLE$ performance comparison between different attention modules and SLAG modules.

| Dataset | Method | F1 (OD) | BLE (OD) | F1 (OC) | BLE (OC) |
|---|---|---|---|---|---|
| Drishti-GS1 | SE+SLAG-CNN | 0.9713 | 5.70 | 0.8796 | 15.56 |
| | CBAM+SLAG-CNN | 0.9724 | 5.40 | 0.8827 | 15.31 |
| | Ours | **0.9791** | **4.20** | **0.8929** | **14.01** |

Figure 3 shows five examples of retinal images and the segmented images corresponding to different attention modules. In order to more intuitively show the segmentation results between the attention modules and the comparison with Ground Truth, we superimpose all the segmentation results on the example retina image and enlarge the OD and OC regions. It can be clearly seen from the segmentation results in Fig. 3 that our method has clearer segmentation

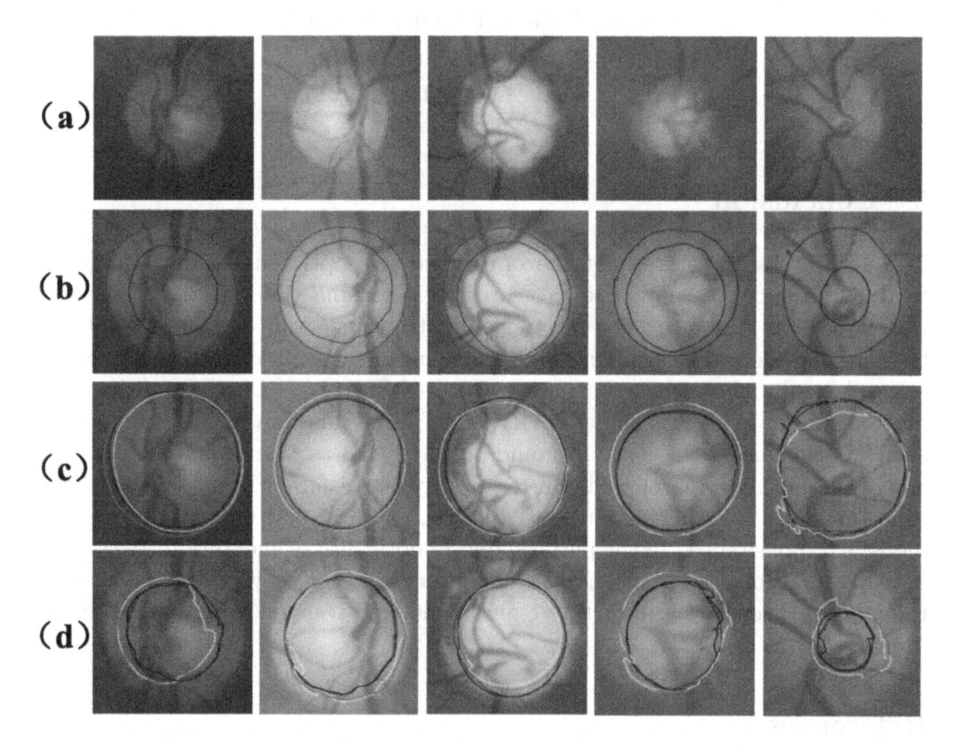


**Fig. 3.** Attention module comparison visualization results. (a) Test images from the Drishti-GS1 dataset; (b) Ground Truth for OD and OC; (c) Ground Truth (blue) and SE+SLAG-CNN (yellow), CBMA+SLAG-CNN (pink) and Ours (black) OD segmentation result comparison graph; (d) Ground Truth (blue) and SE+SLAG-CNN (yellow), CBMA+SLAG-CNN (pink) and Ours (black) OC segmentation result comparison graph. All pictures are pre-processed. (Color figure online)

boundary information. Obviously, the black line representing SLAG is closer to the blue line representing Ground Truth than the other two modules. This proves that on the SLAG-CNN model, the SLAG module has better feature representation capabilities than the SE and CBAM modules.

### 4.2 Does the Model Have Generalization?

In order to verify the generalization ability of the model, we cross-evaluated the model. In other words, the SLAG-CNN model was trained on the Drishti-GS1 dataset, but was tested on the ORIGA dataset at the same time, and the comparison results are shown in Table 3. As expected, SLAG-CNN still achieved the best performance, and the results show that the SLAG-CNN model proposed in this paper has superior generalization ability.

**Table 3.** The $F1$ scores of SLAG and U-Net on the ORIGA dataset.

| Dataset | Method | F1 (OD) | F1 (OC) |
|---|---|---|---|
| ORIGA | U-Net | 0.9590 | 0.9010 |
| | Ours | **0.9841** | **0.9217** |

## 5 Conclusion

The accurate segmentation of the optic disc and optic cup is of great practical significance for helping doctors screen glaucoma. We use the new Skip-Link attention guidance network and polar transformation model SLAG-CNN to jointly segment the optic disc and optic cup in the retinal image. The experimental results show that the performance of SLAG-CNN is better than the most advanced methods such as LARKIFCM, LAC, and HAnet. It is more suitable for segmenting the optic disc and optic cup, so as to automatically screen glaucoma. Future research will aim to improve the accuracy of model segmentation, the correlation between receptive fields, features, and how to improve the model's response speed in the performance of optic disc segmentation. We aim to develop an automated glaucoma screening method that shows lower calculation time, higher accuracy, and further experiments and tests on other retinal image datasets.

## References

1. Singh, V.K., Rashwan, H.A., Saleh, A., et al.: Refuge Challenge 2018-Task 2: Deep Optic Disc and Cup Segmentation in Fundus Images Using U-Net and Multi-scale Feature Matching Networks. arXiv preprint arXiv:1807.11433 (2018)
2. Zhao, R., Liao, W., Zou, B., et al.: Weakly-supervised simultaneous evidence identification and segmentation for automated glaucoma diagnosis. In: Proceedings of the AAAI Conference on Artificial Intelligence, vol. 33, pp. 809–816 (2019)

3. Ding, F., Yang, G., Liu, J., et al.: Hierarchical Attention Networks for Medical Image Segmentation. arXiv preprint arXiv:1911.08777 (2019)
4. Pinz, A., Bernogger, S., Datlinger, P., et al.: Mapping the human retina. IEEE Trans. Med. Imaging **17**(4), 606–619 (1998)
5. Li, H., Chutatape, O.: Automated feature extraction in color retinal images by a model based approach. IEEE Trans. Biomed. Eng. **51**(2), 246–254 (2004)
6. Bhuiyan, A., Kawasaki, R., Wong, T.Y., et al.: A new and efficient method for automatic optic disc detection using geometrical features. In: Dössel, O., Schlegel, W.C. (eds.) World Congress on Medical Physics and Biomedical Engineering, pp. 1131–1134. Springer, Heidelberg (2009). https://doi.org/10.1007/978-3-642-03882-2_301
7. Aquino, A., Gegúndez-Arias, M.E., Marín, D.: Detecting the optic disc boundary in digital fundus images using morphological, edge detection, and feature extraction techniques. IEEE Trans. Med. Imaging **29**(11), 1860–1869 (2010)
8. Roychowdhury, S., Koozekanani, D.D., Kuchinka, S.N., et al.: Optic disc boundary and vessel origin segmentation of fundus images. IEEE J. Biomed. Health Inf. **20**(6), 1562–1574 (2015)
9. Zhou, W., Wu, C., Chen, D., et al.: Automatic microaneurysm detection using the sparse principal component analysis-based unsupervised classification method. IEEE Access **5**, 2563–2572 (2017)
10. Zhou, W., Wu, C., Yi, Y., et al.: Automatic detection of exudates in digital color fundus images using superpixel multi-feature classification. IEEE Access **5**, 17077–17088 (2017)
11. Sudre, C.H., Li, W., Vercauteren, T., Ourselin, S., Jorge Cardoso, M.: Generalised dice overlap as a deep learning loss function for highly unbalanced segmentations. In: Cardoso, M.J., et al. (eds.) DLMIA/ML-CDS -2017. LNCS, vol. 10553, pp. 240–248. Springer, Cham (2017). https://doi.org/10.1007/978-3-319-67558-9_28
12. Long, J., Shelhamer, E., Darrell, T.: Fully convolutional networks for semantic segmentation. In: Proceedings of the IEEE Conference on Computer Vision and Pattern Recognition, pp. 3431–3440 (2015)
13. Kim, J., Tran, L., Chew, E.Y., et al.: Optic disc and cup segmentation for glaucoma characterization using deep learning. In: 2019 IEEE 32nd International Symposium on Computer-Based Medical Systems (CBMS), pp. 489–494. IEEE (2019)
14. Bi, L., Guo, Y., Wang, Q., et al.: Automated Segmentation of the Optic Disk and Cup using Dual-Stage Fully Convolutional Networks. arXiv preprint arXiv:1902.04713 (2019)
15. Shankaranarayana, S.M., Ram, K., Mitra, K., et al.: Fully convolutional networks for monocular retinal depth estimation and optic disc-cup segmentation. IEEE J. Biomed. Health Inf. **23**(4), 1417–1426 (2019)
16. Gao, Y., Yu, X., Wu, C., et al.: Accurate optic disc and cup segmentation from retinal images using a multi-feature based approach for glaucoma assessment. Symmetry **11**(10), 1267 (2019)
17. Gu, Z., et al.: DeepDisc: optic disc segmentation based on atrous convolution and spatial pyramid pooling. In: Stoyanov, D., et al. (eds.) OMIA/COMPAY -2018. LNCS, vol. 11039, pp. 253–260. Springer, Cham (2018). https://doi.org/10.1007/978-3-030-00949-6_30
18. Ronneberger, O., Fischer, P., Brox, T.: U-Net: convolutional networks for biomedical image segmentation. In: Navab, N., Hornegger, J., Wells, W.M., Frangi, A.F. (eds.) MICCAI 2015. LNCS, vol. 9351, pp. 234–241. Springer, Cham (2015). https://doi.org/10.1007/978-3-319-24574-4_28

19. Sevastopolsky, A.: Optic disc and cup segmentation methods for glaucoma detection with modification of U-Net convolutional neural network. Pattern Recogn. Image Anal. **27**(3), 618–624 (2017)
20. Al-Bander, B., Williams, B.M., Al-Nuaimy, W., et al.: Dense fully convolutional segmentation of the optic disc and cup in colour fundus for glaucoma diagnosis. Symmetry **10**(4), 87 (2018)
21. Fu, H., Cheng, J., Xu, Y., et al.: Joint optic disc and cup segmentation based on multi-label deep network and polar transformation. IEEE Trans. Med. Imaging **37**(7), 1597–1605 (2018)
22. Thakur, N., Juneja, M.: Optic disc and optic cup segmentation from retinal images using hybrid approach. Expert Syst. Appl. **127**, 308–322 (2019)
23. Chakravarty, A., Sivaswamy, J.: RACE-net: a recurrent neural network for biomedical image segmentation. IEEE J. Biomed. Health Inf. **23**(3), 1151–1162 (2018)
24. Chen, K., Wang, J., Chen, L.C., et al.: ABC-CNN: an attention based convolutional neural network for visual question answering. arXiv preprint arXiv:1511.05960 (2015)
25. Woo, S., Park, J., Lee, J.Y., et al.: CBAM: Convolutional Block Attention Module (2018)
26. Hu, J., Shen, L., Sun, G.: Squeeze-and-excitation networks. In: Proceedings of the IEEE Conference on Computer Vision and Pattern Recognition, pp. 7132–7141 (2018)
27. Sivaswamy, J., Krishnadas, S.R., Joshi, G.D., et al.: Drishti-GS: retinal image dataset for optic nerve head (ONH) segmentation. In: 2014 IEEE 11th International Symposium on Biomedical Imaging (ISBI), pp. 53–56. IEEE (2014)
28. Ketkar, N.: Deep Learning with Python. Apress, Berkeley (2017)
29. Chakravarty, A., Sivaswamy, J.: Joint optic disc and cup boundary extraction from monocular fundus images. Comput. Methods Programs Biomed. **147**, 51–61 (2017)
30. Wang, L., Yang, S., Yang, S., et al.: Automatic thyroid nodule recognition and diagnosis in ultrasound imaging with the YOLOv2 neural network. World J. Surg. Oncol. **17**(1), 1–9 (2019)
31. Abraham, N., Khan, N.M.: A novel focal tversky loss function with improved attention U-Net for lesion segmentation. In: 2019 IEEE 16th International Symposium on Biomedical Imaging (ISBI 2019), pp. 683–687. IEEE (2019)
32. Kaul, C., Manandhar, S., Pears, N.: FocusNet: An Attention-Based Fully Convolutional Network for Medical Image Segmentation. arXiv preprint arXiv:1902.03091 (2019)
33. Zhang, Z., Yin, F.S., Liu, J., et al.: ORIGA(-light): an online retinal fundus image database for glaucoma analysis and research. In: Engineering in Medicine Biology Society. IEEE (2010)

# LCNet: A Light-Weight Network for Object Counting

Houshun Yu[1] and Li Zhang[1,2](✉)

[1] School of Computer Science and Technology and Joint International Research Laboratory of Machine Learning and Neuromorphic Computing, Soochow University, Suzhou 215006, Jiangsu, China
20195227024@stu.suda.edu.cn, zhangliml@suda.edu.cn

[2] Provincial Key Laboratory for Computer Information Processing Technology, Soochow University, Suzhou 215006, Jiangsu, China

**Abstract.** In recent years, the accuracy of object counting has been greatly improved by applying convolutional neural networks to generate density maps. However, most researchers do not take into consideration the running speed of models but the accuracy of models. The improvement of performance comes at the expense of a large amount of computations, which makes it difficult for models to run on edge devices with a low computing capacity. In this paper, we propose a light-weight but accurate neural network for object counting, called light-weight counting network (LCNet). Our proposed method is a fully convolutional network that contains a small number of parameters and can be trained in an end-to-end way. LCNet adopts a dilated convolution to increase the receptive field and uses a ghost module to further compress the number of network parameters. The performance on the three datasets shows that LCNet achieves the best balance between speed and accuracy. Our method is competent for both sparse and congested scenes. In addition, LCNet can obtain a higher running speed on both GPU and CPU, which is more suitable for edge devices.

**Keywords:** Object counting · Light-weight · Dilated convolution · Ghost module · Convolutional neural network

## 1 Introduction

Object counting has been widely concerned because of its practicality, such as monitoring the crowd density to avoid trampling incidents [2] and obtaining the traffic flow information to thereby control traffic [6]. Most current counting methods are density map-based. These methods generate the density map of its

Supported by the Natural Science Foundation of the Jiangsu Higher Education Institutions of China under Grant No. 19KJA550002, the Six Talent Peak Project of Jiangsu Province of China under Grant No. XYDXX-054, and the Priority Academic Program Development of Jiangsu Higher Education Institutions.


H. Yang et al. (Eds.): ICONIP 2020, LNCS 12532, pp. 411–422, 2020.
https://doi.org/10.1007/978-3-030-63830-6_35

corresponding input image, and obtain the objects number in the input image by summing all the pixels value over the whole density map. Before density map-based methods, researchers also have tried detection-based methods and regression-based methods. The detection-based methods [3,24] first detect the objects in an image, and then obtain the result by counting found objects. For example, Dollár et al. [3] adopted a well-trained classifier that can capture human body features to detect the presence of pedestrians in an image through a sliding window. The final result can be obtained by counting the number of detected pedestrians. Because detection-based methods are prone to lose objects in congested scenes, some researchers proposed methods based on regression [13,16]. Regression-based methods directly map an image to the number of objects contained in this image. The advantage of regression-based methods is that the process of labeling is simple, a label contains only the number of objects in an image. The disadvantage is that they cannot reflect the distribution of the objects in an image, which may result in inaccurate counting.

Since the counting accuracy of detection-based and regression-based methods cannot meet the requirement of real application situations, density map-based methods have been proposed [1,21,27]. By adopting Convolutional Neural Network (CNN), the quality of generated density maps have been greatly improved. Multi-column CNN (MCNN) [27] is a crowd counting network, which composes of three parallel CNN branches. To capture human head features of different scales, the kernel size of filters varies (representing large, medium, and small receptive fields) in different CNN branches. Congested Scene Recognition Network (CSRNet) [11] uses the first ten convolutional layers of pre-trained VGG-16 [20] as the front-end of the model to extract features from the input image. The back-end of CSRNet is a CNN network with dilated convolution. By adopting dilated convolution, CSRNet has achieved the state-of-the-art performance. Due to the success of CSRNet, many later works also use the first ten convolutional layers of pre-trained VGG-16 as the front-end to extract features, and then process the extracted features to generate density maps. For example, Pan-Density Network (PaDNet) [23] inputs the features extracted by the front-end into the several back-end networks trained on images with different density degrees. These density maps are assigned different weights, and then further fused to generate the final density map.

Overall, the accuracy of object counting has been increased with the development of CNNs, but the amount of network parameters, computations, and complexity are also increasing. Heavy computational burden is unfriendly to edge devices, however, few researchers pay attention to the development of light-weight networks for object counting. In addition, most existing light-weight methods are multi-column structure. For instance, top-down feedback CNN (TDF-CNN) [17] contains a bottom-up CNN (BUCNN) with two columns differing in filter sizes to capture features of different scales. TDF-CNN also includes a top-down CNN (TDCNN), which consumes high-level feature maps from BUCNN to generate feedback feature maps. The feedback feature maps are reused by BUCNN to improve the quality of the generated density map.

Another example is Compact CNN (C-CNN) [19] that is based on MCNN [27]. Considering the parameter redundancy of the multi-column structure that causes networks inefficient, C-CNN adopts a solution to reduce redundant parameters in the multi-column structure. The front part of C-CNN contains one three-column convolutional layers with various size filters, while the later part consists of 6 convolutional layers. Filters of different sizes in the front part represent various receptive fields, which can extract features of different sizes. The later part fuses the features extracted from the front part to generate the density map. Compared with MCNN, C-CNN has fewer parameters and runs faster. However, Li et al. [11] showed that the multi-column structure is a less cost-effective way to improve accuracy. In other words, there is room for improvement in the performance of existing light-weight networks.

In this paper, we propose a light-weight network for object counting, called Light-weight Counting Network (LCNet). In order to generate high-quality density maps, LCNet uses dilated convolution to extend the receptive field. Moreover, a ghost module is used to further reduce computations and the number of parameters of the network. Experimental results on three public datasets show that LCNet achieves the optimal balance between accuracy and efficiency.

## 2 Proposed Method

In this section, we introduce the architecture and components of LCNet in detail. Our goal is to achieve lower errors with fewer network parameters. The main idea behind LCNet is to enlarge the receptive field to capture high-level features by deepen a CNN, therefore generating high-quality density maps.

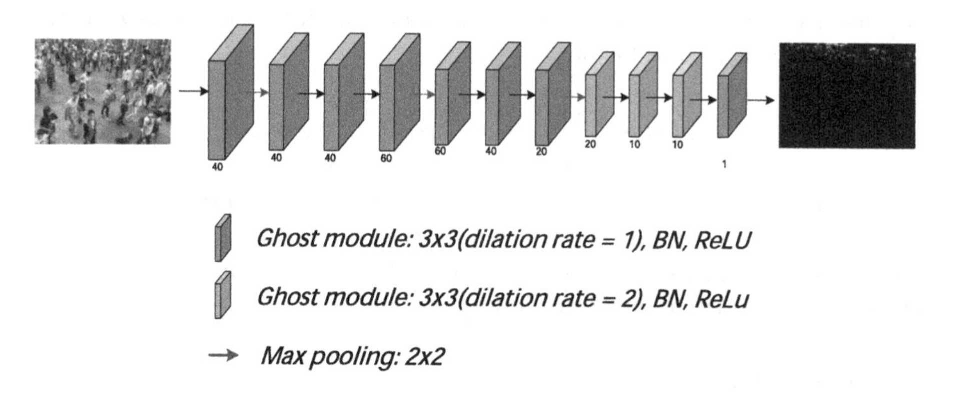


**Fig. 1.** Architecture of LCNet

### 2.1 Architecture

Unlike existing light-weight networks with a multi-column structure, LCNet is a single-column light-weight network with 11 ghost modules [7] (a substitute for a convolutional layer can compress the number of parameters) with a kernel size of $3 \times 3$. Each ghost module is followed by the Batch Normalization (BN) [9] and the ReLU function. To reduce the number of parameters of our method, we adopt the ghost modules instead of traditional convolutional layers to compress the network parameters and reduced the number of filters in each ghost module. The specific network configuration is shown in Fig. 1, where the number of channels is given below the corresponding ghost module.

Generally, a large receptive field may allow the network to extract high-level features, which would result in achieving the goal of generating high-quality density maps. Although deepening a network can also increase the receptive field, the number of parameters would be increased accordingly. Inspired by CSRNet [11], we adopt the dilated convolution operation in each ghost module, which can increase the receptive field of the network without increasing the number of parameters. Referring to CSRNet, we set the dilation rate of ghost modules from seventh to ninth to 2. For other layers, the dilation rate is 1, which means the traditional convolution operation.

The important components, dilated convolution and ghost module, are described in the following.

### 2.2 Dilated Convolution

The effect of dilated convolution is to increase the receptive field of the network without increasing the number of parameters, so as to improve the performance of counting. The main idea behind dilated convolution is to inject "zeros" into a convolution kernel to form a sparse convolution kernel and to increase the receptive field, as illustrated in Fig. 2. Dilated convolution has been proved to be effective. Thus it has been widely used in counting tasks. A 2-D dilated convolution is defined as follow:

$$y(i,j) = \sum_{p=1}^{P} \sum_{q=1}^{Q} x(i + r \cdot p, j + r \cdot q) f(p,q) \tag{1}$$

$x(i,j)$ is the value of input signal at $(i,j)$, $y(i,j)$ is the value of the output signal at $(i,j)$, $f(p,q)$ is the value of a filter of width $P$ and height $Q$ at $(p,q)$, and $r$ represents the dilation rate. Note that when $r = 1$, the dilated convolution is equivalent to the traditional convolution.

### 2.3 Ghost Module

The reason why CNN is powerful is the redundancy in feature maps generated by convolutional layers. Thus, some of these feature maps could be similar to each other, which means that part of feature maps could be generated in cheaper ways instead by the traditional convolution operation. On the basis of the above view, Han et al. [7] proposed a ghost module. In nature, the ghost module applies a series of linear transformations with cheaper cost to generate ghost feature maps based on a set of intrinsic feature maps.

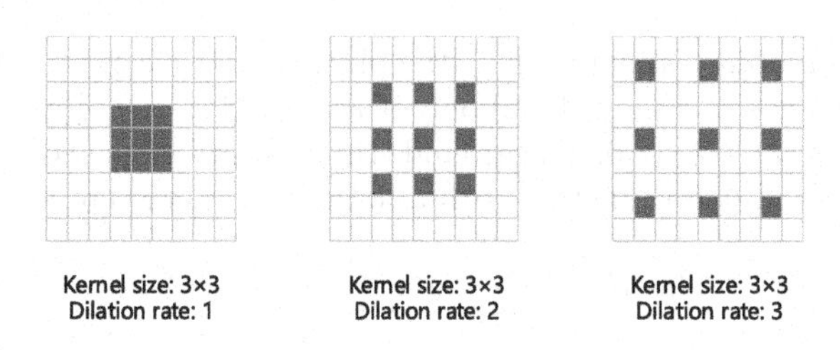


**Fig. 2.** $3 \times 3$ convolution kernels with 1, 2, and 3 dilation rate.

Given an input data $X \in \mathbb{R}^{w \times h \times c}$ for the ghost module, where $w$ and $h$ are the height and width of the input data, respectively, and $c$ stands for the number of channels. Let $s$ be the number of linear operations. To obtain $t$ feature maps, a ghost module first generates $k$ intrinsic feature maps $Y \in \mathbb{R}^{w \times h \times k}$ using a primary convolutional layer, where $k = \lfloor t/s \rfloor$ and $\lfloor \cdot \rfloor$ is a function of rounding the element to the nearest integer towards minus infinity. Namely,

$$Y = X * F + \rho \tag{2}$$

where $*$ denotes the convolution operation, $F \in \mathbb{R}^{c \times P \times Q \times k}$ are the utilized filters, and $\rho$ represents the bias term. Then, the ghost module applies a series of cheap linear operations on each intrinsic feature map in $Y$ to generate the rest $t - k$ ghost features according to:

$$y_{ij}^{g} = \Phi_{i,j}\left(y_i\right), \qquad i = 1, \ldots, k,\ j = 1, \ldots, s-1 \tag{3}$$

where $y_i \in \mathbb{R}^{w \times h}$ is the $i$-th intrinsic feature map in $Y$, $y_{ij}^{g} \in \mathbb{R}^{w \times h}$ is the ghost feature map with respect to $y_i$, and $\Phi_{i,j}$ refers to the $j$-th linear operation on $y_i$. For $j = s$, $\Phi_{i,s}$ is an identity mapping, or $y_{is}^{g} = y_{is}$. The goal of adopting the identity mapping is to preserve intrinsic feature maps. Therefore, each $y_i$ can have one or more ghost feature maps $\left\{y_{ij}^{g}\right\}_{j=1}^{s}$.

By utilizing Eq. (3), we can obtain $t - k$ ghost feature maps, or say $t$ feature maps in total. Because the linear operations cost much less computation, we can generate feature maps in a cheap way. We empirically set $s = 2$ in LCNet, which means the feature maps generated by the ghost module contain half intrinsic feature maps and half ghost feature maps.

# 3 Training Method

## 3.1 Ground Truth Generation

It is known that the annotation information of the input images is center point coordinates of objects. We should first transform the coordinate point annotations into the ground truth density maps. The main idea is to blur each object annotation using a Gaussian kernel [10].

If there is an object annotation at the pixel $(i, j)$, we can represent it with a delta function $\delta(x-i, x'-j)$. If and only if $x = i$ and $x' = j$, $\delta(x-i, x'-j) = 1$; otherwise, $\delta(x-i, x'-j) = 0$. The ground truth density map $D$ can be generated by convolving each delta function with a normalized Gaussian kernel $G_\sigma$:

$$D(x, x') = \sum_{(i,j)\in S} \delta(x-i, x'-j) * G_\sigma \tag{4}$$

where $S$ is the set of annotated points in the input image. Note that the integral of density map $D$ is equal to the object count in the corresponding input image.

### 3.2 Evaluation Metrics

The most commonly used metrics for object counting include Mean Absolute Error (MAE) and Mean Squared Error (MSE), which are defined as follow:

$$MAE = \frac{1}{N}\sum_{i=1}^{N}|\hat{C}_i - C_i| \tag{5}$$

and

$$MSE = \sqrt{\frac{1}{N}\sum_{i=1}^{N}\left(\hat{C}_i - C_i\right)^2} \tag{6}$$

where $N$ is the number of test images, $C_i$ is the count in the $i$-th test image and $\hat{C}_i$ refers to the estimated count for the $i$-th test image that is defined as:

$$\hat{C}_i = \sum_{i=1}^{w'}\sum_{j=1}^{h'} \hat{D}(i, j) \tag{7}$$

where $\hat{D}(i, j)$ is the pixel value at $(i, j)$ in the estimated density map, $w'$ and $h'$ represent the height and width of the estimated density map, respectively. Generally, MAE indicates the accuracy of the result, and MSE represents the robustness. Because MSE is sensitive to outlier samples, the MSE value may be large when the model does not perform well on some samples.

### 3.3 Training Details

Patches with 9/16 size of standard images are first cropped at random, and then they are randomly horizontally flipped for data augmentation. The network parameters are initialized by a Gaussian distribution of zero mean and 0.01 standard deviation. The Adam optimizer with learning rate at $10^{-4}$ is used to train our model. In addition, ground truth density maps are multiplied by a label normalization factor of 1000 according to the way of data preprocess in [4], which helps to accelerate the convergence process and obtain even lower estimation errors. Our experiments are implemented based on the Pytorch framework. The

Euclidean distance is used to measure the difference between the ground truth density map and its estimated density map. Let $\Theta$ represent the parameters of the network. Given the batch of training data $(X_i, D_i)_{i=1}^{N'}$, the loss function can be defined as follow:

$$L(\Theta) = \frac{1}{2N'} \sum_{i=1}^{N'} ||\hat{D}(X_i; \Theta) - D_i||_2^2 \tag{8}$$

where $N'$ represents the number of a training batch and $\hat{D}(X_i; \Theta)$ is the estimated density map.

## 4 Experiments

In this section, we first evaluate and compare our proposed method to the previous state-of-the-art methods on three counting datasets. Then, we compare the speed of different light-weight networks on the same laptop. Finally, we analyze the influence of the number of convolutional layers and number of filters in convolutional layers on the network performance through experimental results.

### 4.1 Accuracy Comparison

First, we show accuracy comparison of LCNet and some state-of-the-art methods on three counting datasets: ShanghaiTech dataset [27], CARPK and PUCPR+ [8].

**ShanghaiTech Dataset.** It is a crowd counting dataset which consists of two parts. Samples in Part_A are randomly crawled from the Internet, samples in Part_B are fixed-size images taken from the streets of Shanghai. 300 images in Part_A are used for training and the remaining 182 images for testing. Part_B contains a total of 716 images, of which 400 images are used for training and 316 images are used for testing. The annotations are the coordinates of the center of human head in target images. The challenges of crowd counting are severe occlusion and the distortion of human head size caused by perspective in congested scenes.

We compare our method with three light-weight networks: MCNN [27], TDF-CNN [17] and C-CNN [19], and report the results in the upper part of Table 1, where the bold value is the lowest among compared light-weight networks. From this part, we can see that LCNet achieves the lowest MAE on both Part_A and Part_B with the smallest parameter number, which indicates that LCNet is capable of object counting in congested scenes. Some samples of Part_B are shown in Fig. 3.

We also compare the proposed LCNet with some large networks: CP-CNN [22], SaCNN [26], ACSCP [18], CSRNet [11] and SFCN [25], and list results in the bottom of Table 1, where the bold value is the lowest among compared large networks. Obviously, not all large networks are superior to our light-weight

**Table 1.** Comparison of networks on the ShanghaiTech dataset.

| Method | Part_A | | Part_B | | #Parameter |
|---|---|---|---|---|---|
| | MAE | MSE | MAE | MSE | |
| MCNN (2016) [27] | 110.2 | 173.2 | 26.4 | 41.3 | 0.15M |
| TDF-CNN (2018) [17] | 97.5 | 145.1 | 20.7 | 32.8 | 0.13M |
| C-CNN (2020) [19] | 88.1 | **141.7** | 14.9 | **22.1** | 0.073M |
| LCNet (Ours) | **87.0** | 143.3 | **13.9** | 22.4 | **0.062M** |
| CP-CNN (2017) [22] | 73.6 | 106.4 | 20.1 | 30.1 | 68.40M |
| SaCNN (2018) [26] | 86.8 | 139.2 | 16.2 | 25.8 | 24.06M |
| ACSCP (2018) [18] | 75.7 | **102.7** | 17.2 | 27.4 | **5.10M** |
| CSRNet (2018) [11] | 68.3 | 115.0 | 10.6 | 16.0 | 16.26M |
| SFCN (2019) [25] | **64.8** | 107.5 | **7.6** | **13.0** | 38.60M |

network even those networks have a large parameter quality. Both CSRNet and SFCN show their strength in the network performance, but the number of their parameters is at least 260 times more than ours.

**Fig. 3.** Some examples from ShanghaiTech dataset Part_B, where the first row shows original images in the test set, the second row contains ground truth density maps, and the third row shows density maps generated by LCNet.

**CARPK and PUCPR+ Datasets.** Both CARPK and PUCPR+ are vehicle counting datasets collected from parking lots, which provide a bounding box annotation for each vehicle [8]. The CARPK dataset contains 989 training images and 459 test images with a total of 89,777 vehicles. All images in this dataset were taken by the drone over the parking lots. The training images are from three different scenes, and the test images are from the fourth scene. The images in PUCPR+ were taken by fixed cameras in high story buildings. This dataset contains 100 training images and 25 test images with a total of 16,930 annotated vehicles.

We compare our method with five detection-based works reported by Goldman et al. [5]: Faster R-CNN [15], YOLO [14], LPN Counting [8], RetinaNet [12], Goldman et al. [5], and two density map-based methods: MCNN and C-CNN. Table 2 shows the detail results, where the lowest values among detection-based methods and density map-based ones are in bold, respectively. The training and test images of the CARPK dataset are from different scenes, which results in a mediocre performance of the density map-based approaches on the CARPK dataset. This exposes the shortcoming of density map-based approaches that are sensitive to scenes and environmental changing. Nevertheless, our proposed approach still obtains the lowest MAE among the light-weight density map-based networks. For the PUCPR+ dataset, the density map-based approaches show their advantage. LCNet achieves the best performance on both MAE and MSE that surpasses other light-weight networks and is close to that of previous state-of-the-art detection-based method of Goldman et al. with only 0.062M parameters.

**Table 2.** Comparison of networks on CARPK and PUCPR+ datasets.

| Method | CARPK | | PUCPR+ | | #Parameter |
|---|---|---|---|---|---|
| | MAE | MSE | MAE | MSE | |
| Faster R-CNN (2015) [15] | 24.32 | 37.62 | 39.88 | 47.67 | 29.67M |
| YOLO (2016) [14] | 48.89 | 57.55 | 156.00 | 200.42 | 119.83M |
| LPN Counting (2017) [8] | 23.80 | 36.79 | 22.76 | 34.46 | **10.24M** |
| RetinaNe t(2018) [12] | 16.62 | 22.30 | 24.58 | 33.12 | 36.33M |
| Goldman et al. (2019) [5] | **6.77** | **8.52** | **7.16** | **12.00** | 38.19M |
| MCNN (2016) | 15.70 | **18.68** | 24.78 | 31.63 | 0.15M |
| C-CNN (2020) | 34.50 | 37.64 | 16.92 | 20.00 | 0.073M |
| LCNet (Ours) | **15.37** | 21.35 | **9.59** | **12.23** | **0.062M** |

## 4.2 Speed Comparison

In this subsection, we compare our proposed LCNet with several other light-weight networks in the performance of running time [17,19,27]. All speed tests are carried out under the same environment in the same laptop with CPU (Intel I5-4210H@2.9 Ghz) and GPU (GeForce GTX 960M). The test samples are tensors of size $[1, 3, 512, 512]$. A frame rate is defined as the number of frame per second (fps) that is used to measure the speed of the networks.

The test results are shown in Table 3. It can be seen from this table that our LCNet runs fastest among compared methods when both GPU and CPU are involved in computing, followed by C-CNN. With CPU alone, owing to the ghost module, LCNet is 66% faster than C-CNN. The advantage of LCNet in speed is obvious when running on CPU. The main reason is that GPU runs so fast to waste time on waiting data transmission when the network is very small. Such experimental results indicate that our method is better suited for deploying on edge devices.

**Table 3.** Speed comparison using frame rate (fps).

| Method | GPU | CPU |
|---|---|---|
| TDF-CNN (2018) [17] | 23.52 | 0.20 |
| MCNN (2016) [27] | 27.39 | 0.21 |
| C-CNN (2020) [19] | 31.84 | 0.30 |
| LCNet (Ours) | **32.12** | **0.50** |

### 4.3 Analysis of Network Architecture

In this subsection, we analyze the influence of the number of convolution layers and filters on the network performance. Ignoring the effect of dilated convolution, LCNet here adopts 11 ghost modules with the traditional convolution. For comparison, we reduce it to a shallow network only with 7 ghost modules with the traditional convolution. The shallower network consists of $GBR(48,3) - M - GBR(64,3) - GBR(64,3) - GBR(48,3) - M - GBR(48,3) - M - GBR(24,3) - GBR(1,3)$, where $GBR(m,n)$ means a ghost module with $m$ filters whose size is $n \times n$ and followed by BN and ReLU, and $M$ represents a max-pooling layer.

Experimental results are shown in Table 4, where the lowest value among compared methods is in bold. Obviously, LCNet (a deeper network) is more accurate than the shallower network with fewer parameters on the ShanghaiTech dataset Part_B. Therefore, on the premise of limited parameters, we consider to deepen the network and reduce the number of filters in each ghost module in the design of our method.

**Table 4.** Comparison between LCNet and the shallower network.

| Method | MAE | MSE | #Parameter |
|---|---|---|---|
| Shallower network | 17.64 | 25.13 | 0.064M |
| LCNet (deeper network) | **15.59** | **25.00** | **0.062M** |

## 5 Conclusion

In order to further improve the accuracy of light-weight networks, a novel light-weight network LCNet is proposed in this paper. The introduction of dilated convolution can increase the receptive fields of the network, thus enabling the network to generate high-quality density maps. Using ghost modules further compresses the number of network parameters and speeds up the network. The results on three datasets show that our method achieves the optimal balance between accuracy and speed within the light-weight networks. According to the experiments of speed comparison, LCNet is suited for deploying on edge devices because LCNet 0.50 fps when running on CPU, that is much better than other compared methods.

## References

1. Boominathan, L., Kruthiventi, S.S.S., Babu, R.V.: Crowdnet: a deep convolutional network for dense crowd counting. In: Proceedings of the 24th ACM International Conference on Multimedia, pp. 640–644 (2016)
2. Cheng, Z., Li, J., Dai, Q., Wu, X., Hauptmann, A.G.: Learning spatial awareness to improve crowd counting. In: Proceedings of the IEEE International Conference on Computer Vision, pp. 6151–6160 (2019)
3. Dollár, P., Wojek, C., Schiele, B., Perona, P.: Pedestrian detection: an evaluation of the state of the art. IEEE Trans. Pattern Anal. Mach. Intell. **34**(4), 743–761 (2012)
4. Gao, J., Lin, W., Zhao, B., Wang, D., Gao, C., Wen, J.: $C^3$ framework: an open-source pytorch code for crowd counting. arXiv preprint arXiv:1907.02724 (2019)
5. Goldman, E., Herzig, R., Eisenschtat, A., Goldberger, J., Hassner, T.: Precise detection in densely packed scenes. In: Proceedings of the IEEE Conference on Computer Vision and Pattern Recognition, pp. 5227–5236 (2019)
6. Guerrero-Gómez-Olmedo, R., Torre-Jiménez, B., López-Sastre, R.J., Maldonado-Bascón, S., Oñoro-Rubio, D.: Extremely overlapping vehicle counting. In: Iberian Conference on Pattern Recognition and Image Analysis (2015)
7. Han, K., Wang, Y., Tian, Q., Guo, J., Xu, C., Xu, C.: Ghostnet: more features from cheap operations. In: Proceedings of the IEEE/CVF Conference on Computer Vision and Pattern Recognition, pp. 1580–1589 (2020)
8. Hsieh, M., Lin, Y., Hsu, W.H.: Drone-based object counting by spatially regularized regional proposal network. In: Proceedings of the IEEE International Conference on Computer Vision, pp. 4165–4173 (2017)
9. Ioffe, S., Szegedy, C.: Batch normalization: accelerating deep network training by reducing internal covariate shift. arXiv preprint arXiv:1502.03167 (2015)
10. Lempitsky, V.S., Zisserman, A.: Learning to count objects in images. In: Advances in Neural Information Processing Systems, pp. 1324–1332 (2010)
11. Li, Y., Zhang, X., Chen, D.: CSRNet: dilated convolutional neural networks for understanding the highly congested scenes. In: CVPR, pp. 1091–1100 (2018)
12. Lin, T., Goyal, P., Girshick, R.B., He, K., Dollár, P.: Focal loss for dense object detection. In: Proceedings of the IEEE International Conference on Computer Vision, pp. 2999–3007 (2017)

13. Rahnemoonfar, M., Sheppard, C.: Deep count: fruit counting based on deep simulated learning. Sensors **17**(4), 905 (2017)
14. Redmon, J., Divvala, S.K., Girshick, R.B., Farhadi, A.: You only look once: unified, real-time object detection. In: Proceedings of the IEEE Conference on Computer Vision and Pattern Recognition, pp. 779–788 (2016)
15. Ren, S., He, K., Girshick, R.B., Sun, J.: Faster R-CNN: towards real-time object detection with region proposal networks. In: Advances in Neural Information Processing Systems, pp. 91–99 (2015)
16. Ryan, D., Denman, S., Fookes, C., Sridharan, S.: Crowd counting using multiple local features. In: 2009 Digital Image Computing: Techniques and Applications, pp. 81–88 (2009)
17. Sam, D.B., Babu, R.V.: Top-down feedback for crowd counting convolutional neural network. In: Thirty-Second AAAI Conference on Artificial Intelligence, pp. 7323–7330 (2018)
18. Shen, Z., Xu, Y., Ni, B., Wang, M., Hu, J., Yang, X.: Crowd counting via adversarial cross-scale consistency pursuit. In: Proceedings of the IEEE Conference on Computer Vision and Pattern Recognition, pp. 5245–5254 (2018)
19. Shi, X., Li, X., Wu, C., Kong, S., Yang, J., He, L.: A real-time deep network for crowd counting. In: ICASSP 2020–2020 IEEE International Conference on Acoustics, Speech and Signal Processing (ICASSP), pp. 2328–2332 (2020)
20. Simonyan, K., Zisserman, A.: Very deep convolutional networks for large-scale image recognition. arXiv preprint arXiv:1409.1556 (2014)
21. Sindagi, V.A., Patel, V.M.: CNN-based cascaded multi-task learning of high-level prior and density estimation for crowd counting. In: 14th IEEE International Conference on Advanced Video and Signal Based Surveillance, pp. 1–6 (2017)
22. Sindagi, V.A., Patel, V.M.: Generating high-quality crowd density maps using contextual pyramid CNNs. In: Proceedings of the IEEE International Conference on Computer Vision, pp. 1879–1888 (2017)
23. Tian, Y., Lei, Y., Zhang, J., Wang, J.Z.: PaDNet: pan-density crowd counting. IEEE Trans. Image Process. **29**, 2714–2727 (2020)
24. Topkaya, I.S., Erdogan, H., Porikli, F.M.: Counting people by clustering person detector outputs. In: IEEE International Conference on Advanced Video and Signal Based Surveillance, pp. 313–318 (2014)
25. Wang, Q., Gao, J., Lin, W., Yuan, Y.: Learning from synthetic data for crowd counting in the wild. In: Proceedings of IEEE Conference on Computer Vision and Pattern Recognition, pp. 8198–8207 (2019)
26. Zhang, L., Shi, M., Chen, Q.: Crowd counting via scale-adaptive convolutional neural network. In: 2018 IEEE Winter Conference on Applications of Computer Vision, pp. 1113–1121 (2018)
27. Zhang, Y., Zhou, D., Chen, S., Gao, S., Ma, Y.: Single-image crowd counting via multi-column convolutional neural network. In: Proceedings of the IEEE Conference on Computer Vision and Pattern Recognition, pp. 589–597 (2016)

# Low-Dose CT Image Blind Denoising with Graph Convolutional Networks

Kecheng Chen[1], Xiaorong Pu[1(✉)], Yazhou Ren[1], Hang Qiu[1], Haoliang Li[2], and Jiayu Sun[3]

[1] School of Computer Science and Engineering, University of Electronic Science and Technology of China, Chengdu, China
puxiaor@uestc.edu.cn

[2] Rapid-Rich Object Search Lab, Nanyang Technological University, Singapore, Singapore

[3] West China Hospital, Sichuan University, Chengdu, China

**Abstract.** Convolutional Neural Networks (CNNs) have been widely applied to the Low-Dose Computed Tomography (LDCT) image denoising problem. While most existing methods aim to explore the local self-similarity of the synthetic noisy CT image by injecting Poisson noise to the clean data, we argue that it may not be optimal as the noise of real-world LDCT image can be quite different compared with synthetic noise (e.g., Poisson noise). To address these issues, instead of manually distorting the clean CT to construct paired training set, we estimate the noise distribution over the real-world LDCT images firstly and then generate noise samples through Generative Adversarial Network (GAN) such that a paired LDCT image dataset can be constructed. To explore the non-local self-similarity of LDCT images, Graph Convolutional Layers (GCLs) is utilized to obtain the non-local patterns of LDCT images. Experiments were performed using real-world LDCT image dataset and the proposed method achieves much better performance compared with other approaches with respect to both quantitative and visual results.

**Keywords:** CT · Graph Convolutional Layers · Noise modeling · GAN

## 1 Introduction

Among various Computed Tomography (CT) techniques, Low-Dose CT (LDCT) examination has become a trend in past decades due to the danger of cumulative ionizing radiation caused by the increasing number of CT examinations. However, compared with the Normal-Dose CT (NDCT) images, the quality of LDCT images (e.g., the details and anatomical structure) can be significantly degraded,

This work was supported in part by Science and Technology Plan Foundation of Sichuan Science and Technology Department (No. 2020YFS0119), National Natural Science Foundation of China (No. 61806043), and China Postdoctoral Science Foundation (No. 2016M602674).


H. Yang et al. (Eds.): ICONIP 2020, LNCS 12532, pp. 423–435, 2020.
https://doi.org/10.1007/978-3-030-63830-6_36

which brings negative impacts to the radiologist's judgment. In order to address this issue, there exist a lot of works based on the noise reduction of LDCT images. These methods can be categorized as sinogram filtration based methods, iterative reconstruction based methods and post-processing based methods. The sinogram filtration based methods directly apply filters (e.g., statistics-based nonlinear filters [19], bilateral filtering [12]) for projection data processing. The iterative reconstruction based methods, including the Total Variation (TV) based [18] and dictionary learning based [4], aim to transform the data from projection domain and image domain, which can further be used for noise reduction of LDCT images. However, the huge time-consumption of data transformation impedes the practical application of aforementioned methods.

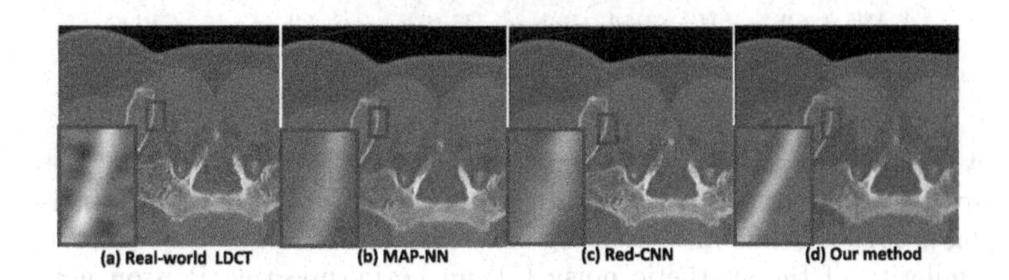


**Fig. 1.** An example of real-world LDCT image denoising. (a) the original LDCT image. (b) denoising results by using MAP-NN [15]. (c) denoising results by using Red-CNN [1]. (d) results of our proposed method. Zoom in for better view. (Color figure online)

Thanks to the development of deep learning, researchers have found that applying post-processing through deep convolutional neural networks (CNNs) based LDCT denoising methods can achieve better performance compared with non-deep learning-based methods. To the best of our knowledge, [2] first proposed to adopt the deep CNNs to process LDCT image. Following [2], some other CNNs-based models have also been introduced, such as residual encoder-decoder (Red-CNN) [1], modularized deep neural network (MAP-NN) [15] and GAN-based methods [14]. However, the existing deep learning-based LDCT denoising methods suffer from two limitations:

- Existing deep learning-based LDCT denoising methods [7,14,15] rely on synthetic LDCT images by injecting Poisson noise to the clean data, such that data pair can be constructed for training purpose. However, the real-world noise can be much more complex, directly applying trained model on synthetic data may not be able to generalize well to the real-world samples. One example is shown in Fig. 1, where we found that the details of the edge can be blurred when we trained with synthetic data and tested with real-world data.
- Existing deep learning-based LDCT denoising methods use CNNs [1,2,14,15], which can only explore the local self-similarity of LDCT images. However, it is found that the non-local self-similarity can also benefit LDCT image denoising problem. Non-local means (NLM) [11] and blocks matching 3D

filtering (BM3D) based methods [10] achieved prominent denoising performances. Such non-local self-similarity information may not be able to be directly learned through CNNs as the convolutional layer processes data only for its receptive field [17].

In this paper, we propose a novel LDCT image denoising framework to tackle the aforementioned problems. Our framework consists of two stages. We first propose to leverage GAN to implicitly learn the real-world LDCT noise such that we can construct the clean and noisy data paired by further leveraging the extra NDCT images and the generated noise through the trained GAN. Secondly, to explore the non-local self-similarity of LDCT images, we propose a novel model based on Graph Convolutional Layers (GCLs) by treating the synthetic LDCT images as input and the NDCT images as ground-truth. In particular, every CT voxel is treated as a vertex and can be represented as a latent feature through a multi-scale feature extraction layer. Thus, the non-local self-similarities can be learned based on the CT voxel in feature space, which can be further adopted to improve the effectiveness of LDCT image denoising.

The contributions of this work are two-fold:

- Instead of simply injecting Poisson noise to construct training data pair, we firstly estimate the noise distribution over the real-world LDCT images and then generate noise samples through GAN. We find that such formulation can deal with real-world noise in LDCT images effectively.
- We propose a novel framework by leveraging the advantage of Graph Convolutional Layer (GCL) together with convolutional neural network, such that both local and non-local self-similarity can be explored.

## 2 Proposed Method

### 2.1 Noise Modeling for LDCT Image

**LDCT Image Noise Extraction.** To extract reliable noise information, we need the relatively smooth patch (the region where the internal parts are very similar) of images, where the background is weak and noise information dominants. For natural images, the smooth patch can be obtained by measuring the similarity of mean and variance between a candidate patch and its sub-patch [3]. Unlike natural images, most regions of the LDCT image naturally have similar mean except the regions of high tissue density [18]. Thus, we propose to extract noise from the relatively smooth patch of LDCT images by leveraging the intrinsic property of the LDCT image and the basic noise extraction methodology. Under this condition, given an LDCT image patch $\mathbf{H}_k$, as well as its $i$-th sub-patch $\mathbf{h}_{k,i}$, where $\mathbf{h}_{k,i}$ is a sub-region on $\mathbf{H}_k$. In our algorithm, the smooth patch of LDCT image must meet a predefined constraint given as

$$|Variance(\mathbf{h}_{k,i}) - Variance(\mathbf{H}_k)| \leq \beta \cdot Variance(\mathbf{H}_k), \tag{1}$$

where $Variance(\cdot)$ denotes the "variance" operator. $\beta$ is a hyper-parameter predefined and ranges from zero to one. For each $i$, if the constraint is satisfied, $\mathbf{H}_k$ will be regarded as a smooth patch. We group the smooth patches which meet the constraint as $\mathbb{H}$, denoted as $\mathbb{H} = \{\mathbf{H}_1, \mathbf{H}_2, ..., \mathbf{H}_K\}$. Under the assumption that the expectation of the noise distribution is zero, the noise information of smooth patch $\mathbf{H}_j$ can be approximated as

$$\mathbf{N}_j = \mathbf{H}_j - Mean(\mathbf{H}_j), \quad j = 1, 2, ..., K \tag{2}$$

The noise patches can be grouped as $\mathbb{N} = \{\mathbf{N}_1, \mathbf{N}_2, ..., \mathbf{N}_K\}$. In general, the imaging areas of CT have two intrinsic properties, including a large area of low or middle tissue density and high resolution. Thus, a large number of eligible smooth areas can be obtained to further generate noise patches for noise modeling purpose through GAN, which will be presented below.

**Noise Modeling with GAN.** Despite a LDCT image can generate a relatively large number of noise patches, the number of available LDCT images is still limited in real world. Besides, the diversity of noise samples also affects the performance of data-driven methods. We then propose to leverage GAN to model the distribution of LDCT image noise patterns, as it can learn the complex distribution over objective samples and outputs a distribution close to objective samples [8]. In our work, the set $\mathbb{N}$ contains all the objective samples. We adopt WGAN-GP [8] for noise modeling purpose, where the objective is given as

$$\mathcal{L}_{GAN} = \mathbb{E}_{\tilde{x}\sim\mathbb{P}_g}[D(\tilde{x})] - \mathbb{E}_{n\sim\mathbb{P}_n}[D(n)] + \lambda\mathbb{E}_{\hat{x}\sim\mathbb{P}_{\hat{x}}}[(\|\nabla_{\hat{x}}D(\hat{x})\|^2 - 1)], \tag{3}$$

where $\mathbb{P}_g$ is the generator distribution, and $\mathbb{P}_n$ is the distribution among extracted noise patch set which satisfies the constraint introduced in previous sub-section. $\mathbb{P}_{\hat{x}}$ is defined as a distribution sampling uniformly along straight lines between pairs of points sampled from $\mathbb{P}_n$ and $\mathbb{P}_g$. The input of GAN follows uniform distribution $U[-1, 1]$. We refer the reader to [8] for more details.

After noise modeling, we can obtain more noise blocks and further use them to construct training data pair by leveraging extra NDCT images. To be more specific, by denoting a clean NDCT image as $\mathbf{I}$ and a random generated noise pattern through GAN as $\mathbf{N}$, we can obtain the training sample pair as $\{(\mathbf{I} + \mathbf{N}, \mathbf{N})\}$.

## 2.2 Low-Dose CT Graph-Convolutional Network

We now introduce our proposed framework for LDCT image denoising. Our framework consists of four main components. The overall framework is shown in Fig. 2.

**Sobel Edge Extraction Layer.** Though CNNs-based methods have been proved to be effective for LDCT image denoising problem, it has also been found that reconstructed images turn out to be blurry on the edge of Region Of Interest (ROI). One explanation is that the general objective function of Mean Square

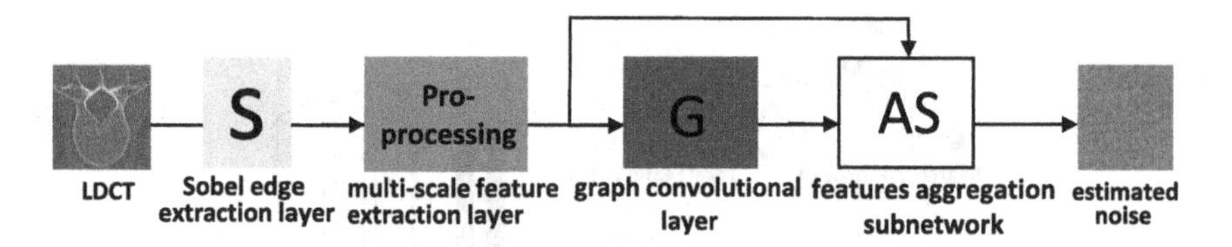


**Fig. 2.** The architecture of the proposed model. The architecture details of feature aggregation subnetwork is introduced in the Sect. 2.2.

Error (MSE) aims to average pixel-based errors, which tends to cause blurring artifacts [14]. Some works aim to address this limitation by optimizing objective function, but the improvement is still limited [20]. Though CNNs can automatically learn the implicit prior of the input image, recent researches have been shown that the effectiveness of image enhancement can be improved by using CNNs with some explicit prior [21]. Thus, we propose to adopt the prior with the edge of a given LDCT image input, which can be obtained through a Sobel edge extraction layer. We expect that the blurring artifact can be mitigated by doing so.

Specifically, the widely used Sobel edge extraction filters are adopted [6]. Then, the channel-wise concatenation process is conducted such that different features can be flexibly selected by subsequent networks [6]. The synthetic LDCT image $\mathbf{I}+\mathbf{N}$ can be concatenated with its filtered output $G(\mathbf{I}+\mathbf{N})$ in channel-wise manner, where $G$ denotes the sobel edge extractor in both vertical and horizontal direction.

**Multi-scale Feature Extraction Layer.** In this paper, a multi-scale feature extraction layer, mainly following a module of parallel branches CNNs in [5], is embedded into our network framework as a role of pro-processing. The effectiveness of this layer has two sides:

- Firstly, multi-scale image features can be extracted by means of convolutional filters with different sizes, it is proven to be beneficial for post-processing [17], especially in the denoising task [5,17]. To be specific, the sizes of parallel convolutional filters are $3 \times 3$, $5 \times 5$, and $7 \times 7$, respectively.
- Secondly, through this CNNs-based module, low-level LDCT image features are abstracted into high-level representation of feature space, that can be further imposed for extracting non-local features by graph convolution.

**Graph Convolutional Layer.** As non-local self-similarity can benefit LDCT image denoising problem. Thus, we propose to adopt graph convolutional layer to explore the non-local self-similarity information. The input will be treated as the high-level representation of feature space and will be further used for aggregation purpose.

It is worth noting that the voxels of CT image has been abstracted into feature space. To simplify the description, we still call a abstracted feature of CT image as a voxel. For every voxel of a LDCT image, a $K$-regular graph $G$ can

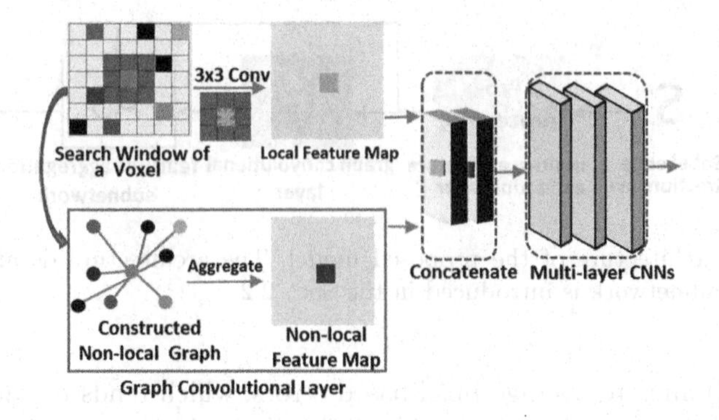


**Fig. 3.** The local and non-local features aggregation subnetwork. To be more intuitive, the Graph Convolutional Layer, which do not belong to this subnetwork, is also shown in here. The output of Graph Convolutional Layer is used by this proposed subnetwork.

be constructed through a $M \times M$ search window (the top left sub-figure in Fig. 3 illustrates the search window of a green voxel) by treating each voxel as a vertex. Let $G = (V, E)$, where $V$ is the set of vertices and $E$ denotes the edge which captures the similarity between vertices pair. In the $l$-th graph convolutional layer, the $i$-th vertex $s_i^l$ can be represented by mapping its corresponding voxel to a feature vector $\mathbf{s}_i^l \to \mathbb{R}^{d_l}$ ($d_l$ is the dimension of the feature vector) through a CNNs-based module. For every vertex $s_i^l$, the L1 norm distances between $s_i^l$ and its non-local voxels are computed within the search window. To explore the non-local self-similarity information, we do not consider the similarity between two vertices if they are adjacent. By denoting the adjacent vertices (a.k.a. local voxels, which are shown as the red voxels in the top left sub-figure of Fig. 3) of the $i$-th vertex as $LV_i$, we can compute the similarity by

$$\mathcal{L}(s_i^l, s_j^l)_{j \in N^l(i)} = \|\mathbf{s}_i^l - \mathbf{s}_j^l\|_1, \tag{4}$$

where $N^l(i) = \{P\} - \{LV_i\}$ denotes the non-local voxels of vertex $i$, and $\{P\}$ denotes the overall voxels of vertex $i$. $\mathbf{s}_j^l$ denotes the feature vector of non-local voxels $s_j^l$. We assign a label to the edge if the L1 norm distance between the $i$-th vertex and its $j$-th non-local vertex is among the $(K-1)$-th, where we denote the edge label as $z^{l,j \to i}$ (the lower left sub-figure in Fig. 3 illustrates a constructed graph for the green voxel). We then adopt the dynamic edge-conditioned convolution (ECC) [16] to aggregate the non-local self-similarity features based on vertex $i$, which can be given as

$$\mathbf{s}_i^{l+1} = \sum_{j \in N^l(i)} \frac{F^l(z^{l,j \to i}, w^l)\mathbf{s}_i^j}{|N^l(i)|} + \mathbf{b}^l = \sum_{j \in N^l(i)} \frac{\Theta_{j,i}^l \mathbf{s}_i^j}{|N^l(i)|} + \mathbf{b}^l, \tag{5}$$

where $F^l$ denotes the output of a network parameterized by $w^l$ which is used to dynamically produce the weight matrix $\Theta_{j,i}^l$ for different edge labels $z^{l,j \to i}$, and $\mathbf{b}^l$ is a learnable bias. The optimization details can be found in [16]. In our work, we adopt a multi-layer perception network for $F^l$.

**Adaptive Local and Non-local Features Aggregation Subnetwork.** Besides the non-local self-similarity information, local self-similarity information can also benefit LDCT image denoising. Thus, we propose a feature aggregation subnetwork to aggregate the learned non-local self-similarity information together with the local one. To be more specific, we first leverage a general $3 \times 3$ convolution operation to obtain the local feature map, which represents the local information of every voxel. The channel-wise concatenation is then used to concatenate the local feature map and the non-local feature map. Note that the non-local feature map is the non-local information obtained from graph convolutional layer. Finally, we adopt the plain CNNs to learn the aggregation manner of local and non-local features, aiming to achieve more adaptive output of approximated noise for aggregation and denoising purpose, where the framework is shown in Fig. 3.

By denoting the local feature of the input as $\mathbf{s}^L$ and the last layer of graph convolutional layer as $\mathbf{s}$, the aggregation process can be denoted in a high level as:

$$\mathbf{Y} = F(Concatenate(\mathbf{s}, \mathbf{s}^L)), \tag{6}$$

where $Concatenate(\cdot, \cdot)$ denotes the channel-wise concatenation operator among two inputs, $F(\cdot)$ denotes a simple three-layer fully convolutional network based on residue module. For each convolution layer, the number of feature channel in the middle layers is set as 32, the size of convolution kernel is $3 \times 3$ with leaky-Relu as the activation function. Instead of estimating the reconstructed LDCT image, we propose to estimate the synthetic noise as it can lead to better convergence and more robust reconstruction performance [21]. The reconstructed image can thus be represented as $\hat{\mathbf{I}} = \mathbf{I} + \mathbf{N} - \mathbf{Y}$.

Similar to other literatures in the field of LDCT image denoising, we impose the MSE loss as the objective to guide the training, which can be represented as $\mathcal{L} = ||\hat{\mathbf{I}} - \mathbf{I}||_2^2$.

# 3 Experiments

## 3.1 Datasets and Experimental Settings

**Datasets and Setting Details.** As there are no groundtruth available for real-world LDCT images, we therefore utilize the LVS CT Image Dataset [9], which includes 7850 real-world low-dose lumbar vertebra CT images in the size of 1024× 1024 without normal-dose counterparts, for noise extraction purpose. Some LVS data samples are shown in Fig. 4(a) to 4(c). We utilize FUMPE[1] dataset, which consists of NDCT images for pulmonary embolism of 35 different subjects with 5000 normal-dose images in the size of $512 \times 512$, for real-world data simulation purpose.

[1] https://figshare.com/collections/FUMPE/4107803.

Regarding the parameters at LDCT image noise extraction, we refer to the settings of natural images in [3], where the size of patch, the size of sub-patch, and $\beta$ are set to 64, 16, 0.25, respectively. We follow [5] to adopt the architecture for the parallel CNNs-based module to conduct pro-processing. For the non-local self-similarity information extraction, we mainly adopt the architecture proposed in [17] by setting the search window size as $65 \times 65$, aiming to basically cover the extracted patch. $K$ is set to 16. We will discuss the impact of $K$ in our ablation study. The Adam algorithm is adopted for optimization purpose with the learning rate starting from $10^{-4}$ and then fixed to $10^{-5}$ when the training error stops decreasing.

**Baseline Methods.** We consider to adopt the data-driven LDCT image denoising models, which recently achieve state-of-the-art denoising performance on synthetic dataset. Some non-data-driven LDCT image denoising models are also considered for comparison. The details of each baseline methods are summarized below.

- **Red-CNN** [1], **CPCE** [14], and **MAP-NN** [15]: These methods are based on the convolutional encoder-decoder structure for LDCT image denoising. MAP-NN is built by stacking multiple CPCEs. All of above-mentioned methods do not adopt specific measures to face blind denoising issue. Instead, they utilize the generalization of the model to handle the real-world LDCT. The settings follow their original papers.
- **CT-BM3D** [10], **DLP-CT** [4]: These are non-deep-learning methods for LDCT image denoising. CT-BM3D naturally leverages the non-local information between patches, while DLP-CT uses a fast dictionary learning method for LDCT image denoising.

In addition, some previous literatures [2,14] often verify the adaptivity of their proposed models compared with natural image denoising models, such as NL-CID [11], FFDNet [21], and TV-L1 [13]. Thus, we also test these methods in our experiments, where NL-CID designs a non-local regularization operator to explore the non-local information of natural image.

**Evaluation Metrics.** We consider to use various metrics based on image quality assessment for evaluation purpose. As we have groundtruth available for synthetic LDCT images, we follow the benchmarks by adopting the Peak Signal-to-Noise Ratio (PSNR), Structural Similarity Index (SSIM) and Root Mean Square Error (RMSE) for evaluation. For real-world LDCT images, as there is no groundtruth available, we use one of the most popular no-reference based image quality assessment metrics, i.e., Contrast-to-Noise Ratio (CNR), for evaluation.

### 3.2 Results on Synthetic LDCT Images

We first evaluate our proposed method on synthetic LDCT images in terms of quantitative and visual quality results. The results are shown in Table 1. We have the following observations:

- Firstly, under the condition of paired training data, general natural image denoising models (e.g., FFDNet, NL-CID) based on deep learning can also achieve acceptable performances. However, compared with our proposed LDCT-GCN, the lower quantitative results of NL-CLD prove that the usage of a non-local special structure (graph convolutional layer in this paper) is very necessary.
- Secondly, our proposed LDCT-GCN outperforms other CNNs-based LDCT denoising methods such as MAP-NN, Red-CNN and CPCE, which further indicates the importance by introducing graph convolutional layer for LDCT image denoising problem. CT-BM3D achieves the best result in terms of RMSE. However, it may lead to blurry artifact as claimed in [1].

**Table 1.** The PSNR, SSIM and RMSE results of different models. B/N-B denote the Blind denoising based methods and Non-Blind based methods, respectively. The higher the better for PSNR and SSIM, the smaller the better for RMSE.

| Methods | TV-L1 | NL-CID | FFDNet | CT-BM3D | MAP-NN | Red-CNN | CPCE | DLP-CT | **Ours** |
|---|---|---|---|---|---|---|---|---|---|
| B/N-B | B | N-B | N-B | B | N-B | N-B | N-B | B | B |
| PSNR | 21.62 | 22.76 | 24.77 | 23.12 | 23.54 | 24.79 | 21.65 | 23.61 | **25.16** |
| SSIM | 0.43 | 0.57 | 0.77 | 0.75 | 0.71 | 0.74 | 0.65 | 0.72 | **0.85** |
| RMSE | 7.59 | 7.58 | 6.04 | **5.40** | 6.48 | 6.84 | 6.48 | 6.89 | 6.21 |

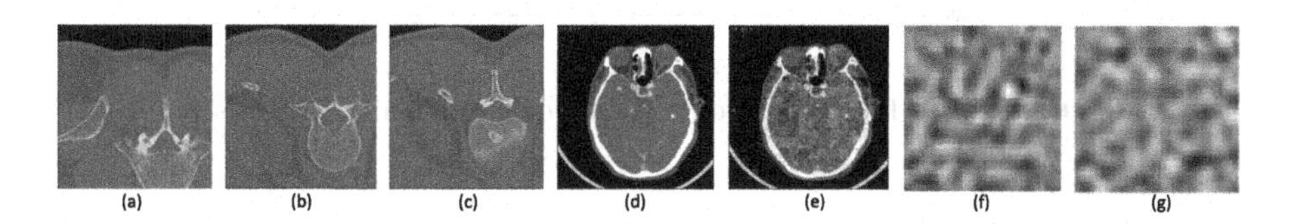


**Fig. 4.** (a) to (c): the blind test samples, every sub-figure is real-world lumbar vertebra LDCT image, (d) the NDCT image sample on FUMPE dataset, (e) the synthetic LDCT image on the foundation of (d), (f) real-world noise which extracted from real LDCT image, (g) generated noise by GAN. Zoom in for better view.

We further conduct comparison based on visual quality. Some results are shown in Fig. 5. As we can see, although all methods can reduce the noise to some extent, the reconstructed image by using our proposed method is sharper with less unexpected artifacts (please refer to the red box and green box in Fig. 5 for more details).

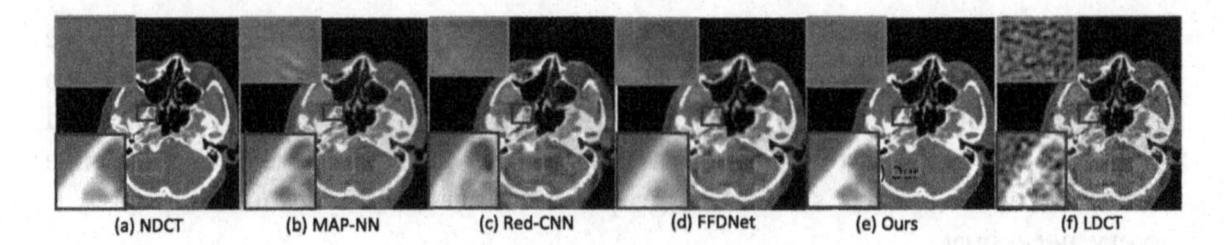


**Fig. 5.** Comparison on FUMPE in the evaluation of synthetic LDCT image. Zoom in for better view. (Color figure online)

**Table 2.** The PSNR, SSIM and RMSE results of different $K$ value setting among feature aggregation. The higher the better for PSNR and SSIM, the smaller the better for RMSE.

| $K$ | 0 | 4 | 8 | 16 | 20 |
|---|---|---|---|---|---|
| PSNR | 25.01 | 25.44 | 25.49 | **25.80** | 24.89 |
| SSIM | 0.72 | 0.76 | 0.73 | **0.77** | 0.68 |
| RMSE | 6.31 | 5.52 | 5.66 | **5.33** | 6.38 |

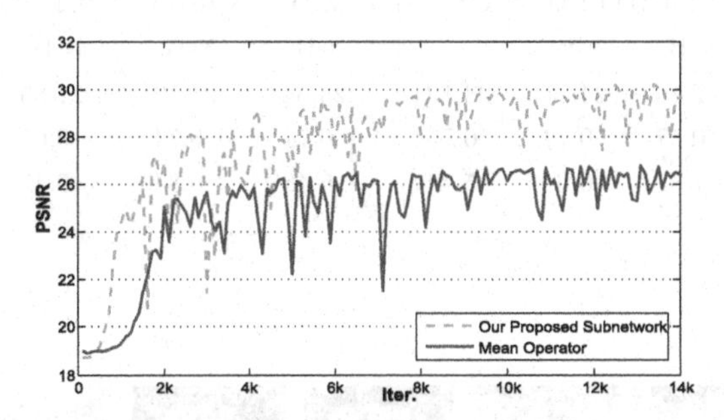


**Fig. 6.** The PSNR results of different models as training progressed.

### 3.3 Ablation Study

**Noise Modeling Analysis.** We are interested in the visual quality of synthetic LDCT images. Figure 4(e) reports the synthetic LDCT image samples by adding the generated noise from GAN, where the clean NDCT images (e.g., Fig. 4(d)) are from FUMPE dataset. We can notice that the texture of synthetic LDCT images is similar with the real-world LDCT images shown in Fig. 4(a) to 4(c). Our proposed method can benefit from this property to handle unknown real-world LDCT image. We further visualize the noise generated by our proposed framework. The visualization results are shown in Fig. 4(g).

**Graph Convolutional Layer Analysis.** There is a key parameter involved in our graph convolutional layer: $K$. As aforementioned, $K$ is used to determine how many non-local voxels are regarded as the vertices in a constructed graph, aiming to explore the non-local self-similarity information. To test its influence, we conduct the experiments on synthetic LDCT images. Table 2 shows the quantitative results by varying $K$. We have the following observations: Firstly, the results of $K = 16$ can achieve better performance compared with $K = 4$ and $K = 8$, which indicates that the introduction of relatively more non-local information in a certain range is beneficial to LDCT image denoising problem. However, the results of $K = 20$ suffer from degradation phenomenon, which indicates that too much non-local information may not be always helpful. Secondly, we observe that the results degrade dramatically when $K = 0$, which indicates that non-local information contribute to the final performance.

**Features Aggregation Subnetwork Analysis.** Local and non-local features aggregation subnetwork aims to achieve more adaptive denoising performance. To prove its effectiveness, we replace this subnetwork with a mean operator for comparison. Figure 6 shows that the results of adaptive subnetwork can achieve significantly better performance as the training progressed.

### 3.4 Results on Real-World LDCT Images

Finally, we conduct experiments on real-world LDCT images by using CNR for evaluation. The results are shown in Table 3. We can observe that our proposed method can effectively handle the real-world LDCT image denoising problem. We also conduct comparison in terms of visual quality in Fig. 7. As we can see, the result by using CT-BM3D turns out to be over-smooth and heavily quantized. Both DLP-CT and Red-CNN have a poor noise suppression effect especially inside the ROI. Our proposed method can lead to a more realistic

**Table 3.** The CNR results of different compared methods by considering real-world noise denoising tasks with LVS dataset. The smaller the better for CNR.

| Methods | MAP-NN | Red-CNN | DLP-CT | CT-BM3D | Our Method |
|---|---|---|---|---|---|
| CNR | 2.55 | 2.53 | 1.45 | 1.31 | **1.23** |

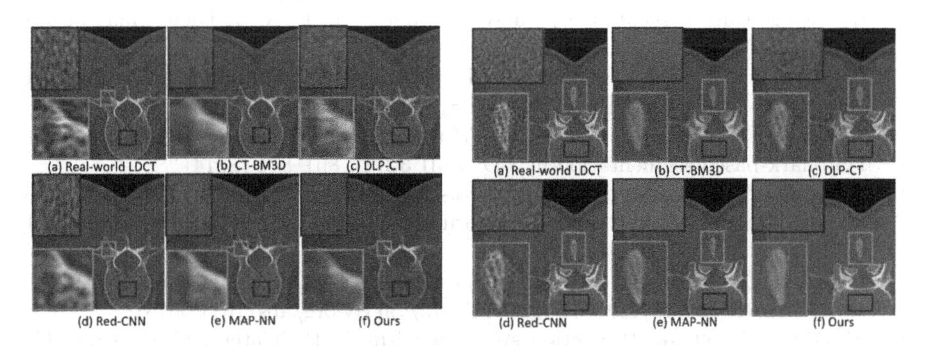


**Fig. 7.** Visual quality comparison on LVS in the evaluation of real-world LDCT image denoising. Zoom in for better view (Color figure online)

textures (see the red box) compared with other methods under real-world noise condition, and can lead to more natural transition of details compared with other methods (see the green box), which is favoured by radiologists.

## 4 Conclusion and Future Work

We first propose to utilize GAN to learn noise distribution and construct data pair for training. Then, graph convolutional layer is adopted to explore the non-local self-similarity of LDCT image, which is typically ignored by existing CNNs-based LDCT denoising methods. Extensive experiments conducted on both synthetic and real-world LDCT images show that our proposed method achieves much better performance compared with state-of-the-art methods. An intuitive advantage is that the proposed framework is more adaptive to clinical situation, as the blind denoising demand always exists in practice. In the future, we will consider more radiologist's favors into the LDCT denoising scheme.

## References

1. Chen, H., et al.: Low-dose CT with a residual encoder-decoder convolutional neural network. TMI **36**(12), 2524–2535 (2017)
2. Chen, H., et al.: Low-dose CT via convolutional neural network. Biomed. Opt. Express **8**(2), 679–694 (2017)
3. Chen, J., Chen, J., Chao, H., Yang, M.: Image blind denoising with generative adversarial network based noise modeling. In: CVPR, pp. 3155–3164 (2018)
4. Chen, Y., et al.: Improving abdomen tumor low-dose CT images using a fast dictionary learning based processing. Phys. Med. Biol. **58**(16), 5803 (2013)
5. Divakar, N., Venkatesh Babu, R.: Image denoising via CNNs: an adversarial approach. In: CVPRW, July 2017
6. Du, C., Wang, Y., Wang, C.: Selective feature connection mechanism: concatenating multi-layer CNN features with a feature selector. Pattern Recogn. Lett. **129**, 108–114 (2018)
7. Gholizadeh-Ansari, M., Alirezaie, J., Babyn, P.: Deep learning for low-dose CT denoising using perceptual loss and edge detection layer. J. Digit. Imaging **33**, 1–12 (2019)
8. Gulrajani, I., Ahmed, F., Arjovsky, M., Dumoulin, V., Courville, A.C.: Improved training of Wasserstein GANs. In: NIPS, pp. 5767–5777 (2017)
9. Ibragimov, B., Likar, B., Pernuš, F., Vrtovec, T.: Shape representation for efficient landmark-based segmentation in 3-D. TMI **33**(4), 861–874 (2014)
10. Kang, D., et al.: Image denoising of low-radiation dose coronary CT angiography by an adaptive block-matching 3D algorithm. In: MIIP, vol. 8669, p. 86692G. ISOP (2013)
11. Lefkimmiatis, S.: Non-local color image denoising with convolutional neural networks. CoRR abs/1611.06757 (2016). http://arxiv.org/abs/1611.06757
12. Manduca, A., et al.: Projection space denoising with bilateral filtering and CT noise modeling for dose reduction in CT. Med. Phys. **36**(11), 4911–4919 (2009)
13. Qin, Z., Goldfarb, D., Ma, S.: An alternating direction method for total variation denoising. Optim. Meth. Softw. **30**(3), 594–615 (2015)

14. Shan, H., et al.: 3-D convolutional encoder-decoder network for low-dose CT via transfer learning from a 2-D trained network. TMI **37**(6), 1522–1534 (2018)
15. Shan, H., et al.: Competitive performance of a modularized deep neural network compared to commercial algorithms for low-dose CT image reconstruction. Nat. Mach. Intell. **1**(6), 269 (2019)
16. Simonovsky, M., Komodakis, N.: Dynamic edge-conditioned filters in convolutional neural networks on graphs. In: CVPR, pp. 3693–3702 (2017)
17. Valsesia, D., Fracastoro, G., Magli, E.: Deep graph-convolutional image denoising. arXiv preprint arXiv:1907.08448 (2019)
18. Varshowsaz, M., Goorang, S., Ehsani, S., Azizi, Z., Rahimian, S.: Comparison of tissue density in hounsfield units in computed tomography and cone beam computed tomography. J. Dentis. (Tehran, Iran) **13**(2), 108 (2016)
19. Wang, J., Lu, H., Li, T., Liang, Z.: Sinogram noise reduction for low-dose CT by statistics-based nonlinear filters. In: MIIP, vol. 5747, pp. 2058–2066. International Society for Optics and Photonics (2005)
20. Yang, Q., et al.: Low-dose CT image denoising using a generative adversarial network with Wasserstein distance and perceptual loss. TMI **37**(6), 1348–1357 (2018)
21. Zhang, K., Zuo, W., Zhang, L.: FFDNet: toward a fast and flexible solution for CNN based image denoising. TIP (2018). https://doi.org/10.1109/TIP.2018.2839891

# Multi Object Tracking for Similar Instances: A Hybrid Architecture

Áron Fóthi(✉), Kinga B. Faragó, László Kopácsi, Zoltán Á. Milacski, Viktor Varga, and András Lőrincz

Department of Artificial Intelligence, Faculty of Informatics, Eötvös Loránd University, Budapest, Hungary
{fa2,faragokinga,kopacsi,miztaai,vv,lorincz}@inf.elte.hu

**Abstract.** Tracking and segmentation of moving objects in videos continues to be the central problem in the separation and prediction of concurrent episodes and situation understanding. Along with critical issues such as collision avoidance, tracking and segmentation have numerous applications in other disciplines, including medicine research. To infer the potential side effects of a given treatment, behaviour analysis of laboratory animals should be performed, which can be achieved via tracking. This presents a difficult task due to the special circumstances, such as the highly similar shape and the unpredictable movement of the subject creatures, but a precise solution would accelerate research by eliminating the need of manual supervision. To this end, we propose Cluster R-CNN, a deep architecture that uses clustering to segment object instances in a given image and track them across subsequent frames. We show that pairwise clustering coupled with a recurrent unit successfully extends Mask R-CNN to a model capable of tracking and segmenting highly similar moving and occluded objects, providing proper results even in certain cases where related networks fail. In addition to theoretical background and reasoning, our work also features experiments on a unique rat tracking data set, with quantitative results to compare the aforementioned model with other architectures. The proposed Cluster R-CNN serves as a baseline for future work towards achieving an automatic monitoring tool for biomedical research.

**Keywords:** Instance segmentation · Tracking · Constrained clustering · Gated recurrent unit · Deep neural networks

## 1 Introduction

Multi-Object Tracking (MOT) is a video processing problem with numerous possible applications in practice. Detecting humans and providing their bounding boxes can be used in video surveillance in order to perform various tasks without supervision, such as monitoring crowd density or recognising suspicious behaviour. The Multi-Object Tracking and Segmentation (MOTS) problem expands on this idea by requiring the detected objects to be segmented.


H. Yang et al. (Eds.): ICONIP 2020, LNCS 12532, pp. 436–447, 2020.
https://doi.org/10.1007/978-3-030-63830-6_37

Naturally, it presents a harder problem, while also making it more difficult to label data manually for training purposes. However, precise models allow for finer observations due to the pixel-wise segmentation provided for each object of interest.

The task becomes harder when tracking interest is shifted from humans animals, especially if the subject creatures are very similar to each other. One can find multiple additional purposes for tracking and segmentation. Regarding medicine research for example, where animal trials are commonplace, automated tracking can aid in solving technical and ethical complexities. Behaviour analysis of laboratory mice and rats is traditionally done manually, either in real time or from a video recording (see Fig. 1), which is a highly monotonous task. RFIDs can eliminate this obstacle [3], but the operation has its own complications and minimising the pain and discomfort of these animals is also of utmost importance. Without means of automation, this also requires continuous monitoring, once again imposing the disadvantages mentioned above. The deep network architecture presented in this paper[1] serves as a first step towards a multidisciplinary project, in which domain specific knowledge in neuroscience could be combined with artificial intelligence to perform as many of these tasks as possible without human supervision.

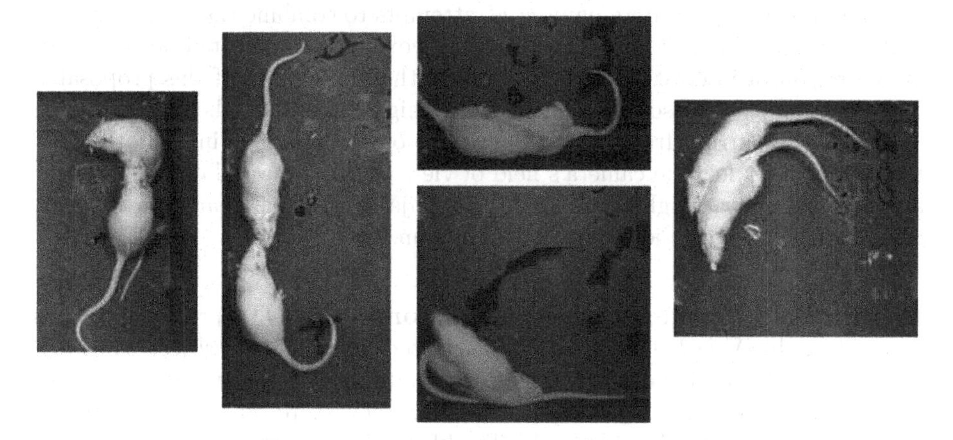

**Fig. 1.** Examples of rat actions to monitor. From left to right: sniffing, head-to-head, mounting/being mounted (the distinction is important from a given rat's point of view), side by side.

In experiments on social behaviours some of the animals may be medicated and/or genetically manipulated making their respective tracking and motion analysis mandatory. The problem is difficult for a variety of reasons. For one, the animals have to be tracked with meticulous precision to yield accurate results for such a vital research. Losing track of the rats invalidates any following tracking

[1] https://github.com/aronfothi/mask_cluster_rcnn.

attempts, as they have no means of visual distinction (marking the animals might influence their behaviour). Frequent mounting and occlusion scenarios, coupled with unpredictable movements make this task even harder.

The structure of our work is as follows. In the next section, we review related architectures and methods (Sect. 2). This is followed by a description of our contributions to the aforementioned rat tracking problem (Sect. 3), implementation details (Sect. 4) and experimental results (Sect. 5.2). Finally, the main observations and expansion possibilities are discussed (Sect. 6), and a summary of our work is provided (Sect. 7).

## 2 Related Work

**Multi-Object Tracking.** As mentioned before, MOT expects the model to track the bounding boxes of multiple objects simultaneously through frames of a video. Most existing tracker architectures follow the two-step *tracking-by-detection* paradigm:

1. Detect objects on each frame independently,
2. Link the appropriate detections together to form the trajectories.

The Tracktor [1] by Bergmann *et al.* attempts to combine these into a single, unified process. In their case, the bounding boxes of the detected objects serve as the region of interest (RoI) proposals for the next frame. If this proposal is accepted by the regressor with a sufficiently high score, then the corresponding object on the current frame matches the one on the previous image, otherwise the object likely left the camera's field of view. Newly detected objects without a previous proposal might actually be past objects that reappeared on the feed, which is tested by a re-identification algorithm.

**Multi-Object Tracking and Segmentation.** The MOTS problem [10] was first coined by Voigtlaender *et al.* as an extension of Multi-Object Tracking. Along with the problem specification, the authors present two data sets for the task based on existing ones, a novel accuracy score for evaluation and a baseline architecture, TrackR-CNN. TrackR-CNN uses Mask R-CNN [4] as a backbone, followed by employing 3D convolutions on the feature maps to attain temporal context. The architecture is further complemented by an *association head*, which creates embedding vectors (referred to as 'association vectors') for each detection. The distance between two given vectors from different frames determines whether they belong to the same instance (and therefore trajectory) or not.

Inspired by the new task and data sets, Porzi *et al.* published a paper [9] with several key contributions. One of these refers to a semi-automatic way of generating training data, mending the shortcoming of KITTI MOTS, which contains only a modest amount of annotations. They also propose MOTSNet, which resembles the previously mentioned TrackR-CNN in a sense that a Mask

R-CNN backbone is amended with an additional head working with embedding vectors. However, MOTSNet makes use of the mask in the tracking head, in addition to the features of the feature pyramid network (FPN). The tracking head utilises a batch hard triplet loss, which is added to the losses of the other heads.

**Clustering for Instance Segmentation.** As opposed to proposal-based methods for instance segmentation, proposal-free methods label pixels without bounding box information. This usually involves learning a pixel-level representation, followed by clustering to obtain pixel groups corresponding to instances. Hsu *et al.* [6] solve the problem in a single forward pass, using a Fully Convolutional Network (FCN) to learn a clustering mechanism based on pixel pair relations. The FCN outputs probability distributions for each pixel as a form of soft clustering, after which the authors use Kullback-Leibler divergence as a distance-like metric. The proposed architecture is fairly fast, performs well, and further supports an unlimited number of instances through a graph colouring inspired extension.

**Temporal Extension of Mask R-CNN.** There have been numerous attempts at injecting temporal information into Mask R-CNN-based architectures. For example, Siamese networks work with input pairs which are later combined into a joint branch. SiamFC [2] takes a crop of the target object and a larger image centered on its last position estimate, and processes them with the same convolutional networks before calculating their cross-correlation. Wang *et al.* propose SiamMask [11], which introduces a binary segmentation head to output masks for each frame.

Incorporating recurrent layers is a more intuitive approach in this regard. Porzi *et al.* [9] employ an RNN in their study, but note that it fails to make a considerable impact on accuracy.

# 3 Methods

## 3.1 Annotation and Augmentation

Due to the slow process of pixel-level human annotation, MOTS problems are usually lacking labelled data. Therefore, we looked for methods to automatically generate training data. We considered two cases: frames with and without contact between the animals. Due to the cage design it is simple to differentiate rats from the background (the floor), segmentation of the separate animals is fairly straightforward with traditional computer vision algorithms. As for the former group (featuring rats close to one another), generating these samples can be achieved through augmentation, offering various approaches. First, one of the rats is removed and replaced with the background, followed by cloning and translating the other:

- **Rotation:** The cloned creature is placed on an arc near the original.
- **Following:** The remaining rat is duplicated with a slight spatial and temporal offset, essentially following the original instance
- **Crossing:** In addition to moving the duplicate, it is also flipped horizontally giving rise to crossing paths.
- **Mixing:** Rat trajectories from different times are combined to form a rich set of overlaps

### 3.2 Segmentation by Clustering

MaskTrack R-CNN [13] approaches the video instance segmentation problem by expanding Mask R-CNN with a tracking head and a memory queue to keep track of objects throughout the video (see, Fig. 2). Assigning instance IDs is aided by bounding box information, but ignores segmentation results. As a result, coinciding bounding boxes reveal a weakness in this architecture, which is also mentioned in the paper. This issue is also prominent in our task, as tail curls may produce completely overlapping bounding boxes even without any real occlusion.

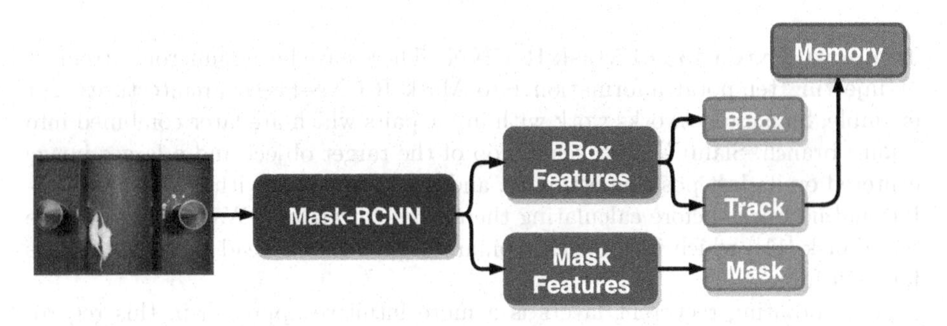


**Fig. 2.** MaskTrack R-CNN architecture. A tracking head with a memory queue was added to the Mask R-CNN architecture to handle the video instance segmentation problem

We propose a clustering head to tackle the aforementioned problem, inspired by [6]. We integrate their proposal-free clustering method into the proposal-based Mask R-CNN to improve accuracy. Instead of using Kullback-Leibler divergence with instance indexes, we opt for a labelless implementation with constrained clustering likelihood (CCL).

Our contributions to the aforementioned clustering methods are as follows: our network takes the RoI proposals, selects pixels from each and compares them to get their similarity (see Fig. 3). The clustering head also clusters the RoIs, and because a single RoI could contain more than one rat, the cluster mask output is appended to the feature map. This step is similar to the method applied in MOTSNet [9] but, rather, masking the background, we concatenate it to the feature map. In addition to examining positive and negative pairs in a given frame, the same comparison can be done between pixels in different frames to

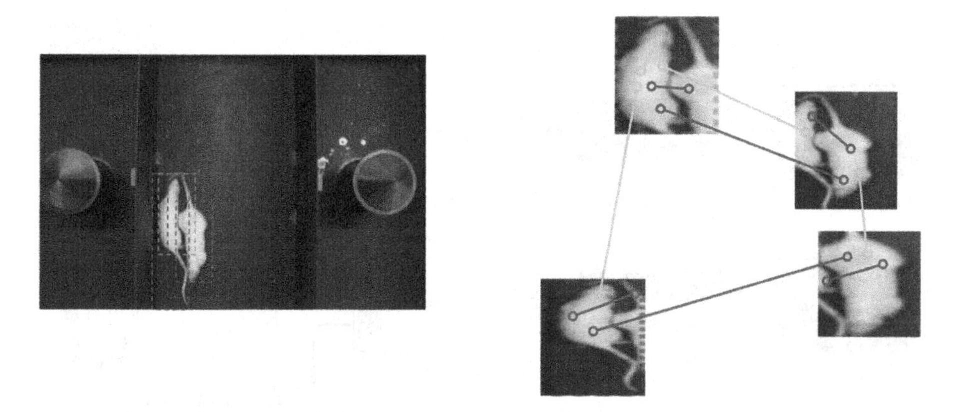

**Fig. 3.** Mask R-CNN proposes multiple Regions of Interest, which are then resized and aligned. Our custom clustering head selects a given number (20 in our experiments) of pixels from each RoI, which are then compared as seen in this image. Positive and negative pairs are identified inside each region and also across different RoIs.

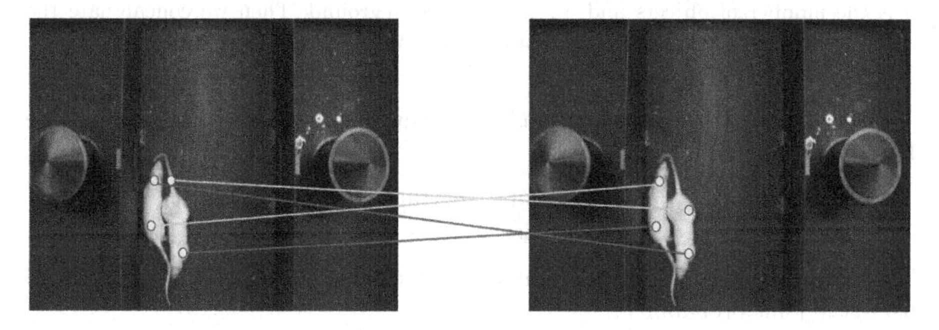

**Fig. 4.** Constrained Clustering Likelihood extended in time. Positive pairs are shown in green, while negative ones are coloured red. (Color figure online)

extend the process through time for tracking purposes (as can be seen in Fig. 4). This approach mends the aforementioned shortcoming of MaskTrack R-CNN, as shown in Fig. 5.

**Formulation.** The clustering method used here has been suggested by [5]. Let $X = \{x_1, \ldots, x_n\}$ be a set of pixels sampled from the RoI proposals throughout the video. Each of these samples belong to a single cluster $c \in \{1, .., k\}$. The clustering head $f_\theta(.)$ calculates the distribution that a given pixel $x_i$ belongs to some cluster $c$:

$$f_\theta(x_i) = [p_{i1}, p_{i2}, \ldots, p_{ik}] \quad \left( \sum_{c=1}^{k} p_{ic} \geq 1 \right),$$

where $P(y_i = c|x_i) = P(y_i = c) = p_{ic}$.

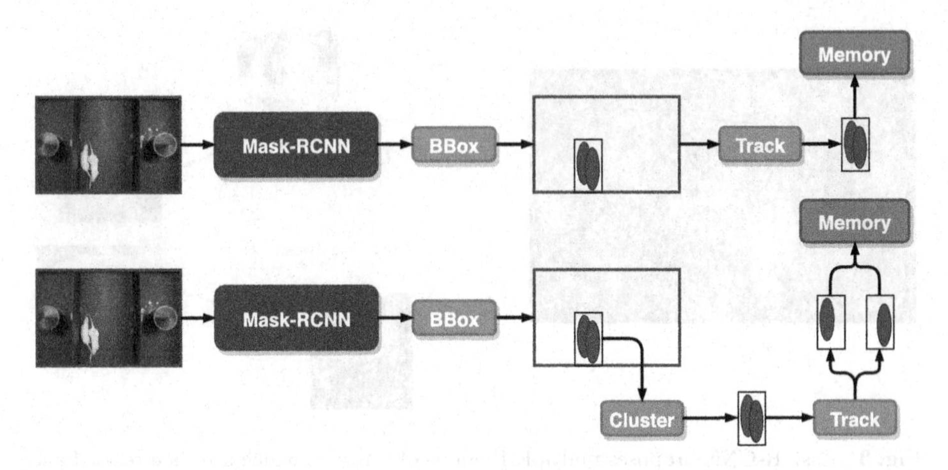


**Fig. 5.** Fixing MaskTrack R-CNN's weakness by clustering. We solve the coinciding bounding boxes problem by adding a clustering head that makes it possible to separate joint segmentation. The clustering head groups each pixel into $k+1$ categories, where $k$ is the number of objects and $+1$ is for the background. Then we concatenate the resulting clusters to the corresponding feature map as in Mask Scoring R-CNN [7].

We say that two samples are similar, if they belong to the same instance (see Fig. 3):

$$R(x_i, x_j) = \begin{cases} 1, \text{ if } x_i \text{ and } x_j \text{ belong to the same cluster} \\ 0, \text{ otherwise} \end{cases}$$

The probability of two pixels belonging to the same cluster can be determined using the joint probability

$$P(y_i = y_j) = \sum_{c=1}^{k} P(y_i = c, y_j = c) = f_\theta(x_i)^T f_\theta(x_j).$$

Likewise the probability that they are dissimilar is given by

$$P(y_i \neq y_j) = 1 - P(y_i = y_j).$$

In this way, the loss for all pairs of samples can be calculated by taking the negative logarithm of the CCL [5]:

$$\mathcal{L}_{pair} = -\frac{1}{|T|} \sum_{(i,j)\in T} \log\left[R(x_i, x_j) \cdot P(y_i = y_j) + (1 - R(x_i, x_j)) \cdot P(y_i \neq y_j)\right],$$

where $T = \{(i,j)\}_{\forall i,j \in [1,..,n]}$ is the set of all possible pairs and $|T|$ denotes the cardinality of $T$.

Following [6], we introduce an additional cluster for the background because of its imbalanced nature. For the background error $\mathcal{L}_{bg}$ we use binary cross-entropy loss as well. Therefore the final loss of the clustering head can be written as

$$\mathcal{L} = \mathcal{L}_{pair} + \mathcal{L}_{bg}.$$

### 3.3 Error Detection

Although we strive for a robust architecture, flawless models are hard, if not impossible to design in such a task. With this remark in mind, we provide two different ways of detecting errors during evaluation. These are beneficial for researchers and other users, due to the ability to mark intervals that need supervision, and those that should be correct. Error may occur:

1. when pixel regions with different cluster IDs neighbor each other closely, or
2. when the number of recognized and tracked objects is less than the actual number of instances, it also indicates an error. This is relevant in the case of MaskTrack R-CNN as discussed in Subsect. 3.2.

### 3.4 Architecture Overview

We extend the Mask R-CNN [4] architecture (Fig. 6). The input frame is fed to a backbone, with its last feature map serving as input for the recurrent layer, a convolutional gated recurrent unit (ConvGRU) [8,12]. The hidden representations are then passed to the neck network (FPN) along with the backbone features. The activation of the mask head is changed to softmax and a CCL loss is added on top of its result. We refer to this as *cluster head.* CCL enforces pixels from the same instance to have strongly similar representations, and through backpropagation the recurrent layer can make use of this information. This combination is thus used to extend instance segmentation in time.

## 4 Implementation

In our experiments, we used a ResNet-50 backbone with an FPN neck, as in the original Mask R-CNN architecture [4]. The RNN's length was set to 10, corresponding to a 10-frame window. This may be reduced during inference to avoid discrepancy between clusters near the beginning and the end of the window being undesirable according to our observations. Training was performed on two RTX TITAN GPUs, with each processing 4 images at a time for 12 epochs. We used Stochastic Gradient Descent and two distinct learning rates: $2 \cdot 10^{-2}$ for auto-annotated samples and $5 \cdot 10^{-3}$ for human-annotated ones.

## 5 Experiments and Results

We evaluate our model on two rat tracking data sets. The first one contains generated samples (refer to Sect. 3.1) similar to the training data. Cluster R-CNN performed exceptionally well on these videos, which is in line with our expectations. The other one measures the robustness of the system on more challenging samples filtered out during the training data preparation.

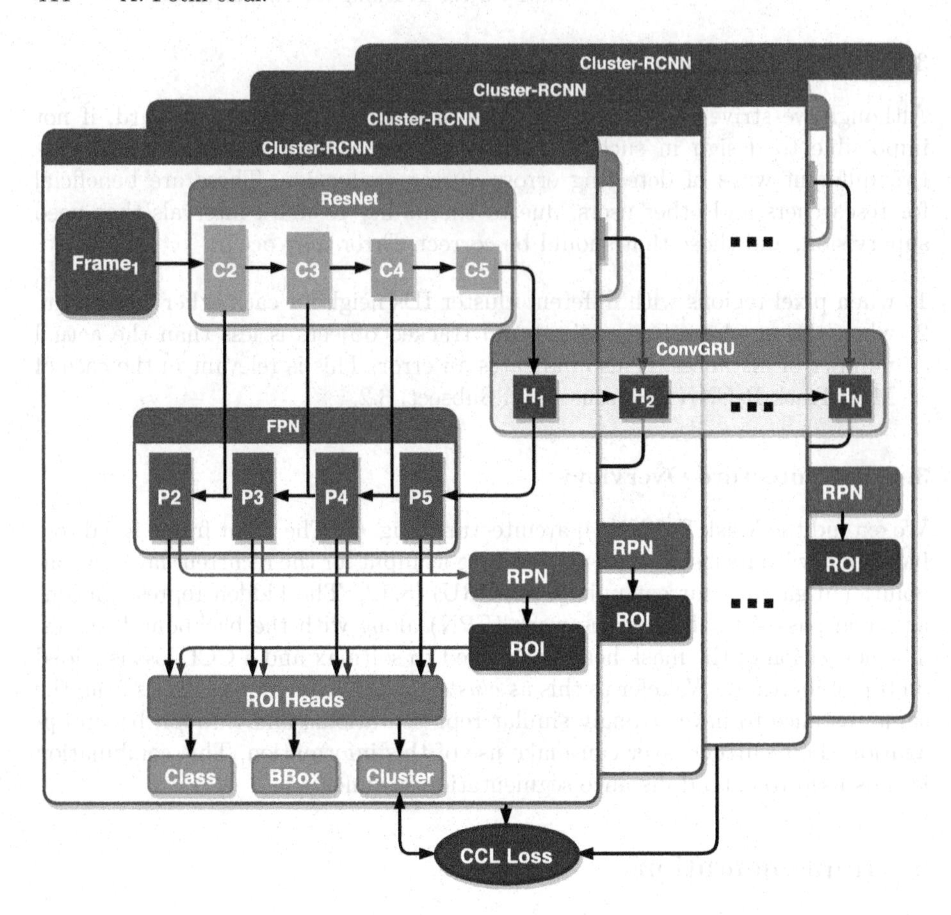


**Fig. 6.** Cluster R-CNN network architecture. For temporal extension a ConvGRU layer was added to the Mask R-CNN architecture and the mask head was replaced by a cluster head with CCL loss to handle coinciding bounding boxes.

## 5.1 Segmentation Quality

To quantify the segmentation quality of the methods we use the evaluation metrics introduced in the Youtube-VIS [13] benchmark. (See Table 1 for results.) MaskTrack R-CNN has higher average precision ($AP$) compared to our proposed architecture, which can be contributed to the activation used in their segmentation head network. MaskTrack R-CNN is using sigmoid, while Cluster R-CNN is using softmax, and as it was shown in [4], for segmentation tasks the former one provides better results. On the other hand, our method has higher average recall ($AR$), which could be justified by the augmentation mechanism (Sect. 3.1) used to train our model (Fig. 7).

**Table 1.** Qualitative results on the generated validation set using metrics introduced in the Youtube-VIS [13] benchmark. The best results are highlighted in bold. Higher values are better.

| **Method** | $AP$ | $AP_{50}$ | $AP_{75}$ | $AR_1$ | $AR_{10}$ |
|---|---|---|---|---|---|
| MaskTrack R-CNN | 0.392 | 0.699 | 0.414 | 0.245 | 0.446 |
| Cluster R-CNN | **0.667** | **0.932** | **0.839** | **0.369** | **0.754** |

**Fig. 7.** From left to right: original video frame, MaskTrack R-CNN and Cluster R-CNN output. This figure presents the most notable case - also described in Subsect. 3.2 - where our architecture corrects the error of MaskTrack R-CNN. Cluster R-CNN successfully tracks and segments both rats despite their coinciding bounding boxes.

### 5.2 Robustness

Our goal is to get robust tracking architecture, for this end we aim to detect and correct errors of MaskTrack R-CNN with our proposed method. To quantify robustness of each tracking method we use the total number of track switches. Due to missing ground truth annotations, we conducted the evaluation with human participants in a controlled environment using fourteen 2000 frame-long test videos. Given randomly sampled tracking results from each experiment, the participants had to count the number of track switches (when detected object masks are swapped or lost) on each video.

The results can be seen in Table 2. MaskTrack R-CNN achieved 9 switches, while our proposed method performed with a reasonable 27 switches total without any post-processing steps. These errors are mostly coming from the incorrect re-initialization of the ConvGRU layer, which can be easily filtered out using Hungarian matching algorithm or it could be mitigated by a bigger window size. By combining our method with MaskTrack R-CNN, we can reduce the total number of switches to 5, which is 56% improvement compared to with what we started.

## 6 Discussion

In the previous section, we focused on tracking accuracy over IoU values. Object masks are also subject to a few noticeable errors in the aforementioned occlusion cases, but the emphasis should always be on flawless tracking. Consider, for example, the mounting action mentioned in the introduction, which can still be

**Table 2.** Results on the test videos. We measure the robustness of the tracking algorithms using the number of switches. The combined method significantly outperforms the individual methods.

| Method | # of switches |
|---|---|
| MaskTrack R-CNN | 9 |
| Cluster R-CNN | 30 |
| Combination | 3 |

identified with imperfect segmentation, but a tracking mistake will reverse all future interactions (which are generally non-symmetrical).

As stated earlier, precision is vital in this task, and so improving it is a priority over steps to incorporate behaviour recognition. In order to achieve this, one could improve the model or provide more annotated data for harder cases. We consider both options for future work. Tracking multiple body parts of the animals (head, body, tail) would also be valuable in differentiating the rats during mounting scenarios.

## 7 Conclusions

In this paper, we approached the tracking and segmentation problem from a biological standpoint, operating on a unique data set to process videos of laboratory rats. We examined existing solutions to related tasks, drawing inspiration from several models and designing the final architecture with the problem's nature in mind. The resulting Cluster R-CNN proves the concept with convincing performance in a variety of situations, albeit suffering from the lack of labeled data in more complex interactions. Our model significantly reduces tracking error frequency when coupled with MaskTrack R-CNN, yielding a potent combination that further reduces the amount of manual intervention needed. This work serves as a foundation for future projects, while also providing valuable insight into certain problems and possible solutions in the field.

**Acknowledgements.** We thank Bence D. Szalay for his careful help in composing this paper; his work was supported by the Hungarian Government and co-financed by the European Social Fund (EFOP-3.6.3-VEKOP-16-2017-00002: EFOP-3.6.3-VEKOP-16-2017-00002, Integrated Program for Training New Generation of Scientists in the Fields of Computer Science). We also thank Árpád Dobolyi and Dávid Keller for providing the database. ÁF and AL were supported by the ELTE Institutional Excellence Program of the National Research, Development and Innovation Office (NKFIH-1157-8/2019-DT) and by the Thematic Excellence Programme (Project no. ED_18-1-2019-0030 titled Application-specific highly reliable IT solutions) of the National Research, Development and Innovation Fund of Hungary, respectively.

## References

1. Bergmann, P., Meinhardt, T., Leal-Taixe, L.: Tracking without bells and whistles. In: Proceedings of the IEEE International Conference on Computer Vision, pp. 941–951 (2019)
2. Bertinetto, L., Valmadre, J., Henriques, J.F., Vedaldi, A., Torr, P.H.S.: Fully-convolutional Siamese networks for object tracking. In: Hua, G., Jégou, H. (eds.) ECCV 2016. LNCS, vol. 9914, pp. 850–865. Springer, Cham (2016). https://doi.org/10.1007/978-3-319-48881-3_56
3. de Chaumont, F., et al.: Real-time analysis of the behaviour of groups of mice via a depth-sensing camera and machine learning. Nat. Biomed. Eng. **3**(11), 930–942 (2019)
4. He, K., Gkioxari, G., Dollár, P., Girshick, R.: Mask R-CNN. In: Proceedings of the IEEE International Conference on Computer Vision, pp. 2961–2969 (2017)
5. Hsu, Y.C., Lv, Z., Schlosser, J., Odom, P., Kira, Z.: A probabilistic constrained clustering for transfer learning and image category discovery. arXiv preprint arXiv:1806.11078 (2018)
6. Hsu, Y.C., Xu, Z., Kira, Z., Huang, J.: Learning to cluster for proposal-free instance segmentation. In: 2018 International Joint Conference on Neural Networks (IJCNN), pp. 1–8. IEEE (2018)
7. Huang, Z., Huang, L., Gong, Y., Huang, C., Wang, X.: Mask scoring R-CNN. In: Proceedings of the IEEE Conference on Computer Vision and Pattern Recognition, pp. 6409–6418 (2019)
8. Jung, M., Tani, J.: Adaptive detrending for accelerating the training of convolutional recurrent neural networks. In: Proceedings of the 28th Annual Conference of the Japanese Neural Network Society, pp. 48–49 (2018)
9. Porzi, L., Hofinger, M., Ruiz, I., Serrat, J., Bulò, S.R., Kontschieder, P.: Learning multi-object tracking and segmentation from automatic annotations. arXiv preprint arXiv:1912.02096 (2019)
10. Voigtlaender, P., et al.: Mots: multi-object tracking and segmentation. In: Proceedings of the IEEE Conference on Computer Vision and Pattern Recognition, pp. 7942–7951 (2019)
11. Wang, Q., Zhang, L., Bertinetto, L., Hu, W., Torr, P.H.: Fast online object tracking and segmentation: a unifying approach. In: Proceedings of the IEEE Conference on Computer Vision and Pattern Recognition, pp. 1328–1338 (2019)
12. Xingjian, S., Chen, Z., Wang, H., Yeung, D.Y., Wong, W.K., Woo, W.c.: Convolutional LSTM network: a machine learning approach for precipitation nowcasting. In: Advances in Neural Information Processing Systems, pp. 802–810 (2015)
13. Yang, L., Fan, Y., Xu, N.: Video instance segmentation. In: Proceedings of the IEEE International Conference on Computer Vision, pp. 5188–5197 (2019)

# Multiple Sclerosis Lesion Filling Using a Non-lesion Attention Based Convolutional Network

Hao Xiong[1](✉), Chaoyue Wang[2], Michael Barnett[3], and Chenyu Wang[3]

[1] Australian Institute of Health Innovation, Macquarie University, Sydney, Australia
hao.xiong@mq.edu.au

[2] School of Computer Science, The University of Sydney, Sydney, Australia

[3] Brain and Mind Centre, The University of Sydney, Sydney, Australia

**Abstract.** Multiple sclerosis (MS) is an inflammatory demyelinating disease of the central nervous system (CNS) that results in focal injury to the grey and white matter. The presence of white matter lesions biases morphometric analyses such as registration, individual longitudinal measurements and tissue segmentation for brain volume measurements. Lesion-inpainting with intensities derived from surrounding healthy tissue represents one approach to alleviate such problems. However, existing methods fill lesions based on texture information derived from local surrounding tissue, often leading to inconsistent inpainting and the generation of artifacts such as intensity discrepancy and blurriness. Based on these observations, we propose a non-lesion attention network (NLAN) that integrates an elaborately designed network with non-lesion attention modules and a designed loss function. The non-lesion attention module is exploited to capture long range dependencies between the lesion area and remaining normal-appearing brain regions, and also eliminates the impact of other lesions on local lesion filling. Meanwhile, the designed loss function ensures that high-quality output can be generated. As a result, this method generates inpainted regions that appear more realistic; more importantly, quantitative morphometric analyses incorporating our NLAN demonstrate superiority of this technique of existing state-of-the-art lesion filling methods.

**Keywords:** Multiple sclerosis · MS Lesion Filling · Non-lesion attention

## 1 Introduction

Multiple sclerosis (MS) is an immune-mediated disease that results in progressive damage to the central nervous system (CNS). MS is characterised by focal inflammatory demyelinating lesions in both the grey and white matter, formation of which is may be accompanied by acute episodes of neurological dysfunction

H. Xiong and C. Wang—Equal contribution.


H. Yang et al. (Eds.): ICONIP 2020, LNCS 12532, pp. 448–460, 2020.
https://doi.org/10.1007/978-3-030-63830-6_38

or relapses. In the longer term, progressive axonal loss and gliosis, both within lesions and normal appearing white matter (NAWM), results in brain atrophy and the accumulation of physical and cognitive disability.

Accordingly, morphological measurements derived from magnetic resonance imaging (MRI) scans are extensively utilized to monitor disease progression [7]. The link between brain volume loss and the evolution of motor and cognitive disability in MS is well established. Estimation of brain atrophy is therefore considered an important surrogate for, and predictor of, clinical disability.

The measurement of brain substructures, such as white matter, gray matter and cerebrospinal fluid (CSF), requires brain tissue segmentation on MRI images [3,16,27]. Although the T1 intensity of white matter lesions may vary according to the severity of tissue injury, similarity with grey matter intensities may result in erroneous substructure volume measurements [1,5,8,19]. Misclassification of MS lesions therefore generates biased white and grey matter volume estimations, and necessitates the development of effective means to address the impact of lesions on morphological analysis.

Imaging processing methods currently use filling algorithms to inpaint lesions with normal appearing white matter-like intensities. Both local and global white matter lesion inpainting methods have been proposed [19]. Local inpainting methods base filling algorithms on surrounding tissue, whilst global methods employ an average intensity derived from the whole brain white matter. More recent models [3] utilize a normal tissue intensity distribution. By sampling the distribution, the lesion is filled with the most probable white matter intensity. Based on this approach, Magon *et al.* [15] proposed slice by slice inpainting of the whole brain and Valverde *et al.* [22] performed lesion filling with the mean intensity of two periplaque NAWM voxels.

Liu *et al.* [14] have recently described a model that performs inpainting on irregular holes and achieves remarkable inpainting effects on natural images. Unlike arbitrary objects, which are morphologically heterogeneous, the human brain consists of defined, reproducible structures such as WM, GM and CSF. When designing deep learning frameworks for natural image ROI inpainting, very few assumptions can be made. By contrast, the relative uniformity of brain structure and appearance between individuals permits the integration of a deep learning framework with a non-lesion attention module to facilitate brain lesion inpainting.

Therefore, we propose a deep learning model with the following distinguishing features: 1) to the best of our knowledge, we are first to apply deep learning algorithm to MS lesion filling issue. The model does not make any assumptions that the lesion area is/should be filled with white matter like intensity; 2) We design the non-lesion attention module and associated loss function to alleviate the effects of lesions on eventual inpainting and thus synthesize realistic healthy tissue like voxels, and simultaneously taking global brain information into account. The model integrates global brain information based on the consistency of brain structure, rather than using neighbourhood texture as the key factor to determine the filled intensity; 3) The algorithm is able to maintain normal tissue architecture rather than simply filling lesion mask ROIs with white matter like intensities. Specifically, when a lesion mask that erroneously overlaps with normal tissues (such as CSF) is supplied as an input, our algorithm is able to inpaint the 'lesion' ROI with appropriately disposed normal tissue intensities. Finally, synthetic lesions can be generated by non-linear co-registration of real FLAIR image derived MS lesion masks with healthy subject MRI T1 images to validate the technique and provide a ready-made ground truth for comparative experiments. Herein, we demonstrate the effectiveness and efficiency of MS lesion inpainting using these methods.

## 2 Related Work

Traditional approaches [3,15,19,22] assume that lesions should, without exception, be filled with white matter like intensities. This hypothesis is flawed, as lesions may occur within grey matter or overlap white and adjacent grey matter. Rather than filling lesions with white matter like intensity only, Battaglini *et al.* [1] inpainted lesions with either white matter or grey matter using a histogram derived from peri-lesional white and grey matter, resulting in improved blending with neighbourhood normal appearing tissues. In the method described by Guizard *et al.* [9], lesions are first pre-filled with the median of intensity from surrounding normal-appearing tissues, and the most similar patches calculated using only the nearby regions. The same group subsequently proposed an iterative, concentric patch-based filling approach to preserve local anatomical information [10]. Similarly, Prados *et al.* [18] use neighbourhood patches to fill lesions with the most appropriate intensity; and exploit a minimal kernel-based convolution to achieve better inpainting effects.

In the general (non-medical) computer vision community, various inpainting techniques have been proposed to remove objects, texts or scratches from images using the remaining information in the image. Examples include (1) an "onion-peel" strategy that fills regions of interest (ROI) from an outermost to innermost concentric ring with reference to available patches [6]; (2) an exemplar-based method [4] that fills specific regions by directly copying similar patches extracted from the whole image; and (3) synthesis of ROI intensities with textures from matched patches by applying a Non-Local Means algorithm [2,23]. Advances in deep learning have significantly enhanced inpainting effects. Initial deep learning based inpainting frameworks, in which networks are trained to fill ROI contents based on its surroundings, have been described by a number of groups [11,13,17,25]. These techniques suffer from both pixel-wise reconstruction loss and adversarial loss, and tend to generate artifacts between inpainted and neighbourhood areas. The addition of more convolutions, which integrates distant areas into inpainting algorithms, has been applied [11,26] to mitigate the appearance discrepancies introduced by deep learning; and further optimization undertaken [24] to minimize the difference between filled regions and their neighbourhood. However, increasing convolutional layers and applying optimization are computationally intensive and are substantially more time consuming.

# 3 Method

We first explain concepts of the non-lesion attention module, then introduce the network architecture and implementation, and lastly explain the devised loss functions for T1 lesion inpainting.

## 3.1 Non-Lesion Atttention Module

Traditional convolution operations focus on the cues within the ROI neighbourhood and thus ignore global information. Even with a large receptive field, convolution operations are still based on local information. In contrast, the non-lesion attention module aims to perform lesion filling by taking long distance cues into consideration, while trying to minimize the effects of lesion area in the feature map. As a result, the output of our method only relies on non-lesion healthy values rather than lesions.

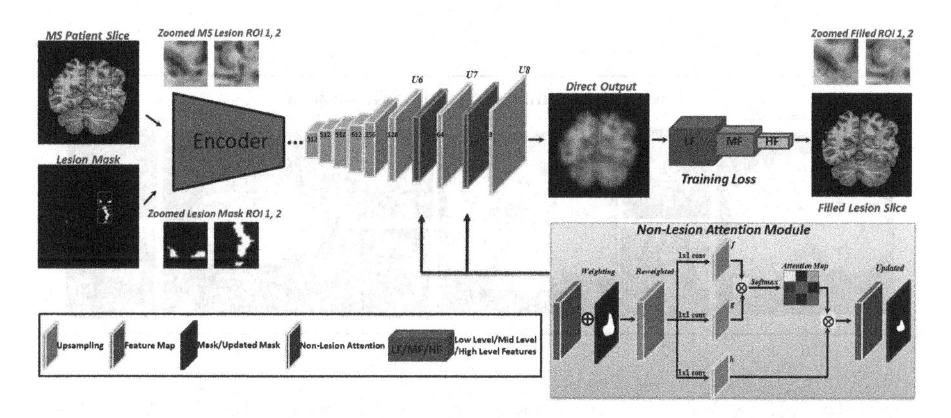


**Fig. 1.** Network architecture. The network takes both the T1 brain image and associated lesion mask as inputs, and then outputs inpainted results. Two non-lesion attention modules are inserted after U6 and U7, respectively.

Given a specific location within brain slice, the non-lesion attention module will reweight the input feature map based on the given masks, and then compute the responses between this location and its counterparts within the whole image, exploiting the relative homogeneity and internal correlation of brain structure. Suppose $X$ refers to the feature values or intensity of the MRI slice, $M$ is the binary mask representing zero and one as lesion and normal brain tissue areas, respectively. The feature map re-weighting function is defined as:

$$x = \begin{cases} W^T(X \odot M)\frac{sum(\mathbf{1})}{sum(M)} + b, & \text{if} \quad sum(M) > 0 \\ 0, & \text{otherwise} \end{cases}, \tag{1}$$

where $W$ and $b$ are convolutional weights and bias respectively. Meanwhile, $\mathbf{1}$ has the same size as $M$ and are filled with all ones.

Given re-weighted feature maps $x \in R^{C \times N}$ from Eq. 1, the non-lesion attention module then transforms $x$ into three feature spaces $f$, $g$ and $h$, where $f(x) = W_f x$,$g(x) = W_g x$ and $h(x) = W_h x$. Here, $W_f$, $W_g$ and $W_h$ are weight matrices to be learned and such linear operations can be implemented as $1 \times 1$ convolutions.

The module subsequently requires $f(x)$ and $g(x)$ for attention map calculation. More specifically, the attention map can be obtained by:

$$a_{ji} = softmax(f(x_i)^T g(x_j)), \tag{2}$$

Here, $a_{ji}$ is namely the softmax operation along the dimension $j$. Evidently, the softmax operation here measures the similarity between $i^{th}$ and $j^{th}$ locations in x. The attention form is finally expressed as:

$$s = ah(x). \tag{3}$$

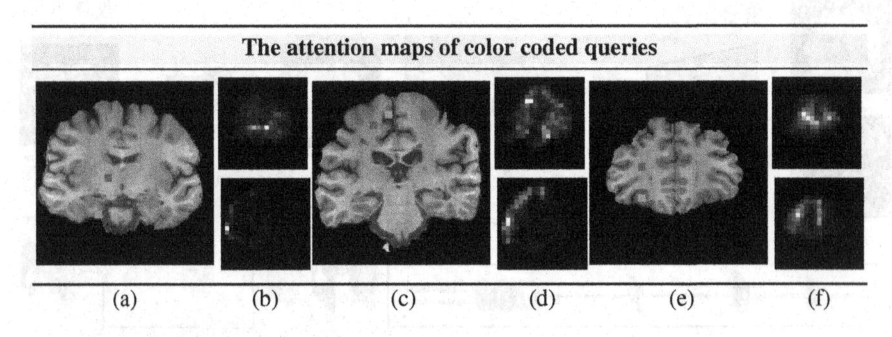


**Fig. 2.** **(a)(c)(e):** Brain slices with query locations **(b)(d)(f):** attention maps of red, green dot queries are shown from top to bottom for each image **Red Dot:** query on white matter or thalamus **Green Dot:** query on grey matter. Note that the whitest regions in the grayscale attention map are best correlated to the queries and vice versa. Clearly, the whitest regions within attention maps are appropriately located in regions that correspond to the respective queries. The attended areas all are within the global domain rather than a local region. (Color figure online)

Concurrently, mask $M$ is updated as follows:

$$m = \begin{cases} 1, & \text{if} \quad sum(M) > 0 \\ 0, & \text{otherwise} \end{cases}. \tag{4}$$

This is to say that the attention module computes a response at a location by attending all positions and taking their weighted average in an embedding space. Subsequently, the attention module is incorporated into a non-local block as:

$$y = W_s s + x, \tag{5}$$

where $y$ refers to the non-local block and $W_s$ is the weight matrix of the $1 \times 1$ convolution to be learned. Initially, $W_s$ is set as zero and thus $y$ is exactly $x$, which means that the inpainting task only relies on neighbourhood cues initially. After gradually updating $W_s$, more non-local evidence is considered in the training and future inpainting.

Here, $m$ is the updated mask. As shown in (a)(c)(e) of Fig. 2, the red, green dots represent query locations within the white matter, grey matter respectively. In (b)(d)(f), the corresponding attention maps of the color coded query dots are displayed. Note that the size of each attention map here is $64 \times 64$ (see Sect. 3.2 for further details), while the size of input MRI slice is $256 \times 256$. Meanwhile, the most-attended regions are the whitest areas in the attention map and vice versa. Though the size of the attention map and input slice are discrepant, the attended area approximates its expected anatomic counterpart. For example, attended regions for grey matter queries are generally located at the periphery of the brain in areas that approximate the location of the cortex.

### 3.2 Network Design and Implementation

Our network is based on the Unet-like architecture (shown in Fig. 1), and both the encoder and decoder in the network has 8 layers. The feature maps and binary masks from the encoder stage are concatenated by skip links into the decoder stage. Furthermore, the size of the binary mask is exactly matched with the respective input brain slice and will be updated in the following encoder and decoder stages. The values of weight matrix and bias for the binary mask are initially set as 1 and 0 respectively, and kept fixed during training. Likewise, the values of weight matrices and bias for $f$, $g$ and $h$ are initially set as 1 and 0, respectively, but only bias is fixed during training (Fig. reffig:example).

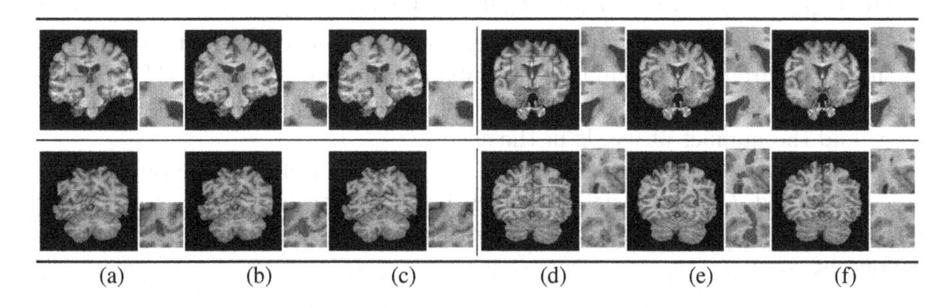


**Fig. 3.** Lesion inpainting examples using non-lesion attention network. **(a)(d):** coronal MS images from 4 patient **(b)(e):** superimposed lesion mask (purple) derived from FLAIR images (not shown) **(c)(f):** inpainted results. (Color figure online)

However, the integration of a non-lesion attention module into the network is computationally intensive. In order to facilitate computational efficiency, the number of channels with feature spaces $f$, $g$ and $h$ is therefore reduced to half of the number of channels of $x$ in our method. Meanwhile, in Eq. 5 the number of channels of $s$ can be increased with $W_s$ so as to match the number of channels of $x$. Additionally, a max pooling layer is applied after $g$ and $h$ to speed computation. In our case, the stride of max pooling layer is 2. In the experimental work presented, the following principles are applied with respect to the integration of non-lesion attention modules: 1) Non-lesion attention modules should be inserted in the decoder stage, where they are closer to the output of the network. 2) The more non-lesion attention modules embedded within the network, the more accurate the inpainting effects are.

Based on Eq. 5, it is optimal to insert two non-lesion modules directly after $U6$ and $U7$ in Fig. 1, respectively. Though some techniques are applied to speed up computation, the maximum number of added non-lesion attention modules is 2. $U8$ is physically the closest layer to the output layer but this layer's feature maps are relatively larger than those relating to $U6$ and $U7$.

Therefore, computational efficiency is improved by not inserting a non-lesion attention module after $U8$, while not sacrificing model accuracy at same time. The training set contains approximately 50800 slices; the initial learning rate is set as 0.00005 and the batch size as 10. The learning rate for fine-tuning is set as 0.00001. There are approximately 125 and 10 epochs for training and fine-tuning respectively.

### 3.3 Loss Function

The output of the above network $I_{output}$ is fed into the following loss function for more accurate per-pixel reconstruction:

$$L_{total} = L_{valid} + 6L_{hole} + 15L_{perceptual} + 0.1L_{tv}, \tag{6}$$

where $L_{valid} = \frac{1}{N_{I_{gt}}}\|M \odot (I_{out} - I_{gt})\|_1$ and $L_{hole} = \frac{1}{N_{I_{gt}}}\|(1 - M) \odot (I_{out} - I_{gt})\|_1$ are the loss terms aiming to enhance per-pixel reconstruction accuracy of the whole brain slice including any inpainted region. Here, $I_{gt}$, $M$ and $N_{I_{gt}}$ respectively represent ground truth slice, binary lesion mask (0 represents lesion area) and the number of voxels in the slice.

In addition to the direct output ($I_{out}$) for pixel wise reconstruction, $L_{perceptual}$ and $L_{style}$ are defined to ensure that the inpainted areas have a more realistic and natural texture and structure. The definition of $L_{perceptual}$ is as follows:

$$L_{perceptual} = \sum_{p=0}^{P-1} \frac{\|\psi_p^{I_{out}} - \psi_p^{I_{gt}}\|_1}{N_p^{I_{gt}}} + \sum_{p=0}^{P-1} \frac{\|\psi_p^{I_{comp}} - \psi_p^{I_{gt}}\|_1}{N_p^{I_{gt}}}. \tag{7}$$

Here, $I_{comp}$ is actually $I_{out}$ but with non-lesion area set to ground truth. $N_p^{I_{gt}}$ is the number of elements in $\psi_p^{I_{gt}}$. The inclusion of $L_{perceptual}$ accepts feature vectors by projecting $I_{out}$, $I_{gt}$ and $I_{comp}$ into a high level feature space $\psi_p^{I_*}$ with a pretrained VGG-16 network [20]. The layers pool1, pool2, pool3 in VGG-16 are utilized for feature extraction in our work. Furthermore, since the VGG-16 network requires an image of at least 236 × 236, all brain slices in our model are padded to size 256 × 256 for training and testing. Lastly, $L_{tv}$ represents the total variation term for smoothing penalty.

$$L_{tv} = \sum_{(i,j)\in R,(i,j+1)\in R} \frac{\|I_{comp}^{i,j+1} - I_{comp}^{i,j}\|_1}{N_{I_{comp}}} + \sum_{(i,j)\in R,(i+1,j)\in R} \frac{\|I_{comp}^{i+1,j} - I_{comp}^{i,j}\|_1}{N_{I_{comp}}}, \quad (8)$$

where $R$ denotes the region of 1 pixel dilation of the lesion area. $N_{I_{comp}}$ is the number of voxels in $I_{comp}$.

# 4 Results

We compared four lesion inpainting algorithms on T1 images, including 1) the lesion filling tool included in FSL v6.0 [12]. FSL is a comprehensive library of analysis tools for FMRI, MRI and DTI brain imaging data; 2) multi-time-point modality-agnostic lesion filling method (MMLF). To the best of our knowledge, MMLF is the most recently described lesion filling method in the medical image field. It is publically available (http://niftyweb.cs.ucl.ac.uk/program.php?p=FILLING); and 3) partial convolutions(PC) based inpainting method. PC based inpainting method is considered state-of-the art in the computer vision community; and 4) our proposed non-lesion attention network (NLAN). None of these methods rely on lesion area information to generate inpainted voxels. Therefore, given synthetic lesion masks, the values of synthetic lesion area were set as zeros before inpainting.

## 4.1 Data

Three different MRI datasets were used in training and testing stages.

1. IXI dataset: a publically available collection of 665 MR images from normal, healthy subjects ([1]). All brain volumes were intensity normalized with mean and variance before training and testing. Synthetic lesions, generated by non-linear co-registration with MS lesion masks (see below), were applied to the 50 selected testing volumes and regarded as ground truth for subsequent evaluation.
2. The second dataset contains MRI scans from 50 patients with RRMS and is primarily exploited for deriving lesion masks for subsequent generation of synthetic lesions on T1 images from the IXI healthy control testing dataset.
3. The third dataset consists of MRI scans from 20 patients with relapsing-remitting multiple sclerosis (RRMS). T1-weighted sequences were acquired on a 3T GE Discovery MR750 scanner(GE Medical Systems, Milwaukee, WI).

[1] https://brain-development.org/ixi-dataset/.

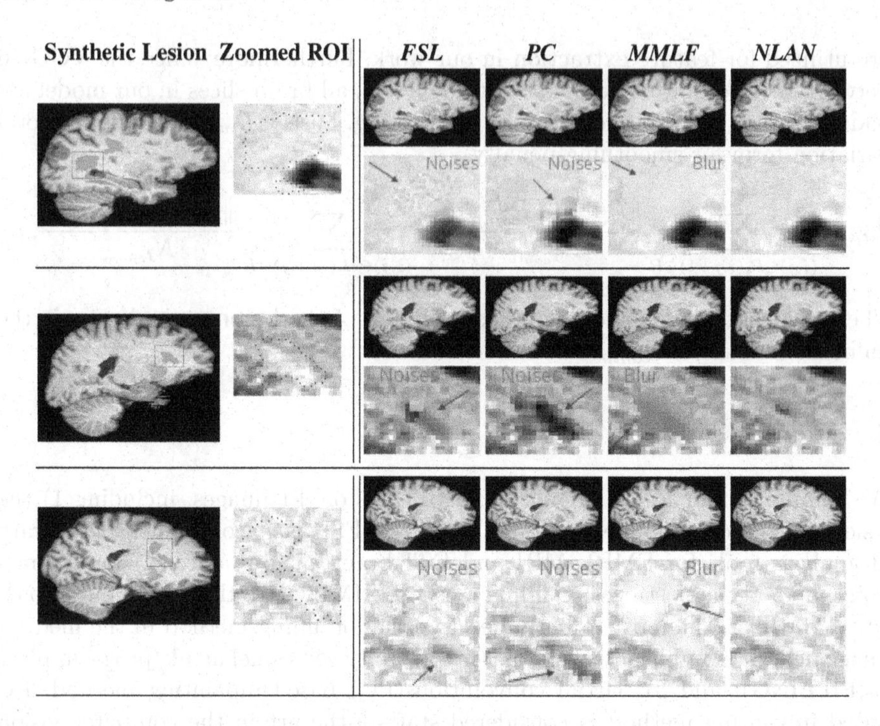


**Fig. 4.** The sagittal and zoomed views of the T1 slice with embedded **synthetic lesions**. Note the noisy samples generated by FSL and PC methods, the blurred boundaries obtained by MMLF method.

## 4.2 Qualitative Evaluation

In this experiment, white matter lesion masks derived from MS FLAIR images were used to create and inpaint synthetic lesions generated on non-linearly coregistered T1 images from healthy controls. Since visible T1 hypointensities are usually a subset of T2 (FLAIR) hyperintensities, the FLAIR-derived manually labelled lesion mask may involve isointense regions on matching MS T1 images. Non-linear co-registration of MS lesion masks and healthy control T1 images is also likely to produce synthetic lesions that erroneously overlap with tissues other than white matter. Synthetic lesions were generated on T1 brain images from 50 healthy controls and inpainted with FSL, MMLF, PC and NLAN, respectively. Figure 4 provides examples of visual comparison of our inpainting method with existing methods. Moreover, Fig. 5 illustrates the inpainitng effects with and without the non-lesion attention module.

## 4.3 Quantitative Evaluation

Since MRI scans acquired from patients with MS have no lesion-region ground truth, most methods [10,18,22] have employed quantitative analysis using synthetically generated lesions in MRI slices. In our case, the original, pre-lesional tissue ROI intensities from 50 healthy subjects were regarded as the ground truth for artificial lesion generation and comparison. We employed several quantitative approaches to facilitate comparison of the MMLF, FSL, PC inpainting techniques, and our own method. The mean square error (MSE) was adopted to measure the difference between the inpainted and ground truth intensities; a smaller MSE value signifies more realistic filled intensities. As a primary goal of lesion inpainting is improve the accuracy of brain substructure volume measurement, we also evaluated WM, GM, and CSF volumes using SIENAX on the native healthy control T1 images (ground truth) and images with synthetic lesions inpainted with each technique. Closer approximation to the ground truth volumes therefore denotes a more accurate inpainting technique. The results of quantitative assessment are shown in Table 1. It showed that the proposed method generated the smallest MSE and therefore the most realistic inpainting result of all four techniques. Moreover, ME and MAE values show that the estimation of GM and WM volumes using SIENAX analysis of T1 brain volumes inpainted by our method are most accurate.

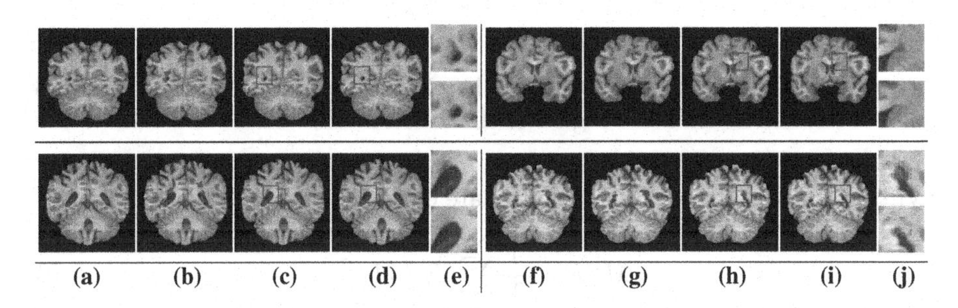


**Fig. 5.** Effect of the non-lesion attention module on **MS lesion** inpainting. (a),(f) MS data. (b),(g) Lesion mask generated using a semi-automated thresholdhing technique by an expert neuroimaging analyst using coregistered FLAIR images. Inpainting results without (c),(h) and with (d),(i) non-lesion attention module. Lesion inpainting is affected by proximity to the lateral ventricles without the non-lesion attention module (top rows in (e),(j)). In contrast, lesion-filling results in this region are cleaner and more realistic with the non-lesion attention module (bottom rows in (e),(j)).

**Synthetic Lesion Generation.** The process of lesion generation is as follows: 1) Whole brain tissue was extracted from all images with the Brain Extraction Tool (BET). Note that the tool BET is part of FSL. 2) MS brain volumes were firstly affine and then non-rigidly registered to healthy brain volumes. 3) Binary lesion masks derived from MS FLAIR images were registered to the healthy T1 images with estimated non rigid transformations. (the CSF mask from SIENAX was used to avoid overlap of the lesion mask with CSF) 4) Synthetic lesion masks generated on healthy subject T1 images were constrained within the relevant whole brain mask derived in Step 1. Lesion inpainting is not dependent on the

**Table 1.** Quantitative evaluation results. 5140 slices with synthetic lesions were used for evaluation. The first row shows mean square error (MSE) (SD) in comparison with ground truth. Rows 2–4 show mean error (ME) (SD). All methods tend to overestimate GM and CSF volumes while underestimating WM volume. Rows 5–7 show mean absolute error (MAE) (SD). In general, our method outperforms other methods with MSE, ME and MAE. Note that this analysis is not an evaluation of the accuracy of each method. Rather, the analysis reflects the variability in MSE, MAE, ME from case to case. Furthermore, the intensity value of each brain volume was normalized between 0 and 1 for testing. Tissue volumes of WM, GM and CSF are represented in voxels.

| | FSL | PC | MMLF | NLAN |
|---|---|---|---|---|
| MSE | 0.000915 (*0.00389*) | 0.001647 (*0.00602*) | 0.000520 (*0.00180*) | **0.000482** (*0.00119*) |
| ME(GM) | 0.0323 (*0.0462*) | 0.0394 (*0.0540*) | 0.0317 (*0.0363*) | **0.0274** (*0.0367*) |
| ME(WM) | −0.0526 (*0.0734*) | −0.0638 (*0.0928*) | −0.0658 (*0.0721*) | **−0.0503** (*0.0791*) |
| ME(CSF) | 0.0095 (*0.0342*) | **0.0007** (*0.0310*) | 0.0230 (*0.0322*) | 0.0423 (*0.0730*) |
| MAE(GM) | 189.023 (*232.728*) | 227.735 (*274.173*) | 171.854 (*191.598*) | **150.383** (*189.463*) |
| MAE(WM) | 230.260 (*287.786*) | 281.297 (*368.709*) | 270.596 (*290.473*) | **217.103** (*308.794*) |
| MAE(CSF) | **58.5919** (*88.608*) | 60.615 (*76.809*) | 73.296 (*85.930*) | 130.388 (*195.675*) |

intensities of the lesion area to be filled; therefore, none of the techniques generate noise of any kind within the region of the lesion mask.

**Synthesisation Error.** The mean square error is defined as:

$$MSE = \frac{1}{N}\sum_{i}^{n}(I_T - I_0)^2, \tag{9}$$

where $n$ is the number of testing volumes and $N$ is the number of lesion voxels. $I_T$ and $I_0$ refer to the intensities of ground truth volume and inpainted volume, respectively.

**WM, GM, CSF Volume Error.** SIENAX-based [21] segmentation was performed on healthy control T1 brain volumes, yielding ground truth WM, GM and CSF volumes respectively. Following insertion of synthetic lesions into T1 brain slices and lesion inpainting by each method, SIENAX-derived WM, GM and CSF volumes were re-estimated. To measure the impact of lesion inpainting using each method, we measured the similarity of substructure volumes derived from T1-inpainted slices to those derived from native (ground truth) images by calculating both the mean error (ME) and mean absolute error (MAE). The mean error is defined as:

$$ME = 100\frac{V_T - V_0}{V_0}, \tag{10}$$

where $V_0$ and $V_T$ are ground truth volume and the volume estimated after inpainting with different methods respectively.

## 5 Conclusion

We proposed a new deep learning model for realistic lesion inpainting in MRI slices by exploiting the non-lesion attention module. Rather than considering the neighbourhood texture for lesion filling, our method inpaints lesion areas with the most plausible intensities by observing the structure and texture information contained within the whole image. To achieve this, the non-lesion attention module compares similarities between the features within lesion regions and remaining brain areas. Although some existing methods take into account non-local information, they are limited by algorithms that search for similar patches within a constrained bounding box, rather than the whole domain. As such, these methods prove inferior to those described in this work. Moreover, integrating the non-lesion attention module under the framework of deep learning generates an effective algorithm that outperforms traditional non-local based methods and can be incorporated into image analysis pipelines for more accurate quantitative assessment of brain substructure volumes.

## References


1. Battaglini, M., Jenkinson, M., Stefano, N.D.: Evaluating and reducing the impact of white matter lesions on brain volume measurements. Hum. Brain Mapp. **33**, 2062–2071 (2014)
2. Buades, A., Coll, B., Morel, J.M.: A non-local algorithm for image denoising. In: Computer Vision and Pattern Recognition, pp. 60–65 (2005)
3. Chard, D., Jackson, J., Miller, D., Wheeler-Kingshott, C.: Reducing the impact of white matter lesions on automated measures of brain gray and white matter volumes. J. Magn. Reson. Imaging. **32**, 223–228 (2010). https://doi.org/10.1002/jmri.22214
4. Criminisi, A., Perez, P., Toyama, K.: Region filling and object removal by exemplar-based image inpainting. IEEE Trans. Image Process **13**, 1200–1212 (2004)
5. Diez, Y., et al.: Intensity based methods for brain MRI longitudinal registration. A study on multiple sclerosis patients. Neuroinformatics **12**, 365–379 (2014)
6. Efros, A.A., Leung, T.K.: Texture synthesis by non-parametric sampling. In: International Conference on Computer Vision (1999)
7. Ganiler, O., et al.: A subtraction pipeline for automatic detection of new appearing multiple sclerosis lesions in longitudinal studies. Neuroradiology **56**(5), 363–374 (2014). https://doi.org/10.1007/s00234-014-1343-1
8. Gelineau-Morel, R., Tomassini, V., Jenkinson, M., Johansen-Berg, H., Matthews, P., Palace, J.: The effect of hypointense white matter lesions on automated gray matter segmentation in multiple sclerosis. Hum. Brain Mapp. **33**, 2802–2814 (2012)
9. Guizard, N., Nakamura, K., Coupe, P., Arnold, D., Collins, D.: Non-local MS MRI lesion inpainting method for image processing. In: The endMS Conference (2013)
10. Guizard, N., et al.: Non-local means inpainting of MS lesions in longitudinal image processing. Front. Neurosci. **456** (2015)
11. Iizuka, S., Simo-Serra, E., Ishikawa, H.: Globally and locally consistent image completion. ACM Trans. Graph. (TOG) **36**(4), 107 (2017)
12. Jenkinson, M., Beckmann, C., Behrens, T., Woolrich, M., Smith, S.: FSL. NeuroImage **62**, 782–790 (2012)

13. Li, Y., Liu, S., Yang, J., Yang, M.H.: Generative face completion. arXiv preprint arXiv:1704.05838 (2017)
14. Liu, G., Reda, F.A., Shih, K.J., Wang, T.C., Tao, A., Catanzaro, B.: Image inpainting for irregular holes using partial convolutions. arXiv preprint arXiv:1804.07723 (2018)
15. Magon, S., et al.: White matter lesion filling improves the accuracy of cortical thickness measurements in multiple sclerosis patients: a longitudinal study. BMC Neurosci. **106** (2014)
16. Nakamura, K., Fisher, E.: Segmentation of brain magnetic resonance images for measurement of gray matter atrophy in multiple sclerosis patients. Neuroimage **44**, 769–776 (2009)
17. Pathak, D., Krahenbuhl, P., Donahue, J., Darrell, T., Efros, A.A.: Context encoders: feature learning by inpainting. In: IEEE Conference on Computer Vision and Pattern Recognition, pp. 2536–2544 (2016)
18. Prados, F., et al.: A multi-time-point modality-agnostic patch-based method for lesion filling in multiple sclerosis. Neuroimage **139**, 376–384 (2016)
19. Sdika, M., Pelletier, D.: Nonrigid registration of multiple sclerosis brain images using lesion inpainting for morphometry or lesion mapping. Hum. Brain Mapp. **30**, 1060–1067 (2008)
20. Simonyan, K., Zisserman, A.: Very deep convolutional networks for large-scale image recognition. arXiv preprint arXiv:1409.1556 (2014)
21. Smith, S., et al.: Accurate, robust and automated longitudinal and cross-sectional brain change analysis. NeuroImage **17**(1), 479–489 (2002)
22. Valverde, S., Oliver, A., Lladó, X.: A white matter lesion-filling approach to improve brain tissue volume measurements. Neuroimage Clin. **6**, 86–92 (2014)
23. Wong, A., Orchard, J.: A nonlocal-means approach to exemplar-based inpainting. In: International Conference on Image Processing (2008)
24. Yang, C., Lu, X., Lin, Z., Shechtman, E., Wang, O., Li, H.: High-resolution image inpainting using multi-scale neural patch synthesis. arXiv preprint arXiv:1611.09969 (2016)
25. Yeh, R., Chen, C., Lim, T.Y., Hasegawa-Johnson, M., Do, M.N.: Semantic image inpainting with perceptual and contextual losses. arXiv preprint arXiv:1607.07539 (2016)
26. Yu, F., Koltun, V.: Multi-scale context aggregation by dilated convolutions. arXiv preprint arXiv:1511.07122 (2015)
27. Zijdenbos, A., Forghani, R., Evans, A.: Automatic quantification of MS lesions in 3D MRI brain data sets: validation of INSECT. In: Wells, W.M., Colchester, A., Delp, S. (eds.) MICCAI 1998. LNCS, vol. 1496, pp. 439–448. Springer, Heidelberg (1998). https://doi.org/10.1007/BFb0056229

# Multi-scale Object Detection in Optical Remote Sensing Images Using Atrous Feature Pyramid Network

Mei Yu[1,2,3], Minyutong Cheng[1,2,3], Han Jiang[4], Jining Shen[1,2,3], Ruiguo Yu[1,2,3], Xiang Ying[1,2,3], Jie Gao[1,2,3], and Xuewei Li[1,2,3](✉)

[1] College of Intelligence and Computing, Tianjin University, Tianjin 300354, China
{yumei,cmyt,jiningshen,rgyu,xiang.ying,gaojie,lixuewei}@tju.edu.cn
[2] Tianjin Key Laboratory of Cognitive Computing and Application, Tianjin 300354, China
[3] Tianjin Key Laboratory of Advanced Networking, Tianjin 300354, China
[4] Laboratory of OpenBayes Machine Intelligence Lab, Beijing 100020, China
hahn@openbayes.com

**Abstract.** Multi-scale feature representations are widely exploited by both the one-stage object detectors and the two-stage object detectors in recent years. Object instances in optical remote sensing images change greatly in scale, orientation, and shape. Therefore, the detection of multi-scale objects, especially densely small objects, is a challenging problem in optical remote sensing images. To alleviate the problem arising from scale variation across object instances, we propose an Atrous Feature Pyramid Network (AFPN) to construct more effective feature pyramids network. The proposed network is an enhanced Faster R-CNN with a novel feature fusing module called Atrous Feature Fusing Module (AFFM). The multi-level features generated by the feature fusion module will be fed to detectors respectively to predict the detection results. The proposed network can not only detect objects more accurately in optical remote sensing images, but also improve the detection accuracy of natural images. Our network can achieve 96.89% mAP on NWPU VHR-10 dataset, 96.75% on RSOD dataset, which ensures the generalization ability of the method on larger and harder datasets.

**Keywords:** Optical remote sensing images · Object detection · Multi-scale feature representations

## 1 Introduction

Object detection is an important task in computer vision. In recent years, with the introduction of deep convolution neural network (DCNN), object detection has made remarkable progress in speed and accuracy. Object detection in optical remote sensing images has broad application prospects, ranging from military applications, urban planning to environmental management [24]. However, it also


H. Yang et al. (Eds.): ICONIP 2020, LNCS 12532, pp. 461–472, 2020.
https://doi.org/10.1007/978-3-030-63830-6_39

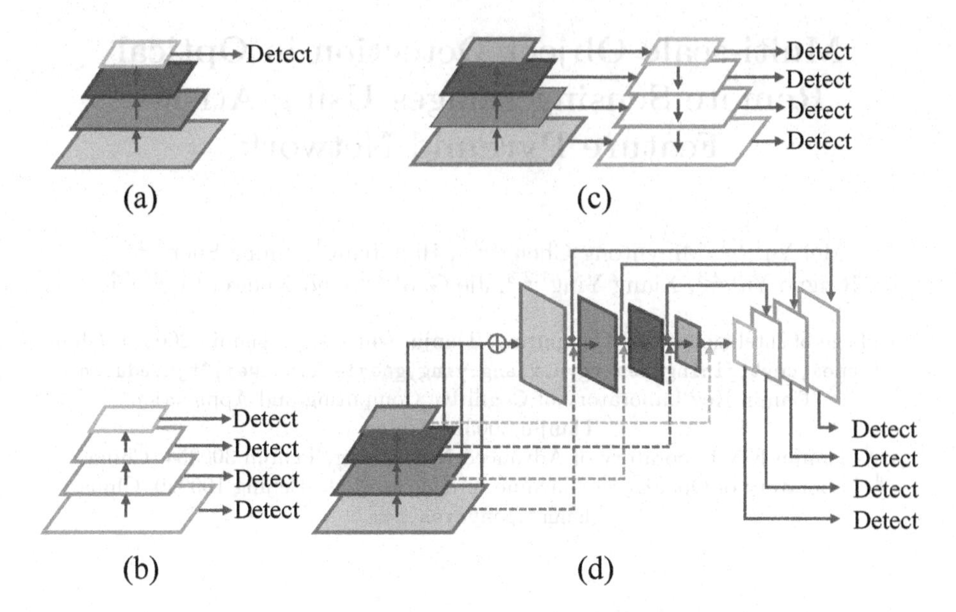


**Fig. 1.** Introductions of different feature pyramid module. (a) This method uses the top-level features in the feature extraction network for object detection. (b) The feature pyramid architecture is constructed by extracting the multi-level features of the backbone network. (c) Use the top-down pathways and lateral connections to feature fusing. (d) Our proposed feature fusion module and feature pyramid construction method.

faces many challenges, which hinder the development and application of object detection methods in the remote sensing field. Multi-scale object detection is one of them. In various tasks of computer vision, there are many ways to explore multi-scale feature representation.

The image pyramid method adjusts the input image to different scales and runs detectors on these images, which is less efficient. Figure 1(a) applies the top-level feature maps from feature extraction network for object detection, which is used by Fast R-CNN [7], Faster R-CNN [22], YOLO [20,21], R-FCN [4] and so on, but they do not perform well for multi-scale objects, especially small objects. Figure 1(b) builds the structure of feature pyramid by extracting multi-level features from backbone network, which is employed by SSD [17] algorithm. Because shallow features lack high-level semantic information, these methods do not work well for the detection of small objects. Figure 1(c) adopts the structure of top-down pathway and lateral connections for building high-level semantic feature maps at all scales, which is used in FPN [14], DSSD [5], YOLOv3 [21] and so on. However, the semantic information of multi-level features is not enough to satisfy the needs of object detection in optical remote sensing images. In order to fully express the multi-scale features, we propose a novel multi-level feature pyramid module to optimize the FPN structure. As shown in Fig. 1(d), we produce a single high-resolution feature with strong contextual semantic information by concatenating and fusing features from different layers. Then we take advantage of multi-layer downsampling and multi-branch atrous convolutional block to generate new multi-level pyramid features, which are fed to Fast R-CNN prediction head to produce the final detection results.

Due to the large spatial resolution of the sensor, the background of optical remote sensing images is extremely complex compared to the natural images and the scale variations of object instances are huge. In order to solve problems mentioned above, this paper proposes a new object detection network based on Faster R-CNN [22] algorithm, which is called Atrous Feature Pyramid network (AFPN). In AFPN, a novel feature fusion module is proposed, called Atrous Feature Fusing Module (AFFM), as shown in Fig. 1(d). We concatenate multi-level features to generate a single high-level feature map of a fine resolution (Multi-level Fusion Feature). Then we utilize convolution kernels with different sizes and multiple atrous rates to construct the Atrous Lateral Connection Block (ALCB). For extracting more representative multi-scale features, we integrate the features generated by ALCB into the features created by multi-layer downsampling of Multi-level Fusion Feature. The feature pyramid structure constructed by AFFM can significantly improve the detection performance of multi-scale objects. Experiments show that our method is superior to most state-of-the-art object detection algorithms in the detection accuracy. The main contributions of this paper are summarized as follows:

- In order to solve problems of scale variations of object instances, especially small objects detection in complex background, we propose a new region-based multi-level feature pyramid network (AFPN). AFPN can integrate high-level semantic information into low-level features, so as to significantly enhance the expressiveness of the feature pyramid network.
- We propose a novel feature fusion module, which is called AFFM. Due to the introduction of the multi-branch atrous convolution, this method can expand the receptive field of features and capture the multi-scale contextual information during feature fusion.
- According to experiments on NWPU VHR-10 and RSOD, we proved that AFPN has a significant improvement over conventional Faster R-CNN. It has relatively strong generalization ability and works better than most state-of-the-art methods, such as FPN, R-FCN, SSD, DSSD, YOLOv3 and so on.

## 2 Related Work

### 2.1 CNN-Based Object Detection Algorithms

In recent years, researchers have proposed a series of object detection algorithms based on DCNN. For example, R-CNN uses Selective Search [25] to generate the region proposals and inputs these region proposals into the detector for classification and bounding box regression. For input images with different resolutions, SPPNet [10] adopts spatial pyramid pool layer (SPP) to reuse feature extraction module, which greatly reduces the calculation and improves the speed. Fast R-CNN [7] introduces ROI pooling layer and fully connected layer to classify the object, which further improved the accuracy and efficiency of detection. Faster R-CNN [22] proposes a Region Proposal Network (RPN) to replace the selective search methods. By eliminating the time-consuming selective search, Faster R-CNN greatly improves the speed of the detection network.

Compared to region-based two-stage detectors, one-stage detectors have a significant increase in speed. YOLO [20,21] divides the input image into $S \times S$ grids to predict the categories and locations of objects for each cell. SSD [17] is another efficient one-stage object detector. It uses multi-level feature maps of different scales to enhance detection performance of detectors. DSSD [5] changes the backbone network of SSD for enhancing the ability of feature extraction and utilize deconvolution layer to generate multi-level feature pyramid networks.

### 2.2 Multi-scale Feature Representations

There are many works to exploit multi-level features fusion methods in the field of computer vision. Both Fast R-CNN and Faster R-CNN [22] initially use the top-level features for object detection. Because the high-level feature maps have lost a lot of detail information, they cannot work well on small objects.

Compared with the image pyramid method, the feature pyramid requires less memory and computational cost while maintaining high accuracy and computational efficiency. For improving the detection accuracy, HyperNet [12], ION [1] and Parsenet [18] realize multi-scale feature fusion by concatenate multi-scale features from multiple layers. SSD [17] utilizes multi-scale features and several additional convolutional layers for different scale object instances prediction. FPN [14] adopts nearest neighbor interpolation and lateral connections to spread high-level semantic information to the lower layer gradually. This structure can be regarded as a lightweight decoder. M2Det [26] proves that multiple FPN structures stack can effectively obtain multi-scale features, which significantly improves the detection performance of object detection. By using neural architecture search method, NAS-FPN [6] constructs a novel feature pyramid architecture which consists of a combination of top-down and bottom-up connections.

### 2.3 Atrous Convolution

Atrous convolution [11] can increase the receptive field of the convolution kernel while maintaining the resolution of the feature maps, and capture multi-scale contextual semantic information. At present, atrous convolution has been widely used in semantic segmentation and object detection task. In Deeplabv3 [2], atrous convolution is introduced into the Atrous Spatial Pyramid Pooling (ASPP) module to improve the segmentation accuracy of multi-scale objects. The ASPP module mainly includes cascade ASPP and parallel ASPP structures, which can capture multi-scale contextual information.

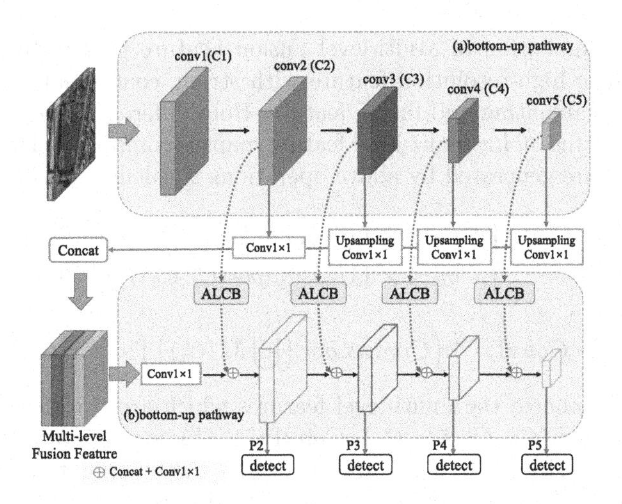


**Fig. 2.** An overview of the proposed Atrous Feature Fusing Module (AFFM) for building AFPN. AFFM utilizes multi-level features of backbone and Atrous Lateral Connection Block (ALCB) to extract features from the input image.

In the field of object detection, DetNet [13] uses atrous convolution in the backbone network to maintain high spatial resolution in deep layers of DCNN. Based on the frame of SSD algorithm [17], RFBNet [16] introduces the RFB module into object detector. The convolution kernels with different sizes and multiple atrous rates are used to enhance the contextual information of the feature map within RFB module.

# 3 Method

## 3.1 Atrous Feature Fusing Module

As shown in Fig. 2, AFFM contains three modules, namely (a) bottom-up pathway, (b) bottom-up pathway and Atrous Lateral Connection Block (ALCB). In the (a) bottom-up pathway, we select the last convolutional outputs from conv2, conv3, conv4, and conv5 stages of backbone network as the basic feature maps. And we express the basic feature maps as $\{C_2, C_3, C_4, C_5\}$, which have strides of $\{4, 8, 16, 32\}$ with respect to the input image. All feature fusion operations later in this paper are based on these basic features. In AFFM, $C_2$ reduces the feature dimension to 256 and conduct cross-channel feature fusion by $1\times1$ convolution operation. We abbreviate $1 \times 1$ convolution operation to Conv$1 \times 1$. For $\{C_3, C_4, C_5\}$, we adjust these feature maps to the size of $C_2$ by bilinear interpolation upsampling, and reduce the number of channels to 256 by Conv$1 \times 1$. The feature maps obtained from above operations are concatenated to obtain the Multi-Level Fusion Feature as shown in Fig. 2. Then we apply Conv$1 \times 1$

to reduce the dimension of Multi-level Fusion Feature to 256. In this way, we product a single high-resolution feature with strong contextual semantic information by concatenating and fusing features from different layers. And the first stage of feature fusion for multi-level feature maps is completed. The single high-resolution feature generated by above operations is calculated as follow:

$$M\left(C_i\right)=\begin{cases}Conv1\times 1\left(C_i\right), i=2\\ Conv1\times 1\left(Upsample\left(C_i, C_2\right)\right), i=3,4,5\end{cases} \tag{1}$$

$$F_1=Conv1\times 1\left(Concat\_opr\left(\bigcup M\left(C_i\right)\right)\right), i=2,3,4,5 \tag{2}$$

where $M\left(C_i\right)$ denotes the multi-level features which are transformed from the base feature maps $\{C_2, C_3, C_4, C_5\}$, $Conv1\times 1\left(C_i\right)$ denotes convolutional operation with kernel size $(1,1)$. $Upsample\left(C_i, C_2\right)$ is the function to resize $C_i$ to size of $C_2$ by bilinear interpolation upsampling. $F_1$ denotes a single high-resolution feature with strong contextual semantic information which is generated by the first stage of feature fusion. $Concat\_opr\left(\bigcup M\left(C_i\right)\right)$ is the feature fusion function, which concatenates multiple features into single feature.

In Fig. 2(b) bottom-up pathway illustrates the process of constructing feature pyramids from bottom up. This additional bottom-up pathway can spread detail information of low-level features to high-level layers efficiently. Multi-level feature maps are carried out by fusing multi-level features generated by some downsampling blocks with the multi-scale features captured by ALCB. In this way, (b) bottom-up pathway can generate multi-level features to construct feature pyramids with powerful multi-scale representation. We represent these feature maps as $\{P_2, P_3, P_4, P_5\}$ respectively, which is corresponding to $\{C_2, C_3, C_4, C_5\}$. Specifically, the multi-level feature maps generated by downsampling and ALCB are combined by a concatenation operation. After that, $\{P_2, P_3, P_4, P_5\}$ is obtained by reducing the dimension to 256 using Conv1 $\times$ 1 operation. These maps are computed as Eq. (3).

$$P_{\mathrm{i}}=\begin{cases}\mathrm{Conv1}\times 1\left(Concat\_opr\left(F_1, ALCB\left(C_i\right)\right)\right), i=2\\ \mathrm{Conv1}\times 1\left(Concat\_opr\left(Upsample\left(P_{i-1}, Ci\right), ALCB\left(Ci\right)\right)\right), i=3,4,5\\ Conv3\times 3\left(P_{i-1}\right), i=6\end{cases} \tag{3}$$

In Eq. (3), $P_{\mathrm{i}}$ denotes the multi-level features which are fed to network for predicting results. The $ALCB\left(C_i\right)$ is the function of multi-branch atrous convolution operation with different size of convolutional kernel and atrous rate. $Conv3\times 3\left(P_{i-1}\right)$ denotes convolutional operation with kernel size $(3,3)$ and stride size of 2, which is a downsampling function. During the experiments, we perform additional downsampling for $P_5$ to produce $P_6$, which is not shown in Fig. 2. Like FPN [14], all levels of feature maps share RPN, classifiers and regressors from the Faster R-CNN [22] model.

### 3.2 Atrous Lateral Connection Block

The lateral connections of FPN [14] only use Conv1 × 1 to reduce dimensionality of feature maps extracted from the backbone. We propose Atrous Lateral Connection Block (ALCB) to capture multi-scale context semantic information by multi-branch convolution with multi-size kernel and different atrous rates. The structure of Atrous Lateral Connection Block (ALCB) is shown in Fig. 3, and it is a multi-branch convolutional block.

Based on the atrous convolution module from DeepLabv3 and RFBNet, we propose a 4-branch convolution block. The above three branches adopt the bottleneck structure and use Conv1 × 1 to perform cross-channel features fusion and dimensionality reduction of feature maps. In addition, the above three branches take convolution kernel with different sizes, namely Conv1 × 1, Conv3 × 3 and Conv5 × 5. In this way, it can increase the receptive fields of multiple branches separately. The Conv3 × 3 with different atrous rates are added after these three branches in order to encode object features at multiple scales. Global Average Pooling branch aims to increase the global information, which can enhance expressiveness of local features. Finally, the feature maps generated by each branch are concatenated together. Compared with the lateral connection in FPN [14], the feature maps generated by ALCB have larger receptive field and powerful multi-scale feature representations.

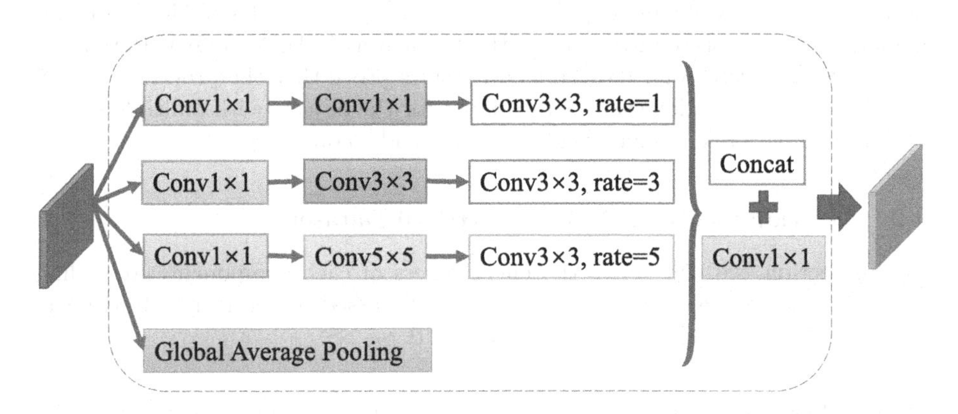


**Fig. 3.** Structure details of Atrous Lateral Connection Block (ALCB).

## 4 Experiments

In this section, we conduct experiments on NWPU VHR-10 [3] and RSOD [19]. Firstly, we compare the experimental results of our AFPN with state-of-the-art methods on NWPU VHR-10 dataset. Next, we aim to further validate the method proposed thorough ablation experiments. Finally, experiments on RSOD indicate that the proposed approach has a good performance and can be generalized easily.

**Table 1.** Comparison of various methods on NWPU VHR-10 dataset.

| Method | Backbone | Storage tank | Airplane | Tennis court | Baseball diamond | Bridge | Vehicle | Basketball court | Harbor | Ground track field | Ship | mAP (%) |
|---|---|---|---|---|---|---|---|---|---|---|---|---|
| Faster R-CNN | ResNet-101 | 67.69 | 99.35 | 91.71 | 93.88 | 68.71 | 81.8 | 88.57 | 82.29 | 98.57 | **96.09** | 86.87 |
| FPN | ResNet-101 | 87.16 | **100** | 96.93 | 97.7 | 77.37 | 91.13 | 93.05 | 93.97 | 98.43 | 94.26 | 93.00 |
| SSD | VGG | 71.21 | 90.60 | 87.16 | 97.10 | **97.56** | 39.05 | 70.18 | 88.21 | 99.70 | 82.44 | 82.32 |
| YOLOV3 | Darknet-53 | 93.39 | 99.99 | 94.92 | **99.03** | 92.85 | 73.56 | 75.25 | 92.87 | **99.88** | 91.07 | 91.28 |
| RetinaNet | ResNet-101 | 85.99 | 90.9 | 90.79 | 89.29 | 72.09 | 86.88 | 90.35 | 85.67 | 97.07 | 89.32 | 87.84 |
| NAS-FPN | ResNet-101 | 85.93 | **100** | **100** | 98.03 | 83.05 | 88.41 | 90.12 | 87.96 | 94.87 | 90.08 | 91.84 |
| **AFPN (Ours)** | ResNet-101 | **93.62** | **100** | 97.71 | 98.05 | 97.28 | **96.8** | **94.33** | **96.7** | 99.35 | 95.04 | **96.89** |

## 4.1 Results on NWPU VHR-10 Dataset

We compare the results of our network with state-of-the-art methods on NWPU VHR-10 dataset. We have implemented Faster R-CNN [22], FPN [14], SSD [17], YOLOv3 [21], RetinaNet [15] and NAS-FPN methods and test these methods on NWPU VHR-10 dataset. As can be seen from Table 1, our AFPN achieves the mAP of 96.89% on the NWPU VHR-10 dataset, which is about 10% higher than the original Faster R-CNN. In addition, our results are also superior to FPN based on Faster R-CNN by roughly 3.8%. In Table 1, the results of AFPN are also higher than most state-of-the-art methods, such as SSD, YOLOv3, RetinaNet and NAS-FPN detectors. Experimental results prove that the proposed AFPN greatly improves the detection performance for the objects with multiple scales, such as storage tank, vehicle, bridge, harbor and so on.

## 4.2 Ablation Study on NWPU VHR-10 Dataset

In this section, we will verify the effectiveness of each component. According to Table 2, the following two aspects will be discussed respectively: Multi-Level Feature Detection and The Effective of ALCB.

**Table 2.** Effectiveness of various designs for building feature pyramid on NWPU VHR-10 dataset.

| Method | Multi-level | ALCB | mAP(%) |
|---|---|---|---|
| (a) Faster R-CNN (Baseline) | – | – | 86.87 |
| (b) Faster R-CNN with multi-level | ✓ | – | 90.34 |
| (c) Faster R-CNN with ALCB | ✓ | ✓ | 93.80 |

**Multi-level Feature Detection.** As shown in Table 2, Faster R-CNN (Table 2(a)) reaches the mAP of 86.87% on NWPU-VHR 10 dataset. Based on Faster R-CNN [22], we build a multi-level feature detector, which is denoted as Faster R-CNN with multi-level (Table 2(b)). This method feeds the basic feature maps ($\{C_2, C_3, C_4, C_5\}$) from the backbone network directly into the Faster R-CNN detection network. We construct feature pyramid by using the basic features, which is similar to Fig. 1(c). The Faster R-CNN with multi-level (Table 2(b)) method increases the mAP from 86.87% to 90.34%, which is an increase of approximately 3.5%. This shows that the detection of multi-level features can significantly improve the detection performance compared to the object detection using single-scale features.

**The Effectiveness of ALCB.** Based on Table 2(b) method, we apply ALCB to the multi-level features from backbone network. We denote this method as Faster R-CNN with ALCB (Table 2(c)). We set all atrous rates of the three branches to $\{1, 3, 5\}$ respectively. As seen in Table 2, the Faster R-CNN with ALCB method increases the mAP to 93.80%. This indicates that the multi-branch structure in ALCB can enhance the expressiveness of the feature pyramid structure and improve the detection performance.

**Table 3.** Comparison of different detection methods on RSOD dataset.

| Method | Backbone | Overpass | Oiltank | Playground | Aircraft | mAP |
|---|---|---|---|---|---|---|
| Faster R-CNN | ResNet-101 | 88.62 | 98.11 | 97.81 | 83.54 | 77.02 |
| FPN | ResNet-101 | 96.90 | **99.18** | 98.16 | 91.93 | 96.54 |
| SSD | VGG | 90.21 | 90.72 | 98.58 | 71.89 | 87.85 |
| YOLOV3 | Darknet-53 | 96.94 | 98.91 | **99.65** | 88.38 | 95.97 |
| RetinaNet | ResNet-101 | 88.00 | 95.04 | 98.18 | 80.81 | 90.51 |
| NAS-FPN | ResNet-101 | 83.68 | 91.15 | 96.96 | 88.20 | 90.00 |
| **AFPN (Ours)** | ResNet-101 | **97.07** | 98.90 | 98.96 | **92.07** | **96.75** |

## 4.3 Results on RSOD

We conduct experiments on RSOD optical remote sensing dataset. We compare our results with the existing methods, and the final results are shown in Table 3. Our AFPN method can reach 96.75% mAP, which is about 20% higher than the traditional Faster R-CNN. In Table 3, FPN method can achieve 96.54% mAP, much higher than the Fast R-CNN, but still slightly lower than our method. For the one-stage detectors, SSD, RetinaNet and YOLOv3 achieve the mAP of 87.85%, 90.51% and 95.97% on RSOD, respectively. As shown in Table 3, the proposed network also outperforms the latest NAS-FPN detector. Finally, we show the renderings of the model on NWPU VHR-10, RSOD and DOTA [23] in Fig. 4.

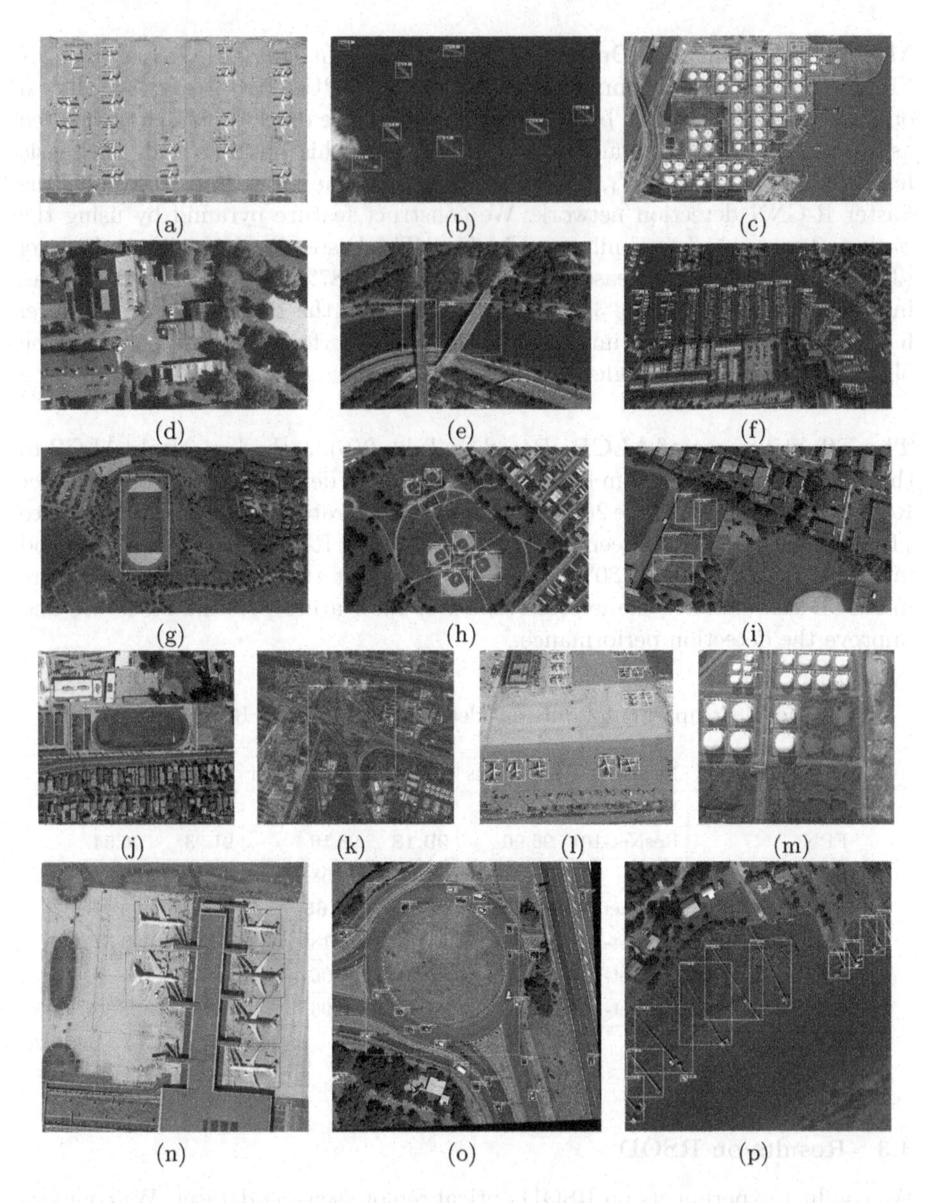


**Fig. 4.** Some examples of detection results for the NWPU VHR-10, RSOD and DOTA datasets. (a–i) images show some results on the NWPU VHR-10 dataset and (j–m) images show some results on the RSOD dataset. In addition, (n–p) images show part of detection examples for DOTA dataset.

## 5 Conclusion

In this work, a novel method called Atrous Feature Pyramid Network (AFPN) is proposed for multi-scale objects, especially small objects detection in optical remote sensing images. By using Atrous Feature Fusing Module (AFFM), we construct a new feature pyramid network with the strong representational power and high robustness to scale variation. In AFFM, we use the multi-branch atrous convolution module to make full use of image feature information. AFPN can achieve 96.89% mAP on NWPU VHR-10 dataset and 96.75% on RSOD dataset. The results show that AFPN can not only improve detection performance of remote sensing images, but also perform well on natural images. Experiments indicate that our method has better generalization performance. Because AFPN is based on two-stage detectors, it is slightly less efficient than the one-stage approaches. In the future, we will try to apply the proposed structure to one-stage detectors for improving the accuracy of the one-stage object detectors while maintaining a high speed of inference.

**Acknowledgment.** This work is supported by National Natural Science Foundation of China (Grant No. 61976155).

## References


1. Bell, S., Lawrence Zitnick, C., Bala, K., Girshick, R.: Inside-outside net: detecting objects in context with skip pooling and recurrent neural networks. In: Proceedings of the IEEE Conference on Computer Vision and Pattern Recognition, pp. 2874–2883 (2016)
2. Chen, L.-C., Papandreou, G., Schroff, F., Adam, H.: Rethinking atrous convolution for semantic image segmentation. arXiv preprint arXiv:1706.05587 (2017)
3. Cheng, G., Han, J.: A survey on object detection in optical remote sensing images. ISPRS J. Photogr. Remote Sens. **117**, 11–28 (2016)
4. Dai, J., Li, Y., He, K., Sun, J.: R-FCN: object detection via region-based fully convolutional networks. In: Advances in Neural Information Processing Systems, pp. 379–387 (2016)
5. Fu, C.-Y., Liu, W., Ranga, A., Tyagi, A., Berg, A.C.: DSSD: deconvolutional single shot detector. arXiv preprint arXiv:1701.06659 (2017)
6. Ghiasi, G., Lin, T.-Y., Le, Q.V.: NAS-FPN: learning scalable feature pyramid architecture for object detection. In: Proceedings of the IEEE Conference on Computer Vision and Pattern Recognition, pp. 7036–7045 (2019)
7. Girshick, R.: Fast R-CNN. In: Proceedings of the IEEE International Conference on Computer Vision, pp. 1440–1448 (2015)
8. Girshick, R., Donahue, J., Darrell, T., Malik, J.: Rich feature hierarchies for accurate object detection and semantic segmentation. In: Proceedings of the IEEE Conference on Computer Vision and Pattern Recognition, pp. 580–587 (2014)
9. He, K., Gkioxari, G., Dollár, P., Girshick, R.: Mask R-CNN. In: Proceedings of the IEEE International Conference on Computer Vision, pp. 2961–2969 (2017)
10. He, K., Zhang, X., Ren, S., Sun, J.: Spatial pyramid pooling in deep convolutional networks for visual recognition. IEEE Trans. Pattern Anal. Mach. Intell. **37**(9), 1904–1916 (2015)
11. Holschneider, M., Kronland-Martinet, R., Morlet, J., Tchamitchian, P.: A real-time algorithm for signal analysis with the help of the wavelet transform. In: Combes, J.M., Grossmann, A., Tchamitchian, P. (eds.) Wavelets, pp. 286–297. Springer, Heidelberg (1990). https://doi.org/10.1007/978-3-642-75988-8_28

12. Kong, T., Yao, A., Chen, Y., Sun, F.: HyperNet: towards accurate region proposal generation and joint object detection. In: Proceedings of the IEEE Conference on Computer Vision and Pattern Recognition, pp. 845–853 (2016)
13. Li, Z., Peng, C., Yu, G., Zhang, X., Deng, Y., Sun, J.: DetNet: a backbone network for object detection. arXiv preprint arXiv:1804.06215 (2018)
14. Lin, T.-Y., Dollár, P., Girshick, R. He, K., Hariharan, B., Belongie, S.: Feature pyramid networks for object detection. In: Proceedings of the IEEE Conference on Computer Vision and Pattern Recognition, pp. 2117–2125 (2017)
15. Lin, T.-Y., Goyal, P., Girshick, R., He, K., Dollár, P.: Focal loss for dense object detection. In: Proceedings of the IEEE International Conference on Computer Vision, pp. 2980–2988 (2017)
16. Liu, S., Huang, D., et al.: Receptive field block net for accurate and fast object detection. In: Proceedings of the European Conference on Computer Vision (ECCV), pp. 385–400 (2018)
17. Liu, W., et al.: SSD: single shot multibox detector. In: Leibe, B., Matas, J., Sebe, N., Welling, M. (eds.) ECCV 2016. LNCS, vol. 9905, pp. 21–37. Springer, Cham (2016). https://doi.org/10.1007/978-3-319-46448-0_2
18. Liu, W., Rabinovich, A., Berg, A.C.: ParseNet: looking wider to see better. arXiv preprint arXiv:1506.04579 (2015)
19. Long, Y., Gong, Y., Xiao, Z., Liu, Q.: Accurate object localization in remote sensing images based on convolutional neural networks. IEEE Trans. Geosci. Remote Sens. **55**(5), 2486–2498 (2017)
20. Redmon, J., Divvala, S., Girshick, R., Farhadi, A.: You only look once: unified, real-time object detection. In Proceedings of the IEEE Conference on Computer Vision and Pattern Recognition, pp. 779–788 (2016)
21. Redmon, J., Farhadi,A.: YOLOv3: an incremental improvement. arXiv preprint arXiv:1804.02767 (2018)
22. Ren, S., He, K., Girshick, R., Sun, J.: Faster R-CNN: towards real-time object detection with region proposal networks. In: Advances in Neural Information Processing Systems, pp. 91–99 (2015)
23. Xia, G.-S., et al.: DOTA: a large-scale dataset for object detection in aerial images. In: Proceedings of the IEEE Conference on Computer Vision and Pattern Recognition, pp. 3974–3983 (2018)
24. Tayara, H., Chong, K.: Object detection in very high-resolution aerial images using one-stage densely connected feature pyramid network. Sensors **18**(10), 3341 (2018)
25. Uijlings, J.R., Van De Sande, K.E., Gevers, T., Smeulders, A.W.: Selective search for object recognition. Int. J. Comput. Vision **104**(2), 154–171 (2013)
26. Zhao, Q., et al.: M2Det: a single-shot object detector based on multi-level feature pyramid network. arXiv preprint arXiv:1811.04533 (2018)

# Object Tracking with Multi-sample Correlation Filters

Qinyi Tian(✉)

Samuel Ginn College of Engineering, Auburn University, Auburn, AL 36849, USA
qzt0006@auburn.edu

**Abstract.** Due to the real-time tracking and location accuracy of the kernel correlation filtering (KCF) algorithm, this method is now widely used in object tracking tasks. However, whether KCF or its improved algorithm, the filter parameter training is achieved by the feature of the current frame, in other words, the training sample is single. If the samples of multiple frames in the previous sequence can be integrated to the filter parameters training, the trained filter parameters should be more reliable. In this paper, we propose an object tracking algorithm based on multi-sample kernel correlation filtering (MSKCF). Meanwhile, In order to select better samples, the average peak correlation energy (APCE) is introduced to measure the stability of tracking effect, and is applied as weight of sample. The frames with higher APCE value are chosen as multi-sample, and then are used to train filter parameters. Experimental results show that the tracking effect of the proposed method is superior to compared state-of-the-art algorithms.

**Keywords:** Object tracking · Kernel correlation filtering (KCF) · Multi-sample · Average peak correlation energy (APCE)

## 1 Introduction

Object tracking is an important part of computer vision technology, and plays an irreplaceable role in acquiring video information and understanding video semantic content. Object tracking can be classified into two types, generative model and discriminative model. The generative tracking algorithm uses the generative model to describe the apparent characteristics of an object, and then search the candidate object to minimize the reconstruction error. Typical algorithms include sparse coding [1,2], online density estimation [3,4], etc. Generative model focuses on the description of the object, ignores the background information, and is prone to drift when the object changes dramatically or is occluded. The discriminative tracking algorithm can complete the tracking task by constructing classifier to distinguish the object from the background. Various machine learning algorithms have been applied to discriminative model, such as online multiple instance learning (MIL) [5], support vector machine (SVM) [6], ensemble learning [7], random forest classifier [8], etc.


H. Yang et al. (Eds.): ICONIP 2020, LNCS 12532, pp. 473–485, 2020.
https://doi.org/10.1007/978-3-030-63830-6_40

In recent years, object tracking algorithms based on correlation filtering (CF) are widely and deeply researched. The main steps of object tracking based on CF are described as follows [9]. First, according to the initial position information of the object, the feature of the object is extracted. In order to avoid the interference of the image features around the object, the cosine function is used to smooth the feature information. Then, the CF parameter (filter parameter) is obtained by training the feature information of the object. Finally, the filter parameter is correlated with the next frame to get the response map. The position with maximum value of the response map corresponds to the center position of the object. In the process of tracking, the filter parameters of correlation filter are updated. Typical works based on CF include minimum output sum of squared error filter (MOSSE) [9], kernel correlation filtering (KCF) [10], discriminative scale space tracking (DSST) [11], spatially regularized correlation filters (SRDCF) [12], large margin tracking method with circulate feature maps (LMCF) [13], etc.

Deep learning method makes use of the advantages of big data, and achieves the goal of automatically learning the object features by training neural network. In addition, transfer learning greatly reduces the requirement for training sample and improves the performance of the tracking algorithm [14]. The obtained features by deep learning method are more robust than the traditional manual features. The low layer features extracted by convolutional neural network have high resolution, more location information and more semantic information. Combining hierarchical convolutional features with CF, some better tracking results are obtained [15,16]. Siamese networks is trained offline based on the big data set, where the previous frame and current frame of video are deal with CF operation to determine the position of the object to be tracked [17]. Based on Siamese networks, the candidate area network and end-to-end training on a large number of video sequence sets are conducted [18–20], these methods greatly improved the speed and effect of object tracking.

KCF based filter parameter training usually is implemented by the feature of the current frame, such as Ref. [10], Ref. [21], etc. In addition, some of variants of KCF use previous frame as constraint to improve tracking effect [22]. In order to improve the robust of tracking, a multi-sample learning is introduced into KCF in this paper, where the filter parameter of the previous frames are adaptively integrated into the filter parameter training of current frame. In other words, the previous frames are also fused into the parameter training of current frame, which is called multi-sample kernel correlation filtering (MSKCF) method. The main contributions of this paper are as follows:

(1) A multi-sample learning method based on KCF is constructed, whereby a strong parameter training model is achieved by integrating the trained filter parameter of previous frames and adjusting the weight of each filter parameter adaptively.
(2) Average peak correlation energy (APCE) is introduced, which is helpful in selecting the filter parameter weight of previous frames with better tracking

performance, so that the trained filter parameter of the current frame are more robust.

(3) Through quantitative comparison, experimental results indicates that our tracker, MSKCF outperforms other state-of-the-art trackers in challenging sequences. The proposed method is shown in Fig. 1, where the filter parameter ($\mathbf{w}$) of current frame is formed by integrating multi-sample.

The remainder of this paper is organized as follows. In Sect. 2, the related work is briefly introduced. Sect. 3 discusses ridge regression, correlation filter, and kernel correlation filter. Sect. 4 derives a novel multi-sample KCF and implementation details. In Sect. 5, experimental results and comparison with state-of-the-art methods are described. Sect. 6 summarizes our work.

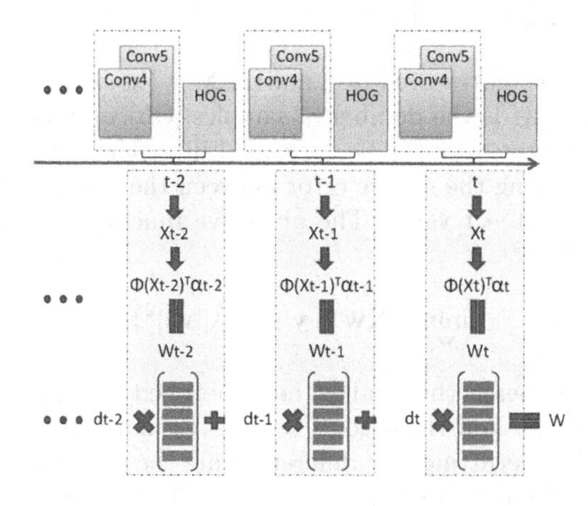


**Fig. 1.** Generation of filter parameter of current frame in MSKCF.

## 2 Related Work

Recently, CF algorithm is widely used in object tracking, and many tracking algorithms based on CF have been proposed. In order to combine different features in KCF, a multi-kernel correlation filter (MKCF) [21] is derived to further improve the tracking performance. The spatial-temporal regularized correlation filters (STRCF) [22] adds spatial and temporal regular terms to discriminative correlation filter (DCF), and constructs a convex objective function, which is optimized by alternating direction multiplier algorithm (ADMM). For alleviating the negative mutual interference of different kernels, Ref. [23] modified MKCF (MKCFup) to reformulate the multi-kernel learning (MKL) version of CF objective function with its upper bound. Multi-template scale-adaptive kernelized correlation filters [24] updates the scale of the tracker by maximizing over the posterior distribution of a grid of scales and achieves better tracking results.

Multi-cue correlation filters [25] proposes a filtering method based on multi cue framework, which combines different weak trackers, each of which is regarded as a cue, and each cue can track the object independently. In this method, multi-threaded divergence is used to ensure the reliability of object tracking. These related works implement filter parameter training by current frame or the constraint of previous frame. Different from above related works, we proposed a multi-sample kernel correlation filtering (MSKCF) algorithm to improve tracking effect, which is able to fully take advantage of previous samples to achieve accurate tracking.

# 3 CF Tracking

## 3.1 Ridge Regression and CF

Training sample set is represented by a matrix $\mathbf{X} = \{\mathbf{x}_1; \mathbf{x}_2; ...; \mathbf{x}_M\}^T$ where each row is a sample and $M$ is the number of samples. Object value of regression are represented by a vector $\mathbf{y} = \{y_1, y_2, ..., y_M\}$. Filter parameter is $\mathbf{w}$, and it can be solved by minimizing the square error between the linear regression function $f(\mathbf{x}_i) = \mathbf{w}^T\mathbf{x}_i$ and object value. The objective function of ridge regression is represented as follow:

$$\min_{\mathbf{w}}\{||\mathbf{X}\mathbf{w} - \mathbf{y}||^2 + \lambda||\mathbf{w}||^2\} \tag{1}$$

where the first term means the smallest error between the linear regression function and object value, and the second term is regularization parameter which prevents the model overfitting. By differentiating the derivative and setting the derivative to zero, the closed solution of the Eq. (1) is obtained:

$$\mathbf{w} = \left(\mathbf{X}^H\mathbf{X} + \lambda\right)^{-1}\mathbf{X}^H\mathbf{y} \tag{2}$$

where $\mathbf{X}^H$ is complex conjugate transposed matrix of $\mathbf{X}$, and $(\cdot)^{-1}$ represents the operation for matrix inversion. Due to the large number of samples and the high complexity of matrix inversion operation, it is difficult to meet the real-time of tracking. Therefore, the characteristics of cyclic shifts and cyclic matrices are used to improve the computational efficiency of Eq. (1).

In CF algorithm, all training samples are obtained from the cyclic shift of the object sample (base sample $\mathbf{x}$). Each training sample is cyclically shifted to obtain a cyclic matrix $\mathbf{X}$ in which base sample $\mathbf{x} = \mathbf{x}_1$ and $\mathbf{y}$ is Gaussian label. The cyclic matrix can be diagonalized by the discrete Fourier transform, according to which the sample matrix is converted into a diagonal matrix for calculation, which can greatly speed up the calculation between the matrices. Let $\hat{\mathbf{x}} = \mathcal{F}(\mathbf{x})$ is discrete Fourier transform of $\mathbf{x}$, then Eq. (2) is transformed into the frequency domain through the Fourier transform, and its complex form is as follow:

$$\hat{\mathbf{w}} = \frac{\hat{\mathbf{x}} \odot \hat{\mathbf{y}}}{\hat{\mathbf{x}}^* \odot \hat{\mathbf{x}} + \lambda} \tag{3}$$

where $(\cdot)^*$ indicates complex conjugation operation and $\odot$ is dot product operation. Further, the filter parameter in Eq. (3) is obtained by inverse Fourier transform $\mathbf{w} = \mathcal{F}^{-1}(\hat{\mathbf{w}})$, where $\mathcal{F}^{-1}(\cdot)$ represents the inverse discrete Fourier transform.

### 3.2 Ridge Regression and KCF

When the sample is linearly inseparable, the sample can be projected into a high-dimensional feature space by introducing a kernel function, so that the better classified result can be obtained. Let $\phi(\cdot)$ be kernel function, in terms of the representer theorem [26], the filter parameter $\mathbf{w}$ can be obtained as a linear combination of all mapped training sample, i.e., $\mathbf{w} = \sum_{i=1}^{M} \alpha^i \phi(\mathbf{x}_i)$, where $\bar{\boldsymbol{\alpha}} = \left[\alpha^1, \alpha^2, ..., \alpha^M\right]^T$ is combination parameter and is a column vector of $M \times 1$. We substitute the $\mathbf{w}$ into the least squares loss function (Eq. (1)) to obtain the following objective function:

$$\min_{\bar{\alpha}} \sum_{j=1}^{M} || \sum_{i=1}^{M} (\alpha^i \phi(\mathbf{x}_i))^T \phi(\mathbf{x}_j) - \mathbf{y}_j ||^2 + \lambda || \sum_{i=1}^{M} (\alpha^i \phi(\mathbf{x}_i)) ||^2 \tag{4}$$

Normally, when the number of samples are large enough, the linear kernel function is selected. When the number of sample is small or we do not know what kernel function to use, the Gaussian kernel function $\mathbf{K}(\mathbf{x}_i, \mathbf{x}_j)$ is preferred. Gaussian kernel has strong spatial locality and learning ability. We use the kernel function to further simplify the Eq. (4) to obtain the kernel least square error objective function:

$$\min_{\bar{\alpha}} ||\mathbf{K}(\mathbf{X}, \mathbf{X})\bar{\boldsymbol{\alpha}} - \mathbf{y}||^2 + \lambda \bar{\boldsymbol{\alpha}}^T \mathbf{K}(\mathbf{X}, \mathbf{X})\bar{\boldsymbol{\alpha}} \tag{5}$$

We construct all $\mathbf{K}(\mathbf{x}_i, \mathbf{x}_j)(i, j = 1, 2, ..., M)$ into a $M \times M$ kernel function matrix $\mathbf{K}(\mathbf{X}, \mathbf{X})$. Similarly, in order to improve the calculation efficiency, the kernel function matrix is diagonalized by the discrete Fourier transform to perform a dot product operation, and then the inverse Fourier transform is used to obtain a response graph $\mathbf{R}$. The position with the maximum value in the response map is predicted as the current object position.

## 4 Multi-sample KCF (MSKCF)

### 4.1 Objective Function

The $i-th$ sample set is presented as $\mathbf{X}_i = \{\mathbf{x}_{i,j} | j = 1, 2, ..., M\}$ and $\phi(\mathbf{X}_i) = [\phi(\mathbf{x}_{i,1}), \phi(\mathbf{x}_{i,1}), ..., \phi(\mathbf{x}_{i,1})]^T$. $\{\mathbf{x}_{i,j} | j = 2, 3, ..., M\}$ are formed by cyclically shifting the base sample $\mathbf{x}_{i,1}$. $\mathbf{w}_i = \phi(\mathbf{X}_i)^T$ is filter parameter of the $i-th$ sample which is transformed using kernel function, and $\boldsymbol{\alpha}_i$ is combination parameter of the sample sets $\mathbf{X}_i$. Considering the correlation between the previous frames and the current frames, we propose a multi-sample correlation filter. Based on multi-sample learning, the filter parameter of current frame (the $t-th$ frame) is represented as $\mathbf{w}_t = \sum_{i=1}^{t} d_i \mathbf{w}_i = \sum_{i=1}^{t} d_i \phi(\mathbf{X}_i)^T \boldsymbol{\alpha}_i$. Objective function of MSKCF is constructed as follow:

$$J(\boldsymbol{\alpha}_t) = ||\phi(\mathbf{X}_t)\mathbf{w}_t - \mathbf{y}||^2 + \lambda||\mathbf{w}_t||^2 \tag{6}$$

Substituting $\mathbf{w}_t$ in Eq. (6) with filter parameter of previous frames, we obtain the unconstrained optimization problem:

$$J(\boldsymbol{\alpha}_t) = ||\phi(\mathbf{X}_t)\sum_{i=1}^{t}(d_i\phi(\mathbf{X}_i)^T\boldsymbol{\alpha}_i) - \mathbf{y}||^2 + \lambda||\sum_{i=1}^{t}(d_i\phi(\mathbf{X}_i)^T\boldsymbol{\alpha}_i)||^2 \tag{7}$$

where $i$ is the number of frame. $d_i$ is weight of filtering parameter $\mathbf{w}_i$ and is determined with APCE value, $\sum_{i=1}^{t} d_i = 1$. Commonly, as the filter parameter of different frame has different robustness, filter parameter should be provided with different weights $d_i$. Let $\mathbf{K}_{ij} = \phi(\mathbf{X}_i)\phi(\mathbf{X}_j)$, then the objective function is transformed into following form:

$$J(\boldsymbol{\alpha}_t) = ||\sum_{i=1}^{t}(d_i\mathbf{K}_{ti}\boldsymbol{\alpha}_i) - \mathbf{y}||^2 + \lambda\sum_{i=1}^{t}\sum_{j=1}^{t}(d_i\boldsymbol{\alpha}_i^T\mathbf{K}_{ij}\boldsymbol{\alpha}_j d_j) \tag{8}$$

which is an unconstrained quadratic programming problem w.r.t. $\boldsymbol{\alpha}_t$,

$$\begin{aligned} minJ(\boldsymbol{\alpha}_t) = \min_{\boldsymbol{\alpha}_t} \boldsymbol{\alpha}_t^T\mathbf{K}_{tt}\mathbf{K}_{tt}\boldsymbol{\alpha}_t d_t^2 + 2\sum_{i=1}^{t-1}(d_i\boldsymbol{\alpha}_i^T\mathbf{K}_{ti}\mathbf{K}_{tt}\boldsymbol{\alpha}_t d_t) \\ - 2\mathbf{y}^T\sum_{i=1}^{t}(d_i\mathbf{K}_{ti}\boldsymbol{\alpha}_i) + \lambda(\boldsymbol{\alpha}_t^T\mathbf{K}_{tt}\boldsymbol{\alpha}_t d_t^2) + 2\lambda\sum_{i=1}^{t-1}(d_i\boldsymbol{\alpha}_i^T\mathbf{K}_{it}\boldsymbol{\alpha}_t d_t) + const \end{aligned} \tag{9}$$

The training samples are cyclical shift of base sample. Therefore, the optimization of $\boldsymbol{\alpha}_t$ can be speeded up by means of the fast Fourier transform (FFT) pair, $\mathcal{F}$ and $\mathcal{F}^{-1}$. Because the samples are circulant, $\mathbf{K}_{ij}$ are circulant [10,27]. The inverses and the sum of circulant matrices are circulant [28]. Then the calculation of Eq. (9) can be accelerated as follows:

$$\mathcal{F}(\boldsymbol{\alpha}_t) = \frac{\mathcal{F}(\mathbf{y}) - (\frac{\lambda}{\mathcal{F}(\mathbf{K}_{tt})+\sigma} + 1)\sum_{i=1}^{t-1}(d_i\mathcal{F}(\mathbf{K}_{ti}) \odot \mathcal{F}(\boldsymbol{\alpha}_i))}{d_t(\mathcal{F}(\mathbf{K}_{tt}) + \lambda\mathbf{I})} \tag{10}$$

where $\mathbf{I}$ is an $M \times M$ identity matrix. When the next frame is inputted, response map $\mathbf{R}$ of object is obtained by $\mathbf{R} = \mathcal{F}^{-1}\sum_{i=1}^{t-1}(d_i\hat{\mathbf{K}}_{ti} \odot \hat{\boldsymbol{\alpha}}_i)$.

### 4.2 Weight of Filter Parameter

In practical application, it is found that when the tracking effect is not ideal or even occurs tracking drift, multi-peaks will appear in the response map. This paper will introduce the multi-peak detection [13,29] to verify the tracking effect. We use the average peak correlation energy (APCE) as tracking confidence indicator to judge whether the object is well tracked, and determine whether reserving the filter parameter of current frame as part of multi-sample. The APCE is formulated as follow:

$$E = \frac{|\mathbf{R}_{max} - \mathbf{R}_{min}|}{mean \sum_{i,j} (\mathbf{R}_{i,j} - \mathbf{R}_{min})} \tag{11}$$

where $\mathbf{R}_{max}$ represents the maximum value of the response graph $\mathbf{R}$, $\mathbf{R}_{min}$ represents the minimum value of the response graph $\mathbf{R}$, $mean(\cdot)$ represents the mean function, $\mathbf{R}_{i,j}$ represents the value of the $i-th$ row and $j-th$ column of the response graph $\mathbf{R}$. If $E_i$ is the APCE value of the $i-th$ frame, the weight of filter parameter $d_i$ is represented as $d_i = \frac{E_i}{\sum_j E_j}$.

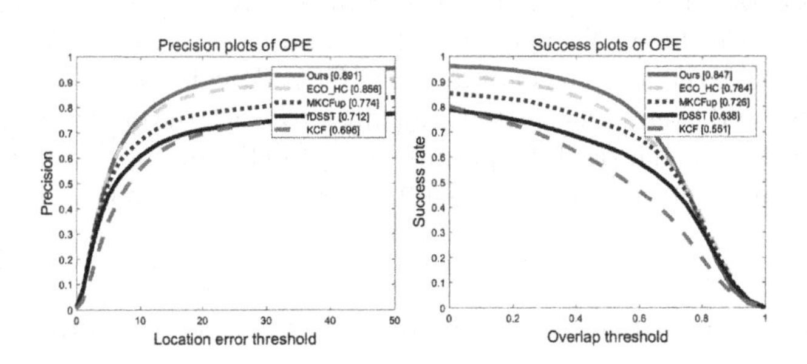


**Fig. 2.** The precision and success plots of OPE. (a) the precision plots of OPE, (b) the success plots of OPE.

## 5 Experiments

### 5.1 Experimental Data and Evaluation Index

Online object tracking benckmark (OTB2015) [30] is used to evaluate the performance of the proposed method in experiments. The benchmark includes 100 video sequences, including various challenges, such as occlusion, deformation, illumination change, etc. We use two evaluation metrics to validate the algorithm performance: the precision and success rate [30]. The precision rate is shown by precision plot where the distance precision is depicted over a range of thresholds. The success rate is shown by the success plot, and the plot indicates the overlap precision based on the bounding box. The success plot shows the percentage of successful frames where the bounding box overlap exceeds a threshold varied from 0 to 1. Both precision plot and success plot of the one-pass evaluation (OPE) are implemented in our experiments.

VOT2016 [31] consists of 60 sequences, based on which expected average overlap (EAO) takes both accuracy and robustness into account and is used to evaluate the overall performance of a tracking algorithm. The robustness of a tracker is evaluated by the failure times. When the overlap ratio between the predicted area and the ground truth becomes zero, a failure occures. The accuracy of a tracker is evaluated by the average overlap ratio between the prediction and the ground truths. Based on these two metrics, EAO is used for overall performance ranking. The proposed MSKCF are compared with recent state-of-the-art algorithms, which include the winner of the VOT2016 challenge.

### 5.2 Compared Algorithms

In experiments on OTB2015, the following four tracking algorithms are selected for comparison: KCF [32], fDSST [33], ECO _ HC [34], MKCFup [23]. KCF algorithm is a tracking algorithm using gradient feature HOG. C-COT [35] algorithm is a deep learning tracking algorithm using continuous convolution operator, and ECO_HC algorithm is a simpler and faster tracking algorithm by reducing dimensions on the basis of C-COT algorithm. MKCFup introduces the MKL into KCF based on multi-kernel correlation filter (MKCF). fDSST uses a two-dimensional position filter and a one-dimensional scale filter to track the object, and uses FFT to speed up the calculation. In experiments on VOT2016, for ensuring a fair and unbiased comparison, the original results provided by the VOT committee are used.

### 5.3 Experimental Environment and Parameter Configuration

The proposed method is implemented in MATLAB. The experiments are performed on a PC Intel Core i7 CPU and 64GB RAM. In experiments, we select three previous frames as multi-sample, that is, current frame and two previous frames. The feature of each frame is constructed by HOG, Conv4 and Conv5 of VGG16 shown in Fig. 1.

### 5.4 Experimental Results and Analysis

(1) OTB Dataset

The overall performance of the algorithm is evaluated quantitatively. The precision plots of OPE are shown in Fig. 2(a). In general, our method achieves the best results and outperforms ECO_HC by 3.5% and MKCFup by 11.8%. The success plots of OPE are shown in Fig. 2(b). Generally, the proposed method achieves the best performance on the success plots of OPE. When the overlap threshold is larger, the success rate of our tracker is comparative. However, when the overlap threshold is smaller, the success rate of our tracker is significantly

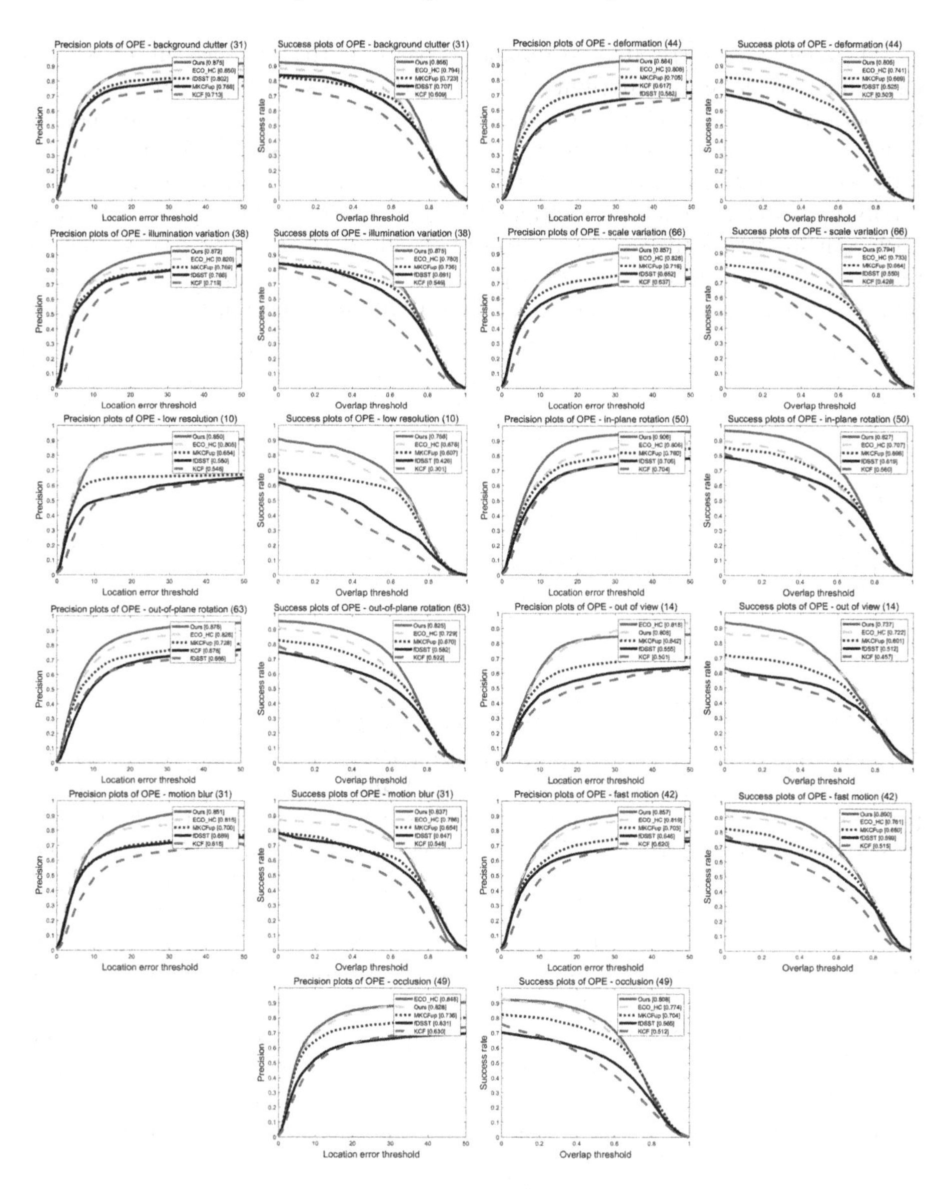


**Fig. 3.** The attribute based success plots and precision plots of OPE.

higher than the compared methods. The robustness of the proposed method lies in the construction of the multi-sample KCF which enhances the proposed tracking model. Overall, the proposed algorithm has achieved excellent performance in benchmark OTB2015, and runs at 18FPS.

The analysis of attribute performance is based on the sequence subsets which have different dominant attributes in OTB2015. These subsets help analyzing the performance of trackers for different challenging aspect. The experimental results is shown in Fig. 3. In general, our method achieves the best and comparable performance in all the subsets. For the precision plots of OPE, in 9 of 11 attribute subsets, our method obtains the best results. For the success plots of OPE, our method achieves the best results in all of 11 attribute subsets. These results indicate the robustness of the proposed method, which is benefit from the multi-sample learning. For the out of view attribute and occlusion, our method

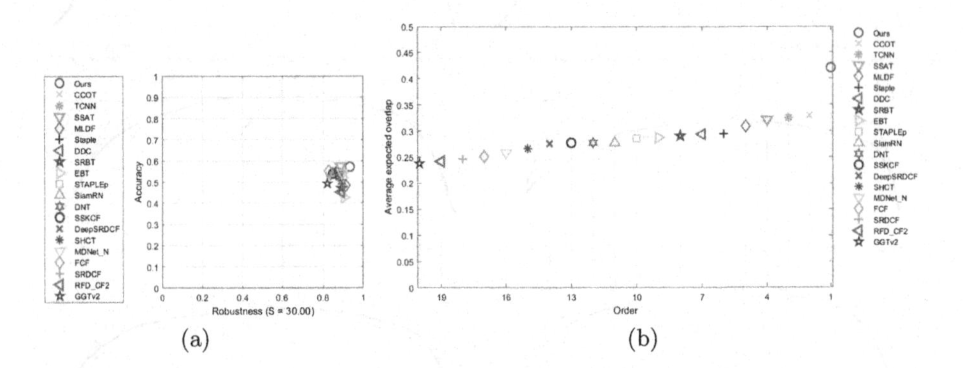


**Fig. 4.** Comparisons on VOT2016. (a) Robustness-accuracy ranking plots under the baseline experiments on the VOT2016 dataset. Trackers closer to upper right corner perform better. (b) Expected overlap scores for baseline. The value (right) indicates better performance. Our MSKCF significantly outperforms the baseline and other method.

ranks second in the precision plots, and ranks first in the success plots. When the local information is lost, the multi-sample method cannot compensate for these missing information, which reduce the discriminative ability of the tracker and lead to little tracking drift. However, compared with other algorithms, other attribution performance of our method is significantly improved which further verifies that multi-sample fusion of MSKCF method is superior to the performance of other compared method and makes tracking effect more stable, robust and reliable.

(2) VOT Dataset

VOT2016 dataset is used to evaluate the proposed MSKCF. We compare the method with all trackers submitted to the both datasets. Two metrics are used to rank all the trackers: accuracy (based on the overlap ratio with ground truth bounding box) and robustness (based on the number of tracking failures).

Figure 4(a) shows accuracy-robustness (AR) ranking plots under the baseline experiments on the VOT2016. In the baseline validations, our method performs best in terms of accuracy and robustness scores (the upper right corner). Figure 4(b) shows the EAO of different trackers. Our MSKCF achieves the best results (EAO score 0.4204), significantly outperforming the other methods. Overall, we can see that the proposed MSKCF achieves the best results.

## 6 Conclusion

In this paper, we propose an efficient multi-sample kernel correlation filter (MSKCF) tracking algorithm based on multi-sample fusion. The proposed method focuses on diversity of training samples of KCF. Experimental results compared with state-of-the-art KCF algorithms demonstrate the effectiveness and robustness of the proposed algorithm.

## References


1. Mei, X., Ling, H.: Robust visual tracking using l1 minimization. In: Proceedings of ICCV, pp. 1436–1443. IEEE, Kyoto (2009)
2. Wang, D., Lu, H., Yang, M.: Least soft-threshold squares tracking. In: Proceedings of CVPR, pp. 2371–2378. IEEE, Portland (2013)
3. Comaniciu, D., Ramesh, V., Meer, P.: Real-time tracking of non-rigid objects using mean shift. In: Proceedings of CVPR, pp. 142–149, IEEE, Hilton Head Island (2000)
4. Achanta, R., Susstrunk, S.: Saliency detection using maximum symmetric surround. In: Proceedings of ICIP, pp. 2653–2656. IEEE, Hong Kong (2010)
5. Babenko, B., Yang, M., Belongie, S.: Visual tracking with online multiple instance learning. In: Proceedings of CVPR, pp. 983–990. IEEE, Miami (2009)
6. Hare, S., Golodetz, S., Saffari, A., et al.: Struck: structured output tracking with kernels. IEEE Trans. Pattern Anal. Mach. Intell. **38**(10), 2096–2109 (2016)
7. Avidan, S.: Ensemble tracking. IEEE Trans. Pattern Anal. Mach. Intell. **29**(2), 261–271 (2007)
8. Kalal, Z., Matas, J., Mikolajczyk, K.: P-N learning: bootstrapping binary classifiers by structural constraints. In: Proceedings of CVPR, pp. 49–56. IEEE, San Francisco (2010)
9. Bolme, D.S., Beveridge, J.R., Draper, B.A., Lui, Y.M.: Visual object tracking using adaptive correlation filters. In: Proceedings of CVPR, pp. 2544–2550. IEEE, San Francisco (2010)
10. Henriques, J.F., Caseiro, R., Martins, P., Batista, J.: High-speed tracking with kernelized correlation filters. IEEE Trans. Pattern Anal. Mach. Intell. **37**(3), 583–596 (2015)
11. Martin, D., Danelljan, G.H., Fahad, S.K., Michael, F.: Discriminative scale space tracking. IEEE Trans. Pattern Anal. Mach. Intell. **39**(8), 1561–1575 (2017)
12. Martin, D., Gustav, H., Fahad, S. K., Michael, F.: Learning spatially regularized correlation filters for visual tracking. In: Proceedings of ICCV, pp. 4310–4318. IEEE, Santiago (2015)
13. Wang, M., Liu, Y., Huang, Z.: Large margin object tracking with circulant feature maps. In: Proceedings of CVPR, pp. 4800–4808. IEEE, Honolulu (2017)

14. Wang, N., Yeung, D. Y.: Learning a deep compact image representation for visual tracking. In: Proceedings of NIPS, Nevada, USA, pp. 809–817 (2013)
15. Wang, L., Ouyang, W., Wang, X., Lu, H.: Visual tracking with fully convolutional networks. In: Proceedings of ICCV, pp. 3119–3127. IEEE, Santiago (2015)
16. Danelljan, M., Robinson, A., Shahbaz Khan, F., Felsberg, M.: Beyond correlation filters: learning continuous convolution operators for visual tracking. In: Leibe, B., Matas, J., Sebe, N., Welling, M. (eds.) ECCV 2016. LNCS, vol. 9909, pp. 472–488. Springer, Cham (2016). https://doi.org/10.1007/978-3-319-46454-1_29
17. Bertinetto, L., Valmadre, J., Henriques, J.F., Vedaldi, A., Torr, P.H.S.: Fully-convolutional siamese networks for object tracking. In: Hua, G., Jégou, H. (eds.) ECCV 2016. LNCS, vol. 9914, pp. 850–865. Springer, Cham (2016). https://doi.org/10.1007/978-3-319-48881-3_56
18. Valmadre, J., Bertinetto, L., Henriques, J., Vedaldi, A., Torr, P. H.: End-to-end representation learning for correlation filter based tracking. In: Proceedings of CVPR, pp. 5000–5008. IEEE, Honolulu (2017)
19. Li, B., Wu, W., Zhu, Z., Hu, X.: High performance visual tracking with Siamese region proposal network. In: Proceedings of CVPR, pp. 8971–8980. IEEE, Salt Lake City (2018)
20. Wang, Q., Zhang, L., Bertinetto, L., Hu, W., Torr, P. H.: Fast online object tracking and segmentation: a unifying approach. In: Proceedings of CVPR, pp. 1328–1338 (2019)
21. Tang, M., Feng, J.: Multi-kernel correlation filter for visual tracking. In: Proceedings of ICCV, pp. 3038–3046. IEEE, Santiago (2015)
22. Li, F., Tian, C., Zuo, W., Zhang, L., Yang, M.: Learning spatial-temporal regularized correlation filters for visual tracking. In: Proceedings of CVPR, pp. 4904–4913. IEEE, Salt Lake City (2018)
23. Tang, M., Yu, B., Zhang, F., Wang, J.: High-speed tracking with multi-kernel correlation filters. In: Proceedings of CVPR, pp. 4874–4883. IEEE, Salt Lake City (2018)
24. Bibi, A., Ghanem, B.: Multi-template scale-adaptive kernelized correlation filters. In: Proceedings of ICCV, pp. 613–620. IEEE, Santiago (2015)
25. Wang, N., Zhow, W., Tian, Q., Hong, R., Wang, M., Li, H.: Multi-cue correlation filters for robust visual tracking. In: Proceedings of CVPR, pp. 4844–4853. IEEE, Salt Lake City (2018)
26. Scholkopf, B., Smola, A.J.: Learning with kernels - support vector machines, regularization, optimization, and beyond. MIT Press, Cambridge (2001)
27. Ma, C., Huang, J., Yang, X., Yang, M.: Hierarchical convolutional features for visual tracking. In: Proceedings of CVPR, pp. 3074–3082. IEEE, Santiago (2015)
28. Nam, H., Han, B.: Learning multi-domain convolutional neural networks for visual tracking. In: Proceedings of CVPR, pp. 4293–4302. IEEE, Las Vegas (2016)
29. Ma, Y., Yuan, C., Gao, P., Wang, F.: Efficient multi-level correlating for visual tracking. In: Jawahar, C.V., Li, H., Mori, G., Schindler, K. (eds.) ACCV 2018. LNCS, vol. 11365, pp. 452–465. Springer, Cham (2019). https://doi.org/10.1007/978-3-030-20873-8_29
30. Wu, Y., Lim, J., Yang, J.: Object tracking benchmark. IEEE Trans. Pattern Anal. Mach. Intell. **37**(9), 1834–1848 (2015)
31. Kristan, M., et al.: The visual object tracking VOT2016 challenge results. In: Hua, G., Jégou, H. (eds.) ECCV 2016. LNCS, vol. 9914, pp. 777–823. Springer, Cham (2016). https://doi.org/10.1007/978-3-319-48881-3_54

32. Henriques, J.F., Caseiro, R., Martins, P., Batista, J.: Exploiting the circulant structure of tracking-by-detection with kernels. In: Fitzgibbon, A., Lazebnik, S., Perona, P., Sato, Y., Schmid, C. (eds.) ECCV 2012. LNCS, vol. 7575, pp. 702–715. Springer, Heidelberg (2012). https://doi.org/10.1007/978-3-642-33765-9_50
33. Danelljan, M., Hhger, G., Khan, F.S., Felsberg, M.: Discriminative scale space tracking. IEEE Trans. Pattern Anal. Mach. Intell. **39**(8), 1561–1575 (2017)
34. Danelljan, M., Bhat, G., Khan, F. S., Felsberg, M.: ECO: Efficient convolution operators for tracking. In: Proceedings of CVPR, pp. 6931–6939. IEEE, Honolulu (2017)
35. Danelljan, M., Robinson, A., Shahbaz Khan, F., Felsberg, M.: Beyond correlation filters: learning continuous convolution operators for visual tracking. In: Leibe, B., Matas, J., Sebe, N., Welling, M. (eds.) ECCV 2016. LNCS, vol. 9909, pp. 472–488. Springer, Cham (2016). https://doi.org/10.1007/978-3-319-46454-1_29

# Real-Time Gesture Classification System Based on Dynamic Vision Sensor

Xiaofan Chen, Jian Wang, Limeng Zhang, Shasha Guo, Lianhua Qu, and Lei Wang(✉)

College of Computer Science and Technology, National University of Defense Technology, Changsha, China
jefferychenxf@gmail.com, wangjian.scrutiny@gmail.com, zhanglimeng13@126.com, lianhuaqu@163.com, {guoshasha13,leiwang}@nudt.edu.cn

**Abstract.** A biologically inspired event camera being able to produce more than 500 pictures per second [1], has been proposed in recent years. Event cameras can achieve profound efficiency in addressing many drawbacks of traditional cameras, for example redundant data and low frame rate during classification. In this paper, we apply a Celex IV DVS camera to fabricate a four-class hand gesture dataset for the first time. Meanwhile, we propose a real-time workflow for reconstructing Celex event data into intensity images while implementing gesture classification with a proposed LeNet-based [2] network and keyframe detection method. More than 30 fps has been achieved with our proposed workflow on a laptop. Compared to the state-of-art work [3] with an accuracy of 99.3%, our proposed network achieves a competent accuracy of 99.75%.

**Keywords:** Dynamic vision sensor (DVS) · LeNet · Keyframe

## 1 Introduction

Computer vision grows rapidly with a profoundly wide range of applications in many fields. Basically, some core tasks such as detection and tracking with frame-based digital cameras have been gradually faced with bottlenecks in terms of high-speed motion and demanding lighting conditions. However, the mentioned drawbacks can be perfectly compensated, with dynamic vision sensor cameras which are featured with high time resolution, high dynamic range, and low power consumption [4]. These biologically inspired cameras record brightness change in the scene, if a certain threshold is reached [5]. Therefore, much of the redundant background data in frame-based cameras will be avoided with event cameras. The sparse and highly temporal correlated event data make event cameras quite efficient on resource-limited platforms. Convolutional neural networks (CNNs) are currently the state-of-art techniques for processing images in computer vision. Normally it can realize high performance with a proper training dataset.


H. Yang et al. (Eds.): ICONIP 2020, LNCS 12532, pp. 486–497, 2020.
https://doi.org/10.1007/978-3-030-63830-6_41

Therefore, there have been many studies applying event cameras with CNNs in recognition tasks. In [6], IBM implements a system to recognize hand gestures in real-time from events streamed by a DVS128 camera [7]. Anadapted PointNet is proposed for real-time gesture recognition in [8], and the work [9] carries out a real-time hand gesture interface. More recently, using a LeNet based network, Lungu et al. [10,11] complete a hand symbol recognition system that can quickly be trained to incrementally learn new symbols recorded with a DVS camera. Besides, there are some studies on object tracking with DVS cameras such as [12–14], with relatively low-resolution event cameras. Nevertheless, when it comes to tracking and recognizing certain small objects like hand gesture or licenseplatewhen certain object is relatively far from the camera, the performance could be quite daunting.

Public DVS gesture datasets include the IBM Gesture Dataset [6] and IITM Gesture Dataset [7]. These two datasets both contain dynamic gesture information recorded from DVS128, but with blurred hand contour due to low camera resolution of 128 × 128 as well as the quick movement.

A lot of DVS gesture recognition work tends to focus on gestures involvingupper limbs [6,7], so low resolutionevent cameraslike DVS128 can provide sufficient eventdataforrelatively coarse gestures contour. As to our knowledge, there is few related work based on anew type of improved DVS camera called Celex [1]. Celex is a highresolutionand even faster DVS camera [3] to capture moreobjects' information, which means better features canbe exploited for detailed hand gesture recognition. In order to involve recognition of gestures in a relatively long distance from DVS, we choose Celex camera as our signal receiver, and also design a Celex DVS gesture dataset.

In this paper, we mainly explore the design of an efficient event data reconstruction and recognition system for hand gesture using only Celex camera. Given the adaptability and lightweight structure of LeNet-5 [2], we propose a network classifier based on LeNet model. We also adopt the idea of key frame detection to improve the real-time performance.

The main contributions of this paper includes: (1) By setting three different illumination conditions, we record a four-class hand gesture DVS dataset with 10 volunteers involved, amounting to 18 bin files per person; (2) We design a real-time pipelined dvs-based intensity image reconstruction and classification system, with event accumulation reconstruction method and our proposed LeNet-based network; (3) By combining Oriented FAST and Rotated BRIEF (ORB) algorithm [15] and a thresholding method, we design a real-time keyframe detection algorithm to effectively boost the classification efficiency, thus improving the overall real-time performance.

## 2 Background

### 2.1 Celex Camera Data

Dynamic vision sensors work asynchronously, recording event data according to Address Event Representation (AER) Protocol [16]. The sensor structure is inspired by that parallel nerve fibers carry continuous-time digital impulses. Celex pixel records event as (X, Y, A ,T) which will be referred as pixel event. (X,Y) is the spatial information of activated pixels. "A" stands for logarithmic gray level value of activated pixels [17], and "T" for activation time.

### 2.2 Event Data Reconstruction

For Celex pixels, the (X, Y, A ,T) is obtained by shifting andcombining the information of three different raw events in bin files, namely the column, row and special events [17], all 4 bytes in length. From row or column events, we can derive Y and T information or X and A respectively. Special events is used to update count cycle for T. Weadopt the accumulation method [18] to obtain binary images, because binary imageswould be enough for contour feature extraction and onlyspatial information is needed.

## 3 Related Work

### 3.1 Gesture Recognition Based on Traditional Camera

There are various approaches to handle traditional gesture recognition [19], from hidden Markov chains [20] to machine learning algorithms. Pavlovic et al. [19] propose a method based on Hidden Markov Models (HMMs) for dynamic gesture trajectory modeling and recognition. A static sign language recognition system using Edge Oriented Histogram (EOH) features and multiclass SVM is proposed in [21]. Convolutional neural networks (CNNs) have been used in gesture recognition and achieved good results [22]. However, these approaches require the signer to wear color tape or the background with restricted color. It reveals that these systems can not well cope with complex and changing scene settings [21].

### 3.2 Gesture Recognition Based on DVS Camera

There has been some work on matching DVS data and traditional CNNs. Simple intensity images can be formulated by accumulating the events along temporal axis [18,23]. In Work [17,24] the latest temporal or illumination information are saved. These algorithms can provide very clear contour features using moderate events. Optical flow methods [25,26] can help derive more features for CNN classification, while with relatively high computation cost. Amir et al. [6] adopt a hardware temporal filter cascade and neurons with stochastic delay to generate spatio-temporal input for CNN. In this way, processing delay is profoundly decreased with spatial-temperal correlation, compared to above methods.

Ahn et al. [27] propose delivery point based method to extract the hand region. Lee et al. [28] propose a four DoF method for fast and efficient hand motion estimation, which makes good use of edge pattern of DVS data. These methods also involve much computation and not suitable for real-time processing.

Lee et al. [9] design a real-time hand gesture GUI control system, in which SNN and various detection algorithms are applied to achieve robust recognition performance, with average classification delay around 120 ms and competent accuracy of 96.49%. More recently, Lungu et al. [10] combine the iCaRL incremental learning algorithm with CNN to boost the retraining efficiency. Lungu et al. [3] implement a CNN containing 5 convolutional layers using jAER SDK, achieving 99.3% accuracy with 70 ms for one recognition. Above DNN approaches can ensure high classification accuracy while remain low real-time performance.

## 4 Method Overview

The overall real-time workflow demo is shown in Fig. 1, and is organized in 5 parts as shown in Fig. 2: (1) new bin files where event data recored from Celex cameras, according to AER protocal, are obtained by the processing board attached to sensor board; (2) pixel events are then extracted frombin files in the form of (X, Y, A, T); (3) in processing 1, when there are enough pixel events, one binary image is reconstructed; (4) the images are then constantly displayed at around 30 fps; (5) meanwhile, in processing 2, keyframe detector selects keyframe by continuously processing the latest reconstructed image; (6) after detection, the classifier will process the keyframe and update label string for displayed images. The following sections will illustrate several key approaches applied in this system.

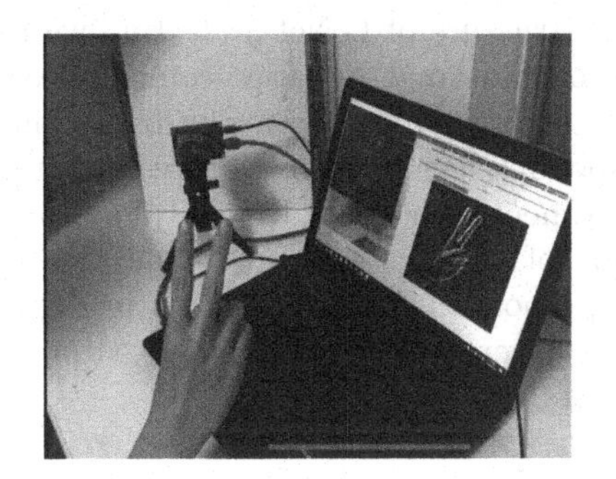

Fig. 1. The real-time workflow demo

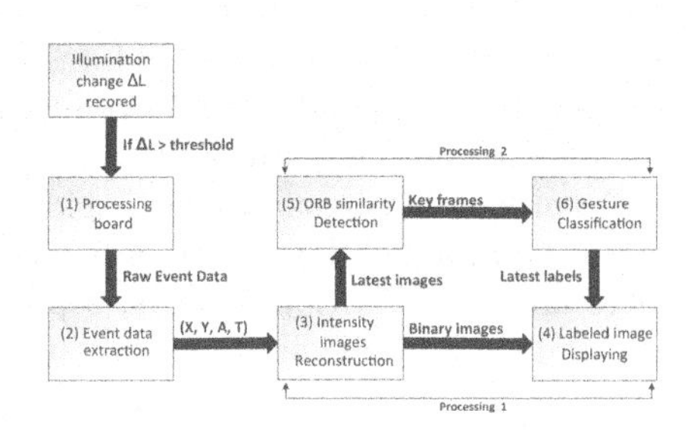


Fig. 2. The workflow of real-time reconstruction and classification.

### 4.1 Data Preprocessing

Liu et al. [29] illustrate the significant background noise due to small illumination and temperature variation. Therefore, we test three different denoising methods. Using reconstruction of 8k, 16k and 24k events/frame as shown in Fig. 6, with different denoising methods, three datasets are set to measure classification accuracy. Each class contains 50k images. We applied time filter [30] to obtain the

TimeFilter dataset. Meanwhile $9 \times 9$ Guassian kernal and Canny algorithm [31] are used to generate GuasCanny dataset. Besides, HeatMap dataset is formulated with heatmap-based filtering.

**Table 1.** The impact of denoising on classification accuracy and delay

| Dataset | Training accuracy | Testing accuracy | Delay per frame |
|---|---|---|---|
| HeatMap [32] | 99.33% | 97.14% | 71.57 ms |
| TimeFilter [30] | 99.20% | 99.26% | 93.14 ms |
| GuasCanny [31] | 99.44% | 99.68% | 57.68 ms |

Considering real-time requirement of more than 30 fps and classification accuracy, it is proper to use dataset of GuasCanny type which causes least computation cost, as shown in Table 1. The main reason could be that GuasCanny contains better contour features than the other two. In fact, images with less noise could be probably distributed unevenly within training and testing datasets due to improper splitting of data for each bin file, further leading to testing error slightly lower than training error.

### 4.2 Proposed Network

For simplicity, we firstly adopt the 8-layer structure of LeNet-5 [2]. By investigating the impact of input image size and convolutional operation configuration, proper parameters can be selected. Inspired by the cascaded convolution-pooling structure in [3], we find more convolution-pooling layers means feature extracted in a more fine-grained way. Following experiments are set for these explorations.

By proper hyper-parameter exploration of network setting and increase of convolution-pooling layers, we derive the proposed network as shown in Fig. 3. There are 6 convolutional layers, 6 pooling layers and 1 fully connected layer. The training and testing accuracy can reach 98%.

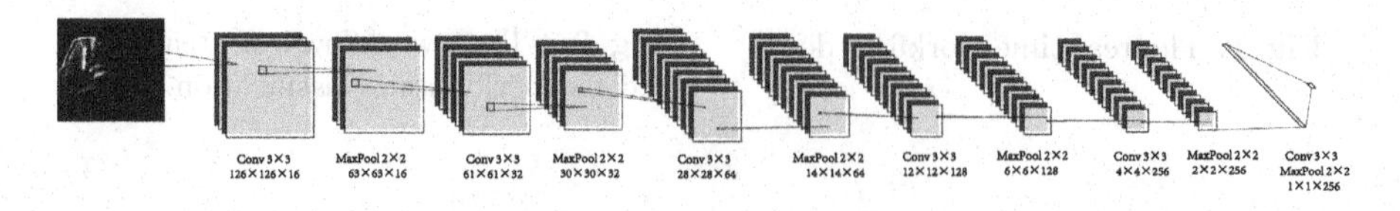


**Fig. 3.** The architecture of our present network which has 5 convolutional layers, 5 pooling layers and 1 fully-connected layer.

**Algorithm 1.** ORB-based key frame detection algorithm

```
1: Fun_orb // function based on Oriented FAST and Rotated BRIEF algorithm
2: kf_threshold, rec_threshold // threshold for determing and triggering classification
   for key frames
3: INPUT Image_new, kf_threshold, rec_threshold
4: OUTPUT Label
5: while New Binary Images Generated do
6:    if Process Initilization Done then
7:       descriptor_cur = Fun_orb(Image_new)
8:       if similarity(descriptor_cur, descriptor_pre) > kf_threshold then
9:          Image_pre = Image_new
10:      else
11:         kf_num = kf_num + 1
12:         if kf_num >= rec_threshold then
13:            Label = Classifier(Image_new)
14:            kf_num = 0
15:         end if
16:      end if
17:   else
18:      Image_pre = Image_new
19:   end if
20: end while
```

### 4.3 ORB-Based Keyframe Detection

Combining ORB algorithm [15] with a thresholding method, we are able to detect keyframes especially when there is a gesture alternation. Firstly, we use FAST algorithm [33] to extract feature points of latest newly reconstructed binary images, and further apply BRIEF algorithm [34] to obtain description for these feature points. Then,we compare the newly detected image with a reference image.The current frame is considered abnormalif certain threshold (Th1)is met. Current frame will be set as reference frame until thenumber of abnormal frames reaches another threshold (Th2),as illustrated in the following Algorithm 1. Th2 is set to alleviate the impact of random blank frames. After Th2 is met, current image will be sent to our proposed network for label updating.

## 5 Dataset

Since there is no open-accessed hand gesture dataset recorded by Celex camera, we designed a four-class dataset. In order to keep our recorded data consistent to the realsituation, we set up three illumination conditions namelythe dark, lamp and natural lighting, with the camera event threshold set at 40 [17], as well asrandom background settings in the same meeting room. Ten people are involvedto record gestures with only hand and arms in the field. Every bin file contains one gesture at a specific setting. Eventually, 18 bin files are generated for one person. The background class is recorded in three illumination settings seperately each for 18 min. The raw data isacquired by recording both hands with different sides facing the camera, while rotating and shifting the hand. We basically reconstruct the data based on the event-accumulated method [18]. The images with 8k, 16k and 24k events/frame are shown in Fig. 6. The total number of each gesture is around 1M. The first 80% and rest 20% of recorded bin files are set as training and testing data respectively. We split the training data into training and validation set randomly by the ratio of 7:1. The dataset details have been uploaded in github[1].

## 6 Experiment

### 6.1 Implementation

The proposed network is simulated on the 1.2.0 Tensorflow platform with Python 3.6.10. Our real-time workflow is running in Inter(R) Core(TM) i5-6300HQ CPU with 8 GB DDR4. The network training is implemented on a cloud server with 4 TESLA-V100 16G GPU. After comparing the specification [1] of Celex cameras, we adopt the Celex IV with resolution of 768 × 640.

### 6.2 Performance

**Network Accuracy:** By setting different the input image size and kernel size, we obtain five preliminary experiments from E1 to E5 in Table 2. We find these layer-fixed networks with 3 × 3 convolution kernel and 128 × 128 input size perform well. By changing the number of convolutional-pooling layers, we obtain the proposed network as shown in Fig. 3.

The proposed dataset is processed with above GuasCanny denoising method. First 80% and rest 20% of the dataset are for training and testing. Table 2 reveals that our proposed network outperforms the other two, with the accuracy of 99.13% and 98.67% respectively. The network is trained from scratch for 20,000 iterations with learning rate set to 0.001. The batch size is 32 for all datasets.

In order to evaluate the generalization of proposed network, comparison experiment between gesture recognition system by [6] and ours is of necessity. The former comprises 5 convolutional layers and 5 pooling layers, which we refer as benchmark [6]. For simplicity, we adopt dataset ROSHAMBO17 in [6]. First 90% and rest 10% of the ROSHAMBO17 data are allocated for training and testing. We adjust the proposed network by removing first pooling layer and resize input data from 128 × 128 to 64 × 64. Input for original LeNet-5 is resized from 128 × 128 to 32 × 32 as well. 25 training epochs were performed in Tensowflow requiring 2 h on TESLA-V100 GPU. Compared to the benchmark, results in Table 3 show that our proposed network performs slightly better by 0.449%. **System Real-time Accuracy:** We further sample another 10 groups of combined bin and label files, numbered from 0 to 9 in Fig. 8. We obtain the real-time accuracy results for fixed reconstruction setting (12k events/frame)

[1] https://github.com/Adnios/GestureRecognitionDVS.

and two keyframe thresholds. As showed in Fig. 8, average accuracy remains over 93% and framerate above 30 fps. The accuracy loss mainly comes from the input initialization and gesture switch intervals.

**Table 2.** Classifier parameters exploration. Accuracy is reported in percentage.

| Experiment (num of layers) | Input (W x H) | Conv layer configuration | Conv layer configuration | Test acc | Train acc |
|---|---|---|---|---|---|
| E1(7) | 32 × 32 | 5 × 5 (stride = 1) | 5 × 5 (stride = 1) | 85.49% | 80.31% |
| E2 (7) | 227 × 227 | 5 × 5 (stride = 1) | 5 × 5 (stride = 1) | 92.21% | 94.21% |
| E3(7) | 128 × 128 | 5 × 5 (stride = 1) | 5 × 5 (stride = 1) | 93.16% | 95.78% |
| E4(7) | 128 × 128 | 11 × 11 (stride = 3) | 3 × 3 (stride = 1) | 94.87% | 97.90% |
| E5(7) | 227 × 227 | 11 × 11 (stride = 3) | 3 × 3 (stride = 1) | 93.21% | 99.40% |
| E6(13) | 128 × 128 | 3 × 3 (stride = 1) | 3 × 3 (stride = 1) | 98.67% | 99.13% |

**Table 3.** The classification accuracy on ROSHAMBO17 dataset

| Model | Testing accuracy |
|---|---|
| Network in [6] | 99.3% |
| Proposed network | 99.749% |
| LeNet-5 | 58.844% |

### 6.3 Reconstruction and Keyframe Parameters Exploration

To improve framerate, it is necessary to explore keframe and reconstruction parameters. We randomly select 4 to 6 bin files from 16 in total as sequential testing inputs. Meanwhile the corresponding label files are concatenated as reference label for real-time accuracy measurement. Considering image integrity and framerate, we basically conduct testing within the range of 8k to 14k. By analyzing the number of similar feature description between two frames, we find the number of similar feature points is almost below 15 between different classes. But around 25 for the same type. When Th2 is more than 3, less than 3% images will be detected as keyframe. Therefore, two threshold values are set to be 10/20/30/40 and 2/3 respectively. From the above results, we can find

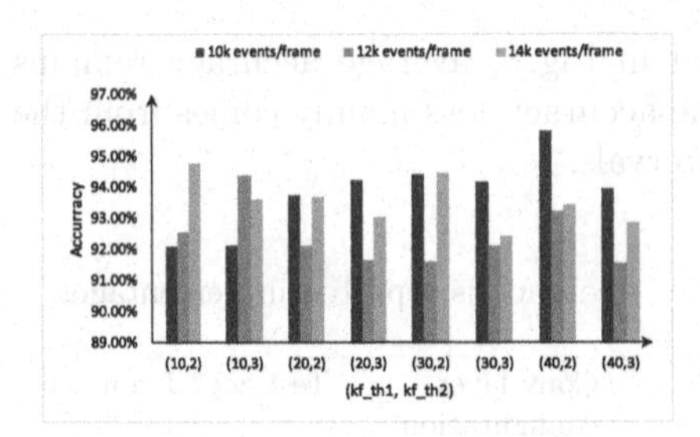


**Fig. 4.** The average accuracy results

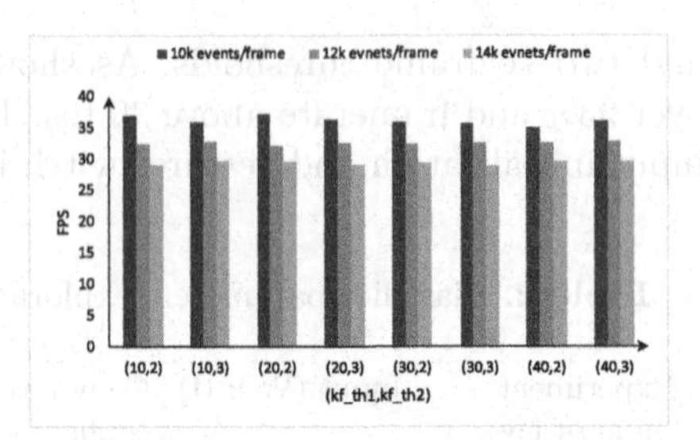


**Fig. 5.** The average Fps results

that framerate is not sensitive to the keyframe parameters shown in Fig. 5, while accuracy changes show no clear discrimination for these parameters in Fig. 4.

Since keyframe amount is influenced by the above three parameters simultaneously, we consider joint influence of both factors and define the following equation to describe classification effectiveness.

$$\begin{aligned} KF_{effective} = & \frac{Accu_{avg}}{P(kf)} \\ & + Accu_{judge\_const} * \; (Accu_{avg} - Accu_{th}) \\ & + Fps_{judge\_const} * \; (Fps_{avg} - Fps_{th}) \end{aligned} \tag{1}$$

where, the avg stands for average result of measured metrics, Accu for accuracy result, P(kf) for the percentage of keyframes among reconstructed images, and judge_const denoting importance constant of accuracy and fps. The second and third terms are to adjust the keyframe effectiveness. We set these two constants at 2 and 1 respectively, since framerate is less correlated to keyframe process. According to Fig. 7, we decide the proper parameters should be 10k and (10, 3).

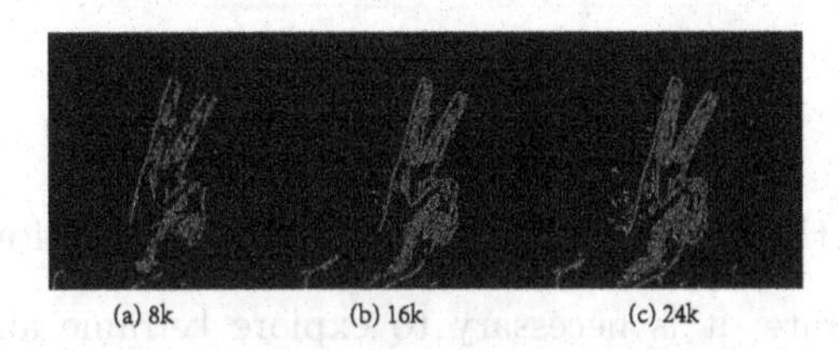


**Fig. 6.** Reconstructing the data from Celex camera with 8k, 16k and 24k events/frame

### 6.4 Ablation

We further investigate the impact on real-time performance without key-frame method. Using no key-frame detection means that every latest reconstructed images will be send to classifier. Thus the computation could be demanding.

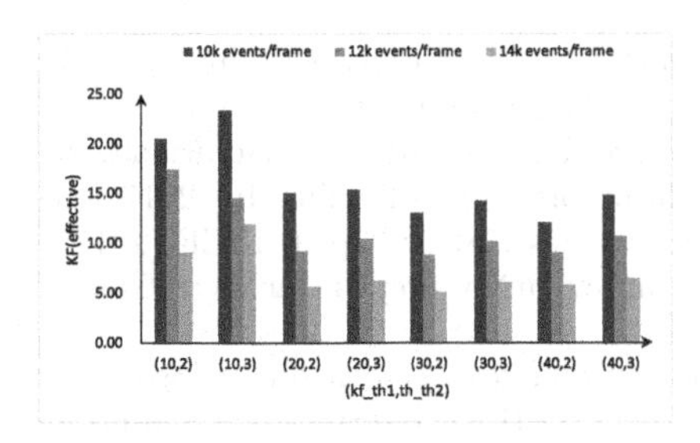

**Fig. 7.** The effective KF results

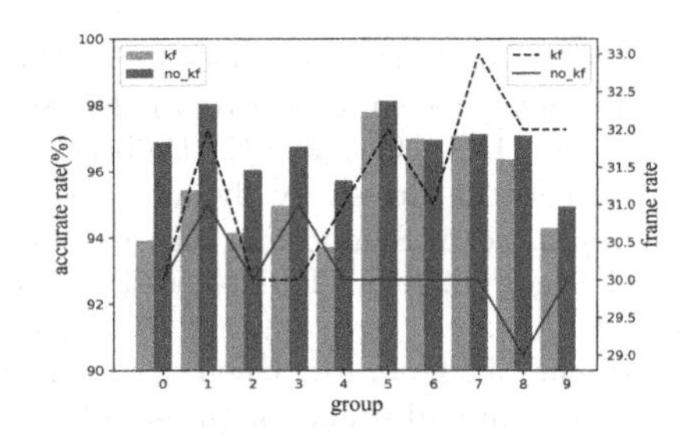


**Fig. 8.** The ablation test for keyframe method.

We feed the system with same testing data along with label files. The results are shown in Fig. 8. The framerate increases by around 2 fps, while the accuracy decreases by around 1%. The main reason should be the similar processing time of keyframe detection (around 0.05 s/frame) and image classification (around 0.10 s/frame). And the percentage of images processed by above algorithm are 60% and 30% respectively on average when reconstructing images of 12k events/frame. Therefore, delay is very much the same whether applying keyframe or not, and keyframe detection may cause label updating delayed slightly, thus a real-time accuracy decrease. However, if given a more complex network for intricate datasets, our keyframe method should boost the performance of both accuracy and framerate, since effective classification becomes dominant in this case.

## 7 Conclusion

In this paper, we present a real-time reconstruction and classification system for event data. The evaluation results demonstrate that our proposed network is more suitable than the benchmark model for DVS gesture recognition. By adapting some off-the-shelf feature detection and description algorithms, we successfully achieve competent accuracy while remaining low displaying latency. In the future, we will improve the keyframe detection and reconstruction algorithm for forming event images of better quality in real time. By implementing more complex DNNs, the proposed system along with Celex Camera should be able to finish recognition tasks with competent accuracy under more intricate situations.

**Acknowledgments.** This work is founded by National Key R&D Program of China [grant numbers 2018YFB2202603].

## References

1. Chen, S., Guo, M.: Live demonstration: CeleX-V: a 1m pixel multi-mode event-based sensor. In: Proceedings of the IEEE Conference on Computer Vision and Pattern Recognition Workshops (2019)

2. LeCun, Y., Bottou, L., Bengio, Y., Haffner, P.: Gradient-based learning applied to document recognition. Proc. IEEE **86**(11), 2278–2324 (1998)
3. Lungu, I.-A., Corradi, F., Delbrück, T.: Live demonstration: convolutional neural network driven by dynamic vision sensor playing roshambo. In: 2017 IEEE International Symposium on Circuits and Systems (ISCAS), p. 1. IEEE (2017)
4. Gallego, G., et al.: Event-based vision: a survey. arXiv preprint arXiv:1904.08405 (2019)
5. Huang, J., Guo, M., Chen, S.: A dynamic vision sensor with direct logarithmic output and full-frame picture-on-demand. In: 2017 IEEE International Symposium on Circuits and Systems (ISCAS), pp. 1–4. IEEE (2017)
6. Amir, A., et al.: A low power, fully event-based gesture recognition system. In: Proceedings of the IEEE Conference on Computer Vision and Pattern Recognition, pp. 7243–7252 (2017)
7. Baby, S.A., Vinod, B., Chinni, C., Mitra, K.: Dynamic vision sensors for human activity recognition. In: 2017 4th IAPR Asian Conference on Pattern Recognition (ACPR), pp. 316–321. IEEE (2017)
8. Wang, Q., Zhang, Y., Yuan, J., Lu, Y.: Space-time event clouds for gesture recognition: From RGB cameras to event cameras. In: 2019 IEEE Winter Conference on Applications of Computer Vision (WACV), pp. 1826–1835. IEEE (2019)
9. Lee, J.H., et al.: Real-time gesture interface based on event-driven processing from stereo silicon retinas. IEEE Trans. Neural Netw. Learn. Syst. **25**(12), 2250–2263 (2014)
10. Lungu, I.A., Liu, S.-C., Delbruck, T.: Fast event-driven incremental learning of hand symbols. In: 2019 IEEE International Conference on Artificial Intelligence Circuits and Systems (AICAS), pp. 25–28. IEEE (2019)
11. Lungu, I.A., Liu, S.-C., Delbruck, T.: Incremental learning of hand symbols using event-based cameras. IEEE J. Emerg. Sel. Top. Circ. Syst. **9**(4), 690–696 (2019)
12. Zong, X., Xiao, P., Wen, S.: An event camera tracking based on MLS surface fitting algorithm. In: 2018 Chinese Control And Decision Conference (CCDC), pp. 5001–5005. IEEE (2018)
13. Zhu, Q., Triesch, J., Shi, B.E.: Retinal slip estimation and object tracking with an active event camera. In: 2020 2nd IEEE International Conference on Artificial Intelligence Circuits and Systems (AICAS), pp. 59–63 (2020)
14. Glover, A., Bartolozzi, C.: Robust visual tracking with a freely-moving event camera. In: 2017 IEEE/RSJ International Conference on Intelligent Robots and Systems (IROS), pp. 3769–3776. IEEE (2017)
15. Rublee, E., Rabaud, V., Konolige, K., Bradski, G.: ORB: an efficient alternative to SIFT or SURF. In: 2011 International Conference on Computer Vision, pp. 2564–2571. IEEE (2011)
16. Sivilotti, M.: Wiring considerations in analog VLSI systems with application to field-programmable networks. Ph.D. Thesis, California Institute of Technology, Pasadena, CA (1991)
17. CelePixel: Celepixel technology. [EB/OL]. https://github.com/CelePixel/CeleX4-OpalKelly/tree/master/Documentation Accessed 15 Apr 2020
18. Moeys, D.P., et al.: Steering a predator robot using a mixed frame/event-driven convolutional neural network. In: 2016 Second International Conference on Event-based Control, Communication, and Signal Processing (EBCCSP), pp. 1–8. IEEE (2016)
19. Pavlovic, V.I., Sharma, R., Huang, T.S.: Visual interpretation of hand gestures for human-computer interaction: a review. IEEE Trans. Pattern Anal. Mach. Intell. **19**(7), 677–695 (1997)

20. Rabiner, L.R.: A tutorial on hidden Markov models and selected applications in speech recognition. Proc. IEEE **77**(2), 257–286 (1989)
21. Nagarajan, S., Subashini, T.S.: Static hand gesture recognition for sign language alphabets using edge oriented histogram and multi class SVM. Int. J. Comput. Appl. **82**(4), 2013
22. Cheng, W., Sun, Y., Li, G., Jiang, G., Liu, H.: Jointly network: a network based on cnn and rbm for gesture recognition. Neural Comput. Appl. **31**(1), 309–323 (2019)
23. Ye, C.: Learning of Dense Optical Flow, Motion and Depth, from Sparse Event Cameras. Ph.D. thesis (2019)
24. Zhu, A.Z., Yuan, L., Chaney, K., Daniilidis, K.: EV-FlowNet: self-supervised optical flow estimation for event-based cameras. arXiv preprint arXiv:1802.06898 (2018)
25. Bardow, P., Davison, A.J., Leutenegger, S.: Simultaneous optical flow and intensity estimation from an event camera. In: Proceedings of the IEEE Conference on Computer Vision and Pattern Recognition, pp. 884–892 (2016)
26. Brosch, T., Tschechne, S., Neumann, H.: On event-based optical flow detection. Front. Neurosci. **9**, 137 (2015)
27. Ahn, E.Y., Lee, J.H., Mullen, T., Yen, J.: Dynamic vision sensor camera based bare hand gesture recognition. In: 2011 IEEE Symposium On Computational Intelligence For Multimedia, Signal And Vision Processing, pp. 52–59. IEEE (2011)
28. Lee, J.H., et al.: Real-time motion estimation based on event-based vision sensor. In: 2014 IEEE International Conference on Image Processing (ICIP), pp. 204–208. IEEE (2014)
29. Liu, H., Brandli, C., Li, C., Liu, S.-C., Delbruck, T.: Design of a spatiotemporal correlation filter for event-based sensors. In: 2015 IEEE International Symposium on Circuits and Systems (ISCAS), pp. 722–725. IEEE (2015)
30. Guo, S., et al.: A noise filter for dynamic vision sensors based on global space and time information. arXiv preprint arXiv:2004.04079 (2020)
31. Canny, J.: A computational approach to edge detection. IEEE Trans. Pattern Anal. Mach. Intell. **6**, 679–698 (1986)
32. Zhang, Y.: Celexmatlabtoolbox. [EB/OL]. https://github.com/yucicheung/CelexMatlabToolbox Accessed 15 Apr 2020
33. Viswanathan, D.G.: Features from accelerated segment test (FAST). Homepages. Inf. Ed. Ac. UK (2009)
34. Calonder, M., Lepetit, V., Strecha, C., Fua, P.: BRIEF: binary robust independent elementary features. In: Daniilidis, K., Maragos, P., Paragios, N. (eds.) ECCV 2010. LNCS, vol. 6314, pp. 778–792. Springer, Heidelberg (2010). https://doi.org/10.1007/978-3-642-15561-1_56

# Res2U-Net: Image Inpainting via Multi-scale Backbone and Channel Attention

Hao Yang and Ying Yu(✉)

School of Information Science and Engineering, Yunnan University, Kunming 650500, China
haoyang@mail.ynu.edu.cn, yuying.mail@163.com

**Abstract.** Most Deep learning-based inpainting approaches cannot effectively perceive and present image information at different scales. More often than not, they adopt spatial attention to utilize information on the image background and ignore the effect of channel attention. Hence, they usually produce blurred and poor-quality restored images. In this paper, we propose a novel Res2U-Net backbone architecture to solve these problems. Both encoder and decoder layers of our Res2U-Net employ multi-scale residual structures, which can respectively extract and express multi-scale features of images. Moreover, we modify the network by using the channel attention and introduce a dilated multi-scale channel-attention block that is embedded into the skip-connection layers of our Res2U-Net. This network block can take advantage of low-level features of the encoder layers in our inpainting network. Experiments conducted on the CelebA-HQ and Paris StreetView datasets demonstrate that our Res2U-Net architecture achieves superior performance and outperforms the state-of-the-art inpainting approaches in both qualitative and quantitative aspects.

**Keywords:** Image inpainting · Deep learning · Multi scale · Channel attention

## 1 Introduction

Image inpainting techniques aim to synthesize visually realistic and semantically correct alternatives for missing or damaged areas in an image. Researchers and engineers use them in such applications as undesired objects removal, occluded regions completion and human faces edition, etc. Numerous literature has proposed various inpainting approaches that can divide into traditional ones and learning-based ones. However, all traditional approaches cannot handle complex structures and novel contents in non-repetitive patterns, which is due to they fail to capture high-level semantics of damaged images.

Recently, learning-based approaches of image inpainting have overcome the limitations of traditional ones. A specific structure of generator called U-net has


H. Yang et al. (Eds.): ICONIP 2020, LNCS 12532, pp. 498–508, 2020.
https://doi.org/10.1007/978-3-030-63830-6_42

been proposed by Ronneberger et al. [13], which is initially utilized in image segmentation. After that, Liu et al. [10] and Yan et al. [15] employed U-net structures in the area of image inpainting. U-net structures have the potentials for the image inpainting task. However, they require some modifications and improvements. Liu et al. [10] replaced vanilla convolutions with partial convolutions and in the U-net structure. Their work effectively utilized valid pixels of original images and achieved high-quality inpainting results. Yan et al. [15] add a special shift-connection layer in the decoder of U-net. Although these works have made encouraging results in image inpainting, they cannot effectively utilize the skip connection of U-Net to capture the low-level image features.

For image inpainting tasks, handling the correlation of information between distant locations in an image is very important. For example, when a photo of a bespectacled face is obscured by a large mask, the main frame of the glasses is missing, and only the arms of the glasses are visible. In this case, only with strong long-range dependencies, the inpainting network can represent image information at different scales, and generate the reasonable semantic contents of the facial features and glasses. There have been many previous attempts to capture the long-range dependencies of the impaired image. Yu et al. [16] designed a contextual attention to model the long-range dependencies, but it needs to build a refinement network to enable end-to-end training. Zheng et al. [19] introduced a short+long term attention layer to enhance the appearance consistency of the impaired image at the cost of increased computational time.

In this paper, we propose a novel U-Net based generator backbone called Res2U-Net for image inpainting. Our goal is to enhance the capture ability of the backbone network for the long-range dependencies, which helps to enhance multi-scale extraction and expression capabilities at a more granular level, thus makes the inpainting results more realistic in details. Inspired by the work of Res2Net [2], we embed hierarchical residual structures into the encoder and decoder layers of the U-Net. This structural modification can increase the size of receptive fields and the number of output feature scales, and hence make the proposed model have more robust multi-scale potentials in feature extraction and expression. To utilize low-level features from encoder layers effectively, we design a dilated multi-scale channel-attention (DMSCA) block and embed it into our Res2U-Net. This block can weight all the multi-scale channel features from the skip-connections according to the relative importance of these features.

We conduct qualitative and quantitative experiments on the Paris StreetView [1] and CelebA-HQ [9] datasets to evaluate the proposed method. Experimental results show that our method can generate better inpainting results as compared to the state-of-the-art approaches. Our contributions are summarized as follows:

- We propose a novel backbone of the multi-scale generator for inpainting tasks, which can effectively capture the long-range dependencies of images.
- We design a dilated multi-scale channel-attention block and embed it into our generator, which improves the utilization of low-level features in the encoder.

- Experiments show that our method and can generate plausible content and delicate details for the damaged images. To verify the effectiveness of each proposed module, we also provide ablation studies.

## 2 Related Work

Learning-based inpainting approaches repair the damaged images by encoding them into high-level features, which can be utilized to guide the generation of missing image regions. As a pioneer work, Context Encoder [12] attempts to employ an encoder-decoder CNN structure in the image inpainting model, and train this structure through the reconstruction loss and the adversarial loss. Iizuka et al. [5] introduced a global discriminator and a local discriminator that are able to generate semantically coherent structures and locally realistic details. In order to repair irregular regions, Liu et al. [10] adopted a kind of partial convolution that only depends on the valid pixels of damaged images. Their work can effectively reduce artifacts in the inpainting results. After that, Yu et al. [17] designed a so-called gated convolution to automatically update the irregular masks. It obtains relatively good inpainting results for irregular regions.

The attention mechanisms can enable a network to effectively allocate available computational resources. According to the dimensionality of a specific application, it can be divided into the spatial attention (SA) and the channel attention (CA). A number of studies [11,15,16] have applied the spatial attention mechanism in their inpainting networks. They consider spatial contextual features to be the references for repairing damaged image regions. Yu et al. [16] proposed a contextual attention layer that copies the most similar features of background patches to fill the missing regions. Yan et al. [15] designed a shift-connection layer that calculates the relationship between the encoder features (from known regions) and the missing contents (from associated decoder layers). To build the correlation among deep features of the missing regions, Liu et al. [11] insert a novel coherent semantic attention (CSA) layer in their network. This operation encourages the spatial semantic coherency of the damaged image regions.

## 3 Proposed Approach

### 3.1 Overview of the Proposed Network

In this paper, we propose a one-stage deep inpainting network mainly based on the multi-scale channel attention and the multi-scale residual backbone. This proposed network is referred to as MS-CARS. Its architecture is illustrated in Fig. 1. Res2U-Net is the backbone of the generator, which is a new U-Net like network. To utilize the low-level features of encoder layers, we embed our proposed dilated multi-scale channel-attention blocks in the outmost two skip-connections. The size of their two feature maps are $64 \times 64$ and $128 \times 128$, respectively. Moreover, we insert a shift-connection layer of the Shift-Net [15] into our MS-CARS network, which enables the restored contents to have sharp

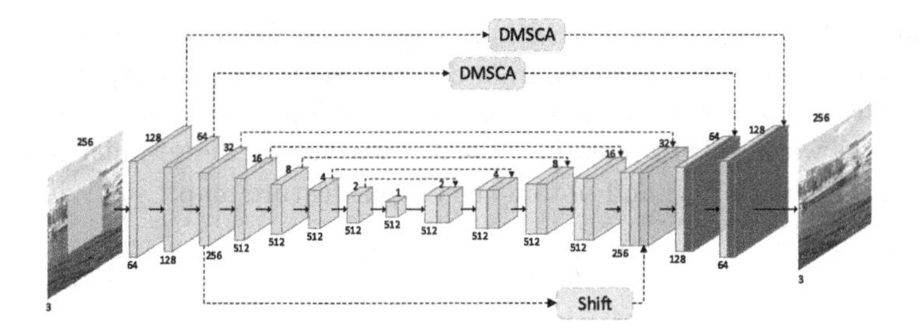


**Fig. 1.** The architecture of MS-CARS network. We add dilated multi-scale channel-attention (DMSCA) block at outmost two skip-connection layers, and adopt the shift-connection layer of Shift-Net.

structures. We adopt the fully convolutional discriminative networks of pix2pix [7] as our discriminator. To make the results have clearer textures and plausible semantics, we combine the pixel reconstruction loss ($L_{rec}$), content loss ($L_{cont}$), style loss ($L_{sty}$), total variation loss ($L_{tv}$) [8] and adversarial loss ($L_{adv}$) [3] into the joint training loss ($L$) of our model, which is defined as

$$L = \lambda_{rec} L_{rec} + \lambda_{cont} L_{cont} + \lambda_{sty} L_{sty} + \lambda_{tv} L_{tv} + \lambda_{adv} L_{adv}. \tag{1}$$

where $\lambda_{rec}$, $\lambda_{cont}$, $\lambda_{sty}$, $\lambda_{tv}$ and $\lambda_{adv}$ are the corresponding tradeoff parameters.

### 3.2 Res2U-Net Architecture

Following Gao et al.'s study on the Res2Net [2], we embed hierarchical residual modules in the U-Net architecture. In this paper, we refer to our proposed U-Net like architecture as Res2U-Net, and use it as the backbone network of our generator. The architecture of our Res2U-Net consists of the res2-encoder, the res2-decoder and the skip-connection layers. The res2-encoder is a down-sampling part of Res2U-Net, which is capable of extracting multi-scale image features. The basic block of res2-encoder is shown in Fig. 2(a). In the downsampling process, $F_{in}$ denotes the feature maps from the previous layer. We send $F_{in}$ into a $1 \times 1$ vanilla convolution adaptively adjust its channel weights. To increase the number of receptive fields at different scales, we averagely divide the weighted features into 4 groups along the channel dimension, and send them into a hierarchical residual connection structure. This process can be formulated as

$$y_i = \begin{cases} x_i & i = 1 \\ W_i \times x_i & i = 2 \\ W_i \times (y_{i-1} + x_i) & i = 3, 4 \end{cases} \tag{2}$$

where $x_i \, (i = 1, 2, 3, 4)$ are the weighted features of $F_{in}$, $y_i \, (i = 1, 2, 3, 4)$ are the output features of the hierarchical structure, and $W_i \, (i = 2, 3, 4)$ denote the weights of 4 groups of convolution. These convolutions have multi-scale receptive fields, which are beneficial to the res2-encoder to aggregate the spatial information from different scales. After that, we send $y_i \, (i = 1, 2, 3, 4)$ into a $4 \times 4$

convolution that is followed by a SE block [4]. These operations are performed to downsample and fuse the features from different scales. Their output is denoted as $F_{se}$. Then, we utilize another $4 \times 4$ convolution to downsample $F_{in}$, and denote the downsampled $F_{in}$ as $F_{in}^{'}$. Finally, we employ the residual connection to combine $F_{in}^{'}$ with $F_{se}$, and obtain the output feature map:

$$F_{out} = F_{se} + F_{in}^{'}. \tag{3}$$

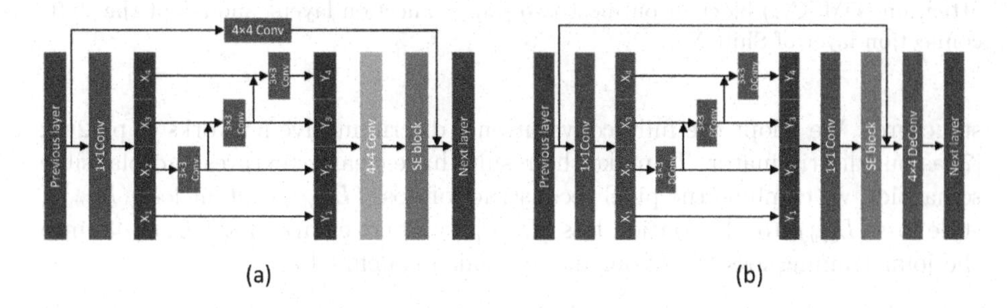


**Fig. 2.** (a) The basic block of res2-encoder and (b) The basic block of res2-decoder.

The res2-decoder is the upsampling part of Res2U-Net, which generates image contents with different scale features. Figure 2(b) shows the basic block of res2-decoder. The major difference between Fig. 2(a) and Fig. 2(b) is that we remove all downsampling convolutions, and replace the four parallel convolutions with dilated convolutions in Fig. 2(b). This modification helps to further increase the size of receptive fields. In addition, a transposed convolution is added to the outmost layer of the basic block. By virtue of all these modified structures, our method can obtain larger scale receptive fields and better multi-scale representation ability. This means that our proposed model can successfully capture the correlation information from distant locations in the occluded image, and finally generates semantically seasonable inpainted images. We denote the original image as Y, the mask as M, and Res2U-Net as G. The output of our model can be defined as G(Y⊙(1-M)), then the final prediction can describe as

$$\hat{\mathrm{Y}} = \mathrm{Y} \odot (1 - \mathrm{M}) + G(Y \odot (1 - \mathrm{M})) \odot \mathrm{M}. \tag{4}$$

where $\hat{\mathrm{Y}}$ denotes the final inpainting results, $\odot$ is the Hadamard product operator.

### 3.3 Dilated Multi-scale Channel-Attention Block

To effectively exploit low-level image features in the encoder layers, we propose a dilated multi-scale channel-attention (DMSCA) block that is employed in the

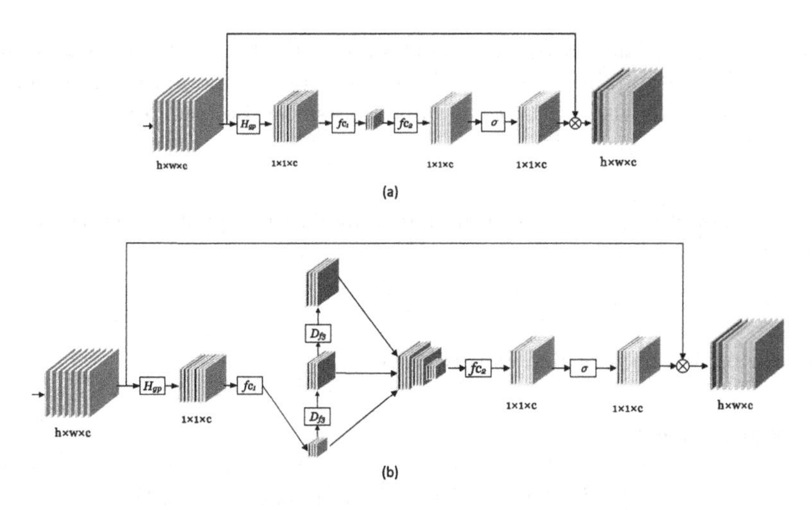


**Fig. 3.** (a) The structure of channel-attention mechanism (b) The structure of our proposed dilated multi-scale channel-attention (DMSCA).

outmost two skip-connection layers of our Res2U-Net. We denote the input feature maps as $x_c \in \mathbb{R}^{H\times W\times C}$ and the output feature maps as $\hat{x_c} \in \mathbb{R}^{H\times W\times C}$, which means both $x_c$ and $\hat{x_c}$ have $C$ features in size of $H\times W$. The global average pooling layer extracts the channel-wise global spatial statistics $z_c \in \mathbb{R}^C$, which are obtained by shrinking $z_c$ through spatial dimensions $H \times W$:

$$z_c = H_{GP}\left(x_c\right) = \frac{1}{h\times w}\sum_{i=1}^{h}\sum_{j=1}^{w} f_c\left(i,j\right) \tag{5}$$

where $f_c\left(i,j\right)$ is the value at position $(i,j)$ of $c-th$ feature map. Note that Channel Attention (CA) [18] employs a gating mechanism (see Fig. 3(a)) to capture the interrelationship between these channels from the global average pooling layer. The gating process of CA can be formulated as

$$r_1 = \delta\left(f_{c1}\left(z_c\right)\right) \tag{6}$$

where $r_1$ denotes their output, $\delta\left(\cdot\right)$ denotes the ReLU function, and $f_{c1}\left(\cdot\right)$ is a fully connected layer. To extract better multi-scale features, we add two dilated $3\times 3$ convolutions after $f_{c1}\left(\cdot\right)$, which is denoted by $D_{f_3}\left(\cdot\right)$. The outputs of dilated convolutions are computed by

$$r_3 = \delta\left(D_{f_3}\left(r_1\right)\right) \tag{7}$$

$$r_5 = \delta\left(D_{f_3}\left(r_3\right)\right) \tag{8}$$

where $r_3$ and $r_5$ are representations of different scales (see Fig. 3(b)). As studied in Inception-V2 [6], large convolution kernels such as $5\times 5$ filter can be divide into two $3\times 3$ filters. We adopt this strategy to reduce parameters as well as the computational time. Then, we concatenate $r_1$,$r_3$ and $r_5$ as a whole, and feed them into the upsampling layer $f_{c_2}(\cdot)$ with a sigmoid function so as to get the multi-scale channel-wise statistics $s$. This process can be formulated as

$$s = \sigma\left(f_{c_2}\left([r_1, r_3, r_5]\right)\right) \tag{9}$$

where $\sigma(\cdot)$ denotes the sigmoid function. Finally, we exploit the statistics $s$ to rescale the input feature map as

$$\hat{x_c} = s_c \cdot x_c \tag{10}$$

where $\hat{x_c}$ denotes the rescaled feature map.

# 4 Experiments

We evaluate our method on Paris StreetView [1] and CelebA-HQ [9] datasets. For Paris StreetView, we adopt their 14900 training images and 100 test images. The CelebA-HQ including 28000 images for training and 2000 images for testing. All images are with the size of 256 × 256 px, and their center mask has a size of 128 × 128 px. Our network is optimized by the Adam alogrithm. The tradeoff parameters of joint loss are set as $\lambda_{rec} = 1.0$, $\lambda_{cont} = 1.0$, $\lambda_{sty} = 10.0$, $\lambda_{tv} = 0.01$ and $\lambda_{adv} = 0.2$. All experiments were implemented on NVIDIA 1080Ti GPU and Intel Core i7-6850K 3.60 GHz CPU with Pytorch. We compare our method with three state-of-the-art approaches: GMCNN [14], PIC [19] and Shift-Net [15].

## 4.1 Quantitative Comparisons

In order to evaluate the inpainting performance, we utilize Peak Signal-to-Noise Ratio (PSNR), Structural Similarity (SSIM) and Fréchet Inception Distance (FID) as the metrics to assess the inpainted results. Note that larger PSNR and SSIM denote better performance, whereas larger FID means worse restoration results. The quantitative results in the test dataset of CelebA-HQ and Paris StreetView are shown in Table 1. It can be seen that our Res2U-Net achieves the best performance of PSNR, SSIM and FID on both two datasets.

## 4.2 Qualitative Comparisons

Figure 4 shows the qualitative comparisons of our method with the other 3 state-of-the-art approaches. The red box in the figure indicates that our proposed method can produce clearer contextual contents with the fine details. Specifically, in the first and second rows of Fig. 4, we can see that the structure and texture of our generated glassses are more realistic than other methods. This demonstrates the strong ability of our method for the capturing long-range dependencies in the repaired images. The arches and windows in the fifth and sixth rows also support this view. For details such as pupils and blinds in Fig. 4, our inpainting results are more satisfactory and visual-pleasing, which shows that our method has a good generation of the small-scale details of the repaired images.

**Table 1.** Center mask numerical comparison on CelebA-HQ and Paris StreetView.

| Method | | Shift-Net | GMCNN | PIC | Our Method |
|---|---|---|---|---|---|
| CelebA-HQ | PSNR | 26.51 | 26.18 | 24.61 | **26.86** |
| | SSIM | 0.8932 | 0.8927 | 0.8728 | **0.9034** |
| | FID | 6.947 | 7.665 | 6.781 | **5.85** |
| Paris StreetView | PSNR | 24.98 | 24.10 | 23.72 | **25.59** |
| | SSIM | 0.8425 | 0.8316 | 0.8294 | **0.8528** |
| | FID | 46.23 | 47.60 | 45.05 | **43.95** |

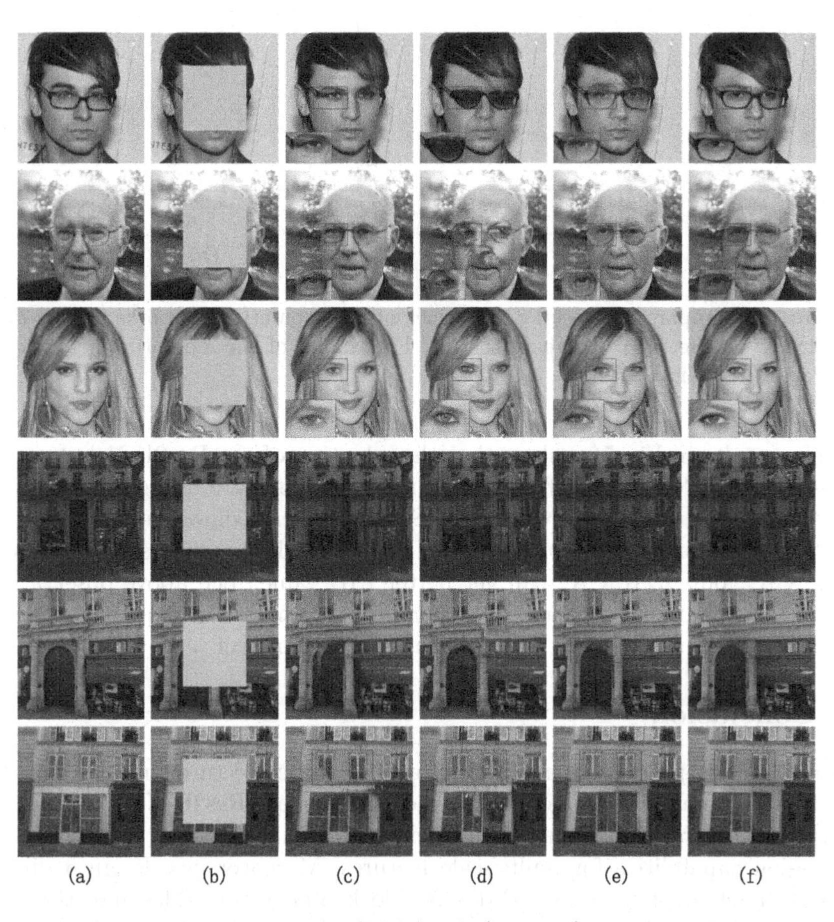


**Fig. 4.** Qualitative comparisons on CelebA-HQ (1–3 row) and Paris StreetView (4–6 row), from left to right: (a) Ground Truth (b) input, (c) GMCNN, (d) PIC, (e) Shift-Net and (f) ours.

### 4.3 Ablation Study

To verify the effectiveness of each proposed module in our network, we conduct ablation study on Paris StreetView dataset and caompare the performs of our network variants and baseline. All of these networks are trained with the same settings for fair comparison. In Fig. 5, "CA" denotes the standard channel attention, "DMSCA" is our dilated multi-scale channel-attention. "Res2U" denotes our Res2U-Net, "Full Method" combine the baseline with "DMSCA" and "Res2U".

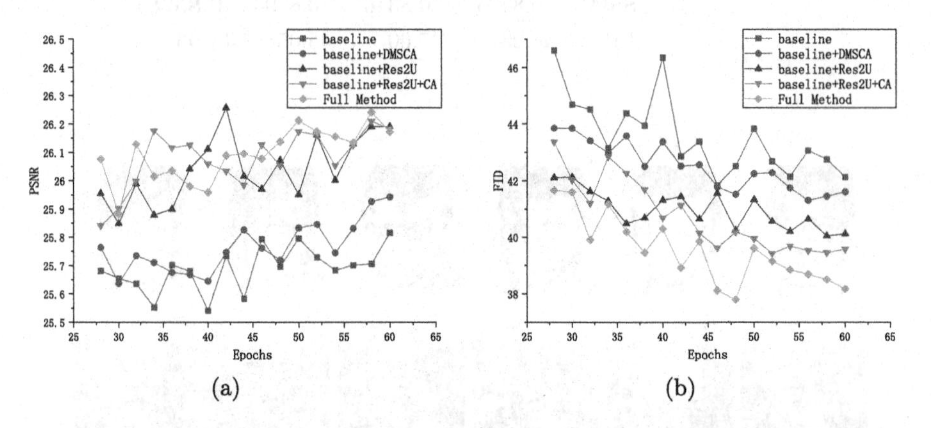


**Fig. 5.** (a): Compare the test PSNR values of the inpainting networks on the Paris StreetView dataset. (b): Compare the test FID values of the inpainting networks on the Paris StreetView dataset.

The results in Fig. 5(a) show that the adoption of our Res2U-Net structure increase the PSNR values of inpainted images obviously. Adding "DMSCA" to Res2U-Net will further increase the PSNR value. As shown in Fig. 5(b), our Res2U-Net significantly reduces the FID value of the repair result. Compared with "CA", our "DMSCA" can get better FID value. The above observations indicate that our proposed modules significantly improve the quality of image inpainting both at the pixel level and the perception level.

## 5 Conclusion

In this paper we propose an innovative Res2U-Net architecture for image inpainting. By using hierarchical and residual mechanisms, our Res2U-Net enhances the capture ability of long-range dependencies, and achieves better extraction and expression capabilities for multi-scale features. Moreover, we design a dilated multi-scale channel-attention (DMSCA) block that is embedded into the skip-connection layers. The DMSCA block can utilize low-level features of the encoder more effectively. Experiments show that our method outperforms other state-of-the-arts with both qualitative and quantitative evaluation. Our method can obtain fine details and plausible semantic structures.

**Acknowledgements.** This work was supported by the National Natural Science Foundation of China (Grant No. 61263048) and, by the Applied Basic Research Project of Yunnan Province (Grant No. 2018FB102), and by the Young and Middle-Aged Backbone Teachers' Cultivation Plan of Yunnan University (Grant No. XT412003).

## References

1. Doersch, C., Singh, S., Gupta, A., Sivic, J., Efros, A.A.: What makes Paris look like Paris? ACM Trans. Graph. **31**(4) (2012)
2. Gao, S.H., Cheng, M.M., Zhao, K., Zhang, X.Y., Yang, M.H., Torr, P.: Res2Net: a new multi-scale backbone architecture. arXiv e-prints arXiv:1904.01169 (Apr 2019)
3. Goodfellow, I., et al.: Generative adversarial nets. In: Advances in Neural Information Processing Systems, pp. 2672–2680 (2014)
4. Hu, J., Shen, L., Sun, G.: Squeeze-and-excitation networks. In: The IEEE Conference on Computer Vision and Pattern Recognition (CVPR), pp. 7132–7141, June 2018
5. Iizuka, S., Simo-Serra, E., Ishikawa, H.: Globally and locally consistent image completion. ACM Trans. Graph. **36**(4), 1–14 (2017)
6. Ioffe, S., Szegedy, C.: Batch normalization: accelerating deep network training by reducing internal covariate shift. arXiv e-prints arXiv:1502.03167 (2015)
7. Isola, P., Zhu, J.Y., Zhou, T., Efros, A.A.: Image-to-image translation with conditional adversarial networks. arXiv e-prints arXiv:1611.07004 (2016)
8. Johnson, J., Alahi, A., Fei-Fei, L.: Perceptual losses for real-time style transfer and super-resolution. In: Leibe, B., Matas, J., Sebe, N., Welling, M. (eds.) ECCV 2016. LNCS, vol. 9906, pp. 694–711. Springer, Cham (2016). https://doi.org/10.1007/978-3-319-46475-6_43
9. Karras, T., Aila, T., Laine, S., Lehtinen, J.: Progressive growing of GANs for improved quality, stability, and variation. arXiv e-prints arXiv:1710.10196 (2017)
10. Liu, G., Reda, F.A., Shih, K.J., Wang, T.C., Tao, A., Catanzaro, B.: Image inpainting for irregular holes using partial convolutions. arXiv e-prints arXiv:1804.07723 (2018)
11. Liu, H., Jiang, B., Xiao, Y., Yang, C.: Coherent semantic attention for image inpainting. arXiv e-prints arXiv:1905.12384 (2019)
12. Pathak, D., Krahenbuhl, P., Donahue, J., Darrell, T., Efros, A.A.: Context encoders: feature learning by inpainting. In: IEEE Conference on Computer Vision and Pattern Recognition, pp. 2536–2544. IEEE, New York (2016)
13. Ronneberger, O., Fischer, P., Brox, T.: U-net: convolutional networks for biomedical image segmentation. arXiv e-prints arXiv:1505.04597 (2015)
14. Wang, Y., Tao, X., Qi, X., Shen, X., Jia, J.: Image inpainting via generative multi-column convolutional neural networks. In: Advances in Neural Information Processing Systems, pp. 331–340 (2018)
15. Yan, Z., Li, X., Li, M., Zuo, W., Shan, S.: Shift-net: image inpainting via deep feature rearrangement. arXiv e-prints arXiv:1801.09392 (2018)
16. Yu, J., Lin, Z., Yang, J., Shen, X., Lu, X., Huang, T.S.: Generative image inpainting with contextual attention. In: The IEEE Conference on Computer Vision and Pattern Recognition (CVPR), pp. 5505–5514, June 2018
17. Yu, J., Lin, Z., Yang, J., Shen, X., Lu, X., Huang, T.S.: Free-form image inpainting with gated convolution. In: The IEEE International Conference on Computer Vision (ICCV), pp. 4470–4479, October 2019

18. Zhang, Y., Li, K., Li, K., Wang, L., Zhong, B., Fu, Y.: Image super-resolution using very deep residual channel attention networks. In: The European Conference on Computer Vision (ECCV), pp. 286–301, September 2018
19. Zheng, C., Cham, T.J., Cai, J.: Pluralistic image completion. In: The IEEE Conference on Computer Vision and Pattern Recognition (CVPR), pp. 1438–1447, June 2019

# Residual Spatial Attention Network for Retinal Vessel Segmentation

Changlu Guo[1,2](✉), Márton Szemenyei[2], Yugen Yi[3](✉), Wei Zhou[4], and Haodong Bian[5]

[1] Eötvös Loránd University, Budapest, Hungary
clguo.ai@gmail.com
[2] Budapest University of Technology and Economics, Budapest, Hungary
[3] Jiangxi Normal University, Nanchang, China
yiyg510@jxnu.edu.cn
[4] Chinese Academy of Science, Shengyang, China
[5] Qinghai University, Xining, China
https://github.com/clguo/RSAN

**Abstract.** Reliable segmentation of retinal vessels can be employed as a way of monitoring and diagnosing certain diseases, such as diabetes and hypertension, as they affect the retinal vascular structure. In this work, we propose the Residual Spatial Attention Network (RSAN) for retinal vessel segmentation. RSAN employs a modified residual block structure that integrates DropBlock, which can not only be utilized to construct deep networks to extract more complex vascular features, but can also effectively alleviate the overfitting. Moreover, in order to further improve the representation capability of the network, based on this modified residual block, we introduce the spatial attention (SA) and propose the Residual Spatial Attention Block (RSAB) to build RSAN. We adopt the public DRIVE and CHASE DB1 color fundus image datasets to evaluate the proposed RSAN. Experiments show that the modified residual structure and the spatial attention are effective in this work, and our proposed RSAN achieves the state-of-the-art performance.

**Keywords:** Retinal vessel segmentation · Residual block · DropBlock · Spatial attention

## 1 Introduction

Retinal images contain rich contextual structures, such as retinal vascular structures that can provide important clinical information for the diagnosis of diseases such as diabetes and hypertension. Therefore, the accuracy of retinal blood

This work is supported by the China Scholarship Council, the Stipendium Hungaricum Scholarship, the National Natural Science Foundation of China under Grants 62062040 and 61672150, the Chinese Postdoctoral Science Foundation 2019M661117, and Hungarian Government and co-financed by the European Social Fund (EFOP-3.6.3-VEKOP-16-2017-00001).


H. Yang et al. (Eds.): ICONIP 2020, LNCS 12532, pp. 509–519, 2020.
https://doi.org/10.1007/978-3-030-63830-6_43

vessel segmentation can be used as an important indicator for the diagnosis of related diseases. However, manual segmentation of retinal blood vessels is a time-consuming task, so we are working to find a way to automatically segment retinal blood vessels.

In recent years, convolutional neural network (CNN) based methods have shown strong performance in automatically segmenting retinal blood vessels. In particular, Ronneberger et al. [1] proposed the famous U-Net that combines coarse features with fine features through "skip connections" to have superior performance in the field of medical image processing. Therefore, numerous retinal vessel segmentation methods are based on U-Net, for example, Zhuang et al. [2] proposed a chain of multiple U-Nets (LadderNet), which includes multiple pairs of encoder-decoder branches. Wu et al. [3] reported the multi-scale network followed network (MS-NFN) for retinal vessel segmentation and each sub-model contains two identical U-Net models. Then Wang et al. [4] proposed the Dual Encoding U-Net (DEU-Net) that remarkably enhances networks capability of segmenting retinal vessels in an end-to-end and pixel-to-pixel way. Although these U-Net-based methods have achieved excellent performance, they all ignore the inter-spatial relationship between features, which is important for retinal vessel segmentation because the distinction between vascular and non-vascular regions in the retinal fundus image is not very obvious, especially for thin blood vessels. To address this problem, we introduce spatial attention (SA) in this work because it can learn where is able to effectively emphasize or suppress and refine intermediate features [5].

In this paper, we propose a new Residual Spatial Attention Network (RSAN) for segmentation of retinal vessels in retinal fundus images. Specifically, inspired by the success of DropBlock [6] and residual network [7], we add DropBlock to the pre-activation residual block [8], which can be used to build a deep network to obtain deeper vascular features. Then, based on the previous discussion, we integrate the SA into this modified residual block and propose a Residual Spatial Attention Block (RSAB). Finally, combined with the advantage that "skip connection" in U-Net is able to save more structural information, the original convolution unit of U-Net is replaced by the modified residual block and RSAB to form the proposed RSAN. Through comparison experiments, we observe that the use of DropBlock can improve the performance. Then, after the introduction of SA, that is, using the proposed RSAN for retinal vessel segmentation, our performance surpasses other existing state-of-the-art methods.

## 2 Related Work

### 2.1 U-Net

In 2015, Ronneberger et al. [1] proposed a U-shaped fully convolutional network for medical image segmentation called U-Net, which has a typical symmetrical codec network structure. U-Net has an obvious advantage that it can make good use of GPU memory. This advantage is mainly related to extraction of image features at multiple image scales. U-Net transfers the feature maps obtained in the encoding stage to the corresponding decoding stage, and merges the feature maps of different stages through "skip connection" to merge coarse and fine-level dense predictions.

### 2.2 ResNet

He et al. [7] observed that when deeper networks begin to converge, there will be a degradation problem: as the network deepens, the accuracy quickly degrades after reaching saturation. In other words, simply deepening the network can hinder training. To overcome these problems, the residual network proposed by He et al. shows significantly improved training characteristics, allowing the network depth to be previously unachievable. The residual network consists of some stacked residual blocks, and each residual block can be illustrated as a routine form:

$$\begin{aligned} y_i &= F(x_i, w_i) + h(x_i) \\ x_{i+1} &= \sigma(y_i) \end{aligned} \tag{1}$$

where $x_i$ and $x_j$ represent the input and output of the current residual block, $\sigma(y_i)$ is an activation function, $F(\bullet)$ is the residual function, and $h(x_i)$ is an identity mapping function, typically $h(x_i) = x_i$.

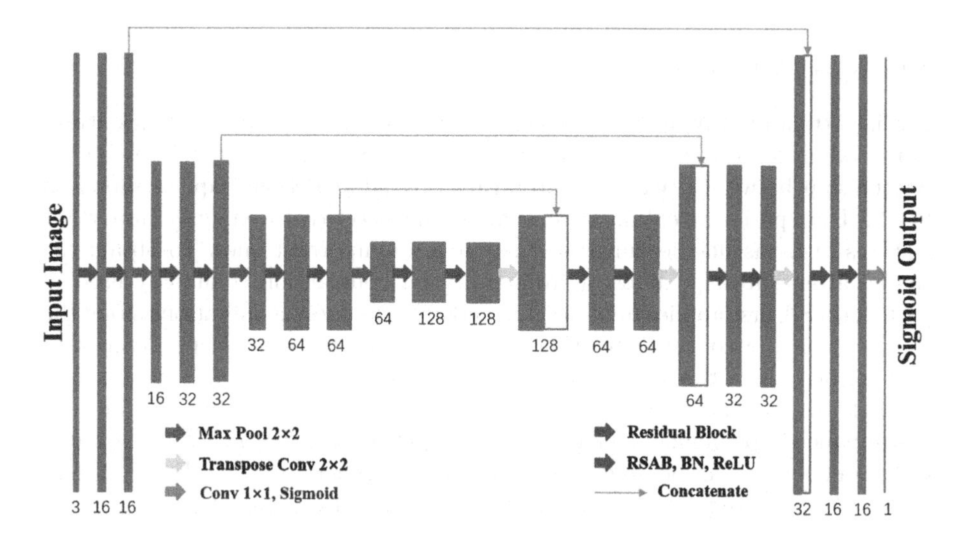


**Fig. 1.** Diagram of the proposed RSAN.

## 3 Method

Figure 1 illustrates the proposed Residual Spatial Attention Network (RSAN) with a typical encoder-decoder architecture. RSAN consists of three encoder blocks (left side) and three decoder blocks (right side) that are connected by a concatenate operation. Each encoder block and decoder block contain a pre-activation residual block with DropBlock, a Residual Spatial Attention Block (RSAB), a Batch Normalization (BN) layer, and a Rectified Linear Unit (ReLU). In the encoder, the max pooling with a pooling size of 2 is utilized for down-sampling, so that the size of the image after each RSAB is halved, which is beneficial to reduce the computational complexity and save training time. The decoder block and the encoder block is similar, except that the former uses a $2 \times 2$ transposed convolution for upsampling instead of the pooling layer. The last layer utilizes a $1 \times 1$ convolution followed by a Sigmoid activation function to obtain the required feature map.

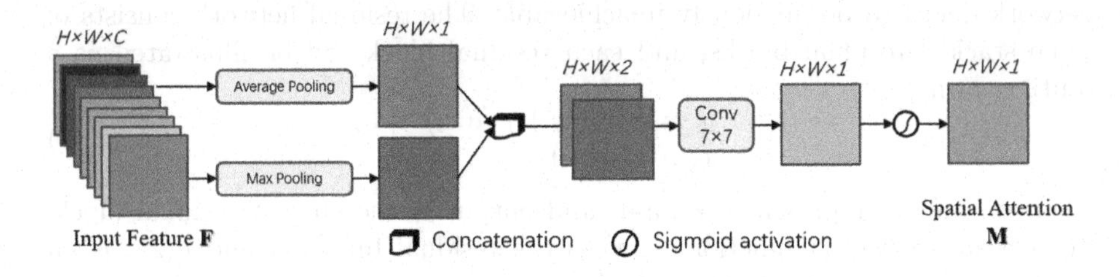


**Fig. 2.** Diagram of the spatial attention.

## 3.1 Spatial Attention

Spatial Attention (SA) was introduced as a part of the convolutional block attention module for classification and detection [5]. SA employs the inter-spatial relationship between features to produce a spatial attention map, as shown in Fig. 2. The spatial attention map enables the network to enhance important features (e.g. vascular features) and suppress unimportant ones. To obtain the spatial attention map, different from the $1 \times 1$ convolution commonly used in past work, SA first applies max-pooling and average-pooling operations along the channel axis and concatenates them to produce an efficient feature descriptor. The reason behind this is that the amount of SA parameters will be very small. A single SA module contains only 98 parameters, but it can bring significant performance improvements. Generally, the input feature $F \in R^{H \times W \times C}$ through the channel-wise max-pooling and average-poling generate $F_{mp} \in R^{H \times W \times 1}$ and $F_{ap} \in R^{H \times W \times 1}$, respectively, e.g., at the $i$-th pixel in $F_{mp}$ and $F_{ap}$:

$$F_{mp}^{i} = Max(P^{(i,c)}), 0 < c < C, 0 < i < H \times W \tag{2}$$

$$F_{ap}^{i} = \frac{1}{C} \sum_{c=1}^{C} (P^{(i,c)}), 0 < c < C, 0 < i < H \times W \tag{3}$$

where $Max(\cdot)$ obtain the maximum number, $P^{(i,c)}$ represents the pixel value of the $i$-th pixel at the $c$-th channel, and $H$, $W$, and $C$ denote the height, width, and the number of channels for the input feature $F$, respectively. Then a convolutional layer followed by a Sigmoid activation function on the concatenated feature descriptor which is utilized to produce a spatial attention map $M(F) \in R^{H \times W \times 1}$. Briefly, the spatial attention map is calculated as:

$$M(F) = \sigma(f^{7\times7}([F_{mp}; F_{ap}])) \tag{4}$$

where $f^{7\times7}(\cdot)$ denotes a convolution operation with a kernel size of 7 and $\sigma(\cdot)$ represents the Sigmoid function.

### 3.2 Modified Residual Block

In this work, shallow networks may limit the network's ability to extract the vascular features required [9]. We argue that building deeper neural networks can learn more complex vascular features, but He et al. [7] observed that simply increasing the number of network layers may hinder training, and degradation problems may occur. In order to address the above problems, He et al. [7] proposed the residual network (ResNet) achieving a wide influence in the field of computer vision. Furthermore, He et al. [8] discussed in detail the effects of the residual block consisting of multiple combinations of ReLU activation, Batch normalization (BN), and convolutional layers, and proposed a pre-activation residual block, as shown in Fig. 3(b). We utilize this pre-activation residual block to replace the original convolutional unit of U-Net shown in Fig. 3(a), and call this modified network as "Backbone".

In addition, Ghiasi et al. [6] proposed DropBlock, a structured variant of dropout, and also proved its effectiveness in convolutional networks, then SD-Unet [10] and DRNet [11] showed that DropBlock can effectively prevent overfitting problems in fully convolutional networks (FCNs). Inspired by the above work, we introduce DropBlock in the pre-activation residual block, as shown in Fig. 3(c). If the numbers of input and output channels are different, we employ $1 \times 1$ convolution to compress or expand the number of channels.

### 3.3 Residual Spatial Attention Block

Spatial Attention automatically learns the importance of each feature spatial through learning, and uses the obtained importance to emphasize features or suppress features that are not important to the current retinal vessel segmentation task. Combining the previous discussion and inspired by the successful application of the convolutional block attention module in classification and detection, we integrate SA into the modified residual block shown in Fig. 3(c) and propose the Residual Spatial Attention Block (RSAB). The structure of RSAB is shown in Fig. 4, and we argue that the introduction of SA can make full use of the inter-spatial relationship between features to improve the network's representation capability, and moreover, the integration of DropBlock

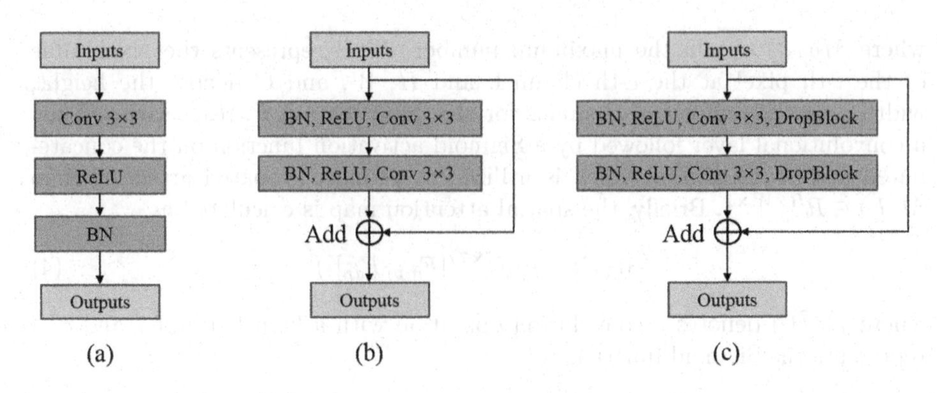


**Fig. 3.** (a) Convolutional unit of U-Net, (b) pre-activation residual block, (c) pre-activation residual block with DropBlock.

and pre-activation residual block is effective without worrying about overfitting or degradation problems, even for small sample datasets such as retinal fundus image datasets.

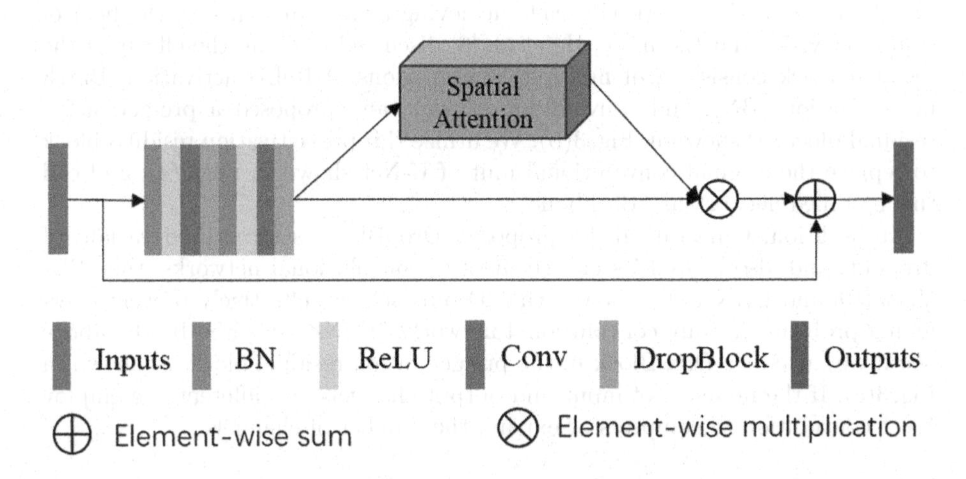


**Fig. 4.** Diagram of the proposed RSAB.

## 4 Experiments

### 4.1 Datasets

We employ DRIVE and CHASE DB1 retinal image datasets to evaluate the proposed RSAN. The DRIVE dataset includes 40 color fundus images, of which 20 are officially designated for training and 20 for testing. The CHASE DB1 contains 28 retinal fundus images. Although there is no initial division of training and testing sets, usually the first 20 are used as the training set, and the remaining 8 are used as the testing set [3,12]. The resolution of each image in DRIVE and CHASE DB1 is $565 \times 584$ and $999 \times 960$ respectively. In order to fit our network, we resize each image in DRIVE and CHASE DB1 to $592 \times 592$ and $1008 \times 1008$ by padding it with zero in four mar-gins, but in the process of evaluating, we crop the segmentation results to the initial resolution. The manual segmented binary vessel maps of both datasets provided by human experts can be applied as the ground truth.

### 4.2 Evaluation Metrics

To evaluate our model quantitatively, we compare the segmentation results with the corresponding ground truth and divide the comparison results of each pixel into true positive (TP), false positive (FP), false negative (FN), and true negative (TN). Then, the sensitivity (SEN), specificity (SPE), F1-score (F1), and accuracy (ACC) are adopted to evaluate the performance of the model. To further evaluate the performance of our model, we also utilize the Area Under the ROC Curve (AUC). If the value of AUC is 1, it means perfect segmentation.

### 4.3 Implementation Details

We compare the performance of Backbone, Backbone+DropBlock and the proposed RSAN on DRIVE and CHASE DB1. All three models are trained from scratch using the training sets and tested in the testing sets. In order to observe whether the current training model is overfitting, we randomly select two images as the validation set from the training set of both datasets. For both datasets, we utilize the Adam optimizer to optimize all models with binary cross entropy as the loss function. During the training of DRIVE, we set the batch size to 2. RSAN first trains 150 epochs with the learning rate of $1 \times 10^{-3}$, and the last 50 epochs with $1 \times 10^{-4}$. For CHASE DB1, the batch size is 1, and a total of 150 epochs are trained, of which the first 100 epochs with a learning rate of $1 \times 10^{-3}$ and the last 50 epochs with $1 \times 10^{-4}$.

For the setting of DropBlock in RSAN, the size of block to be dropped for all datasets is set to 7. To reach the best performance, we set the probability of keeping a unit for DRIVE and CHASE DB1 to 0.85 and 0.78, respectively. In the experiments, Backbone+DropBlock and RSAN have the same configuration. For Backbone, We observed serious overfitting problems, so we use the results obtained from its best training epochs.

### 4.4 Results

Figure 5 displays some examples of two color fundus images from the DRIVE and CHASE DB1 datasets, segmentation results performed by Backbone, Backbone+DropBlock and RSAN, and the corresponding ground truth. The segmentation results illustrate that RSAN can predict most thick and thin blood vessels (pointed by red and green arrows) comparing with Backbone and Backbone+DropBlock. In particular, RSAN has a clearer segmentation result for thin blood vessels, and can retain more detailed vascular space structure. In addition, we quantitatively compare the performance of Backbone, Backbone+DropBlock and the proposed RSAN on the DRIVE, CHASE DB1 datasets, as displayed in Tables 1 and 2. From the results in these table, we can get several notable observations: First, Backbone+DropBlock has better performance than the Backbone, which shows that the strategy of using the DropBlock to regularize the network is effective. Second, the SEN, F1, ACC, and AUC of RSAN on the two datasets are higher than Backbone+DropBlock about 2.41%/1.56%, 1.12%/1.21%, 0.14%/0.15%, and 0.33%/0.23%, respectively. It proves that the introduction of spatial attention can improve the performance of the network in retinal vessel segmentation task. At last, our proposed RSAN has the best segmentation performance overall, which means that RSAN is an effective method for retinal vessel segmentation.

**Table 1.** Experimental results on DRIVE. (*The results is obtain from [14])

| Datasets | DRIVE | | | | |
|---|---|---|---|---|---|
| Metrics | SEN | SPE | F1 | ACC | AUC |
| U-Net [5]* | 0.7537 | 0.9820 | 0.8142 | 0.9531 | 0.9755 |
| Backbone | 0.7851 | 0.9826 | 0.7985 | 0.9653 | 0.9762 |
| Backbone+DropBlock | 0.7908 | **0.9847** | 0.8110 | 0.9677 | 0.9822 |
| **RSAN** | **0.8149** | 0.9839 | **0.8222** | **0.9691** | **0.9855** |

**Table 2.** Experimental results on CHASE DB1. (*The results is obtain from [14])

| Datasets | CHASE DB1 | | | | |
|---|---|---|---|---|---|
| Metrics | SEN | SPE | F1 | ACC | AUC |
| U-Net [5] * | 0.8288 | 0.9701 | 0.7783 | 0.9578 | 0.9772 |
| Backbone | 0.7843 | **0.9844** | 0.7781 | 0.9718 | 0.9805 |
| Backbone+DropBlock | 0.8330 | 0.9830 | 0.7990 | 0.9736 | 0.9871 |
| **RSAN** | **0.8486** | 0.9836 | **0.8111** | **0.9751** | **0.9894** |

Finally, we compare our proposed RSAN with several existing state-of-the-art methods. We summarize the release year of the different methods and their performance on DRIVE and CHASE DB1, as shown in Tables 3 and 4, respectively. From the results in these tables, our proposed RSAN achieves the best performance on both datasets. In detail, on the DRIVE and CHASE DB1, our

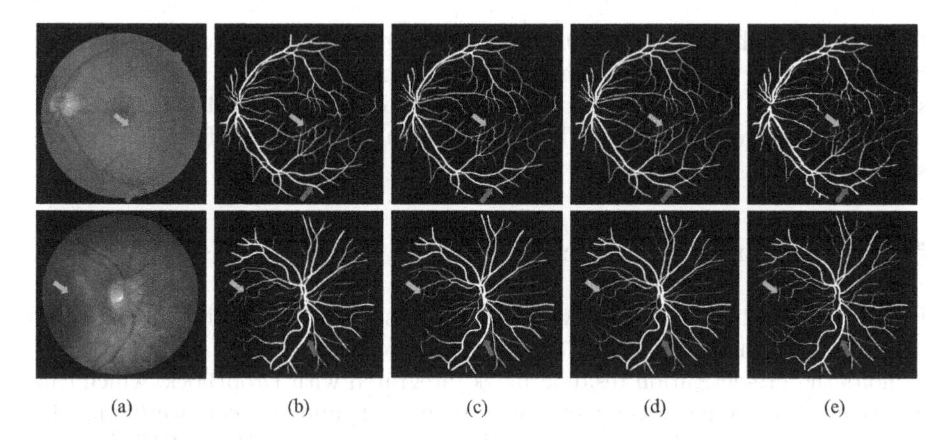


**Fig. 5.** Row 1 is for DRIVE dataset. Row 2 is for CHASE DB1 dataset. (a) Color fundus images, (b) segmentation results of Backbone, (c) segmentation results of Backbone+DropBlock, (d) segmentation results of RSAN, (e) corresponding ground truths.

**Table 3.** Results of RSAN and other methods on DRIVE dataset

| Methods | Year | SEN | SPE | F1 | ACC | AUC |
|---|---|---|---|---|---|---|
| Orlando et al. [12] | 2017 | 0.7897 | 0.9684 | 0.7857 | 0.9454 | 0.9506 |
| Yan et al. [13] | 2018 | 0.7653 | 0.9818 | N.A | 0.9542 | 0.9752 |
| R2U-Net [14] | 2018 | 0.7799 | 0.9813 | 0.8171 | 0.9556 | 0.9784 |
| LadderNet [2] | 2018 | 0.7856 | 0.9810 | 0.8202 | 0.9561 | 0.9793 |
| MS-NFN [3] | 2018 | 0.7844 | 0.9819 | N.A | 0.9567 | 0.9807 |
| DEU-Net [4] | 2019 | 0.7940 | 0.9816 | **0.8270** | 0.9567 | 0.9772 |
| Vessel-Net [15] | 2019 | 0.8038 | 0.9802 | N.A | 0.9578 | 0.9821 |
| **RSAN** | **2020** | **0.8149** | **0.9839** | 0.8222 | **0.9691** | **0.9855** |

**Table 4.** Results of RSAN and other methods on CHASE DB1 dataset.

| Methods | Year | SEN | SPE | F1 | ACC | AUC |
|---|---|---|---|---|---|---|
| Orlando et al. [12] | 2017 | 0.7277 | 0.9712 | 0.7332 | 0.9458 | 0.9524 |
| Yan et al. [13] | 2018 | 0.7633 | 0.9809 | N.A | 0.9610 | 0.9781 |
| R2U-Net [14] | 2018 | 0.7756 | 0.9820 | 0.7928 | 0.9634 | 0.9815 |
| LadderNet [2] | 2018 | 0.7978 | 0.9818 | 0.8031 | 0.9656 | 0.9839 |
| MS-NFN [3] | 2018 | 0.7538 | **0.9847** | N.A | 0.9637 | 0.9825 |
| DEU-Net [4] | 2019 | 0.8074 | 0.9821 | 0.8037 | 0.9661 | 0.9812 |
| Vessel-Net [15] | 2019 | 0.8132 | 0.9814 | N.A | 0.9661 | 0.9860 |
| **RSAN** | **2020** | **0.8486** | 0.9836 | **0.8111** | **0.9751** | **0.9894** |

RSAN has the highest AUC (0.34%/0.34% higher than the best before), the highest accuracy (1.13%/0.90% higher than the best before) and the highest sensitivity. Besides, F1 and specificity are comparable in general. The above results mean that our method achieves the state-of-the-art performance for retinal vessel segmentation.

## 5 Discussion and Conclusion

Residual Spatial Attention Network (RSAN) is presented in this paper to be utilized to accurately segment retinal blood vessel in fundus images. RSAN exploits the pre-activation residual block integrated with DropBlock, which can effectively extract more complex vascular features and prevent overfitting. In addition, the newly designed Residual Spatial Attention Block (RSAB) significantly improves the network's representation capability via introducing the spatial attention mechanism. We evaluate the RSAN on two datasets, including DRIVE and CHASE DB1, and the results indicate that RSAN reaches the state-of-the-art performance. The improvement of RSAN's performance in retinal blood vessel segmentation, especially the improvement of AUC and ACC indicators, is of great significance in the clinical detection of early eye-related diseases. It is worth mentioning that RSAN has not yet considered the connectivity of blood vessels, which is also the focus of our next work.

## References

1. Ronneberger, O., Fischer, P., Brox, T.: U-net: convolutional networks for biomedical image segmentation. In: Navab, N., Hornegger, J., Wells, W.M., Frangi, A.F. (eds.) MICCAI 2015. LNCS, vol. 9351, pp. 234–241. Springer, Cham (2015). https://doi.org/10.1007/978-3-319-24574-4_28
2. Zhuang, J.: LadderNet: multi-path networks based on U-net for medical image segmentation. arXiv preprint arXiv:1810.07810 (2018)
3. Wu, Y., Xia, Y., Song, Y., Zhang, Y., Cai, W.: Multiscale network followed network model for retinal vessel segmentation. In: Frangi, A.F., Schnabel, J.A., Davatzikos, C., Alberola-López, C., Fichtinger, G. (eds.) MICCAI 2018. LNCS, vol. 11071, pp. 119–126. Springer, Cham (2018). https://doi.org/10.1007/978-3-030-00934-2_14
4. Wang, B., Qiu, S., He, H.: Dual encoding U-net for retinal vessel segmentation. In: Shen, D., et al. (eds.) MICCAI 2019. LNCS, vol. 11764, pp. 84–92. Springer, Cham (2019). https://doi.org/10.1007/978-3-030-32239-7_10
5. Woo, S., Park, J., Lee, J.-Y., Kweon, I.S.: CBAM: convolutional block attention module. In: Ferrari, V., Hebert, M., Sminchisescu, C., Weiss, Y. (eds.) ECCV 2018. LNCS, vol. 11211, pp. 3–19. Springer, Cham (2018). https://doi.org/10.1007/978-3-030-01234-2_1
6. Ghiasi, G., Lin, T.-Y., Le, Q.V.: DropBlock: a regularization method for convolutional networks. In: Neural Information Processing Systems (2018)
7. He, K., Zhang, X., Ren, S., Sun, J.: Deep residual learning for image recognition. In: CVPR 2016, pp. 770–778 (2016)

8. He, K., Zhang, X., Ren, S., Sun, J.: Identity mappings in deep residual networks. In: Leibe, B., Matas, J., Sebe, N., Welling, M. (eds.) ECCV 2016. LNCS, vol. 9908, pp. 630–645. Springer, Cham (2016). https://doi.org/10.1007/978-3-319-46493-0_38
9. Li, D., Dharmawan, D.A., Ng, B.P., Rahardja, S.: Residual U-net for retinal vessel segmentation. In: 2019 IEEE International Conference on Image Processing (ICIP), Taipei, Taiwan, pp. 1425–1429 (2019)
10. Guo, C., Szemenyei, M., Pei, Y., Yi, Y., Zhou, W.: SD-Unet: a structured dropout u-net for retinal vessel segmentation. In: 2019 IEEE 19th International Conference on Bioinformatics and Bioengineering (BIBE), Athens, Greece, pp. 439–444 (2019)
11. Guo, C., Szemenyei, M., Yi, Y., Xue, Y., Zhou, W., Li, Y.: Dense residual network for retinal vessel segmentation. In: ICASSP 2020–2020 IEEE International Conference on Acoustics, Speech and Signal Processing (ICASSP), Barcelona, Spain, pp. 1374–1378 (2020)
12. Orlando, J.I., Prokofyeva, E., Blaschko, M.B.: A discriminatively trained fully connected conditional random field model for blood vessel segmentation in fundus images. IEEE Trans. Biomed. Eng. **64**(1), 16–27 (2017)
13. Yan, Z., Yang, X., Cheng, K.T.: Joint segment-level and pixel-wise losses for deep learning based retinal vessel segmentation. IEEE Trans. Biomed. Eng. **65**(9), 1912–1923 (2018)
14. Alom, M.Z., Hasan, M., Yakopcic, C., Taha, T.M., Asari, V.K.: Recurrent residual convolutional neural network based on U-net (R2U-Net) for medical image segmentation. arXiv preprint arXiv:1802.06955 (2018)
15. Wu, Y., et al.: Vessel-net: retinal vessel segmentation under multi-path supervision. In: Shen, D., et al. (eds.) MICCAI 2019. LNCS, vol. 11764, pp. 264–272. Springer, Cham (2019). https://doi.org/10.1007/978-3-030-32239-7_30

# REXUP: I REason, I EXtract, I UPdate with Structured Compositional Reasoning for Visual Question Answering

Siwen Luo, Soyeon Caren Han(✉), Kaiyuan Sun, and Josiah Poon

School of Computer Science, The University of Sydney,
1 Cleveland Street, Sydney, NSW 2006, Australia
{siwen.luo,caren.han,kaiyuan.sun,josiah.poon}@sydney.edu.au

**Abstract.** Visual Question Answering (VQA) is a challenging multi-modal task that requires not only the semantic understanding of images and questions, but also the sound perception of a step-by-step reasoning process that would lead to the correct answer. So far, most successful attempts in VQA have been focused on only one aspect; either the interaction of visual pixel features of images and word features of questions, or the reasoning process of answering the question of an image with simple objects. In this paper, we propose a deep reasoning VQA model (REXUP- REason, EXtract, and UPdate) with explicit visual structure-aware textual information, and it works well in capturing step-by-step reasoning process and detecting complex object-relationships in photo-realistic images. REXUP consists of two branches, image object-oriented and scene graph-oriented, which jointly works with the super-diagonal fusion compositional attention networks. We evaluate REXUP on the benchmark GQA dataset and conduct extensive ablation studies to explore the reasons behind REXUP's effectiveness. Our best model significantly outperforms the previous state-of-the-art, which delivers 92.7% on the validation set, and 73.1% on the test-dev set. Our code is available at: https://github.com/usydnlp/REXUP/.

**Keywords:** GQA · Scene graph · Visual Question Answering

## 1 Introduction

Vision-and-language reasoning requires the understanding and integration of visual contents and language semantics and cross-modal alignments. Visual Question Answering (VQA) [2] is a popular vision-and-language reasoning task, which requires the model to predict correct answers to given natural language questions based on their corresponding images. Substantial past works proposed VQA models that focused on analysing objects in photo-realistic images but worked well only for simple object detection and yes/no questions [14,17,25].

S. Luo and S. C. Han—Both authors are first author.


H. Yang et al. (Eds.): ICONIP 2020, LNCS 12532, pp. 520–532, 2020.
https://doi.org/10.1007/978-3-030-63830-6_44

To overcome this simple nature and improve the reasoning abilities of VQA models, the Clever dataset [13] was introduced with compositional questions and synthetic images, and several models [9,20] were proposed and focused on models' inferential abilities. The state-of-the-art model on the Clevr dataset is the compositional attention network (CAN) [11], which generates reasoning steps attending over both images and language-based question words. However, the Clevr dataset is specifically designed to evaluate reasoning capabilities of a VQA model. Objects in the Clevr dataset images are only in three different shapes and four different spatial relationships, which results in simple image patterns. Therefore, a high accuracy on Clevr dataset hardly prove a high object detection and analysis abilities in photo-realistic images, nor the distinguished reasoning abilities of a VQA model. To overcome the limitations of VQA and Clevr [2,7], the GQA dataset [12] includes photo-realistic images with over 1.7 K different kinds of objects and 300 relationships. GQA provides diverse types of answers for open-ended questions to prevent models from memorizing answer patterns and examine the understanding of both images and questions for answer prediction.

The state-of-the-art models on the Clevr and VQA dataset suffered large performance reductions when evaluated on GQA [1,11,19]. Most VQA works focus on the interaction between visual pixel features extracted from images and question features while such interaction does not reflect the underlying structural relationships between objects in images. Hence, the complex relationships between objects in real images are hard to learn. Inspired by this motivation, we proposed REXUP (REason, EXtract, UPdate) network to capture step-by-step reasoning process and detect the complex object-relationships in photo-realistic images with the scene graph features. A scene graph is a graph representation of objects, attributes of objects and relationships between objects where objects that have relations are connected via edge in the graph.

The REXUP network consists of two parallel branches where the image object features and scene graph features are respectively guided by questions in an iterative manner, constructing a sequence of reasoning steps with REXUP cells for answer prediction. A super-diagonal fusion is also introduced for a stronger interaction between object features and question embeddings. The branch that processes scene graph features captures the underlying structural relationship of objects, and will be integrated with the features processed in another branch for final answer prediction. Our model is evaluated on the GQA dataset and we used the official GQA scene graph annotations during training. To encode the scene graph features, we extracted the textual information from the scene graph and used Glove embeddings to encode the extracted textual words in order to capture the semantic information contained in the scene graph. In the experiments, our REXUP network achieved the state-of-the-art performance on the GQA dataset with complex photo realistic images in deep reasoning question answering task.

## 2 Related Work and Contribution

We explore research trends in diverse visual question answering models, including fusion-based, computational attention-based, and graph-based VQA models.

**Fusion-Based VQA.** Fusion is a common technique applied in many VQA works to integrate language and image features into a joint embedding for answer prediction. There are various types of fusion strategies for multi-modalities including simple concatenation and summation. For example, [22] concatenated question and object features together and pass the joint vectors to a bidirectional GRU for further processes. However, the recent bilinear fusion methods are more effective at capturing higher level of interactions between different modalities and have less parameters in calculation. For example, based on the tensor decomposition proposed in [3], [4] proposed a block-term decomposition of the projection tensor in bilinear fusion. [5] applied this block-term fusion in their proposed MuRel networks, where sequences of MuRel cells are stacked together to fuse visual features and question features together.

**Computational Attention-Based VQA.** Apart from fusion techniques, attention mechanisms are also commonly applied in VQA for the integration of multi-modal features. Such attention mechanisms include soft attention mechanism like [1,11] using softmax to generate attention weights over object regions and question words, self attention mechanism like [18,24] that applied dot products on features of each mode, and co-attention mechanisms like in [6,16] using linguistic features to guide attentions of visual features or vice versa.

**Graph Representations in VQA.** In recent years, more works have been proposed to integrate graph representations of images in VQA model. [19] proposed a question specific graph-based model where objects are identified and connected with each other if their relationships are implied in the given question. There are also works use scene graph in VQA like we did. [21] integrates scene graphs together with functional programs for explainable reasoning steps. [8] claimed only partial image scene graphs are effective for answer prediction and proposed a selective system to choose the most important path in a scene graph and use the most probable destination node features to predict an answer. However, these works did not apply their models on GQA.

**REXUP's Contribution.** In this work, we move away from the classical attention and traditional fusion network, which have been widely used in simple photo-realistic VQA tasks and focus mainly on the interaction between visual pixel features from an image and question embeddings. Instead, we focus on proposing a deeper reasoning solution in visual-and-language analysis, as well as complex object-relationship detection in complex photo-realistic images. We propose a new deep reasoning VQA model that can be worked well on complex images by processing both image objects features and scene graph features and integrating those with super-diagonal fusion compositional attention networks.

# 3 Methodology

The REXUP network contains two parallel branches, object-oriented branch and scene-graph oriented branch, shown in Fig. 1a. Each branch contains a sequence of REXUP cells where each cell operates for one reasoning step for the answer

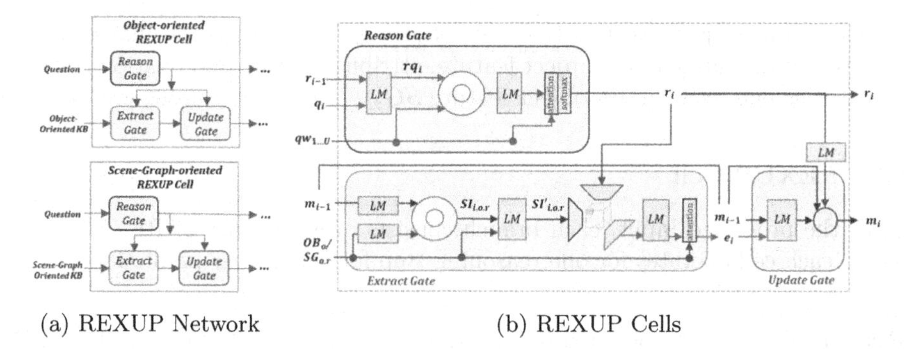


(a) REXUP Network (b) REXUP Cells

**Fig. 1. REXUP Network and REXUP cell.** (a) The REXUP network includes two parallel branches, object-oriented *(top)* and scene graph-oriented *(bottom)*. (b) A REXUP cell contains reason, extract, and update gate which conduct multiple compositional reasoning and super-diagonal fusion process

prediction. As shown in Fig. 1b, each REXUP cell includes a reason, an extract and an update gate. At each reasoning step, the reason gate identifies the most important words in the question and generates a current reasoning state with distributed attention weights over each word in the question. This reasoning state is fed into the extract gate and guides to capture the important objects in the knowledge base, retrieving information that contains the distributed attention weights over objects in the knowledge base. The update gate takes the reasoning state and information from extract gate to generate the current memory state.

### 3.1 Input Representation

Both Object-oriented branch and Scene graph-oriented branch take question and knowledge base as inputs; image object-oriented knowledge base (OKB) and scene-graph-oriented knowledge base (SGKB). For a question $q \in Q$ with maximum $U$ words, contextual words are encoded via a pre-trained 300$d$ Glove embedding and passed into bi-LSTM to generate a sequence of hidden states $qw_{1...U}$ with $d$ dimension for question contextual words representation. The question is encoded by the concatenation of the last backward and forward hidden states, $\overleftarrow{qw_1}$ and $\overrightarrow{qw_U}$. Object features are extracted from a pre-trained Fast-RCNN model, each image contains at most 100 regions represented by a 2048$d$ object feature. For each $o_{th}$ object in an image, linear transformation converts the object features with its corresponding coordinates to a 512$d$ object region embedding. The SGKB is the matrix of scene graph objects each of which is in 900 dimensions after concatenating with their corresponding attribute and relation features. To encode the scene graph object features, all the objects names, their attributes and relations in the scene graph are initialized as 300$d$ Glove embedding. For each object's attributes, we take the average of those attributes features $A$. For each object's relations, we first average each relation feature $r_s \in R$ and the subject feature $o_j \in O$ that it is linked to, and then average all such relation-subject features that this object $o_n \in O$ has as its final relation feature. We concatenate the object feature, attribute feature and relation feature together as one scene graph object feature $SG_{o,r}$ of the whole scene graph.

### 3.2 REXUP Cell

With the processed input, each branch consists of a sequence of REXUP cells where each cell operates for one reasoning step for the answer prediction.

**Reason Gate.** At each reasoning step, the reason gate in each REXUP cell $i = 1, ..., P$ takes the question feature $q$, the sequence of question words $qw_1, qw_2, ..., qw_U$ and the previous reasoning state $r_{i-1}$ as inputs. Before being passed to the reason gate, each question $q$ is processed through a linear transformation $q_i = W_i^{d \times 2d} q + b_i^d$ to encode the positional-aware question embedding $q_i$ with $d$ dimension in the current cell. A linear transformation is then processed on the concatenation of $q_i$ and the previous reasoning state $r_{i-1}$,

$$rq_i = W^{d \times 2d} \left[ r_{i-1}, q_i \right] + b^d \tag{1}$$

in order to integrate the attended information at the previous reasoning step into the question embedding at the current reasoning step.

Then an element-wise multiplication between $rq_i$ and each question word $qw_u$, where $u = 1, 2, ..., U$, is conducted to transfer the information in previous reasoning state into each question word at the current reasoning step, the result of which will be processed through a linear transformation, yielding a sequence of new question word representations $ra_{i,1}, ..., ra_{i,u}$ containing the information obtained in previous reasoning state. A softmax is then applied to yield the distribution of attention scores $rv_{i,1}, ..., rv_{i,u}$ over question words $qw_1, ..., qw_u$.

$$ra_{i,u} = W^{1 \times d} (rq_i \odot qw_u) + b \tag{2}$$

$$rv_{i,u} = softmax(ra_{i,u}) \tag{3}$$

$$r_i = \sum_{u=1}^{U} rv_{i,u} \cdot qw_u \tag{4}$$

The multiplications of each $rv_{i,u}$ and question word $qw_u$ are summed together and generates the current reasoning state $r_i$ that implies the attended information of a question at current reasoning step.

**Extract Gate.** The extract gate takes the current reasoning state $r_i$, previous memory state $m_{i-1}$ and the knowledge base features as inputs. For the OKB branch, knowledge base features are the object region features $OB_o$, and for the SGKB branch, knowledge base features are the scene graph features $SG_{o,r}$. For each object in the knowledge base, we first multiplied its feature representation with the previous memory state to integrate the memorized information at the previous reasoning step into the knowledge base at the current reasoning step, the result of which is then concatenated with the input knowledge base features and projected into $d$ dimensions by a linear transformation. This interaction $SI'_{i,o,r}$ generates the knowledge base features that contains the attended information memorized at the previous reasoning step as well as the yet unattended information of knowledge base at current reasoning step. The process of the extract gate in the SGKB branch can be shown in the following equations, where the interaction $SI'_{i,o,r}$ contains the semantic information extracted from the object-oriented scene graph.

$$SI_{i,o,r} = \left[W_m^{d \times d} m_{i-1} + b_m^d\right] \odot \left[W_s^{d \times d} SG_{o,r} + b_s^d\right] \tag{5}$$

We then make $SI'_{i,o,r}$ interact with $r_i$ to let the attended question words guide the extract gate to detect important objects of knowledge base at the current reasoning step. In the SGKB branch, such integration is completed through a simple multiplication as shown in (7).

$$SI'_{i,o,r} = W^{d \times 2d}\left[SI_{i,o,r}, SG_{o,r}\right] + b^d \tag{6}$$

$$ea_{i,o,r} = W^{d \times d}(r_i \odot SI'_{i,o,r}) + b^d \tag{7}$$

However, in OKB branch, $SG_{o,r}$ in Eq. (5) and (6) is replaced with the object region features $OB_o$, and generated interaction $I'_{i,o}$, which will be integrated with $r_i$ through a super-diagonal fusion [4] as stated in Eq. (8), where $\theta$ is a parameter to be trained. Super-diagonal fusion projects two vectors into one vector with $d$ dimension through a projection tensor that would be decomposed into three different matrices during calculation in order to decrease the computational costs while boosting a stronger interaction between input vectors. The resulted $F_{r_i,I'_{i,o}}$ is passed via a linear transformation to generate $ea_{i,o}$.

$$F_{r_i,I'_{i,o}} = SD(r_i, I'_{i,o}; \theta) \quad \text{and} \quad ea_{i,o} = W^{d \times d} F_{r_i,I'_{i,o}} + b^d \qquad (8) \text{ and } (9)$$

Similar to the process in the reason gate, $ea_{i,o,r}$ and $ea_{i,o}$ are then processed by softmax to get the distribution of attention weights for each object in the knowledge base. The multiplications of each $ea_{i,o,r}/ea_{i,o}$ and knowledge base $SG_{o,r}/OB_o$ are summed together to yield the extracted information $e_i$.

$$ev_{i,o,r} = softmax(ea_{i,o,r}) \quad \text{and} \quad ev_{i,o} = softmax(ea_{i,o}) \tag{10}$$

$$e_i = \sum_{o=1}^{O} ev_{i,o,r} \cdot SG_{o,r} \quad \text{and} \quad e_i = \sum_{o=1}^{O} ev_{i,o} \cdot OB_o \tag{11}$$

**Update Gate.** We apply a linear transformation to the concatenation of the extracted information $e_i$ and previous memory state $m_{i-1}$ to get $m_i^{prev}$.

$$m_i^{prev} = W^{d \times 2d}\left[e_i, m_{i-1}\right] + b^d \tag{12}$$

$$m_i = \sigma(r'_i) m_{i-1} + (1 - \sigma(r'_i)) m'_i \tag{13}$$

To reduce redundant reasoning steps for short questions, we applied sigmoid function upon $m_i^{prev}$ and $r_i'$, where $r_i' = W^{1 \times d} r_i + b^1$, to generate the final memory state $m_i$.

The final memory states generated in the OKB branch and SGKB branch respectively are concatenated together as the ultimate memory state $m_P$ for overall $P$ reasoning steps. $m_P$ is then integrated with the question sentence embedding $q$ for answer prediction. In this work, we set $P = 4$.

# 4 Evaluation

## 4.1 Evaluation Setup

**Dataset.** Our main research aim is proposing a new VQA model that provides not only complex object-relationship detection capability, but also deep reasoning ability. Hence, we chose the GQA that covers 1) complex object-relationship: 113,018 photo-realistic images and 22,669,678 questions of five different types, including *Choose, Logical, Compare, Verify and Query,* and 2) deep reasoning tasks: over 85% of questions with 2 or 3 reasoning steps and 8% of questions with 4+ reasoning steps. The GQA is also annotated with scene graphs extracted from the Visual Genome [15] and functional programs that specify reasoning operations for each pair of image and question. The dataset is split into 70% training, 10% validation, 10% test-dev and 10% test set.

**Training Details.** The model is trained on GQA training set for 25 epochs using a 24 GB NVIDIA TITAN RTX GPU with 10.2 CUDA toolkit. The average per-epoch training times and total training times are 7377.31 s and 51.23 h respectively. We set the batch size to 128 and used an Adam optimizer with an initial learning rate of 0.0003.

## 4.2 Performance Comparison

In Table 1, we compare our model to the state-of-the-art models on the validation and test-dev sets of GQA. Since the GQA test-dev set does not provide pre-annotated scene graphs, we used the method proposed in [26] to predict relationships between objects and generate scene graphs from images of GQA test-dev set for the evaluation procedure. However, the quality of the generated scene graphs are not as good as the pre-annotated scene graphs in the GQA validation set, which lead to the decreased performance on test-dev. Nevertheless, our model still achieves the state-of-the-art performance with 92.7% on validation and 73.1% on test-dev. Compared to [1,11,23] that only used the integration between visual pixel features and question embedding through attention mechanism, our model applies super-diagonal fusion for a stronger interaction and also integrates the scene graph features with question embedding, which help to yield much higher performance. Moreover, our model greatly improves over [10], which used the graph representation of objects but concatenated the object features with contextual relational features of objects as the visual features to be integrated with question features through the soft attention. The significant improvement over [10] indicates that the parallel training of OKB and SGKB branch can successfully capture the structural relationships of objects in images.

**Table 1.** State-of-the-art performance comparison on the GQA dataset

| Methods | Val | Test-dev |
|---|---|---|
| CNN+LSTM [12] | 49.2 | – |
| Bottom-Up [1] | 52.2 | – |
| MAC [11] | 57.5 | – |
| LXMERT [23] | 59.8 | 60.0 |
| Single-hop [10] | 62 | 53.8 |
| Single-hop+LCGN [10] | 63.9 | 55.8 |
| **Our model** | **92.7** | **73.1** |

**Table 2.** Results of ablation study on **validation** and **test-dev** set of GQA. 'O' and 'X' refers to the existence and absence of scene-graph oriented knowledge branch (*SGKB*) and super-diagonal (*SD*) fusion applied in object-oriented knowledge branch (*OKB*) branch respectively

| # | OKB | SD | SGKB | Val | Test-dev |
|---|---|---|---|---|---|
| 1 | O | X | X | 62.35 | 56.92 |
| 2 | O | O | X | 63.10 | 57.25 |
| 3 | O | X | O | 90.14 | 72.38 |
| **4** | **O** | **O** | **O** | **92.75** | **73.18** |

### 4.3 Ablation Study

We conducted the ablation study to examine the contribution of each component in our model. As shown in Table 2, integrating object-oriented scene graph features is critical in achieving a better performance on the GQA. Using only OKB branch leads to a significant drop of 29.65% in the validation accuracy and 15.93% in the test-dev accuracy. The significant performance decrease also proves the importance of semantic information of objects' structural relationships in VQA tasks. Moreover, applying the super-diagonal fusion is another key reason of our model's good performance on GQA. We compared performances of models that apply super-diagonal fusion and models that apply element-wise multiplication. The results show that using element-wise multiplication causes a drop of 2.61% on the validation set and 0.8% on the test-dev set. It still shows that the concrete interaction between image features and question features generated by super-diagonal fusion contributes to an improved performance on the GQA.

**Table 3.** Parameter testing with different number of the REXUP cell

| # of cells | Val | Test-dev |
|---|---|---|
| 1 | 90.97 | 72.08 |
| 2 | 90.98 | 72.13 |
| 3 | 92.67 | 72.56 |
| **4** | **92.75** | **73.18** |

### 4.4 Parameter Comparison

Sequences of REXUP cells will lead to sequential reasoning steps for the final answer prediction. The three gates in each cell are designed to follow questions' compositional structures and retrieve question-relevant information from knowledge bases at each step. To reach the ultimate answer, a few reasoning steps should be taken, and less cells are insufficient to extract the relevant knowledge base information for accurate answer prediction, especially for compositional questions with longer length. In order to verify this assumption, we have conducted experiments to examine the model's performances with different numbers of REXUP cells in both branches. The results of different performances are shown in Table 3. From the result, we can see that the prediction accuracy on both validation and test-dev set will gradually increase (90.97% to 92.75% on validation and 72.08% to 73.18% on test-dev) as the cell number increases. After experiment, we conclude that four REXUP cells are best both for clear presentation of reasoning capabilities and a good performance on the GQA.

### 4.5 Interpretation

To have a better insight into the reasoning abilities of our model, we extract the linguistic and visual attention weights our model computes at each reasoning step to visualize corresponding reasoning processes. Taking the first row in Fig. 2 as an example, at the first reasoning step, concrete objects - man's hand and head obtain high visual attention score. When it comes to the second and third reasoning step, linguistic attention focuses on *wearing* and corresponding visual attention focuses on man's shirt and pants. This indicates that our model's abilities in capturing the underlying semantic words of questions as well as detecting relevant objects in image for answer prediction. Moreover, our model's good understanding of both images and questions is also shown when given different questions for a same image. For example, in the second row in Fig. 2, the model successfully captures the ***phone*** in image for the question, but for images of third row in Fig. 2, the ***dog*** is detected instead. We also found that

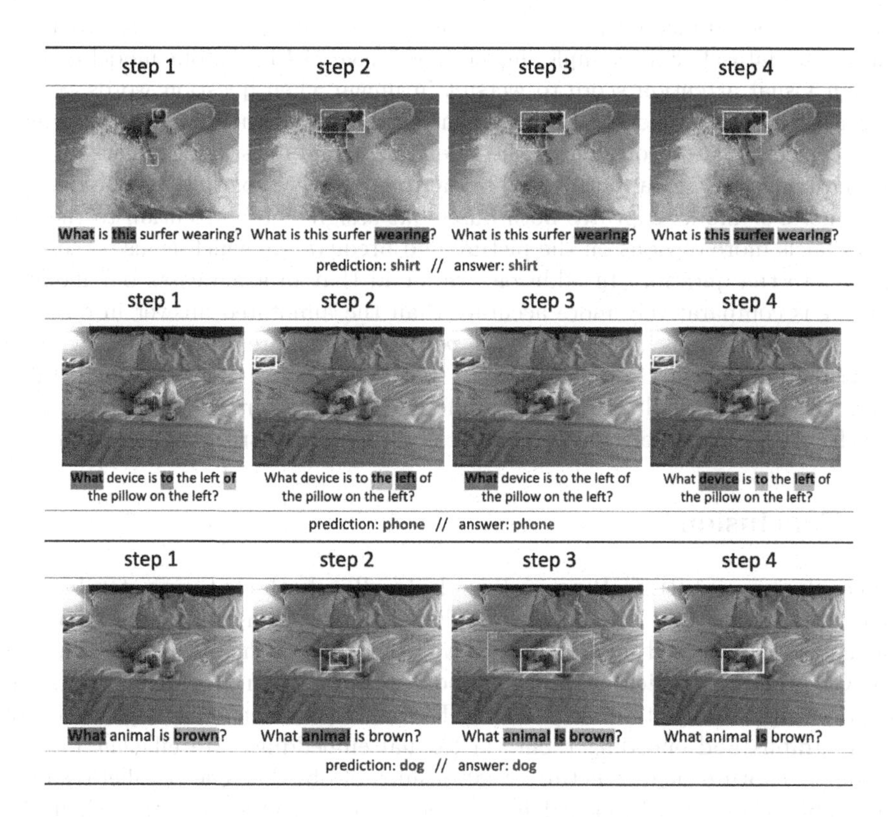


**Fig. 2.** Visualization of important image objects and question words at each reasoning step. Object regions with high attention weights are framed with white bounding boxes. The thicker the frame, the more important the object region is. Question words with high attention weights are colored blue in the question. (Color figure online)

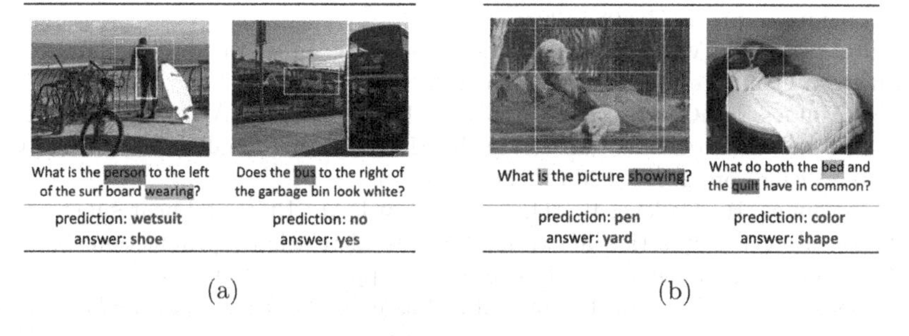


**Fig. 3.** Figure (a) shows examples when the ground truth answer and our prediction are both correct to the given question. Figure (b) shows examples when our prediction is more accurate than the ground truth answer in dataset

sometimes our predicted answer is correct even though it's different from the answer in dataset. For example, in the first image of Fig. 3a, our model assigns a high visual attention score to wetsuit in image when question words *person* and *wearing* are attended. Our model then gives the prediction ***wetsuit***, which is as correct as ***shoe*** considering the given image and question. Similarly, in the second image of Fig. 3a, both white bus and red bus are spatially on the right of garbage. Our model captures both buses but assigns more attention to the red bus that is more obvious on the picture and predicts *no*, which is also a correct answer to the question. In addition, we found that in some cases our model's answer is comparatively more accurate than the annotated answer in dataset. For example, for first image of Fig. 3b, ***pen***, as a small area surrounded by fence to keep animal inside, is more accurate than the annotated answer ***yard***. Likewise, the bed and quilt are actually different in shape but both in white color, which makes our model's answer correct and the ground truth answer incorrect.

## 5 Conclusion

In conclusion, our REXUP network worked well in both capturing step-by-step reasoning process and detecting a complex object-relationship in photo-realistic images. Our proposed model has achieved the state-of-the-art performance on the GQA dataset, which proves the importance of structural and compositional relationships of objects in VQA tasks. Extracting the semantic information of scene graphs and encoding them via textual embeddings are efficient for the model to capture such structural relationships of objects. The parallel training of two branches with object region and scene graph features respectively help the model to develop comprehensive understanding of both images and questions.

## References


1. Anderson, P., et al.: Bottom-up and top-down attention for image captioning and visual question answering. In: IEEE Conference on Computer Vision and Pattern Recognition, pp. 6077–6086 (2018)
2. Antol, S., et al.: VQA: Visual question answering. In: IEEE International Conference On Computer Vision, pp. 2425–2433 (2015)
3. Ben-Younes, H., Cadene, R., Cord, M., Thome, N.: MUTAN: multimodal tucker fusion for visual question answering. In: IEEE International Conference on Computer Vision, pp. 2612–2620 (2017)
4. Ben-Younes, H., Cadene, R., Thome, N., Cord, M.: Block: bilinear superdiagonal fusion for visual question answering and visual relationship detection. In: AAAI Conference on Artificial Intelligence, vol. 33, pp. 8102–8109 (2019)
5. Cadene, R., Ben-Younes, H., Cord, M., Thome, N.: MUREL: multimodal relational reasoning for visual question answering. In: IEEE Conference on Computer Vision and Pattern Recognition, pp. 1989–1998 (2019)
6. Gao, P., et al.: Dynamic fusion with intra- and inter-modality attention flow for visual question answering. In: IEEE Conference on Computer Vision and Pattern Recognition (CVPR) (2019)

7. Goyal, Y., Khot, T., Summers-Stay, D., Batra, D., Parikh, D.: Making the V in VQA matter: elevating the role of image understanding in visual question answering. In: IEEE Conference on Computer Vision and Pattern Recognition, pp. 6904–6913 (2017)
8. Haurilet, M., Roitberg, A., Stiefelhagen, R.: It's not about the journey; it's about the destination: following soft paths under question-guidance for visual reasoning. In: IEEE Conference on Computer Vision and Pattern Recognition, pp. 1930–1939 (2019)
9. Hu, R., Andreas, J., Darrell, T., Saenko, K.: Explainable neural computation via stack neural module networks. In: European Conference on Computer Vision (ECCV), pp. 53–69 (2018)
10. Hu, R., Rohrbach, A., Darrell, T., Saenko, K.: Language-conditioned graph networks for relational reasoning. In: IEEE International Conference on Computer Vision, pp. 10294–10303 (2019)
11. Hudson, D.A., Manning, C.D.: Compositional attention networks for machine reasoning. In: International Conference on Learning Representations (2018)
12. Hudson, D.A., Manning, C.D.: GQA: a new dataset for real-world visual reasoning and compositional question answering. In: IEEE Conference on Computer Vision and Pattern Recognition, pp. 6700–6709 (2019)
13. Johnson, J., Hariharan, B., van der Maaten, L., Fei-Fei, L., Lawrence Zitnick, C., Girshick, R.: CLEVR: a diagnostic dataset for compositional language and elementary visual reasoning. In: IEEE Conference on Computer Vision and Pattern Recognition, pp. 2901–2910 (2017)
14. Kim, J.H., On, K.W., Lim, W., Kim, J., Ha, J.W., Zhang, B.T.: Hadamard product for low-rank bilinear pooling. In: International Conference on Learning Representations (2016)
15. Krishna, R., et al.: Visual genome: connecting language and vision using crowdsourced dense image annotations. Int. J. Comput. Vis. **123**(1), 32–73 (2017)
16. Liu, F., Liu, J., Fang, Z., Hong, R., Lu, H.: Densely connected attention flow for visual question answering. In: 28th International Joint Conference on Artificial Intelligence, pp. 869–875 (2019)
17. Lu, J., Yang, J., Batra, D., Parikh, D.: Hierarchical question-image co-attention for visual question answering. In: Advances in Neural Information Processing Systems, pp. 289–297 (2016)
18. Nguyen, D.K., Okatani, T.: Improved fusion of visual and language representations by dense symmetric co-attention for visual question answering. In: IEEE Conference on Computer Vision and Pattern Recognition, pp. 6087–6096 (2018)
19. Norcliffe-Brown, W., Vafeias, S., Parisot, S.: Learning conditioned graph structures for interpretable visual question answering. In: Advances in Neural Information Processing Systems, pp. 8334–8343 (2018)
20. Perez, E., Strub, F., De Vries, H., Dumoulin, V., Courville, A.: FiLM: visual reasoning with a general conditioning layer. In: Thirty-Second AAAI Conference on Artificial Intelligence (2018)
21. Shi, J., Zhang, H., Li, J.: Explainable and explicit visual reasoning over scene graphs. In: IEEE Conference on Computer Vision and Pattern Recognition, pp. 8376–8384 (2019)
22. Shrestha, R., Kafle, K., Kanan, C.: Answer them all! Toward universal visual question answering models. In: IEEE Conference on Computer Vision and Pattern Recognition, pp. 10472–10481 (2019)

23. Tan, H., Bansal, M.: LXMERT: learning cross-modality encoder representations from transformers. In: 2019 Conference on Empirical Methods in Natural Language Processing and the 9th International Joint Conference on Natural Language Processing (EMNLP-IJCNLP), pp. 5103–5114 (2019)
24. Yu, Z., Yu, J., Cui, Y., Tao, D., Tian, Q.: Deep modular co-attention networks for visual question answering. In: IEEE Conference on Computer Vision and Pattern Recognition, pp. 6281–6290 (2019)
25. Yu, Z., Yu, J., Fan, J., Tao, D.: Multi-modal factorized bilinear pooling with co-attention learning for visual question answering. In: IEEE International Conference on Computer Vision, pp. 1821–1830 (2017)
26. Zellers, R., Yatskar, M., Thomson, S., Choi, Y.: Neural motifs: scene graph parsing with global context. In: Proceedings of the IEEE Conference on Computer Vision and Pattern Recognition, pp. 5831–5840 (2018)

# Simultaneous Inpainting and Colorization via Tensor Completion

Tao Li and Jinwen Ma(✉)

Department of Information Science, School of Mathematical Sciences and LMAM, Peking University, Beijing, China
li_tao@pku.edu.cn, jwma@math.pku.edu.cn

**Abstract.** Both image inpainting and colorization can be considered as estimating certain missing pixel values from a given image, which is still a challenging problem in image processing. In fact, it is a more challenging problem to make image inpainting and colorization simultaneously, that is, given a corrupted gray-scale image and a few color pixels, we try to restore the original color image. In this paper, we propose a novel tensor completion model to solve the problem of simultaneous inpainting and colorization. Moreover, it can be applied to each of inpainting and colorization tasks separately as a special case. Experimental results on test images demonstrate that our proposed model is effective and efficient for simultaneous inpainting and colorization task and outperforms the state-of-the-art inpainting methods.

**Keywords:** Inpainting · Colorization · Tensor

## 1 Introduction

Image inpainting [2,14,20] and colorization [15,23,28] are two fundamental tasks in image processing and computer vision. For image inpainting, a corrupted color image is given, together with a mask indicating the pixels need to be inpainted, and the goal is to restore the image. For image colorization, a gray-scale image is given, together with a few labeled color pixels, and the task is to generate a color image consistent with the gray-scale image and the labeled pixels. It can be easily observed that both inpainting and colorization can be regarded as estimating missing values from a given image, which can be considered as a tensor completion problem.

In fact, tensors [13] have been extensively used in recommender systems, image processing [17,29], signal processing and machine learning [6,21]. For example, in recommender systems, the recorded ratings can be viewed as a 3-order tensor of $user \times item \times time$. In image processing, a color image is a 3-order tensor of $weight \times height \times channel$, and a video can be modeled as a 4-order tensor of $weight \times height \times channel \times time$. Due to technical reasons, tensors in most applications are incomplete. The tensor completion problem is to estimate the missing values in tensors, which is very important for


H. Yang et al. (Eds.): ICONIP 2020, LNCS 12532, pp. 533–543, 2020.
https://doi.org/10.1007/978-3-030-63830-6_45

practical applications. Clearly, image inpainting, and colorization can be easily transformed into tensor completion problems. Since the concept of a tensor is a natural extension of a matrix, various matrix completion techniques [3,19] have been extended to the tensor case [17,29], and the core assumption underlying these methods is the tensors are low-rank.

In practical applications of image processing, there is a very challenging problem that tries to make image inpainting and colorization simultaneously, that is, given a corrupted gray-scale image and a few color pixels, we need to restore the original color image, as illustrated in Fig. 1. In inpainting task, a relatively large proportion of the color image is known and the corresponding gray-scale image is unknown, while in the colorization task the gray-scale image is completely known but only a few color pixels are labeled. Can we restore the color image given a corrupted gray-scale image and a few color pixels? We refer to this task as simultaneous inpainting and colorization, which is obviously far more challenging. In this paper, we solve this simultaneous inpainting and colorization problem via the tensor completion model with constraints and penalties in which the tensor nuclear norm is penalized under the constraint that the tensor is consistent with the given or known information. Extensive experimental results demonstrate the proposed method can solve the simultaneous inpainting and colorization problem efficiently and effectively.

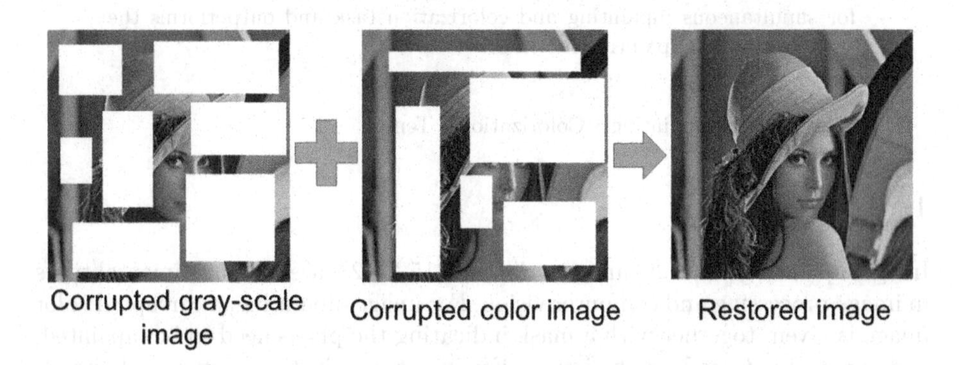


**Fig. 1.** Simultaneous inpainting and colorization problem.

## 2 Related Work

Traditional inpainting methods include anisotropic diffusion model [2], partial differential equation models [20], belief propagation [14] and so on. One seminal work of applying tensor completion techniques in inpainting is [17]. In [29], Zhao et al. proposed the Bayesian CP factorization of tensors with several applications in image processing. Recently, the total variation regularized reweighted low-rank tensor completion method was proposed in [16]. These studies are excellent for the inpainting task, but none of them considers the relationship between inpainting and colorization. In [15], Levin et al. modeled colorization problem as an optimization problem. Wang and Zhang proposed a matrix completion model for colorization in [23]. However, in [23], R/G/B channels were processed separately, thus the inner structure between channels was ignored.

There are also various deep learning based methods for inpainting [25–27] and colorization [4,7,28]. However, most deep learning based algorithms require a large amount of training data and computational costs. Our tensor completion method only requires a single image as input and obtain a satisfying result with several ADMM iterations, which is fast and efficient.

The theory and algorithms of matrix and tensor completion [5,9,10,24] has developed rapidly in recent years. However, most of these works emphasize on the general theoretical aspects, while this paper mainly concerns the application of tensor completion in image processing tasks. The recent work [8,11] apply matrix/tensor models to the background estimation problem, but we mainly consider image inpainting and colorization tasks in this work.

# 3 Proposed Method

## 3.1 Preliminaries and Notations

We only consider 3-order tensors in the following, since a color image can be naturally regarded as a 3-order tensor. We use upper case letters to denote matrices, e.g., $Z$, and use calligraphical letters for tensors, e.g., $\mathcal{X}$. Suppose $\mathcal{X}, \mathcal{Y} \in \mathbb{R}^{I_1 \times I_2 \times I_3}$, then we use $\langle \mathcal{X}, \mathcal{Y} \rangle$ to denote the standard inner product of $\mathcal{X}$ and $\mathcal{Y}$. The Frobenius norm of $\mathcal{X}$ is denoted by $\|\mathcal{X}\|_\mathrm{F}$.

We use $\mathcal{X} \times_3 S \in \mathbb{R}^{I_1 \times I_2 \times J}$ for the mode-3 multiplication between a tensor $\mathcal{X} \in \mathbb{R}^{I_1 \times I_2 \times I_3}$ and a matrix $S \in \mathbb{R}^{J \times I_3}$. More precisely, $(\mathcal{X} \times_3 S)_{i_1,i_2,j} = \sum_{i_3=1}^{I_3} x_{i_1,i_2,i_3} s_{j,i_3}$.

We can unfold a tensor $\mathcal{X} \in \mathbb{R}^{I_1 \times I_2 \times I_3}$ along the $k$-th mode to get a matrix $\mathcal{X}_{(k)}$. The inverse operator of unfolding is represented by $\mathrm{Fold}_k$, and it's obvious that $\mathrm{Fold}_k(\mathcal{X}_{(k)}) = \mathcal{X}$. For a matrix, the nuclear norm is defined as the summation of its singular values. For a tensor $\mathcal{X}$, the nuclear norm is defined as $\|\mathcal{X}\|_* = \sum_{i=1}^{3} \alpha_i \|\mathcal{X}_{(i)}\|_*$, where $\{\alpha_i\}_{i=1}^{3}$ are arbitrary constants satisfying $\sum_{i=1}^{3} \alpha_i = 1, \alpha_i \geq 0$. We always set $\alpha_i = 1/3$ in this paper.

Suppose $X = U\Sigma V'$ is the singular value decomposition of $X \in \mathbb{R}^{m \times n}$, where $\Sigma = \mathrm{Diag}(\sigma_1, \cdots, \sigma_m)$. The singular value thresholding operator is defined as $\mathbf{D}_\tau(X) = U\tilde{\Sigma}V'$, where $\tilde{\Sigma} = \mathrm{Diag}(\max(\sigma_1 - \tau, 0), \cdots, \max(\sigma_m - \tau, 0))$. Given a matrix $\Omega \in \{0,1\}^{m \times n}$, the projection of $X \in \mathbb{R}^{m \times n}$ to $\Omega$ is defined as $\mathbf{P}_\Omega(X) = X \circledast \Omega$, where $\circledast$ denotes Hadamard product (element-wise multiplication). We can also apply $\mathbf{P}_\Omega$ to a tensor $\mathcal{X}$ similarly when $\Omega$ has the same shape as $\mathcal{X}$.

## 3.2 Problem Formulation

Suppose we are given a partially observed gray-scale image $Z \in \mathbb{R}^{m \times n}$ together with a corrupted color image $\mathcal{O} \in \mathbb{R}^{m \times n \times 3}$, the goal is to restore the image

$\mathcal{X} \in \mathbb{R}^{m \times n \times 3}$. For the gray-scale image, the known pixels are indexed by $\Omega_g \in \{0,1\}^{m \times n}$, and the known pixels are indexed by $\Omega_c \in \{0,1\}^{m \times n \times 3}$ for the color image. We define the proportion of known pixels as

$$p_g = \frac{\|\Omega_g\|_{\mathrm{F}}^2}{m \times n}, p_c = \frac{\|\Omega_c\|_{\mathrm{F}}^2}{m \times n \times 3}. \tag{1}$$

In image inpainting task, usually the gray-scale image is unknown, while a majority of the color image pixels are known, i.e., $p_g = 0$ and $p_c$ is large. In image colorization task, the gray-scale image is complete and a few color pixels are known, i.e., $p_g = 1$ and $p_c$ is small. The problem we consider here is far more challenging: $p_g$ and $p_c$ can vary in $(0, 1]$ arbitrarily.

Low-rank assumption is essential for matrix and tensor completion. Following [17], we use nuclear norm $\|\mathcal{X}\|_*$ as a surrogate of the rank of $\mathcal{X}$. Furthermore, we assume the known color pixels are exact, while the known gray-scale pixels may be corrupted by Gaussian noise. Therefore, we restrict $\mathcal{X}$ in the feasible domain $\mathcal{Q} = \{\mathcal{X} \in \mathbb{R}^{m \times n \times 3} | \mathbf{P}_{\Omega_c}(\mathcal{X}) = \mathbf{P}_{\Omega_c}(\mathcal{O})\}$ and penalize the Frobenius norm of $\mathbf{P}_{\Omega_g}(\mathcal{X} \times_3 \lambda - Z)$. Here, $\lambda$ is the row vector consisting of the transformation coefficients from color space to gray-scale space. We work in RGB color space in the following, thus $\lambda = [0.2989, 0.5870, 0.1140]$. The problem of simultaneous inpainting and colorization can be formulated as follows,

$$\begin{aligned} \min \quad & \|\mathcal{X}\|_* + \frac{\mu}{2}\|\mathbf{P}_{\Omega_g}(\mathcal{X} \times_3 \lambda - Z)\|_{\mathrm{F}}^2, \\ \text{s.t.} \quad & \mathbf{P}_{\Omega_c}(\mathcal{X}) = \mathbf{P}_{\Omega_c}(\mathcal{O}). \end{aligned} \tag{2}$$

The parameter $\mu$ may be determined by our prior knowledge on how accurate the given gray-scale image is.

**Algorithm 1.** ADMM based algorithm for simultaneous inpainting and colorization

**Input:** $\Omega_c, \Omega_g, \mathcal{O}, Z$
**Hyper parameters:** $\mu, \rho, \delta$

1: Initialize $\mathcal{X}, \{\mathcal{Y}_i, \mathcal{M}_i\}_{i=1}^3$ by random guess
2: $\mathbf{P}_{\Omega_c}(\mathcal{X}) \leftarrow \mathbf{P}_{\Omega_c}(\mathcal{O})$
3: **while** not converged **do**
4: $\quad \mathcal{M}_i \leftarrow \mathrm{Fold}_i\left[\mathbf{D}_{\alpha_i/\rho}(\mathcal{X}_{(i)} - \frac{1}{\rho}\mathcal{Y}_{(i)})\right], \quad i = 1, 2, 3$
5: $\quad \mathcal{T} = \frac{1}{3}\sum_{i=1}^3(\mathcal{M}_i + \frac{1}{\rho}\mathcal{Y}_i)$
6: $\quad T = \mathcal{T} \times_3 \lambda$
7: $\quad \mathcal{X} = \mathcal{T} - \left(\frac{1}{\|\lambda\|_2^2 + \frac{\rho}{\mu}}\mathbf{P}_{\Omega_g}(T - Z) \times_3 \lambda'\right)$
8: $\quad \mathcal{Y}_i = \mathcal{Y}_i + \delta(\mathcal{M}_i - \mathcal{X}), \quad i = 1, 2, 3$
9: **end while**
10: return $\mathcal{X}$

### 3.3 ADMM Based Algorithm

We apply ADMM algorithm to tackle the optimization problem (2). Introducing auxiliary tensor variables $\{\mathcal{M}_i\}_{i=1}^3$, problem (2) is equivalent to

$$\begin{aligned} \min \quad & \sum_{i=1}^{3} \alpha_i \|\mathcal{M}_{i_{(i)}}\|_* + \frac{\mu}{2}\|\mathbf{P}_{\Omega_g}(\mathcal{X}) \times_3 \lambda - Z\|_{\mathrm{F}}^2, \\ \text{s.t.} \quad & \mathcal{X} = \mathcal{M}_i, \quad i = 1, 2, 3 \quad \text{and} \quad \mathcal{X} \in \mathcal{Q} \end{aligned} \tag{3}$$

The augmented Lagrangian function is:

$$\mathcal{L} = \sum_{i=1}^{3} \left[\alpha_i \|\mathcal{M}_{i_{(i)}}\|_* + \langle \mathcal{Y}_i, \mathcal{M}_i - \mathcal{X}\rangle + \frac{\rho}{2}\|\mathcal{M}_i - \mathcal{X}\|_{\mathrm{F}}^2\right] + \frac{\mu}{2}\|\mathbf{P}_{\Omega_g}(\mathcal{X}) \times_3 \lambda - Z\|_{\mathrm{F}}^2. \tag{4}$$

According to the ADMM algorithm, we can iterate the variables alternately using the following scheme:

$$\mathcal{M}_i^{t+1} = \arg\min \frac{\alpha_i}{\rho}\|\mathcal{M}_{i_{(i)}}^t\|_* + \frac{1}{2}\|\mathcal{M}_i^t - \mathcal{X}^t + \frac{1}{\rho}\mathcal{Y}_i\|_{\mathrm{F}}^2, \tag{5}$$

$$\mathcal{X}^{t+1} = \arg\min_{\mathcal{X}\in\mathcal{Q}} -\langle \sum_{i=1}^{3} \mathcal{Y}_i^t, \mathcal{X}\rangle + \sum_{i=1}^{3} \frac{\rho}{2}\|\mathcal{M}_i^{t+1} - \mathcal{X}\|_{\mathrm{F}}^2 + \frac{\mu}{2}\|\mathbf{P}_{\Omega_g}(\mathcal{X}) \times_3 \lambda - Z\|_{\mathrm{F}}^2 \tag{6}$$

$$\mathcal{Y}_i^{t+1} = \mathcal{Y}_i^t + \delta(\mathcal{M}_i^{t+1} - \mathcal{X}^{t+1}). \tag{7}$$

As indicated in [17], the optimal solution of $\mathcal{M}_i$ is:

$$\mathrm{Fold}_i\left[\mathbf{D}_{\alpha_i/\rho}(\mathcal{X}_{(i)}^t - \frac{1}{\rho}\mathcal{Y}_{(i)}^t)\right]. \tag{8}$$

For $\mathcal{X} \in \mathcal{Q}$, we need to find the minimum of

$$-\langle \sum_{i=1}^{3} \mathcal{Y}_i^t, \mathcal{X}\rangle + \sum_{i=1}^{3} \frac{\rho}{2}\|\mathcal{M}_i^{t+1} - \mathcal{X}\|_{\mathrm{F}}^2 + \frac{\mu}{2}\|\mathbf{P}_{\Omega_g}(\mathcal{X}) \times_3 \lambda - Z\|_{\mathrm{F}}^2, \tag{9}$$

We consider the corresponding unconstrained problem, then project the solution to $\mathcal{Q}$. If $\Omega(i,j) = 0$, then $\mathcal{X}(i,j,k)^{t+1}$ is the minimum of

$$\sum_{l=1}^{3}\sum_{k=1}^{3} \left[\frac{\rho}{2}\left(\mathcal{X}(i,j,k) - \mathcal{M}_l^t(i,j,k)\right)^2 - \mathcal{X}(i,j,k)\mathcal{Y}_l^t(i,j,k)\right]. \tag{10}$$

If $\Omega(i,j) = 1$, then $\mathcal{X}(i,j,k)^{t+1}$ is the minimum of

$$\begin{aligned} &\sum_{l=1}^{3}\sum_{k=1}^{3} \left[\frac{\rho}{2}\left(\mathcal{X}(i,j,k) - \mathcal{M}_l^t(i,j,k)\right)^2 - \mathcal{X}(i,j,k)\mathcal{Y}_l^t(i,j,k)\right] \\ &\qquad + \frac{\mu}{2}\left(\sum_{k=1}^{3} \lambda_k \mathcal{X}(i,j,k) - Z(i,j)\right)^2. \end{aligned} \tag{11}$$

Both (10) and (11) are unconstrained quadratic programming, thus have closed-form solution. The detailed algorithm is given in Algorithm 1. The problem itself is convex, one can hope the general convergence result of ALM also holds for this algorithm. However, the difficult part is the so-called exact-recovery property, which is also very important in Robust PCA. It is not trivial to extend existing exact-recovery results for this model.

# 4 Experiments

## 4.1 Inpainting

Our method applies to image inpainting when a gray-scale image is not given, i.e., $\Omega_g = 0$. To validate the effectiveness of the proposed algorithm, we compare it with four inpainting algorithms. Navier-Stokes equation based inpainting (NS) [1] and fast marching based inpainting (FM) [22] are implemented in OpenCV [12]. We also consider two PDE based methods: Harmonic model (HA) and Mumford Shah model (MS) [18,20]. In this experiment, the mask $\Omega_c$ is generated by randomly selecting $p_c$ percent of the pixels, and $p_c$ varies in $\{20, 50, 80\}$. We use signal-noise-ratio (SNR) to measure the performances of various algorithms and record the running time. All tests are performed on Intel Core i7 2.60 GHz and 8 GB RAM computer. All images in this experiment are resized to $512 \times 820$.

**Table 1.** Comparison with other methods on image inpainting task.

| | | NS | | FM | | HA | | MS | | Proposed | |
|---|---|---|---|---|---|---|---|---|---|---|---|
| | | SNR | Time | SNR | Time | SNR | Time | SNR | Time | SNR | Time |
| Test1 | 20% | 19.3531 | 45.91 | 19.0108 | 47.81 | 19.7814 | 14.07 | **20.0837** | 154.44 | 19.6307 | 65.71 |
| | 50% | 23.7776 | 99.44 | 22.1123 | 109.55 | 23.5746 | 5.72 | 24.285 | 177.25 | **24.9674** | 28.99 |
| | 80% | 29.3645 | 3.69 | 27.1132 | 24.07 | 28.0483 | 3.51 | 29.8671 | 158.71 | **31.2528** | 20.04 |
| Test2 | 20% | 21.0029 | 45.34 | 20.3325 | 44.70 | 21.3168 | 15.46 | **21.8731** | 162.49 | 20.2820 | 121.32 |
| | 50% | 26.7146 | 100.66 | 23.9512 | 96.97 | 26.4812 | 6.04 | 27.7183 | 193.11 | **28.4775** | 39.67 |
| | 80% | 33.2443 | 3.55 | 29.5071 | 23.38 | 31.9467 | 3.29 | 34.4596 | 163.16 | **37.9288** | 22.65 |
| Test3 | 20% | 15.8646 | 44.89 | 15.1005 | 45.07 | 16.1002 | 18.52 | 16.6838 | 167.40 | **17.8480** | 112.32 |
| | 50% | 21.8975 | 99.92 | 18.7328 | 102.69 | 21.476 | 6.77 | 22.8321 | 191.02 | **26.2275** | 37.30 |
| | 80% | 28.8432 | 3.59 | 24.5412 | 22.70 | 27.2031 | 3.32 | 29.9101 | 162.33 | **37.3820** | 25.79 |

The results are shown in Fig. 2 and Table 1. In most cases, the proposed method outperforms other algorithms in terms of SNR. Furthermore, although Mumford Shah model has better SNR occasionally, our algorithm is significantly faster than Mumford Shah model. Therefore, we claim the proposed method applies well to the inpainting task.

## 4.2 Colorization

When $p_g = 1$ and $p_c$ is relatively small, the proposed method applies to colorization problem. We set $p_c = 1\%, 3\%, 5\%$ respectively, and the colorized images

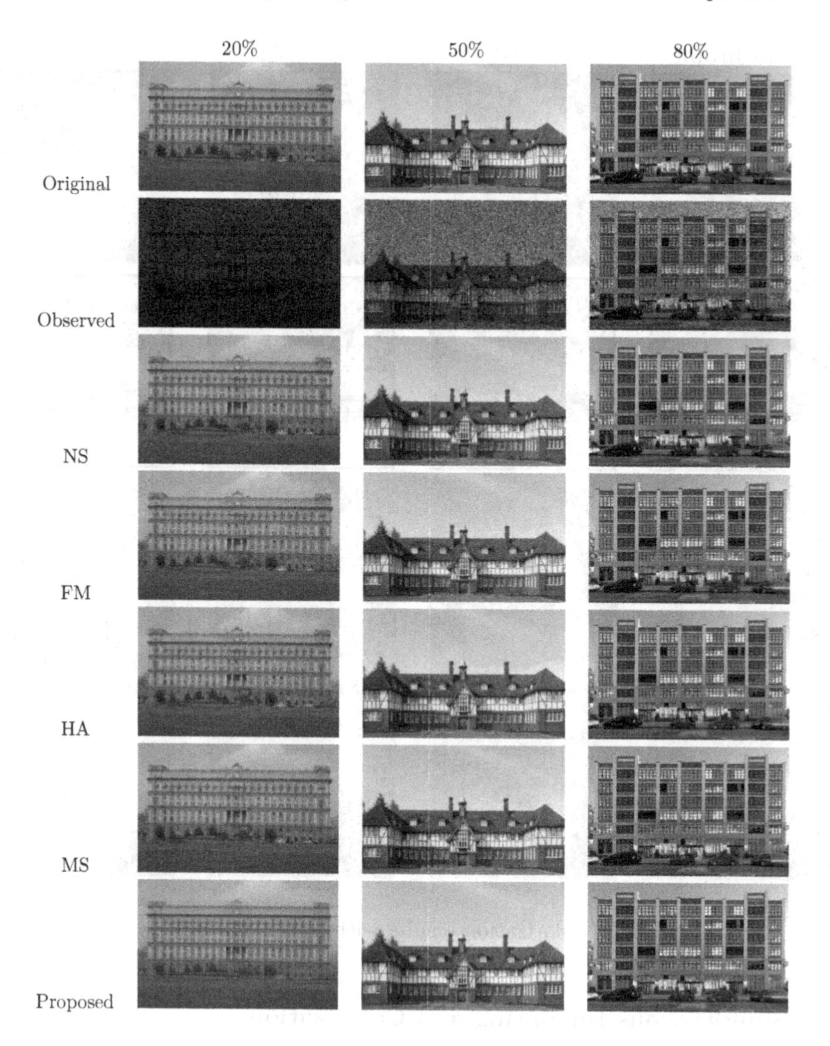


**Fig. 2.** Comparison with other methods on image inpainting task. From top to bottom: test1 with 20% pixels observed, test2 with 50% pixels observed, test3 with 80% pixels observed. Please zoom in to see more details.

are shown in Fig. 3. For experimental purposes, the images we used in this part have corresponding ground-truth color images, and all the color pixels labels are given by the ground-truth image. From Fig. 3, we notice that when $p_c = 1\%$ the generated color image seems a little unrealistic. However, once $p_c \geq 3\%$, the obtained color images are vivid. Note that it's also possible to combine the proposed method with other colorization methods. For example, we can use the local color consistency assumption to increase the proportion of known color pixels as in [23].

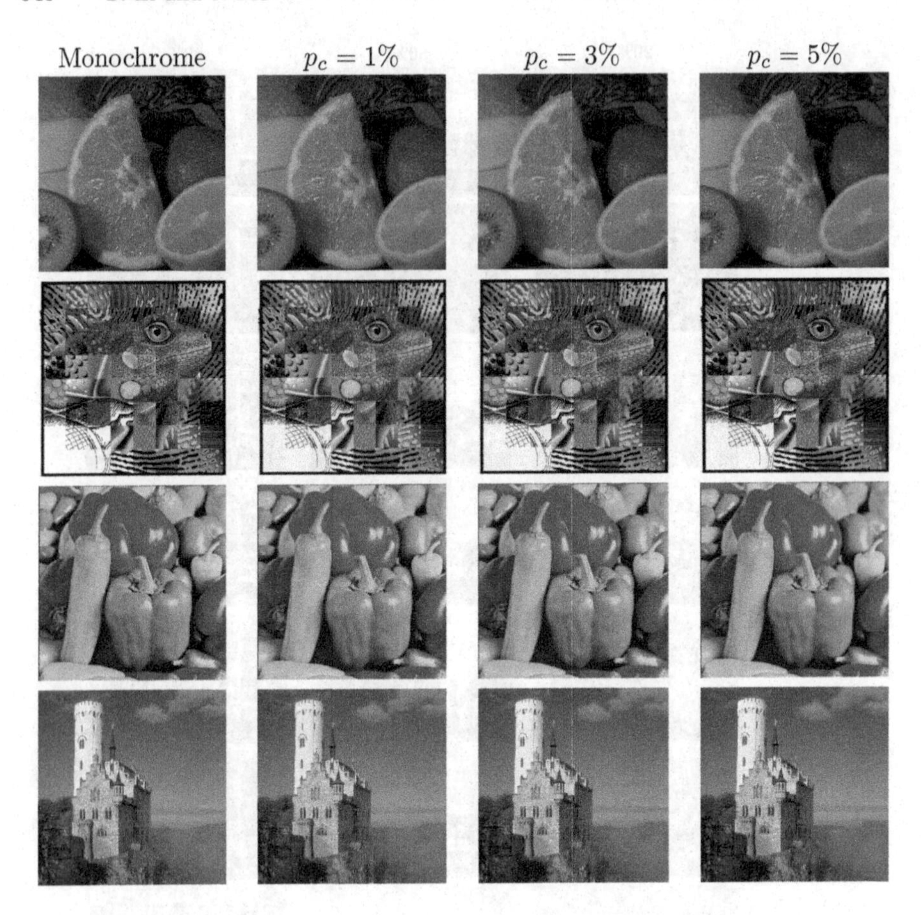


**Fig. 3.** Colorization results with $p_c = 1\%, 3\%, 5\%$.

### 4.3 Simultaneous Inpainting and Colorization

We evaluate the proposed method on the simultaneous inpainting and colorization task in this part. Unlike inpainting and colorization, the known proportions of pixels of a color image and gray-scale image are both small. In Fig. 4, we let $p_c, p_g$ varies in $\{5\%, 10\%, 15\%, 20\%, 25\%\}$ and show the corresponding restored image. It's almost impossible to recover a color image by inpainting techniques if only $\sim$10% of the pixels are available. However, if an extra corrupted (also $\sim$10% known) gray-scale image is provided, we can restore the image with relatively high SNR, as shown in Fig. 4. All the masks are generated by randomly selecting pixels.

We also show the SNR curves with respect to $p_c$ and $p_g$ in Fig. 5. From this figure, we conclude that the more information we have, the more accurate the result is, which is consistent with our intuitions. Besides, we find that color pixels provide more information than gray-pixels. When $p_c$ is fixed, SNR grows slowly as we increase $p_g$. However, when $p_g$ is fixed, SNR grows relatively rapidly with respect to $p_c$.

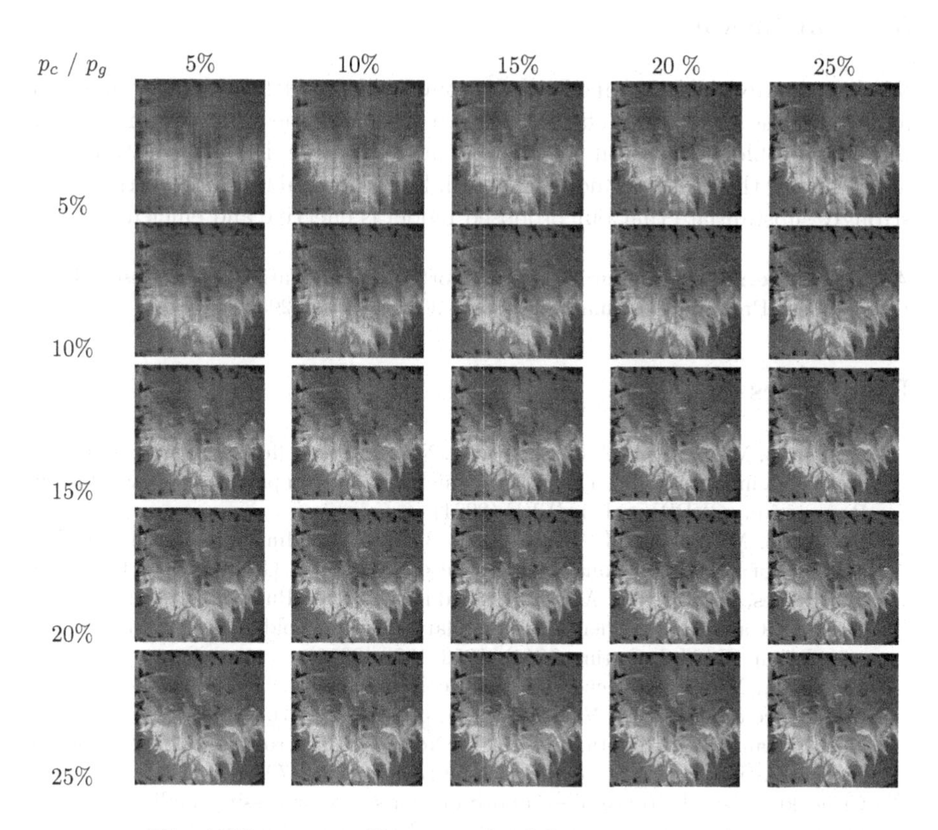


**Fig. 4.** The recovered image under different settings of $p_c, p_g$

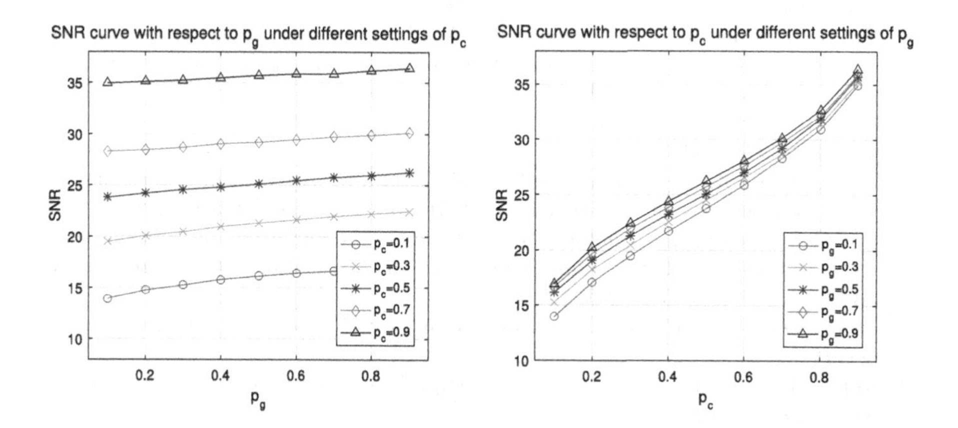


**Fig. 5.** SNR curves with respect to $p_g$ and $p_c$.

## 5 Conclusion

We have established a unified tensor completion model for image inpainting and colorization which can effectively solve the simultaneous inpainting and colorization problem. Based on the ADMM procedure, we derive the optimization algorithm for the proposed model in detail. Experimental results on various tasks strongly substantiate that our proposed model is effective and efficient.

**Acknowledgements.** This work was supported by the National Key Research and Development Program of China under grant 2018AAA0100205.

## References

1. Bertalmio, M., Bertozzi, A.L., Sapiro, G.: Navier-stokes, fluid dynamics, and image and video inpainting. In: The IEEE Conference on Computer Vision and Pattern Recognition (CVPR), vol. 1. IEEE (2001)
2. Bertalmio, M., Sapiro, G., Caselles, V., Ballester, C.: Image inpainting. In: Proceedings of the 27th Annual Conference on Computer Graphics and Interactive Techniques, pp. 417–424. ACM Press/Addison-Wesley Publishing Co. (2000)
3. Cai, J.F., Candès, E.J., Shen, Z.: A singular value thresholding algorithm for matrix completion. SIAM J. Optim. **20**(4), 1956–1982 (2010)
4. Cheng, Z., Yang, Q., Sheng, B.: Deep colorization. In: The IEEE International Conference on Computer Vision (ICCV), pp. 415–423 (2015)
5. Cherapanamjeri, Y., Gupta, K., Jain, P.: Nearly optimal robust matrix completion. In: International Conference on Machine Learning, pp. 797–805 (2017)
6. Cichocki, A., et al.: Tensor decompositions for signal processing applications: from two-way to multiway component analysis. IEEE Signal Process. Mag. **32**(2), 145–163 (2015)
7. Deshpande, A., Lu, J., Yeh, M.C., Chong, M.J., Forsyth, D.A.: Learning diverse image colorization. In: The IEEE Conference on Computer Vision and Pattern Recognition (CVPR), pp. 2877–2885 (2017)
8. Dutta, A., Gong, B., Li, X., Shah, M.: Weighted singular value thresholding and its application to background estimation. arXiv preprint arXiv:1707.00133 (2017)
9. Dutta, A., Hanzely, F., Richtárik, P.: A nonconvex projection method for robust PCA. In: Proceedings of the AAAI Conference on Artificial Intelligence, vol. 33, pp. 1468–1476 (2019)
10. Dutta, A., Li, X.: A fast weighted SVT algorithm. In: 2018 5th International Conference on Systems and Informatics (ICSAI), pp. 1022–1026. IEEE (2018)
11. Dutta, A., Richtárik, P.: Online and batch supervised background estimation via l1 regression. In: 2019 IEEE Winter Conference on Applications of Computer Vision (WACV), pp. 541–550. IEEE (2019)
12. Itseez: Open source computer vision library (2015). https://github.com/itseez/opencv
13. Kolda, T.G., Bader, B.W.: Tensor decompositions and applications. SIAM Rev. **51**(3), 455–500 (2009)
14. Komodakis, N.: Image completion using global optimization. In: The IEEE Conference on Computer Vision and Pattern Recognition (CVPR), vol. 1, pp. 442–452. IEEE (2006)

15. Levin, A., Lischinski, D., Weiss, Y.: Colorization using optimization. ACM Trans. Graph. **23**, 689–694 (2004)
16. Li, L., Jiang, F., Shen, R.: Total variation regularized reweighted low-rank tensor completion for color image inpainting. In: The 25th IEEE International Conference on Image Processing (ICIP), pp. 2152–2156. IEEE (2018)
17. Liu, J., Musialski, P., Wonka, P., Ye, J.: Tensor completion for estimating missing values in visual data. IEEE Trans. Pattern Anal. Mach. Intell. **35**(1), 208–220 (2013)
18. Parisotto, S., Schönlieb, C.: MATLAB/Python codes for the Image Inpainting Problem. GitHub repository, MATLAB Central File Exchange, September 2016
19. Salakhutdinov, R., Mnih, A.: Bayesian probabilistic matrix factorization using Markov chain Monte Carlo. In: Proceedings of the 25th International Conference on Machine Learning (ICML), pp. 880–887. ACM (2008)
20. Schönlieb, C.B.: Partial Differential Equation Methods for Image Inpainting, vol. 29. Cambridge University Press, Cambridge (2015)
21. Sidiropoulos, N.D., De Lathauwer, L., Fu, X., Huang, K., Papalexakis, E.E., Faloutsos, C.: Tensor decomposition for signal processing and machine learning. IEEE Trans. Signal Process. **65**(13), 3551–3582 (2017)
22. Telea, A.: An image inpainting technique based on the fast marching method. J. Graph. Tools **9**(1), 23–34 (2004)
23. Wang, S., Zhang, Z.: Colorization by matrix completion. In: AAAI (2012)
24. Wright, J., Ganesh, A., Rao, S., Peng, Y., Ma, Y.: Robust principal component analysis: exact recovery of corrupted low-rank matrices via convex optimization. In: Advances in Neural Information Processing Systems, pp. 2080–2088 (2009)
25. Xie, J., Xu, L., Chen, E.: Image denoising and inpainting with deep neural networks. In: Advances in Neural Information Processing Systems (NIPS), pp. 341–349 (2012)
26. Yang, C., Lu, X., Lin, Z., Shechtman, E., Wang, O., Li, H.: High-resolution image inpainting using multi-scale neural patch synthesis. In: The IEEE Conference on Computer Vision and Pattern Recognition (CVPR), vol. 1, p. 3 (2017)
27. Yeh, R.A., Chen, C., Lim, T.Y., Schwing, A.G., Hasegawa-Johnson, M., Do, M.N.: Semantic image inpainting with deep generative models. In: The IEEE Conference on Computer Vision and Pattern Recognition (CVPR), vol. 2, p. 4 (2017)
28. Zhang, R., Isola, P., Efros, A.A.: Colorful image colorization. In: Leibe, B., Matas, J., Sebe, N., Welling, M. (eds.) ECCV 2016. LNCS, vol. 9907, pp. 649–666. Springer, Cham (2016). https://doi.org/10.1007/978-3-319-46487-9_40
29. Zhao, Q., Zhang, L., Cichocki, A.: Bayesian CP factorization of incomplete tensors with automatic rank determination. IEEE Trans. Pattern Anal. Mach. Intell. **37**(9), 1751–1763 (2015)

# Take a NAP: Non-Autoregressive Prediction for Pedestrian Trajectories

Hao Xue[1,2(✉)], Du Q. Huynh[1], and Mark Reynolds[1]

[1] The University of Western Australia, Perth, Australia
{du.huynh,mark.reynolds}@uwa.edu.au
[2] RMIT University, Melbourne, Australia
hao.xue@rmit.edu.au

**Abstract.** Pedestrian trajectory prediction is a challenging task as there are three properties of human movement behaviors which need to be addressed, namely, the social influence from other pedestrians, the scene constraints, and the multimodal (multi-route) nature of predictions. Although existing methods have explored these key properties, the prediction process of these methods is autoregressive. This means they can only predict future locations sequentially. In this paper, we present NAP, a non-autoregressive method for trajectory prediction. Our method comprises specifically designed feature encoders and a latent variable generator to handle the three properties above. It also has a future-time-agnostic context generator and a future-time-oriented context generator for non-autoregressive prediction. Through extensive experiments that compare NAP against eleven recent methods, we show that NAP achieves state-of-the-art trajectory prediction performance.

**Keywords:** Trajectory prediction · Non-autoregressive · Motion analysis

## 1 Introduction

Pedestrian trajectory prediction is an important component in a range of applications such as social robots and self-driving vehicles, and plays a key role in understanding human movement behaviors. This task is not trivial due to three key properties in pedestrian trajectory prediction: (i) **social interaction:** People do not always walk alone in public places. Pedestrians often socially interact with others to avoid collisions, walk with friends and keep a certain distance from strangers; (ii) **environmental scene constraints:** Besides social interaction, pedestrians' routes also need to obey scene constraints such as obstacles and building layouts; (iii) **multimodal nature of future prediction:** People can follow different routes as long as these routes are both socially and environmentally acceptable. For example, a person can choose to turn right or turn left to bypass an obstacle.

Recently, researchers have made progress in incorporating these properties into the trajectory prediction process. For example, the Social LSTM model [1]


H. Yang et al. (Eds.): ICONIP 2020, LNCS 12532, pp. 544–556, 2020.
https://doi.org/10.1007/978-3-030-63830-6_46

is one of the methods that can capture social influence information in a local neighborhood around each pedestrian. Based on the generative model GAN [4], the SGAN model proposed by Gupta *et al.* [7] can handle multimodality in the prediction process while also capturing the social influence from other pedestrians in the scene. To deal with the scene constraints, Convolutional Neural Networks (CNNs) are often used to extract scene information in the trajectory prediction network, such as SS-LSTM [22], SoPhie [18], and Social-BiGAT [11].

While other methods like [8,14,19,20,23,24,26] miss one or two aforementioned key properties, SoPhie [18], Liang *et al.* [15], and Social-BiGAT [11] are three typical papers that have taken all three properties into consideration. However, these methods predict the future locations recurrently. There are two main limitations in using autoregression to generate trajectory prediction: (i) the autoregressive prediction process works in a recursive manner and so the prediction errors accumulated from previous time steps are passed to the prediction for the next time step; (ii) the process cannot be parallelized, *i.e.*, predictions must be generated sequentially.

To overcome the above limitations and inspired by the application of non-autoregressive decoder in other areas such as machine translation [5,6] and time series forecasting [21], we propose a novel trajectory prediction method that can predict future trajectories non-autoregressively. We name our method *NAP* (short for *N*on-*A*utoregressive *P*rediction). Our research contributions are threefold: (a) To the best of our knowledge, we are the first to explore non-autoregressive trajectory prediction. The network architecture of NAP includes trainable context generators to ensure that context vectors are available for the non-autoregressive decoder to forecast good quality predictions. The state-of-the-art performance of NAP is demonstrated through the extensive experiments and ablation study conducted. (b) Both the social and scene influences are handled by NAP through specially designed feature encoders. The social influence is captured by social graph features propagated through an LSTM; the scene influence is modeled by a CNN. The effectiveness of these encoders is confirmed from the performance of NAP. (c) Unlike previous work in the literature, NAP tackles multimodal predictions by training a latent variable generator to learn the latent variables of the sampling distribution for each pedestrian's trajectory. The generator is shown to give NAP top performance in multimodal predictions.

## 2 Proposed Method

### 2.1 Problem Definition and System Overview

Pedestrian trajectory prediction is defined as forecasting the future trajectory of the person $i$ given his/her observed trajectory. We assume that trajectories have already been obtained in the format of time sequences of coordinates (*i.e.*, $\mathbf{u}_t^i = (x_t^i, y_t^i) \in \mathbb{R}^2, \forall i$). The lengths of the observed trajectories and predicted trajectories are represented by $T_\text{o}$ and $T_\text{p}$. Thus, considering an observed trajectory $X^i = \{\mathbf{u}_t^i \,|\, t = 1, \cdots, T_\text{o}\}$, our target is to generate the prediction $\hat{Y}^i = \{\hat{\mathbf{u}}_t^i \,|\, t = T_\text{o}+1, \cdots, T_\text{o}+T_\text{p}\}$.

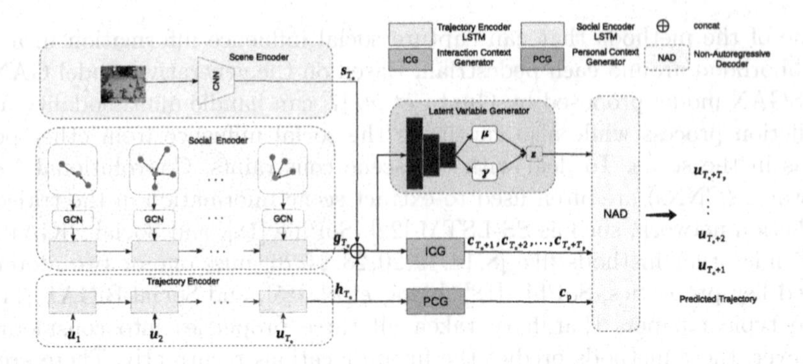


**Fig. 1.** The network architecture of NAP. There are three encoders to extract features, two context generators to generate context vectors for the non-autoregressive decoder (NAD), and a latent variable generator to handle multimodal predictions. The embedding layer and the superscript $i$ are not shown to simplify visualization.

Our proposed NAP comprises four major parts (Fig. 1): (i) three feature encoders for encoding the input information such as observed trajectories and scene images (Sect. 2.2); (ii) two context generators to yield context vectors for prediction (Sect. 2.3); (iii) a latent variable generator for multimodality (Sect. 2.4); (iv) a non-autoregressive decoder for predicting future trajectories (Sect. 2.5). Details of these parts are described in the following subsections.

## 2.2 Feature Encoders

In NAP, there are three feature encoders: a trajectory encoder, to learn the representation of the observed history movement of each pedestrian; a social encoder, to learn the representation of the influence from other pedestrians; and a scene encoder, to learn the representation of the scene features.

**Trajectory Encoder.** The coordinates of the $i^{\text{th}}$ pedestrian in the observation phase ($t = 1, \cdots, T_{\text{o}}$) are firstly embedded into a higher dimensional space through an embedding layer $\phi(\cdot)$. Then, across different time steps, the embedded features are used as inputs of an LSTM layer (denoted by $\text{LSTM}_{\text{ENC}}(\cdot)$) to get the encoded hidden state $\mathbf{h}_t^i$ which captures the observed path information. This trajectory encoding is given by:

$$\mathbf{e}_t^i = \phi(x_t^i, y_t^i; \mathbf{W}_{\text{EMB}}), \tag{1}$$

$$\mathbf{h}_t^i = \text{LSTM}_{\text{ENC}}(\mathbf{h}_{t-1}^i, \mathbf{e}_t^i; \mathbf{W}_{\text{ENC}}), \tag{2}$$

where $\mathbf{W}_{\text{EMB}}$ and $\mathbf{W}_{\text{ENC}}$ are trainable weights of the corresponding layers.

**Social Encoder.** At each time step $t$, NAP captures the social influence on the $i^{\text{th}}$ pedestrian through a graph $\mathcal{G}_t^i = (V_t^i, E_t^i)$. The $i^{\text{th}}$ pedestrian and all other pedestrians $\mathcal{N}_t^{(i)}$ at the same time step are considered as nodes in the set $V_t^i$. Edges linking the $i^{\text{th}}$ pedestrian and pedestrians in $\mathcal{N}_t^{(i)}$ form the edge set $E_t^i$.

We then use a graph convolutional network (GCN) to process these graphs. In the $\ell^{\text{th}}$ graph convolutional layer, the node feature of pedestrian $i$ is aggregated as follows:

$$\mathbf{a}_t^{i,\ell} = \text{ReLU}\left(\mathbf{b}^\ell + \frac{1}{|\mathcal{N}_t^{(i)}|} \sum_{j \in \mathcal{N}_t^{(i)}} \mathbf{W}^\ell \mathbf{a}_t^{j,\ell-1}\right), \tag{3}$$

where $\mathbf{W}^\ell$ and $\mathbf{b}^\ell$ are the weight matrix and bias term. At the first layer, we initialize the node feature $\mathbf{a}_t^{i,0}$ as the location coordinates of the $i^{\text{th}}$ pedestrian, *i.e.*, $\mathbf{a}_t^{i,0} = (x_i^t, y_i^t)$.

The social graph feature $\mathbf{g}_t^i$ (Eq. (4)) is designed to model the surrounding (or *social*) information of pedestrian $i$ at each time step $t$. To compute this feature, the node features, denoted by $\{\mathbf{a}_t^i \,|\, t = 1, \cdots, T_\text{o}\}$, from the final GCN layer across all the time steps in the observation phase are passed through an LSTM layer with trainable weights $\mathbf{W}_\text{SG}$, *i.e.*,

$$\mathbf{g}_t^i = \text{LSTM}_\text{SG}(\mathbf{g}_{t-1}^i, \mathbf{a}_t^i; \mathbf{W}_\text{SG}). \tag{4}$$

**Scene Encoder.** Different from other methods (such as [18,22]) that process each image frame in the observation phase, the scene encoder of NAP takes only image $I_{T_\text{o}}^i$ as input, since the scene encoder focuses on the static information like scene layouts and obstacles. Not only does this save the computation time, but it also supplies the most up-to-date and sufficient scene context before prediction kicks in at $t = T_\text{o} + 1$. We use a CNN to model the scene feature $\mathbf{s}_{T_\text{o}}^i$ as follows:

$$\mathbf{s}_{T_\text{o}}^i = \text{CNN}(I_{T_\text{o}}^i; \mathbf{W}_\text{CNN}). \tag{5}$$

In the CNN$(\cdot)$, we take the merit of the state-of-the-art semantic segmentation architecture DeepLabv3+ [2] and adopt their encoder module to extract semantic features. Similar to SoPhie [18], the encoded semantic features is then passed through a convolutional layer and embedded to the scene feature $\mathbf{s}_i^{T_\text{o}}$ through a single fully connected layer. The encoder module of DeepLabv3+ used in our CNN$(\cdot)$ is initialized with the weight matrix that has been pre-trained on the Cityscapes dataset [3]. The scene encoder (CNN$(\cdot)$) is then trained together with other parts of NAP.

### 2.3 Context Generators

The role of the context generators is to aggregate the outputs of the feature encoders for the downstream decoder for trajectory forecasting. We use two context generators in NAP: (i) a *personal context generator* that is *future-time-agnostic*, as its input is the hidden state $\mathbf{h}_t^i$ computed from the $i^{\text{th}}$ pedestrian's own trajectory only; (ii) an *interaction context generator* that is *future-time-oriented* as its input includes both social graph and scene interaction features also.

**Personal Context Generator (PCG).** We use a Multi-Layer Perceptron (MLP) to model this context generator. The output context vector $\mathbf{c}_{\mathrm{p}}^{i}$ is computed as

$$\mathbf{c}_{\mathrm{p}}^{i} = \mathrm{MLP}_{\mathrm{A}}(\mathbf{h}_{T_{\mathrm{o}}}^{i}; \mathbf{W}_{\mathrm{A}}), \tag{6}$$

where $\mathbf{W}_{\mathrm{A}}$ is the corresponding weight matrix. The context $\mathbf{c}_{\mathrm{p}}^{i}$ captures the "personal" cues such as the pedestrian's preferred walking speed and direction in the observation phase, oblivious of his/her surrounding. This context is *future-time-agnostic* because, without considering the social and scene influences, such personal cues can remain the same for the entire trajectory, *e.g.*, the pedestrian can continue to walk in a straight line or at a fast pace with no penalty when bumping into obstacles or other pedestrians since both the social graph and scene features are not present in the equation.

**Interaction Context Generator (ICG).** This context generator incorporates both the social graph and scene features. These two types of influences allow the context generator to be future-time-oriented, *e.g.*, while a pedestrian can walk at a fast pace in the initial part of his/her trajectory, he/she would need to slow down or detour at a later part of the trajectory in order to avoid other pedestrians or scene obstacles. Similar to PCG, we use an MLP to model ICG but its input, being the concatenation of $\mathbf{h}_{T_{\mathrm{o}}}^{i}$, $\mathbf{g}_{T_{\mathrm{o}}}^{i}$, and $\mathbf{s}_{T_{\mathrm{o}}}^{i}$, contains richer information. The output of ICG comprises different context vectors for different time steps in the prediction phase, as given below:

$$(\mathbf{c}_{T_{\mathrm{o}}+1}^{i}, \mathbf{c}_{T_{\mathrm{o}}+2}^{i}, \cdots, \mathbf{c}_{T_{\mathrm{o}}+T_{\mathrm{p}}}^{i}) = \mathrm{MLP}_{\mathrm{B}}(\mathbf{h}_{T_{\mathrm{o}}}^{i} \oplus \mathbf{g}_{T_{\mathrm{o}}}^{i} \oplus \mathbf{s}_{T_{\mathrm{o}}}^{i}; \mathbf{W}_{\mathrm{B}}), \tag{7}$$

where $\mathbf{W}_{\mathrm{B}}$ is the corresponding weight matrix.

### 2.4 Latent Variable Generator

For multimodal prediction, NAP is designed to generate multiple trajectories through the latent variables $\boldsymbol{\mu}$ and $\boldsymbol{\gamma}$ (see Fig. 1). Although several existing trajectory prediction methods such as [7,9,12,14] also use latent variables to handle multimodality, the latent variables in these methods are either directly sampled from the normal distribution or a multivariate normal distribution conditioned on the observed trajectories. To make our latent variables more aware of the social and scene cues, we design NAP to learn the parameters ($\boldsymbol{\mu}_i$ and $\boldsymbol{\gamma}_i$) of the sampling distribution from the observed trajectories, the social influence, and the scene influence features. To this end, the concatenated feature $\mathbf{h}_{T_{\mathrm{o}}}^{i} \oplus \mathbf{g}_{T_{\mathrm{o}}}^{i} \oplus \mathbf{s}_{T_{\mathrm{o}}}^{i}$ is passed to two different MLPs (Eqs. (8)–(9)) to yield the mean vector $\boldsymbol{\mu}_i$ and logarithmic variance $\boldsymbol{\gamma}_i$ and finally $\mathbf{z}_i$ for the downstream non-autoregressive decoder:

$$\boldsymbol{\mu}_i = \mathrm{MLP}_{\mu}(\mathbf{h}_{T_{\mathrm{o}}}^{i} \oplus \mathbf{g}_{T_{\mathrm{o}}}^{i} \oplus \mathbf{s}_{T_{\mathrm{o}}}^{i}; \mathbf{W}_{\mu}), \tag{8}$$

$$\log \boldsymbol{\sigma}_i^2 \triangleq \boldsymbol{\gamma}_i = \mathrm{MLP}_{\sigma}(\mathbf{h}_{T_{\mathrm{o}}}^{i} \oplus \mathbf{g}_{T_{\mathrm{o}}}^{i} \oplus \mathbf{s}_{T_{\mathrm{o}}}^{i}; \mathbf{W}_{\sigma}), \tag{9}$$

$$\mathbf{z}_i \sim \mathcal{N}(\boldsymbol{\mu}_i, \mathrm{diag}(\boldsymbol{\sigma}_i^2)), \tag{10}$$

where $\mathbf{W}_\mu$ and $\mathbf{W}_\sigma$ are trainable weights of $\text{MLP}_\mu(\cdot)$ and $\text{MLP}_\sigma(\cdot)$. The reparameterization trick [10] is applied to sample the latent variable $\mathbf{z}_i$.

### 2.5 Non-Autoregressive Decoder (NAD)

In the work of Gu *et al.* [5], the authors introduce in their model a component that enhances the inputs passed to the decoder for their machine translation problem. The idea is to help the model learn the internal dependencies (which are absent in their non-autoregressive translator) within a sentence. In the work of Guo *et al.* [6], the authors use a positional module to improve the decoder's ability to perform local reordering. The context generators in NAP play a similar role as these two approaches. In the testing stage, the trained PCG and ICG are able to generate context vectors for new (unseen) observed trajectories to help the decoder improve its time awareness for trajectory prediction. Both the ICG, which generates future-time-oriented contexts $\{\mathbf{c}_t \,|\, T_\text{o}{+}1 \leqslant t \leqslant T_\text{o}{+}T_\text{p}\}$, and the future-time-agnostic PCG are needed in the NAD for generating trajectories.

To make multimodal predictions at time step $t$, the NAD therefore takes the concatenation of the two contexts $\mathbf{c}_\text{p}^i$ and $\mathbf{c}_t^i$ and the latent variable $\mathbf{z}_i$ as inputs, modeled by the MLP below:

$$(\hat{x}_t^i, \hat{y}_t^i) = \text{MLP}_\text{out}(\mathbf{c}_t^i \oplus \mathbf{c}_\text{p}^i \oplus \mathbf{z}_i; \mathbf{W}_\text{out}), \tag{11}$$

where $\text{MLP}_\text{out}(\cdot)$ is the MLP used for predicting the location coordinates. Its parameter $\mathbf{W}_\text{out}$ is shared across all the time steps in the prediction phase. Note that the input passed to the NAD is different for each time step $t$ in the prediction phase as $\mathbf{c}_t^i$ depends on $t$. We can consider that the contexts $\{\mathbf{c}_t^i\}$ function is like the hidden states in the decoder of an LSTM except that they are not recursively defined.

### 2.6 Implementation Details

The embedding layer $\phi$ in Eq. (1) is modeled as a perceptron that outputs 32-dimensional embedded vectors for the input location coordinates. The dimensions of the hidden states of the LSTM layers for both the Trajectory and Social Encoders are 32. The GCN in the Social Encoder is a single graph convolutional layer (*i.e.*, $\ell = 1$ in Eq. (3)). For the ICG, $\text{MLP}_\text{B}$ is a three-layer MLP with ReLU activations. All the other MLPs used in Eqs. (6), (8), (9), and (11) are single-layer MLPs. Except for in Sect. 3.4 where we explore the prediction performance for different prediction lengths, the observed length of input trajectories is 8 time steps ($T_\text{o} = 8$) and the prediction length is 12 time steps ($T_\text{p} = 12$) for all other experiments. We implemented NAP and its variants (Sect. 3.3) using the PyTorch framework in Python. The Adam optimizer was used to train our models with the learning rate set to 0.001 and the mini-batch size to 128.

# 3 Experiments

## 3.1 Datasets and Metrics

We use the popular ETH [17] and UCY [13] datasets, which, altogether, include 5 scenes: ETH, HOTEL, UNIV, ZARA1, and ZARA2. Similar to [18,24], we normalize each pedestrian's coordinates and augment the training data by rotating trajectories. As raw scene images are used as inputs for extracting scene features, we also rotate the input images when input trajectories are rotated. Same as previous work in the literature [1,7,9,18], the leave-one-out strategy is adopted for training and testing. All methods are evaluated based on two standard metrics: the Average Displacement Error (ADE) [17] and the Final Displacement Error (FDE) [1]. Smaller errors indicate better prediction performance.

**Table 1.** The ADEs/FDEs (in meters) of various methods. The settings are: $T_o = 8$ and $T_p = 12$. The results with a † are taken from the authors' papers. The result with ‡ is taken from [7].

| Method | # | ETH & UCY scenes | | | | | |
|---|---|---|---|---|---|---|---|
| | | ETH | HOTEL | UNIV | ZARA1 | ZARA2 | Average |
| Social-LSTM [1]‡ | | 1.09/2.35 | 0.79/1.76 | 0.67/1.40 | 0.47/1.00 | 0.56/1.17 | 0.72/1.54 |
| SGAN 1V-1 [7]† | | 1.13/2.21 | 1.01/2.18 | 0.60/1.28 | 0.42/0.91 | 0.52/1.11 | 0.74/1.54 |
| MX-LSTM [8]† | | – | – | **0.49/1.12** | 0.59/1.31 | 0.35/0.79 | – |
| Nikhil & Morris [16]† | | 1.04/2.07 | 0.59/1.17 | 0.57/1.21 | 0.43/0.90 | 0.34/0.75 | 0.59/1.22 |
| Liang *et al.* [15]† | | 0.88/1.98 | 0.36/0.74 | 0.62/1.32 | 0.42/0.90 | 0.34/0.75 | 0.52/1.14 |
| MATF [25]† | | 1.33/2.49 | 0.51/0.95 | 0.56/1.19 | 0.44/0.93 | 0.34/0.73 | 0.64/1.26 |
| SR-LSTM [24]† | | 0.63/1.25 | 0.37/0.74 | 0.51/1.10 | **0.41**/0.90 | 0.32/0.70 | **0.45**/0.94 |
| STGAT 1V-1 [9]† | | 0.88/1.66 | 0.56/1.15 | 0.52/1.13 | **0.41**/0.91 | **0.31/0.68** | 0.54/1.11 |
| NAP (ours) | | **0.59/1.13** | **0.30/0.51** | 0.59/1.23 | **0.41/0.86** | 0.36/0.72 | **0.45/0.89** |
| SGAN 20V-20 [7]† | ✓ | 0.81/1.52 | 0.72/1.61 | 0.60/1.26 | 0.34/0.69 | 0.42/0.84 | 0.58/1.18 |
| SoPhie [18]† | ✓ | 0.70/1.43 | 0.76/1.67 | 0.54/1.24 | 0.30/0.63 | 0.38/0.78 | 0.54/1.15 |
| Liang *et al.* [15]† | ✓ | 0.73/1.65 | 0.30/0.59 | 0.60/1.27 | 0.38/0.81 | 0.31/0.68 | 0.46/1.00 |
| MATF GAN [25]† | ✓ | 1.01/1.75 | 0.43/0.80 | **0.44/0.91** | 0.26/0.45 | 0.26/0.57 | 0.48/0.90 |
| IDL [14]† | ✓ | 0.59/1.30 | 0.46/0.83 | 0.51/1.27 | **0.22/0.49** | **0.23/0.55** | 0.40/0.89 |
| STGAT 20V-20 [9]† | ✓ | 0.65/1.12 | 0.35/0.66 | 0.52/1.10 | 0.34/0.69 | 0.29/0.60 | 0.43/0.83 |
| Social-BiGAT [11]† | ✓ | 0.69/1.29 | 0.49/1.01 | 0.55/1.32 | 0.30/0.62 | 0.36/0.75 | 0.48/1.00 |
| NAP (ours) | ✓ | **0.53/1.08** | **0.26/0.46** | 0.58/1.22 | 0.30/0.65 | 0.28/0.60 | **0.39/0.80** |

## 3.2 Comparison with Other Methods

We compare NAP against the following state-of-the-art trajectory prediction methods: Social-LSTM [1], SGAN [7], MX-LSTM [8], Nikhil & Morris [16], Liang *et al.* [15], MATF [25], SR-LSTM [24], SoPhie [18], IDL [14], STGAT [9], and Social-BiGAT [11].

In Table 1, all the compared methods are put into two groups depending on whether they generate only one prediction (top half of the table) or multiple predictions (bottom half and indicated by a tick under the # column) for each input observed trajectory. The multimodal predictions being considered is 20. The reported ADEs and FDEs are computed from the best predictions out of the 20. Five methods report both single and multimodal prediction results and so they appear in both halves of the table: SGAN, MATF, STGAT, Liang *et al.*, and NAP.

Our proposed method is able to achieve results on par with the state-of-the-art methods for the single prediction setting. Specifically, NAP has the same smallest average ADE (0.45 m) as SR-LSTM while outperforming all methods on the average FDE (0.89 m). In addition to NAP, SR-LSTM, MX-LSTM, and STGAT 1V-1 also have the best performance on one or more scenes. In the lower half of Table 1, results of multimodal predictions are given and compared. On average, our NAP achieves the smallest ADE of 0.39 m and the smallest FDE of 0.80 m. For each scene, the best performers that achieve the smallest ADE/FDE in the lower half of the table include NAP, IDL, MATF, and GAN. Taken together, these results demonstrate the efficacy of our proposed method in both single and multimodal prediction settings.

### 3.3 Ablation Study

To explore the effectiveness of different contexts working together in our proposed method, we consider four variants of NAP listed below:

- NAP-P: This variant only uses the Personal Context Generator (future-time-agnostic context, the light blue PCG box in Fig. 1), *i.e.*, the interaction context $\mathbf{c}_t^i$ in Eq. (11) is removed. Accordingly, a different MLP taking only $\mathbf{c}_\mathrm{p}^i$ as input is used in the NAD so that it is still capable of returning different outputs at different time steps.
- NAP-ISS: In contrast to the NAP-P above, NAP-ISS disables the personal context and forecasts predictions based on the future-time-oriented interaction context (the pink box in Fig. 1). The personal context $\mathbf{c}_\mathrm{p}^i$ in Eq. (11) is removed so the $\mathrm{MLP}_\mathrm{out}$ in that equation has fewer input neurons. The rest of NAP-ISS is the same as NAP.
- NAP-ISg: In order to further investigate the impact of removing the scene influence, we drop the scene feature $\mathbf{s}_{T_\mathrm{o}}^i$ from the Interaction Context Generator in Eq. (7) to form this variant. That is, the interaction context $\mathbf{c}_t^i$ in NAP-ISS is computed using both the social graph and scene features, whereas the $\mathbf{c}_t^i$ in NAP-ISg is computed using the social graph feature only. The $\mathrm{MLP}_\mathrm{B}$ (Eq. (7)) used in this variant has only input neurons for $\mathbf{h}_{T_\mathrm{o}}^i \oplus \mathbf{g}_{T_\mathrm{o}}^i$.
- NAP-ISc: Similar to NAP-ISg, this variant is designed to investigate the impact of removing the social influence. We drop the social graph feature $\mathbf{g}_{T_\mathrm{o}}^i$ but keep the scene feature $\mathbf{s}_{T_\mathrm{o}}^i$ in Eq. (7) so the context $\mathbf{c}_t^i$ is computed from the scene feature only.

In our ablation study, we compare only the single prediction performance (see Table 2) of these four variants, *i.e.*, the latent variable $\mathbf{z}_i$ for multimodality is removed from Eq. (11) in the experiments. In general, NAP-P, which uses only

**Table 2.** The ADEs / FDEs (in meters) of the four variants for single predictions in the ablation study, with $T_o = 8$ and $T_p = 12$.

| | NAP-P | NAP-ISS | NAP-ISg | NAP-ISc |
|---|---|---|---|---|
| ETH | 0.87/1.63 | 0.66/1.22 | 0.69/1.31 | 0.74/1.52 |
| HOTEL | 0.43/0.77 | 0.34/0.61 | 0.37/0.70 | 0.38/0.73 |
| UNIV | 0.71/1.42 | 0.68/1.37 | 0.68/1.35 | 0.70/1.39 |
| ZARA1 | 0.46/0.95 | 0.45/0.94 | 0.47/0.96 | 0.45/0.95 |
| ZARA2 | 0.44/0.88 | 0.42/0.83 | 0.44/0.84 | 0.44/0.86 |
| Average | 0.58/1.13 | 0.51/0.99 | 0.53/1.03 | 0.54/1.09 |

the personal context (future-time-agnostic), has a poorer performance than the other three variants. This is not unexpected as, without the future-time-oriented context, the NAD is not able to forecast good predictions for different time steps in the prediction phase. Comparing the three interaction context based variants against each other, it is not surprising to see that NAP-ISS outperforms the other two variants due to the presence of both the social graph and scene features. As for NAP-ISg versus NAP-ISc, we observe that NAP-ISg slightly outperforms NAP-ISc. This demonstrates that the social influence is more important than the scene influence. However, it should be noted that the five scenes in the ETH/UCY datasets do not have many obstacles scattered in the pedestrian pathways. The slightly better performance of NAP-ISg confirms that there are more social (pedestrian) interactions than scene interactions in these datasets.

Comparing the results from the four variants in Table 2 and from NAP in Table 1, we observe that NAP outperforms all the four variants. Our ablation study justifies the need for all the contexts to be present in NAP.

**Table 3.** The ADEs/FDEs (in meters) of various methods for different prediction lengths ($T_o = 8$). A method with * indicates that it generates 20 predictions for each input observed trajectory.

| | $T_p = 8$ | $T_p = 12$ | Error increment |
|---|---|---|---|
| Social-LSTM | 0.45/0.91 | 0.72/1.54 | 60.00%/69.23% |
| SGAN 1V-1 | 0.49/1.00 | 0.74/1.54 | 51.02%/54.00% |
| STGAT 1V-1 | 0.39/0.81 | 0.54/1.11 | 38.46%/37.03% |
| NAP (ours) | 0.35/0.67 | 0.45/0.89 | **28.57%/32.84%** |
| SGAN 20V-20* | 0.39/0.78 | 0.58/1.18 | 48.72%/51.28% |
| STGAT 20V-20* | 0.31/0.62 | 0.43/0.83 | 38.71%/33.87% |
| NAP (ours)* | 0.31/0.61 | 0.39/0.80 | **25.81%/31.15%** |

### 3.4 Different Prediction Lengths

In addition to the prediction length setting ($T_\text{p} = 12$ frames, corresponding to 4.8 seconds) used in Tables 1 and 2 and similar to previous work such as SGAN [7] and STGAT [9], we conduct experiments for the prediction length $T_\text{p} = 8$ frames (or 3.2 seconds) to further evaluate the performance of NAP. Table 3 shows the average ADE/FDE results for this prediction length setting. The figures under the '$T_\text{p} = 12$' column are copied from the *Average* column of Table 1. Each error increment (last column) due to the increase of $T_\text{p}$ is calculated as: $(e_{\text{p}12} - e_{\text{p}8})/e_{\text{p}8} \times 100\%$, where $e_{\text{p}12}$ and $e_{\text{p}8}$ are errors (ADEs or FDEs) for $T_\text{p} = 12$ and $T_\text{p} = 8$ of the same method.

As expected, all methods shown in Table 3 have better performance for the shorter prediction length. In the top half of the table, when generating a predicted trajectory for each input, the error increments of Social-LSTM and SGAN 1V-1 are over 50%. Compared to these two methods, STGAT 1V-1 has smaller error increments for both ADE and FDE. For the multimodal predictions (bottom half of the table), STGAT 20V-20 again outperforms SGAN 20V-20.

We observe from Table 3 that NAP consistently outperforms all other methods for both prediction length settings and for both single and multimodal predictions. Furthermore, NAP also has the smallest error increments for both ADE and FDE when $T_\text{p}$ increases. This demonstrates that NAP is more robust in generating long trajectories. The reason is due to the non-autoregressive nature of the decoder, which not only allows the location coordinates at different time steps to be independently forecast but also helps minimize the accumulation of prediction errors when the prediction length increases.

### 3.5 Qualitative Results

Figure 2 illustrates some prediction examples generated by NAP. The observed and ground truth trajectories are shown in yellow and green; the best trajectory of the 20 predictions of each pedestrian is shown in pink. For better visualization, the video frames have been darkened and blurred. These examples cover different movement behaviors of pedestrians. For example, Fig. 2(a) shows two simple straight path scenarios, Fig. 2(b) and (c) show a gentle turning scenario, and Fig. 2(d) shows a more difficult scenario in which an abrupt turn occurs near the end of the observation phase. Although the predicted trajectory (in pink) in Fig. 2(d) does not perfectly overlap with the ground truth trajectory, NAP is still able to correctly predict the trajectory from the late turning cue.

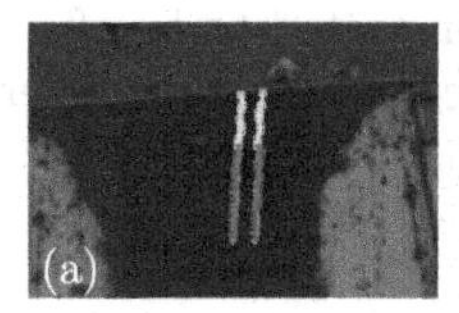

**Fig. 2.** Examples of predicted trajectories (shown in pink) generated by NAP. The observed trajectories and ground truth trajectories are shown in yellow and green. (Color figure online)

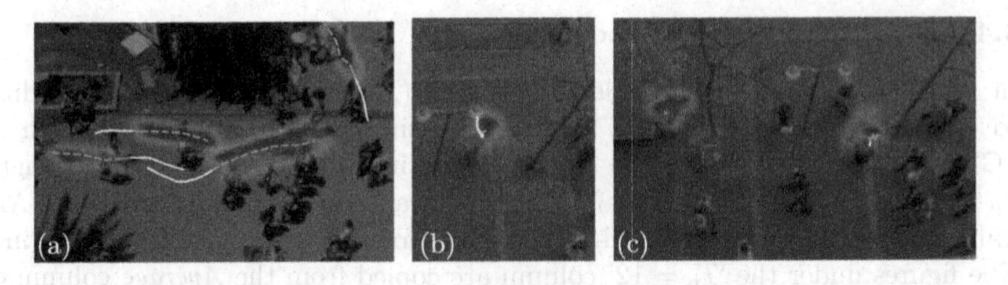


**Fig. 3.** Examples of multiple predicted trajectories shown as heatmaps. The observed trajectories and ground truth trajectories are shown in yellow and green. (Color figure online)

All the 20 predicted trajectories of each pedestrian are displayed as a heatmap in Fig. 3. The generated heatmaps in Fig. 3(a) for these pedestrians show a well coverage of the ground truth trajectories. Two difficult scenarios are shown in Fig. 3(b) and (c). For the pedestrian in Fig. 3(b) and the right pedestrian in Fig. 3(c), each made an abrupt turn at almost the last frame of the observation phase. However, NAP is still able to give good predicted trajectories, as all the plausible paths (including the ground truth trajectories (green)) are well covered by the heatmaps. The left pedestrian in Fig. 3(c) is a stopping scenario. After stopping, the pedestrian can remain still or resume walking in any direction. The generated heatmap shows a good coverage of possible paths; however, it has a small dent in the bottom left hand region due to the presence of a bench there, showing that the pedestrian must bypass the obstacle. This example shows the importance of including scene influence in the method.

## 4 Conclusion

We have presented a novel method called NAP which can handle both social influence and scene influence in the pedestrian trajectory prediction process. NAP captures these influences using the trainable feature encoders in the network. In addition, NAP handles multimodal predictions via a latent variable generator which models the sampling distribution that describes the multiple plausible paths of each pedestrian. Unlike existing trajectory prediction methods, the decoder of NAP is non-autoregressive. NAP is therefore able to forecast predictions for different time steps simultaneously or to forecast only for those time steps that are of interest. From our extensive experiments and ablation study, the context encoders used in NAP have been demonstrated to be effective. Not only does NAP achieve state-of-the-art performance, it also has smaller error increments when the prediction length increases.

## References

1. Alahi, A., Goel, K., Ramanathan, V., Robicquet, A., Fei-Fei, L., Savarese, S.: Social LSTM: human trajectory prediction in crowded spaces. In: CVPR, June 2016
2. Chen, L.C., Zhu, Y., Papandreou, G., Schroff, F., Adam, H.: Encoder-decoder with atrous separable convolution for semantic image segmentation. In: ECCV, September 2018
3. Cordts, M., et al.: The cityscapes dataset for semantic urban scene understanding. In: CVPR (2016)
4. Goodfellow, I., et al.: Generative adversarial networks. In: NeurIPS, pp. 2672–2680 (2014)
5. Gu, J., Bradbury, J., Xiong, C., Li, V.O., Socher, R.: Non-autoregressive neural machine translation. In: ICLR (2018). https://openreview.net/forum?id=B1l8BtlCb
6. Guo, J., Tan, X., He, D., Qin, T., Xu, L., Liu, T.Y.: Non-autoregressive neural machine translation with enhanced decoder input. In: AAAI, vol. 33, pp. 3723–3730 (2019)
7. Gupta, A., Johnson, J., Fei-Fei, L., Savarese, S., Alahi, A.: Social GAN: socially acceptable trajectories with generative adversarial networks. In: CVPR, June 2018
8. Hasan, I., Setti, F., Tsesmelis, T., Del Bue, A., Galasso, F., Cristani, M.: MX-LSTM: mixing tracklets and vislets to jointly forecast trajectories and head poses. In: CVPR, June 2018
9. Huang, Y., Bi, H., Li, Z., Mao, T., Wang, Z.: STGAT: modeling spatial-temporal interactions for human trajectory prediction. In: ICCV, October 2019
10. Kingma, D.P., Welling, M.: Auto-encoding variational bayes. arXiv preprint arXiv:1312.6114 (2013)
11. Kosaraju, V., Sadeghian, A., Martín-Martín, R., Reid, I., Rezatofighi, S.H., Savarese, S.: Social-BiGAT: multimodal trajectory forecasting using bicycle-GAN and graph attention networks. In: NeurIPS (2019)
12. Lee, N., Choi, W., Vernaza, P., Choy, C.B., Torr, P.H.S., Chandraker, M.: DESIRE: distant future prediction in dynamic scenes with interacting agents. In: CVPR (2017)
13. Lerner, A., Chrysanthou, Y., Lischinski, D.: Crowds by example. In: Computer Graphics Forum, vol. 26, pp. 655–664. Wiley Online Library (2007)
14. Li, Y.: Which way are you going? Imitative decision learning for path forecasting in dynamic scenes. In: CVPR, June 2019
15. Liang, J., Jiang, L., Niebles, J.C., Hauptmann, A.G., Fei-Fei, L.: Peeking into the future: predicting future person activities and locations in videos. In: CVPR, June 2019
16. Nikhil, N., Morris, B.T.: Convolutional neural network for trajectory prediction. In: ECCV Workshop, September 2018
17. Pellegrini, S., Ess, A., Schindler, K., Van Gool, L.: You'll never walk alone: modeling social behavior for multi-target tracking. In: ICCV, pp. 261–268 (2009)
18. Sadeghian, A., Kosaraju, V., Sadeghian, A., Hirose, N., Rezatofighi, H., Savarese, S.: SoPhie: an attentive GAN for predicting paths compliant to social and physical constraints. In: CVPR, June 2019
19. Su, H., Zhu, J., Dong, Y., Zhang, B.: Forecast the plausible paths in crowd scenes. In: IJCAI, pp. 2772–2778 (2017)
20. Vemula, A., Muelling, K., Oh, J.: Social attention: modeling attention in human crowds. In: ICRA, pp. 1–7, May 2018

21. Wen, R., Torkkola, K., Narayanaswamy, B., Madeka, D.: A multi-horizon quantile recurrent forecaster. In: NIPS Workshop (2017)
22. Xue, H., Huynh, D.Q., Reynolds, M.: SS-LSTM: a hierarchical LSTM model for pedestrian trajectory prediction. In: WACV, pp. 1186–1194. IEEE (2018)
23. Xue, H., Huynh, D.Q., Reynolds, M.: Pedestrian trajectory prediction using a social pyramid. In: Nayak, A.C., Sharma, A. (eds.) PRICAI 2019. LNCS (LNAI), vol. 11671, pp. 439–453. Springer, Cham (2019). https://doi.org/10.1007/978-3-030-29911-8_34
24. Zhang, P., Ouyang, W., Zhang, P., Xue, J., Zheng, N.: SR-LSTM: state refinement for LSTM towards pedestrian trajectory prediction. In: CVPR, June 2019
25. Zhao, T., et al.: Multi-agent tensor fusion for contextual trajectory prediction. In: CVPR, June 2019
26. Zou, H., Su, H., Song, S., Zhu, J.: Understanding human behaviors in crowds by imitating the decision-making process. In: AAAI (2018)

# Temporal Smoothing for 3D Human Pose Estimation and Localization for Occluded People

M. Véges(✉) and A. Lőrincz

Eötvös Loránd University, Budapest, Hungary
{vegesm,lorincz}@inf.elte.hu

**Abstract.** In multi-person pose estimation actors can be heavily occluded, even become fully invisible behind another person. While temporal methods can still predict a reasonable estimation for a temporarily disappeared pose using past and future frames, they exhibit large errors nevertheless. We present an energy minimization approach to generate smooth, valid trajectories in time, bridging gaps in visibility. We show that it is better than other interpolation based approaches and achieves state of the art results. In addition, we present the synthetic MuCo-Temp dataset, a temporal extension of the MuCo-3DHP dataset. Our code is made publicly available. (https://github.com/vegesm/pose_refinement).

**Keywords:** Human activity recognition · Pose estimation · Temporal · Absolute pose

## 1 Introduction

The task of 3D human pose estimation is to predict the coordinates of certain body joints based on an input image or video. The potential applications are numerous, including augmented reality, sport analytics and physiotherapy. In a multiperson settings, 3D poses may help analyzing the interactions between the actors.

Recent results on the popular Human3.6M database [9] are starting to saturate [10,28] and there is an increasing interest in more natural settings. The standard evaluation protocol in Human3.6M uses hip-relative coordinates. It makes the prediction task easier, as the localization of the pose is not required. However, in multi-person settings the distance between the subjects can be important too. This led to the introduction of absolute pose estimation [19,29], where the relative pose prediction is supplemented with the localization task.

Note that this problem is underdefined when only a single camera provides input. It is impossible to estimate the scale of the real scene based on one image, if semantic information is not available. Therefore, we use a scale-free error metric that calculates the prediction error up to a scalar multiplier. Additionally, our


H. Yang et al. (Eds.): ICONIP 2020, LNCS 12532, pp. 557–568, 2020.
https://doi.org/10.1007/978-3-030-63830-6_47

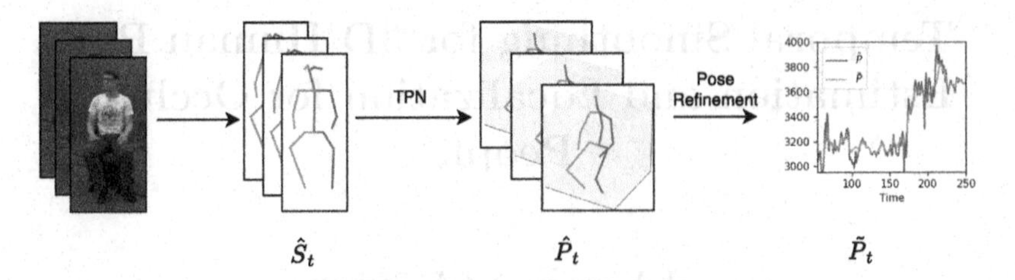


**Fig. 1. Overview of our algorithm.** First 2D pose estimations ($\hat{S}_t$) are generated for all frames. Then the Temporal PoseNet (*TPN*) converts the 2D estimates to initial 3D poses ($\hat{P}_t$). Finally, the pose refinement step smoothes out the trajectories, producing the final 3D pose estimates ($\tilde{P}_t$).

method is temporal, that lets us discern special cases such as jumping. An image-based method does not have enough information to decide, whether the jumping person is closer to the camera or higher above the ground.

Temporal methods can exploit information in neighboring frames. When a person disappears for a short time (for instance someone walks in front of him/her), these methods could fill in the void based on the previous and following frames. Still, the output is often noisy and even a simple linear interpolation yields better results (see Table 2).

To overcome the occlusion problem, we propose two contributions. First, we introduce the MuCo-Temp synthetic dataset. It is a temporal extension of the MuCo-3DHP dataset [18]. The latter database is commonly used as the training set for MuPoTS-3D [18]. The MuPoTS-3D dataset is a multi-person 3D human pose database, containing 20 video sequences. It consists of a test set only so the authors of [18] introduced MuCo-3DHP, that is generated from the single person MPI-INF-3DHP dataset [17]. Each image in MuCo-3DHP is a composition of four poses from MPI-INF-3DHP. This new synthetic dataset contains occlusions typical to multi-person settings but has images only while our method is temporal. Thus, we created the MuCo-Temp dataset, containing videos composited from sequences in the MPI-INF-3DHP database. To keep compatibility, we used the same algorithm as the authors of MuCo-3DHP.

Our second contribution is an energy minimization based smoothing function, targeting specifically those frames where a person became temporarily invisible. It adaptively smoothes the prediction stronger at frames where the pose is occluded and weaker when the pose is visible. This way large noises during heavy occlusions can be filtered without 'over-smoothing' unoccluded frames. The method does not require additional training and can be applied after any temporal pose estimation algorithm.

To summarize, we introduce an approach to predict 3D human poses even when the person is temporarily invisible. It achieves state-of-the-art results on the MuPoTS-3D dataset, showing its efficiency.

## 2 Related Work

**3D Pose Estimation.** Before the widespread usage of deep learning, various approaches were explored for 3D human pose prediction, such as conditional random fields [2] or dictionary based methods [23]. However, recent algorithms are all based on neural networks [15,20,28]. The primary difficulty in 3D human pose estimation is the lack of accurate in the wild datasets. Accurate measurements require a studio setting, such as Human3.6M [9] or MPI-INF-3DHP [17]. To overcome this, several approaches were proposed to use auxiliary datasets. Zhou et al. [33] uses 2D pose datasets with a reprojection loss. Pavlakos et al. [20] employed depthwise ordering of joints as additional supervision signal. An adversarial loss added to the regular regression losses ensure the plausibility of generated 3D poses [6]. It requires no paired 2D-3D data.

More importantly, Martinez et al. [15] introduced a two step prediction approach: first the 2D pose is predicted from the image, then the 3D pose is estimated solely from the 2D joint coordinates. This places the task of handling image features on the 2D pose estimator, for which large, diverse datasets exist [13]. The 2D-to-3D regression part can be a simple feed-forward network [15], one that uses recurrent layers [7,11] or a graph convolutional network [32]. Additionally to employing off-the-shelf algorithms [27], the joint training of the 2D pose estimator and the 2D-to-3D part is possible via soft-argmax [14,28]. The soft-argmax function is a differentiable approximation of the argmax operation that lets gradients flow to the 2D estimator. In our paper we follow the two step approach with a pretrained 2D pose estimator [27].

**Temporal Methods.** While 3D poses can be inferred from an image only, for videos temporal methods provide better performance than simple frame-by-frame approaches. A natural idea is to use recurrent layers over per-frame estimates [8]. However, even with LSTMs, RNN based methods exploit only a small temporal neighborhood of a frame. This can be solved with 1D convolution [21]. To increase the receptive field without adding extra parameters, the authors used dilated convolutions. Additionally, bundle adjustment [1] was used to refine body mesh prediction in time. Our work is closest to [1], however their method smoothes coordinates for visible poses only and does not handle occlusions.

Instead of working directly on joint coordinates, Lin et al. first decomposed the trajectories with discrete cosine transform [12]. In [31] the authors introduced a special motion loss that encourages realistic motion paths. Finally, Cai et al. [3] use a spatio-temporal graph convolutional network. The standard uniform convolution was changed such that the kernels became different for different joints.

**Localization and Pose Estimation.** Most work focuses on relative pose estimation: the joint coordinates are predicted relative to a root joint, usually the hip. This is sufficient for many use cases, however for multi-person images, the locations of the actors might be important too. The joint estimation of location and relative pose is also called absolute pose estimation [29]. In [30], the authors propose a direct regression approach that learns from RGB-D data as well. Moon et al. [19] uses two separate networks to predict the location and the 3D pose.

Additional performance can be gained by using multiple cameras. The authors of [22] predict the joint locations based on many viewpoints. They employ a modified pictorial structure model that recursively increases its resolution around the joint locations, improving accuracy while keeping speed.

**Occlusions.** Occlusions can be grouped in two categories: self-occlusions and occlusions caused by other objects or people. In typical single-pose estimation datasets, only the first type arises. Mehta et al. introduces Occlusion Robust Heatmaps, that store joint location coordinates as a 2D heatmap [18]. Augmenting datasets with synthetic occlusions also helps. Sarandi et al. overlaid rectangles and objects from MS-COCO on individual frames [26]. Cylindrical man models can be used to predict visibility from 3D joint coordinates [5]. The visibility flags combined with 2D joint detection score can form the basis of an effective regularizer.

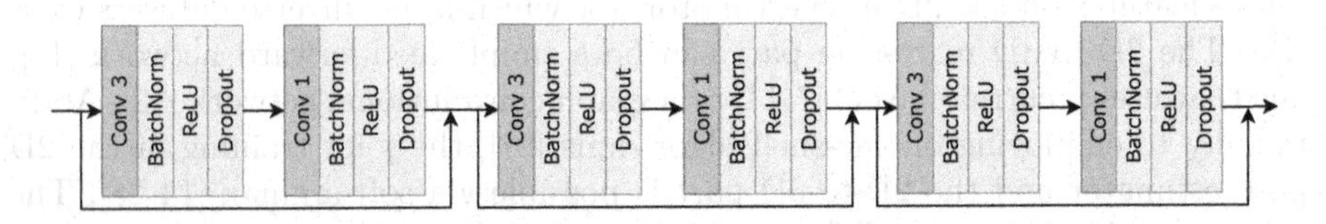


**Fig. 2. Temporal PoseNet architecture.** The network has three residual blocks, each block contains two convolutional layer followed by BatchNorm and Dropout layers. The activation function was ReLU.

## 3 Method

Our algorithm predicts the coordinates of the joints of a person in a camera relative coordinate system, where the origin is at the camera center and the axes are parallel to the camera plane. The estimation of coordinates is separated into two tasks: localization and relative pose estimation, similar to previous work [19,29]. That is, we predict the location of the root joint (the hip) and the location of the other joints relative to the hip. We denote the former $P^{loc}$, the latter $P^{rel}$.

The general outline of our algorithm is as follows (see Fig. 1): first $\hat{S}_{k,t}$, the 2D position of the $k$th person on frame $t$, is predicted using an off-the-shelf algorithm [27]. Then, a temporal pose network (*TPN*) predicts the 3D location and pose of the person ($\hat{P}^{loc}_{k,t}$ and $\hat{P}^{rel}_{k,t}$ respectively), see Sect. 3.1 for details. Finally, the pose refinement step smoothes the predictions of TPN, producing the final predictions $\tilde{P}^{loc}_{k,t}$ and $\tilde{P}^{rel}_{k,t}$ (see Sect. 3.2). The TPN and the pose refinement are run for each person separately. For brevity, we drop the index $k$ in the discussions below.

### 3.1 Temporal PoseNet

The Temporal PoseNet takes the 2D joint coordinates produced by the 2D pose estimator and predicts the 3D location and pose. To be robust against the different cameras in the training and test set, the 2D joints are normalized by the inverse of the camera calibration matrix. Then, for frame $t$ the joint locations are predicted based on a window of size $2w+1$:

$$\left[\hat{P}_t^{loc}; \hat{P}_t^{rel}\right] = f_{TPN}\left([K^{-1}\hat{S}_i]_{i=t-w}^{t+w}\right),$$

where $K$ is the intrinsic camera calibration matrix and $f_{TPN}$ is the Temporal PoseNet.

The network architecture of TPN was inspired by [21], Fig. 2 shows an overview. It is a 1D convolutional network, that has three residual modules. Each module contains a convolution of size 3 followed by a convolution of size 1. The convolutional layers are followed by a Batch Normalization layer, a ReLU activation function and a Dropout layer. The input and the output of the network are normalized by removing the mean and dividing by the standard deviation.

The loss function is the $\ell_1$ loss, to be robust against outliers. The final loss is:

$$L_{TPN} = \sum_t \left\|\hat{P}_t^{loc} - P_t^{loc}\right\|_1 + \left\|\hat{P}_t^{rel} - P_t^{rel}\right\|_1,$$

where $P_t^{loc}$ and $P_t^{rel}$ are the ground-truth location and the relative pose coordinates for frame $t$.

### 3.2 Pose Refinement

While the input of the TPN is a large temporal window of frames, it still makes mistakes, especially during heavy occlusions. We propose an energy optimization based smoothing method that adaptively filters the keypoint trajectories. The objective function is calculated separately for the location and relative pose. It is formulated as:

$$E_{ref}(\tilde{P}) = vE_{pred}(\tilde{P}) + (1-v)\lambda_1 E_{smooth}(\tilde{P}, \tau_1) + \lambda_2 E_{smooth}(\tilde{P}, \tau_2), \qquad (1)$$

where $\tilde{P}$ is is either $\tilde{P}^{loc}$ or $\tilde{P}^{rel}$, indicating whether the objective function optimizes the location or the relative pose; $v$ is the visibility score, $E_{pred}$ is an error function ensuring the smoothed pose is close to the original prediction, and $E_{smooth}$ is a smoothness error.

The parameter $\tau$ of $E_{smooth}$ controls the smoothing frequency. A small value corresponds to a low-pass filter with a high frequency threshold, while a large value corresponds to a lower frequency threshold (see below the definition of $E_{smooth}$). We choose $\tau_1 \gg \tau_2$.

The visibility score $v$ is close to 1 when the pose is visible, and to 0 when the pose is occluded. If the person is fully invisible, $v$ is set to 0. Thus, in Eq. (1), when the pose is visible, only $E_{pred}$ is taken into account and the (stronger, $\tau_1$ parametrized) smoothing is ignored. In other words, when the pose was visible, the objective function does not override the predictions of TPN. On the other hand, when the pose is heavily occluded, only $E_{smooth}$ is active and the optimization ensures that the predicted pose will be smooth.

The prediction error is

$$E_{pred}\left(\tilde{P}\right) = \sum_t \min\left(\left\|\tilde{P}_t - \hat{P}_t\right\|_2^2, m\right),$$

where $\hat{P}_t$ is either the location or the pose predicted by the Temporal PoseNet at frame $t$, and $\tilde{P}_t$ is the target of the optimization. The min function ensures that large outliers do not affect the objective function. This is essential because the predicted pose $\hat{P}$ can be noisy when the person is occluded. $m$ is a fixed parameter.

The smoothing error is a zero velocity loss [1], and $\tau$ controls the scale of the smoothing:

$$E_{smooth}\left(\tilde{P}, \tau\right) = \sum_t \left\|\tilde{P}_t - \tilde{P}_{t-\tau}\right\|_2^2.$$

The error encourages that the pose changes smoothly over time. If $\tau$ is small, then the filtering works in a small window, removing local noise. If $\tau$ is large, the error function looks at a larger timescale.

Finally, the visibility score $v$ is the confidence score predicted by the 2D pose estimator. Since the 2D estimator returns per-joint confidences, these are averaged. Finally, a median filter is applied on $v$.

The total energy function evaluates Eq. (1) both for the location and relative position:

$$E_{total}(\tilde{P}^{loc}) = E_{ref}(\tilde{P}^{loc}) + \lambda_{rel} E_{ref}(\tilde{P}^{rel})$$

where $\lambda_{rel}$ is a weighting parameter.

# 4 Experiments

## 4.1 Datasets

We evaluated our method on the multi-person 3D pose dataset MuPoTS-3D [18]. It contains videos taken in indoor and outdoor environments. Since MuPoTS-3D contains only test sequences and has no training set, commonly the MPI-INF-3DHP [17] and MuCo-3DHP [18] databases serve as the training set. However, MuCo-3DHP contains single frames only, so temporal models can not be trained on it. Therefore, we created the MuCo-Temp synthetic dataset.

**MuCo-Temp Dataset.** The MuCo-3DHP database contains synthetic images, composited from frames of MPI-INF-3DHP. The latter dataset was recorded in a green-screen studio, so segmenting of the actors is easy. Each synthesized frame contains four persons copied on a single image. Optionally, the background can be augmented with arbitrary images.

MuCo-Temp is using the same generation algorithm as MuCo-3DHP but it consists of videos instead of frames. Each video contains 4 person and 2000 frames. In total, we generated 77 videos for training. The validation set was created using the same process. We did not augmented the background, as the 2D pose estimator was already trained on a visually diverse dataset.

Our method was trained on the concatenation of the MPI-INF-3DHP and MuCo-Temp datasets.

### 4.2 Metrics

We calculate various metrics to evaluate the performance of our algorithm. The following list briefly summarizes each:

- **MRPE** or Mean Root Position Error, the average error of the root joint (the hip) [19].
- **MPJPE** or Mean Per Joint Position Error, the mean Euclidean error averaged over all joints and all poses, calculated on *relative* poses [15].
- **3D-PCK** or Percentage of Correct Keypoints, the percentage of joints that are closer than 150 mm to the ground truth. This metric is also calculated on relative poses only.

Since our method uses a single a camera, predicting the poses and locations of people is an underdefined problem. The above metrics do not take into account this, and calculate every error in mm. Therefore, we also include unit-less variants of the above metrics called *N-MRPE* and *N-MPJPE* [24]. The definition of N-MRPE is

$$\min_{s\in\mathbb{R}} \frac{1}{N} \sum_{k,t} \left\| s\tilde{P}_{k,t}^{hip} - P_{k,t}^{hip} \right\|_2^2,$$

where $\tilde{P}_{k,t}^{hip}$ and $P_{k,t}^{hip}$ are the predicted and ground-truth locations of the hip for person $k$ on frame $t$. The total number of poses is $N$. The formula above finds an optimal scaling parameter $s$ to minimize the error. In other words, the scaling of the prediction comes from the ground truth. The scaling constant $s$ is calculated over all people and frames of the video: that is, the size of two predicted skeletons must be correct relative to each other, while the absolute size is still unknown.

MuPoTS has two kinds of pose annotations: universal and normal. The normal coordinates are the joint locations in millimeters. The universal coordinates are like the normal ones, but each person is rescaled from their hip to have a normalized height. Following previous work [19,30], the 3D-PCK metric is calculated on the universal coordinates of MuPoTS-3D while the (N-)MRPE and (N-)MPJPE errors are calculated on the normal coordinates. All the metrics are calculated over each sequence separately, then averaged.

### 4.3 Implementation Details

The half-window size $w$ in the Temporal PoseNet was 40, thus the full input window length was 81 frames. The dropout rate was set to 0.25. The training algorithm was Adam with a learning rate of $10^{-3}$, decayed by a multiplier of 0.95 on each epoch. The network was trained for 80 epochs.

To avoid overfitting, the training set was augmented by scaling the 2D skeletons. When the focal length of the camera remains unchanged, this roughly corresponds to zooming the image. This step is essential, otherwise the TPN may overfit to the $y$ location of joints.

The optimization algorithm minimizing $E_{total}$ was Adam, with a learning rate of $10^{-2}$. The optimization ran for 500 iterations. The temporal timestep $\tau_1$ was 20, $\tau_2$ was 1, and the threshold $m$ was 1. The values of the weighting parameters: $\lambda_{rel} = 0.1$, $\lambda_1 = 0.1$ and $\lambda_2 = 1$.

## 5 Results

First, we compare the results of our model to the state of the art (Table 1). We improve on all metrics, however, this is not a fair comparison, since these models are single-frame algorithms, while ours is a temporal model. Also, MRPE and MPJPE errors can be calculated on detected poses only for these models

**Table 1. Comparison with state-of-the-art.** MPJPE and MRPE errors are in mm. * Non-temporal methods. † Error is calculated on detected frames only.

| | MRPE | MPJPE | 3D-PCK |
|---|---|---|---|
| LCR-Net++ [25]* | – | – | 70.6 |
| Moon et al. [19]* | 277† | – | 81.8 |
| Veges et al. [30]* | 238† | 120† | 78.2 |
| **Ours**† | 225† | 100† | 86.8† |
| **Ours** | **252** | **103** | **85.3** |

To have a fair comparison, we also test against other baselines (Table 2). First, we use a simple linear interpolation for frames when a person is occluded (*Interpolation*). While on relative pose estimation metrics it achieves similar performance as our model, the localization performance is considerably worse (272 vs 252 in MRPE, a 7.4% relative drop).

Additionally, we apply the 1-Euro filter [4] on the interpolated poses (*1-Euro*). This filter was applied in previous work to reduce high-frequency noises [16,34]. It is noticeable, that the filter had no additional effect over the interpolation, the metrics either did not change or marginally worsened. That is, our smoothing procedure performs better than the 1-Euro method.

**Table 2. Comparison with baselines.** *Interpolation* uses simple linear interpolation for unseen poses. *1-Euro* applies a 1-Euro filter on interpolated poses. (N-)MPJPE and MRPE errors are in mm.

| | MRPE | MPJPE | 3D-PCK | N-MRPE | N-MPJPE |
|---|---|---|---|---|---|
| Ours w/o refine | 340 | 107 | 83.4 | 307 | 113 |
| Interpolation | 272 | 103 | 85,1 | 243 | 109 |
| 1-Euro | 273 | 104 | 85.1 | 243 | 109 |
| **Ours** | **252** | **103** | **85.3** | **221** | **108** |

## 5.1 Ablation Studies

We performed an ablation study to confirm that each component of our algorithm contributes positively, the results are presented in Table 3a. The *Baseline* corresponds to the Temporal PoseNet only, trained on MPI-INF-3DHP. Adding MuCo-Temp improves on all performance metrics. Moreover, adding Pose Refinement further decreases errors by a large amount (MRPE goes from 340 to 252, a 25% drop).

**Table 3. Results of ablation studies.** a) Results when components are turned on sequentially. b) Errors calculated on visible poses only.

| a) Performance of components | | | | | |
|---|---|---|---|---|---|
| | MRPE | MPJPE | 3D-PCK | N-MRPE | N-MPJPE |
| Baseline | 372 | 116 | 81,2 | 332 | 122 |
| + MuCo-Temp | 340 | 107 | 83.4 | 307 | 113 |
| + Pose Refinement | 252 | 103 | 85.3 | 221 | 108 |
| b) Results on visible poses | | | | | |
| | MRPE | MPJPE | 3D-PCK | N-MRPE | N-MPJPE |
| Ours w/o refine | 227 | 100 | 86.5 | 191 | 106 |
| Ours | 225 | 100 | 86.6 | 189 | 106 |

We hypothesized that our model improves the estimation of occluded poses. To show that the pose refinement process does not hurt the accuracy of visible joints, we evaluated the model on those poses, where the visibility score reached a threshold (0.1 in our experiments). The results are shown in Table 3b. The addition of the refinement step changed the performance only marginally, indicating that a) our method improves prediction on heavily occluded poses b) does not decrease performance on unoccluded poses.

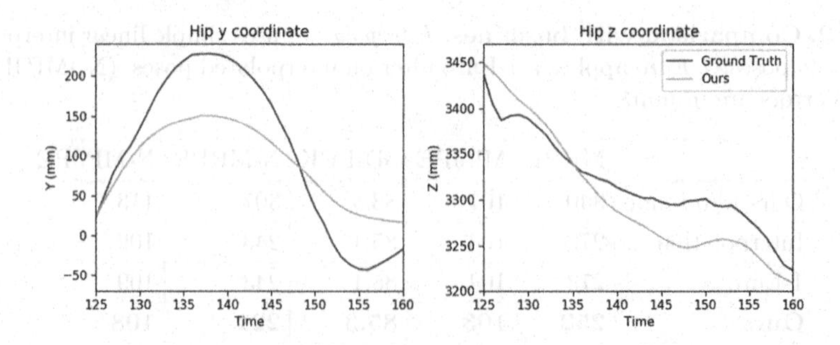


**Fig. 3. Trajectory of hip during a jump.** During a jump, the trajectory follows the ground-truth, showing that the model did not overfit to vertical location. Left plot shows the vertical coordinate of the hip in a camera centered coordinate system. The right plot shows the depth of the joint.

### 5.2 Vertical Location and Depth

One may assume that the TPN overfitted and simply calculates the depth based on the $y$ coordinates. However, this is mitigated by the fact that the training and test set have different camera viewpoints. Moreover, we applied an augmentation that effectively zooms the the cameras, further increasing the diversity of the inputs (see Sect. 4.3). We also show an example sequence in Fig. 3. It contains the trajectory of the hip of a person jumping backward, taken from MuPoTS Sequence 15. The figure demonstrates, that even though the $y$ coordinate of the joint is increasing, the $z$ coordinate still follows the ground truth. If the TPN was overfitted to the $y$ location, we would expect a jump in the trajectory of the $z$ coordinate

## 6 Conclusion

We proposed a pose refinement approach, that corrects predictions of heavily occluded poses. Our method improves both localization and pose estimation performance, achieving state-of-the-art results on the MuPoTS dataset. We demonstrated that it does not impair performance on unoccluded poses. Our algorithm could be further extended by making the refinement process part of the pose estimation network, in an end-to-end fashion. Also, one drawback of our approach is that it does not include tracking, the combination with a tracking algorithm remains future work.

**Acknowledgment.** MV received support from the European Union and co-financed by the European Social Fund (EFOP-3.6.3-16-2017-00002). AL was supported by the National Research, Development and Innovation Fund of Hungary via the Thematic Excellence Programme funding scheme under Project no. ED_18-1-2019-0030 titled Application-specific highly reliable IT solutions.

## References

1. Arnab, A., Doersch, C., Zisserman, A.: Exploiting temporal context for 3D human pose estimation in the wild. In: CVPR (2019)
2. Belagiannis, V., Amin, S., Andriluka, M., Schiele, B., Navab, N., Ilic, S.: 3D pictorial structures for multiple human pose estimation. In: CVPR (2014)
3. Cai, Y., Ge, L., Liu, J., Cai, J., Cham, T.J., Yuan, J., Thalmann, N.M.: Exploiting spatial-temporal relationships for 3D pose estimation via graph convolutional networks. In: ICCV (2019)
4. Casiez, G., Roussel, N., Vogel, D.: 1€ filter: a simple speed-based low-pass filter for noisy input in interactive systems. In: Proceedings of the SIGCHI Conference on Human Factors in Computing Systems, pp. 2527–2530 (2012)
5. Cheng, Y., Yang, B., Wang, B., Yan, W., Tan, R.T.: Occlusion-aware networks for 3D human pose estimation in video. In: ICCV (2019)
6. Drover, D., Rohith, M.V., Chen, C.H., Agrawal, A., Tyagi, A., Huynh, C.P.: Can 3D pose be learned from 2D projections alone? In: ECCV Workshops (2019)
7. Fang, H.S., Xu, Y., Wang, W., Liu, X., Zhu, S.C.: Learning pose grammar to encode human body configuration for 3D pose estimation. In: AAAI (2018)
8. Hossain, M.R.I., Little, J.J.: Exploiting temporal information for 3D pose estimation (2017). arXiv:1711.08585
9. Ionescu, C., Papava, D., Olaru, V., Sminchisescu, C.: Human3.6m: Large scale datasets and predictive methods for 3D human sensing in natural environments. TPAMI **36**(7), 1325–1339 (2014)
10. Kocabas, M., Karagoz, S., Akbas, E.: Self-supervised learning of 3D human pose using multi-view geometry. In: CVPR (2019)
11. Lee, K., Lee, I., Lee, S.: Propagating LSTM: 3D pose estimation based on joint interdependency. In: Ferrari, V., Hebert, M., Sminchisescu, C., Weiss, Y. (eds.) ECCV 2018. LNCS, vol. 11211, pp. 123–141. Springer, Cham (2018). https://doi.org/10.1007/978-3-030-01234-2_8
12. Lin, J., Lee, G.H.: Trajectory space factorization for deep video-based 3D human pose estimation. In: BMVC (2019)
13. Lin, T.-Y., et al.: Microsoft COCO: common objects in context. In: Fleet, D., Pajdla, T., Schiele, B., Tuytelaars, T. (eds.) ECCV 2014. LNCS, vol. 8693, pp. 740–755. Springer, Cham (2014). https://doi.org/10.1007/978-3-319-10602-1_48
14. Luvizon, D.C., Picard, D., Tabia, H.: 2D/3D pose estimation and action recognition using multitask deep learning. In: CVPR (2018)
15. Martinez, J., Hossain, R., Romero, J., Little, J.J.: A simple yet effective baseline for 3D human pose estimation. In: ICCV (2017)
16. Mehta, D., et al.: VNect: real-time 3D human pose estimation with a single RGB camera. ACM Trans. Graph. **36**(4), 44 (2017)
17. Mehta, D., Rhodin, H., Casas, D., Fua, P., Sotnychenko, O., Xu, W., Theobalt, C.: Monocular 3D human pose estimation in the wild using improved CNN supervision. In: 3DV (2017)
18. Mehta, D., et al.: Single-shot multi-person 3D pose estimation from monocular RGB. In: 3DV (2018)
19. Moon, G., Chang, J.Y., Lee, K.M.: Camera distance-aware top-down approach for 3D multi-person pose estimation from a single RGB image. In: ICCV (2019)
20. Pavlakos, G., Zhou, X., Daniilidis, K.: Ordinal depth supervision for 3D human pose estimation. In: CVPR (2018)

21. Pavllo, D., Feichtenhofer, C., Grangier, D., Auli, M.: 3D human pose estimation in video with temporal convolutions and semi-supervised training. In: CVPR (2019)
22. Qiu, H., Wang, C., Wang, J., Wang, N., Zeng, W.: Cross view fusion for 3D human pose estimation. In: ICCV (2019)
23. Ramakrishna, V., Kanade, T., Sheikh, Y.: Reconstructing 3D human pose from 2D image landmarks. In: Fitzgibbon, A., Lazebnik, S., Perona, P., Sato, Y., Schmid, C. (eds.) ECCV 2012. LNCS, vol. 7575, pp. 573–586. Springer, Heidelberg (2012). https://doi.org/10.1007/978-3-642-33765-9_41
24. Rhodin, H., et al.: Learning monocular 3D human pose estimation from multi-view images. In: CVPR (2018)
25. Rogez, G., Weinzaepfel, P., Schmid, C.: LCR-Net++: multi-person 2D and 3D pose detection in natural images. TPAMI **42**, 1146–1161 (2019)
26. Sarandi, I., Linder, T., O. Arras, K., Leibe, B.: Synthetic occlusion augmentation with volumetric heatmaps for the 2018 ECCV PoseTrack challenge on 3D human pose estimation (2018, unpublished). arXiv:1809.04987
27. Sun, K., Xiao, B., Liu, D., Wang, J.: Deep high-resolution representation learning for human pose estimation. In: CVPR (2019)
28. Sun, X., Xiao, B., Liang, S., Wei, Y.: Integral human pose regression. In: ECCV (2018)
29. Veges, M., Lorincz, A.: Absolute human pose estimation with depth prediction network. In: IJCNN (2019)
30. Veges, M., Lorincz, A.: Multi-person absolute 3D human pose estimation with weak depth supervision (2020). arXiv:2004.03989
31. Wang, J., Yan, S., Xiong, Y., Lin, D.: Motion guided 3D pose estimation from videos (2020). arXiv:2004.13985
32. Zhao, L., Peng, X., Tian, Y., Kapadia, M., Metaxas, D.N.: Semantic graph convolutional networks for 3D human pose regression. In: CVPR (2019)
33. Zhou, X., Huang, Q., Sun, X., Xue, X., Wei, Y.: Towards 3D human pose estimation in the wild: a weakly-supervised approach. In: ICCV (2017)
34. Zou, Y., Yang, J., Ceylan, D., Zhang, J., Perazzi, F., Huang, J.: Reducing footskate in human motion reconstruction with ground contact constraints. In: WACV (2020)

# The Dynamic Signature Verification Using population-Based Vertical Partitioning

Marcin Zalasiński[1](✉), Krzysztof Cpałka[1], and Tacjana Niksa-Rynkiewicz[2]

[1] Department of Computational Intelligence, Częstochowa University of Technology, Częstochowa, Poland
{marcin.zalasinski,krzysztof.cpalka}@pcz.pl
[2] Department of Marine Mechatronics, Gdańsk University of Technology, Gdańsk, Poland
tacniksa@pg.edu.pl

**Abstract.** The dynamic signature is an attribute used in behavioral biometrics for verifying the identity of an individual. This attribute, apart from the shape of the signature, also contains information about the dynamics of the signing process described by the signals which tend to change over time. It is possible to process those signals in order to obtain descriptors of the signature characteristic of an individual user. One of the methods used in order to determine such descriptors is based on signals partitioning. In this paper, we propose a new method using a population-based algorithm for determining vertical partitions of the signature and its descriptors. Our method uses a Differential Evolution algorithm for signals partitioning and an authorial one-class fuzzy classifier for verifying the effectiveness of this process. In the simulations, we use a commercial BioSecure DS2 dynamic signature database.

**Keywords:** Dynamic signature · Population-based algorithm · Computational intelligence · Biometrics

## 1 Introduction

The dynamic signature is a biometric attribute which is commonly used for identity verification of an individual. However, this process, which is also referred to in the literature as signature verification, is very difficult to perform. This results from the fact that the signature is characterized by a relatively high intra-class variation and is likely to change over time. Due to this, as the literature shows, there are a number of approaches allowing to determine the most characteristic descriptors of the signature for an individual user (see e.g. [13]). Their use makes the signature verification process more efficient.

The dynamic signature can be acquired by using various digital input devices, e.g. graphic tablets or smartphones, so no expensive sensors are necessary to


H. Yang et al. (Eds.): ICONIP 2020, LNCS 12532, pp. 569–579, 2020.
https://doi.org/10.1007/978-3-030-63830-6_48

obtain this biometric attribute. This kind of signature is described by signals which change over time, e.g. the pen position, pen pressure, instant pen velocity, pen tilt angle, etc. These signals contain a lot of important data which are characteristic of an individual user; however, the information should be properly extracted using appropriate signal processing methods.

### 1.1 Motivation

Typical signals constituting the stylus input in on-line signature identity verification are the stylus velocity, pressure on the tablet surface, its angle to the tablet surface, and their derivative signals as well [13]. Processing of these signals in the context of the dynamic signature verification is not easy. This results, among others, from the fact that the signature verification should take into account the result of an analysis of many available signals describing the signing process. The problem consists in a proper aggregation of the results of this analysis. In the previous works, the authors dealt with this using the division of signals into parts (partitions). The boundary values between these parts were properly calculated with the use of the arithmetic mean [2,3,10]. In this case, it was difficult to process the signatures or their parts that were different from the others. However, this situation is typical because users at the stage of the acquisition of reference signatures create several signatures and some of them may differ from others. In this paper, this problem has been reduced by using a population-based algorithm. The purpose of this algorithm is to divide signatures without the use of the arithmetic mean. This allows us to optimize the division of the signatures into parts, and the expected effect should lead to an increase in the accuracy of the method. In such a procedure of dividing the signatures into parts, one should additionally consider the fact that this procedure must be implemented independently for each user from the database. Therefore, it is needed to ensure that the signature descriptors of individual users differ from one another, but are similar to one another as much as possible in the context of different signatures provided by the same user [8,12,18]. The method proposed in this work meets this very condition.

### 1.2 Novel Elements of the Proposed Approach

In this paper, we propose a new method for dynamic signature signals processing using population-based vertical partitioning. In this method the signature is divided into regions, called partitions, which contain the most characteristic information about the way in which an individual user writes his/her signature. The signature descriptors which are used in the identity verification process are created in the partitions. As previously mentioned, this approach makes it easier to map the specifics of reference signatures of individual users. In this method the signatures differing from the others, or their fragments, do not determine the value of signature descriptors. The method proposed in this paper can be used as a part of the classic approach to the signature verification which we proposed earlier [2] to create more characteristic partitions for an individual and to increase verification efficiency. The method presented in [2] is not focused on signals processing for selecting the most characteristic partitions which are created always in the same way using predefined division points of the signals. The approach offered in this paper consisting in using a population-based algorithm for vertical partitioning has not yet been presented in the literature, thus making this approach a novel one.

### 1.3 Structure of the Paper

This paper is organized into five sections. Section 2 contains a description of the idea of the signature verification using vertical partitioning. Section 3 presents a description of the signal processing method presented in this paper. Section 4 shows the simulation results, and Sect. 5 contains the conclusions.

## 2 Idea of the Signature Verification Using Vertical Partitioning

In this section, we present a general idea of the method for the dynamic signature verification used in this paper. Our method involves processing signature signals and dividing them into partitions characteristic of an individual using a population-based Differential Evolution algorithm, creating signature descriptors in the partitions, and using them by the one-class fuzzy classifier to perform the signature verification. The details of the algorithm are presented below.

The dynamic signature verification system works in two phases – the learning phase (see Fig. 1) and the test (verification) phase. In the learning phase, the user creates a few signatures, which should be pre-processed in order to match their length, rotation, scale and offset. This is realized by commonly known methods (see e.g. [7,10]). For example, matching of the signals' length is performed using the Dynamic Time Warping algorithm [14] which uses different signals $s$ describing the dynamics of the signing process to align the signatures. This alignment is not performed directly on the signals describing the shape so as to avoid losing characteristic features describing the shape of the signature. It is worth noting that many different signals can be used to match the length of the signatures.

Next, the signals of the signature are processed to create $R$ vertical partitions. The number of partitions is dependent on the number of signals $Ns$ used for the alignment of the signatures in the pre-processing phase and it equals $2 \cdot Ns$. This is due to the fact that we can align the signature using different signals describing its dynamics. Partitioning of the signature is realized by selecting division points of the signals describing the signature. In our algorithm, this process is performed using a population-based algorithm. After the selection of division points, each signature is divided into two parts-the initial and final parts of the signing process. The details concerning the partitioning algorithm are presented in Sect. 2.

After creating the partitions, the templates of signature trajectories are created in the partitions. They represent the shape of the genuine signatures of an

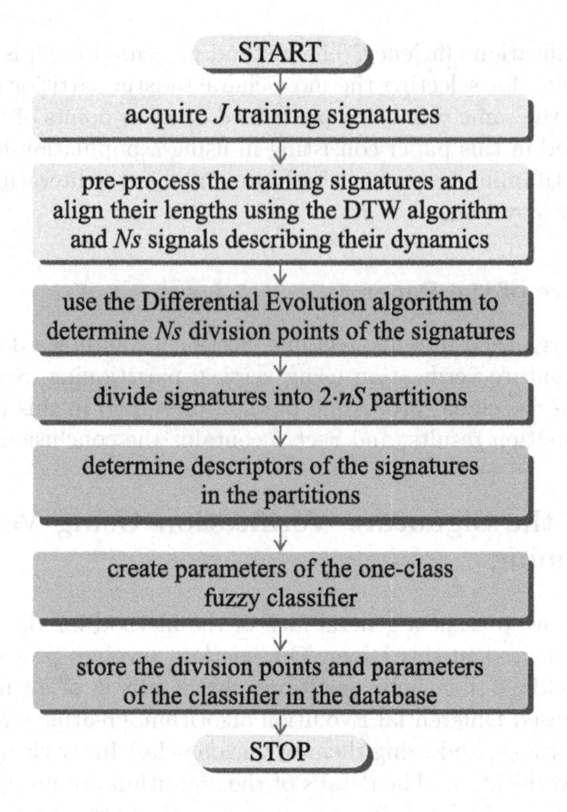


**Fig. 1.** Learning phase of the dynamic signature verification algorithm.

individual user in the created partitions. Templates $\mathbf{t}_{i,r}^{\{s,a\}} = [t_{i,r,k=1}^{\{s,a\}}, t_{i,r,k=2}^{\{s,a\}}, \ldots, t_{i,r,k=K_{i,r}^{\{s\}}}^{\{s,a\}}]$ are also used in the signature verification phase to compare the test signature to the reference signatures ($k$ is the index of a signal sample in the partition, $K_{i,r}^{\{s\}}$ is the number of samples in partition $r$ of user $i$ created after alignment of the signature on the basis of signal $s$, $a$ is the shape trajectory used for creating the template). Their components $t_{i,r,k}^{\{s,a\}}$ are determined as follows:

$$t_{i,r,k}^{\{s,a\}} = \frac{1}{J} \sum_{j=1}^{J} a_{i,j,r,k}^{\{s\}}, \tag{1}$$

where $J$ is the number of the reference signatures created in the learning phase and $a \in \{x, y\}$ represents the value of the trajectory signal of signature $j$ of user $i$ in partition $r$ created on the basis of dynamic signal $s$ at discretization point $k$.

In the next step of the algorithm, signature descriptors $d_{i,j,r}^{\{s,a\}}$ should be created. In the learning phase the descriptors of the reference signatures are used for determining the parameters of the one-class fuzzy classifier (see [3]). In the test phase, they are created from the test signature and they are used as input values of the system used for identity verification. The descriptors of the signatures are determined as follows:

$$d_{i,j,r}^{\{s,a\}} = \sqrt{\sum_{k=1}^{K_{i,r}^{\{s\}}} \left( t_{i,r,k}^{\{s,a\}} - a_{i,j,r,k}^{\{s\}} \right)^2}. \tag{2}$$

After the determination of the descriptors of the reference signatures, which are characteristic of an individual user, a classifier can be trained. Since in the learning phase in the dynamic signature verification problem we do not have any samples representing the signature forgeries, we have decided to use the one-class flexible fuzzy classifier presented in our previous works. A detailed description of the way in which it is determined and how it works can be found, for example, in [3]. In this paper, we have chosen to focus on a new method for the dynamic signature signal processing presented in the next section.

In the test phase (see Fig. 2), one test signature is created, and it should be pre-processed and next partitioned. In the partitions, the test signature descriptors which are used in the identity verification of an individual are created.

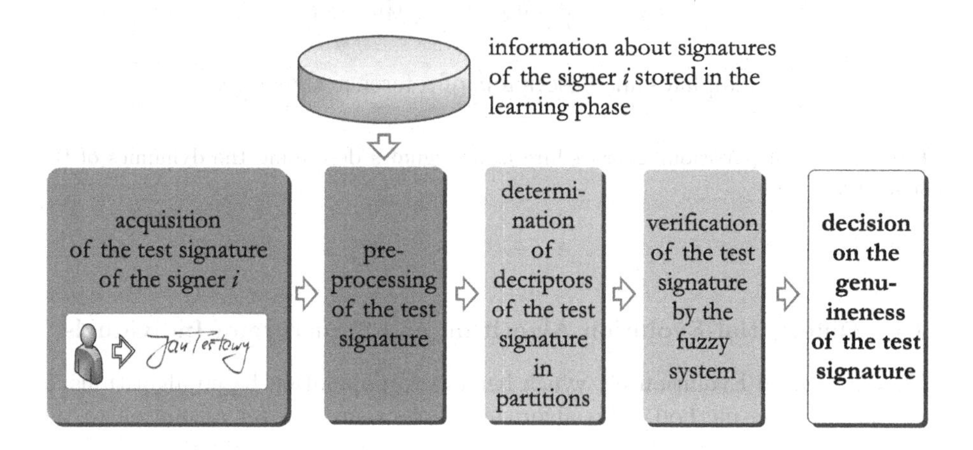


**Fig. 2.** Test phase of the dynamic signature verification algorithm.

## 3 New Method for Signals Partitioning

In this section, we present a new method for signal processing used to create partitions of the signature which are characteristic of an individual user. Our method involves the use of a Differential Evolution population-based algorithm (see Sect. 3.1) for selecting division points values $div_i^{\{s\}}$ used for creating the signature partitions. These values are used to create two vertical partitions of the signature for each considered signal $s$ (e.g. pen velocity or pressure) describing the dynamics of the signature (a general idea of this method is presented in Fig. 3). The membership of the $l$-th sample of the signature of the $i$-th user to the $r$-th partition is described as follows:

$$r_{i,l}^{\{s\}} = \begin{cases} 0 \text{ for } l < div_i^{\{s\}} \\ 1 \text{ for } l \geq div_i^{\{s\}}. \end{cases} \quad (3)$$

The description of the general idea of the Differential Evolution algorithm is presented below. The Differential Evolution algorithm is convenient to use and effective for various applications [4,15,16], but any method based on population can be used instead.

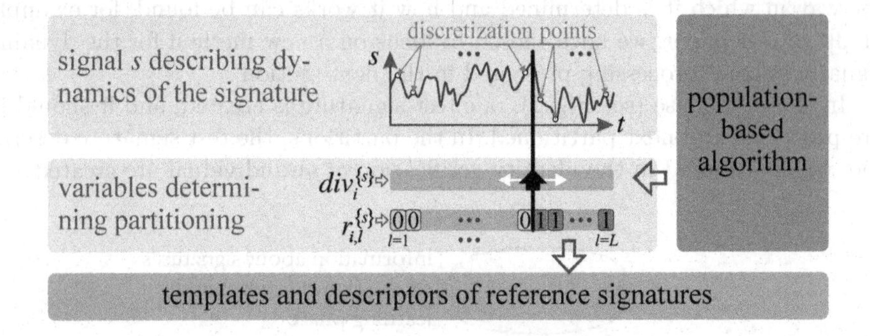


**Fig. 3.** Vertical partitioning procedure using signal $s$ describing the dynamics of the signing process.

### 3.1 Differential Evolution Algorithm and Structure of Individuals

The differential Evolution [6], which belongs to population-based algorithms, is an optimization method. This algorithm starts searching for a solution to the problem by initializing a population which contains $Npop$ individuals. Each of them encodes parameters randomly distributed in the so-called search space. Each individual in the population is evaluated using a properly defined evaluation function. This process is followed by iterative processing of the population, repeated until a specific iteration number is obtained or a satisfactory result is found. In each iteration for each individual, a clone is created, which is modified according to the following formula:

$$X'_{ch,g} = \begin{cases} X_{l,g} + F \cdot (X_{m,g} - X_{n,g}) \text{ for } rand < CR \text{ or } g = R \\ X_{ch,g} \text{ otherwise}, \end{cases} \quad (4)$$

where $\mathbf{X}$ are the individuals, $\mathbf{X}'$ are the individual clones, $ch = 1, ..., Npop$ are the indexes of the individuals, $g = 1, ..., D$ are the indexes of the individuals parameters ($D$ is the number of the parameters being optimized), $F \in \langle 0; 2)$

is the differential weight, $CR \in \langle 0; 1 \rangle$ is the crossover probability, $l, m, n$ are the indexes of the randomly selected individuals from the population (without repetition and different from $ch$), $R$ is a randomly selected integer from set $\{1, D\}$ (thus always one individual parameter is modified), and $rand$ is a random function that returns the values from range $\langle 0; 1 \rangle$. Each clone is repaired (the parameters are narrowed to the search space) and evaluated. If the clone has a better value of the objective function than its parent, then it replaces the parent. After the iteration or interruption of the algorithm, the best solution found is presented.

In the approach proposed in this paper, each individual $\mathbf{X}_{i,ch}$ from the population encodes $Ns$ division points of the signature determined for user $i$ and associated with different signals describing the dynamics of the signature which were used in the signature alignment process. The individual is represented as follows:

$$\begin{aligned} \mathbf{X}_{i,ch} &= \left\{ div_i^{\{s_1\}}, div_i^{\{s_2\}}, \ldots, div_i^{\{s_{Ns}\}} \right\} \\ &= \left\{ X_{i,ch,s_1}, X_{i,ch,s_2}, \ldots, X_{i,ch,s_{Ns}} \right\}, \end{aligned} \tag{5}$$

where $\{s_{1,}, s_{2,}, \ldots, s_{Ns,}\}$ is a set of signals used for the signature alignment and associated with the created partitions.

The purpose of the algorithm is to select the subset with division points which will allow us to create partitions containing the most characteristic descriptors for user $i$. The partitions created using the Differential Evolution algorithm have flexible boundaries and are selected individually for each signer. Due to this, templates and descriptors of the signatures determined in the created partitions should be selected with greater precision.

In the next section, we present the evaluation function created for the problem of the dynamic signature signal processing.

### 3.2 Evaluation Function for the Dynamic Signature Signal Processing

Assessment of the individuals being part of the population is achieved by using a properly designed evaluation function. Its value is calculated taking into account the values of the descriptors from the partitions created using the encoded division points and a specially defined ratio between the number of the discretization points in both partitions created using the same division point. This ratio has been introduced because the number of points in the partition should not be too small. It is determined as follows:

$$Rp_{i,j}^{\{s\}} = \begin{cases} 1 - \frac{K_{i,1}^{\{s\}}}{K_{i,0}^{\{s\}}} \text{for } K_{i,0}^{\{s\}} \geq Kc_{i,1}^{\{s\}} \\ 1 - \frac{K_{i,0}^{\{s\}}}{K_{i,1}^{\{s\}}} \text{ otherwise.} \end{cases} \tag{6}$$

Next, the values of the descriptors representing the similarity of the training signatures to the template should be normalized. This is performed using membership function $\mu\left(d_{i,j,r}^{\{s,a\}}\right)$ defined as follows:

$$\mu\left(d_{i,j,r}^{\{s,a\}}\right)=\left(1+\exp\left(5-2\cdot d_{i,j,r}^{\{s,a\}}\right)\right)^{-1}. \tag{7}$$

Next, the values of the above-mentioned variables should be normalized for each training signature $j$ of user $i$. As the result of normalization, the values of parameters $avgD_{i,j}$ and $avgR_{i,j}$ are obtained. They are defined as follows:

$$\begin{cases} avgD_{i,j}=\frac{1}{4\cdot Ns}\cdot\left(\begin{array}{c}\mu\left(d_{i,j,0}^{\{s_1,x\}}\right)+\mu\left(d_{i,j,1}^{\{s_1,x\}}\right)+\\ +\mu\left(d_{i,j,0}^{\{s_1,y\}}\right)+\mu\left(d_{i,j,1}^{\{s_1,y\}}\right)+\ldots\\ +\mu\left(d_{i,j,0}^{\{s_{Ns},x\}}\right)+\mu\left(d_{i,j,1}^{\{s_{Ns},x\}}\right)+\\ +\mu\left(d_{i,j,0}^{\{s_{Ns},y\}}\right)+\mu\left(d_{i,j,1}^{\{s_{Ns},y\}}\right)\end{array}\right)\\ avgR_{i,j}=\frac{1}{Ns}\cdot\left(Rp_{i,j}^{\{s_1\}}+\ldots+Rp_{i,j}^{\{s_{Ns}\}}\right),\end{cases} \tag{8}$$

where $avgD_{i,j}$ is the average value of descriptors (see Eq. (2)) of signature $j$ and $avgR_{i,j}$ is the average value of parameters $Rp_{i,j}^{\{s\}}$ (see Eq. (6)) for signature $j$.

The value of the evaluation function for individual $\mathbf{X}_{i,ch}$ in population ff $(\mathbf{X}_{i,ch})$ is calculated using the above-mentioned coefficients. Their values are aggregated using weighted algebraic triangular norm $T^*\{\cdot\}$ [1]. The evaluation function can be determined as follows:

$$\begin{array}{l}\mathrm{ff}\left(\mathbf{X}_{i,ch}\right)=T^*\left\{\begin{array}{c}avgD_{i,1},\ldots,avgD_{i,J},avgR_{i,1},\ldots,avgR_{i,J};\\ wD_{i,1},\ldots,wD_{i,J},wR_{i,1},\ldots,wR_{i,J}\end{array}\right\}\\ =1-wD_{i,1}\cdot\left(1-avgD_{i,1}\right)\cdot\ldots\cdot 1-wR_{i,J}\cdot\left(1-avgR_{i,J}\right),\end{array} \tag{9}$$

where t-norm $T^*\{\cdot\}$ is a generalization [1] of the usual two-valued logical conjunction (studied in the classical logic), $wD_{i,j}\in[0,1]$ and $wR_{i,j}\in[0,1]$ are the weights of importance of arguments $avgD_{i,j}$ and $avgR_{i,j}$, which are the algorithm's parameters.

In the method proposed in this paper, the Differential Evolution algorithm tends to maximize the value of the evaluation function. After a certain number of iterations $Nit$, the population-based algorithm stops its operation and returns the determined values of the division points, which are used for signature partitioning.

## 4 Simulation Results

In the simulations, we implemented the proposed method for the dynamic signature signal processing using population-based vertical partitioning in the C#.NET language. Then, we performed identity verification on the basis of the dynamic signature divided into the partitions created with the use of our method. It was realized to evaluate the efficiency of the new method offering a population-based vertical division of the signature. We assumed that identity verification using the partitions created by the new method should work better than in the case of our previous method [2] which creates fixed size vertical partitions. In our simulations, we used the one-class fuzzy classifier which we proposed in [3].

The problem of the dynamic signature verification is specific. Therefore, authors of new algorithms should primarily use databases of signatures in their simulations. This is due to the following reasons: (a) the biometric data describing signatures are sensitive and protected by law. This makes it difficult to obtain, store, and process them. This also makes it difficult to share them with other people for, e.g., testing various verification methods. Using signature databases eliminates this problem because the databases contain completely anonymous data-obtained and made available in accordance with applicable law. The great advantage of signature databases is also the fact that they contain data from many people who differ, for example, in age, education, type of work, etc. (b) biometric data used in tests should contain the so-called skilled forgeries created by professional forgers. It would be difficult to acquire signatures of several dozen users and to prepare forgeries for them. Meanwhile, the signature databases contain such forgeries. (c) using signature databases allows you to reliably compare the effectiveness of signature verification methods, which is a key issue. We have been dealing with the problem of signature verification for several years, so using signature databases allows us to practically verify the effectiveness of the proposed solutions. If we used our own signature databases, presented the results of our experimental research, and wrote that we cannot share the databases, then the results obtained by us could not be recognized.

The simulations were performed using the BioSecure dynamic signature database DS2 [5], which contains signatures of 210 users acquired in two sessions with the use of a graphics tablet. Each session contains 15 genuine signatures and 10 skilled forgeries per person. In the learning phase, we used 5 randomly selected genuine signatures of each signer. During the test phase, we used 10 genuine signatures and 10 so-called skilled forgeries [11] of each signer. The alignment of the signatures and partitioning were realized using two signals describing the dynamics of the signing-velocity and pressure ($Ns = 2$).

We adopted the following values of the Differential Evolution algorithm parameters in the simulations: (a) the number of individuals in population $Npop = 100$, (b) the value of parameter $F$ [17] of the DE algorithm is 0.5, (c) the value of parameter $CR$ [17] of the DE algorithm is 0.9, (d) the value of weights $wD_{i,j}$ is 0.7, (e) the value of weights $wR_{i,j}$ is 0.2, and (f) the number of iterations $Nit$ of the Differential Evolution algorithm is 100.

The simulations were repeated 5 times in accordance with the standard cross-validation procedure. The results of the simulations are presented in Table 1 in the form of FAR (False Acceptance Rate), FRR (False Rejection Rate) and EER (Equal Error Rate) coefficients which are used in the literature to evaluate the effectiveness of biometric methods [11].

**Table 1.** Comparison of the accuracy of the method using on-line signature partitioning with a population-based algorithm to other methods for the dynamic signature verification using the BioSecure database.

| Method | Average FAR | Average FRR | Average error |
|---|---|---|---|
| Methods of other authors [9] | – | – | 3.48%–30.13% |
| Method using fixed vertical partitions [2] | 3.13% | 4.15% | 3.64% |
| **Our method** | **2.80%** | **3.04%** | **2.92%** |

Given the obtained results, we can see that the proposed method has achieved a good accuracy in comparison to the methods proposed by other authors. It means that the selection of characteristic parts of the signature for each user can have a key role in the identity verification process. Moreover, the accuracy of the proposed method is better than the accuracy of the method using fixed size vertical partitions which we proposed in [2]. It confirms that the partitions created in a more flexible way by the population-based signal processing are more characteristic in the case of individual users and can improve verification efficiency.

## 5 Conclusions

In this paper, we have presented the dynamic signature signal processing method using a population-based algorithm in order to perform vertical partitioning. The purpose of the method is to create the most characteristic partitions of the signature for an individual user. These partitions should contain a certain degree of important information about the dynamics of his/her signing process. This information is very useful in the identity verification process, which has been confirmed by our simulations performed using the BioSecure dynamic signature database. The accuracy of the signature verification process with the use of the dynamically created characteristic partitions is better than in the case of using partitions with a fixed size.

Our future plans include, among others, using different population-based algorithms in order to compare their performance in this field, taking into account the changes occurring in signatures over time while determining partitions and creating a new one-class classifier based on possibilities offered by convolutional neural-networks.

**Acknowledgment.** The authors would like to thank reviewers for their helpful comments.

This paper was financed under the program of the Minister of Science and Higher Education under the name 'Regional Initiative of Excellence' in the years 2019–2022, project number 020/RID/2018/19 with the amount of financing PLN 12 000 000.

## References


1. Cpałka, K.: Design of Interpretable Fuzzy Systems. SCI, vol. 684. Springer, Cham (2017). https://doi.org/10.1007/978-3-319-52881-6_9
2. Cpałka, K., Zalasiński, M.: On-line signature verification using vertical signature partitioning. Expert Syst. Appl. **41**, 4170–4180 (2014)
3. Cpałka, K., Zalasiński, M., Rutkowski, L.: New method for the on-line signature verification based on horizontal partitioning. Pattern Recogn. **47**, 2652–2661 (2014)
4. Dawar, D., Ludwig, S.A.: Effect of strategy adaptation on differential evolution in presence and absence of parameter adaptation: an investigation. J. Arti. Intell. Soft Comput. Res. **8**, 211–235 (2018)
5. Homepage of Association BioSecure. https://biosecure.wp.imtbs-tsp.eu. Accessed 26 May 2020
6. Das, S., Suganthan, P.N.: Differential evolution: a survey of the state-of-the-art. IEEE Trans. Evol. Comput. **15**(1), 4–31 (2010)
7. Fierrez, J., Ortega-Garcia, J., Ramos, D., Gonzalez-Rodriguez, J.: HMM-based on-line signature verification: feature extraction and signature modeling. Pattern Recogn. Lett. **28**, 2325–2334 (2007)
8. Galbally, J., Martinez-Diaz, M., Fierrez, J.: Aging in biometrics: an experimental analysis on on-line signature. PLoS One **8**(7), e69897 (2013)
9. Houmani, N., et al.: BioSecure signature evaluation campaign (BSEC'2009): evaluating online signature algorithms depending on the quality of signatures. Pattern Recogn. **45**, 993–1003 (2012)
10. Ibrahim, M.T., Khan, M.A., Alimgeer, K.S., Khan, M.K., Taj, I.A., Guan, L.: Velocity and pressure-based partitions of horizontal and vertical trajectories for on-line signature verification. Pattern Recogn. **43**, 2817–2832 (2010)
11. Jain, A.K., Ross, A.: Introduction to biometrics. In: Jain, A.K., Flynn, P., Ross, A.A. (eds.) Handbook of Biometrics, pp. 1–22. Springer, Heidelberg (2008). https://doi.org/10.1007/978-0-387-71041-9_1
12. Liew, L. H., Lee, B. Y., Wang, Y. C. Intra-class variation representation for on-line signature verification using wavelet and fractal analysis. In: International Conference on Computer and Drone Applications (IConDA), Kuching, pp. 87–91 (2017)
13. Linden, J., Marquis, R., Bozza, S., Taroni, F.: Dynamic signatures: a review of dynamic feature variation and forensic methodology. Forensic Sci. Int. **291**, 216–229 (2018)
14. Müller, M.: Dynamic time warping. In: Müller, M. (ed.) Information Retrieval for Music and Motion, pp. 69–84. Springer, Heidelberg (2007). https://doi.org/10.1007/978-3-540-74048-3_4
15. Nasim, A., Burattini, L., Fateh, M.F., Zameer, A.: Solution of linear and non-linear boundary value problems using population-distributed parallel differential evolution. J. Artif. Intell. Soft Comput. Res. **9**, 205–218 (2019)
16. Opara, K.R., Arabas, J.: Differential evolution: a survey of theoretical analyses. Swarm Evol. Comput. **44**, 546–558 (2019)
17. Pedersen, M.E.H.: Good parameters for differential evolution. Hvass Laboratories Technical Report, vol. HL1002 (2010)
18. Xia, X., Song, X., Luan, F., Zheng, J., Chen, Z., Ma, X.: Discriminative feature selection for on-line signature verification. Pattern Recogn. **74**, 422–433 (2018)

# Triple Attention Network for Clothing Parsing

Ruhan He[1], Ming Cheng[1], Mingfu Xiong[1](✉), Xiao Qin[2], Junping Liu[1], and Xinrong Hu[1]

[1] Engineering Research Center of Hubei Province for Clothing Information, Wuhan Textile University, Wuhan 430200, China
{heruhan,jpliu,hxr}@wtu.edu.cn, cm_jsw@163.com, xmf2013@whu.edu.cn

[2] Department of Computer Science and Software Engineering, Auburn University, Auburn, USA
xqin@auburn.edu

**Abstract.** Clothing parsing has been actively studied in the vision community in recent years. Inspired by the color coherence for clothing and the self-attention mechanism, this paper proposes a Triple Attention Network (TANet) equipped with a color attention module, a position attention module and a channel attention module, to facilitate fine-grained segmentation of clothing images. Concretely, the color attention module is introduced for harvesting color coherence, which selectively aggregates the color feature of clothing. The position attention module and the channel attention module are designed to emphasize the semantic interdependencies in spatial and channel dimensions respectively. The outputs of the three attention modules are incorporated to further improve feature representation which contributes to more precise clothing parsing results. The proposed TANet has achieved 69.54% mIoU - a promising clothing parsing performance on ModaNet, the latest large-scale clothing parsing dataset. Especially, the color attention module is also demonstrated to bring semantic consistency and precision improvement obviously. The source code is made available in the public domain.

**Keywords:** Image processing and computer vision · Clothing parsing · Attention network · Color coherence

## 1 Introduction

Clothing parsing is a special branch of image semantic parsing where the categories are one of the clothing items such as t-shirt, skirt, dress. It has been actively studied in the vision community in recent years [1–6], which is also the foundation of clothing recognition [7], recommendation [8], retrieval [2,7,9], and the like.

Supported by National Natural Science Foundation of China (No. 61170093).


H. Yang et al. (Eds.): ICONIP 2020, LNCS 12532, pp. 580–591, 2020.
https://doi.org/10.1007/978-3-030-63830-6_49

Clothing parsing requires the model has a good contextual recognition ability for the deformations and occlusions. The existing clothing parsing methods have achieved promising performance. Simo-Serra *et al.* [3] proposed a pose-aware *conditional random field* (CRF) to capture the appearance, shape of clothing and structure information of human body, which has achieved significant improvement on Fashionista dataset [9]. Tangseng *et al.* [1] extended the FCN [10] architecture with an *outfit encoder* and *conditional random field* (CRF) to obtain clothing global context information. However, color, which is the first visual impact to people for clothing, is usually ignored by the existing clothing parsing methods. In addition, attention mechanism, which can capture the long-range dependencies, has been successfully applied for various tasks. Especially, channel attention and spatial attention are usually used together [11–14].

In this paper, a novel framework named Triple Attention Network (TANet) for clothing parsing is proposed to make full use of color consistency of clothing and take advantage of attention mechanism, which is illustrated in Fig. 1. The TANet consists of three attention modules, namely color attention module, position module and channel attention module, for fine-grained segmentation of clothing images. It not only learns the feature interdependencies between the channel and spatial dimensions, which are frequently used in previous methods [11–14], but also makes full use of the color coherence. The position attention module and the channel attention module, are designed to emphasize the semantic interdependencies in spatial and channel dimensions respectively. In particular, the color attention module selectively aggregates the color feature of clothing. The outputs of the three attention modules are incorporated to further improve feature representation which contributes to more precise clothing parsing results. It shows that the proposed approach outperforms the existing methods. The main contributions of this paper are summarized as following:

- This study proposes a Triple Attention Network (TANet) which is equipped with a color attention module, a position attention module and a channel attention module for clothing parsing.
- The color correlation is first used to capture the essential features of clothing and has improves the parsing results obviously.
- The proposed method achieves a promising clothing parsing results on ModaNet [15], where all the crucial implementation details are included. The code and trained models are publicly available[1].

## 2 Related Work

### 2.1 Clothing Parsing

Clothing parsing is one of the hot topics in the field of computer vision. It distinguishes itself from general object or scene segmentation problems in that

[1] https://www.github.com/cm-jsw/TANet.

fine-grained categories such as *T-shirt, skirt and dress*, require high-level judgment based on the semantics of clothing and the deforming structure within an image.

Yamaguchi *et al.* first built the Fahionista [9] and then extended it to Paper Doll [2] for clothing paring. However, confusing labels like *top* and *blouse* are the main challenge of clothing parsing. Some conventional methods made use of the Markov Random Field [16] (MRF) and Conditional Random Field [3] (CRF) as an auxiliary tool. Tangseng *et al.* developed a outfit encoder side-branch network based on FCN [10] to learn clothing combinatorial preference.

However, the above methods employ insufficient color information, which is important for clothing. It is extremely challenging to design a model to capture the color information for clothing parsing.

## 2.2 Semantic Segmentation

As a powerful method, deep convolutional neural networks (DCNN) have been used for semantic segmentation in the last few years. FCN [10] is a fundamental work of popular approaches that replaces the last fully connected layers of DCNN by convolutional layers to restore resolution for segmentation. A few FCN-based techniques modifying the network structure have made a great progress in semantic segmentation. Deeplabv3 [17] fuse information of low-level and high-level layers to predict segmentation mask by exploit encoder-decoder structures. CRF-RNN [18] and [1] use *Graph Model* such as CRF for semantic segmentation.

Meanwhile, to enrich feature representation some works integrate context information. PSPNet [19] utilizes pyramid pooling to aggregate contextual information. DANet [11] model the semantic interdependencies in spatial and channel dimensions to adaptively integrate local features with their local dependencies. These methods of semantic segmentation are very useful for clothing parsing.

## 2.3 Attention Model

Attention models, which recently become very popular, are widely used in various tasks. Recently, it is popular to fuse multiple attentions for different tasks. Fan *et al.* [12] proposed a spatial channel parallelism network (SCPNet), in which each channel in the ReID feature pays attention to a given spatial part of the body. Su *et al.* [13] proposed a Spatial, Channel-wise Attention Residual Bottelneck (SCARB), integrating the spatial and channel-wise attention mechanism into the original residual unit. Chen *et al.* [14] propose a network with multi-context attention mechanism into an end-to-end framework for human pose estimation.

In Non-local Network [20], non-local module was proposed to generate attention map in the feature map, thereby furnishing dense contextual information aggregation. Fu *et al.* [11] exploited self-attention mechanism to obtain contextual information in spatial-wise and channel-wise separately. Huang *et al.* proposed CCNet [21] to get dense contextual information in a recurrent way by a effectively and efficiently criss-cross attention module.

Attention models have achieved nonnegligible achievement in many application areas. Inspired by the effectiveness of multiple attention modules, the proposed method takes advantage of the self-attention mechanism in the task of clothing parsing and uses the color information of clothing for better segmentation.

# 3 Approach

In this section, the details of the proposed Triple Attention Network (TANet) tailored for clothing parsing will be elaborated. First of all, a general framework of this network will be presented. Then, the three attention modules are introduced, which capture contextual information in color, spatial and channel aspect. Finally, it will be described how to integrate the contextual information in the framework.

## 3.1 Overall

TANet architecture is depicted in Fig. 1. To enhance feature representation for clothing parsing, this method utilizes multiple attention mechanisms [11–13] with three parallel attention modules (the color attention module, the position attention module and the channel attention module). Especially, the color attention module aims to draw global color context over local features, thus improving feature representations for pixel-level prediction.

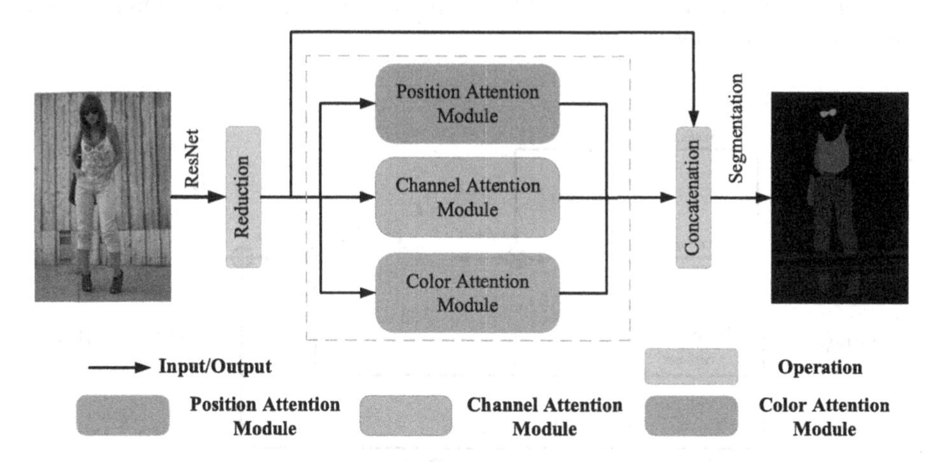


**Fig. 1.** An overview of the proposed Triple Attention Network (TANet) for clothing parsing.

Similar to the previous studies [11,17,21], the proposed method removes the last down-sampling operations and employs dilation convolutions in the subsequent convolutional layers of ResNet [22], thus enlarging size of the output feature map size to the 1/8 of an input image. After that, a convolution layer is applied to obtain the dimension reduction feature map of the previous one.

Then, the dimension reduction feature map, retaining details without adding extra parameters, is fed into the three parallel attention modules (color, position and channel attention module) to generate new feature representations (see more details in Sect. 3.2 and Sect. 3.3).

The prior position module, channel module and the novel color attention module are the self-contained modules which can be augmented into a CNN architecture at any point, and in any number, capturing rich contextual information. In order to aggregate features from these three modules, the proposed method concatenates the modules with the original local representation features. This procedure is followed by one or several convolutional layers with batch normalization and activation for feature fusion.

Finally, the fused features are fed into the segmentation layer to generate a final segmentation map. Simplicity is central to this design and; therefore, this method construct a model that is robust to a wide variety of design choices. The proposed method does not adopt cascading operations, because such operations need large GPU memory.

### 3.2 Color Attention Module

Color information is important for clothing parsing. Therefore, the proposed method introduces a color attention module to capture the color coherence in clothing parsing images, which is beneficial to segment objects. By exploiting the color information, this method is in a position to improve the representation capability of clothing parsing.

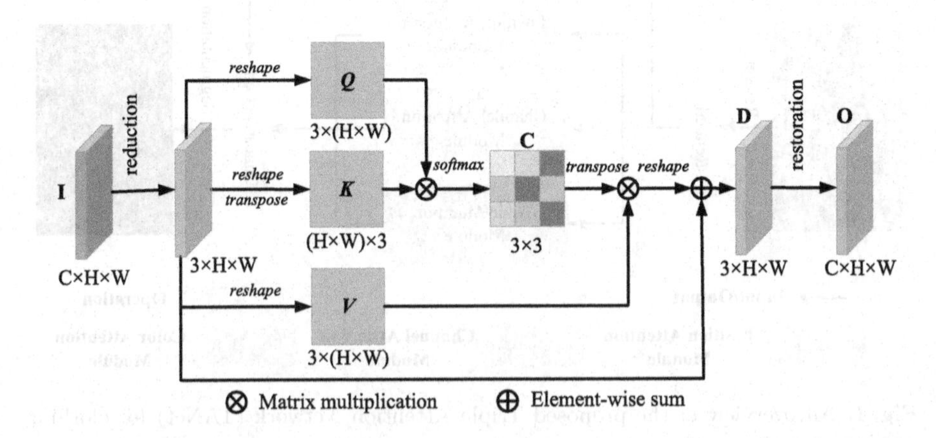


**Fig. 2.** Details of color attention module. (Best viewed in color)

Figure 2 shows the structure of the proposed color attention module. The proposed method first performs a convolution operation for input features to reduce dimension from $I \in \mathbb{R}^{C\times H\times W}$ to $I' \in \mathbb{R}^{3\times H\times W}$, where 3 denotes the color space. Then, this system reshapes $I'$ to $\{Q,V\} \in \mathbb{R}^{3\times N}$ and transpose it to $K \in \mathbb{R}^{N\times 3}$, where $N = H \times W$. Finally, a softmax layer is applied to obtain the color attention map $C \in \mathbb{R}^{3\times 3}$. Thus:

$$C_{j,i} = \frac{exp(Q_i \cdot K_j)}{\sum_{i=1}^{3}(exp(Q_i \cdot K_j))} \tag{1}$$

where $C_{j,i}$ measures the $i^{\text{th}}$ color attribute's impact on $j^{\text{th}}$ color attribute. In addition, a matrix multiplication is performed between the transpose of $C$ and $V$, thereby reshaping the result to $\mathbb{R}^{3\times H\times W}$. Then the result is multiplied by a scale parameter $\gamma$ followed by performing an element-wise sum operation with $I'$ to obtain the output $D \in \mathbb{R}^{3\times H\times W}$:

$$D_j = \gamma \sum_{i=1}^{3}(C_{i,j}V_j') + I_j' \tag{2}$$

where $\gamma$ is learned during the training phase. For consistency, a convolution layer is applied to restore the feature dimension to $O \in \mathbb{R}^{C\times H\times W}$. Therefore, this feature representations, achieving mutual gains, are robust for clothing parsing.

The color attention module improves the previous attention module by adaptively capturing the color information. Experimental results show that this module can improve contour segmentation and reduce misclassification (see more details in Sect. 4.3). Desinging and optimizing connection are out the scope of this paper, so the proposed method opt for the aforementioned simple design.

### 3.3 Position/Channel Attention Module

The position attention module and the channel attention module have been widely used. They are the fundamental modules in many previous works [11–13]. In addition, they have achieved good results in some domains such as semantic segmentation, person re-identification, pose estimation and etc.

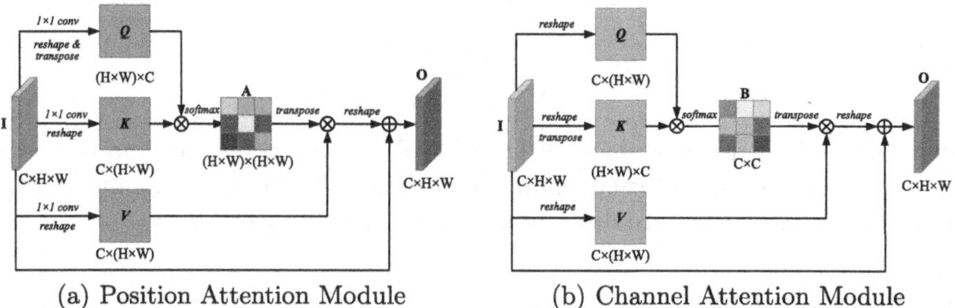


(a) Position Attention Module (b) Channel Attention Module

**Fig. 3.** The details of the position attention module, the channel attention module are illustrated in (a), (b). (Best viewed in color)

Modeling rich contextual relationship over local features is non-trivial in the previous FCN-based method. Furthermore, images used for clothing parsing are diverse on categories, scales and self-occlusion. Hence, the proposed method use the position attention module encoding a wider range of contextual information into local features to enhancing their representation capability. The structure of position attention module is illustrated in Fig. 3(a). The operation implemented in the position attention module is formulated as:

$$O_j = \alpha \sum_{i=1}^{N} (A_{j,i} V_i) + I_j \tag{3}$$

where $\alpha$ is initialized as 0 and gradually learns to assign an appropriate weight, $I \in \mathbb{R}^{C \times H \times W}$ denotes input feature maps, $V \in \mathbb{R}^{C \times N}(N = H \times W)$ denotes the feature maps $I$ performs $1 \times 1$ convolution and reshape operation. $A \in \mathbb{R}^{N \times N}$ is a position attention map and $A_{j,i}$ measures the $i^{\text{th}}$ position's impact on $j^{\text{th}}$ position as follows:

$$A_{j,i} = \frac{exp(Q_i \cdot K_j)}{\sum_{i=1}^{N} (exp(Q_i \cdot K_j))} \tag{4}$$

where $Q \in \mathbb{R}^{N \times C}$ is the feature maps $I$ that performs $1 \times 1$ convolution, reshape and transpose operations, and $K \in \mathbb{R}^{C \times N}$ is the feature maps $I$ that performs $1 \times 1$ convolution and reshape operations. Therefore, the position attention module has a global contextual view that selectively aggregates contexts.

Some previous studies [11–13] suggest that each channel map of high level can be regarded as a class-specific responses, and different semantic response are associated with each other. Therefore, the proposed method applies channel attention module to explicitly model interdependencies between channels for clothing parsing. As illustrated in Fig. 3(b). The output $O \in \mathbb{R}^{C \times H \times W}$ is expressed as:

$$O_j = \beta \sum_{i=1}^{C} (B_{j,i} V_i) + I_j \tag{5}$$

where $\beta$ is a scale parameter which is initialized by 0, $I \in \mathbb{R}^{C \times H \times W}$ is an input features, $B \in \mathbb{R}^{C \times C}$ is the channel attention map, and $V \in \mathbb{R}^{C \times N}(N = H \times W)$ denotes the $I$ perform a reshape operation. Channel attention map $B$ is written as:

$$B_{j,i} = \frac{exp(Q_i \cdot K_j)}{\sum_{i=1}^{C} (exp(Q_i \cdot K_j))} \tag{6}$$

where $Q \in \mathbb{R}^{C \times N}$ is the reshape of $\{I, K\} \in \mathbb{R}^{N \times C}$ is the reshape and transpose of $I$. The final feature of each channel is a weighted sum of all channels, which helps to strengthen feature discriminability of clothing parsing.

## 4 Experiment

To evaluate the proposed method, comprehensive experiments are carried out on the ModaNet dataset [15]. The experimental results demonstrate that TANet achieves promising performance. In the following subsections, the dataset and performance metric are introduced, then the implementation details are presented. Finally, the results on ModaNet dataset are reported.

### 4.1 Dataset and Evaluation Metric

Some datasets are tailored for clothing parsing, such as CFPD (2,682 images) [16], CCP (1,004 images) [6] and Fashionista (685 images) [9], but they are small relatively for deep learning algorithms. So the proposed method chooses ModaNet [15], the latest large-scale clothing parsing dataset, as experimental dataset.

ModaNet, a dataset for clothing fashion parsing, contains 55,176 street images, in which the clothing images are annotated by polygonal mask to 13 categories. The dataset includes 52,377/2,799 images for training and validation purposes. It is noted that there are some invalid data in the dataset[2] [15]. The valid data are randomly divided into 49,620/2,612 (95%/5%) of the *train-validation / test* set.

The proposed method adopts Mean IoU (Mean of class-wise intersection over union), a commonly used evaluation metric for semantic segmentation, for experimental data.

### 4.2 Implementation Details

The proposed method uses the open source pytorch segmentation toolbox [23], and chooses the ImageNet pre-trained ResNet-101 [22] as network backbone (see also [11,19,21]). The model trained with Inplace-ABN [24] to the mean and standard-deviation of BatchNorm across multiple GPUs following the previous work reported in [21]. SGD with mini-batch used for training, momentum and weight decay coefficients are set to 0.9 and 0.0001 respectively. The initial learning rate is 1e−2. Following [11,19,21,25], the proposed method employs a poly learning rate policy where the initial learning rate is multiplied by $1-(\frac{iter}{max\_iter})^{power}$ with $power = 0.9$. In addition, for data augmentation, the proposed method applies flipping horizontally and rotates at a rotation angle of $\pm 10^{\circ}$ randomly. The model is trained 150K iterations (∼24 epochs) on 4 × GTX 1080 GPUs. Batch size is set to 8 due to computational power limitations.

### 4.3 Result on ModaNet Dataset

**Comparisons.** For comparison, some promising semantic segmentation methods are tested on ModaNet. As shown in Table 1, the backbones and training settings in FCN [10], CRFasRNN [18] and Deeplabv3+ [26] are different from the proposed method. Particularly, DANet [11], Deeplabv3 [17], PSPNet [19], CCNet [21] and TANet use the same backbone and training strategies as the proposed

[2] https://github.com/eBay/modanet.

method. The results reveal that the proposed TANet achieves the superb performance. Among these approaches, DANet [11] is the most related work to the proposed method; DANet generates position and channel attention for scene segmentation. This method can achieve good performance with fewer parameters adjunction compared to all parameters.

**Table 1.** Experimental performance on test set across leading competitive models. *PAM* represents Position Attention Module, *CAM* represents Channel Attention Module, *CoAM* represents Color Attention Module.

| Method | Backbone | PAM | CAM | CoAM | mIoU (%) |
|---|---|---|---|---|---|
| FCN-32s [10]† | VGG | – | – | – | 35.36 |
| FCN-16s [10]† | VGG | – | – | – | 36.93 |
| FCN-8s [10]† | VGG | – | – | – | 38.00 |
| CRFasRNN [18]† | VGG | – | – | – | 40.57 |
| Deeplabv3+ [26]† | Xception-65 | – | – | – | 51.14 |
| DANet [11] | ResNet-101 | ✓ | ✓ | – | 68.16 |
| Deeplabv3 [17] | ResNet-101 | – | – | – | 68.57 |
| PSPNet [19] | ResNet-101 | – | – | – | 68.80 |
| CCNet [21] | ResNet-101 | – | – | – | 69.22 |
| TANet | ResNet-101 | ✓ | ✓ | ✓ | **69.55** |

† Result given in [15]

## Ablation Studies

*The Effect of Attention Modules.* As shown in Table 1, compared with the baseline DANet that only uses the position attention module and channel attention module, the proposed method with the additional color attention module yields a results of 69.55% in Mean IoU, which represents an improvement of 1.39%. The results unveil that the color attention module brings great benefit to clothing parsing.

As illustrated in Table 2, experimental results show that TANet is sensitive to small objects such as *belt, boots, footwear*. Meanwhile, Fig. 4 demonstrates that, with the color attention module, a handful of misclassified categories are now correctly classified (*e.g.*, 'footwear' in the left column). Thanks to the usage of color information, TANet can improve contour segmentation and reduce misclassification (in the middle column). Moreover, color features helps in segmenting small objects (such as 'sunglasses' in the right column). The semantic consistency is improved obviously by the virtue of the color attention module.

*Result Analysis.* Fig. 5(a) unravels the successful parsing results, demonstrating that the proposed model is good at capturing contour information for improved segmentation by utilizing color features, which are overlooked by the position attention module and the channel position module. Nevertheless, misclassification occurs when some adjacent objects have similar colors (see Fig. 5(b)).

**Table 2.** Per-class IoU (%) on test set. TANet outperforms existing approaches and achieves 69.55% in Mean IoU.

| Method | mIoU | bg | Bag | Belt | Boots | Footwear | Outer | Dress | Sunglasses | Pants | Top | Shorts | Skirts | Headwear | Scarf&Tie |
|---|---|---|---|---|---|---|---|---|---|---|---|---|---|---|---|
| FCN-32s [10]† | 35.36 | 95.00 | 27.00 | 12.00 | 32.00 | 33.00 | 36.00 | 28.00 | 25.00 | 51.00 | 38.00 | 40.00 | 28.00 | 33.00 | 17.00 |
| FCN-16s [10]† | 36.93 | 96.00 | 26.00 | 19.00 | 32.00 | 38.00 | 35.00 | 25.00 | 37.00 | 51.00 | 38.00 | 40.00 | 23.00 | 41.00 | 16.00 |
| FCN-8s [10]† | 38.00 | 96.00 | 24.00 | 21.00 | 32.00 | 40.00 | 35.00 | 28.00 | 41.00 | 51.00 | 38.00 | 40.00 | 24.00 | 44.00 | 18.00 |
| CRFasRNN [18]† | 40.57 | 96.00 | 30.00 | 18.00 | 41.00 | 39.00 | 43.00 | 32.00 | 36.00 | 56.00 | 40.00 | 44.00 | 26.00 | 45.00 | 22.00 |
| Deeplabv3+ [26]† | 51.14 | 98.00 | 42.00 | 28.00 | 40.00 | 51.00 | 56.00 | 52.00 | 46.00 | 68.00 | 55.00 | 53.00 | 41.00 | 55.00 | 31.00 |
| DANet [11] | 68.16 | 98.11 | 75.77 | 53.47 | 45.29 | 53.94 | 70.89 | 68.99 | 61.92 | 79.34 | 70.56 | 75.01 | 70.49 | 73.40 | 57.10 |
| Deeplabv3 [17] | 68.57 | 98.23 | 76.18 | 53.07 | 43.85 | 51.78 | **73.99** | 69.80 | 61.18 | 80.30 | **72.48** | 75.66 | 72.01 | 72.73 | 58.72 |
| PSPNet [19] | 68.80 | 98.25 | 76.82 | 53.04 | 46.52 | 54.07 | 72.65 | 69.70 | 60.82 | 80.20 | 70.99 | 74.55 | 72.71 | 73.48 | 59.39 |
| CCNet [21] | 69.22 | 98.26 | 76.82 | 55.18 | 44.69 | 53.09 | 73.61 | 69.29 | **62.69** | 80.15 | 72.44 | **76.08** | **73.11** | **74.23** | **59.49** |
| TANet | **69.55** | **98.38** | **77.49** | **55.28** | **50.97** | **56.35** | 73.07 | **70.23** | 61.56 | **81.41** | 72.05 | 73.60 | 71.50 | 73.35 | 58.42 |

† Result given in [15] which keep two significant digits.

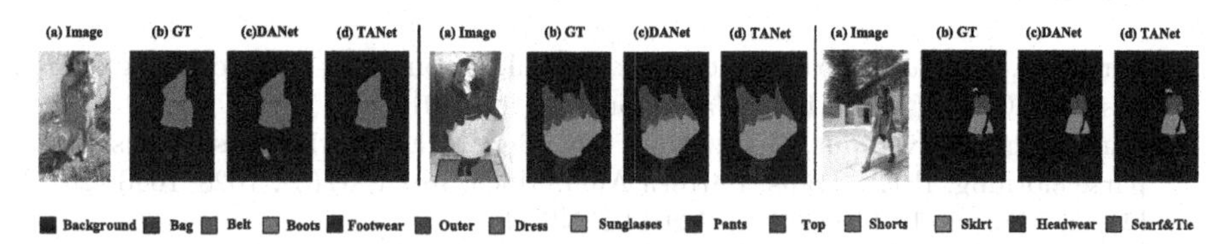


**Fig. 4.** Experimental results on ModaNet. Input image, ground-truth, baseline DANet and TANet. (Best viewed in color)

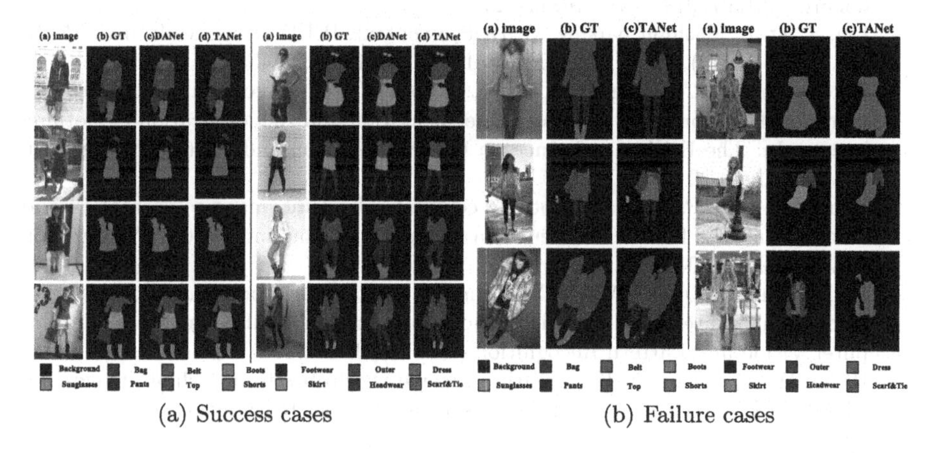


(a) Success cases (b) Failure cases

**Fig. 5.** (a) Visual improvements of the proposed method on ModaNet, TANet produces more accurate and detailed results. (b) In the failure mode, it is difficult for TANet to classify the adjacent objects that have similar colors. (Best viewed in color)

## 5 Conclusion and Future Work

In this paper, a Triple Attention Network (TANet) is proposed to solve the clothing parsing problem. TANet adaptively captures long-range contextual features, which is equipped with the color attention module, the position attention module and the channel attention module. Specially, the proposed method builds a color attention module to utilize color information for improving feature representations. The ablation experiments demonstrate that TANet is able to capture long-range contextual information, thereby offering more precise parsing results. TANet achieves outstanding performance on the ModaNet dataset.

In the future, we intend to refine this model from multiple aspects. For example, we plan to integrate a graph model such as CRF and MRF into the proposed network. We also have a plan to do experiments with a wide range of datasets and apply this method to other sources of data such as depth maps and videos. Also, we expect to study how prior information like human pose estimations and clothing structures affect clothing parsing.

## References

1. Tangseng, P., Wu, Z., Yamaguchi, K.: Looking at outfit to parse clothing. CoRR abs/1703.01386 (2017). http://arxiv.org/abs/1703.01386
2. Yamaguchi, K., Kiapour, M.H., Ortiz, L.E., Berg, T.L.: Retrieving similar styles to parse clothing. IEEE Trans. Pattern Anal. Mach. Intell. **37**(5), 1028–1040 (2015). https://doi.org/10.1109/TPAMI.2014.2353624
3. Simo-Serra, E., Fidler, S., Moreno-Noguer, F., Urtasun, R.: A high performance CRF model for clothes parsing. In: Cremers, D., Reid, I., Saito, H., Yang, M.-H. (eds.) ACCV 2014. LNCS, vol. 9005, pp. 64–81. Springer, Cham (2015). https://doi.org/10.1007/978-3-319-16811-1_5
4. Liu, S., et al.: Fashion parsing with video context. IEEE Trans. Multimed. **17**(8), 1347–1358 (2015). https://doi.org/10.1109/TMM.2015.2443559
5. Ge, Y., Zhang, R., Wang, X., Tang, X., Luo, P.: DeepFashion2: a versatile benchmark for detection, pose estimation, segmentation and re-identfication of clothing images. In: The IEEE Conference on Computer Vision and Pattern Recognition (CVPR), June 2019
6. Yang, W., Luo, P., Lin, L.: Clothing co-parsing by joint image segmentation and labeling. In: The IEEE Conference on Computer Vision and Pattern Recognition (CVPR), June 2014
7. Liu, Z., Luo, P., Qiu, S., Wang, X., Tang, X.: DeepFashion: powering robust clothes recognition and retrieval with rich annotations. In: The IEEE Conference on Computer Vision and Pattern Recognition (CVPR), June 2016
8. Hsiao, W.L., Grauman, K.: Creating capsule wardrobes from fashion images. In: The IEEE Conference on Computer Vision and Pattern Recognition (CVPR), June 2018
9. Yamaguchi, K., Kiapour, M.H., Ortiz, L.E., Berg, T.L.: Parsing clothing in fashion photographs. In: 2012 IEEE Conference on Computer Vision and Pattern Recognition, pp. 3570–3577, June 2012. https://doi.org/10.1109/CVPR.2012.6248101
10. Long, J., Shelhamer, E., Darrell, T.: Fully convolutional networks for semantic segmentation. In: The IEEE Conference on Computer Vision and Pattern Recognition (CVPR), June 2015
11. Fu, J., et al.: Dual attention network for scene segmentation. In: The IEEE Conference on Computer Vision and Pattern Recognition (CVPR), June 2019
12. Fan, X., Luo, H., Zhang, X., He, L., Zhang, C., Jiang, W.: SCPNet: spatial-channel parallelism network for joint holistic and partial person re-identification. In: ACCV (2018)
13. Su, K., Yu, D., Xu, Z., Geng, X., Wang, C.: Multi-person pose estimation with enhanced channel-wise and spatial information. In: The IEEE Conference on Computer Vision and Pattern Recognition (CVPR), June 2019

14. Chu, X., Yang, W., Ouyang, W., Ma, C., Yuille, A.L., Wang, X.: Multi-context attention for human pose estimation. In: The IEEE Conference on Computer Vision and Pattern Recognition (CVPR), July 2017
15. Zheng, S., Yang, F., Kiapour, M.H., Piramuthu, R.: ModaNet: a large-scale street fashion dataset with polygon annotations. In: Proceedings of the 26th ACM International Conference on Multimedia, MM 2018, pp. 1670–1678. ACM, New York (2018). https://doi.org/10.1145/3240508.3240652
16. Liu, S., et al.: Fashion parsing with weak color-category labels. IEEE Trans. Multimed. **16**(1), 253–265 (2014). https://doi.org/10.1109/TMM.2013.2285526
17. Chen, L., Papandreou, G., Schro, F., Adam, H.: Rethinking atrous convolution for semantic image segmentation. CoRR abs/1706.05587 (2017). http://arxiv.org/abs/1706.05587
18. Zheng, S., et al.: Conditional random fields as recurrent neural networks. In: International Conference on Computer Vision (ICCV) (2015)
19. Zhao, H., Shi, J., Qi, X., Wang, X., Jia, J.: Pyramid scene parsing network. In: The IEEE Conference on Computer Vision and Pattern Recognition (CVPR), July 2017
20. Wang, X., Girshick, R., Gupta, A., He, K.: Non-local neural networks. In: The IEEE Conference on Computer Vision and Pattern Recognition (CVPR), June 2018
21. Huang, Z., Wang, X., Huang, L., Huang, C., Wei, Y., Liu, W.: CCNet: criss-cross attention for semantic segmentation. In: The IEEE International Conference on Computer Vision (ICCV), October 2019
22. He, K., Zhang, X., Ren, S., Sun, J.: Deep residual learning for image recognition. In: The IEEE Conference on Computer Vision and Pattern Recognition (CVPR), June 2016
23. Huang, Z., Wei, Y., Wang, X., Liu, W.: A pytorch semantic segmentation toolbox (2018). https://github.com/speedinghzl/pytorch-segmentation-toolbox
24. Bulò, S.R., Porzi, L., Kontschieder, P.: In-place activated batchnorm for memory-optimized training of DNNs. CoRR abs/1712.02616 (2017). http://arxiv.org/abs/1712.02616
25. Chen, L., Papandreou, G., Kokkinos, I., Murphy, K., Yuille, A.L.: DeepLab: semantic image segmentation with deep convolutional nets, atrous convolution, and fully connected CRFs. IEEE Trans. Pattern Anal. Mach. Intell. **40**(4), 834–848 (2018). https://doi.org/10.1109/TPAMI.2017.2699184
26. Chen, L.C., Zhu, Y., Papandreou, G., Schro, F., Adam, H.: Encoder-decoder with atrous separable convolution for semantic image segmentation. In: The European Conference on Computer Vision (ECCV), September 2018

# U-Net Neural Network Optimization Method Based on Deconvolution Algorithm

Shen Li[1], Junhai Xu[1](✉), and Renhai Chen[2](✉)

[1] College of Intelligence and Computing, Tianjin University, Tianjin 300350, China
jhxu@tju.edu.cn

[2] College of Intelligence and Computing, Shenzhen Research Institute of Tianjin University, Tianjin University, Tianjin, China
renhai.chen@tju.edu.cn

**Abstract.** U-net deep neural network has shown good performances in medical image segmentation analysis. Most of the existing works are a single use of upsampling algorithm or deconvolution algorithm in the expansion path, but they are not opposites. In this paper, we proposed a U-net network optimization strategy, in order to use the available annotation samples more effectively. One deconvolution layer and upsampling output layer were added in the splicing process of the high-resolution features in the contraction path, and then the obtained "feature map" was combined with the high-resolution features in the contraction path in the way that broaden the channel. The training data used in the experiment is the pathological section image of prostate tumor. The average Dice scores for models based on our optimization strategy improve from 0.749 to 0.813. It proves that the deconvolution algorithm can extract feature information different from the upsampling algorithm, and the complementarity can achieve a better data enhancement effect.

**Keywords:** Medical image segmentation · Convolutional neural networks · Deconvolution · Optimization strategy

## 1 Introduction

Prostate cancer is the most common malignancy of the male reproductive system and is considered to be one of the major health hazards for middle-aged and elderly men. Prostate cancer accounted for 7% of new cancer cases worldwide in 2018 and there were 359, 000 prostate cancer-related deaths worldwide in the same year [1]. In order to treat prostate cancer, high-energy X-rays in different directions should be used in clinical practice. Therefore, the segmentation of prostate in pathological section image is of great clinical significance for accurate location. However, it is very time-consuming and laborious to provide the description of prostate segmentation manually by physician [2]. Therefore, in


H. Yang et al. (Eds.): ICONIP 2020, LNCS 12532, pp. 592–602, 2020.
https://doi.org/10.1007/978-3-030-63830-6_50

recent years, there are increasing requirements for the development of computer-aided segmentation methods.

Recently, many researches on automatic or semi-automatic methods of prostate segmentation have been proposed [3–6] . But most of the researches are focused on prostate MR images. Relatively few studies have been done with pathological sections. Manu Goyal et al. used FCN series full convolution neural network combined with transfer learning to segment skin cancer pathological section image [7]. The experiment shows that cancer pathological image segmentation still faces the problem of unbalanced data in data set, and it is difficult to increase the types of features extracted by neural network. The acquisition of tumor pathological images is also difficult. Therefore, according to our research, there is no research on prostate tumor pathological section image processing combined with full convolution neural network. This paper is the first to use U-net network to treat pathological sections of prostate tumors, hoping to provide reference for future studies in this field. Because the U-net network needs a small amount of data to achieve good segmentation results [8,9], it is widely used in the field of medical image segmentation.

The classical U-net deep neural network architecture is composed of a contraction path and a symmetric extension path. The U-net network achieves the effect of data enhancement by mixing the upsampling output layer feature information and high-resolution feature information, so that the features of images can be fully extracted. Because of this, the network can conduct end-to-end training with few training images. In recent years, many improved methods for the U-net network have been published. Two studies [10,11] both proposed the idea of using multi-scale convolution to extract global and local information of images. Brosch et al. proposed a jump connection structure,which is added between the first convolution layer and the last upsampling layer of U-net network to achieve the effect of data enhancement [12]. Xiaomeng et al. proposed H-DenseUNet, which optimizes the network performance by combining in-chip and inter-chip features into a mixed feature fusion (HFF) layer [13]. Another method is to replace the upsampling layer with the deconvolution layer in the extended path [14]. These methods are single use of upsampling or deconvolution in the network, although they can also improve the performance of the network, but ignore the complementary relationship between them.

In this paper, We propose an optimization method which combines the upsampling and deconvolution. Compared with existing methods, main contributions of the proposed method are as follows:

- We applied the U-net network to the segmentation of prostate pathological section image.
- Different from the previous optimization algorithm, we propose an optimization method which combines the upsampling and deconvolution in the expansion path of the U-net structure.

## 2 Materials and Methods

### 2.1 Database

The data used in this experiment was pathological images of prostate tumors collected from 90 patients. In order to reduce the influence of the edge part of the image on the segmentation result, we cut the original image according to expert tagged information into many patches. And in general, in each patch, the target area is larger than the background area. In this way, the target region can be segmented as the main body and the efficiency of target information extraction can be improved. After obtaining the patches, we uniformly scale each patch cut to 1024 × 1024. In order to better save the original image information, we choose the bilinear interpolation method for the scaling algorithm [15]. Then, we normalized each patch. We normalized the initial pixel value of each point of the trained image to [0, 1] as the input of the network. The normalized formula we used was:

$$\frac{X - X_{min}}{X_{max} - X_{min}}$$

Called maximum and minimum normalization method.

In the above formula, X represents the pixel value at this point, $X_{min}$ represents the minimum pixel value of the whole image, and $X_{max}$ represents the maximum pixel value of the whole image. Finally, we selected 1000 pieces of normalized patches as the data set for the experiment. We set the ratio of training set and verification set as 0.2. That means, when starting the training 200 images are randomly selected from the whole data set during each round of training as the verification set.

### 2.2 Algorithm Design Mentality

Among the existing algorithms, the upsampling or deconvolution algorithm used in the U-net extension path is single. We think that the two algorithms are totally different from each other in terms of operation mode. As mentioned earlier, the core of U-net network is to achieve data enhancement by splicing the high-resolution output layer in the contraction path with the upsampling output layer. In the process of upsampling, which is determined by the calculation method of upsampling, the calculation method is shown in Fig. 1. The pixel values in the enlarged region are set to the same values. This method enlarges the sensory field of the original image, but the position information of each pixel value in the original image will be weakened. For example, in Fig. 1, the upper-left corner of the upsampled image has the same four pixel values, and we can't determine which pixel in the upsampled image is the real location of the pixel after the upsampled operation.

Unlike the upsampling algorithm, the deconvolution algorithm obtains the output image by convolution inverse operation through the convolution core. In the process of convolution, the value of convolution kernel determines the feature extraction of a certain position in the image. Like a forward convolution,

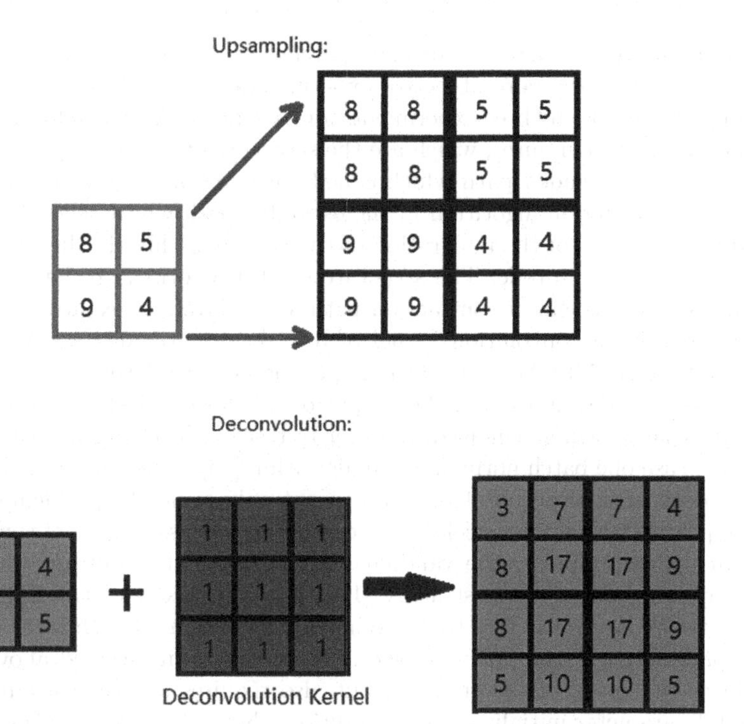


**Fig. 1.** Differences between deconvolution and upsampling Algorithms. The upper part represents the upsampling algorithm, and the lower part represents the deconvolution algorithm. In upsampling algorithm, the pixel values of enlarged images are the same in the certain region, which is impossible to distinguish the location of the truth value. In deconvolution algorithm, the pixel value is obtained by convolution inverse operation through the convolution core. Pixel values are distinct at different locations in the certain region.

deconvolution preserves the position information of the feature in the original image. In theory, the features extracted by these two algorithms are different, so it is not redundant to mix them. Pixel perception field and location information are both very important in the network training. Therefore, this paper proposes a method combining the two algorithms to fuse the extracted different image features. The "mixed features" obtained after fusion not only increase the size of the receptive field, but also include the location information of the pixel. We think that they complement each other can achieve better data enhancement effect and improve the segmentation performance of the network.

### 2.3 Methods

In order to prove our point of view, we designed a comparative experiment. The training data used in the experiment are 800 patches, which are divided into 600 pieces of training data and 200 pieces of verification data in the process of training network. There are also 200 patches as verification data after network training to provide more objective experimental results. The final results were obtained after about three experiments for each network. We build three different U-net network structures, which are the structure of only the up sampling algorithm in the extended path which called Unet-upsample, the structure of only the anti convolution algorithm in the extended path which called Unet-deconv, and the structure of the integration of the two algorithms in the extended path which called Unet-up-de. The structure of all networks is 10 layers, which are composed of 5 layers of contraction path and 5 layers of expansion path. Each layer contains a convolution kernel of $3 \times 3$, also contains a BN layer and a Dropout layer. After the convolution operation of each layer of the network, the characteristic distribution of the output of this layer, which is also the input of the next layer, will be changed. In order to restore the changed feature distribution, we use one batch normalization algorithm [16] at each layer of the network. The batch normalization algorithm can not only avoid the problem of gradient disappearance, but also accelerate the training process of the network.

In our experiment, the random gradient descent algorithm was used for the training optimization strategy [17]. The complexity of training deep neural network lies in the fact that the input distribution of each layer varies with the parameters of the previous layer. The questions mentioned above not only reduces the speed of the training, but also requires a lower learning rate and careful parameter initialization. In addition, the problem also makes it very difficult to train models with saturated nonlinearity. We set the learning rate and learning rate attenuation values in the stochastic gradient descent algorithm as 0.01 and 1e-2 respectively. The maximum number of training rounds is set to 30. This parameter controls the network to train up to 30 times on the data set. In the training process, we set an early stop mechanism. When the loss function on the verification set is no longer reduced for two consecutive rounds, the network training can be stopped to avoid the network overfitting. Furthermore, all we have to do is just to build a two-class model at the pixel level of the image, so we use a binary cross-entropy function as the loss function. The formula of loss function is as follows:

$$L = -[y \log \widehat{y} + (1 - y) \log(1 - \widehat{y})]$$

In the formula, y is the pixel value in mask images and it is the ground_truth of network training. So, the value of y is only 0 or 1. $\widehat{y}$ is the output of the sigmoid function at the end of the network, which symbolizes the probability that the label of the current pixel value is 1. The reason that we choose the binary cross-entropy loss function is that the gradient of the last layer weight is no longer related to the derivative of the activation function. It is only related to the interpolation of the output value and the real value. So, the greater the probability, the more likely the label is to be true. Relatively, the smaller the value of loss function is, the higher the similarity between segmented image and mask is. Besides, like all deep learning problems, we also need to solve the problem about overfitting in the network. The regularization we used is to add a

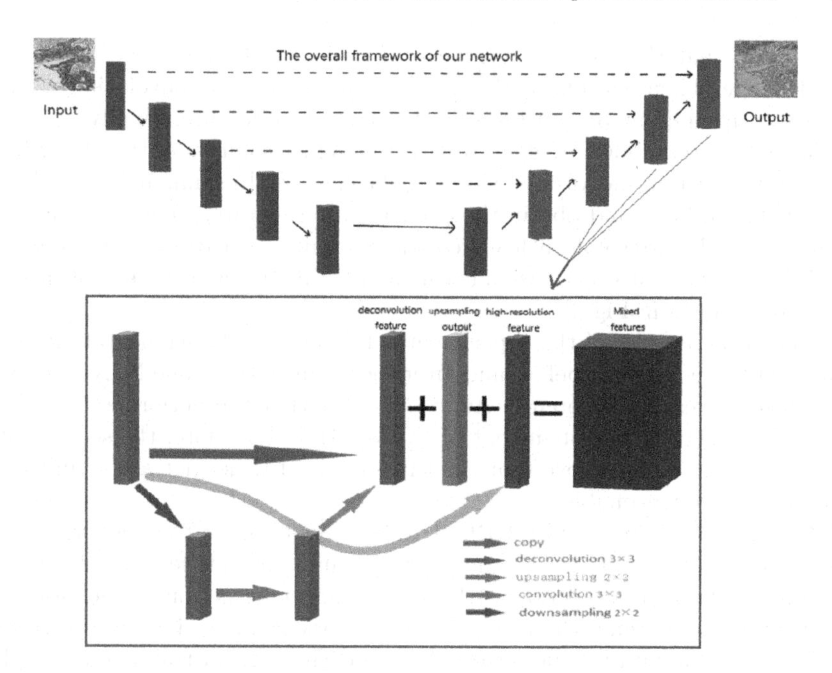


**Fig. 2.** The optimization method proposed in this paper. The optimization method is applied to the red layer of the whole structure. The optimization algorithm of each layer is the same and as shown in the figure below. Through a part of the network layer structure, we describe in details the operation process of the optimization algorithm in this paper. The blue layer represents the output layer obtained after convolution of the previous layer of contraction path, that is, the high-resolution features obtained after convolution. The red layer represents the deconvolution features obtained by deconvolution layer. The purple layer represents the upsampling feature obtained from the upsampling layer. The three features are combined to produce the brown features in the figure. The mixed feature layer channel number is the sum of three feature layer channel numbers. In the proposed network, a deconvolution layer is added to each layer, and each layer performs the same operation as shown in the figure. (Color figure online)

Dropout layer after each convolution layer, in which the parameter is set to 0.3. After each deconvolution layer we also add a Dropout layer, with the parameter 0.4 [18].

Here is a detailed introduction to our proposed structure. The network structure is 10 layers, which is divided into contraction path and expansion path. We use the deconvolution operation to obtain the information about the deconvolution features. Then, deconvolution features are added in the stitching process, and we first perform a deconvolution operation on the high-resolution output layer in the contraction path that needs to be spliced. In the deconvolution process, 0 is padded in the image periphery to extract the image edge information and ensure that the output image size after deconvolution is the same as the input image. Then we spliced the features obtained by deconvolution with the same size upsampling output layer to obtain a mixed feature. Finally, the mixed features obtained above are spliced with the high-resolution output layer in the contraction path in the form of widening the channel. In summary, we added the deconvolution layer and obtain the deconvolution feature. Then the deconvolution feature, high-resolution features and upsampling features were combined. And the network with such fusion features were finally trained. The optimization method is shown in Fig. 2.

The basic purpose of the experiment is to compare the segmentation performance of U-net network before and after the optimization. The FCN-8s network we joined serves as a reference. As we all know that the performance of U-net network must be better than that of FCN-8s [19]. Therefore, the segmentation results of FCN-8s can prove that the three kinds of U-net networks built in the experiment are reasonable.

Our network is trained on CPU. It takes about 5 h for the network to complete one round of training. The number of training rounds through the network training to the highest accuracy (that is, the accuracy no longer increases with the training) represents the time cost of network training. During the training process, a set of data will be returned when each round of training is complete. The data contains the loss function value and the segmentation accuracy value of the network on the verification set, and each value are calculated by the network using the real-time weight. When the loss function on the verification set is lower than that in the previous round, the weight value of the network will be saved. After the training stops, the optimal weights of the final network after training are obtained. According to the comparison results, we calculate the Dice scores of each patch, and then take the average value to get the Dice scores as the evaluation criteria of each network segmentation result. In our work, we regard Dice scores as the main criterion.

## 3 Experimental Results

The Dice score is used to calculate the similarity between two samples. When applied to the field of image segmentation, its meaning becomes the overlap ratio between the segmented image and mask image. According to this definition, the bigger the Dice score, the more similar the image segmented by the classifier is to the ground_truth, and the better the segmentation effect is. In order to compare the differences between the three networks in more details, we summarize the Dice scores, the loss function values and the segmentation accuracy values of the network on the verification set, and the number of training rounds that the network experienced at the end of training into one table in (Table 1).

As shown in Table 1, the loss function of Unet-upsample is 0.6787 and that of Unet-deconv is 0.6532. After optimization, the loss function of the Unet-up-de is 0.5698. This shows that the optimized network has better robustness. The segmentation accuracy of Unet-upsample is 0.6503 and that of Unet-deconv is 0.6574.

**Table 1.** Dice scores and other network training data of the three networks. The Dice is averaged. Other data are generated during the training process, and the other data are averaged through many experiments. FCN-8s is also used as a reference to explain the rationality of the data.

| | Val-loss | Accuracy | **Dice scores** | Rounds |
|---|---|---|---|---|
| FCN-8s | 0.8335 | 0.3929 | 0.5989 | 5 |
| Unet-upsample | 0.6787 | 0.6503 | 0.7495 | 6 |
| Unet-deconv | 0.6532 | 0.6574 | 0.7547 | 5 |
| Unet-up-de | **0.5698** | **0.7173** | **0.8134** | **3** |

The segmentation accuracy of Unet-up-de is 0.7173 which is also the best. It proves that the segmentation accuracy of the optimized network on the verification set is improved. Dice scores as the main evaluation criterion, the average Dice scores of Unet-up-de is 0.8134, and the average Dice scores of Unet-upsample is 0.7495, the average Dice scores of Unet-deconv is 0.7547. The average Dice scores of the optimized network increased by 0.0639, which means that the similarity between the segmented image and ground_truth increases by about 6.4%, which is a great improvement. Besides, we also found that, Unet-up-de achieves a segmentation accuracy of 0.7173 in only three rounds. We can draw a conclusion that the optimized network can achieve a higher accuracy in a shorter time than before. To put it another way, this is equivalent to that the network can use fewer training data to achieve a higher training accuracy. It can prove that our optimization method improves the effect of data enhancement.

## 4 Segmentation Results

Figure 3 shows the segmentation result of the optimized network. In the process of generating the segmented image, we set the background to blue, which represents the normal area. The target area is red, which represents the lesion area. As a result, the lesion cells were stained red in the image. According to the segmentation result, we can observe intuitively that the generated image is very similar to the mask image. Besides, in the unlabeled part, we found that some cells were dyed red, which means that the unlabeled pathological cells were also separated. Therefore, we think that the features learned by neural network through marker information have good generalization, and can segment the scattered pathological cells in the unlabeled area. The results show that learning feature information can achieve better results than expected, which proves the powerful application of neural network in the field of medical imaging.

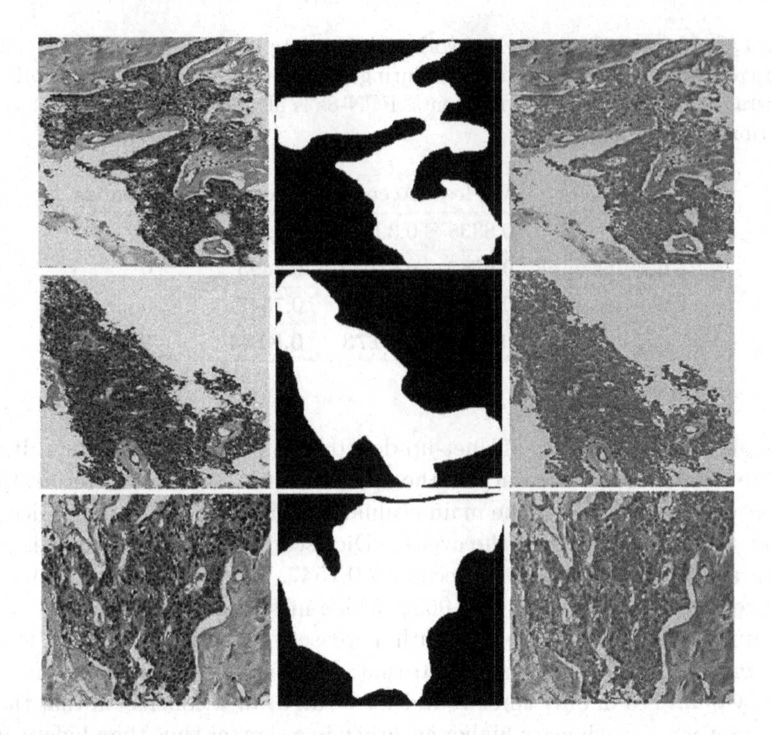

**Fig. 3.** The graphical results are obtained from the optimized network. From left to right, the original image, mask and segmentation result are in order.

## 5 Conclusion

In this paper, we developed an optimization method of U-net neural networks. This is a significant problem, because the development of U-net is closely related to the development of medical image segmentation. We proposed an optimization method based on deconvolution and compared the optimized network with the two network. In addition, we also built the FCN-8s network as an experimental reference, which can make the experimental data obtained by us more convincing. We use average Dice scores as the main evaluation criteria of segmentation effect. The results showed that, the average Dice scores increased by 0.0639 after optimization. Compared with the U-net network, the optimized network needs less training data. We can conclude that the data enhancement effect of our optimization method is higher than that of the two kinds of single U-net network. Finally, according to the experimental results, combined with the upsampling algorithm and deconvolution algorithm, we explain the optimization method. According to the difference of the calculation methods between two algorithm, we suggest that the deconvolution algorithm is different from the upsampling algorithm. So the extracted image features are also different. We take advantage of this difference and combine the two features in the form of stitching, enlarging the receptive field of the feature while retaining the location information, which has been proved to be capable of achieving a better data enhancement performance.

**Acknowledgment.** This work was supported by the National Natural Science Foundation of China (61703302) and partially supported by Shenzhen Science and Technology Foundation (JCYJ20170816093943197).

## References

1. Siegel, R.L., Miller, K.D., Jemal, A.: Cancer statistics. CA Cancer J. Clin. **68**(1), 7–30 (2018)
2. Itakura, K., Hosoi, F.: Background and foreground segmentation in plant images with active contour model and plant image classification using transfer learning. Eco Eng. **30**, 81–85 (2018)
3. Yu, L., Yang, X., Chen, H., Qin, J., Heng, P.A.: Volumetric convnets with mixed residual connections for automated prostate segmentation from 3D MR images, p. 66C72 (2017)
4. Tian, Z., Liu, L., Fei, B.: A supervoxel-based segmentation method for prostateMR images. In: Ourselin, S., Styner, M.A. (eds.) Medical Imaging 2015: Image Processing, vol. 9413, pp. 321–327. SPIE, Bellingham (2015). International Society for Optics and Photonics
5. Guo, Y., Gao, Y., Shen, D.: Deformable MR prostate segmentation via deep feature learning and sparse patch matching. IEEE Trans. Med. Imaging **35**(4), 1077–1089 (2016)
6. Clark, T., Wong, A., Haider, M.A., Khalvati, F.: Fully deep convolutional neural networks for segmentation of the prostate gland in diffusion-weighted MR images. In: International Conference Image Analysis and Recognition, pp. 97–104 (2017)
7. Clark, T., Wong, A., Haider, M.A., Khalvati, F.: Fully deep convolutional neural networks for segmentation of the prostate gland in diffusion-weighted MR images. In: Karray, F., Campilho, A., Cheriet, F. (eds.) ICIAR 2017. LNCS, vol. 10317, pp. 97–104. Springer, Cham (2017). https://doi.org/10.1007/978-3-319-59876-5_12
8. He, H., Yang, D., Wang, S.: Road segmentation of cross-modal remote sensing images using deep segmentation network and transfer learning. Ind. Robot Int. J. **46**, 384–390 (2019)
9. Ronneberger, O., Fischer, P., Brox, T.: U-Net: convolutional networks for biomedical image segmentation. In: Navab, N., Hornegger, J., Wells, W.M., Frangi, A.F. (eds.) MICCAI 2015. LNCS, vol. 9351, pp. 234–241. Springer, Cham (2015). https://doi.org/10.1007/978-3-319-24574-4_28
10. Kamnitsas, K.: Efficient multi-scale 3D CNN with fully connected CRF for accurate brain lesion segmentation. Med. Image Anal. **36**, 61–78 (2017)
11. Ghafoorian, M.: Non-uniform patch sampling with deep convolutional neural networks for white matter hyperintensity segmentation. In: IEEE 2016 IEEE 13th International Symposium on Biomedical Imaging, pp. 1414–1417 (2016)
12. Brosch, T., Tang, L.Y., Yoo, Y., Li, D.K., Traboulsee, A., Tam, R.: Deep 3D convolutional encoder networks with shortcuts for multiscale feature integration applied to multiple sclerosis lesion segmentation. IEEE Trans. Med. Imaging **35**(5), 1229–1239 (2016)

13. Li, X., Chen, H., Qi, X., Dou, Q., Fu, C.W., Heng, P.A.: H-denseunet: hybrid densely connected UNET for liver and tumor segmentation from CT volumes. IEEE Trans. Med. Imaging **37**(12), 2663–2674 (2018)
14. Chang, Y., Song, B., Jung, C., Huang, L.: Automatic segmentation and cardiopathy classification in cardiac mri images based on deep neural networks. In: IEEE International Conference on Acoustics, Speech and Signal Processing (ICASSP), pp. 1020–1024 (2018)
15. Jing, L., Xiong, S., Shihong, W.: An improved bilinear interpolation algorithm of converting standard-definition television images to high-definition television images. In: 2009 WASE International Conference on Information Engineering. IEEE (2009)
16. Ioffe, S., Szegedy, C.: Batch normalization: Accelerating deep network training by reducing internal covariate shift. arXiv: Learning (2015)
17. Mercier, Q., Poirion, F., Désidéri, J.A.: A stochastic multiple gradient descent algorithm. Eur. J. Oper. Res. **271**(3), 808–817 (2018). S0377221718304831
18. Hinton, G.E., Srivastava, N., Krizhevsky, A., Sutskever, I., Salakhutdinov, R.R.: Improving neural networks by preventing co-adaptation of feature detectors. arXiv: Neural and Evolutionary Computing (2012)
19. Long, J., Shelhamer, E., Darrell, T.: Fully convolutional networks for semantic segmentation. IEEE Trans. Pattern Anal. Mach. Intell. **39**(4), 640–651 (2015)

# Unsupervised Tongue Segmentation Using Reference Labels

Kequan Yang[1], Jide Li[1], and Xiaoqiang Li[1,2](✉)

[1] School of Computer Engineering and Science, Shanghai University, Shanghai, China
{yangkequan,iavtvai,xqli}@shu.edu.cn

[2] Shanghai Institute for Advanced Communication and Data Science, Shanghai University, Shanghai, China

**Abstract.** Accurate and robust tongue segmentation is helpful to acquire the result of automatic tongue diagnosis. There are numerous existing methods for tongue segmentation using traditional image processing or deep learning. However, these methods often require complicated preprocessing or to be trained using a large number of manual labels with expensive cost. To overcome these limitations, we propose an end-to-end unsupervised tongue segmentation method using reference labels based on the adversarial approach. Firstly, we use the segmentation network to extract the object mask which is then binarized into reference label. Secondly, the generation network is used to redraw the segmented area by inputing the object mask and noise. Finally, the segmentation network and generation network are controlled by a discriminator and reference labels to refine the segmentation network with best performance. The proposed method only requires unlabeled datasets for training. To the best of our knowledge, this work is the first deep learning approach to achieve unsupervised tongue segmentation. We perform experiments using the proposed method and achieve satisfactory segmentation results on different validation sets.

**Keywords:** Unsupervised tongue segmentation · Reference labels · Adversarial approach

## 1 Introduction

Tongue diagnosis is under a pivotal role in Traditional Chinese Medicine(TCM). Evidence suggests that information of the tongue is the most important factors reflecting physical condition [1]. The traditional methods of tongue diagnosis depend entirely on the experience of the clinician. In addition, it requires a doctor to observe each patient's tongue in person, which will inevitably cause some wrong judgments because of fatigue. So people hope that computer technology can be used to make judgments [2] based on the input tongue images. Since any tongue diagnosis system makes decisions according to different features

Thanks to Shanghai Daosheng Medical Technology Co., Ltd.


H. Yang et al. (Eds.): ICONIP 2020, LNCS 12532, pp. 603–615, 2020.
https://doi.org/10.1007/978-3-030-63830-6_51

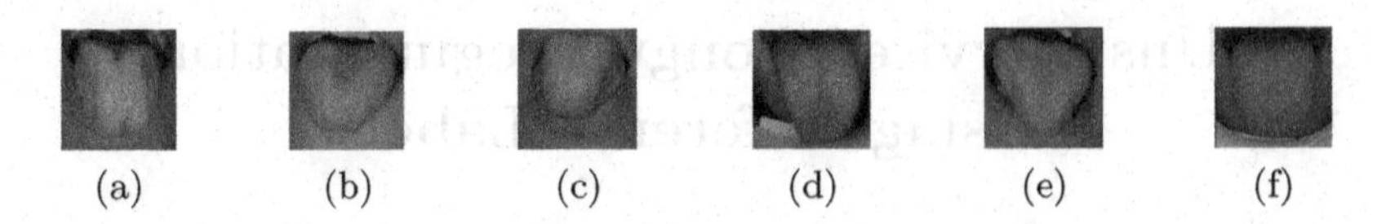


**Fig. 1.** Tongue images of different patients.

of tongue, the quality of diagnostic result is directly related to the segmented tongue from the source image. Therefore, the tongue segmentation is one of the key technologies for automatic tongue diagnosis. The challenges of tongue segmentation are that the characteristics of a tongue are different, such that it is difficult to mine common attributes using conventional geometric and iterative image processing methods [3]. And due to the similar appearance and texture, the tongue body is hard to be segmented from lips and cheeks pixels. Different tongue images are displayed in the Fig. 1.

Until now, a variety of automatic tongue segmentation methods have been put forward as part of the integrated tongue diagnosis system. Most of those methods are based on the traditional image processing technologies, which mainly divided into shape-based methods [4], region-based methods [5,6], and color-based methods [7–9]. These methods are susceptible to the tongue color texture which is similar to the lips and cheeks (see in Fig. 1) and the complicated preprocessing. Therefore, these methods do not achieve satisfactory results.

In recent years, deep learning has developed rapidly. Although deep learning has achieved good performance in classification, object detection, and segmentation, only a few tongue segmentation methods based on deep learning. The latest tongue segmentation methods [10,11] applying the deep conventional neural network (CNN) which has outstanding ability of feature learning and representation outperform some traditional tongue segmentation methods. However these methods suffer from a significant amount of pixel-wise labeled training data limitations, which can be difficult to collect on novel domains. When the tongue images collection environment changes, such as different closed collection environments, tongue images collected by mobile phone etc. It is necessary to relabel the datasets. Such pixel-wise annotation is much more time consuming than box-level or image-level annotation. We follow here a very recent trend that aims at revisiting the unsupervised image segmentation problem with new tools and new ideas from the recent success of deep learning to achieve unsupervised tongue segmentation with box-level dataset.

The rest of this paper is organized as follows. Section 2 introduces the implementation of the proposed method and the various parts of the network. Plentiful experiments including ablation experiment of the proposed method are conducted and discussed in Sect. 3. Finally, we give a conclusion in Sect. 4.

## 2 Method

We propose a novel end-to-end unsupervised tongue segmentation methods using reference labels (see Fig. 2). It relies on the idea that it should be possible to change the textures or colors of the objects without changing the overall distribution of the dataset [12–14]. Following this assumption, our model consists of two parts, a generation part $G_g$ and a discrimination part $D_g$. The generation part $G_g$ is mainly used to synthesize new image, which consists of segmentation network $F$ and generation network $G_F$. This task is trained in the unlabeled box-level RGB images $I \in \mathbb{R}^{W \times H \times C}$. The segmentation network $F$ is guided by an input image: it extracts the object mask and reference labels by given an image. Then the generation network $G_F$ redraws a new object at the same location by inputing the object mask $M_i$ and noise $z$. The discrimination part $D_g$ is used to control the generation part $G_g$: i)The discrimination network $D$ is used to discriminate the distribution difference between the composite images and the original dataset. ii) To avoid invalid segmentation, network $\delta$ determines whether the composite image contains input noise $z$. iii) The reference labels constrain the segmentation network $F$ in a way that minimizes entropy [15]. Through the game between the generation part $G_g$ and the discrimination part $D_g$, a segmentation network $F$ with better results will be obtained gradually.

### 2.1 The Generation Part

The generation part generates new images by cooperating with the segmentation network $F$ and the generation network $G_F$. Pixels in the input image $I$ can be divided into two categories, tongue and background. Let us denote the mask $M_i, i \in \{0, 1\}$ corresponding to object which associates binary value to each pixels in the final image so that $M_0^{x,y} = 1$ if the pixel of coordinate $(x, y)$ belongs to the tongue body. It is easy to know the background mask $M_1 = 1 - M_0$. The mask of the tongue obtained directly from the input image $I$ through the segmentation function $F$. The composite image generation method is described in Fig. 2. There are three main steps in the generation process $G_g$. i) Randomly select an input image in the dataset and use $F(I) \rightarrow M_0$ to calculate the masks. ii) Generate new pixel value $V_i = G_F(z, M_i)$ in the corresponding area according to the masks $M_i$ and the noise $z$. iii) The generated new pixel value $V_i$ and the old pixel value of the input image $I$ are stitched together according to the mask $M_i$ to form a new image $\widetilde{I} = I \odot (1 - M_i) \oplus V_i$.

In this way, we can use the difference between the distribution of the composite image and the datasets to optimize the parameters of the segmentation network, and obtain a segmentation network with the best possible result.

### 2.2 The Discrimination Part

**Discriminator.** The discriminator $D$ takes the real images and the composite images as inputs, which is used to estimate the probability that the current input image belongs to that of real image.

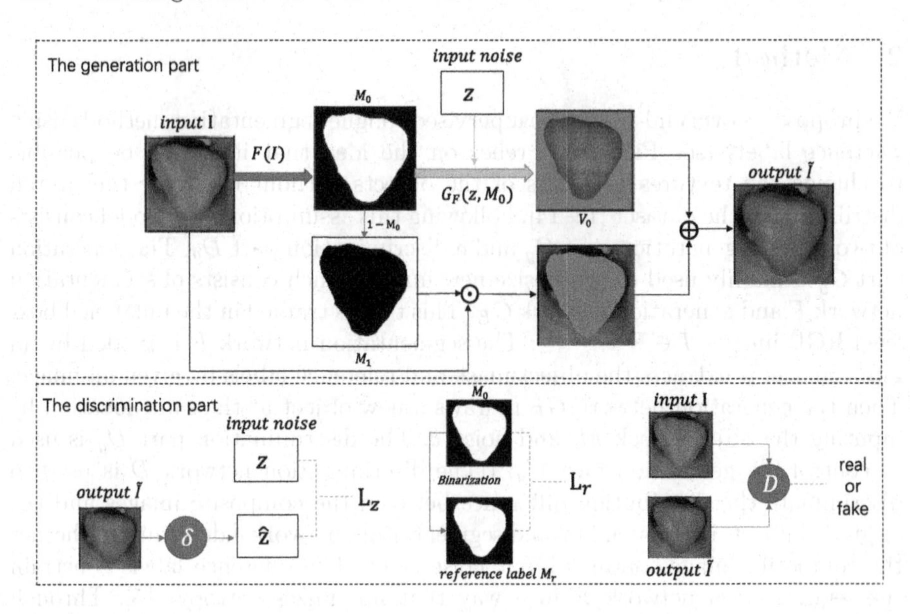


**Fig. 2.** The framework of the proposed method. An image $I$ is input into the segmentation network $F$ to obtain the object mask $M_0$. The object mask $M_0$ and noise $z$ are input into the generation network $G_F$ to obtain the generation foreground $V_0$. The composite image $\widetilde{I}$ is obtained by combining the generated foreground $V_0$ and the real background. It is also possible to combine the generated background and the real foreground by inputing the background mask $M_1$ into the generation network $G_F$. The discrimination network $D$ is used to judge the distribution difference between the composite image $\widetilde{I}$ and the input image $I$. The network $\delta$ is used to determine whether the generated foreground $V_0$ contains the noise $z$. $M_0$ is binarized as reference label $M_r$ constraining the generation part $G_g$.

**The Network $\delta$.** One possible unwanted case in training segmentation network process is that all pixels value of the input image was considered as the background, and results in empty generation foreground $V_0$, i.e. $m_0^{x,y} = 0$ for all $(x, y)$. As introduced in Sect. 2.1, the composite image is generated by inputing mask and noise $z$. Therefore, if this case happened, the input noise $z$ will not appear in the composite images $\widetilde{I}$. Equivalently, if the information of $z$ can be retrieved from the composite images $\widetilde{I}$, then the segmented area is not empty. So we design a network $\delta$ based on SAGAN [14,16] which objective is to infer the value of $z$ given any image $\widetilde{I}$, i.e. $\widehat{z} = \delta(\widetilde{I})$. This information can be achieved by adding a regular term to the loss function of the generator $G_g$. This strategy is similar to the mutual information maximization used in InfoGAN [17]. The corresponding loss function are as follows:

$$\begin{aligned} \widehat{z} &= \delta(\widetilde{I}) \\ L_z &= ||\widehat{z} - z)||_2^2 \end{aligned} \tag{1}$$

**Reference Labels.** Image segmentation is a process of assigning a label to each pixel in the image such that pixels with the same label are connected with respect to some visual or semantic property. The mask $M_i$ obtained by segmentation network $F$ can be considered as pixel-wise probability map the same size as the input. As shown in Fig. 3, although visually the predicted mask is acceptable, this is not conducive to the entire optimization process. The probability values of the foreground and background are too close, which will lead to a high degree of class overlap and an excessively large entropy value. And when this mask is used to synthesize a new image, some information of the original image will be introduced, which will affect the training of the network. As shown in Fig. 4(a), when a soft mask is used as an input to the generation network $G_F$, the generated segmented area will have a lot of interference information.

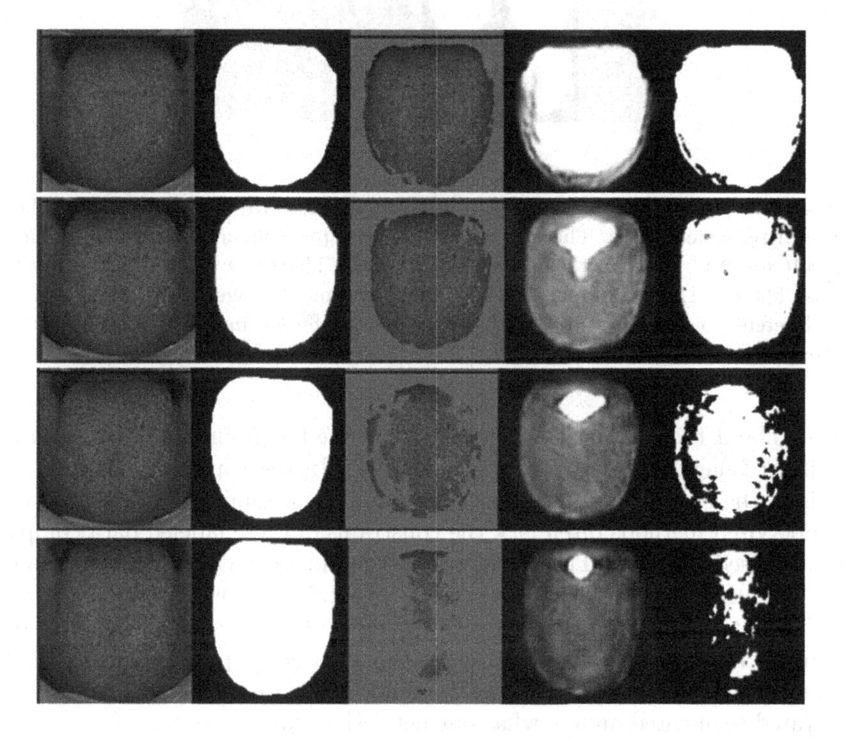

**Fig. 3.** Comparison of segmentation network results under different iterations. From the left column to the right is the input image, ground-truth, tongue segmented by the binary mask, the prediction result soft mask, and the binary mask. From top to bottom, the number of iterations is gradually increasing. Uncertain classification results for some pixels lead to optimization of network in a bad direction.

The results in Fig. 4(a) demonstrate that the discriminator of GAN [18] is not conducive to the optimization of the segmentation network. Following the minimum entropy regularizer [15], when the segmentation network is more

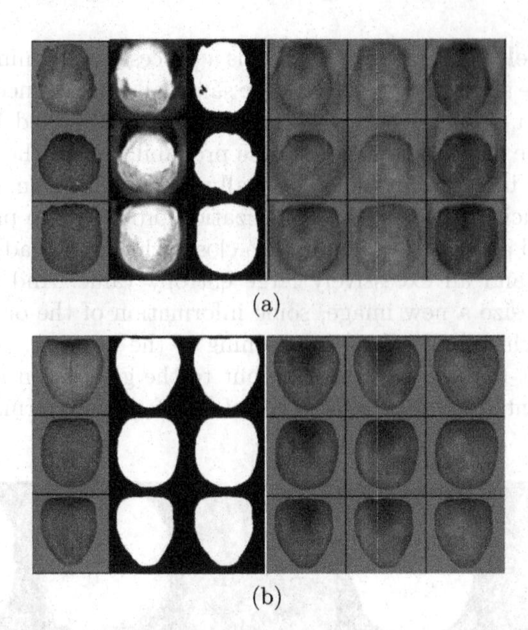

**Fig. 4.** The effect of reference labels on generated pixels. (a) Without reference labels, (b) with reference labels. The first column is the split tongue. The second column is the soft mask obtained by segmentation network. The third column is the reference labels. The last three columns are the generated tongue image $V_0$ using the soft mask and different noise $z$. (a) and (b) are the results of different inputs at the same number of iterations.

certain about the classification probability of the foreground and background in the input images, the entropy is the smallest. In the supervised methods, the labels of the samples are often used to guide the optimization of the network. Unsupervised methods often lack the constrains of these labeled data compared to supervised methods. Therefore, we use the binary mask as labels to emulate the supervised method. The binary mask is obtained from the soft mask which is the results of segmentation network by setting the threshold. The selection of a suitable threshold is introduced in Sect. 3.4. In our paper, the binary mask is named reference label. When using reference labels (as shown in Fig. 4(b)), the generated segmented area is what our network need.

The reference labels obtained by each iteration are used as the ground-truth. The mask ranges from 0 to 1 and is thus trained with binary cross entropy (BCE) loss $L_r$. The corresponding loss function is as follows:

$$L_r = BCE(M, M_r) \tag{2}$$

### 2.3 Loss Function

The proposed framework is optimized in an end-to-end fashion using a multi-task loss. Through the introduced in the previous Sect. 2.1, parameters of the network are mainly optimized by adversarial method. The discriminator $D_g$ is designed to help the generation part $G_g$ generate plausible composite image $\widetilde{I}$. Following [19,20] and leverage hinge loss for stabilizing training, which is given by

$$\begin{aligned} L_{D_g} &= \mathbb{E}[max(0, 1 - D(I))] + \mathbb{E}[max(0, 1 + D(\widetilde{I}))] \\ L_{G_g} &= -\mathbb{E}[D(\widetilde{I})] + \lambda_z * L_z + \lambda_r * L_r \end{aligned} \tag{3}$$

When training $D_g$ by minimizing $L_{D_g}$, $D_g$ is encouraged to produce large scores for real images. While training $G_g$ by minimizing $L_{G_g}$, the composite images are expected to fool $D$ and obtain large scores. The loss function $L_z$ and $L_r$ are introduced in the previous Sect. 2.2. The trade-off parameter $\lambda_r$ and $\lambda_z$ are empirically set as 20 and 5 in our experiments.

# 3 Experiments

In this section, we perform tongue segmentation experiments using the proposed method to demonstrate its effectiveness and efficiency. Two criteria to evaluate the segmentation performance of single tongue image are adopted: the mean Intersection over Union (mIoU) and the mean Pixel Accuracy (mPA). Then we compared the results with some supervised methods in test set.

## 3.1 Dataset and Pretraining

The tongue images are collected from hospitals and communities in multiple batches, and collection equipment is also customized. These datasets contain part of the tongue with lesions. Because of its privacy, it is difficult to obtain a large number of data sets. The size of the collected raw datasets is 3240 * 4320. Training of our network is based on box-level datasets. Each tongue image is sampled to generate image patches which are most likely to contain the tongue body. Then transform patches of different sizes to the size $128 * 128$ required by the network. The dataset has a total of 1185 tongue images. We select 1085 as the training set and 100 masks corresponding to the human hand segmentation as the test set to evaluate the performance of model. This is called Testset1. Another dataset is the tongue image dataset published in the internet by BioHit [21], which is composed of 300 tongue images with labels. Therefore, we use this public data as a test set called TestSet2. TestSet1 and TestSet2 are shown in the Fig. 5.

Using a reference label requires that the segmentation network has certain segmentation capabilities. So we take the first pre-training of the segmentation network $F$ on the Flowers dataset [22], and then fine-tune it on our dataset. Generation network $G_F$, discrimination network $D$ and network $\delta$ use random parameter initialization.

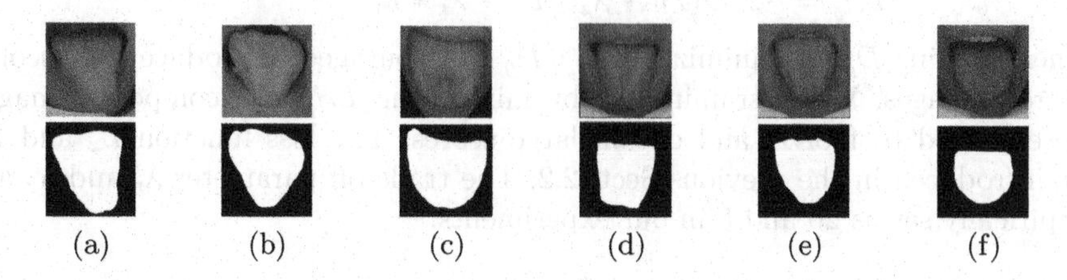


**Fig. 5.** The tongue images and the corresponding artificial segmentation masks. (a)-(c)are the examples in TestSet1, (d)-(f) are the examples in TestSet2.

### 3.2 Implementation Details

The proposed method training mainly depends on the game between the generation part and the discrimination part. One component of the generation part is the segmentation network $F$ which uses PSPNet [23]. The other component is the generation network $G_F$ based on SAGAN [16], which is enhanced by the attention block. The discriminator $D_g$ and network $\delta$ are also based on SAGAN [16]. We sample noise $z$ of size of 32 from $N(0, I_d)$ distribution. Unless specified, our network architecture and hyperparameters are same as [14]. We conduct experiment on a single NVIDIA GTX 1080Ti GPU with the mini-batches of size 12.

### 3.3 Evaluation Methods and Experimental Result

In order to evaluate the segmentation performance of the unsupervised tongue segmentation network, we use two metrics commonly used for segmentation tasks. The mean pixel classification accuracy (mPA) measures the proportion of pixels that have been assigned to the correct region. The mean intersection over union (mIOU) is the ratio of area intersection to union between mask and ground truth. The formulations of mPA and mIOU are shown as following:

$$mPA = (\frac{1}{n_{cl}}) \sum_i \frac{n_{ii}}{t_i} \tag{4}$$

$$mIOU = (\frac{1}{n_{cl}}) \sum_i \frac{n_{ii}}{t_i + \sum_j n_{ji} - n_{ii}} \tag{5}$$

where $n_{cl}$ is the number of the pixel classes, $n_{ij}$ is the number of pixels of class $i$ predicted to belong to class $j$, and $t_i = \sum_j n_{ij}$ is the total number $j$ of pixels of class $i$.

The Fig. 6 shows the visual segmentation results of the model on TestSet1 and TestSet2. And Table 1 shows the evaluation results of the model on the two test sets respectively using the evaluation methods introduced early. And compared with supervised methods. Among them, the Deeplabv3-LBM [24] uses the same private dataset as us to evaluate, SegNet [25], Deeptongue [10], and Tonguenet [11] all use the TestSet2 to evaluate. All methods require a lot of labeled data for training except ours.

The data in Table 1 shows that for private dataset TestSet1, our unsupervised method is completely better than the supervised method. On public dataset TestSet2, although our method is slightly inferior to the latest Tonguenet [11] and other supervised methods, the training process of our model does not require a large number of labeled dataset.

### 3.4 Ablation Studies

**Whether to Use Reference Labels for Comparison.** Through the introduction in Sect. 2.2, reference labels play a vital role in the training process. After adding reference labels to participate in training, the segmentation results have improved significantly, of which mIOU increased by 2% points, and mPA increased by 1% points. The comparison results under different iterations in TestSet1 are shown in the Fig. 7. And impact of reference labels on generated images is shown in Fig. 4.

It can be seen from the Fig. 7 that when the reference labels are used, the curves of mIOU and mPA have relatively small fluctuations, the model is more stable and robust. When iterating to 7k, the mIOU and mPA of the model without reference labels decreased significantly, and the volatility is large.

**Threshold Selection.** However, since the model itself generates targets, they may very well be incorrect. If too much weight is given to the generated targets, the cost of inconsistency outweighs that of misclassification, preventing the learning of new information. In effect, the model suffers from confirmation bias. So how to determine the suitable threshold to improve the quality of targets is very important. Some comparative experiments are conducted on different threshold. The results are shown in the Fig. 8. It is easy to see that the proposed method has the best result when the threshold value is 0.5.

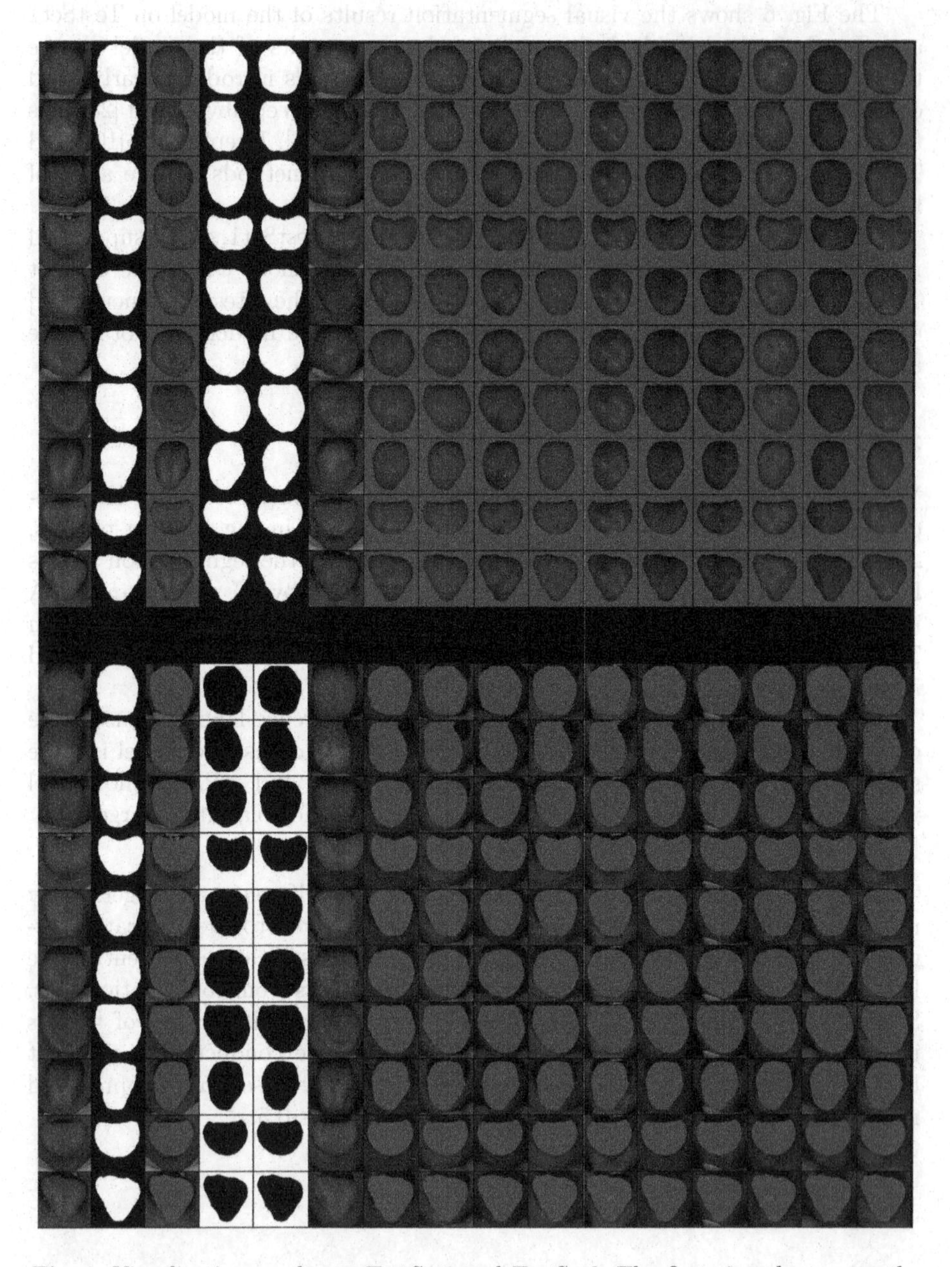

**Fig. 6.** Visualization results on TestSet1 and TestSet2. The first six columns are the input images, ground truth masks, segmentation results, predicted masks, reference labels and composite images. The last ten columns are the images generated based on the predicted masks and noise $z$. The first ten lines are the generated foreground area, and the last ten lines are the generated background area.

**Table 1.** Comparison of our method with other supervised segmentation methods

| Dataset | Method | mPA | mIOU |
|---|---|---|---|
| TestSet1 | Deeplabv3 [26] | 94.1% | 93.7% |
| | Deeplabv3-LBM [24] | 91.4% | 86.2% |
| | Ours | 96.31% | 94.10% |
| TestSet2 | FCN [27] | 95.36% | 96.29% |
| | U-Net [28] | 95.86% | 96.99% |
| | SegNet [25] | 92.88% | 95.16% |
| | Deeptongue [10] | 92.78% | 94.17% |
| | Tonguenet [11] | 98.63% | 97.74% |
| | Ours | 94.31% | 89.63% |

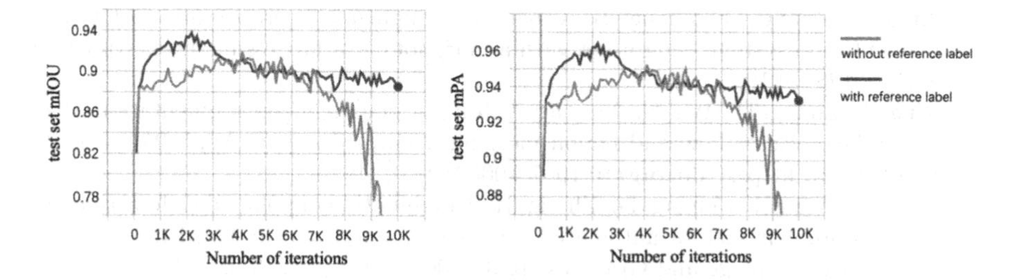


**Fig. 7.** Comparison results of whether to add reference label on TestSet1.

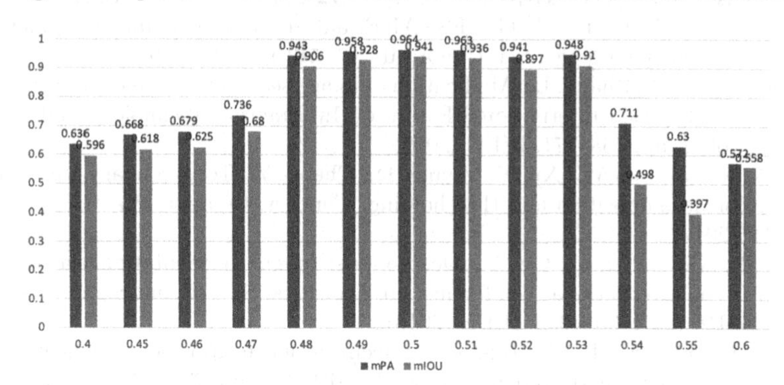


**Fig. 8.** Comparison results of network with different thresholds on TestSet1.

## 4 Conclusion

In this work, we present a novel unsupervised tongue segmentation method using reference labels. We treat the tongue segmentation as an adversarial learning problem to achieve unsupervised learning. The segmentation network and the generation network are used to segment the tongue area and generate new pixels, respectively. Then the discriminator is used to optimize the segmentation network and generation network according to the distribution difference between the composite images and original images. To overcome the invalid segmentation and optimize the results of generating pixels, we propose reference label to constrain the segmentation network. Extensive experimental results on our dataset TestSet1 and public dataset TestSet2 demonstrate that the proposed method has comparable performance with supervised methods.

## References


1. Pang, B., Zhang, D., Wang, K.: Tongue image analysis for appendicitis diagnosis. Inf. Sci. **175**, 160–176 (2005)
2. Zhang, H.Z., Wang, K.Q., Zhang, D., Pang, B., Huang, B.: Computer aided tongue diagnosis system. In: 2005 IEEE Engineering in Medicine and Biology 27th Annual Conference, pp. 6754–6757. IEEE (2006)
3. Zhou, J., Zhang, Q., Zhang, B., Chen, X.: Tonguenet: a precise and fast tongue segmentation system using u-net with a morphological processing layer. Appl. Sci. **9**(15), 3128 (2019)
4. Ning, J., Zhang, D., Wu, C., Yue, F.: Automatic tongue image segmentation based on gradient vector flow and region merging. Neural Comput. Appl. **21**(8), 1819–1826 (2012). https://doi.org/10.1007/s00521-010-0484-3
5. Zuo, W., Wang, K., Zhang, D., Zhang, H.: Combination of polar edge detection and active contour model for automated tongue segmentation. In: Third International Conference on Image and Graphics (ICIG 2004), pp. 270–273. IEEE (2004)
6. Kebin, W., Zhang, D.: Robust tongue segmentation by fusing region-based and edge-based approaches. Expert Syst. Appl. **42**(21), 8027–8038 (2015)
7. Shi, M., Li, G., Li, F.: $C^2$ G $^2$ FSNAKE: automatic tongue image segmentation utilizing prior knowledge. Sci. China Inf. Sci. **56**(9), 1–14 (2013)
8. Zhai, X., Lu, H., Zhang, L.: Application of image segmentation technique in tongue diagnosis. In: 2009 International Forum on Information Technology and Applications, vol. 2, pp. 768–771. IEEE (2009)
9. Li, Z., Yu, Z., Liu, W., Xu, Y., Zhang, D., Cheng, Y.: Tongue image segmentation via color decomposition and thresholding. Concurr. Comput. Pract. Exp. **31**(23), e4662 (2019)
10. Lin, B., Xle, J., Li, C., Qu, Y.: Deeptongue: tongue segmentation via resnet. In: 2018 IEEE International Conference on Acoustics, Speech and Signal Processing (ICASSP), pp. 1035–1039. IEEE (2018)
11. Zhou, C., Fan, H., Li, Z.: Tonguenet: accurate localization and segmentation for tongue images using deep neural networks. IEEE Access **7**, 148779–148789 (2019)
12. Dai, S., Li, X., Wang, L., Wu, P., Tong, W., Chen, Y.: Learning segmentation masks with the independence prior. In: Proceedings of the AAAI Conference on Artificial Intelligence, vol. 33, pp. 3429–3436 (2019)
13. Remez, T., Huang, J., Brown, M.: Learning to segment via cut-and-paste. In: Proceedings of the European Conference on Computer Vision (ECCV), pp. 37–52 (2018)
14. Chen, M., Artieres, T., Denoyer, L.: Unsupervised object segmentation by redrawing. In: Advances in Neural Information Processing Systems, pp. 12705–12716 (2019)
15. Grandvalet, Y., Bengio, Y.: Semi-supervised learning by entropy minimization. In: Advances in Neural Information Processing Systems, pp. 529–536 (2005)

16. Zhang, H., Goodfellow, I., Metaxas, D., Odena, A.: Self-attention generative adversarial networks. arXiv preprint arXiv:1805.08318 (2018)
17. Chen, X., Duan, Y., Houthooft, R., Schulman, J., Sutskever, I., Abbeel, P.: Infogan: interpretable representation learning by information maximizing generative adversarial nets. In: Advances in Neural Information Processing Systems, pp. 2172–2180 (2016)
18. Goodfellow, I., et al.: Generative adversarial nets. In: Advances in Neural Information Processing Systems, pp. 2672–2680 (2014)
19. Miyato, T., Kataoka, T., Koyama, M., Yoshida, Y.: Spectral normalization for generative adversarial networks, February 2018
20. Cong, W.: Deep image harmonization via domain verification. In: CVPR. Dovenet (2020)
21. BioHit. TongueImageDataset (2014). https://github.com/BioHit/TongeImageDataset
22. Nilsback, M., Zisserman, A.: Automated flower classification over a large number of classes. In: 2008 Sixth Indian Conference on Computer Vision, Graphics Image Processing, pp. 722–729, December 2008
23. Zhao, H., Shi, J., Qi, X., Wang, X., Jia, J.: Pyramid scene parsing network. In: Proceedings of the IEEE Conference on Computer Vision and Pattern Recognition, pp. 2881–2890 (2017)
24. Xue, Y., Li, X., Wu, P., Li, J., Wang, L., Tong, W.: Automated tongue segmentation in Chinese medicine based on deep learning. In: Cheng, L., Leung, A.C.S., Ozawa, S. (eds.) ICONIP 2018. LNCS, vol. 11307, pp. 542–553. Springer, Cham (2018). https://doi.org/10.1007/978-3-030-04239-4_49
25. Badrinarayanan, V., Kendall, A., Cipolla, R.: Segnet: a deep convolutional encoder-decoder architecture for image segmentation. IEEE Trans. Pattern Anal. Mach. Intell. **39**(12), 2481–2495 (2017)
26. Chen, L.-C., Papandreou, G., Schroff, F., Adam, H.: Rethinking atrous convolution for semantic image segmentation. arXiv preprint arXiv:1706.05587 (2017)
27. Long, J., Shelhamer, E., Darrell, T.: Fully convolutional networks for semantic segmentation. In: Proceedings of the IEEE Conference on Computer Vision and Pattern Recognition, pp. 3431–3440 (2015)
28. Ronneberger, O., Fischer, P., Brox, T.: U-Net: convolutional networks for biomedical image segmentation. In: Navab, N., Hornegger, J., Wells, W.M., Frangi, A.F. (eds.) MICCAI 2015. LNCS, vol. 9351, pp. 234–241. Springer, Cham (2015). https://doi.org/10.1007/978-3-319-24574-4_28

# Video-Interfaced Human Motion Capture Data Retrieval Based on the Normalized Motion Energy Image Representation

Wei Li[1], Yan Huang[1](✉), and Jingliang Peng[2,3]

[1] School of Software, Shandong University, Jinan, China
{wli,yan.h}@sdu.edu.cn

[2] Shandong Provincial Key Laboratory of Network Based Intelligent Computing, University of Jinan, Jinan 250022, China
ise_pengjl@ujn.edu.cn

[3] School of Information Science and Engineering, University of Jinan, Jinan 250022, China

**Abstract.** To retrieve expected clips from human motion capture (MoCap) databases for effective reuse with a user-friendly interface is a challenging task. In this work, we propose an effective video-based human MoCap data retrieval scheme, which lets the user act in front of a video camera to specify the query and searches for similar motion clips in the MoCap database. Specifically, we propose a novel normalized motion energy image (NMEI) representation to bridge the representational gap between video clips and MoCap clips. Then, the discriminative feature of each NMEI is extracted by computing its local augmented Gabor features, constructing its identity vector in a learned variability space, and making a subspace projection by linear discriminative analysis. Finally, effective similarity metric between the extracted features of any two NMEIs is designed. Experimental results demonstrate the promising performance of the proposed video-based MoCap data retrieval approach.

**Keywords:** Human motion capture · Video · Retrieval · NMEI

## 1 Introduction

Motion capture (MoCap) has been increasingly employed in many fields including interactive virtual reality, film production, 3D animation and so forth. With the explosive growth of MoCap data volume, it is important to retrieve right clips from MoCap databases for efficient reuse.

Many algorithms have been proposed for content-based MoCap data retrieval. They adopt different modalities of query, including MoCap clip (*e.g.*, [15]), hand-drawn sketch (*e.g.*, [24]), puppet motion (*e.g.*, [16]) and Kinect skeleton motion (*e.g.*, [10]). While good results have been demonstrated for these modalities, they have inherent disadvantages. MoCap clips are often not easy to obtain due to the high device cost and complex capturing process. The quality and style of hand-drawn sketches influence the retrieval performance.


H. Yang et al. (Eds.): ICONIP 2020, LNCS 12532, pp. 616–627, 2020.
https://doi.org/10.1007/978-3-030-63830-6_52

The puppet device is not easy to construct and perform. The Kinect-based methods usually restrict the user at a limited range of distances and orientations for good results. Compared with the above-mentioned modalities, video clip appears to be a natural, easy and affordable one. As such, we adopt video clips as queries for MoCap data retrieval in this work. In this work, we focus on this application scenario: the user performs in front of a monocular video camera to specify the query motion, and MoCap clips of similar motions are returned from the MoCap database as the result. For this purpose, we propose an effective video-based MoCap data retrieval method whose key contributions include:

- A novel and discriminative human motion descriptor, which is formed by computing and projecting the augmented Gabor tensor features of the normalized motion energy image (NMEI) for each video or MoCap clip.
- Integration of the similarity metric based on probabilistic linear discriminative analysis (PLDA) and the proposed human motion descriptor to make an effective retrieval system.
- A gallery dataset of human MoCap clips and a query dataset of human action videos, which may be published for research uses.

## 2 Related Work

For the retrieval of MoCap data, different types of data have been used as queries, which include MoCap clip (*e.g.*, [15]), hand-drawn sketch (*e.g.*, [24]), puppet motion (*e.g.*, [16]) and Kinect skeleton motion (*e.g.*, [10]). Compared with them, video clip appears to be more user-friendly, as the data may be easily acquired by recording a user's performance in front of a camera. Nevertheless, it is harder to bridge the representational gap between video clips and MoCap clips and, probably because of which, there has been little research published except for the work by Gupta *et al.* [8] to retrieve MoCap data with video clips as query. They make frame-by-frame alignment of the query video clip and MoCap clips and adopt the dense trajectories [23] for data description, making an effective but time-consuming approach.

Video-based action recognition is related to video-based MoCap data retrieval in that both need to effectively describe motions in videos. Recognizing human actions from video clips is an important research topic in computer vision and, in this context, various methods have been proposed to describe human motions in video clips. The binary motion energy image (MEI) and motion history image (MHI) representations are proposed in the reference [2]. MEI represents where motion has occurred in an image sequence, while MHI is a scalar-valued image where intensity is a function of the motion's recency. The space-time shape feature is proposed in the reference [1] for human action recognition. The template-based method by a maximum average correlation height (MACH) filter is proposed in the reference [18] for recognizing human actions. Besides, gait energy image (GEI) based method is used [13] for individual gait recognition.

Many subspace analysis techniques have been proposed to obtain discriminative feature descriptors for classification, retrieval *etc.* in various contexts.

For instance, identity vector (i-vector) [3,6] was proposed for speaker recognition and tensor subspace learning method [26] was used for human gait recognition.

Convolutional neural network (CNN) based methods have achieved great successes in various fields [20,21,25], and they were recently utilized for human action recognition as well [4,7,12,14,22]. Feichtenhofer *et al.* [4] proposed a convolutional two-stream network fusion scheme for video action recognition. In [14], a decomposition method for 3D convolutional kennels was proposed by modeling spatial and temporal information separately. Li *et al.* [12] encoded spatio-temporal features collaboratively by imposing a weight-sharing constraint on parameters for action recognition. Tran *et al.* [22] proposed a channel separated convolutional network (CSN) for video classification. Girdhar *et al.* [7] proposed an action transformer model for recognizing and localizing human actions in video clips. Generally, CNN based methods extract effective features, but have relatively complex models and demand high hardware configurations. Besides, while the methods mentioned above are related to our work, they focus on video data recognition and analysis. The retrieval across the two different data modalities (video and MoCap data) has drawn little attention so far. Hence, we are motivated to develop a compact solution for video-based MoCap data retrieval.

## 3 Overview

We propose a novel scheme for human MoCap data retrieval given a human video clip from a monocular view as query, whose flowchart is shown in Fig. 1 and explained in the following. We make a simple 3D human model and use each MoCap clip in the database to drive the motion of this model. For each MoCap frame in a clip, we render the posed model to several views. At each view, we construct a normalized motion energy image (NMEI) for the clip from the binarized human silhouettes of all the rendered images at that view. For each video clip, we obtain the human silhouette in each frame by background subtraction and construct the NMEI of the video clip from all the frames' binary human silhouettes. Thereafter, we represent each NMEI as a set of local augmented Gabor features. A total variability subspace is learned from the augmented Gabor tensor descriptors of the NMEIs of all the MoCap clips in the database and, based on which, the identity vector is extracted for each NMEI of the MoCap clips

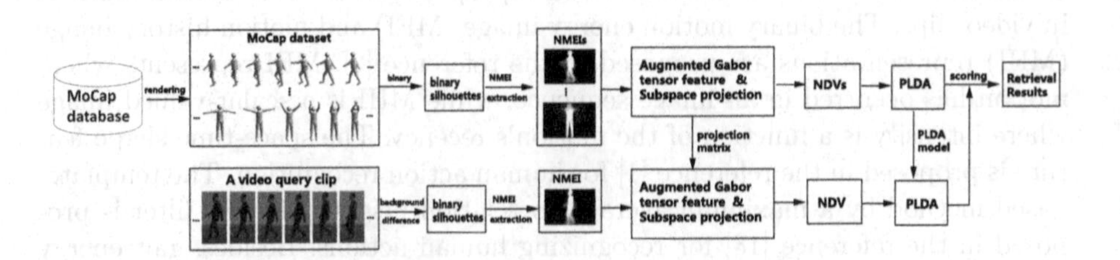


**Fig. 1.** Flowchart of the proposed video-based human MoCap data retrieval scheme.

in the database and the query video clips, respectively. Further, the linear discriminant analysis (LDA) [5] is conducted to find a discriminative subspace for the identity vectors, leading to the NMEI-based discriminative vector (NDV) for each MoCap clip at each specific view and each query video clip. Finally, probabilistic linear discriminant analysis (PLDA) [17] is used to measure the similarity between NDVs and a subset of the MoCap clips in the database whose NDVs are the closest to the query's are returned as the result. Components of the proposed scheme are detailed in Sect. 4.

# 4 Algorithm

## 4.1 Normalized Motion Energy Image

For each MoCap frame, we make a simple 3D human model by fitting the head with a sphere and each of the other bones with a cylinder for the skeleton as posed in the current frame. Then we render the posed 3D model by orthographic projection onto four views (*i.e.*, Front, Back, Left and Right) and binarize each rendered image to get the human silhouettes. By doing such, we obtain four binarized image sequences for each MoCap clip at four views, respectively. Inspired by the GEI representation for gait recognition [9], we propose to represent each image sequence with a normalized motion energy image. Specifically, we compute the bounding box of each silhouette and normalize it to a fixed height. The horizontal centroid of each silhouette is used as the horizontal center for frame alignment. Finally, we compute the NMEI by

$$\mathbf{M}(x,y) = \frac{1}{N}\sum_{t=1}^{N}\mathbf{I}_t(x,y), \tag{1}$$

where $N$ is the number of frames in the MoCap clip and $\mathbf{I}_t$ is the $t$-th normalized silhouette image. Exemplar normalized silhouette images and NMEI for a MoCap sequence are shown in Fig. 2. Note that the traditional MEI [2] is binary and not normalized. Therefore, it records less information about the temporal variation of human posture and is more sensitive to global and local speed variations.

The computation of NMEI for a video clip is similar except that the binary silhouette for each frame is obtained via background subtraction. Exemplar normalized silhouette images and NMEI for a video clip are also shown in Fig. 2.

Based on the NMEI representation, we compute the similarity of a video query to each of the four rendered sequences of a MoCap clip, and use the highest similarity score to measure the resemblance between the query and the MoCap clip.

## 4.2 Augmented Gabor Feature Extraction

We extract augmented Gabor features for each NMEI for robust and discriminative description. As the work [26] that extracts Gabor features on GEIs for

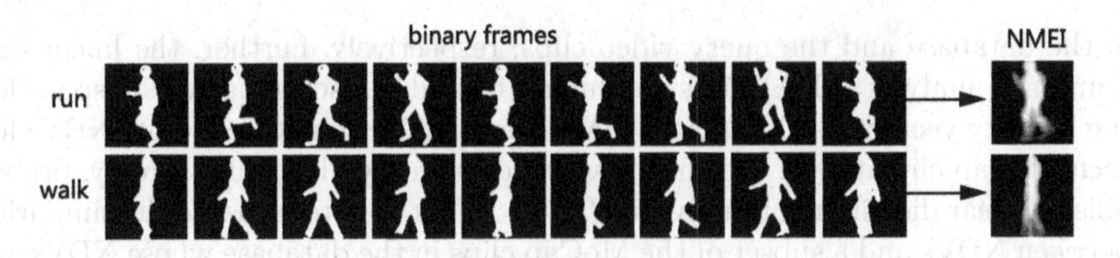


**Fig. 2.** The normalized silhouettes in one action clip (left) and the corresponding NMEI (right) for MoCap data (the top row) and video data (the bottom row).

gait recognition, we define a family of Gabor kernel functions for one given pixel $p(x, y)$ as

$$\psi_{\tau,\upsilon}(p) = \frac{\| \gamma_{\tau,\upsilon} \|^2}{\delta^2} e^{-\frac{\|\gamma_{\tau,\upsilon}\|^2 \|p\|^2}{2\delta^2}} [e^{i\gamma_{\tau,\upsilon} \cdot p} - e^{-\frac{\delta^2}{2}}] \tag{2}$$

where $\gamma_{\tau,\upsilon}$ determines the scale and the orientation of the Gabor kernel function. Following the work [26], we set $\delta = 2\pi$, $\tau = \{0, 1, 2, 3, 4\}$ and $\upsilon = \{0, 1, 2, 3, 4, 5, 6, 7\}$, leading to 40 Gabor kernel functions. For enhanced representation, we further add the pixel's 2D coordinates and intensity information into the local feature representation. Specifically, assuming that $g_i$ is the 40-D Gabor feature vector from the 40 Gabor-filtered images at the $i$-th pixel of an NMEI and $(x_i,y_i)$ and $I(x_i, y_i)$ are the 2D coordinates and intensity of the $i$-th pixel, respectively, the augmented Gabor feature is defined as $q_i = [g_i, x_i, y_i, I(x_i, y_i)] \in R^\rho$ ($\rho = 43$ in this work). For an $H \times W$ NMEI, we describe it by a set of local augmented Gabor features, *i.e.*, $\{q_i|_{i=1}^N\}$, where $N = H \times W$.

### 4.3 Discriminative Vector Extraction

Based on all the Gabor tensor descriptors of the MoCap samples in the database, we firstly learn a model of subspace projections, and then apply this model to the Gabor tensor descriptor of any action sample, either video or MoCap, to construct its discriminative vector (*i.e.*, NDV), as detailed below.

We learn a universal background model (UBM), essentially a Gaussian mixture model (GMM), with the augmented Gabor tensor descriptors of all the MoCap samples. The UBM can be described as a parameter set $\Omega = \{m_k, \mu_k, \Sigma_k|_{k=1}^K\}$ where $K$ is the total number of Gaussian components and $m_k \in R$, $\mu_k \in R^\rho$ and $\Sigma_k \in R^{\rho\times\rho}$ are the weight, the mean vector, and the covariance matrix of the $k$-th Gaussian component, respectively. Further, in order to obtain a fixed-length i-vector $\eta \in R^D$ for each sample, we learn a total variability space $T \in R^{K\rho \times D}$ by factor analysis. Specifically, the new UBM supervector $M \in R^{K\times\rho}$ for each sample is defined by

$$M = \mu + T\omega \tag{3}$$

where $\mu \in R^{K\times\rho}$ is the original UBM supervector, $T$ is a rectangular matrix of low rank and $\omega$ is a random vector having a standard normal distribution

$N(0, I)$. The total factor $\omega$ is a hidden variable, which can be defined by its posterior distribution conditioned to the Baum-Welch statistics. The i-vector $\eta$ is a standard normally distributed latent variable obtained as the maximum a posteriori (MAP) point estimate of $\omega$. We refer the reader to the detailed description of i-vector in the reference [3]. Further, we apply linear discriminative analysis (LDA) [5] on the i-vectors of the MoCap samples to find a lower and more discriminative subspace. Finally, projecting the i-vector of each action sample into this subspace, we obtain its NDV.

### 4.4 Similarity Metric by PLDA

Probabilistic linear discriminative analysis (PLDA) [6,11,17] is used to evaluate the similarity of two samples for its generality and robustness. Since the NDVs are of sufficiently low dimension (*i.e.*, 15 for our experiments), we follow the modified PLDA data generation process proposed in [11]. Specifically, we model the data generation process for each NDV sample $x$ by

$$x = \mu + Fh + \epsilon. \tag{4}$$

It can be seen that the model contains two components: 1) the signal component $\mu + Fh$ which describes between-class variation, and 2) the noise component $\epsilon$. The term $\mu$ represents the mean of all the NDVs of the MoCap action samples. The columns of $F$ provide a basis for the between-class subspace and the term $h$ is a latent vector having a standard normal distribution. The residual noise term $\epsilon$ is defined to be Gaussian with diagonal covariance $\Sigma$.

There are two phases to utilize the PLDA model. In the training phase, we learn the parameters $\theta = \{\mu, F, \Sigma\}$ from all the MoCap samples' NDVs using the expectation maximization (EM) algorithm as in [17]. In the scoring phase, if the video query sample $\eta_1$ and one MoCap sample $\eta_2$ belong to the same action class, they are expected to have the same latent variable $h$. We test two alternative hypotheses: $H_s$ that both $\eta_1$ and $\eta_2$ share the same latent variable $h$, or $H_d$ that the two samples were generated using different latent variables $h_1$ and $h_2$. The similarity score can be computed as the log-likelihood-ratio as

$$score = log \frac{P(\eta_1, \eta_2 | H_s)}{P(\eta_1 | H_d) P(\eta_2 | H_d)} \tag{5}$$

## 5 Experiments

In this section, we conduct comprehensive experiments to evaluate the performance of the proposed video-based MoCap data retrieval scheme.

### 5.1 Platform and Implementation

All the proposed NDVP method and the benchmark methods were implemented in Matlab. We run the Matlab code on a laptop with Intel Core 2, 2.10-GHz CPU, 24 GB RAM and 64-bit Win7 operating system. We make use of the MSR Identity Toolkit [19] for i-vector and PLDA implementation.

**Table 1.** MAP statistics of MHPN, MEPN, NPN, NDVN and NDVP.

| Action | MAP statistics of the five methods | | | | |
|---|---|---|---|---|---|
| (#) name | MHPN | MEPN | NPN | NDVN | NDVP |
| (1) Calisthenics movements | 0.2151 | 0.2071 | 0.5848 | 0.7043 | 0.7272 |
| (2) Arms waving | 0.6928 | 0.7005 | 0.4930 | 0.7497 | 0.7093 |
| (3) Phone answering | 0.6139 | 0.6558 | 0.1929 | 0.6550 | 0.6436 |
| (4) Jump | 0.2765 | 0.3408 | 0.3829 | 0.4234 | 0.5811 |
| (5) Punch in horse stance | 0.1449 | 0.1622 | 0.3820 | 0.7115 | 0.8689 |
| (6) Basketball bouncing | 0.1153 | 0.1480 | 0.2519 | 0.6262 | 0.7692 |
| (7) Front raise | 0.4629 | 0.6583 | 0.2357 | 0.5650 | 0.4279 |
| (8) Walking in circles | 0.3759 | 0.3426 | 0.7247 | 0.5258 | 0.7411 |
| (9) Normal running | 0.1769 | 0.5359 | 0.2834 | 0.5135 | 0.7151 |
| (10) Step | 0.0995 | 0.1496 | 0.3534 | 0.2875 | 0.2893 |
| (11) Basket shot | 0.2647 | 0.3457 | 0.3512 | 0.8429 | 0.8740 |
| (12) Bending down | 0.5728 | 0.6760 | 0.6889 | 0.7454 | 0.8271 |
| (13) Turning in circles | 0.3999 | 0.2232 | 0.8763 | 0.8226 | 0.8937 |
| (14) Parade step | 0.1443 | 0.2120 | 0.5293 | 0.7251 | 0.7407 |
| (15) Normal walking | 0.0643 | 0.1678 | 0.3932 | 0.4715 | 0.5421 |
| (16) Sitting down | 0.1375 | 0.2709 | 0.4408 | 0.6290 | 0.6824 |
| Average MAP | 0.2973 | 0.3623 | 0.4590 | 0.6249 | 0.6895 |

### 5.2 Databases and Performance Metrics

We make a human MoCap database with 16 daily action classes. Four males and one female of various body shapes are employed to capture these motion clips. Each actor or actress performs each motion 4 times, resulting in a total number of 320 clips in the MoCap database.

We also make a video query database with the same 16 daily action classes. Three males and one female of various body shapes are employed to capture these video action clips by a monocular camera. Each actor or actress performs each motion 10 times at each of the four viewpoints (*i.e.*, Front, Back, Left and Right), resulting in a total number of 2,560 clips in the video query database.

The performance metrics we use include mean average precision (MAP), precision-recall curve (P-R curve) and precision at $n$ (P@$n$) as often used in the general field of information retrieval.

### 5.3 Benchmark Algorithms

We test and compare five algorithms, as described below.

- MHI [2] with subspace projection by PCA for feature description and nearest neighbor (NN) for similarity metric (MHPN);

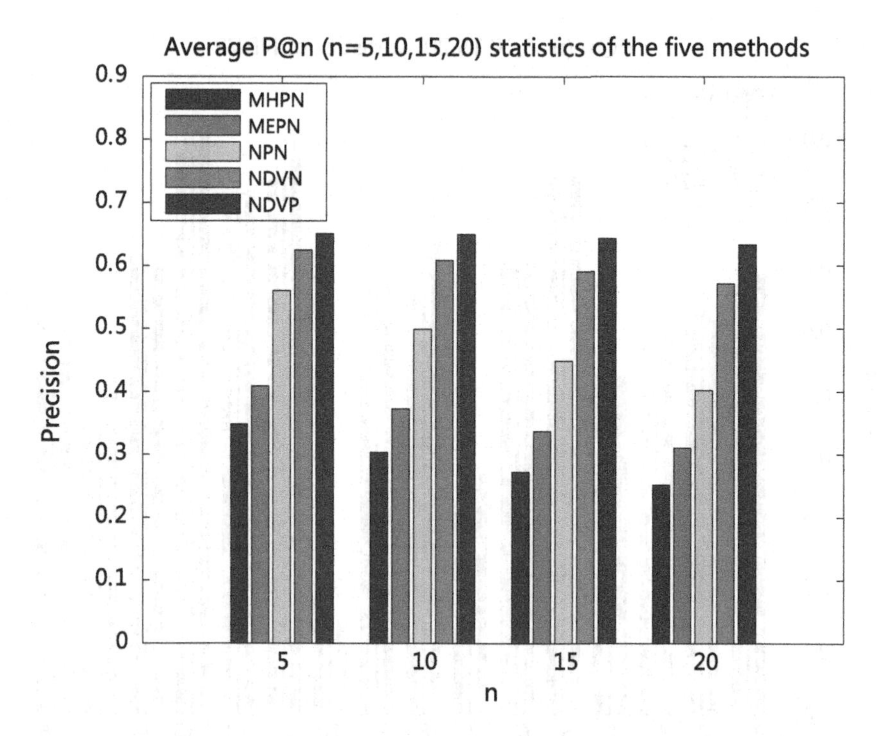


**Fig. 3.** Average P@$n$ statistics of the five methods over all the motion classes.

- MEI [2] with subspace projection by PCA for feature description and NN for similarity metric (MEPN);
- NMEI with subspace projection by PCA for feature description and NN for similarity metric (NPN);
- NMEI based discriminative vector (NDV) for feature description and NN for similarity metric (NDVN);
- NMEI based discriminative vector (NDV) for feature description and PLDA for similarity metric (NDVP).

Note that we do not compare with the work [8] as it focuses on extracting similar (to the video query) sub-MoCap-sequences through effective but time-consuming frame-to-frame alignment, while our work targets at quick search of similar whole MoCap clips based on overall motion characteristics.

### 5.4 MAP Statistics

MAP statistics of the five methods are presented in Table 1 for comparison. From Table 1 we have the following observations: 1) NPN outperforms MHPN and MEPN, showing a better discriminating power of NMEI than MHI and MEI, 2) NDVN outperforms NPN, showing a better discriminating power of NDV than NMEI, and 3) NDVP outperforms NDVN, showing the better performance of PLDA than NN for the similarity measurement.

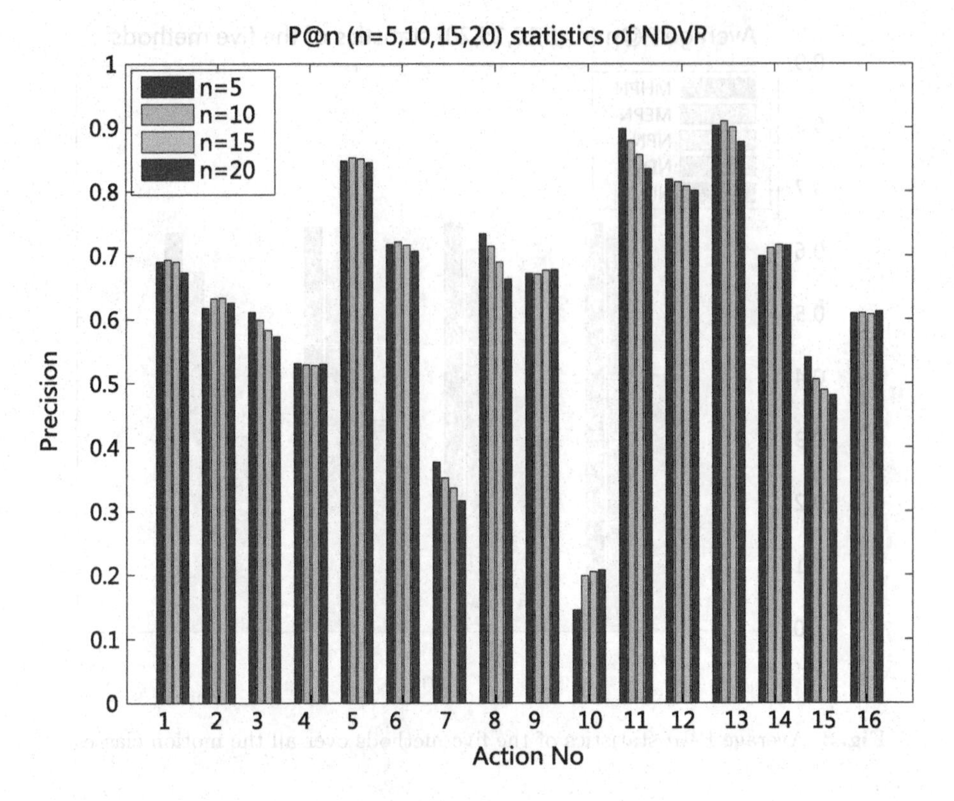


**Fig. 4.** P@$n$ statistics of NDVP for all the action classes.

## 5.5 P@n Statistics and P-R Curves

In Fig. 3, we plot for each method its average P@n (n = 5, 10, 15, 20) statistics over all the action classes, which again shows better performance of NDVP over the others. For a complete reference, we also plot the P@n (n = 5,10,15,20) statistics and the P-R curves of NDVP for all the action classes in Fig. 4 and Fig. 5.

## 5.6 Computing Efficiency

In our experiments, for NDVN and NDVP, we set the GMM to have 128 components, the total variability space to have 150 dimensions and the LDA to reduce to C-1 (C is the number of action classes) dimensions; for MHPN, MEPN and NPN, we set the PCA to preserve 99% of the original data's energy, and each of MHI, MEI and NMEI to have a size of $100 \times 150$.

Under these settings, the image pre-processing tasks (*e.g.*, background subtraction and NMEI construction) can be conducted offline at a very high efficiency. The time for learning the UBM model, the total variability space and the LDA model for NDVN and NDVP are 143.80 s, 289.63 s and 0.05 s, respectively.

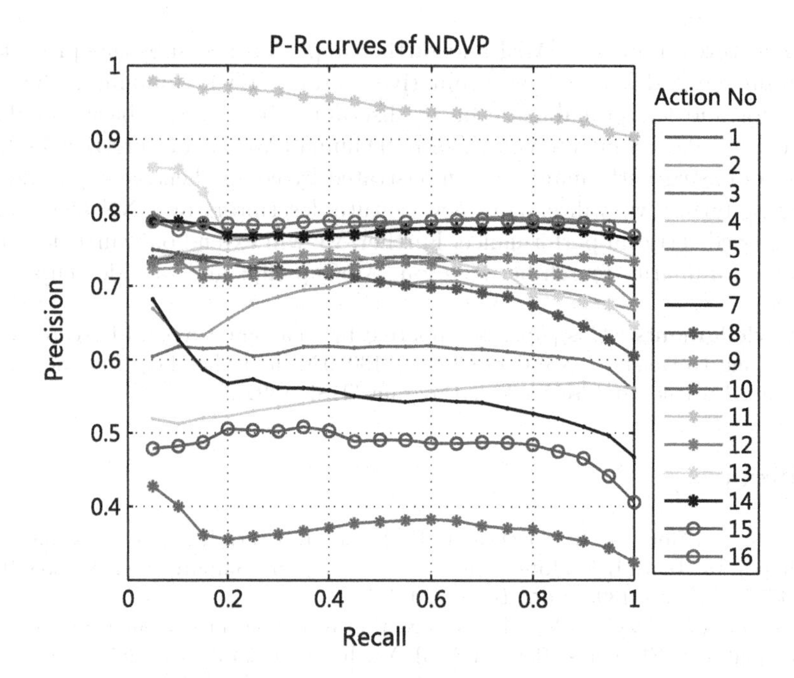


**Fig. 5.** P-R curves of NDVP for all the action classes.

Besides, it takes 0.11 s for learning the PLDA model for NDVP. The augmented Gabor feature extraction takes 0.06s for each NMEI sample on the average.

When evaluating a retrieval scheme's computing efficiency, we focus more on the average online query time. As such, we report the average running time of a query excluding the MHI, MEI or NMEI construction and the model learning parts for the five methods in Table 2. From Table 2, we observe that MHPN, MEPN and NPN achieve the highest computing efficiencies due to their simple feature extraction processes and similarity metrics. NDVN and NDVP have similar and lower computing efficiencies, but still achieve less than 0.1s per online query, which is quite fast.

**Table 2.** Averaged online query time

| Methods | MHPN | MEPN | NPN | NDVN | NDVP |
|---|---|---|---|---|---|
| Average time | 0.024 s | 0.024 s | 0.024 s | 0.087 s | 0.084 s |

## 6 Conclusion and Future Work

A novel and efficient video-based 3D human MoCap data retrieval method is proposed in this work. We construct a normalized motion energy image (NMEI) to represent the motion in each MoCap or video clip, extract augmented Gabor tensor features from the NMEI and further apply learned subspace projections to obtain the NMEI-based discriminative vector (NDV). Combining the NDV descriptor and the probabilistic linear discriminative analysis based similarity metric, we make a complete video-based 3D human MoCap data retrieval scheme whose promising performance is demonstrated by comprehensive experiments.

In the future, we will explore new feature descriptors by CNN technologies for improved retrieval performance. Further, we will extend our current method for partial similarity search between long MoCap clips and/or video clips.

**Acknowledgments.** This paper is supported by "The Fundamental Research Funds of Shandong University", NSFC of China (61907026) and Shandong Province Higher Educational Science and Technology Program (J18KA392).

## References

1. Blank, M., Gorelick, L., Shechtman, E., Irani, M., Basri, R.: Actions as space-time shapes. In: Tenth IEEE International Conference on Computer Vision, 2005. ICCV 2005, vol. 2, pp. 1395–1402. IEEE (2005)
2. Bobick, A.F., Davis, J.W.: The recognition of human movement using temporal templates. IEEE Trans. Pattern Anal. Mach. Intell. **23**(3), 257–267 (2001)
3. Dehak, N., Kenny, P.J., Dehak, R., Dumouchel, P., Ouellet, P.: Front-end factor analysis for speaker verification. IEEE Trans. Audio Speech Lang. Process. **19**(4), 788–798 (2011)
4. Feichtenhofer, C., Pinz, A., Zisserman, A.: Convolutional two-stream network fusion for video action recognition, pp. 1933–1941 (2016)
5. Fukunaga, K.: Introduction to Statistical Pattern Recognition. Academic press, Cambridge (2013)
6. Garcia-Romero, D., Espy-Wilson, C.Y.: Analysis of i-vector length normalization in speaker recognition systems. In: Interspeech, pp. 249–252 (2011)
7. Girdhar, R., Carreira, J.J., Doersch, C., Zisserman, A.: Video action transformer network. In: IEEE Conference on Computer Vision and Pattern Recognition (CVPR) (2019)
8. Gupta, A., He, J., Martinez, J., Little, J.J., Woodham, R.J.: Efficient video-based retrieval of human motion with flexible alignment. In: 2016 IEEE Winter Conference on Applications of Computer Vision (WACV), pp. 1–9. IEEE (2016)
9. Han, J., Bhanu, B.: Individual recognition using gait energy image. IEEE Trans. Pattern Anal. Mach. Intell. **28**(2), 316–322 (2006)
10. Kapadia, M., Chiang, I.k., Thomas, T., Badler, N.I., Kider Jr, J.T., et al.: Efficient motion retrieval in large motion databases. In: Proceedings of the ACM SIGGRAPH Symposium on Interactive 3D Graphics and Games, pp. 19–28. ACM (2013)
11. Kenny, P.: Bayesian speaker verification with heavy-tailed priors. In: Odyssey, p. 14 (2010)
12. Li, C., Zhong, Q., Xie, D., Pu, S.: Collaborative spatio-temporal feature learning for video action recognition. In: International Conference on Computer Vision and Pattern Recognition (2019)
13. Li, W., Kuo, C.C.J., Peng, J.: Gait recognition via GEI subspace projections and collaborative representation classification. Neurocomputing **275**, 1932–1945 (2018)

14. Luo, C., Yuille, A.L.: Grouped spatial-temporal aggregation for efficient action recognition, pp. 5512–5521 (2019)
15. Müller, M., Röder, T., Clausen, M.: Efficient content-based retrieval of motion capture data. In: ACM Transactions on Graphics (TOG), vol. 24, pp. 677–685. ACM (2005)
16. Numaguchi, N., Nakazawa, A., Shiratori, T., Hodgins, J.K.: A puppet interface for retrieval of motion capture data. In: Proceedings of the 2011 ACM SIGGRAPH/Eurographics Symposium on Computer Animation, pp. 157–166. ACM (2011)
17. Prince, S.J., Elder, J.H.: Probabilistic linear discriminant analysis for inferences about identity. In: 2007 IEEE 11th International Conference on Computer Vision, pp. 1–8. IEEE (2007)
18. Rodriguez, M.D., Ahmed, J., Shah, M.: Action mach a spatio-temporal maximum average correlation height filter for action recognition. In: IEEE Conference on Computer Vision and Pattern Recognition, 2008. CVPR 2008, pp. 1–8. IEEE (2008)
19. Sadjadi, S.O., Slaney, M., Heck, L.: MSR identity toolbox-a Matlab toolbox for speaker recognition research. Microsoft Research, Conversational Systems Research Center (CSRC) (2013)
20. Song, X., et al.: Channel attention based iterative residual learning for depth map super-resolution. In: The IEEE Conference on Computer Vision and Pattern Recognition (CVPR) (2020)
21. Song, X., et al.: Apollocar3D: a large 3D car instance understanding benchmark for autonomous driving. In: The IEEE Conference on Computer Vision and Pattern Recognition (CVPR) (2019)
22. Tran, D., Wang, H., Feiszli, M., Torresani, L.: Video classification with channel-separated convolutional networks. In: International Conference on Computer Vision and Pattern Recognition, pp. 5552–5561 (2019)
23. Wang, H., Schmid, C.: Action recognition with improved trajectories. In: Proceedings of the IEEE International Conference on Computer Vision, pp. 3551–3558 (2013)
24. Xiao, J., Tang, Z., Feng, Y., Xiao, Z.: Sketch-based human motion retrieval via selected 2D geometric posture descriptor. Sig. Process. **113**, 1–8 (2015)
25. Xibin Song, Y.D., Qin, X.: Deeply supervised depth map super-resolution as novel view synthesis. IEEE Trans. Circuits Syst. Video Technol. **29**(8), 2323–2336 (2019)
26. Xu, D., Huang, Y., Zeng, Z., Xu, X.: Human gait recognition using patch distribution feature and locality-constrained group sparse representation. In: IEEE Transactions on Image Processing, pp. 316–326 (2012)

# WC2FEst-Net: Wavelet-Based Coarse-to-Fine Head Pose Estimation from a Single Image

Zhen Li[1], Wei Li[1], Zifei Jiang[1], Peng Jiang[2], Xueqing Li[1], Yan Huang[1(✉)], and Jingliang Peng[3,4]

[1] School of Software, Shandong University, Jinan, China
jensleesd@gmail.com, jiangzifei@mail.sdu.edu.cn
{wli,xqli,yan.h}@sdu.edu.cn

[2] School of Qilu Transportation, Shandong University, Jinan, China
sdujump@gmail.com

[3] Shandong Provincial Key Laboratory of Network Based Intelligent Computing, University of Jinan, Jinan 250022, China

[4] School of Information Science and Engineering, University of Jinan, Jinan 250022, China
ise_pengjl@ujn.edu.cn

**Abstract.** This paper proposes a novel head pose estimation scheme that is based on image and wavelets input and conducts a coarse to fine regression. As wavelets provide low-level shape abstractions, we add them as extra channels to the input to help the neural network to make better estimation and converge. We design a coarse-to-fine regression framework that makes coarse-grained head pose classification followed by fine-grained angles estimation. This framework helps alleviate the influence of biased training sample distribution, and combines segment-wise mappings to form a better global fitting. Further, multiple streams are used in the neural network to extract a rich feature set for robust and accurate regression. Experiments show that the proposed method outperforms the state-of-the-art methods of the same type for the head pose estimation task.

**Keywords:** Head pose estimation · Wavelet transform · Coarse-to-fine · Deep learning · Convolutional neural network

## 1 Introduction

There is a growing demand for automatic and effective head pose estimation in many important applications, such as face recognition, driver attention estimation and human-computer interaction. However, head pose estimation still faces many challenges, such as variations in identity and facial expression and

Z. Li, W. Li—Both authors have contributed equally to this work.


H. Yang et al. (Eds.): ICONIP 2020, LNCS 12532, pp. 628–640, 2020.
https://doi.org/10.1007/978-3-030-63830-6_53

occlusion in imaging, making it a difficult task to automatically and robustly estimate the head pose from an input image.

Many methods have been proposed for head pose estimation, such as landmark-based approaches [27]. Nevertheless, detection of facial landmarks itself is a complex problem and the results are sensitive to noise, lighting and occlusion in the images. As convolutional neural network (CNN) techniques achieved great successes in various fields, e.g., face detection [20], autonomous driving [24], geological analysis [14,15], time series data mining [8–10], they were recently utilized for head pose estimation as well. For instance, Ranjan *et al.* [20] proposed to use a deep multi-task scheme for head pose estimation; Yang *et al.* [26] proposed to use two-streams network with attention-based module for head pose estimation. More works along this line are reviewed in Sect. 2. Though these works make a direct end-to-end estimation of head poses, they have paid little particular attention to address the issue of biased training sample distribution, and they use purely the raw images as input.

In this work, we also make end-to-end head pose estimation using CNN techniques. In particular, we propose a coarse-to-fine regression framework to counteract biased training sample distribution, add image wavelets to the input and use multiple streams in the network for a deeper exploitation of features. We call the proposed wavelet-based coarse-to-fine regression network WC2FEst-Net for short. While the outstanding performance of WC2FEst-Net is fully demonstrated in Sect. 4, we show in Fig. 1 several examples of head pose estimation results on different head pose datasets.

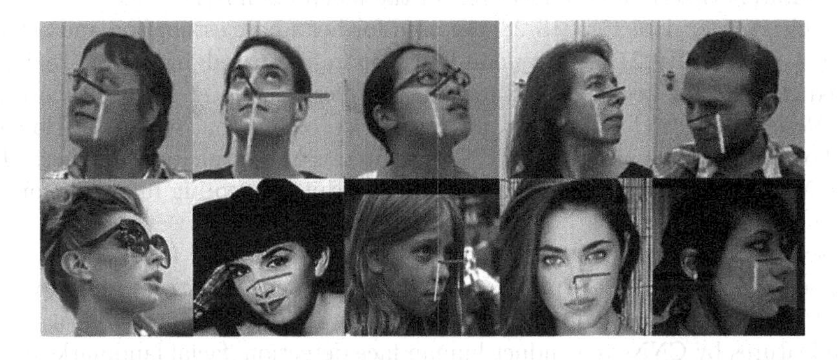

**Fig. 1.** Examples of head pose estimation on BIWI [4] and AFLW2000 datasets [12].

The key contributions of this work are listed below.

- A novel framework of coarse-to-fine regression is proposed for head pose estimation. The coarse-grained classification followed by fine-grained regression alleviates biased training and forms a better approximation to the global mapping function for accurate head pose estimation.

- A higher level shape abstraction is provided besides the raw pixel image by introducing wavelets information into the input, which provides extra guidance for the estimation and helps the convergence of the network.
- Multiple streams are used with the fine-grained stagewise feature aggregation to extract and utilize a richer feature set for robust estimation.
- Comprehensive experiments are made to evaluate the proposed scheme on multiple public head pose datasets, showing its outstanding performance over the state-of-the-arts.

## 2 Related Work

Some works use the depth image as input to conduct head pose estimation. Fanelli *et al.* [5] use the discriminant random regression forest to estimate the head pose of the depth image. Meyer *et al.* [18] propose to register a three-dimensional (3D) deformable model in a depth image and gradually redefining the registration over time. Nevertheless, depth cameras are expensive, limiting its wide application. Most head pose estimation methods [3,17,20,22,25] directly utilize RGB images as input for head pose estimation.

Recently facial landmark detectors [13,28] based on the neural network have been used for facial modeling and analysis. With this progress, landmark-based head pose estimation methods have achieved promising performance. However, landmark-based methods usually require manually labeling the ground-truth, which is a labor-intensive task. For low-resolution head pose images, it is difficult to accurately determine the locations of the facial landmarks.

Multiple face detectors [13,28] are used for head pose estimation and achieve greater success [19,21]. Then, non-linear manifold-based methods are introduced [1,7,25] for head pose estimation. Besides, Chang *et al.* [3] propose the FacePoseNet method by a CNN for head pose estimation, while the Hopenet method [22] exploits a multi-loss CNN architecture to estimate head poses. The FSA-Net method [26] utilizes the attention module by scoring functions for feature aggregation.

Several approaches [13,20] use the multi-task framework for head pose estimation. Kumar *et al.* [13] propose a modified GoogleNet structure to predict facial landmarks and head pose jointly. Ranjan *et al.* [20] propose to learn common features by CNNs to conduct human face detection, facial landmarks detection, gender estimation and head pose estimations. These methods have shown that learning multiple related tasks together could improve the accuracy of head pose estimation. However, multi-task methods may lead to high cost for computing multiple tasks in parallel. Besides, multi-task methods also require more kinds of training data labels to train the modules for different tasks, constraining its wide application.

# 3 Method

## 3.1 Overview of WC2FEst-Net

The high-level architecture of the proposed WC2FEst-Net is shown in Fig. 2. As shown in Fig. 2, the RGB image and its wavelets are used as input for *Module C* that classifies the head pose according to its yaw, pitch and roll angles, respectively. To be specific, for each type of rotation angle (yaw, pitch or roll), the whole gamut is divided into classes, i.e., regions with small overlaps, and an input image is classified based on the yaw, pitch and roll angles, respectively, of the head pose in it. Then, based on the classification, the RGB image and its wavelets are sent to the corresponding fine-grained yaw, pitch, and roll regression networks (*Module* $Y_i$, *Module* $P_i$, and *Module* $R_i$, $i = 1, 2, 3$) to estimate the three angles, respectively.

All the modules in Fig. 2 have the same network structure as shown in Fig. 4 and explained in Sects. 3.3 and 3.4. *Module C* differs in that the output head pose vector contains the class labels rather than the specific values of the three angles. Also note that each fine-grained regression network outputs a head pose vector but only one component (i.e., yaw for *Module* $Y_i$, pitch for *Module* $P_i$, and roll for *Module* $R_i$) is used to form the final result.

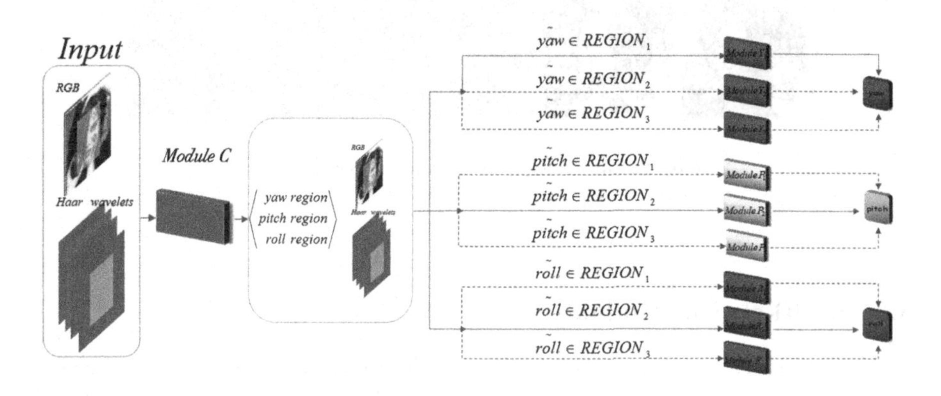


**Fig. 2.** Architecture of the proposed WC2FEst-Net scheme.

## 3.2 Wavelet Transform

Our method is based on wavelet transform, which decomposes an image into a sequence of wavelet sub-bands of the same size. The wavelet transform function is defined as follows:

$$WT_f\left(a, b, c\right) = \frac{1}{\sqrt{\alpha}} \iint f\left(x, y\right) \phi_x \left(\frac{x-b}{a}\right) \phi_y \left(\frac{y-c}{a}\right) \, dxdy \qquad (1)$$

The parameters $a$, $b$ and $c$ are continuous variables. If the translation parameters of one-dimensional wavelet transform along the $x$ and $y$ directions are $b$ and $c$ respectively, then the $x$-direction wavelet function can be expressed as $\phi_x(\frac{x-b}{a})dx$. And the wavelet function in $y$-direction can be expressed as $\phi_y(\frac{y-c}{a})dy$. $\alpha > 0$ is called the scale factor, whose function is to scale the basic wavelet $\Phi$ function.

Supposing that $f(x, y)$ is an image, its wavelet transform is equivalent to filtering $f(x, y)$ along the $x$-direction and the $y$-direction respectively through different one-dimensional filters, details of which can be found in the reference [16].

After one pass of wavelet transform, the original image is transformed to four wavelet sub-bands. The low-frequency sub-band keeps the overall appearance while the high-frequency sub-bands capture the edge information of the image. To be specific, the original image is transformed into four sub-bands $cA, cH, cV, cD$ which represent low frequency component, horizontal detail component, vertical detail component and diagonal detail component, respectively. Figure 3 shows an example of Haar wavelet transformation, where the first sub-figure shows the original color image, and the rest sub-figures show the four wavelet sub-bands.

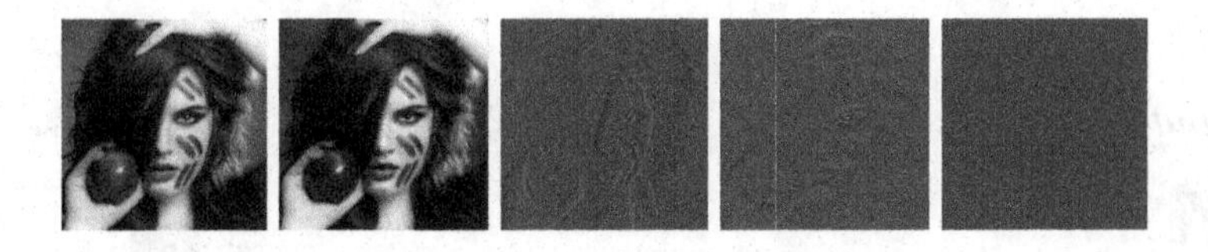

**Fig. 3.** Illustration of wavelet decomposition.

### 3.3 Multi-stream Module

As shown in Fig. 2, all the component modules of WC2FEst-Net share the same network structure, which is shown in Fig. 4. As shown in Fig. 4, the module structure contains three streams and each stream has $S$ ($S = 3$ in our design) stages to generate feature maps. In order to increase the variety of extracted features, three streams are constructed from different building blocks, $Block_L$, $Block_M$ and $Block_R$, which are defined as

$$\begin{aligned} Block_L(c) &\equiv \{SepConv2D(3 \times 3, c) - BN - ReLU\}, \\ Block_M(c) &\equiv \{SepConv2D(3 \times 3, c) - BN - Sigmoid\}, \\ Block_R(c) &\equiv \{SepConv2D(3 \times 3, c) - BN - Tanh\}, \end{aligned}$$

where $SepConv2D$ is separable 2D convolution, $BN$ is batch normalization and $c$ denotes channel count. The detailed structure of the proposed multi-stream module is shown in Table 1.

**Table 1.** Multi-stream module. $B$ means *Block*, $AP$ means AvgPool, $MP$ means MaxPool.Multi-stream module. $B$ means *Block*, $AP$ means AvgPool, $MP$ means MaxPool.

| $S_L$ | $B_L(16)$ | $AP(2\times 2)$ | $B_L(32)$ | $B_L(32)$ | $AP(2\times 2)$ | $B_L(64)$ | $B_L(64)$ | $AP(2\times 2)$ | $B_L(128)$ | $B_L(128)$ |
|---|---|---|---|---|---|---|---|---|---|---|
| $S_M$ | $B_M(16)$ | $MP(2\times 2)$ | $B_M(32)$ | $B_M(32)$ | $MP(2\times 2)$ | $B_M(64)$ | $B_M(64)$ | $MP(2\times 2)$ | $B_M(128)$ | $B_M(128)$ |
| $S_R$ | $B_R(16)$ | $AP(2\times 2)$ | $B_R(32)$ | $B_R(32)$ | $AP(2\times 2)$ | $B_R(64)$ | $B_R(64)$ | $AP(2\times 2)$ | $B_R(128)$ | $B_R(128)$ |

Each stream extracts a feature map at one stage, as shown in Fig. 4. For the $i$-th stage, the extracted feature maps are fused together by a stage fusion process. Specifically, it makes element-wise multiplication of the output feature maps from $l\_layer_i$ and $m\_layer_i$ to generate a feature map, $feat\_lm_i$ Then, it makes element-wise addition of $feat\_lm_i$ and the output feature map from $r\_layer_i$ to generate $feat\_lmr_i$. Finally, $c$ 1×1 convolutions and average pooling are made on $feat\_lmr_i$ to obtain a $w \times h \times c$ feature map $Feat_i$ for the $i$-th stage. The width $w$, height $h$ and channel count $c$ of the final feature map is set to 8, 8 and 64, respectively.

Note that, compared with the FSA-Net method [26], we add an extra stream $S_M$ in our scheme to effectively extract more features.

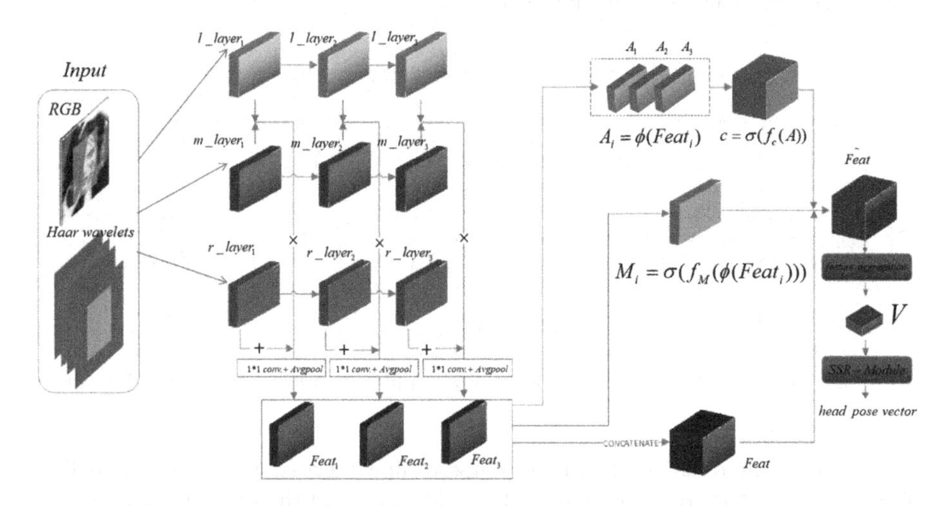


**Fig. 4.** Component module structure with the multi-stream module and the fine-grained structure feature aggregation.

### 3.4 Fine-Grained Structure Feature Aggregation

Now that we have obtained the combined feature map $Feat = (Feat_1, ...Feat_S)$, we conduct the same fine-grained structure feature aggregation as done in the FSA-Net method [26]. Specifically, for each feature map $Feat_i$, we compute its attention map $A_i$ by $A_i(m,n) = \Phi(Feat_i(m,n))$ where $m$, $n$ denote each pixel position in $Feat_i$. Three options are used to implement the function $\Phi(\cdot)$, which are $1 \times 1$ convolution, variance across channels, and uniform. With the extracted feature maps $Feat_i$ and their attention maps $A_i$, fine-grained structure mapping is used to extract a set of representative features $\widetilde{Feat}$. Further, feature aggregation is conducted on $\widetilde{Feat}$ and SSR-Module is then used to derive the final head pose vector.

### 3.5 Coarse-to-Fine Regression

In our tests with the FSA-Net method [26], we observe that the estimation errors tend to be smaller for central poses while larger for poses towards the extremes. For instance, we train the FSA-Net with 70% samples and test it with the rest 30% of the AFLW2000 dataset. The error-angle curves for pitch, yaw and roll estimations are shown in Fig. 6, where the horizontal axis marks the ground-truth angles and the vertical axis marks the corresponding estimation errors. These curves show smaller/larger errors for smaller/larger poses. Further, we plot the statistics of the labeled head pose angles of the AFLW2000 dataset in Fig. 5, which shows uneven distributions of angles with more samples around the center while less samples toward the extremes. Based on the above observations, we project that biased training sample distributions lead to (at least partially) worse estimation accuracy for larger poses.

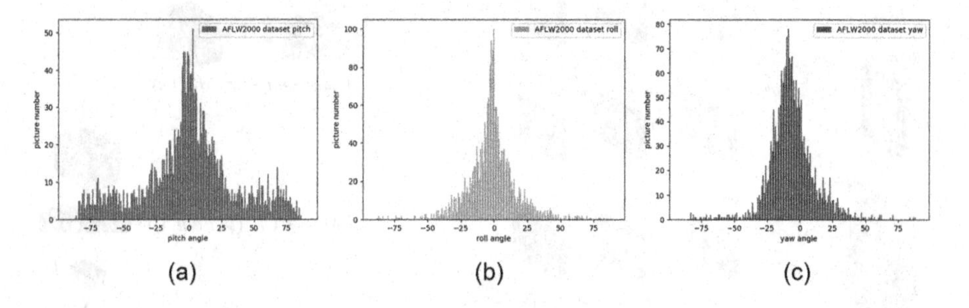


**Fig. 5.** The angle distributions of yaw, pitch, and roll in the AFLW2000 dataset.

This motivated us to design a coarse-to-fine regression scheme. The first stage estimates the coarse region of the head pose, and the second stage derives the angle values using the corresponding fine-grained regression networks. Each fine-grained regression network is specifically trained using samples falling inside the corresponding region. Using specifically trained multiple regression networks instead of one globally trained network, we address the issue of biased sample distribution and form a better approximation to the mapping function.

The proposed coarse-to-fine regression framework consists of $3 \times M + 1$ modules which are $Module\ C$, $Module\ Y_m$, $Module\ P_m$ and $Module\ R_m$, $m = 1, ..., M$. The parameter $M$ is the number of coarse angle regions and $m$ denotes the region index. We empirically set $M$ to 3 with $REGION_1 = [-90^o, -20^o]$, $REGION_2 = [-30^o, 30^o]$ and $REGION_3 = [20^o, 90^o]$.

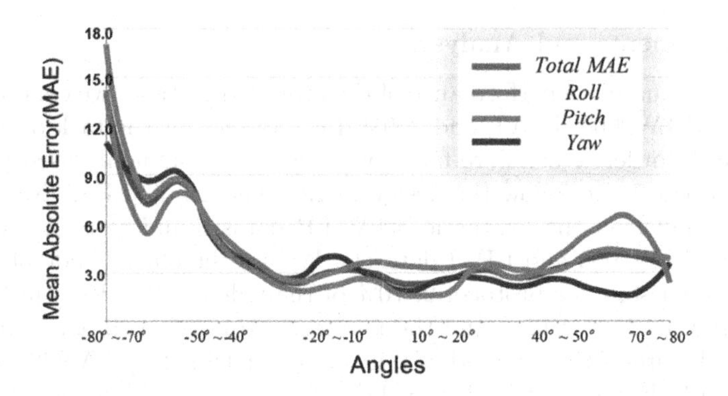


**Fig. 6.** The MAE distributions of yaw, pitch, and roll angle estimations in the AFLW2000 dataset.

# 4 Experiments

We use Keras with Tensorflow backend to implement the proposed WC2FEst-Net method. All the experiments are performed on Ubuntu 16.0 system with an Intel i5 CPU and a GTX1080Ti GPU. The epochs and batch size are set to 90 and 16 respectively to train the network by the Adam optimizer. Regarding the computing efficiency, it takes about 3 ms on the average for our proposed method to estimate the head pose of each image.

## 4.1 Datasets

To make a comprehensive comparison, we conduct experiments on multiple public head pose datasets: AFLW2000 [12], BIWI [4] and 300W-LP [23], which contain precise pose annotations, numerous identities, different lighting conditions, expressions and complex backgrounds.

**AFLW2000:** This dataset is derived from the first 2000 identities of the in-the-wild AFLW dataset [12], which has been re-annotated with 68 3D landmarks points. The faces in the dataset have large pose variations with various illumination conditions and expressions.

**BIWI:** This dataset is collected in a laboratory setting by recording RGB-D videos of different subjects across different head poses by a Kinect v2 device. It contains 24 videos of 20 subjects and roughly 15,000 frames. The rotations are $\pm 75^o$ for yaw, $\pm 60^o$ for pitch and $\pm 50^o$ for roll.

**300W-LP:** This dataset is derived from the 300 W dataset [23] which unifies several datasets for face alignment through 68 landmarks. Zhu *et al.* [29] use face profiling with 3D image meshing to generate 61,225 samples across large poses, which are further expanded to 122,450 samples with flipping.

### 4.2 Experiments and Analysis

In order to evaluate the performance of the proposed method, we conduct experiments on AFLW2000, BIWI, and 300W-LP datasets for the task of head pose estimation. In order to compare fairly, we follow two commonly used protocols.

For protocol 1, we follow the design of Hopenet [22] and FSA-Net [26] that trains the model on the synthetic 300W-LP dataset and tests it on the two real-world AFLW2000 and BIWI datasets. We use the official code of FSA-Net [26] for comparison. For protocol 2, 70% of the videos (16 videos) in the BIWI dataset are used for training and the rest for testing. We compare our method with several state-of-the-art methods, including 3DDFA [28], FAN [2], Dlib [11], KEPLER [13], Hopenet [22], Martin [17], VGG16 [6] and FSA-Net [26].

The estimation results following protocol 1 are shown in Table 2. The estimation is a challenging task due to the large differences between the training and testing data. The training dataset is synthetic, while the testing datasets are from the real world. The mean absolute error (MAE) is used as the evaluation metric. WEst-Net represents the network shown in Fig. 4. In other words, we use the single network in Fig. 4 to complete the head pose estimation without using the coarse-to-fine regression. From Table 2, we observe that the landmark-free methods (Hopenet, FSA-Net, WEst-Net and WC2FEst-Net) perform better than landmark-based ones on both AFLW2000 and BIWI datasets. The proposed WEst-Net method achieves good performance because of the utilization of wavelet information and multi-streams structure. With the coarse-to-fine regression architecture further introduced, the proposed WC2FEst-Net method achieves the best performance.

**Table 2.** Results under protocol 1. All the methods are trained on 300W-LP dataset, and the first group of MAE is evaluated on AFLW2000 dataset, while the second group of MAE is evaluated on BIWI dataset.

| Method | MAE (AFLW2000) | | | | MAE (BIWI) | | | |
|---|---|---|---|---|---|---|---|---|
| | Total | Yaw | Pitch | Roll | Total | Yaw | Pitch | Roll |
| 3DDFA [28] | 7.52 | 5.63 | 8.71 | 8.22 | 18.50 | 32.50 | 14.80 | 8.20 |
| Dlib(68 points) [11] | 16.20 | 22.90 | 14.10 | 11.60 | 14.80 | 17.50 | 15.32 | 11.58 |
| FAN(12 points) [2] | 10.25 | 7.31 | 13.50 | 9.94 | 8.47 | 7.84 | 8.63 | 8.94 |
| KEPLER [13] | – | – | – | – | 13.5 | 9.01 | 15.42 | 16.07 |
| Hopenet($\alpha = 2$) [22] | 6.21 | 6.04 | 6.72 | 5.87 | 5.62 | 5.31 | 6.60 | 4.95 |
| Hopenet($\alpha = 1$) [22] | 6.49 | 6.47 | 6.80 | 6.20 | 5.20 | 4.85 | 6.61 | 4.14 |
| FSA-Net(Fusion) [26] | 5.31 | 4.35 | 5.83 | 5.75 | 4.17 | 4.25 | 5.24 | 3.02 |
| WEst-Net | 5.22 | 4.34 | 5.70 | 5.62 | 4.04 | 4.33 | 4.97 | 2.82 |
| WC2FEst-Net | **5.01** | **4.20** | **5.49** | **5.32** | **3.90** | **4.20** | **4.95** | **2.55** |

Table 3 reports the performance on the BIWI dataset with different input modalities. The BIWI dataset provides information from multiple modalities such as depth or temporal information to improve performance. The proposed WC2FEst-Net method achieves comparable performance with Martin(RGB + Depth) and VGG16 + RNN(RGB + Time) without the help of depth map and time frames. It is worth mentioning that both WEst-Net and WC2FEst-Net outperform the RGB-based VGG16 and FSA-Net methods.

### 4.3 Discussion

In this section, we reveal the ablation study for key components in the proposed WC2FEst-Net, which promote the FSA-Net method [26] mainly by: 1) wavelet input, 2) added stream, and 3) coarse-to-fine regression architecture. For the ablation study, we compare the performance of the following methods: 1) FSA-Net [26], 2) FSA-C2F-Net (coarse-to-fine regression using FSA-Net as component module), 3) WEst-Net (shown in Fig. 4), and 4) WC2FEst-Net (coarse-to-fine regression using WEst-Net as component module).

Figure 7 shows the effectiveness of the components in general, though a single component may not always promote the performance in all cases.

**Table 3.** Results on BIWI dataset under protocol 2.

| Method | MAE | | | |
|---|---|---|---|---|
| | Total | Yaw | Pitch | Roll |
| Martin(RGB + Depth) [17] | **2.96** | 3.62 | **2.66** | **2.60** |
| VGG16 + RNN(RGB + Time) [6] | 3.15 | 3.21 | 3.57 | 2.67 |
| VGG16(RGB) [6] | 3.73 | 3.96 | 3.80 | 3.43 |
| FSA-Net(Fusion)(RGB) [26] | 3.85 | 3.23 | 4.56 | 3.76 |
| WEst-Net | 3.48 | 3.22 | 3.80 | 3.42 |
| WC2FEst-Net | 3.22 | **2.92** | 3.43 | 3.31 |

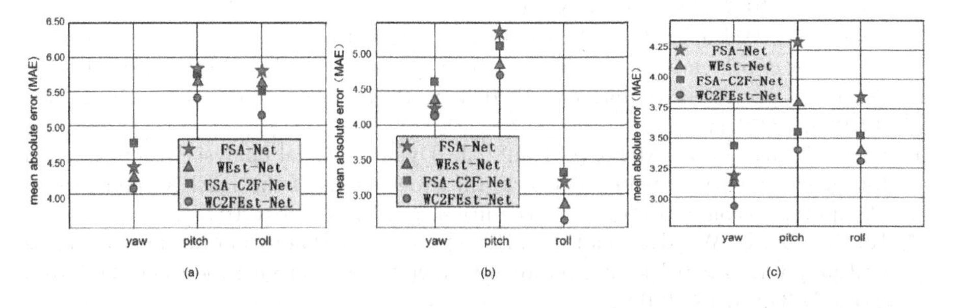


**Fig. 7.** Estimation performance of various methods for the ablation study. Part (a) shows the results on AFLW2000 dataset under protocol 1. Part (b) shows the results on BIWI dataset under protocol 1. Part (c) shows the results on BIWI dataset under protocol 2.

## 5 Conclusion

In this paper, we have proposed a wavelet-based coarse-to-fine regression method for head pose estimation. As wavelet information provides a higher level of shape abstraction, we input it with the original color image to guide the neural network for effective feature extraction. Further, we propose a coarse-to-fine regression architecture to deal with biased training sample distribution and make a better mapping approximation. In addition, multi-stream module and fine-grained structure feature aggregation method are exploited to extract much more effective pose features. Experiments show that the proposed method outperforms the state-of-the-art methods for the head pose estimation task. In the future, we will extend the current framework to estimate the head poses in a video sequence, exploiting the inter-frame correlations for improved precision of estimation.

**Acknowledgments.** This paper is supported by "The Fundamental Research Funds of Shandong University", NSFC of China (61907026) and Shandong Province Higher Educational Science and Technology Program (J18KA392).

## References

1. Balasubramanian, V.N., Ye, J., Panchanathan, S.: Biased manifold embedding: a framework for person-independent head pose estimation. In: 2007 IEEE Conference on Computer Vision and Pattern Recognition, pp. 1–7. IEEE (2007)
2. Bulat, A., Tzimiropoulos, G.: How far are we from solving the 2D & 3D face alignment problem? (and a dataset of 230,000 3D facial landmarks). In: Proceedings of the IEEE International Conference on Computer Vision, pp. 1021–1030 (2017)
3. Chang, F.J., Tuan Tran, A., Hassner, T., Masi, I., Nevatia, R., Medioni, G.: Faceposenet: making a case for landmark-free face alignment. In: Proceedings of the IEEE International Conference on Computer Vision Workshops, pp. 1599–1608 (2017)
4. Fanelli, G., Dantone, M., Gall, J., Fossati, A., Van Gool, L.: Random forests for real time 3D face analysis. Int. J. Comput. Vis. **101**(3), 437–458 (2013). https://doi.org/10.1007/s11263-012-0549-0
5. Fanelli, G., Weise, T., Gall, J., Van Gool, L.: Real time head pose estimation from consumer depth cameras. In: Mester, R., Felsberg, M. (eds.) DAGM 2011. LNCS, vol. 6835, pp. 101–110. Springer, Heidelberg (2011). https://doi.org/10.1007/978-3-642-23123-0_11
6. Gu, J., Yang, X., De Mello, S., Kautz, J.: Dynamic facial analysis: from Bayesian filtering to recurrent neural network. In: Proceedings of the IEEE Conference on Computer Vision and Pattern Recognition, pp. 1548–1557 (2017)
7. Hu, N., Huang, W., Ranganath, S.: Head pose estimation by non-linear embedding and mapping. In: IEEE International Conference on Image Processing 2005. vol. 2, pp. II-342. IEEE (2005)
8. Hu, Y., Ji, C., Zhang, Q., Chen, L., Zhan, P., Li, X.: A novel multi-resolution representation for time series sensor data analysis. Soft. Comput. **24**(14), 10535–10560 (2019). https://doi.org/10.1007/s00500-019-04562-7
9. Hu, Y., Ren, P., Luo, W., Zhan, P., Li, X.: Multi-resolution representation with recurrent neural networks application for streaming time series in IoT. Comput. Netw. **152**, 114–132 (2019)

10. Hu, Y., Zhan, P., Xu, Y., Zhao, J., Li, Y., Li, X.: Temporal representation learning for time series classification. Neural Comput. Appl. **32**, 1–13 (2020). https://doi.org/10.1007/s00521-020-05179-w
11. Kazemi, V., Sullivan, J.: One millisecond face alignment with an ensemble of regression trees. In: Proceedings of the IEEE Conference on Computer Vision and Pattern Recognition, pp. 1867–1874 (2014)
12. Koestinger, M., Wohlhart, P., Roth, P.M., Bischof, H.: Annotated facial landmarks in the wild: a large-scale, real-world database for facial landmark localization. In: 2011 IEEE International Conference on Computer Vision Workshops (ICCV workshops), pp. 2144–2151. IEEE (2011)
13. Kumar, A., Alavi, A., Chellappa, R.: Kepler: keypoint and pose estimation of unconstrained faces by learning efficient H-CNN regressors. In: 2017 12th IEEE International Conference on Automatic Face & Gesture Recognition (FG 2017), pp. 258–265. IEEE (2017)
14. Li, S., et al.: Deep-learning inversion of seismic data. IEEE Trans. Geosci. Remote Sens. **58**(3), 2135–2149 (2020)
15. Liu, B., et al.: Deep learning inversion of electrical resistivity data. IEEE Trans. Geosci. Remote Sensing **58**, 1–14 (2020). https://doi.org/10.1109/TGRS.2020.2969040
16. Mallat, S.: A theory for multiresolution signal decomposition: the wavelet representation. IEEE Trans. Pattern Anal. Mach. Intell. **11**(7), 674–693 (1989)
17. Martin, M., Van De Camp, F., Stiefelhagen, R.: Real time head model creation and head pose estimation on consumer depth cameras. In: 2014 2nd International Conference on 3D Vision, vol. 1, pp. 641–648. IEEE (2014)
18. Meyer, G.P., Gupta, S., Frosio, I., Reddy, D., Kautz, J.: Robust model-based 3D head pose estimation. In: Proceedings of the IEEE International Conference on Computer Vision, pp. 3649–3657 (2015)
19. Osuna, E., Freund, R., Girosit, F.: Training support vector machines: an application to face detection. In: Proceedings of IEEE Computer Society Conference on Computer Vision and Pattern Recognition, pp. 130–136. IEEE (1997)
20. Ranjan, R., Patel, V.M., Chellappa, R.: Hyperface: a deep multi-task learning framework for face detection, landmark localization, pose estimation, and gender recognition. IEEE Trans. Pattern Anal. Mach. Intell. **41**(1), 121–135 (2017)
21. Rowley, H.A., Baluja, S., Kanade, T.: Neural network-based face detection. IEEE Trans. Pattern Anal. Mach. Intell. **20**(1), 23–38 (1998)
22. Ruiz, N., Chong, E., Rehg, J.M.: Fine-grained head pose estimation without keypoints. In: Proceedings of the IEEE Conference on Computer Vision and Pattern Recognition Workshops, pp. 2074–2083 (2018)
23. Sagonas, C., Tzimiropoulos, G., Zafeiriou, S., Pantic, M.: 300 faces in-the-wild challenge: the first facial landmark localization challenge. In: Proceedings of the IEEE International Conference on Computer Vision Workshops, pp. 397–403 (2013)
24. Song, X., et al.: Apollocar3D: a large 3D car instance understanding benchmark for autonomous driving. In: The IEEE Conference on Computer Vision and Pattern Recognition, CVPR, June 2019
25. Srinivasan, S., Boyer, K.L.: Head pose estimation using view based eigenspaces. In: Object Recognition Supported by User Interaction for Service Robots, vol. 4, pp. 302–305. IEEE (2002)
26. Yang, T.Y., Chen, Y.T., Lin, Y.Y., Chuang, Y.Y.: FSA-NET: learning fine-grained structure aggregation for head pose estimation from a single image. In: Proceedings of the IEEE Conference on Computer Vision and Pattern Recognition, pp. 1087–1096 (2019)

27. Zhu, X., Ramanan, D.: Face detection, pose estimation, and landmark localization in the wild. In: 2012 IEEE Conference on Computer Vision and Pattern Recognition, pp. 2879–2886. IEEE (2012)
28. Zhu, X., Lei, Z., Liu, X., Shi, H., Li, S.Z.: Face alignment across large poses: a 3D solution. In: Proceedings of the IEEE Conference on Computer Vision and Pattern Recognition, pp. 146–155 (2016)
29. Zhu, X., Liu, X., Lei, Z., Li, S.Z.: Face alignment in full pose range: a 3D total solution. IEEE Trans. Pattern Anal. Mach. Intell. **41**, 78–92 (2019)

# Natural Language Processing

# A Memory-Based Sentence Split and Rephrase Model with Multi-task Training

Xiaoning Fan, Yiding Liu, Gongshen Liu(✉), and Bo Su(✉)

School of Electronic Information and Electrical Engineering, Shanghai Jiao Tong University, Shanghai 200240, China
{fxn627,danyliu,lydlovehdq,lgshen,subo}@sjtu.edu.cn

**Abstract.** The task of sentence split and rephrase refers to breaking down a complex sentence into some simple sentences with the same semantic information, which is a basic preprocess method for simplification in many natural language processing (NLP) fields. Previous works mainly focus on applying conventional sequence-to-sequence models into this task, which fails to capture relations between entities and lacks memory of the decoded parts, and thus causes duplication of generated subsequences and confuses the relationship between subjects and objects. In this paper, we introduce a memory-based Transformer model with multi-task training to improve the accuracy of the sentence information obtained by the encoder. To enrich the semantic representation of the model, we further incorporated a conditional Variational Autoencoder (VAE) component to our model. Through experiments on the WebSplit-v1.0 benchmark dataset, results show that our proposed model outperforms other state-of-the-art baselines from both BLEU and human evaluations.

**Keywords:** Split and rephrase · Memory gate · Multi-task · VAE

## 1 Introduction

Sentence split and rephrase is essentially a sentence rewriting problem, which splits a given complex sentence into several simple sentences and rephrases them into correct forms. This task is firstly introduced by [18]. Through splitting and rephrasing, a complex sentence can be converted to a more concise version, which is shorter and easier to understand. With these beneficial properties, the sentence split and rephrase problem has been gradually drawn attention from both NLP fields and social services. In statistical machine translation systems, especially relationship extraction and semantic similarity computation, splitting and rephrasing as a preprocessing step can effectively improve the performance due to the intelligibility of simpler sentences. In real world, the conversion from complex sentences to shorter ones helps people catch the meanings faster and easier, especially those who have difficulties in reading, such as aphasia patients, language learners and children.


H. Yang et al. (Eds.): ICONIP 2020, LNCS 12532, pp. 643–654, 2020.
https://doi.org/10.1007/978-3-030-63830-6_54

Previous studies introduce the sequence-to-sequence model [23] to address the problem of splitting and rephrasing sentences. By leveraging stacked LSTM, [21] improves the original sequence-to-sequence framework. Later, [18] first proposed the split and rephrase dataset from WEBNLG. Besides, to give a benchmark of split and rephrase task, the author applies five basic models: a probabilistic, semantic-based model [17], a basic sequence-to-sequence approach, a multi-source sequence-to-sequence approach with sets of RDF triples (A triples of subjects, objects and the relationship between them) given by training data, a split sequence-to-sequence model and a multi-source split sequence-to-sequence model. [1] presents a new data-split form and a copy-augmented model inspired by the copy mechanism from the abstractive summarization task. Another method based on the Wikipedia edit history also improves the model's ability to split and rephrase [3].

Some of the models mentioned above lack the consideration of the facts in the dataset, like the RDF triples. Consequently, the generated sentences may misrepresent or even miss the facts in the original sentence. Other methods fail to take into account the order of the generated simpler sentences, which may confuse the model. To handle these problems, a fact-aware framework with permutation invariant training is proposed by [8]. This framework trains a fact classifier from the RDF triples and computes the minimal loss on the different orders of generated sentences. However, these two components only work in the training step and are removed in the decoding phase, causing a gap between training and decoding. Further more, the number of permutations is always 6 according to the hypothesis that there are often 3 short sentences generated which is not consistent with the situation in the dataset. Since VAE was proposed [12], several following works proved its ability to increase the diversity of generated sentences. [9] introduces a conditional VAE architecture that improves the ability of the traditional sequence-to-sequence model.

In this paper, we introduce a novel memory-based sentence split and rephrase model with multi-task training. To encourage the variations of the simple sentences generated by the model, a conditional VAE part is incorporated between the encoder and decoder. We further designed a multi-task training method to employ the RDF triples in the dataset. Thus, in our proposed method, the model needs to generate the correct simple sentences and extract the reasonable RDF triples at the same time. This component will help the model focus on the facts in the given complex sentences and avoid missing relations between the subjects and objects of the generated sentences. A memory gate layer is added between the encoder and decoder to decide what information has been decoded which is not important in the after steps. This structure helps the model only focus on the remaining semantic information of the given sentence and avoid the gap between training and decoding in the permutation training method. We summarized the main contributions of this paper as follows.

- We implement a memory gate mechanism between the encoder and decoder of the transformer to help the model only focusing on the remaining semantic information and avoiding generating repetitive phrases.

- We propose a multi-task training mechanism to enable the model not only to split and rephrase sentences but also to identify the facts in the complex sentence.
- We introduce a conditional VAE architecture into the traditional transformer model to generate sentences of rich representations by combining the latent code (Fig. 1).

## 2 Preliminaries

In this section, we introduce the VAE structure in our work. Variational auto-encoder [12] assumes observed data follows a latent distribution called "code" $z$. As an encoder-decoder framework, VAE can obtain the latent distribution of data and reconstruct the input from the features. The encoder generates a posterior distribution $q_\phi(z|x)$, which is supposed to be close to the prior distribution $p_\theta(z)$. These two distributions are usually presumed some parametrized distribution, like a Gaussian distribution. As the prior $p_z$ is assumed to be a standard normal distribution $\mathcal{N}(0, I)$, the posterior $q_\phi(z|x)$ follows a normal distribution $\mathcal{N}(\mu(x), \sigma^2(x))$.

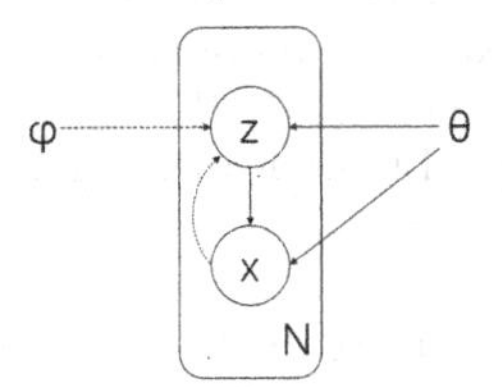


**Fig. 1.** An illustration of variational Auto-Encoder.

To reconstruct input $x$ from code $z$, the VAE contains another distribution $p_\theta(x|z)$ as a decoder. The joint probability function of VAE model can be represented as:

$$p_\theta(x, z) = p_\theta(x|z)p(z) \tag{1}$$

the $p(z)$ is approximated by $q_\phi(z|x)$. The parameters of the encoder $\phi$ and decoder $\theta$ are learned from a neural network that is trained by maximizing the following objective:

$$\mathcal{L}(\theta, \phi; x) = \mathbb{E}_{q_\phi(z|x)}[\log_{p_\theta}(x|z)] - KL(q_\phi(z|x)||p(z)) \tag{2}$$

The first term is the reconstruction error. KL divergence, in the second term, forces the latent code $z$ to be close to prior $p(z)$. VAE has been proved by lots of previous works to be useful in generation tasks [4,9].

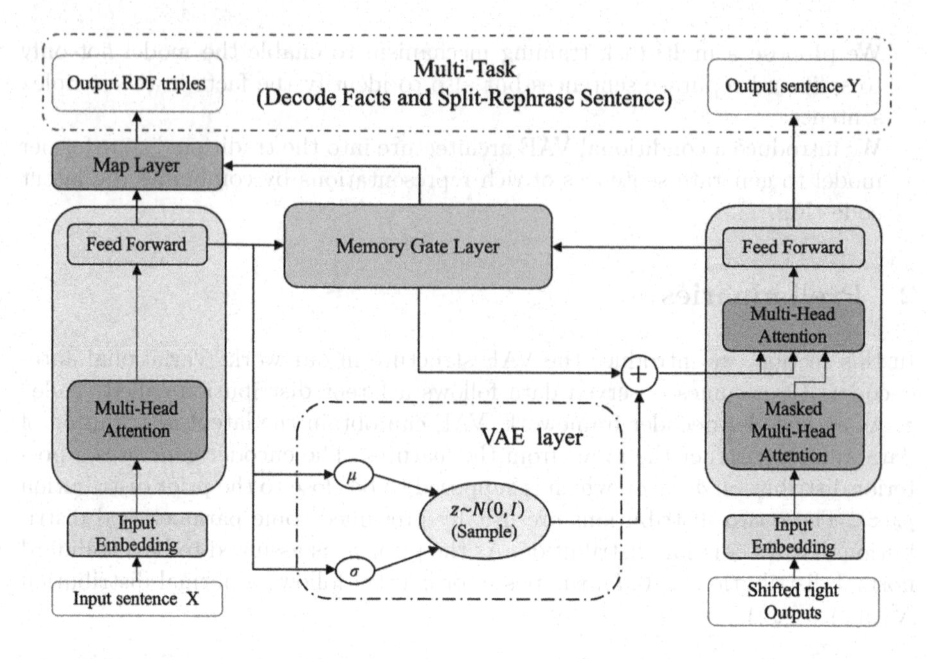


**Fig. 2.** An Overview of our model. The memory gate layer determines the information to generate in the next steps based on representations of the encoder and generated outputs. The map layer extracts the facts (RDF triples) from the input and the memory. A conditional VAE is incorporated to enrich the representations.

To avoid KL divergency converging to 0, we employ a KL term annealing method which is a weighted function of the KL divergence. The KL weight is computed as [15]:

$$KL_{weight} = \frac{1}{1 + e^{-k*step+b}} \tag{3}$$

where *step* is the training step and $k$, $b$ are hyperparameters. The total loss is defined as:

$$\begin{aligned} \mathcal{L}_{\mathcal{G}}(\theta, \phi; x) = & -KL_{weight} * KL(q_\phi(z|x)||p(z)) \\ & + \mathbb{E}_{q_\phi(z|x)}[\log_{p_\theta}(x|z)]. \end{aligned} \tag{4}$$

## 3 Methodology

As mentioned above, we introduce a memory-based sentence split and rephrase model with a multi-task training method. The training data contains pairs of original complex sentences $S_c$, their matched simpler ones $S_s$, and the RDF triples $S_f$, namely facts. The $X_c$ and $X_s$ refer to the vector representations of $S_c$ and $S_s$. In our methodology, we do not use any pre-trained word embeddings or extra corpus to pre-train our model. The word embeddings are the combination of positional encoding and original embeddings. The Fig. 2 illustrates an overview of our proposed model.

### 3.1 Memory-Based Transformer

We use the transformer model [24] as our backbone model. Following the original transformer, we apply 6 stacked layers as the encoder part which contains a multi-head attention layer and a feed-forward neural network. The multi-head attention is composed of the self-attention mechanism, which is defined as follows:

$$\text{Attention}(Q, K, V) = \text{softmax}(\frac{QK^T}{\sqrt{d_k}})V \tag{5}$$

Multi-head attention is computed as:

$$\begin{aligned} \text{MultiHead}(Q, K, V) &= \text{Concat}(\text{head}_i, ..., \text{head}_h)W^O \\ \text{where head}_i &= \text{Attention}(QW_i^Q, KW_i^K, VW_i^V) \end{aligned} \tag{6}$$

The $W_i^Q \in \mathbb{R}^{d_{model} \times d_k}$, $W_i^K \in \mathbb{R}^{d_{model} \times d_k}$, $W_i^V \in \mathbb{R}^{d_{model} \times d_v}$ and $W_i^O \in \mathbb{R}^{hd_v \times d_{model}}$ are the parameter matrices of linear projections.

In the split and rephrase task, one complex sentence always has several corresponding short sentences. Due to the diverse orders of $S_s$, the traditional transformer model is easily confused by the various outputs of one input. So we add the memory gate layer to filter out part of information that has been already decoded. Thus, our model can focus on the information that awaits further decoding.

The memory gate layer takes the vector representations generated by encoder $E_o$ and decoder $D_o$ as inputs and outputs new vector representations $M_o$. The memory gate layer can be divided into a transmit layer, a forget layer and a fuse gate. The transmit layer decides which part of the information from the encoder will be transmitted to the decoder. The forget layer takes the decoder output as the input and judges which part of the information has been decoded and is not valuable no longer. They are defined as:

$$T_o = max((E_o W_{t_1} + b_t), 0) \tag{7}$$

$$F_o = max((D_o W_{f_1} + b_f), 0) \tag{8}$$

The fuse layer takes $T_o$ and $F_o$ as inputs and computes the remaining semantic information of future steps, based on the feed-forward layer:

$$M_o = max(0, T_o W_{t_2} + F_o W_{f_2} + b_{c_1})W_c + b_{c_2} \tag{9}$$

where $W_{t_1}, b_t, w_{f_1}, b_f, W_{t_2}, W_{f_2}, b_{c_1}, W_c$ and $b_{c_2}$ refer to the parameters of linear projections. Through the memory gate mechanism, our model can automatically focus on the sentence information of the undecoded part. The decoder is adaptive to the situation that one complex sentence has different splitting and rephrasing versions, because the memory gate can tell the remaining information from the vector representations of the encoder. Different from [8], our model follows the same way in the training and decoding phase, which can discriminate the gap question in the permutation training method.

### 3.2 Multi-task Training

One of the benefits of splitting and rephrasing is that it helps extract relationships between objectives from simple sentences. The WebSplit-v1.0 dataset provides a pair of RDF triples for every complex sentence. Due to these properties, we design a multi-task training method to make the model focus on the facts (RDF triples) in given sentences. [8] implements a fact classifier to use the facts information. However, the classifier and the sequence-to-sequence model are independent in decoding phase, which may cause the only the facts encoder and classifier can learn the knowledge of the relationship of entities while a sequence-to-sequence model benefits little from such architecture.

In our work, we define a fact generator that enables the model to extract the facts from the encoding information and memory vectors. The memory gate layer can directly obtain feedback from this part, which will strongly enhance the information extraction ability of our model. Furthermore, the generation task is usually more difficult than the classification task, which requires a higher learning ability of our model to learn more information from given inputs. The map layer takes the encoder output $E_o$ and memory gate layer outputs $M_o$ as inputs, and output the relations of facts in complex sentences. We also use a two-layer feed-forward layer to handle the map layer. After that, a CNN layer is introduced to extract the facts relations and try to reconstruct the given RDF triples. The loss $L_M$ is cross-entropy loss.

### 3.3 Conditional VAE

To encourage the variations of splitting and rephrasing, we apply a conditional VAE between the encoder and the decoder. The VAE cell in our model takes the output of the encoder as inputs and generates a normal distribution of $\mathcal{N}(\mu, \sigma^2)$ by linear projection. The latent code $z$ is sampled from this distribution. Such sampling operations will bring randomness to decoder so that the decoder can be adaptive to the different orders of given simple sentences. Reparameterization trick [12] is a common trick in VAE to overcome the non-continuous and non-differentiable problem in the sampling operations. Using the trained $\mu$ and $\sigma$, we compute the code $z$ by

$$z = \mu + \sqrt{\sigma}\beta \tag{10}$$

where $\beta$ is a sample from the standard normal distribution $\beta \sim \mathcal{N}(0, I)$. The sampled $z$ will be added to the memory gate output to enrich the diversity of the representations. The decoder side of our model is similar to the original transformer which is composed of stacked layers of a feed-forward layer and two multi-head attention layers. Different from the traditional transformer, we change the $Q$ and $K$ value with a combination of the additional VAE part and memory gate. Without directly connecting between encoder and decoder, the new $Q$ and $K$ is computed as:

$$Q = M_o + z \tag{11}$$

$$K = M_o + z \tag{12}$$

In the decoding phase, the VAE part will be partially ignored and the latent code $z$ is sampled from a standard normal distribution $\mathcal{N}(0, I)$. Combining the above 3 parts, the loss of our model is computed as follows:

$$L = \lambda L_G + (1 - \lambda) L_M \tag{13}$$

where $L_G$ is the generating loss as defined in Sect. 2 and $L_M$ if the mapping loss of extract facts. $\lambda$ is the hyper-parameter to control the weights for the two tasks.

# 4 Experiments

In this section, we introduce the WebSplit-v1.0 dataset, experimental setup, evaluation metrics and our results.

## 4.1 Dataset

We use WebSplit-v1.0 dataset [18] to evaluate the performance of our model. This dataset contains a number of pairs of complex sentences and their corresponding split and rephrased ones. For each point of data, the dataset provides a group of RDF triples. In this corpus, a complex sentence usually has several simple sentences. The basic statistics are shown in Table 1.

**Table 1.** Dataset statistics

| Dataset | Train | Val | Test |
|---|---|---|---|
| Distinct (complex sentences) | 16K | 1K | 1K |
| Total | 1300K | 40K | 40K |

**Table 2.** Results on the WebSplit-v1.0 dataset. TRANS + MEMO + MULTI-TASK + VAE represents the memory-based VAE transformer model with a multi-task training method introduced by us.

| Model | BLEU | #S/C | #T/S |
|---|---|---|---|
| Reference | – | 2.5 | 10.9 |
| SOURCE | 57.2 | 1.0 | 20.5 |
| SPLIT HALF | 55.7 | 2.0 | 10.8 |
| LSTM(AG18) [1] | 25.5 | 2.3 | 11.8 |
| LSTM(BOTHA-WEBSPLIT) [3] | 30.5 | 2.0 | 8.8 |
| LSTM + FASE + PIT [8] | 37.3 | 2.4 | 9.9 |
| DISSIM [19] | 59.7 | 2.7 | 9.2 |
| BOTHA-BOTH [3] | 60.1 | 2.0 | 11 |
| TRANSFORMER [24] | 69.2 | 2.3 | 10.1 |
| TRANS + FASE + PIT [8] | 70.1 | 2.4 | 9.8 |
| TRANS + MEMO + MULTI-TASK + VAE(OURS) | **71.2** | 2.5 | 10.3 |

### 4.2 Baseline

We compare our model with several baselines of the WebSplit-v1.0 dataset to show the effectiveness of our model. Reference is the standard splitting and rephrasing sentences given by the corpus. SPLIT HALF means directly splitting a sentence into two equal-length token sequences and appending a period to the first one. LSTM(AG18) is the basic LSTM used in [1]. LSTM(BOTHA-WEBSPLIT) refers to the LSTM only that is only trained by the WebSplit-v1.0 dataset [3]. LSTM + FASE + PIT is the fact-aware LSTM model with a permutation invariant training method. DISSIM indicates the model which transforms complex sentences into a semantic hierarchy. BOTHA-BOTH is also introduced by [3], which is an LSTM model trained by WebSplit-v1.0 dataset and Wikipedia edit history. TRANSFORMER is the basic transformer model in [24]. TRANS+FASE+PIT is similar to the LSTM+FASE+PIT with the change of base sequence-to-sequence model [8].

To evaluate the performance of models, following previous works, we use the BLEU metric [20]. The $\#S/C$ stands for the ratio of the number of simple sentences and the number of complex sentences. The $\#T/S$ means the average number of the tokens in splitting and rephrasing sentences. Higher BLEU score is better. The metrics of the $\#S/C$ and $\#T/S$ are also referencing indexes, whose value are closer to the ones in the corpus reflecting better results.

### 4.3 Setup

The transformer part in our model is similar to the original model in [24]. A stack of 6-layer transformer with hidden layers and the word embedding dimension $h = 512$ is built as the basic framework. We use Adam [11] to train our model

**Table 3.** Results of our ablation study.

| Model | BLEU | #S/C | #T/S |
|---|---|---|---|
| TRANSFORMER | 69.2 | 2.3 | 10.1 |
| TRANS + MEMO | 70.3 | 2.4 | 10.4 |
| TRANS + MULTI-TASK | 69.8 | 2.5 | 10.1 |
| TRANS + VAE | 69.6 | 2.3 | 11.0 |
| TRANS + MEMO + MULTI-TASK | 70.8 | 2.4 | 10.3 |
| TRANS + MEMO + VAE | 70.6 | 2.4 | 10.7 |
| TRANS + MULTI-TASK + VAE | 70.3 | 2.5 | 10.4 |
| TRANS + MEMO + MULTI-TASK + VAE | 71.2 | 2.5 | 10.3 |

with an initial learning rate $\alpha = 0.0004$, momentum parameters $\beta_1 = 0.9$ and $\beta_2 = 0.998$ respectively. We also employ the label smoothing value $\epsilon = 0.1$.

In the memory-gate layers, the dimensions of the parameters in the transmit layer, the forget layer and the fuse layer are all set to 1024 to ensure the performance of our model. In the map layer, we use two 512-dimension layers to map the important facts in complex sentences. In the VAE cell, the dimension of latent code $z$ is set to 256. For KL term annealing, the parameters $k$ and $b$ used in KL weight are set to 0.0025 and 2500 respectively. To keep the balance of the two tasks (decoding splitting and rephrasing sentences and extracting the RDF triplets), we choose $\lambda$ as 0.6 to control the weights of generating loss and mapping loss.

### 4.4 Results

We present the experiment results in Table 2. Through comparing several baselines, we show that our work outperforms other state-of-the-art models. The combination of memory gate and VAE architecture with the multi-task training method improves a large margin in the BLEU metric about 1.1 points more than the state-of-the-art result. Both the $\#S/C$ and $\#T/S$ values of our model are also close to the given reference index, which means our model can locate the correct segmentation position and rewrite the splitting subsequences into proper sentences (Table 4).

In Table 3, we did an ablation study to observe the effects of every part in our model. The memory gate layer, the VAE and the map layer are connected in the final version of our model, so we can not remove one of them directly. In the ablation study, we try to maintain the remaining structure as possible as we can. As testing the memory gate transformer, we simply drop the map layer and the VAE cell, so that the output of the memory gate layer will be delivered to the second encoder-decoder multi-head attention layer in the decoder. To evaluate the function of the map layer, we only use the encoder's output to extract the facts in complex sentences. The VAE cell is operated in the same way. The

**Table 4.** Results of human evaluation on a random sample of 50 sentences on the WebSplit-v1.0.

| Model | G | M | S | AVG |
|---|---|---|---|---|
| Reference | 4.92 | 4.64 | 1.67 | 3.74 |
| BOTHA [3] | 3.95 | 3.77 | 0.76 | 2.83 |
| DISSIM [19] | 4.27 | 3.65 | 1.1 | 3.01 |
| FASE + PIT [8] | 4.76 | 4.39 | 1.69 | 3.61 |
| MEMO + MULTI-TASK + VAE | **4.79** | **4.42** | 1.68 | **3.63** |

results show that each part of our model has a positive impact on enhancing the model's ability of splitting and rephrasing.

Human evaluation is also conducted to confirm our results. Five well-educated volunteers with strong English background are asked to score on a subset of 50 randomly sampled sentences which is generated by each model. Each volunteer will evaluate sentences from three aspects: grammatically(G), meaning preservation(M) and structural simplicity(S). The guidelines are described in the Appendix A in [19] in details. Considering all three aspects, our model achieves highest average score among several baselines.

## 5 Related Work

Sentence Split and Rephrase task is firstly introduced by [18], which is different from the previous sentence rewriting tasks, like: sentence compression [5,13], sentence fusion [6,16], sentence paraphrasing [2] and sentence simplification [22,26]. In [18], authors introduce the WeSplit corpus which contains tuples mapping complex sentences to their simple ones. A series of sequence-to-sequence models [23] have been implemented to show the difficulties of this task. Traditional sequence-to-sequence model are composed of the LSTM [10]. In [21], a stacked LSTM framework with deeper layers is proposed to improve the performance with deeper layers. To evaluate the Sentence Split and Rephrase task more accurately, [9] presents a new train-development-test split distribution and a neural model augmented with a copy mechanism [7]. [3] proposes a sequence-to-sequence model combining the information from Wikipedia edit history. Different from the data-driven ways, [19] adopts a recursive method with hand-crafted transformation rules to simplify sentences. However, the conventional sequence-to-sequence model lacks the consideration about the facts in complex sentences and is easily confused by the order variance of the simple sentences. A fact-aware sentence splitting and rephrasing method with permutation invariant training is introduced by [8] to relief the two questions mentioned above. Permutation invariant training is spirited by [14,25] in speech fields.

Transformer model is an advanced sequence-to-sequence model and achieves state-of-the-art performance in many NLP tasks [24], which is the backbone framework of our model. We add the variational auto-encoder cell (VAE) [12] to encourage model to generate more various and fluent simple sentences. In [9], a conditional VAE LSTM network is introduced to address the paraphrase task, which is similar to our model. [15] also designs a Transformer-based VAE to the sentence generation task.

## 6 Conclusion

In this paper, we introduce a novel memory-based transformer model with combining multi-task training method and VAE architecture for the Sentence Split and Rephrase task. The memory gate mechanism addresses the confusion question caused by the different orders of splitting sentences. The facts information in the dataset is efficiently used by the multi-task training method. The VAE cell is introduced to improve the diversity of representations of the semantic vector. Through experiments over several baselines on WebSplit-v1.0 dataset, results show that our model outperforms the state-of-the-art work and the ablation study also shows the positive affect of each part in our proposed model.

**Acknowledgement.** This research work has been funded by the National Natural Science Foundation of China (Grant No.61772337, U1736207) and the National Key R&D Program of China (2018YFC0832004).

## References

1. Aharoni, R., Goldberg, Y.: Split and rephrase: better evaluation and stronger baselines. In: Gurevych, I., Miyao, Y. (eds.) Proceedings of the 56th Annual Meeting of the Association for Computational Linguistics, ACL 2018, Melbourne, Australia, 15–20 July 2018, Volume 2: Short Papers, pp. 719–724. Association for Computational Linguistics (2018). https://doi.org/10.18653/v1/P18-2114. https://www.aclweb.org/anthology/P18-2114/
2. Barzilay, R., McKeown, K.R.: Extracting paraphrases from a parallel corpus. In: Association for Computational Linguistic, 39th Annual Meeting and 10th Conference of the European Chapter, Proceedings of the Conference, 9–11 July 2001, Toulouse, France, pp. 50–57. Morgan Kaufmann Publishers (2001). https://doi.org/10.3115/1073012.1073020. https://www.aclweb.org/anthology/P01-1008/
3. Botha, J.A., Faruqui, M., Alex, J., Baldridge, J., Das, D.: Learning to split and rephrase from Wikipedia edit history. In: EMNLP, pp. 732–737. Association for Computational Linguistics (2018)
4. Bowman, S.R., Vilnis, L., Vinyals, O., Dai, A.M., Józefowicz, R., Bengio, S.: Generating sentences from a continuous space. In: CoNLL, pp. 10–21. ACL (2016)
5. Cohn, T., Lapata, M.: Sentence compression beyond word deletion. In: COLING, pp. 137–144 (2008)
6. Filippova, K.: Multi-sentence compression: finding shortest paths in word graphs. In: COLING, pp. 322–330. Tsinghua University Press (2010)
7. Gu, J., Lu, Z., Li, H., Li, V.O.K.: Incorporating copying mechanism in sequence-to-sequence learning. In: ACL(1). The Association for Computer Linguistics (2016)
8. Guo, Y., Ge, T., Wei, F.: Fact-aware sentence split and rephrase with permutation invariant training. CoRR abs/2001.11383 (2020)

9. Gupta, A., Agarwal, A., Singh, P., Rai, P.: A deep generative framework for paraphrase generation. In: AAAI, pp. 5149–5156. AAAI Press (2018)
10. Hochreiter, S., Schmidhuber, J.: Long short-term memory. Neural Comput. **9**(8), 1735–1780 (1997)
11. Kingma, D.P., Ba, J.: Adam: a method for stochastic optimization. In: ICLR (Poster) (2015)
12. Kingma, D.P., Welling, M.: Auto-encoding variational Bayes. In: Bengio, Y., LeCun, Y. (eds.) 2nd International Conference on Learning Representations, ICLR 2014, Banff, AB, Canada, 14–16 April 2014, Conference Track Proceedings (2014). http://arxiv.org/abs/1312.6114
13. Knight, K., Marcu, D.: Statistics-based summarization - step one: Sentence compression. In: AAAI/IAAI, pp. 703–710. AAAI Press/The MIT Press (2000)
14. Kolbaek, M., Yu, D., Tan, Z., Jensen, J.: Multitalker speech separation with utterance-level permutation invariant training of deep recurrent neural networks. IEEE ACM Trans. Audio Speech Lang. Process. **25**(10), 1901–1913 (2017)
15. Liu, D., Liu, G.: A transformer-based variational autoencoder for sentence generation. In: International Joint Conference on Neural Networks, IJCNN 2019 Budapest, Hungary, 14–19 July 2019, pp. 1–7. IEEE (2019). https://doi.org/10.1109/IJCNN.2019.8852155
16. McKeown, K.R., Rosenthal, S., Thadani, K., Moore, C.: Time-efficient creation of an accurate sentence fusion corpus. In: HLT-NAACL, pp. 317–320. The Association for Computational Linguistics (2010)
17. Narayan, S., Gardent, C.: Hybrid simplification using deep semantics and machine translation. In: ACL(1), pp. 435–445. The Association for Computer Linguistics (2014)
18. Narayan, S., Gardent, C., Cohen, S.B., Shimorina, A.: Split and rephrase. In: EMNLP, pp. 606–616. Association for Computational Linguistics (2017)
19. Niklaus, C., Cetto, M., Freitas, A., Handschuh, S.: Transforming complex sentences into a semantic hierarchy. In: ACL(1), pp. 3415–3427. Association for Computational Linguistics (2019)
20. Papineni, K., Roukos, S., Ward, T., Zhu, W.: Bleu: a method for automatic evaluation of machine translation. In: ACL, pp. 311–318. ACL (2002)
21. Prakash, A., et al.: Neural paraphrase generation with stacked residual LSTM networks. In: COLING, pp. 2923–2934. ACL (2016)
22. Siddharthan, A., Nenkova, A., McKeown, K.R.: Syntactic simplification for improving content selection in multi-document summarization. In: COLING (2004)
23. Sutskever, I., Vinyals, O., Le, Q.V.: Sequence to sequence learning with neural networks. In: NIPS, pp. 3104–3112 (2014)
24. Vaswani, A., et al.: Attention is all you need. In: NIPS, pp. 5998–6008 (2017)
25. Yu, D., Kolbæk, M., Tan, Z., Jensen, J.: Permutation invariant training of deep models for speaker-independent multi-talker speech separation. In: ICASSP, pp. 241–245. IEEE (2017)
26. Zhu, Z., Bernhard, D., Gurevych, I.: A monolingual tree-based translation model for sentence simplification. In: COLING, pp. 1353–1361. Tsinghua University Press (2010)

# A Neural Framework for English-Hindi Cross-Lingual Natural Language Inference

Tanik Saikh[1(✉)], Arkadipta De[2], Dibyanayan Bandyopadhyay[3], Baban Gain[3], and Asif Ekbal[1]

[1] Indian Institute of Technology Patna, Patna, India
{1821cs08,asif}@iitp.ac.in
[2] Indian Institute of Technology Hyderabad, Hyderabad, India
ai20mtech14002@iith.ac.in
[3] Government College of Engineering and Textile Technology, Berhampore, India
{dibyanayan.bandyopadhyay,baban.gain}@gcettb.ac.in

**Abstract.** Recognizing Textual Entailment (RTE) between two pieces of texts is a very crucial problem in Natural Language Processing (NLP), and it adds further challenges when involving two different languages, i.e. in cross-lingual scenario. The paucity of a large volume of datasets for this problem has become the key bottleneck of nourishing research in this line. In this paper, we provide a deep neural framework for cross-lingual textual entailment involving English and Hindi. As there are no large dataset available for this task, we first create this by translating the premises and hypotheses pairs of Stanford Natural Language Inference (SNLI) (https://nlp.stanford.edu/projects/snli/) dataset into Hindi. We develop a Bidirectional Encoder Representations for Transformers (BERT) based baseline on this newly created dataset. We perform experiments in both mono-lingual and cross-lingual settings. For the mono-lingual setting, we obtain the accuracy scores of 83% and 72% for English and Hindi languages, respectively. In the cross-lingual setting, we obtain the accuracy scores of 69% and 72% for English-Hindi and Hindi-English language pairs, respectively. We hope this dataset can serve as valuable resource for research and evaluation of Cross Lingual Textual Entailment (CLTE) models.

**Keywords:** Cross-lingual textual entailment · English-Hindi CLTE dataset · SNLI · Deep learning

## 1 Introduction

Combating with language variability is one of the utmost challenging tasks in NLP. Textual Entailment (TE) [4,11] or Natural Language Inference (NLI) [1] is considered to be the trusted testing ground for this language variability. We use these two terms (TE and NLI) interchangeably throughout this paper. Modeling

T. Saikh and A. De—Equal Contribution.


H. Yang et al. (Eds.): ICONIP 2020, LNCS 12532, pp. 655–667, 2020.
https://doi.org/10.1007/978-3-030-63830-6_55

such language variability is the main challenging task for the NLP researchers. This is essentially an important mean to combat various tasks that include (broadly saying) information retrieval, semantic parsing and commonsense reasoning. Given two pieces of texts, one being the Premise (P) and another one is Hypothesis (H), the system has to decide:

- Whether $H$ is the logical consequence of $P$ or not.
- The $H$ is true in every circumstance (possible world) in which $P$ is true.

For example, the $P$: *I left the restaurant satisfactorily* entails $H$: *I had good food.* So a human reading $P$ would most likely infer that $H$ is true, i.e. $P$ entails the $H$. It is an interesting problem and has been well studied over the years in monolingual settings (i.e. both P and H is in the same language), especially involving English language. In this paper, we create a framework for Cross-lingual Textual Entailment (CLTE) involving low-resource language like Hindi. In a cross-lingual setting, it is assumed that both P and H are in different languages. A challenge on CLTE was launched in [12]. This problem had come up with its application scenario in Content Synchronization in SemEval-2012 [13] and SemEval-2013 [14]. The number of teams participated in SemEval-2012 and SemEval-2013 were 10 and 6, respectively. Such a less representations in both of these shared tasks exhibit the challenges of CLTE. The use of deep learning for an investigation into the problem of TE, in general, and cross-lingual entailment, in particular, has been limited due to the lack of a sufficiently large amount of P–H pairs. However, the availability of two large scale datasets, namely SNLI [1] and Multi-Genre Natural Language Inference (MultiNLI) [27] have made it possible to leverage the advantages of deep learning techniques and thereby, have introduced several new models [2,10,16,19,26].

### 1.1 Motivation

Cross-linguality adds a dimension of the TE recognition problem that has not been explored significantly. Monolingual textual entailment recognition systems have been used in several NLP applications like Question Answering (QA) [7], Novelty detection [21], Evaluation of Machine Translation systems [15], Fake Information Detection [20], and Summarization [24] etc. However, due to the absence of potential CLTE dataset, similar improvements have not been achieved yet in cross-lingual applications. This task aims at promoting research to fill this gap. Research towards this direction can now be taken due to the recent advances in other related areas, especially in Neural Machine Translation (NMT), and with the availability of the following: (i). a substantial amount of parallel and comparable corpora in many languages, (ii). open-source software to compute word-alignments from the parallel corpora, and (iii). open source software to set-up strong machine translation (MT) baseline systems. These resources can be effectively used for building systems for Cross-lingual and Multi-lingual TE. In our work, we focus on building TE systems involving English and Hindi. Hindi is the most widely spoken language in India, and in terms of native speakers, it ranks 5th all over in the world.

### 1.2 Contribution

The contributions of our current work are two-folds:

- We construct almost 45K inference annotated P–H (i.e. Premise-Hypothesis) pairs. We translate such pairs of SNLI [1] from English to Hindi using Google's English-Hindi Neural Machine Translation (NMT) System [28]. This Google's NMT has competitive BLEU score (on WMT'14 test set) to state-of-the-art Machine Translation (MT) Systems. This dataset will provide opportunities for Cross-lingual Textual Entailment research and development.
- We develop a baseline model based on the state-of-the-art deep neural framework for CLTE involving English and Hindi. This would open up new research avenues for CLTE.

The rest of the paper is arranged as follows. Section 2 presents a literature survey. The dataset preparation is described in Sect. 3. A qualitative evaluation of our baseline system on the proposed dataset is elaborated in Sect. 4. Results obtained and comparison with existing state-of-the-art systems are shown in Sect. 5. This Section is followed by an error analysis in Sect. 6. Sect. 7 summarises the paper with directions to the future research avenues.

## 2 Related Work

TE has been extensively studied before the pre-deep learning era [4,11,22,23]. With the recent advancement of deep learning techniques, researchers have also started to explore these techniques for TE [1,2,16]. However, as already mentioned, the CLTE [3] is of very recent interest to the community. The work of [3] is considered to be a potential work of TE in a cross-lingual setting. The work presented the Cross-lingual Natural Language Inference (XNLI) benchmark dataset comprising of 14 languages apart from English. The work of [25] presented XLDA, (i.e. cross-lingual data augmentation), where a segment of the input text is translated to another language in order to improve the training/inference of any NLP module. This reported to have improved the performance of 14 tested languages of XNLI.

## 3 Data Preparation

The MultiNLI Corpus is extended to 15 languages, including Hindi [3] and called the dataset as XNLI, that comprises of multiple domains. The SNLI dataset was constructed from the Flickr30k corpus [29], where the authors considered captions of an image as the premises and asked the CrowdWorkers' (Amazon Mechanical Turk) to make three hypotheses which entail or contradict with respect to the premise; or sometimes could be neutral with the premise. Our dataset is an extension (in terms of translated version) of this SNLI corpus, and is limited to only one domain. Keeping the labels (entail/contradict/neutral) same, we perform translation of each pair of Hypothesis and Premise. We use

Google's Neural Machine Translation System (GNMT) [28] for this purpose. We check the outputs manually, and observe the outputs to be of good quality with acceptable fluency and adequacy. We employ three translators who are having good knowledge in English and Hindi, and are with an age group of 20–30. They were asked to make three reference translations for randomly picked up 1000 samples. Three translators independently rated the translated outputs of Premise and Hypothesis based on adequacy and fluency (in the scale of 0–4, where 0 and 4 indicate very poor and very rich translation, respectively. The values in-between indicate the translation quality to lie in-between). We compute the inter-annotator agreement ratio in terms of kappa coefficient [6]. It was found to be 0.81 which is considered to be good as per [9]. This yields the English-Hindi and Hindi-English NLI dataset and coin it as EH-XNLI (English-Hindi Cross Lingual Natural Language Inference) Corpus[1]. The distribution for training, development and test sets are shown in Table 1. The format of the dataset is as in Table 2. Apart from these, we compute some key statistics about the data that are shown in Table 3.

**Table 1.** Distribution (train/dev/test splits) of the dataset

| Dataset | Training | Development | Testing |
|---|---|---|---|
| | 36000 | 5000 | 4000 |

**Table 2.** Data format: sample

| Column | Description |
|---|---|
| ID | Sentence identification number |
| Premise | English premise |
| Hypothesis | English hypothesis |
| Hindi_Premise | Translated Hindi premise |
| Hindi_Hypothesis | Translated Hindi hypothesis |
| Label | Neutral, contradiction, entailment |

[1] http://www.iitp.ac.in/~ai-nlp-ml/resources.html.

**Table 3.** Various statistics of the dataset. Here, Eng: English; Hin: Hindi; Dev: development

| Dataset | Average words per sentence | | | | Word count | | Vocab count | |
|---|---|---|---|---|---|---|---|---|
| | Premise | | Hypothesis | | Eng | Hin | Eng | Hin |
| | Eng | Hin | Eng | Hin | | | | |
| Training | 12 | 14 | 7 | 8 | 733717 | 811075 | 12202 | 12409 |
| Dev | 13 | 14 | 7 | 8 | 102923 | 113502 | 4272 | 4493 |
| Test | 12 | 14 | 7 | 8 | 81368 | 89774 | 5380 | 5488 |

# 4 Model for Evaluation

We develop the following baseline model and evaluate on the newly created dataset.

## 4.1 Model

This model is based on Bidirectional Encoder Representations for Transformers (BERT) [5]. There has been a trend of increasing the complexity of neural network architecture in this era of deep learning. There are many works that could be found that made use of more and more complex architectures in order to capture the similarity between the two sentences. However, a few recent language studies like [5,17] showed progress on many popular NLP tasks like Reading Comprehension on the widely used datasets like SQuAD [18] and RACE [8]. This is a simple model pre-trained on a large text corpus that takes the concatenation of passages and questions without modeling any direct interaction between the passage and question that can perform well. We foster this concept into our CLTE task for our baseline model. The layers of this model are as follows: *Multilingual Cased BERT Embedding Layer*, *Single Layered Feed- Forward Network*, and the *Softmax Classification Layer.*

***Multilingual Cased BERT Embedding Layer:*** The input to this layer is a sentence pair (P-H pair) like: $P = \{x_1, x_2, x_3....x_N\}$ and $H = \{y_1, y_2, y_3....y_N\}$. These two input sentences are separated by a special *[SEP]* token. Along with separating the sentences in a sentence pair, we append this special *[SEP]* token to the end of the sequence. The special token *[CLS]* is used at the front of the sequence. So mathematically, it can be expressed as:

$$[CLS]x_1, x_2, x_3....x_N[SEP]y_1, y_2, y_3....y_N[SEP].$$

BERT also uses **Segment Embeddings** to differentiate the premise from the hypothesis. These are simply two embeddings (for segments **A** and **B**) that BERT learned, and it adds to the token embeddings before feeding them into the input layer. The two sentences in the sentence pair are in different languages i.e. English and Hindi or vice-versa. We conduct two experiments using this model.

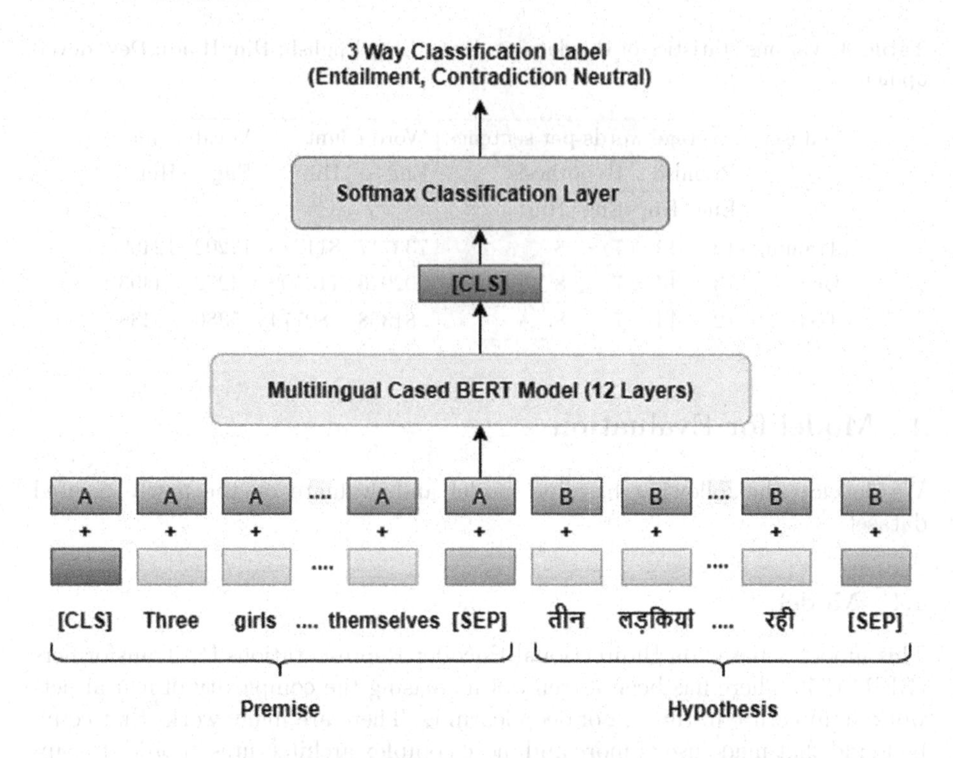


**Fig. 1.** Architecture of the proposed baseline model

Both the experiments correspond to using different language pairs as input (e.g., Hindi Premise - English Hypothesis and English Premise - Hindi Hypothesis). As the input to our system is pair of sentence of two different languages, we use multilingual **(Multilingual-Cased-BERT-Base)** version of it.

***Single Layer Feed Forward Layer:*** In order to obtain a fixed-dimensional representation of the input sequence (here a sentence pair for classification), we take the final hidden state (i.e., the output of the Transformer) for the first token, which corresponds to the special *[CLS]* embedding. We obtain this vector (denoted as **C**) with a dimension of $\mathbf{1XH}$ (H = 768). The only new parameters added during fine-tuning are for this feed forward layer with dimension $K \times H$, where **K** is the number of classes (i.e. Entailment, Contradiction and Neutral in this task). We further denote the parameter of this layer as $W_1$. So in this layer, the output Y is as follows:

$$Y = W_1 \times C + b \tag{1}$$

Where $\times$ denotes the dot product between the weight matrix and the context vector C and b is a bias term.

***Softmax Classification Layer:*** The output of the previous layer is fed into the final layer with soft-max activation function. The label probabilities K are computed with a standard softmax, $\mathbf{P} = \mathbf{softmax(O)}$. Essentially, P has a dimension of $1 \times K$, where K is the number of labels. All the parameters of BERT and **W** (the parameters of the previous layer) are fine-tuned jointly to maximize the log-probability of the correct label. So the final layer with soft-max activation function layer output (O) would be:

$$O = f(W_f \times Y + b_f) \quad (2)$$

where f is a soft-max function, Y is the output obtained from the previous feed forward layer, $W_f$ and $b_f$ represent weight and bias, respectively, of this final layer. The architecture of this model is shown in Fig. 1.

## 5 Experiments and Results

BERT is basically a language model pre-trained on Wikipedia and Books Corpus. We fine-tune the model for our downstream task using the training set of our dataset. We also perform task specific hyper-parameter tuning for our model using the development set. For tuning, several experiments were conducted. Each model is run by four different input combinations *viz.*: (i). English Premise and Hindi Hypothesis; (ii). Hindi Premise and English Hypothesis; (iii). English Premise and Hindi Hypothesis and (iv). Hindi Premise and Hindi Hypothesis. Thus, on this model, in total four experiments are performed. The results obtained by the baseline model are shown in Table 5. This table shows the class-wise (i.e. for Entailment, Contradiction and Neutral) precision, recall and F1 score. Evaluation results of monolingual experiments are shown in Table 4.

We provide comparisons in Table 6 with the existing model [3]. The paper presented multiple models including a machine translation based baseline. From the table, we can see that our proposed models outperform all the existing state-of-the-arts in terms of accuracy by a considerable margin. The proposed model is based on BERT which is robust in capturing semantic similarity between the two pieces of text snippets. The most simple way of doing CLTE is to rely on the translation system. Translate the hypotheses into Premises' language and then simply run entailment model to classify. In Table 6, TRANSLATE TRAIN refers to the system where training data (English) is translated into the target language (Hindi) during training, and in TRANSLATE TEST, the model is trained with original data (English) and input sentences (Hindi) are translated to the training language (English) at the time of testing. The third one (Evaluation of XNLI multilingual sentence encoders (in-domain)) refers to the another approach that relies on language universal embedding of texts and builds multilingual classification models on top of these embedding. X-BILSTM-last and X-BILSTM-max are trained on NLI, where the former one takes the last representation and the later one takes the maximum among all the representations. In contrast, X-CBOW encoder is pre-trained universal sentence embedding that takes the average of word embeddings. It evaluates transfer learning.

**Table 4.** Different results on monolingual settings: Hindi–Hindi and English–English language pairs

| Language Pair: Hindi-Hindi | | | | Language Pair: English–English | | |
|---|---|---|---|---|---|---|
| Classwise | Precision | Recall | F1 score | Precision | Recall | F1 score |
| Entailment | 0.76 | 0.74 | 0.75 | 0.84 | 0.86 | 0.85 |
| Contradiction | 0.71 | 0.76 | 0.73 | 0.88 | 0.84 | 0.86 |
| Neutral | 0.69 | 0.66 | 0.67 | 0.78 | 0.80 | 0.79 |
| Overall | | | | Overall | | |
| Macro avg | 0.54 | 0.54 | 0.54 | 0.62 | 0.62 | 0.62 |
| Weighted avg | 0.72 | 0.72 | 0.72 | 0.83 | 0.83 | 0.83 |
| **Accuracy: 0.72** | | | | **Accuracy: 0.83** | | |

**Table 5.** Different results on English–Hindi and Hindi–English cross-lingual settings

| Language Pair: English Hindi | | | | Language Pair: Hindi English | | |
|---|---|---|---|---|---|---|
| ClassWise | Precision | Recall | F1 score | Precision | Recall | F1 score |
| Entailment | 0.70 | 0.74 | 0.72 | 0.73 | 0.75 | 0.74 |
| Contradiction | 0.72 | 0.79 | 0.71 | 0.77 | 0.73 | 0.75 |
| Neutral | 0.66 | 0.65 | 0.66 | 0.68 | 0.69 | 0.68 |
| Overall | | | | Overall | | |
| macro avg | 0.52 | 0.52 | 0.52 | 0.54 | 0.54 | 0.54 |
| weighted avg | 0.69 | 0.69 | 0.69 | 0.72 | 0.72 | 0.72 |
| **Accuracy: 0.69** | | | | **Accuracy: 0.72** | | |

**Table 6.** Comparison of our results with existing state-of-the-art results. Here, SOTA: State-of-the-art

| Models | | Accuracy (%) |
|---|---|---|
| Proposed | BERT(Hindi–English) | **72** |
| | BERT (English–Hindi) | **69** |
| SOTA (XNLI [3]) | Machine translation baselines (TRANSLATE TRAIN) | |
| | BiLSTM-last | 61.3 |
| | BiLSTM-max | 62.1 |
| | Machine translation baselines (TRANSLATE TEST) | |
| | BiLSTM-last | 61.8 |
| | BiLSTM-max | 64.2 |
| | Evaluation of XNLI multilingual sentence encoders (in-domain) | |
| | X-BiLSTM-last | 62.8 |
| | X-BiLSTM-max | 64.1 |
| | Evaluation of pretrained multilingual sentence encoders (transfer learning) | |
| | X-CBOW | 56.3 |

## 6 Error Analysis

Evaluation results are presented in Table 4 and Table 5. For testing we use a manually created ground truths, where no translation is used. We extract the wrongly classified example pairs for both monolingual and cross-lingual experimental settings and perform a detailed error analysis for each of them. We take class-wise statistics of such instances. It is found that in both the cross-lingual settings i.e. English-Hindi and Hindi-English, maximum number of example pairs having *Neutral* class have been wrongly classified followed by *Contradiction* and *Entailment* class. We can argue that the model is efficient in handling *Entailment* and *Contradiction* cases compared to the *Neutral* cases. It demands further investigation. Later, we carefully examine those error examples. Our observations could be summarised as follows:

- Length difference between the two sentences is high, which could be one of the reasons of system failure.
- Translation system failed to produce the fluent and adequate translations of ambiguous sentences. Such an example is the following: *The camera-man shot the policeman with a gun.* The sentence is having lexical ambiguity, like: "shoot" refers to firing shot or taking a photograph; and structural ambiguity like: policeman with a gun and shot with a gun. It resulted wrong prediction in the Entailment decision.
  It is also observed that there are improper translation of some words. As an example, the correct translation of the word *Football* is फ़ुटबॉल (Footbaanl), but the translation system yields the translated word गेंद (*Gend*) in context of the sentence, which means ball but not football and thus it gives a sense of contradiction in the sentence pair.
- The system is weak to understand out-of-vocabulary (OOV) noun phrases (denotes the phrases that are not typically found in traditional dictionaries and that are used as spoken language) and this is the reason it misinterprets them as being different than their equivalent traditional phrases having the same meaning. As for example *on coffee* is a noun phrase equivalent to the traditional phrase *drinking coffee*, despite having the same meaning the system understands them as different, leading to the overall change in the perceived meaning of the sentence pair.
- The system performs poorly when there are named entities (NEs) in the sentence pair. This problem is observed for both mono-lingual and cross-lingual experimental settings. As for example, *Denim* and *Jeans* are the same, but the system does not capture this. The system again performs poorly when there are multi-word expressions in the sentence. As an example, *Sushi Bar* and *Hot Dog Stand* are not alike but the system does not capture the contradictory meanings.
- The system seems to fail when there is a complex causal relationship between the premise and hypothesis sentence pairs. Due to the existence of such causal relationships, the system fails to capture these and wrongly classifies the relation between the sentence pairs. As for example consider the Table 7 below:

**Table 7.** Example of causal relationship between the sentences from **EH-XNLI** dataset

| **Premise** | **Hypothesis** |
|---|---|
| एक आदमी हरे भरे घास काटने की मशीन पर एक बड़े मैदान में घास काट रहा | **A man rides a lawn mower** |
| शहर का नक्शा देख रहा लड़का | **Someone is trying to find directions** |

## 7 Conclusion

In this paper, we have presented a novel dataset for English–Hindi CLTE. The data was constructed by translating the original SNLI corpus into Hindi. This dataset offers the research community to a new dimension of TE research i.e. CLTE involving a low-resource language like Hindi. We have implemented a BERT based baseline model and ran on our dataset. We obtained the competitive results to the state-of-the-art. We hope this dataset will stimulate the development of more advanced CLTE models. Our future line of research would be as follows: • Enrichment of this dataset up to more than 100k for efficient training of the deep learning models. • Would like to create this kind of datasets for other languages too including low resource Indian languages. • More complex deep learning based models for CLTE.

**Acknowledgements.** We would like to acknowledge "Elsevier Centre of Excellence for Natural Language Processing" at Indian Institute of Technology Patna for partial support of the research work carried out in this paper. Asif Ekbal gratefully acknowledges Visvesvaraya Young Faculty Research Fellowship Award. We also acknowledge the annotators for manually checking the translated outputs.

## References

1. Bowman, S.R., Angeli, G., Potts, C., Manning, C.D.: A large annotated corpus for learning natural language inference. In: Proceedings of the 2015 Conference on Empirical Methods in Natural Language Processing, pp. 632–642, Lisbon, Portugal. Association for Computational Linguistics, September 2015. https://doi.org/10.18653/v1/D15-1075. https://www.aclweb.org/anthology/D15-1075
2. Chen, Q., Zhu, X., Ling, Z.H., Wei, S., Jiang, H., Inkpen, D.: Enhanced LSTM for natural language inference. In: Proceedings of the 55th Annual Meeting of the Association for Computational Linguistics (Volume 1: Long Papers), pp. 1657–1668, Vancouver, Canada. Association for Computational Linguistics, July 2017. https://doi.org/10.18653/v1/P17-1152. https://www.aclweb.org/anthology/P17-1152
3. Conneau, A., et al.: XNLI: evaluating cross-lingual sentence representations. In: Proceedings of the 2018 Conference on Empirical Methods in Natural Language Processing, Brussels, Belgium, pp. 2475–2485. Association for Computational Linguistics, October–November 2018. https://doi.org/10.18653/v1/D18-1269. https://www.aclweb.org/anthology/D18-1269

4. Dagan, I., Glickman, O., Magnini, B.: The PASCAL Recognising Textual Entailment Challenge. In: Quiñonero-Candela, J., Dagan, I., Magnini, B., d'Alché-Buc, F. (eds.) MLCW 2005. LNCS (LNAI), vol. 3944, pp. 177–190. Springer, Heidelberg (2006). https://doi.org/10.1007/11736790_9
5. Devlin, J., Chang, M.W., Lee, K., Toutanova, K.: BERT: pre-training of deep bidirectional transformers for language understanding. In: Proceedings of the 2019 Conference of the North American Chapter of the Association for Computational Linguistics: Human Language Technologies, Volume 1 (Long and Short Papers), Minneapolis, Minnesota, pp. 4171–4186. Association for Computational Linguistics, June 2019. https://doi.org/10.18653/v1/N19-1423. https://www.aclweb.org/anthology/N19-1423
6. Fleiss, J.L.: Measuring nominal scale agreement among many raters. Psychol. Bull. **76**(5), 378 (1971)
7. Harabagiu, S., Hickl, A.: Methods for using textual entailment in open-domain question answering. In: Proceedings of the 21st International Conference on Computational Linguistics and the 44th Annual Meeting of the Association for Computational Linguistics, pp. 905–912. Association for Computational Linguistics (2006)
8. Lai, G., Xie, Q., Liu, H., Yang, Y., Hovy, E.: RACE: large-scale reading comprehension dataset from examinations. In: Proceedings of the 2017 Conference on Empirical Methods in Natural Language Processing, Copenhagen, Denmark, pp. 785–794. Association for Computational Linguistics, September 2017. https://doi.org/10.18653/v1/D17-1082. https://www.aclweb.org/anthology/D17-1082
9. Landis, J.R., Koch, G.G.: The measurement of observer agreement for categorical data. Biometrics **33**, 159–174 (1977)
10. Liu, X., He, P., Chen, W., Gao, J.: Multi-task deep neural networks for natural language understanding. In: Proceedings of the 57th Annual Meeting of the Association for Computational Linguistics, Florence, Italy, pp. 4487–4496. Association for Computational Linguistics, July 2019. https://doi.org/10.18653/v1/P19-1441. https://www.aclweb.org/anthology/P19-1441
11. MacCartney, B., Grenager, T., de Marneffe, M.C., Cer, D., Manning, C.D.: Learning to recognize features of valid textual entailments. In: Proceedings of the Human Language Technology Conference of the NAACL, Main Conference, New York City, USA, pp. 41–48. Association for Computational Linguistics, June 2006. https://www.aclweb.org/anthology/N06-1006
12. Mehdad, Y., Negri, M., Federico, M.: Towards cross-lingual textual entailment. In: Human Language Technologies: The 2010 Annual Conference of the North American Chapter of the Association for Computational Linguistics, Los Angeles, California, pp. 321–324. Association for Computational Linguistics, June 2010. https://www.aclweb.org/anthology/N10-1045
13. Negri, M., Marchetti, A., Mehdad, Y., Bentivogli, L., Giampiccolo, D.: Semeval-2012 task 8: cross-lingual textual entailment for content synchronization. In: *SEM 2012: The First Joint Conference on Lexical and Computational Semantics - Volume 1: Proceedings of the main conference and the shared task, and Volume 2: Proceedings of the Sixth International Workshop on Semantic Evaluation (SemEval 2012), Montréal, Canada, 7–8 June 2012, pp. 399–407. Association for Computational Linguistics (2012). https://www.aclweb.org/anthology/S12-1053

14. Negri, M., Marchetti, A., Mehdad, Y., Bentivogli, L., Giampiccolo, D.: Semeval-2013 task 8: cross-lingual textual entailment for content synchronization. In: Second Joint Conference on Lexical and Computational Semantics (*SEM), Volume 2: Proceedings of the Seventh International Workshop on Semantic Evaluation (SemEval 2013), Atlanta, Georgia, USA, pp. 25–33. Association for Computational Linguistics, June 2013. https://www.aclweb.org/anthology/S13-2005
15. Padó, S., Galley, M., Jurafsky, D., Manning, C.D.: Robust machine translation evaluation with entailment features. In: Proceedings of the Joint Conference of the 47th Annual Meeting of the ACL and the 4th International Joint Conference on Natural Language Processing of the AFNLP, Suntec, Singapore, pp. 297–305. Association for Computational Linguistics, August 2009. https://www.aclweb.org/anthology/P09-1034
16. Parikh, A., Täckström, O., Das, D., Uszkoreit, J.: A decomposable attention model for natural language inference. In: Proceedings of the 2016 Conference on Empirical Methods in Natural Language Processing, Austin, Texas, pp. 2249–2255. Association for Computational Linguistics, November 2016. https://doi.org/10.18653/v1/D16-1244, https://www.aclweb.org/anthology/D16-1244
17. Radford, A., Narasimhan, K., Salimans, T., Sutskever, I.: Improving language understanding by generative pre-training (2018). https://s3-us-west-2.amazonaws.com/openai-assets/researchcovers/languageunsupervised/languageunderstandingpaper.pdf
18. Rajpurkar, P., Zhang, J., Lopyrev, K., Liang, P.: SQuAD: 100,000+ questions for machine comprehension of text. In: Proceedings of the 2016 Conference on Empirical Methods in Natural Language Processing, Austin, Texas, pp. 2383–2392. Association for Computational Linguistics, November 2016. https://doi.org/10.18653/v1/D16-1264. https://www.aclweb.org/anthology/D16-1264
19. Rocktäschel, T., Grefenstette, E., Hermann, K.M., Kociský, T., Blunsom, P.: Reasoning about entailment with neural attention. In: 4th International Conference on Learning Representations, ICLR 2016, San Juan, Puerto Rico, 2–4 May 2016. Conference Track Proceedings (2016)
20. Saikh, T., Anand, A., Ekbal, A., Bhattacharyya, P.: A novel approach towards fake news detection: deep learning augmented with textual entailment features. In: Métais, E., Meziane, F., Vadera, S., Sugumaran, V., Saraee, M. (eds.) NLDB 2019. LNCS, vol. 11608, pp. 345–358. Springer, Cham (2019). https://doi.org/10.1007/978-3-030-23281-8_30
21. Saikh, T., Ghosal, T., Ekbal, A., Bhattacharyya, P.: document level novelty detection: textual entailment lends a helping hand. In: Proceedings of the 14th International Conference on Natural Language Processing (ICON-2017), Kolkata, India, pp. 131–140. NLP Association of India, December 2017. http://www.aclweb.org/anthology/W/W17/W17-7517
22. Saikh, T., Naskar, S.K., Ekbal, A., Bandyopadhyay, S.: Textual entailment using machine translation evaluation metrics. In: Gelbukh, A. (ed.) CICLing 2017. LNCS, vol. 10761, pp. 317–328. Springer, Cham (2018). https://doi.org/10.1007/978-3-319-77113-7_25
23. Saikh, T., Naskar, S.K., Giri, C., Bandyopadhyay, S.: Textual entailment using different similarity metrics. In: Gelbukh, A. (ed.) CICLing 2015. LNCS, vol. 9041, pp. 491–501. Springer, Cham (2015). https://doi.org/10.1007/978-3-319-18111-0_37
24. Saini, N., Saha, S., Bhattacharyya, P., Tuteja, H.: Textual entailment-based figure summarization for biomedical articles. ACM Trans. Multimedia Comput. Commun. Appl. (TOMM) **16**(1s), 1–24 (2020)

25. Singh, J., McCann, B., Keskar, N.S., Xiong, C., Socher, R.: XLDA: cross-lingual data augmentation for natural language inference and question answering. arXiv preprint arXiv:1905.11471 (2019)
26. Wang, S., Jiang, J.: Learning natural language inference with LSTM. In: Proceedings of the 2016 Conference of the North American Chapter of the Association for Computational Linguistics: Human Language Technologies, San Diego, California, pp. 1442–1451. Association for Computational Linguistics, June 2016. https://doi.org/10.18653/v1/N16-1170. https://www.aclweb.org/anthology/N16-1170
27. Williams, A., Nangia, N., Bowman, S.: A broad-coverage challenge corpus for sentence understanding through inference. In: Proceedings of the 2018 Conference of the North American Chapter of the Association for Computational Linguistics: Human Language Technologies, Volume 1 (Long Papers), pp. 1112–1122. Association for Computational Linguistics (2018). http://aclweb.org/anthology/N18-1101
28. Wu, Y., Schuster, M., et al.: Google's neural machine translation system: bridging the gap between human and machine translation. arXiv preprint arXiv:1609.08144 (2016)
29. Young, P., Lai, A., Hodosh, M., Hockenmaier, J.: From image descriptions to visual denotations: new similarity metrics for semantic inference over event descriptions. Trans. Assoc. Comput. Linguist. **2**, 67–78 (2014). https://doi.org/10.1162/tacla000166

# A Token-Wise CNN-Based Method for Sentence Compression

Weiwei Hou[1(✉)], Hanna Suominen[1,2,3(✉)], Piotr Koniusz[1,2(✉)], Sabrina Caldwell[1(✉)], and Tom Gedeon[1(✉)]

[1] The Australian National University, Canberra, Australia
{weiwei.hou,hanna.suominen,sabrina.caldwell,tom.gedeon}@anu.edu.au
[2] Data61/CSIRO, Eveleigh, Australia
piotr.koniusz@data61.csiro.au
[3] University of Turku, Turku, Finland

**Abstract.** Sentence compression is a *Natural Language Processing* (NLP) task aimed at shortening original sentences and preserving their key information. Its applications can benefit many fields *e.g.*, one can build tools for language education. However, current methods are largely based on *Recurrent Neural Network* (RNN) models which suffer from poor processing speed. To address this issue, in this paper, we propose a token-wise*Convolutional Neural Network*, a CNN-based model along with pre-trained *Bidirectional Encoder Representations from Transformers* (BERT) features for deletion-based sentence compression. We also compare our model with RNN-based models and fine-tuned BERT. Although one of the RNN-based models outperforms marginally other models given the same input, our CNN-based model was ten times faster than the RNN-based approach.

**Keywords:** Neural networks · NLP · Application

## 1 Introduction

Deletion-based sentence compression refers to the task of extracting key information from a sentence by deleting some of its words. It is often used as an initial step for generating document summaries or as an intermediary step in information retrieval and machine translation. Its applications may also benefit many fields such as e-Learning and language education.

Recent studies focus on adapting methods based on neural networks to solve deletion-based sentence compression as a sequential binary classification problem (see Sect. 2). *Recurrent Neural Networks* (RNNs) are one of the most popular network architectures that handle sequential tasks. Models such as *Gated Recurrent Units* (GRUs), *Long-Short Term Memory* (LSTM) networks, and BiLSTM *Bidirectional LSTMs* (BiLSTMs) are found to be suited for this task. However,

W. Hou—This work is supported by the Australian National University.


H. Yang et al. (Eds.): ICONIP 2020, LNCS 12532, pp. 668–679, 2020.
https://doi.org/10.1007/978-3-030-63830-6_56

training these RNN-based models can be time consuming. Applications with poor response speed are the cause of negative user experience.

In contrast, *Convolutional Neural Networks* (CNNs) outperform RNNs in their training speed and reportedly have a similar or better performance than RNN-based models in many tasks [1,12]. They are widely applied to tasks such as object recognition in Computer Vision. In *Natural Language Processing* (NLP), CNNs have been studied for document summarization, sentiment analysis, and sentence classification, among others. However, the majority of these methods concern the sentence or document level.

In this paper, we apply CNNs to sentence compression at the token level. However, CNNs are weaker at capturing sequential information. To circumvent this issue, we train our model with pre-trained *Bidirectional Encoder Representations from Transformers* (BERT) [4] features. In addition, we also compare our model performance against RNN and BERT fine-tuned models. We test the performance in both the correctness and efficiency.

## 2 Related Work

### 2.1 Recurrent Neural Networks

RNN-based approaches are widely applied to sequential problems such as machine translation, sentence parsing, and image captioning. Inspired by these core NLP tasks, [6] concatenated input sequence and its labels by a key word 'GO' as an input to a sequence-to-sequence framework. The goal was to predict sequence labels for each word succeeding 'GO'. Their network architecture is composed of three layers of LSTMs. On top of the LSTM layers is a softmax layer that produces final outputs. Their method takes on input only word embeddings in the form of the vector representations.

Subsequently, authors of [11] and [13] discovered that adding syntactic information improves the performance. Both works included the *Part-of-Speech* (POS) tags, dependency type, and word embeddings. The results showed significant accuracy improvements. In addition, instead of concatenating embedding sequences and labels, both these studies used hidden vectors to predict labels. The difference between these two methods is that approach [11] uses a framework with three layers of BiLSTMs while approach [13] has a more complex architecture which includes a layer of *Bi-directional RNNs* (BiRNNs) and a layer of Graph Neural Networks. A year later, authors of [14] proposed an approach based on reinforcement learning which includes a BiRNN-based policy network containing two types of actions – REMOVE and RETAIN, where words marked as REMOVE are deleted to obtain a sequence of predicted compression. Such a sequence is then fed into a syntax-based evaluation step that examines the predicted compression according to two rewards. The first reward concerns the fluency of generated sentence against a syntax-based language model. The second reward is based on the comparison of the generated sentence compression rate with an average compression rate. Their method works well on both large unlabeled and labeled datasets. However, such a model is difficult to train.

### 2.2 Bidirectional Encoder Representations from Transformers

BERT is a language representation model which takes the word position into account to capture a word with its context [4]. Unlike *Global Vectors for Word Representation* (GloVe), Word2Vec, and many other context independent word embeddings, BERT not only embeds semantic information but it also captures the structural information in a sequence. In addition, BERT enables bidirectional prediction. It uses a *Masked Language Model* mechanism that randomly masks a certain percentage of tokens in an input sequence. The objective is to use both the preceding and succeeding context to predict the masked token.

Apart from providing pre-trained language representations, BERT can be fine-tuned on related tasks. It has reportedly achieved the state-of-the-art performance in nine NLP tasks [4]. BERT is said to have also the ability to capture high-level linguistic knowledge (*e.g.*, semantics, syntax, or grammar) [3,7,8].

To the best of our knowledge, our is the first work using BERT for deletion-based sentence compression tasks. We compare pre-trained BERT layers with word embeddings, POS embeddings, and dependency embeddings given the same network architecture to explore whether BERT is able to capture complex syntactic information.

## 3 Method

We define the deletion-based sentence compression task as a token level segmentation task. Specifically, we have given a sequence of $s = \{w_1, w_2, w_3, \cdots, w_i\}$ as an original sentence, where $i$ is the number of tokens in this sequence, for $s$, and we have a corresponding sequence of the mask $y = \{y_1, y_2, y_3, \cdots, y_i\}$, where $y_i \in \{0, 1\}$ is the ground truth label of $w_i$. Moreover, by zero (one) we mean that a token needs to be deleted (retained) from the original sequence. The goal is to train a model to predict whether $w_i$ in sequence $s$ should be deleted or retained.

### 3.1 Network Architectures

Our approach is largely based on U-Net [10], with some differences. The network was originally designed for pixel-level image segmentation tasks. It is a fast and lightweight network which shows extraordinary performance in many image segmentation tasks. The reason we choose U-Net as our base network is that our task is a token level binary classification. Regular CNN cannot capture various levels of coarseness of information and "expand" it back to the token level for segmentation. In addition, we believe that the max-pooling operation can be seen as realizing "compression".

We adapted the original U-Net network architecture to train a model for a text-based task. As Fig. 1 shows, we assume we have word $w_i$ in sequence $s$ where each $w_i$ has associated with it a $j$-dimensional vector $w_i = \{e_{i,1}, e_{i,2}, e_{i,3}, \cdots, e_{i,j}\}$. This setting yields a matrix of size $i \times j$ which forms the input to the first layer of our network.

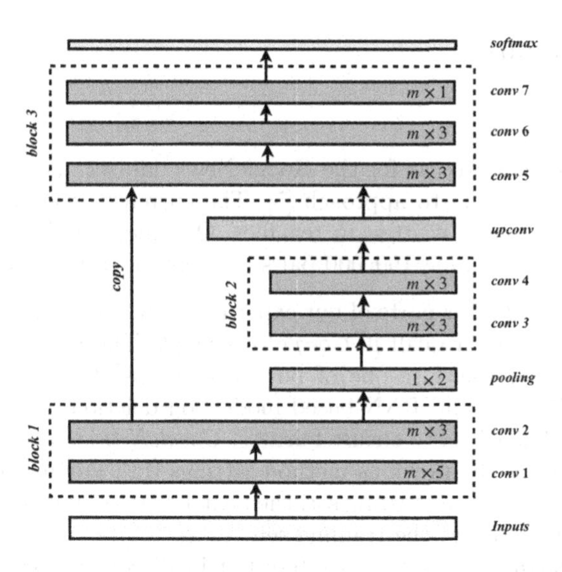


**Fig. 1.** Graphical illustration of a CNN-based network architecture. Convolutional layers are denoted by *conv* 1, *conv* 2, *conv* 3, ... ,*conv* 7.

Figure 1 shows (from the bottom to the top) that the network architecture contains three main blocks. The first block has two standard convolution layers (conv 1 and conv 2). The conv 1 layer has a $m \times 5$ kernel; $m$ is determined by the number of input channels (see Sect. 4) and conv 2 has a $m \times 3$.

Moreover, between the first and second block, there is a $1 \times 2$ pooling layer (pool1) with the stride size of 2 for down-sampling. The pooling output is fed to the second block which contains two $m \times 3$ convolution layers (conv3, conv4). After the above operations, the output is halved in size.

In order to map the output into a token level, we up-sample (deconv) the output from the second block to double back the output size and then the output is concatenated with the previous output from the first block. This operation allows the network to remember "early" features. The third block includes two regular $m \times 3$ convolutions (conv5, conv6) and one $m \times 1$ convolutions (conv7). All above convolutions are followed by a *Rectified Linear Unit* (ReLU).

Lastly, we pass the convolution output $x = \{x_1, x_2, x_3, \cdots, x_i\}$ though a token-wise softmax layer to predict the mask $y_i$ as follows:

$$P(y_i|x_i) = softmax(Wx_i), \quad (1)$$

where $W$ is a weight matrix to be learned.

We introduce two main modifications to the original network architecture [10]. The first one is that instead of using the same kernel sizes, our network uses a mix of kernel sizes in order to capture multi-gram information. The second change is that we reduced the depth of the network due to the size of our sentences being much smaller than image sizes from the original work. Hence, empirically, a shallower network tends to perform better.

# 4 Experiments

## 4.1 Data

For our experiments, we use for the GoogleNews dataset, an automatically collected dataset according to approach [5]. The currently released data contains 200,000 sentence pairs, written in ten files. One file includes 10,000 pairs for evaluation while remaining 190,000 pairs are contained in other files.

***GoogleNewsSmall.*** For parts of our experiments, we use only the 10,000 pairs based evaluation set. We call it *GoogleNewsSmall*. The reason that we choose *GoogleNewsSmall* is because one of purposes of this study is to compare the performance of CNN- and RNN-based models on deletion-based sentence compression tasks. Therefore, to ensure fairness, *GoogleNewsSmall* includes exactly the same dataset as the previous method settings [11]. Furthermore, we use the first 1,000 sentences as the testing set, the next 1,000 sentences as the validation set, and the remainder as the training set.
***GoogleNewsLarge.*** We are also interested in discovering the impact of the training data size on performance. Therefore, in the second setting, we include the entire 200,000 sentence pairs in our experiments. We denote this setting as *GoogleNewsLarge*. For testing and validation, we use the setting already described above. The remaining 198,000 pairs are used as the training set.

## 4.2 Experimental Setup

***BiLSTM + Emb.*** In this experiment, we use the same network settings as the one described in [11], which includes a base model architecture with three BiLSTM layers and a dropout layer between them. The input to this network is contains word embeddings only. The purpose of this experiment is to train a base-line RNN-based network.

***CNN + Emb.*** In this experiment, we use our method proposed in Sect. 3.1. For input, we again use word embeddings as the only input to ensure the comparison with RNNs in the same testbed.
***BiLSTM + BERT.*** In this experiment, we adapt the setting of **BiLSTM + Emb** by replacing the word embeddings with the last layer of pre-trained BERT outputs. We also record the training time, in seconds, from the point of starting training to full convergence.

***CNN + Multilayer-BERT.*** This experiment is composed of a group of sub-experiments. We use the same network architecture as in Sect. 4.2. We firstly extract the last four BERT layers ($-1$, $-2$, $-3$, $-4$) from a pre-trained model. Then, we run the experiments by feeding the network with four different input layer settings: (i) input with the layer $-1$, (ii) input with the layer $-1$, and $-2$, (iii) input with the layer $-1$, $-2$, and $-3$), and (iv) input with all last four layers. We record the training time of experiment with the layer $-1$ as input, from the point of start of training to the full converge.

### 4.3 Experiments on Different Network Settings

This experiment aims to test the model performance w.r.t. reducing network layers. In the first sub-experiment, we remove one convolution layer from each block (conv 2, 4, and 5 layers from the original settings) in order to test the impact of convolution layers. In the second sub-experiment, we remove pooling, conv 4, 5 of block 2, and upsampling layers. The purpose of this experiment is to compare the performance of a stack of convolution layers and our model.

### 4.4 Experiments w.r.t. the Training Size

This experiment evaluates the performance w.r.t. different sizes of training data. We vary the training size between 8,000 and 19,800. We divide the training progress into ten steps. At each step, the training size increases by 20,000. We detail this experiment in Sect. 4.2 by only testing on the "last four layers" setting.

The data is labeled as described in approach [5]. Word embeddings are initialised by GloVe 100-dimensional pre-trained embeddings [9]. BERT representations are extracted from the 12/768-uncased-base pre-trained model[1]. All experiments are conducted on a single GPU (Nvidia GeForce GTX Titan XP 12 GB). For CNN-based networks, all input sequences have a fixed length of 64[2]. The number of input channels is and layers are the same as in BERT.

### 4.5 Quantitative Evaluations

To evaluate the performance, we use the accuracy and F1 measure. F1 scores are derived from precision and recall values [2] where the precision is defined as the percentage of predicted labels (1 = retained words) that match ground truth labels, and recall is defined as the percentage of labeled retained words in ground truth that overlap with the predictions.

Regarding the training efficiency, we evaluate it as follows. Firstly, we compare both the F1 scores and accuracy of CNN and RNN-based model as a function of each recorded time point. Secondly, we compare the training time of two models.

### 4.6 Perception-Based Evaluations

In order to test the readability of outputs and their relevance to the original inputs, we asked two native English speakers to manually evaluate outputs from the **CNN-Multilayer-BERT** and **BiLSTM-Emb** (baseline). The inputs are the top 200 sentences from the test set. Evaluation methods follow approach [6].

---

[1] https://github.com/google-research/bert.

[2] We concluded that it is the best input size after calculating the mean and max length of all sentences in the entire dataset, and evaluating the efficiency of extracting BERT features.

**Table 1.** Results given different experimental settings for models trained on *Google- NewsSmall.*

| Method | F1 |
|---|---|
| BiLSTM + Emb | 0.74 |
| BiLSTM + BERT | 0.79 |
| BiLSTM + SynFeat [11] | 0.80 |
| LK-GNN + Emb [13] | 0.80 |
| LK-GNN + All Features [13] | **0.82** |
| CNN + Emb | 0.72 |
| CNN + Multilayer-BERT | 0.80 |

**Table 2.** Results w.r.t. using different layers of BERT features (trained on *GoogleNewsSmall*).

| BERT layers | F1 | Accuracy |
|---|---|---|
| Only −1 | 0.78 | 0.80 |
| −1, −2 | 0.79 | 0.81 |
| −1, −2, −3 | 0.80 | 0.81 |
| −1, −2, −3, −4 | **0.80** | **0.82** |

**Table 3.** Results given different network settings (trained on *GoogleNewsSmall*).

| Network Setup | F1 |
|---|---|
| Remove conv 2, 4, and 5 | 0.797 |
| Remove pooling, conv 4,5 and upsampling | 0.786 |
| Original setting (channel size conv4 = 128, conv6 = 256) | 0.805 |

## 5 Results

We report the accuracy of different experimental settings in Table 1. We note that **LK-GNN + All Features** achieves the best results. **CNN + Multilayer-BERT** performs the same as **BiLSTM + SynFeat** and **LK-GNN + Emb** (0.80 = 80%). Next, we compare the performance of different models with the same input settings. When model inputs are word embeddings, the table shows that the RNN-based model outperforms the CNN-based one. The results reflect our assumption that CNNs are weaker in capturing sequential information compared to RNNs. However, the performance gap between **CNN+Emb** and the **BiLSTM+Emb** is not significant. Therefore, CNN-based models are a reasonable choice.

Looking at experiments that use the same network architecture but different input settings, we notice that models with BERT used as the input have significantly better performance than the models with Emb inputs. When comparing BERT to add-on syntactic features, **BiLSTM + BERT** slightly under-performs **BiLSTM + SynFeat** (*ie.*, 1% lower on F1). This implies that BERT captures both syntactic and semantic information. Moreover, when we test multiple layers of BERT features, we can see that **CNN + Multilayer-BERT** performs the same as the model with add-on syntactic features. Therefore, multiple layer BERT features enhance learning ability of the model. To further investigate BERT, Table 2 shows the impact of using different layers of BERT features on the model performance. The model trained with last four BERT output layers

**Table 4.** Perception-based Evaluations (trained on *GoogleNewsSmall*).

| Method | Readability | Informativeness |
|---|---|---|
| BiLSTM + Emb | 4.0 | 3.4 |
| CNN + Multilayer-BERT | 4.3 | 3.6 |

achieves best results (F1 = 0.80, Accuracy = 0.82). However, the model that only uses the last layer of BERT under-performs by 2% (F1 score) and by 1% accuracy. We notice a trend that increasing the number of BERT layers will result in a slightly better performance. This supports our hypothesis that using multiple layers of BERT features improves the performance. Note that we tested multiple layers of BERT features on our CNN based model but not on RNN-based models because concatenating features of four BERT output layers results in an extremely long input (each layer of BERT has a hidden size equal $4 \times 768$). Thus, the computation cost would be very expensive. Moreover, Table 4 shows results for the perception-based evaluations of our model compared to the baseline. As one can see, the model trained with BERT scores better on readability and informativeness.

**Table 5.** Training time of CNN- *vs*. LSTM-based models. The first layer of BERT output is used as input to the model.

| Run time(s) | CNN + BERT | | LSTM + BERT | |
|---|---|---|---|---|
| | F1 | Accuracy | F1 | Accuracy |
| 16 | 0.60 | 0.73 | 0.60 | 0.66 |
| 64 | 0.75 | 0.73 | 0.50 | 0.64 |
| 120 | 0.766 | 0.79 | 0.59 | 0.67 |
| 210 | 0.78 | 0.80 | 0.50 | 0.69 |
| 720 | – | – | 0.70 | 0.76 |
| 1095 | – | – | 0.73 | 0.78 |
| 1483 | – | – | 0.75 | 0.78 |
| 1863 | – | – | 0.75 | 0.79 |
| 2239 | – | – | 0.78 | 0.80 |
| 2622 | – | – | 0.79 | 0.79 |
| 3303 | – | – | 0.80 | 0.81 |

The comparisons are demonstrated in Table 5. As we expected, the CNN-based model converged faster when compared to the BiLSTM-based model. It reached its best accuracy performance (F1 = 0.78, Accuracy = 0.81) after 210 s. However, while reaching the same F1 score, the BiLSTM-based model needs 10 times longer training time. To outperform the CNN-based models in terms of

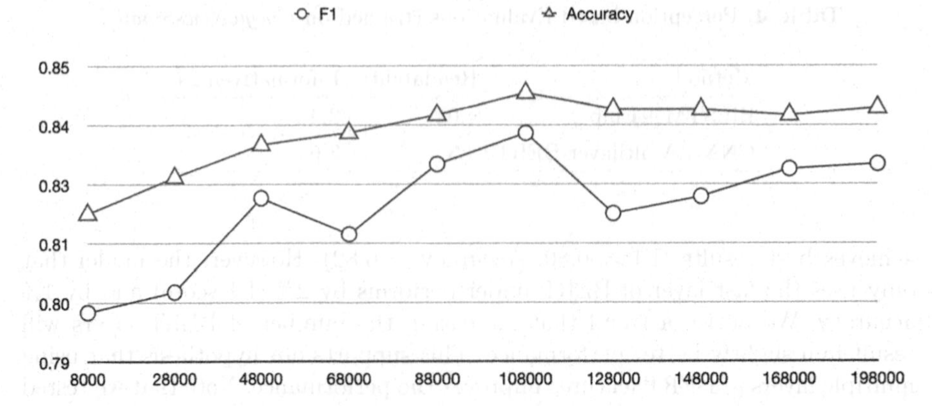


**Fig. 2.** Performance of CNN + Multilayer-BERT (last four layers used) in terms of F1 and accuracy. Experiments are conducted on *GoogleNewsLarge*.

both the F1 and accuracy, the BiLSTM model took 15 times more time to train. Therefore, our model significantly outperforms BiLSTM models if F1, accuracy and time are taken into account. Table 3 shows the F1 scores for different network settings. As we can see, by removing three convolutional layers, the model slightly under-performs the original model. However, when the pooling, conv 4, conv 5 and upsampling were removed, the results are nearly 2% lower compared to the original settings. This result supports our assumption that pooling help extract compressed information.

In addition, we also investigated the impact of the training data size on the CNN-based model. Figure 2 shows the F1 scores and accuracy of our CNN-BERT model (**CNN + Multilayer-BERT**) w.r.t. different size of training data. We observed that by increasing the training size to approximately 100, 000 pairs, the model performs reaches over 0.84 in both the F1 score and accuracy. For more training data, the results show no further improvement and when we increase the training size to be equal to the size of full training set, the results drop to 0.83 (F1 score). We believe that this is caused by the noise in the dataset as authors of approach [14] note that this dataset is automatically labeled based on syntactic-tree-pruning method. Noise can be introduced by syntactic tree parsing errors. Approach [14] scores 0.84 (F1 score) on their LSTM models with the same training data as in our experiments. However, under the same settings, a model trained on 2 million training pairs reported 0.82 (F1 score) [6]. Authors of approach [14] also augured that this training data may contain errors caused by syntactic tree parsing errors during data annotation. However, we do not evaluate the quality of the data. Our results indicate that the effective training size equals 100, 000.

## 6 Discussion

Section 5 showed that, the RNN-based models slightly outperform CNN-based models in their correctness; on average, their F1 scores were 2% higher. However, the CNN-based models performed over ten times faster. In addition, to improve the results of CNN-based models, we adopted BERT features as our networks inputs. The results showed that the model with four layers of BERT features achieved equal performance compared to approach with add-on syntactic features [11] given training size equal 8,000. It implies that multiple layers of BERT capture both the syntactic and semantic information. We argue that the CNN with multiple layers of BERT features was quite a reasonable setting. Since each layer of BERT features has a vector size of 678, concatenating multiple BERT layers as inputs for RNN-based models is computationally prohibitive.

In addition, we also tested the impact of the training data size on our **CNN + Multilayer-BERT** model given four distinct feature settings of BERT, and we found that the F1 scores do not improve further when the training date size reaches approximately 100,000 pairs. We believe that such a result is caused by the noise in the dataset. We believe that such a noise was introduced during the data collecting and labeling process. Similar observations were reported by authors of approaches [13,14]. In contrast to previous works, we report what is a reasonable trade-off in terms of the dataset size.

In this paper, we did not directly compare our results with the state-of-the-art model [14] for two main reasons. Firstly, our work mainly focused on comparing the performance on different base model settings. Authors of approach [14] proposed a reinforcement learning method, implementing bidirectional RNNs as the base model of the policy network, and this setting is quite similar to approach [11]. Secondly, we tested methods that do not include any domain specific knowledge. One of reward rules in their method uses scoring the sentence compression rates. Since the data was generated by predefined rules, adding such a rule could improve the performance. Although we did not directly compared our results with theirs, we reported results given the training size of 100,000. Our method reaches 84% (F1 score) which is equal to their implemented LSTM model, and only about 1% lower than their reinforcement-based method. We believe if the base model was replaced with our CNN model, the final accuracy would be similar while enjoying faster training.

## 7 Conclusions

In this paper, we studied the performance of CNN-based models for the deletion-based sentence compression task. We first tested the correctness results against the most commonly implemented RNN-based models as well as fine-tuned BERT. Subsequently, we examined the training efficiency of both models. We also compared the results when using a pre-trained BERT language representation as an input to the models with classical word-embeddings and/or other add-on syntactic information.

Our results show that the CNN-based model requires much less training time than the RNN-based model. In addition, the pre-trained BERT language representation highlighted its ability to capture deeper information compared to classical word embedding models. BERT could also serve as a replacement of manually introduced add-on syntactic information. Finally, we observed that increasing the size of training data beyond certain point does not improve the performance further.

Our approach can potentially reduce the cost of building sentence compression applications such as language education tools. Our approach saves computational resources which promotes interactive applications while preserving their accuracy. In the future, we will use our model as a backbone in a reading assistant tool supporting university *English as a Second Language* (ESL) students in their reading activities. We will also continue to study sentence compression with the focus on approaches that can customise the output.

## References

1. Bai, S., Kolter, Z.J., Koltun, V.: An empirical evaluation of generic convolutional and recurrent networks for sequence modeling. https://arxiv.org/abs/1803.01271 (2018)
2. Chinchor, N.: MUC-4 evaluation metrics. In Proceedings of the 4th Conference on Message Understanding, pp. 22–29 (1992)
3. Clark, K., Khandelwal, U., Levy, O., Mannning, C.D.: What Does BERT look at? An analysis of BERT's attention. https://arxiv.org/abs/1906.04341 (2019)
4. Devlin, J., Chang, M.W., Lee, K., Toutanova, K.: BERT: pre-training of deep bidirectional transformers for language understanding. https://arxiv.org/abs/1810.04805 (2019)
5. Filippova, K., Altun, Y.: Overcoming the lack of parallel data in sentence compression. In Proceedings of the 2013 Conference on Empirical Methods in Natural Language Processing, Seattle, Washington, pp. 1481–1491. ACL (2013)
6. Filippova, K., Alfonseca, E., Colmenares, C. A., Kaiser, L., Vinyals, O.: Sentence compression by deletion with LSTMs. In Proceedings of the 2015 Conference on Empirical Methods in Natural Language Processing, Lisbon, Portugal. ACL (2015)
7. Goldberg, Y.: Assessing BERT's syntactic abilities (2019). http://u.cs.biu.ac.il/yogo/bert-syntax.pdf
8. Lie, C., Andrede, A.D., Osama, M.: Exploring multilingual syntactic sentence representations. https://arxiv.org/abs/1910.11768 (2019)
9. Pennington, J., Socher, R., Manning, C.D.: Glove: global vectors for word representation. In Proceedings of the Conference on Empirical Methods in Natural Language Processing, Doha, Qatar. ACL (2014)
10. Ronneberger, O., Fischer, P., Brox, T.: U-Net: convolutional networks for biomedical image segmentation. In: Navab, N., Hornegger, J., Wells, W.M., Frangi, A.F. (eds.) MICCAI 2015. LNCS, vol. 9351, pp. 234–241. Springer, Cham (2015). https://doi.org/10.1007/978-3-319-24574-4_28
11. Wang, L., Jiang, J., Chieu, H.L., Ong, C.H., Song, D., Liao, L.: Can syntax help?. Improving an LSTM-based sentence compression model for new domains. ACL, Vancouver, Canada (2017)

12. Yin, W., Kann, K., Yu, M., Schütze, H.: Comparative study of CNN and RNN for natural language processing. https://arxiv.org/abs/1702.01923 (2017)
13. Zhao, Y., Senuma, H., Shen, X., Aizawa, A.: Gated neural network for sentence compression using linguistic knowledge. In: Frasincar, F., Ittoo, A., Nguyen, L.M., Métais, E. (eds.) NLDB 2017. LNCS, vol. 10260, pp. 480–491. Springer, Cham (2017). https://doi.org/10.1007/978-3-319-59569-6_56
14. Zhao, Y., Luo, Z., Aizawa, A.: A language model based evaluator for sentence compression. In Proceedings of the 56th Annual Meeting of the Association for Computational Linguistics, Melbourne, Australia, ACL pp 170–175. (2018)

# Automatic Parameter Selection of Granual Self-organizing Map for Microblog Summarization

Naveen Saini(✉), Sriparna Saha, Sahil Mansoori, and Pushpak Bhattacharyya

Indian Institute of Technology Patna, Dayalpur Daulatpur, Bihar, India
{naveen.pcs16,sriparna,smansoori.cs15,pb}@iitp.ac.in

**Abstract.** In this paper, a neural-network-based unsupervised classification technique is proposed for summarizing a set of tweets where informative tweets are selected based on their importance. The approach works in two stages: in the first stage, the concept of a self-organizing map (SOM) is utilized to reduce the number of tweets. In the second stage, a granular self-organizing map (GSOM), which is a 2-layer feed-forward neural network utilizing the fuzzy rough set theory for its training, is considered for clustering the reduced set of tweets. Then, a fixed length summary is generated by selecting tweets from the obtained clusters. GSOM is having a set of parameters; proper selection of these parameter values influences the performance. Therefore an evolutionary optimization technique is utilized for the selection of the optimal parameter combinations. We have evaluated the efficacy of the proposed approach on four disaster-related microblog datasets. Results obtained clearly illustrate that our proposed method outperforms the state-of-the-art methods.

**Keywords:** Microblogs · Self-organizing map (SOM) · Granular SOM · Genetic algorithm · Extractive summarization · Parameter selection

## 1 Introduction

Nowadays, social networking and microblogging sites such as *Tumblr*, *Twitter*, have become popular due to its usage by a large number of users [1,2]. These services have changed the way people think, live, and communicate. Due to their ease of use, users are allowed to share and discuss any interesting topics such as political issues, recommendations, education, among others. Especially in the case of a disaster event scenario, microblogging sites become the important source of getting the current status of the affected location. Most of the posted tweets are conversational, including emotions and sympathy related; only 3.6% of the tweets convey relevant information. These tweets can be relevant to the disaster management authority for immediate decision making, but reading all such tweets is a very tedious task. Therefore, there is indeed a need to develop


H. Yang et al. (Eds.): ICONIP 2020, LNCS 12532, pp. 680–692, 2020.
https://doi.org/10.1007/978-3-030-63830-6_57

a technique named as multi-tweet summarization or microblog summarization that can extract the informative tweets from these relevant tweets.

***Related Work:*** In the literature, various algorithms are proposed for microblog summarization. Some examples of well-known algorithms are, LexRank [3], LUHN [4], cluster-rank [5], LSA [6], SumDSDR [7], MEAD [8], SumBasic [9], FreqSum [9]. The comparison between these algorithm is provided in the paper [10] on disaster-related tweets. To the best of our knowledge, there are three most recent methods in the literature which are COWTS [11], EnGraphSumm [12] and MOOTweetSumm [13]. COWTS utilizes the concept of integer linear programming to maximize the coverage of tweets having maximum number of content (numerals, verbs and noun) words. EnGraphSumm makes use of ensembling of different existing algorithms, to generate a single final summary. MOOTweet-Summ is the most recent method utilizing the concept of multi-objective evolutionary algorithms and simultaneously optimizes different aspects of the tweets. *EnGraphSum* method generates the final summary after considering the summaries generated by different existing algorithms like LUHN, LexRank, among others. The method, *MOOTweetSumm* [13] uses the multi-objective binary differential evolution (MOBDE) for the simultaneous optimization of various perspectives of the tweets. Note that MOBDE is computationally expensive due to the evaluation of different perspectives over the different iterations of the iterative procedure of MOBDE. *COWTS* extracts the tweets having maximum coverage of content words (verbs, nouns, and numerals) using the Integer Linear Programming.

After conducting a thorough literature survey, it has been found that neural network based architecture was never explored for solving the microblog summarization task. In [14], convolution neural network is considered, but for solving opinion based microblog summarization task which is different from our task. Moreover, CNN makes use of annotated data for its training and it is computationally expensive due to the involvement of hidden layers, convolution and pooling operations. In this direction, the use of self-organizing Map (SOM) [15] seems to be viable which is a special type of 2-layer (input and output) neural network and benefited from the advantage of unsupervised training (more details discussed in Sect. 2). Moreover, it has been successfully applied in solving problems of different area including video summarization, document/web-page clustering, data collection, wireless network, among others.

***Theme of the Paper:*** In this paper, we have posed the problem of microblog summarization as a partitioning/clustering problem where a given set of tweets is partitioned into several groups based on some similarity/dissimilarity measure. To represent the tweets, the well know word2vec [16] model is used. And for cluster analysis, we have utilized SOM. But in the paper by Ray et al. [17], SOM is not able to handle overlapping class boundaries. An example of the same is shown in Fig. 1 taken from the paper [17]. In part (a) of this figure, overlapping patterns are shown belonging to different classes. In Fig. 1(b), clusters obtained using SOM are shown; the input patterns wrongly assigned to different clusters are also shown. This drawback was removed by a granular self-organizing map

[17], which utilizes the fuzzy rough set theory [18] to assign patterns to different clusters. Note that the working principle of GSOM is the same as that of SOM. The part (c) of Fig. 1, shows the efficacy of GSOM in handling overlapping classes. But, GSOM takes more time than SOM due to the involvement of the fuzzy concept, and convergence speed is low (shown in Figure-1 of the supplementary sheet of GSOM paper [17] available at https://avatharamg.webs.com/GSOM-UFRFS.pdf). Therefore, to take advantage of both architectures, i.e., the higher convergence rate of SOM and capability of GSOM in handling overlapping patterns, in the current paper, we build a sequence of two 2-layer neural networks. For the first network, we utilize SOM, where the mapping of input tweets to different neurons (or also known as clusters) is performed, and then, representative tweets are selected from each cluster to get the reduced number of input tweets. Here GSOM is used as the second architecture where firstly, clustering using the reduced number of tweets is performed, and then, the summary is generated. The clusters obtained using GSOM are ranked and high-scoring tweets from each cluster are selected to generate the summary.

In GSOM, various parameters like the number of training iterations, the number of clusters (neurons), learning rate, neighborhood radius, are crucial; the usage of proper values of these parameters extremely influences the performance of GSOM. Identifying the optimal values of these parameters involves a thorough sensitivity analysis which is highly time-complex. Therefore in the current study, these parameter values are automatically selected utilizing the search capability of well-known genetic algorithm [19] which is a meta-heuristic based optimization strategy.

***Key-Contributions:*** The key-contributions of the paper are listed below:

1. The task of microblog summarization, which is having a great social impact, is solved in an unsupervised way. For this purpose, fusion of two 2-layer neural networks, i.e., SOM and GSOM, is performed. First one helps in reducing the number of tweets, while, second one is utilized for summary generation.
2. We have shown that optimal values of different parameters of GSOM selected in an automatic way utilizing the evolutionary procedure help in improving the performance.
3. The proposed method is generalized in nature and can be applicable in developing other summarization systems. We have shown the potentiality of our generalized framework on four disaster-related microblog data sets.

The rest of the paper is structured as follows: Section 2 discusses background knowledge. Problem formulation is provided in Sect. 3. The proposed methodology and experimental setup are discussed in Sect. 4 and 5, respectively. The discussion of results is covered in Sect. 6. Finally, Sect. 7 concludes the paper.

## 2 Background Knowledge

In this section we have discussed about some basic concepts which are used in the current paper.

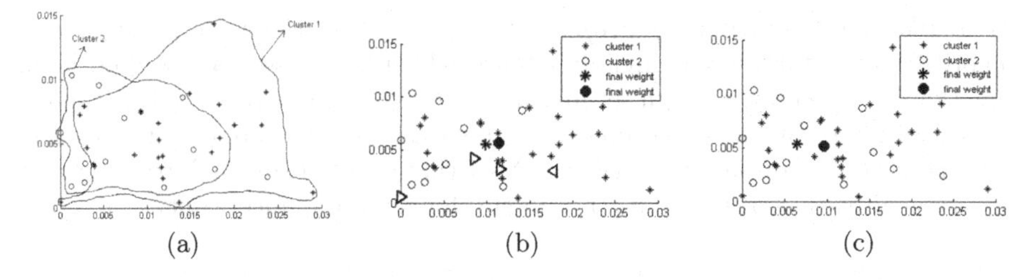


**Fig. 1.** Cluster analysis of SOM and GSOM [17]. Here, (a): Illustration of overlapping patterns belonging to actual clusters using 2-D plot; (b): Output clusters obtained after applying SOM with misclassified patterns. Here ▷ denotes the input patterns wrongly assigned to cluster 2 and ◁ denotes the input patterns wrongly assigned to cluster 1.; (c) Output clusters obtained after application of GSOM where no misclassification occurs.

***Self-organizing Map:*** SOM [15,20] is a well-known 2-layer special type of artificial feed-forward neural network and simulates the principle of feature mapping of input patterns/instances in the cerebral cortex of human brain. Noted that as opposed to other neural-networks like recurrent neural-network, convolution neural networks, it includes no hidden layer and performs its training in an unsupervised way. Each $u^th$ neuron (also referred as cluster) in the output grid, has two associated vectors: (a) position vector: $z^u = (z_1^u, z_2^u)$; (b) a weight vector: $w^u = (w_1^u.w_2^u, \ldots, w_n^u)$, having same length as that of the input vector (Lat say n-dimensional).

Its training is performed using the competitive and cooperative learning among neurons. At $t^{th}$ iteration, when an input is fed to the neuron's grid, all neurons fight with each other to become the winning neuron, but only one becomes the winning neuron (having the shortest distance from the input instance). And, then, *Gaussian function* is used to identify the neighboring neurons of the winning neuron using their position vectors (mathematically shown by Eq. 1).

$$h_{w,j}(t) = \exp\big(\frac{-d^2(z^w, z^j)}{2\sigma^2(t)}\big) \tag{1}$$

where, $w$ is the winning neuron index and $j$ is the index of neighboring neuron, $d$ is the distance between winning neuron and neighboring neuron, $t$ is the iteration number. Here, the neighboring neuron will be a neuron which has distance lesser than neighborhood radius ($\sigma$) at $t^{th}$ iteration. Thereafter, weight vectors of the winning neuron along with the neighboring neurons are updated so that they become close to the presented input instance. It is important to note that neighborhood radius and the learning rate keep on decreasing with the increase in the number of iterations in the iterative procedure of SOM. Main drawback of SOM arises due to the overlapping patterns belonging to different clusters [17] (also shown in Fig. 1) which was overcome by granular self-organizing map [17].

***Granular Self-organizing Map:*** GSOM [17] follows the same principle and algorithm as SOM with a single difference which is: it uses the neighborhood function based on fuzzy rough set instead of *Gaussian function.* The Eq. 2 shows the definition of this modified function. This was done to overcome the problem raised due to the overlapping patterns. Note that effectiveness of fuzzy rough set theory is already shown in the [21] and [18] for handling uncertainty.

$$h_{w,j}(t) = \exp^{-\left(\frac{L_{wj}^{avg} - B_{wj}^{avg}}{2}\right)} \tag{2}$$

Here, $w$ is the winning neuron index and $w = 1, 2, \ldots, K$ (number of neurons in GSOM); $B_{wj}^{avg}(A)$ is the boundary region of set A and represented by $U_{wj}^{avg}(A) - L_{wj}^{avg}(A)$, where, $U_{wj}(A)$ and $L_{wj}(A)$ denote the membership values belonging to upper and lower approximations of the patterns in set A obtained using rough set theory; $U_{wj}^{avg}(A)$ and $L_{wj}^{avg}(A)$ are the average of all membership values in upper and lower approximations of set A. For more details, one can refer to [17].

## 3 Problem Formulation

We have divided the task of microblog summarization in two phases as described below. Consider a dataset/event D consisting of $\mathcal{N}$ tweets, $\mathcal{D} = \{T_1, T_2, \ldots, T_{\mathcal{N}}\}$.

***First Phase:*** In the first phase, $\mathcal{K}$ representative tweets denoted as $\mathcal{R}_p^s = \{\mathcal{T}_{p1}, \mathcal{T}_{p2}, \ldots, \mathcal{T}_{p\mathcal{K}}\}$, are selected after application of SOM, where, $\mathcal{K}$ ($< \mathcal{N}$) is the number of neurons in SOM; each neuron is called as cluster. If $M_i$ is the number of mapping tweets in $ith$ cluster, then representative tweet, $\mathcal{T}_{pi}$, of the same cluster is calculated as

$$\mathcal{T}_{pi} = \operatorname*{argmin}_{j=1}^{M_i} \frac{\sum_{m=1, j\neq m}^{M_i} \mathcal{D}(T_j^i, T_m^i)}{\mathcal{M}_i - 1} \tag{3}$$

where, $i = \{1, 2, \ldots, \mathcal{K}\}$, $T_j^i$ and $T_m^i$ are the $jth$ and $mth$ tweet mapping to $ith$ cluster, $\mathcal{D}$ is the distance between two tweets. This step is performed as the pre-processing step, i.e., to reduce the number of tweets. Further, these tweets are used for GSOM training. The concept of single-objective genetic algorithm [19] is utilized for selecting optimal parameter values of GSOM.

***Second Phase:*** In this phase, our aim is to find a subset of tweets, $T \in \mathcal{R}_p^s$, utilizing the clusters obtained after application of GSOM, such that

$$max(\sum_{i=1}^{\mathcal{K}} \sum_{k=1}^{q} score(w_{k,T_{pi}})) \tag{4}$$

$$\text{such that} \quad \sum_{i=1}^{\mathcal{K}} Z_i = L \quad and \quad Z_i = \begin{cases} 1, & \text{if } T_{pi} \in \mathcal{R}_p^s \\ 0, & \text{otherwise} \end{cases} \tag{5}$$

where, $L$ is the maximum number of tweets in the summary, $q$ is the number of words in a particular tweet, $score(w_{k,T_{pi}})$ is the tf-idf score of the $kth$ word of $ith$ representative tweet, $T_{pi}$. To calculate tf-idf score, each tweet is considered as a bag of words, each word having it's own tf-idf score. Thus, tf-idf vector of a tweet $T_{pi}$ can be represented as

$$V_{T_{pi}} = [w_{1T_{pi}}, w_{2T_{pi}}, w_{3T_{pi}}, \ldots\ldots, w_{qT_{pi}}] \quad (6)$$

where

$$w_{k,T_{pi}} = tf_{(k,T_{pi})} \cdot \left(1 + \log \frac{1 + \mathcal{N}}{1 + \{t' \in \mathcal{D} | w_{k,T_{pi}} \in t'\}}\right) \quad (7)$$

and '$tf''_{k,T_{pi}}$ is calculated by counting the number of occurrences of $kth$ word in the same tweet, $T_{pi}$, $\mathcal{N}$ is the total number of tweets. Thus, summation of tf-idf scores of different words in the tweet is considered as the tweet tf-idf score.

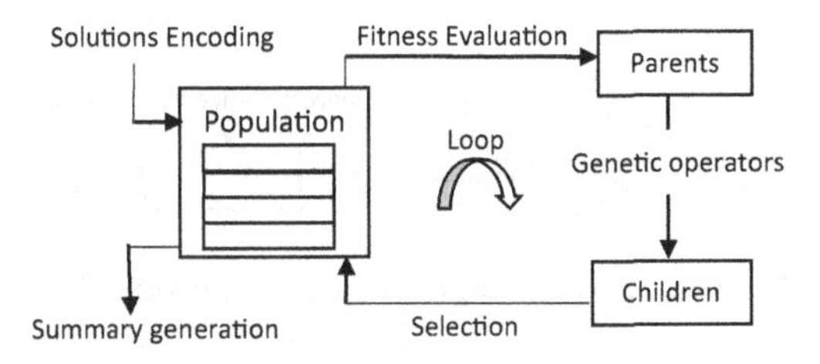


**Fig. 2.** Flowchart of genetic algorithm.

# 4 Methodology

For solving the microblog summarization task, we have utilized the neural-network-based clustering for which we have used the self-organizing map (SOM) and granular self-organizing map (GSOM). In the first step, SOM is trained using the tweet vectors of the tweets belonging to the original dataset. After training, each neuron of SOM acts as a cluster and tweets which are similar in semantic space are assigned to that neuron. The representative tweets ($\mathcal{R}_p^s$) from different clusters are selected using Eq. 3. This step acts as a sample selection algorithm; reducing the number of tweets.

Now, in the first phase of our algorithm, GSOM training is performed using these representative tweets, $\mathcal{R}_p^s$. Similar to SOM, the similar tweets (belonging to $\mathcal{R}_p^s$) are mapped to the same neuron in GSOM, and thus, each neuron can be considered as a different cluster. The training algorithm of GSOM is described in ref. [17]. Note that GSOM requires a set of input parameters like the number of neurons/clusters, the number of training iterations ($\#Iter$), learning rate ($\eta_0$), neighborhood radius ($\sigma_0$), which must be initialized before training. Therefore, to select the appropriate values of these parameters, we have utilized the concept of genetic algorithm [19] which is a population based evolutionary algorithm. The evolution process of genetic algorithm is shown in Fig. 2 and described below in relation with GSOM.

*1. Population Initialization:* The GA starts from a set of candidate solutions and each solution encodes the possible values of these parameters. For a solution, these parameters are randomly selected from their possible ranges written below: (a) learning rate ($\eta_0$): 0 to 1; (b) number of training iterations ($\#Iter$): 10 to 100; (c) number of neurons ($K$): 2 to $\sqrt{\mathcal{R}_p^s}$; (d) neighborhood radius ($\sigma_0$): Maximum number of neurons in row or column of output layer. The possible ranges of the shown parameters are suggested in the paper [17]. GSOM training will be performed in each solution using the associated parameters and tweet vectors of the tweets belonging to $\mathcal{R}_p^s$. An example of solution encoding is shown in Fig. 3 where four parameters are encoded, each having possible range.

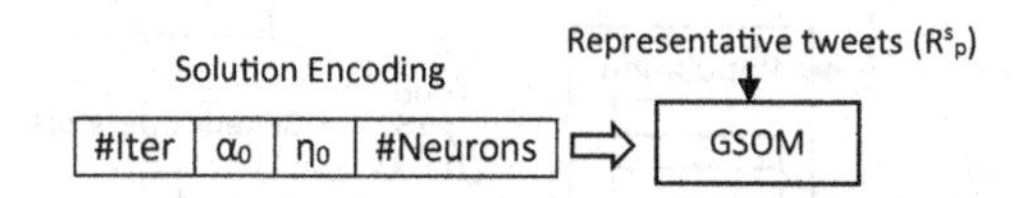


**Fig. 3.** An example of solution encoding.

*2. Fitness Evaluation:* For each solution, firstly the parameter values are extracted and GSOM is trained using these associated parameters. Neurons in GSOM represent various clusters. Therefore, to evaluate the quality of these clusters, we have used an internal cluster validity index namely, $\beta$-index [22]. It is the ratio of total variation and within cluster variation, and is defined as

$$\beta = \frac{\sum_{k=1}^{K}\sum_{j=1}^{M_i}(T_{ij}^v - \bar{C})}{\sum_{i=1}^{K}\sum_{j=1}^{M_i}(T_{ij}^v - C_i)} \tag{8}$$

where, $K$ is the number of neurons/clusters in GSOM, $\bar{C}$ is the average of all representative vectors, $C_i$ is the cluster center of $ith$ cluster/neuron, $T_{ij}^v$ is the tweet-vector of $jth$ tweet in $ith$ cluster. Thus, for each solution, $\beta$-index is evaluated. For good quality clusters, high values of $\beta$ are desired.

*3. Parent Selection:* Two solutions are randomly selected, and then, one with a high $\beta$ value will be the first parent. Repeat the same process until the second parent is decided. If population size is $PopSize$, then $PopSize$ pairs of parents are selected to generate new children/offsprings.

*4. Genetic Operators:* It includes crossover and mutation. First, we will perform crossover and then mutation. For crossover, a single-point crossover scheme is used, which further uses crossover probability ($CR$). If for a parent, $CR$ is greater than random probability generated then, components of both the parents are exchanged starting from the crossover point. Thus, for each pair of parents, two new solutions will be generated. The mutation is to make a small change in the new solution generated. For example, if the number of iterations in the new solution is 10, then it may be increased or decreased by 1. Similarly for (a) $\eta_0$, we can add or subtract 0.05; (b) for $K$, we can add/subtract 1. These changes are done based on some random probability. If the probability is less than 0.5, then we can add; otherwise, subtract. After generating a new solution, GSOM will be trained using the mentioned parameters followed by fitness function evaluation.

***5. Selection: Survival of the Fitness:*** In this step, all the new solutions are merged with the population, and then, they are sorted based on their fitness values. Only the top-best solutions (equals to population size) are selected to proceed for the next generation. Thus, the process of parent selection, genetic operators, and then selection continues until we reach the maximum number of generations.

***6. Summary Generation:*** In this step, the summary is formulated corresponding to the single-best solution selected from the final population generated at the end of the execution. GSOM provides clusters; therefore, from each cluster, relevant tweets are picked up until we get the desired length of the summary. To select the relevant tweets, there exist many measures like selecting tweet having (a) highest tf-idf score [13]; (b) maximum length [12], (c) maximum BM25 score [12], among others. Out of these, the second one is given higher importance than others [13]; therefore, in this paper, the same feature is used to select the tweet from each cluster. But, before this, the clusters should be ranked to select the sentences from them in a rank-wise manner. The following steps are executed for summary generation.

- Firstly, document center is calculated by averaging the tweet vectors as $\bar{C} = \sum_{i=1}^{|\mathcal{R}_p^s|} \mathcal{T}_{pi}^v / |\mathcal{R}_p^s|$.
- Compute distance of each cluster center with respect to document center as $\mathbb{D}_i = \| \mathcal{C}_i - \bar{C} \|$, where, $i$: cluster center index; $\mathcal{C}_i$: $ith$ cluster center; $\bar{C}$: dataset cluster center; $\mathcal{T}_{pi}^v$ is the tweet vector of $ith$ representative tweet belonging to set $\mathcal{R}_p^s$ (obtained using SOM).
- Cluster having maximum distance is assigned a higher priority, i.e., rank-1, assuming that it is different from other clusters. Similarly, ranks are assigned to other clusters based on distances computed in the above step.
- tf-idf scores of different tweets in the cluster are computed (tf-idf score of a tweet is calculated by summing up the tf-idf scores of different words in the tweet.
- Select high-scoring tweet from each cluster according to the ranks (rank-1 is the higher rank) until the desired length of the summary is reached.

# 5 Experimental Setup

## 5.1 Datasets

For experimentation, we use four datasets related to various disaster events. These datasets are (a) Sandy Hook elementary school shooting in the USA; (b) Typhoon Hangupit in the Philippines; (c) Flood in Uttarakhand, India; and (d) Bomb blasts in Hyderabad, India. These datasets have 2080, 1461, 2069, and 1413 number of tweets, respectively and provided with gold summaries [12] containing 37, 41, 34, and, 33 number of tweets, respectively. It is important to note that before utilizing these datasets, we have performed various pre-processing steps including removal of (a) stop words; (b) special characters; (c) URLs; (d) user mentions; and (e) hashtags. Moreover, all the words in the tweets are converted to lower case.

## 5.2 Evaluation Measure

For evaluating the performance of our generated summary, we have used ROUGE-N score which measures the overlapping words between our and gold summary. Here, the value of N is taken as 1, 2, and L (longest common subsequence) to provide ROUGE-1, ROUGE-2, and ROUGE-L, respectively. Higher the value of ROUGE score, more closer our summary is with respect to the gold summary.

## 5.3 Parameter Settings

To get the reduced number of tweets using SOM, the number of neurons for Sandyhook, Hagupit, Ukflood, Hblast datasets are considered as 400, 225, 144, and, 289, respectively, and same number of tweets are given as input to GSOM. The neurons in both the components are arranged in 1D. For SOM, the values of $\sigma_0$, $\eta_0$, and $\#Iter$ are considered as 1, 0.5, 100, respectively. These parameters are selected after conducting a thorough sensitivity analysis. While for GSOM, its parameters are adaptively selected using the genetic algorithm. In order to get a tweet vector (utilized in phase-1 of the algorithm), we have performed averaging of word vectors present in the tweet. These word vectors are obtained using a pre-trained *word2vec* model on large corpora of 53 million tweets available at http://crisisnlp.qcri.org/lrec2016/lrec2016.html. For GSOM training, we have utilized the code available at http://avatharamg.webs.com/software-code. Experiments are executed using Intel Core-i3 4th generation processor having Windows 10, CPU 1.7 GHz and 4 GB of RAM. For genetic algorithm, the number of generations and crossover probability are taken as 10 and 0.8, respectively.

### 5.4 Comparative Methods

For comparison, we have used the existing well-known algorithms including LSA [6], Lex-Rank [3], LUHN [4], MEAD [8], SumBasic [9], SumDSR [7]. In addition to these, some recent approaches namely, *EnGraphSumm* [12] and *MOOTweet-Summ* [13], are also utilized for comparison. In *EnGraphSumm*, there are many variants, but we have used top four versions, namely, VecSim-Community-maxSumTFIDF, VecSim-ConComp-MaxDeg, and VecSim-ConComp-MaxLen, VecSim-ConComp-maxSumTFIDF. To establish the potentiality of using GSOM over SOM in the second phase of the approach, we have also executed experiments after replacing GSOM by SOM, i.e., the reduced number of tweets obtained using the first SOM are given as input to the second SOM and then, the summary is generated. Along these approaches, some other approaches like, Lex-Rank [3], LSA [6], LUHN [4], SumBasic [9], MEAD [8] SumDSR [7], are also taken into account for the purpose of comparison.

## 6 Discussion of Results

In Table 1, the results obtained by our proposed approach are shown over individual datasets in terms of Rouge-1, Rouge-2, and Rouge-L measures. In this Table, the results obtained utilizing SOM in place of GSOM in the second phase, are also shown. As can be observed, GSOM performs better than SOM in terms of ROUGE scores for summary generation. As existing methods report the average results of all datasets; therefore, we have shown the average Rouge scores in comparison with the existing methods in Table 2. From Table 2, it is evident that our proposed method achieves gains of 9.7% and 2.8% over the best existing methods, i.e., MOOTweetSumm, in terms of Rouge-2 and Rouge-L measure, respectively. Note that the *MOOTweetSumm* method was based on multi-objective optimization, where different quality measures are simultaneously optimized to improve the quality of the summary while our method is based on a 2-layer unsupervised neural network. Even the method *COWTS*, which was based on the concept of integer linear programming, lacks behind by 93% and 12% by our proposed

**Table 1.** Results attained by our proposed architecture over individual datasets. Here, $N1$: Reduced number of tweets obtained using SOM. Bold entries indicate the best results.

| Dataset | N1 | GSOM | | | SOM | | |
|---|---|---|---|---|---|---|---|
| | | Rouge-1 | Rouge-2 | Rouge-L | Rouge-1 | Rouge-2 | Rouge-L |
| UkFlood | 144 | 0.5533 | 0.3471 | 0.5465 | 0.5125 | 0.2920 | 0.5057 |
| Sandyhook | 400 | 0.5427 | 0.3422 | 0.5274 | 0.5091 | 0.3404 | 0.5000 |
| HBlast | 289 | 0.5247 | 0.3956 | 0.5247 | 0.4659 | 0.3277 | 0.4635 |
| Hagupit | 225 | 0.4133 | 0.2985 | 0.4016 | 0.3821 | 0.2442 | 0.3684 |
| **Average** | **265** | **0.5085** | **0.3458** | **0.5000** | 0.4674 | 0.3011 | 0.4594 |

**Table 2.** Comparison of our ROUGE scores in comparison with state-of-the-art methods. The symbol † indicates that results are statistically significant at 5% significant level. The p-values obtained are shown in fourth and fifth column.

| Approach | Rouge-2 | Rouge-L | P-values | |
|---|---|---|---|---|
| Proposed | **0.3458**† | **0.5000**† | – | – |
| MOOTweetSumm | 0.3150 | 0.4860 | 1.094E−034 | 2.568E−017 |
| VecSim–ConComp–MaxLen | 0.1940 | 0.4506 | 7.660E−279 | 1.727E−073 |
| VecSim–ConComp–MaxDeg | 0.1919 | 0.4457 | 6.838E−283 | 3.733E−083 |
| VecSim–Community–maxSumTFIDF | 0.1898 | 0.4591 | 6.443E−287 | 1.599E−057 |
| VecSim–ConComp–maxSumTFIDF | 0.1886 | 0.4600 | 3.305E−289 | 6.672E−056 |
| ClusterRank (CR) | 0.0859 | 0.2684 | 0.00 | 0.00 |
| COWTS (CW) | 0.1790 | 0.4454 | 3.003E−307 | 9.358E−084 |
| FreqSum (FS) | 0.1473 | 0.3602 | 0.00 | 2.058E−262 |
| Lex-Rank (LR) | 0.0489 | 0.1525 | 0.00 | 0.00 |
| LSA (LS) | 0.1599 | 0.4234 | 0.00 | 9.994E−130 |
| LUHN (LH) | 0.1650 | 0.4015 | 0.00 | 7.159E−177 |
| Mead (MD) | 0.1172 | 0.3709 | 0.00 | 5.807E−241 |
| SumBasic (SB) | 0.1012 | 0.3289 | 0.00 | 1.007E−321 |
| SumDSDR (SM) | 0.0985 | 0.2602 | 0.00 | 0.00 |

method. Statistical significance t-test is also conducted at a significance level of 5% to validate our results. This t-test provides p-value. Lesser is the p-value; more significant is the obtained improvements. The p-values are also shown in the Table 2. It has been found that the obtained p-values are less than 5% and thus, evident that the improvements obtained are statistically significant.

The optimal values of different parameters for GSOM including the number of iterations, neighborhood radius, learning rate, the number of neurons/clusters considered, for different datasets are shwon below. These values are obtained using our genetic algorithm based procedure and the results shown in Table 1 are obtained using these parameter values for GSOM training. The quality of clusters is evaluated using $\beta$-index whose value is also shown along with these parameters values (fifth one): (a) UkFlood: 51, 11, 0.50, 11, 1.07; (b) Sandyhook: 26, 17, 0.06, 17, 1.04; (c) HBlast: 89, 5, 0.19, 5, 1.01; (d) Hagupit: 32, 5, 0.13, 1.02.

## 7 Conclusion

In the current paper, we have proposed a neural-network-based technique for solving the microblog summarization task. The advantage of the used neural network, i.e., Granular self-organizing Map, is two-fold: (a) no supervised information is required for training; (b) it is just a 2-layer neural network without any hidden layer and thus, requires less computation time. To select its optimal values of parameters, the concept of a genetic algorithm is utilized where solutions encode the possible values of parameters and evolve during the iterative procedure. Efficacy of the proposed approach is tested on four disaster-related datasets, and we obtain 9.7% and 2.8% improvements over the best existing method in the literature, in terms of Rouge-2 and Rouge-L score, respectively. In the future, we would like to evaluate our proposed approach for solving the task of multilingual microblog summarization where tweets available are in different languages. Moreover, we would also like to investigate the effect of BERT model in generating context-aware tweet vectors in our model.

**Acknowledgement.** Sriparna Saha would like to acknowledge the support of SERB WOMEN IN EXCELLENCE AWARD 2018 (SB/WEA-07/2017) of the Department of Science and Technology for carrying out this research.

## References

1. Rudra, K., Goyal, P., Ganguly, N., Mitra, P., Imran, M.: Identifying sub-events and summarizing disaster-related information from microblogs. In: The 41st International ACM SIGIR Conference on Research & Development in Information Retrieval, pp. 265–274 (2018)
2. Saini, N., Kumar, S., Saha, S., Bhattacharyya, P.: Mining graph-based features in multi-objective framework for microblog summarization. In: IEEE Congress on Evolutionary Computation (CEC), pp. 1–8. IEEE (2020)
3. Erkan, G., Radev, D.R.: LexRank: graph-based lexical centrality as salience in text summarization. J. Artif. Intell. Res. **22**, 457–479 (2004)
4. Luhn, H.P.: The automatic creation of literature abstracts. IBM J. Res. Dev. **2**(2), 159–165 (1958)
5. Garg, N., Favre, B., Reidhammer, K., Hakkani-Tür, D.: ClusterRank: a graph based method for meeting summarization. In: Tenth Annual Conference of the International Speech Communication Association (2009)
6. Blei, D.M., Ng, A.Y., Jordan, M.I.: Latent Dirichlet allocation. J. Mach. Learn. Res. **3**, 993–1022 (2003)
7. He, Z., et al.: Document summarization based on data reconstruction. In: Twenty-Sixth AAAI Conference on Artificial Intelligence, pp. 620–626 (2012)
8. Radev, D.R., Jing, H., Styś, M., Tam, D.: Centroid-based summarization of multiple documents. Inf. Process. Manag. **40**(6), 919–938 (2004)
9. Nenkova, A., Vanderwende, L.: The impact of frequency on summarization. Microsoft Research, Redmond, Washington, Technical report MSR-TR-2005 101 (2005)
10. Dutta, S., Chandra, V., Mehra, K., Ghatak, S., Das, A.K., Ghosh, S.: Summarizing microblogs during emergency events: a comparison of extractive summarization algorithms. In: Abraham, A., Dutta, P., Mandal, J., Bhattacharya, A., Dutta, S. (eds.) Emerging Technologies in Data Mining and Information Security, pp. 859–872. Springer, Singapore (2019). https://doi.org/10.1007/978-981-13-1498-8_76
11. Rudra, K., Ghosh, S., Ganguly, N., Goyal, P., Ghosh, S.: Extracting situational information from microblogs during disaster events: a classification-summarization approach. In: Proceedings of the 24th ACM International on Conference on Information and Knowledge Management, pp. 583–592 (2015)
12. Dutta, S., Chandra, V., Mehra, K., Das, A.K., Chakraborty, T., Ghosh, S.: Ensemble algorithms for microblog summarization. IEEE Intell. Syst. **33**(3), 4–14 (2018)

13. Saini, N., Saha, S., Bhattacharyya, P.: Multiobjective-based approach for microblog summarization. IEEE Trans. Comput. Soc. Syst. **6**(6), 1219–1231 (2019)
14. Li, Q., Jin, Z., Wang, C., Zeng, D.D.: Mining opinion summarizations using convolutional neural networks in Chinese microblogging systems. Knowl. Based Syst. **107**, 289–300 (2016)
15. Kohonen, T.: The self-organizing map. Neurocomputing **21**(1), 1–6 (1998)
16. Mikolov, T., Chen, K., Corrado, G., Dean, J.: Efficient estimation of word representations in vector space. arXiv preprint arXiv:1301.3781 (2013)
17. Ray, S.S., Ganivada, A., Pal, S.K.: A granular self-organizing map for clustering and gene selection in microarray data. IEEE Trans. Neural Netw. Learn. Syst. **27**(9), 1890–1906 (2016)
18. Ganivada, A., Ray, S.S., Pal, S.K.: Fuzzy rough granular self-organizing map and fuzzy rough entropy. Theor. Comput. Sci. **466**, 37–63 (2012)
19. Deb, K., Pratap, A., Agarwal, S., Meyarivan, T.: A fast and elitist multiobjective genetic algorithm: NSGA-II. IEEE Trans. Evol. Comput. **6**(2), 182–197 (2002)
20. Saini, N., Saha, S., Bhattacharyya, P.: Cascaded SOM: an improved technique for automatic email classification. In: International Joint Conference on Neural Networks (IJCNN), pp. 1–8. IEEE (2018)
21. Ganivada, A., Dutta, S., Pal, S.K.: Fuzzy rough granular neural networks, fuzzy granules, and classification. Theor. Comput. Sci. **412**(42), 5834–5853 (2011)
22. Pal, S.K., Ghosh, A., Shankar, B.U.: Segmentation of remotely sensed images with fuzzy thresholding, and quantitative evaluation. Int. J. Remote Sens. **21**(11), 2269–2300 (2000)

# CARU: A Content-Adaptive Recurrent Unit for the Transition of Hidden State in NLP

Ka-Hou Chan[1,2(✉)], Wei Ke[1,2], and Sio-Kei Im[2]

[1] School of Applied Sciences, Macao Polytechnic Institute, Macao, China
{chankahou,wke}@ipm.edu.mo
[2] Macao Polytechnic Institute, Macao, China
marcusim@ipm.edu.mo

**Abstract.** This article introduces a novel RNN unit inspired by GRU, namely the Content-Adaptive Recurrent Unit (CARU). The design of CARU contains all the features of GRU but requires fewer training parameters. We make use of the concept of weights in our design to analyze the transition of hidden states. At the same time, we also describe how the content adaptive gate handles the received words and alleviates the long-term dependence problem. As a result, the unit can improve the accuracy of the experiments, and the results show that CARU not only has better performance than GRU, but also produces faster training. Moreover, the proposed unit is general and can be applied to all RNN related neural network models.

**Keywords:** Recurrent neural network · Gate recurrent unit · Long-Short Term Memory · Content-adaptive · Natural Language Processing

## 1 Introduction

With the rapid development of the Internet and social networks, people can easily view and share information with each others to express their thoughts across the digital networks. Unlike face-to-face conversations, the presence of intermediary media provides a way to collect those typed sentences and text which can be further analyzed in the Natural Language Processing (NLP) tasks and human behavior research. However, the structure of social network data is complex and diverse, and usually contains some cultural information related to the living environment. How to query and confirm relevant and meaningful information from the perceptual data (within image and language) has important research significance.

For the natural language in sentences analysis, there is unstructured characteristics of sequence data. A language expression has diversity and irregularity, and the complexity of semantics makes it difficult for computers to extract the real intention and implication of the expression. NLP researchers aim to gather


H. Yang et al. (Eds.): ICONIP 2020, LNCS 12532, pp. 693–703, 2020.
https://doi.org/10.1007/978-3-030-63830-6_58

knowledge about how human beings understand and use languages in order to develop appropriate methods and techniques to allow computer systems to understand and interpret natural languages to perform further tasks. However, traditional NLP methods require a lot of feature extraction and expression in a sentence [13,28]. How to learn the meaning of perceptual data to determine the morphological structure and the grammar of an entire sentence, to express the natural information of text, and to improve the performance of the emotional classification are problems to be solved in text analysis. However, the development of cross-domain (linguistics and computer science) methods will always lag behind the evolution of modern languages.

In recent years, with the development of machine learning for NLP [7], through the neural network models, expressions can be studied through a large number of training parameters without manual design. This is convenient in practice with theory and evidence [1,23]. The neural networks can approximate any unknown expressions through a combination of nonlinear functions and specific layers [21], and these NLP problems can be well studied in the context of recurrent neural networks (RNNs) [14,16,26]. Exploring the RNN models, it can be found that there are some layers being used repeatedly, with each layer consisting of various RNN units. The Elman network [6] is also known as a simple recurrent networks model whose units perform linear operations. Specifically, the sequence of tensor $\left\{v_{m\times 1}^{(1)}, v_{m\times 1}^{(2)}, \dots, v_{m\times 1}^{(l)}\right\}$, each of size $(m \times 1)$ as features, is input to the RNN, then produces the last hidden state $h_{n\times 1}^{(l)}$, and the hidden state results:

$$h_{n\times 1}^{(t+1)} = \boldsymbol{w}_{n\times n} h_{n\times 1}^{(t)} + \boldsymbol{w}_{n\times m} v_{m\times 1}^{(t)} \tag{1}$$

where $h_{n\times 1}^{(t)}$ is the $t$-th hidden state of size $(n \times 1)$. $\boldsymbol{w} = [\boldsymbol{w}_{n\times n} | \boldsymbol{w}_{n\times m}]$ is the weight for training use[1]. The input tensors can be projected onto

$$\left\{x_{n\times 1}^{(1)}, x_{n\times 1}^{(2)}, \dots, x_{n\times 1}^{(l)}\right\},$$

when $x_{n\times 1} = \boldsymbol{w}_{n\times m} v_{m\times 1}$. Therefore, the expansion form of (1) becomes

$$h_{n\times 1}^{(t+1)} = \boldsymbol{w}_{n\times n}^{\boldsymbol{t}} x_{n\times 1}^{(1)} + \boldsymbol{w}_{n\times n}^{\boldsymbol{t-1}} x_{n\times 1}^{(2)} + \cdots + \boldsymbol{w}_{n\times n}^{\boldsymbol{0}} x_{n\times 1}^{(t)} \tag{2}$$

where $\boldsymbol{w}_{n\times n}^{\boldsymbol{0}}$ is an identity matrix $I_n$. (2) can be seen as a system with radix $\boldsymbol{w}_{n\times n}$, which is a non-integer square matrix. The current hidden state $h^{(t)}$ can uniquely correspond to the sequence of the received tensors $v^{(\leq t)}$, implying their feature and order, thus RNNs can overcome the dynamic length data, allowing one to only consider the last hidden state.

[1] For the complete hidden state, (1) should include the bias parameter as $h_{n\times 1}^{(t+1)} = \boldsymbol{w}_{n\times n} h_{n\times 1}^{(t)} + \boldsymbol{b}_{n\times 1} + \boldsymbol{w}_{n\times m} v_{m\times 1}^{(t)} + \boldsymbol{b}_{n\times 1}$, and followed by a non-linear activation function $\tanh\left(h^{(t+1)}\right)$ that is to prevent divergence during training. To facilitate derivation, we ignore them in this section.

### 1.1 Related Work

In recent years, RNN has become the mainstream model for NLP tasks. It can be used to represent language expressions with different cultures through unified embedded vectors. Analysis of these word vectors through neural network models can obtain higher-level information. However, the (2) reflects that each input tensor has the same weight, but this feature conflicts with the importance of various words, since a word may have other meanings or different importance in various combinations. Especially the punctuation, preposition and postposition is often less important than other words, and the main information will always be diluted if the above model is used. Fortunately, this problem can be significantly alleviated by the Long-Short-Term-Memory (LSTM) networks [8,9,11] and its variant, known as the Gated Recurrent Unit (GRU) [3]. These are designed to avoid the long-term dependence of less important information. Instead of overwriting previous hidden states, they use various gate units to adjust the ratio of the updated state to alleviate word weighting conflicts. Later, some simplified gated units have been proposed but the performance has not been improved [10]. For the simple NLP problem, LSTM-like method can achieve better accuracy. Their hidden states can be used in classification problems for further purposes, such as parts-of-speech tagging [17,25] and sentiment analysis [20,27].

In addition, the last hidden state can also be regarded as an encoding result, which can be further decoded by another radix. In this case, the coding result must contain all words in the source sentence, but the weights are difference [12]. The dynamic coming along with the ratio of word weighting is well compatible by the Seq2Seq model [22]. Its primary components are one encoder network and one decoder network (both of them are RNNs), and the decoder reverses the encoding process [2]. Furthermore, ideas from RNN architectures that improve the ease of training, such as that the decoder accepts all hidden states which stores the entire context and allows the decoder to look at the input sequence selectively as with the attention mechanisms [24]. In view of these architectures, researchers can propose novel RNN units for various tasks. [4] introduces an Update Gate RNN (UGRNN) that is similar to the Minimal Gated Unit (MGU) [29] which is directly inspired by the GRU. Of course, their variants may work well in some cases, but there is no consensus on the best and comprehensive LSTM-like architecture.

In this work, we propose a new RNN unit inspired by GRU, that contains the linear update unit like (2) and the transition ratio of the hidden state will be completed by interpolation. In particular, we intend to remove the reset gate from the GRU and replace its function with our proposed adaptive gate. In our architecture, the GRU update gate and the proposed adaptive gate form the content adaptive gates, thus we name the proposed unit as the Content-Adaptive Recurrent Unit (CARU). Meanwhile, the proposed method has the smallest possible number of training parameters (equivalent to MGU) in GRU-like units, and has met the requirements of various aspects as much as possible.

## 2 Content-Adaptive Recurrent Unit

For the context of perceptual data, human beings can capture the meaning or patterns immediately if they have learned it before. Corresponding to NLP, for a standard sentence, its subject, verb and object should be paid more attention before considering other auxiliary information. This process can be seen as a weighted assignment of words in a sentence, which is related to the tagging task in NLP. However, weighting by using tagging is not suitable for non-standardized sequence data, such as phrases or casual chats. In order to overcome this weighting problem, we first analyze the architecture of GRU:

$$r^{(t)} = \sigma\left(\boldsymbol{w}_{\boldsymbol{hr}} h^{(t)} + \boldsymbol{b}_{\boldsymbol{hr}} + \boldsymbol{w}_{\boldsymbol{vr}} v^{(t)} + \boldsymbol{b}_{\boldsymbol{vr}}\right) \tag{3a}$$

$$n^{(t)} = \tanh\left(r^{(t)} \odot \left(\boldsymbol{w}_{\boldsymbol{hn}} h^{(t)} + \boldsymbol{b}_{\boldsymbol{hn}}\right) + \boldsymbol{w}_{\boldsymbol{vn}} v^{(t)} + \boldsymbol{b}_{\boldsymbol{vn}}\right) \tag{3b}$$

$$z^{(t)} = \sigma\left(\boldsymbol{w}_{\boldsymbol{hz}} h^{(t)} + \boldsymbol{b}_{\boldsymbol{hz}} + \boldsymbol{w}_{\boldsymbol{vz}} v^{(t)} + \boldsymbol{b}_{\boldsymbol{vz}}\right) \tag{3c}$$

$$h^{(t+1)} = \left(1 - z^{(t)}\right) \odot n^{(t)} + z^{(t)} \odot h^{(t)} \tag{3d}$$

where $r^{(t)}$ and $z^{(t)}$ is the result of reset gate and update gate, $n^{(t)}$ is a variant of the next hidden state. It presents that the current hidden state $h^{(t)}$ is processed by the Hadamard operation ($\odot$) with $r^{(t)}$ in (3b), that controls the amount of history used to update the candidate states. It would effectively become only dependent on the current word $v^{(t)}$ when $r^{(t)}$ is close to zero, thus the GRU reset gate has achieved the dynamic weighting of previous hidden states. Furthermore, the result of the GRU update gate in (3d) makes the gradual transition of the hidden states instead of overwriting previous states.

Nevertheless in (3a) and (3b), the GRU allows the result of the reset gate to weaken the hidden state, thereby alleviating the dependence on long-term content, but this approach can further dilutes the information in the entire sentence. It is still challenging for compound-complex sentence like relative-clauses. Therefore, our design does not intend to directly interfere with any current hidden state, as shown below briefly:

$$x^{(t)} = \boldsymbol{w}_{\boldsymbol{vn}} v^{(t)} + \boldsymbol{b}_{\boldsymbol{vn}} \tag{4a}$$

$$n^{(t)} = \tanh\left(\left(\boldsymbol{w}_{\boldsymbol{hn}} h^{(t)} + \boldsymbol{b}_{\boldsymbol{hn}}\right) + x^{(t)}\right) \tag{4b}$$

$$z^{(t)} = \sigma\left(\boldsymbol{w}_{\boldsymbol{hz}} h^{(t)} + \boldsymbol{b}_{\boldsymbol{hz}} + \boldsymbol{w}_{\boldsymbol{vz}} v^{(t)} + \boldsymbol{b}_{\boldsymbol{vz}}\right) \tag{4c}$$

$$l^{(t)} = \sigma\left(x^{(t)}\right) \odot z^{(t)} \tag{4d}$$

$$h^{(t+1)} = \left(1 - l^{(t)}\right) \odot h^{(t)} + l^{(t)} \odot n^{(t)} \tag{4e}$$

(4a) It first project the current word $v^{(t)}$ into $x^{(t)}$ as the input feature. This result would be used in the next hidden state and passed to the proposed content-adaptive gate.

(4b) The reset gate has been taken out. It just combines the parameters related to $h^{(t)}$ and $x^{(t)}$ to produce a new hidden state $n^{(t)}$. So far, it is the same as a simple RNN unit like (1).
(4c) It is the same as the update gate in GRU and is used to the transition of the hidden state.
(4d) There is Hadamard operator to combine the update gate with the weight of current feature. We name this gate the content-adaptive gate, which will influence the amount of gradual transition, rather than diluting the current hidden state (See more details below).
(4e) The next hidden state is combined with $h^{(t)}$ and $n^{(t)}$.

According to (4a) to (4e), the complete diagram of our proposed architecture is shown in Fig. 1.

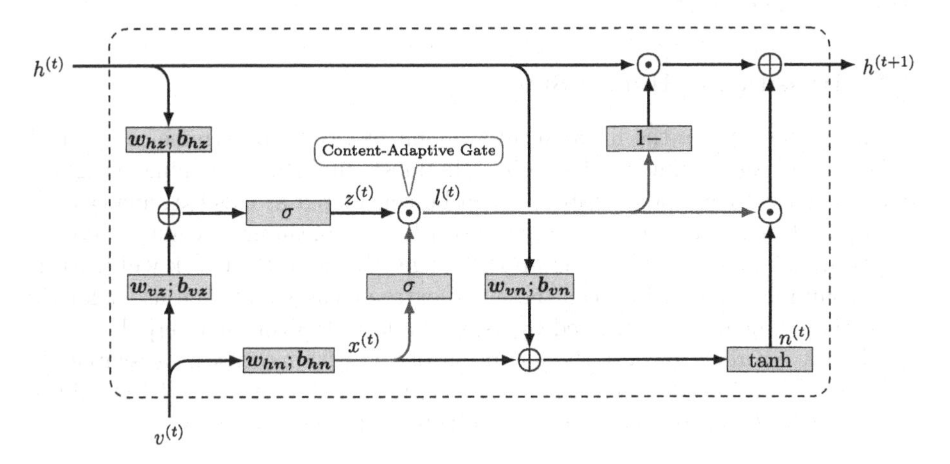


**Fig. 1.** The proposed CARU architecture. The direction of data flow is indicated by arrows, involved functions are indicated by yellow rectangles, and various gates (operations) are indicated by blue circles. Our approach had drawn in red color corresponding to (4a) and (4d). (Color figure online)

As illustrated in Fig. 1, functions like $(\boldsymbol{w}; \boldsymbol{b})$ is the linear function like (4a), and there are four sets of linear function training parameters. The tanh and $\sigma$ are the sigmoid activation function within range $(-1.0, +1.0)$ and $(0.0, 1.0)$ respectively, where tanh prevents divergence during training, and the result of $\sigma$ is used as the reference amount in the following operation.

### 2.1 Data Flow of Hidden State and Weight

As mentioned in (3b) from GRU, the hidden state is weighted by $r^{(t)}$ according to the current content. Inspired by this, our method further weights the hidden state according to the current word, and introduces a content-adaptive gate instead of using the reset gate to alleviate the dependence on long-term content. Thus we dispatch these processes into three trunk of data flows:

**content-state** It produces a new hidden state $n^{(t)}$ achieved by (4b), this part is equivalent to simple recurrent networks.

**word-weight** It produces the weight $\sigma\left(x^{(t)}\right)$ of the current word, it has the capability like a GRU reset gate but is only based on the current word instead of the entire content. More specifically, it can be considered as the tagging task that connects the relation between the weight and parts-of-speech.

**content-weight** It produces the weight $z^{(t)}$ of the current content, the form is the same as a GRU update gate but with the purpose to overcome the long-term dependence.

Comparing to GRU, we do not intend to process those data flow like (3b), but dispatch the word-weight to the proposed gate and multiply it by the content-weight. In such a way, the content-adaptive gate takes into account of both the word and the content.

## 2.2 Transition of Hidden State

As mentioned above, the linear recurrent unit as (2) is an extreme case, which reflects that each input word also has the most important meaning (weight is one), thus the new hidden state $n^{(t)}$ can be considered as this extreme case. In the opposite extreme case, the input word has no meaning (weight is zero) in the entire sentence. Once we clamp the range, the neural network will learn to find a suitable amount between these two extreme cases and then use it for the transition. This can be achieved by using the last step (linear interpolation) in GRU, but in order to meet the condition of the extreme cases, we switch the $n^{(t)}$ and $h^{(t)}$ from (3d) to (4e). For instance, the word-weight $\sigma\left(x^{(t)}\right)$ should be close to zero if the current word is a "full stop", implied as:

$$
\begin{aligned}
\sigma\left(x^{(t)}\right) \to 0 \implies & \sigma\left(x^{(t)}\right) \odot z^{(t)} = l^{(t)} \to 0 \\
\implies & \left(1 - l^{(t)}\right) \odot h^{(t)} + l^{(t)} \odot n^{(t)} = h^{(t+1)} \to h^{(t)}
\end{aligned}
$$

where no matter how the content-weight $z^{(t)}$ is, the result of the content-adaptive gate $l^{(t)}$ will be close to zero, meaning that the next hidden state $h^{(t+1)}$ will also tend to the current one $h^{(t)}$. It can be seen that low-weight words have less influence to the hidden state. Similarly, the long-term dependence problem can also be alleviated by the implication of the content-weight $z^{(t)}$.

Overall speaking, the proposed CARU can be summarized in two parts: The first part involves all the training parameters (used for linear function $(\boldsymbol{w};\boldsymbol{b})$), and focuses on the calculation of the state and weight, which can be done independently. The second part focuses on the transition of the hidden state, starting from the content-adaptive gate, it combines all the obtained weights and the current state then produces the next hidden state directly. Since CARU contains the factor and concept of word tagging, it can perform well in the work of NLP tasks.

# 3 Experimental Results

In this section, we evaluate the performance of CARU for NLP tasks. There are IMDB [15] dataset and Multi30k [5] dataset for sentiment analysis and machine translation problems, respectively. Because the proposed unit has fewer parameters than GRU, the training period requires less time. We use a simple model as the baseline architecture (no additional layers or activation functions). In order to compare the proposed CARU with GRU and MGU, we have implemented our design and re-implemented the MGU unit by PyTorch, which is a scientific deep learning framework [18]. All our experiments were conducted on an NVIDIA GeForce GTX 1080 with 8.0 GB of video memory. We evaluated the performance on the same environment with and without the use of the proposed unit, and with the same configuration and the same number of neural nodes. We only changed the RNN unit without applying other optimization techniques.

## 3.1 Sentiment Analysis

The first experiment is about sentiment analysis of IMDB movie reviews, that contains 25,000 movie reviews for training, another 25,000 for testing. Our task is to determine the positive and negative comments according to the content of the paragraph. Since the sequence length provided is not fixed and the maximum length is of thousands, it can be handled by a simple recurrent model. In this way, we apply various RNN units to compare their accuracy on the test set. Moreover, we also make use of a pre-trained word embedding [19] in order to convert words to vectors as input feature of the RNN unit, and all units process a size of 100 input and 256 output. The batch size is 100 (with shuffling), and we use the Adam optimizer and initialize the learning rate to 0.001. Similar to other classification problems, we apply fully connect layer to the last hidden state to classify movie reviews as positive or negative according to the probability calculated from a *Softmax* function, also we cooperate with cross-entropy function during training.

As indicated in Fig. 2, we find out the test accuracy after 100 epochs. Our proposed unit has improved the performance. In addition, we compare the training time of these units. We know that three sets of linear function parameters are required in GRU, but CARU and MGU requires two sets. Therefore we can roughly estimate that the time required for GRU is 1.5 times that of ours. Meanwhile, the convergence of our proposed unit is faster than the other units. It also performed with better stability in all experiments, and the difference of results is very small. This advantage in the training time and parameters is most obvious for the machine translation tasks.

## 3.2 Neural Machine Translation

The second experiment is the neural machine translation of Multi30k, that contains 30,000 images originally described in English, a collection of professionally translated German sentences, and another of crowdsourced German descriptions. Beyond other NLP tasks, there is no established answer in translation

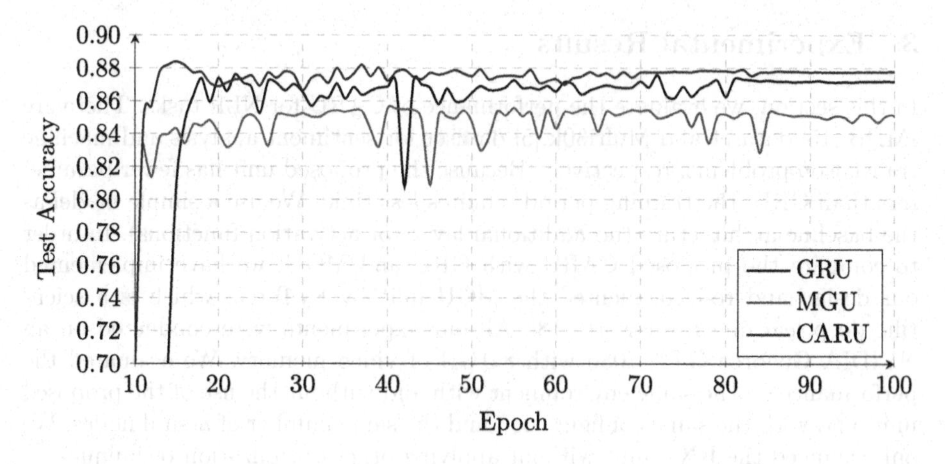


**Fig. 2.** Test accuracy rates for IMDB test set.

problem. It requires comprehension and representation of the source and target languages, respectively. For this task, we use the Seq2Seq model that can meet these requirements well, which is a type of Encoder-Decoder model using RNN. In fact, its Encoder and Decoder are two independent simple recursive models, which can directly correspond to comprehension and representation. Therefore, we can compare the performance of translation by using various RNN units (English to German, and reversed). Because there are two different languages, a huge vocabulary is required for training, all RNN units must increase the size to 256 input and 2048 output, other settings are the same as previous.

**Table 1.** BLEU-4 scores of translation results on the Multi30k development set.

| | GRU | MGU | CARU |
|---|---|---|---|
| *German to English* | $0.2793 \pm 0.0074$ | $0.2651 \pm 0.0173$ | $0.3019 \pm 0.0022$ |
| *English to German* | $0.2852 \pm 0.0042$ | $0.2892 \pm 0.0070$ | $0.3008 \pm 0.0020$ |

**Table 2.** BLEU-4 scores of translation results on the Multi30k test set.

| | GRU | MGU | CARU |
|---|---|---|---|
| *German to English* | $0.2921 \pm 0.0033$ | $0.2849 \pm 0.0184$ | $0.3199 \pm 0.0041$ |
| *English to German* | $0.3020 \pm 0.0033$ | $0.3123 \pm 0.0062$ | $0.3257 \pm 0.0015$ |

As indicated in Table 1 and Table 2, the scores by the CARU is found to improve the overall translation performance in terms of BLEU-4 scores with multiple runs. The most obvious improvement is German to English, by a gap of 0.03. As expected, both of CARU and MGU took about 20 to 25 min to complete a training epoch, and GRU took more than half an hour for this. Furthermore, no matter which RNN unit we use during training, the convergence often stagnates after completing several epochs. Whenever we encounter this issue, we will halve the learning rate and the convergence will start to improve again.

In all the experiments, our proposed unit performs well in sentence-based NLP tasks, but has no advantage in handling paragraphs or articles. The reason for this problem is that weighted words are active throughout the sequence, but when entering the entire paragraph, the active-weighted words may mislead other unrelated sentences. Fortunately, this issue can be alleviated by the sentence tokenization during pre-processing. In addition, our experiment also requires more pre-processing and post-processing, the reason for that is less relevant and we are not going into the detail. Please find the complete source code on following this link:

`github.com/ChanKaHou/CARU`

## 4 Conclusion

In this article, we propose a new RNN unit for NLP tasks. The proposed Content-Adaptive Recurrent Unit (CARU) clearly illustrates the content and weighted data flow and the gates used, making the model easier to design and understand. It requires fewer training parameters than GRU, which means it requires less memory and less training time. We compared CARU with MGU and GRU on two datasets that process sequence data in various NLP tasks. CARU achieved better results and faster training time than others.

However, all RNN units so far can only take into account the current and previous contents of sequence data. If we want to consider the following content to achieve complete sentence understanding, it can only be achieved by the network model, which is beyond the capabilities of the RNN unit. For the future work, the proposed unit has great potential for further improvement and analysis, such as answering questions and even being used in other non-NLP tasks or any neural network involving RNN units.

**Acknowledgment.** The article is part of the research project funded by The Science and Technology Development Fund, Macau SAR (File no. 0001/2018/AFJ).

## References

1. Bianchini, M., Scarselli, F.: On the complexity of neural network classifiers: a comparison between shallow and deep architectures. IEEE Trans. Neural Netw. Learn. Syst. **25**(8), 1553–1565 (2014)
2. Cho, K., van Merrienboer, B., Bahdanau, D., Bengio, Y.: On the properties of neural machine translation: encoder-decoder approaches. In: SSST@EMNLP, pp. 103–111. Association for Computational Linguistics (2014)

3. Cho, K., et al.: Learning phrase representations using RNN encoder-decoder for statistical machine translation. In: EMNLP, pp. 1724–1734. ACL (2014)
4. Collins, J., Sohl-Dickstein, J., Sussillo, D.: Capacity and trainability in recurrent neural networks. In: ICLR (Poster). OpenReview.net (2017)
5. Elliott, D., Frank, S., Sima'an, K., Specia, L.: Multi30k: multilingual English-German image descriptions. In: VL@ACL. The Association for Computer Linguistics (2016)
6. Elman, J.L.: Finding structure in time. Cogn. Sci. **14**(2), 179–211 (1990)
7. François, T., Miltsakaki, E.: Do NLP and machine learning improve traditional readability formulas? In: PITR@NAACL-HLT, pp. 49–57. Association for Computational Linguistics (2012)
8. Gers, F.A., Schmidhuber, J.: LSTM recurrent networks learn simple context-free and context-sensitive languages. IEEE Trans. Neural Netw. **12**(6), 1333–1340 (2001)
9. Gers, F.A., Schmidhuber, J., Cummins, F.A.: Learning to forget: continual prediction with LSTM. Neural Comput. **12**(10), 2451–2471 (2000)
10. Heck, J.C., Salem, F.M.: Simplified minimal gated unit variations for recurrent neural networks. In: MWSCAS, pp. 1593–1596. IEEE (2017)
11. Hochreiter, S., Schmidhuber, J.: Long short-term memory. Neural Comput. **9**(8), 1735–1780 (1997)
12. Kalchbrenner, N., Blunsom, P.: Recurrent continuous translation models. In: EMNLP, pp. 1700–1709. ACL (2013)
13. Kim, S., Seo, H., Rim, H.: Information retrieval using word senses: root sense tagging approach. In: SIGIR, pp. 258–265. ACM (2004)
14. Lopez, M.M., Kalita, J.: Deep learning applied to NLP. CoRR abs/1703.03091 (2017)
15. Maas, A.L., Daly, R.E., Pham, P.T., Huang, D., Ng, A.Y., Potts, C.: Learning word vectors for sentiment analysis. In: ACL, pp. 142–150. The Association for Computer Linguistics (2011)
16. Mikolov, T., Kombrink, S., Burget, L., Cernocký, J., Khudanpur, S.: Extensions of recurrent neural network language model. In: ICASSP, pp. 5528–5531. IEEE (2011)
17. Nguyen, D.Q., Dras, M., Johnson, M.: A novel neural network model for joint POS tagging and graph-based dependency parsing. In: CoNLL Shared Task (2), pp. 134–142. Association for Computational Linguistics (2017)
18. Paszke, A., et al.: Automatic differentiation in Pytorch (2017)
19. Pennington, J., Socher, R., Manning, C.D.: Glove: global vectors for word representation. In: EMNLP, pp. 1532–1543. ACL (2014)
20. Poria, S., Cambria, E., Gelbukh, A.F.: Aspect extraction for opinion mining with a deep convolutional neural network. Knowl. Based Syst. **108**, 42–49 (2016)
21. Rumelhart, D.E., Hinton, G.E., Williams, R.J.: Learning representations by back-propagating errors. Nature **323**(6088), 533–536 (1986). https://doi.org/10.1038/323533a0
22. Sutskever, I., Vinyals, O., Le, Q.V.: Sequence to sequence learning with neural networks. In: NIPS, pp. 3104–3112 (2014)
23. Szegedy, C., et al.: Going deeper with convolutions. In: CVPR, pp. 1–9. IEEE Computer Society (2015)
24. Vaswani, A., et al.: Attention is all you need. In: NIPS, pp. 5998–6008 (2017)
25. Wang, P., Qian, Y., Soong, F.K., He, L., Zhao, H.: A unified tagging solution: bidirectional LSTM recurrent neural network with word embedding. CoRR abs/1511.00215 (2015)

26. Werbos, P.J.: Generalization of backpropagation with application to a recurrent gas market model. Neural Netw. **1**(4), 339–356 (1988)
27. Zhang, L., Wang, S., Liu, B.: Deep learning for sentiment analysis: a survey. Wiley Interdiscip. Rev. Data Min. Knowl. Discov. **8**(4) (2018)
28. Zhang, Y., Clark, S.: Joint word segmentation and POS tagging using a single perceptron. In: ACL, pp. 888–896. The Association for Computer Linguistics (2008)
29. Zhou, G.-B., Wu, J., Zhang, C.-L., Zhou, Z.-H.: Minimal gated unit for recurrent neural networks. Int. J. Autom. Comput. **13**(3), 226–234 (2016). https://doi.org/10.1007/s11633-016-1006-2

# Coarse-to-Fine Attention Network via Opinion Approximate Representation for Aspect-Level Sentiment Classification

Wei Chen[1], Wenxin Yu[2](✉), Gang He[3], Ning Jiang[2], and Gang He[2]

[1] College of Computer Science, Chongqing University, Chongqing, China
cwei_01@163.com
[2] Southwest University of Science and Technology, Mianyang, China
yuwenxin@swust.edu.cn
[3] Xidian University, Xian, Shanxi, China

**Abstract.** Aspect-level sentiment classification aims to determine the sentiment polarity of the aspect that occurs in the sentence. Some prior methods regard the aspect as the attention goal and learn the association between aspect and context directly. Although improved results are achieved, the pure representation of aspect can not effectively reflect the differences of multiple sentences and will limit the improvement of sentiment classification. To address this issue, we propose a coarse-to-fine attention network via opinion approximate representation, which first extracts the neighboring words of aspect to approximate the opinion representation that contains the descriptive information of the aspect. Moreover, we design the coarse-to-fine attention module to complete the interaction of useful information from the sentence and word level, the former learns the rough representation of the context, while the latter learns the precise representation of opinion. The experimental results on three public datasets show our model achieves the state-of-the-art performance.

**Keywords:** Natural Language Processing · Sentiment classification · Context representation · Attention network

## 1 Introduction

Aspect level sentiment classification is a fundamental task in Natural Language Processing (NLP) [11,19,23], it aims to infer the sentiment polarity of sentence with respect to the aspects, which includes positive, negative and neutral. For example, there is a sentence "delicious food but the service was bad", the polarity of service is negative while the polarity of food is positive.

Since the sentiment polarity classification of an aspect needs to consider both the aspect and the context, the key is to characterize the relationship between the aspect and the context [24]. Some previous works extract aspect-related features to build sentiment classifiers in traditional machine learning methods [8], or


H. Yang et al. (Eds.): ICONIP 2020, LNCS 12532, pp. 704–715, 2020.
https://doi.org/10.1007/978-3-030-63830-6_59

to establish aspect-specific recursive structure used for the input in Recursive Neural Network methods [1,13,19]. However, these models treat the aspect as the attention goal directly, and cannot efficiently identify the difference of the same aspect existing multiple sentences. For example, there are two sentences, "delicious food, i like it." and "the food is too salty". The aspect word "food" appears in two sentences, but the sentiment polarities are opposite, if we use "food" as one of the interactive objects directly, it may cause insufficient use of aspect information.

From the perspective of language expression, we observe that when people express their sentiment towards a specific aspect, they tend to use adjectives or adverbs to describe the aspect and express their inner feelings, in addition, these descriptive words always appear near the aspect words. Consider the sentence "delicious food, but the service was bad", the aspect word is "food". If we want to judge the sentiment information, according to the previous solution, first we should establish the relationship between aspect and the context, then get the opinion information word "delicious", and lastly obtain the sentiment polarity through the opinion word. Although obtain the sentiment polarity, the pure aspect representation will limit the full use of contextual information, furthermore, the process of judging polarity is a little complicated, and it is easy to get wrong results.

Attention mechanism, is an effective method and successfully applied to machine translation [12] and reading comprehension [6]. Some works exploit this advantage and achieve superior performance for aspect-level sentiment classification, i.e. [1,11,13,22], but there are still some disadvantages shared by previous methods. For the aspect that contains multiple words, they simply use the average pooling to represent the aspect semantic, take the aspect "dessert food" as an example, intuitively, the word "food" is more import for the judgment of sentiment polarity than word "dessert", hence the aspect representations are not accurate enough and may cause the loss of the semantic information in this way. In addition, they calculate single attention scores of context/aspect to aspect/context, which is not fully complete the interaction of useful information.

Motivated by the above intuition, we propose a coarse-to-fine attention network via opinion approximate representation, we first extract the opinion approximate representation that contains the descriptive information of the aspect, take the sentence "most importantly, it is reasonably priced." as the example, the aspect is "priced", if we take a word before and after it (means k equals 1), so the opinion becomes "reasonably priced". In the proposed model, the coarse-to-fine attention module is to complete the interaction of useful information from sentence and word level, the former learns the rough representation of the context, while the latter learns the precise representation of opinion. The experimental results show our model can effectively predict the polarity of the given aspect sentence and reach the highest level.

## 2 Related Work

Aspect-level sentiment analysis is designed to determine the sentiment polarity of the sentence for the given aspect. Traditional approaches to solve the problem are to manually design the features and most of them focus on building sentiment classifiers with features [8]. However, the results highly depend on the quality of these features and the feature engineering is labor intensive.

Later, neural networks have been employed to learn useful features automatically, achieving better results on this task. [19] introduced the TD-LSTM approach which learns the feature representation from the leftmost and rightmost sides of the sentence. Most of the neural network models often suffer from the semantic mismatching problems.

Currently, attention mechanism has been successfully applied to the aspect-level sentiment analysis. [22] proposed the attention based LSTM with aspect embedding (ATAE-LSTM), which firstly applied attention mechanism to aspect level sentiment analysis by simply concatenating the aspect vector into the sentence hidden representations and achieving good performance. [20] developed a deep memory network based on a multi-hop attention mechanism (Mem-Net), which introduced the position information into the hidden layer. [13] proposed the interactive attention networks (IAN) to interactively learn attention weights in the contexts and targets. But the above methods calculate single attention scores of context/aspect to aspect/context, which is not fully complete the interaction of useful information.

Some researchers began to attempt introduction opinion into this researches [7,21]. But their work is implicitly expressing the relationship between the context and opinion. Our work firstly shows the connections between context and opinion in the aspect-level classification.

Compared to the above methods, we first extract the opinion approximate representation that contains the descriptive information of the aspect. and then automatically extract important parts of context. Moreover, this is the first work to explore the opinion-context interactions.

## 3 Model

In this section, we present the overall architecture of our model in Fig. 1. It mainly consists of the contextual module, coarse-to-fine attention module and classification module.

### 3.1 Contextual Module

**Embedding Layer.** The embedding layer contains of three components: word embedding, position embedding and binary indicator embedding.

Given a sentence $w = [w_1, w_2 ... w_i ... w_n]$ where $n$ is the sentence length and the opinion approximate representation (abbreviated as opinion) $t = [t_i, t_{i+1} ... t_{i+m-1}]$ where $m$ is the opinion length. Each word would be mapped

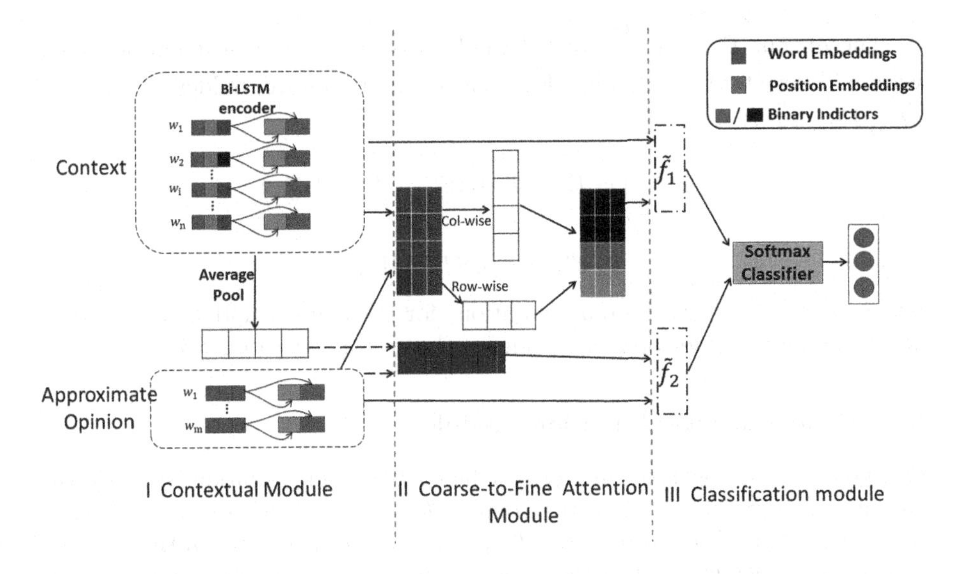


**Fig. 1.** The framework of coarse-to-fine attention network.

from its one hot vector into its dense embedding vector by the unsupervised method such as Glove [18] or CBOW [16], which outputs $[v_1; v_2; ...; v_n] \in \mathrm{R}^{\mathrm{n} \times \mathrm{d}_1}$ and $[v_i; v_{i+1}; ...; v_{i+m-1}] \in \mathrm{R}^{\mathrm{m} \times \mathrm{d}_1}$ as two embedding matrixes $M^{v \times d_1}$ where $d_1$ is the dimension of word embedding and $v$ is the vocabulary size.

The closer the word is to the aspect, the more important the word is for judging the sentiment polarity. Based upon this, we use the relative distance of each word from aspect word as its index and suppose the index's length equal to the length of the context. If a word occurs in the aspect, then the index will be marked as zero. For example, given a sentence "not only was the food outstanding but the little perks were great.", and the aspect word is "food", so the position index sequence is represented as "4, 3, 2, 1, 0, 1, 2, 3, 4, 5, 6, 7". And its corresponding position embedding are obtained by looking up a position embedding matrix $N \in \mathrm{R}^{\mathrm{n} \times \mathrm{d}_2}$, which is randomly initialized, and updated during the training process, the $d_2$ is the dimension of the position embedding and $n$ is length of the context.

Similar to position embedding, we use a binary feature to indicate whether each word is part of an aspect or not, each binary indicator is mapped to a vector representation using a randomly initialized embedding table.

**BiLSTM Layes.** RNN has already demonstrated its superior performance on sequences modeling, which can capture the long distance semantic information.

The Bi-directional Long Short-Term Memory (BiLSTM) is then used to capture the contextual information for each word. For the context, we combine the three parts of the embedding as input to the BiLSTM.

The output $E = [\overrightarrow{E}, \overleftarrow{E}]$ of BiLSTM encoder is the concatenation of forward hidden state $\overrightarrow{E} = (\overrightarrow{E}_1, \overrightarrow{E}_2 \ldots \overrightarrow{E}_j)$ and backward hidden state $\overleftarrow{E} = (\overleftarrow{E}_1, \overleftarrow{E}_2 \ldots \overleftarrow{E}_j)$:

$$E(x^c) = BiLSTM(W^c_{emb}; W^c_{pos}; W^c_{bin}) \quad (1)$$

$$E(x^o) = BiLSTM(W^o_{emb}) \quad (2)$$

where $E(x^c) \in R^{2d}$ is the representations for the context and $E(x^o) \in R^{2d}$ is the representations for opinion, ';' means the vectors concatenation.

### 3.2 Coarse-to-Fine Attention Module

Our goal is to strength the aspect-specific context representation, when getting the hidden representations of the context $E(x^c) = [e(w_1), e(w_2), \ldots, e(w_n)]$ and the opinion $E(x^o) = [e(t_1), e(t_2), \ldots, e(t_m)]$ respectively, in the proposed coarse-to-fine attention algorithm, one to learn the rough representation of the context in sentence-level, another to capture the correlation between context and opinion in word-level and obtain the final feature representation. The implementation of the coarse-to-fine attention algorithm will be described in detail in the following subsections.

**Coarse Attention Algorithm.** Because we use the neighboring words of the aspect word as the interaction object, the opinion word is composed of multiple words, obviously, each word contributes differently to the judgment of polarity. To steer the attention weights on the opinion words, first we use the average pooling mechanism to get the context average vector $e(w_{mean})$, and then compute the weights for each word in the opinion phrase:

$$S[e(t_i), e(w_{mean})] = e(t_i) \cdot w_1 \cdot e(w_{mean}) \quad (3)$$

$$\alpha_i = \frac{\exp S[e(t_i), e(w_{mean})]}{\sum_{k=1}^{m} \exp S[e(t_k), e(w_{mean})]} \quad (4)$$

where the $S$ function calculates the importance score of each context word to the opinion word and $t_i$ represents the opinion word. $w_1 \in R^{2d \times 2d}$ is a weight matrix that can be trained and $d$ is the dimension of hidden layer, and $\alpha_i \in R^{m \times 1}$, $m$ is the length of opinion phrase.

We compute the final weighted combination of opinion vector $m_c \in R^{2d}$ as follows:

$$m_c = \sum_{i=1}^{m} \alpha_i \cdot E(x^o) \quad (5)$$

**Fine Attention Algorithm.** When getting the hidden semantic representations of the context and the opinion respectively, we use the dot product to obtain the pairwise correlation matrix $M$ as follows:

$$M = E\left(x^{c}\right)^{T} \cdot E\left(x^{o}\right) \in R^{n \times m} \tag{6}$$

where each element $M_{i,j}$ in the correlation matrix refers to the relevance between the $w_i$ in the context and the $t_j$ in the opinion.

The attention mechanism is used to automatically determine the importance of each word, rather than using simple heuristics such as summation or averaging to focus these words' attention on the final attention. Inspired by the AOA module in question answering [3]. Firstly, we perform the softmax operation on each line to obtain the attention scores of the opinion to the context:

$$\rho_{i,j} = \frac{\exp\left(M_{i,j}\right)}{\sum_{i} \exp\left(M_{i,j}\right)} \tag{7}$$

Similarly, with the softmax operation on each column, we can also obtain the attention vector of each context word to the opinion, which is calculated as:

$$\sigma_{i,j} = \frac{\exp\left(M_{i,j}\right)}{\sum_{j} \exp\left(M_{i,j}\right)} \tag{8}$$

Since the contribution of each word in the opinion to the context is different, so we design following mechanism to accurately obtain the opinion information representation using the $\rho$ and $\sigma$ and the final context feature representation $m_f \in R^{2d}$:

$$\gamma_i = \frac{\sum_{j=1}^{m}(\rho_{i,j} \odot \sigma_{i,j})}{m} \tag{9}$$

$$m_f = \sum_{i=1}^{n} E\left(x^{c}\right)^{T} \cdot \gamma_i \tag{10}$$

where $n$ is the sentence length, and $\gamma_i \in R^{n \times 1}$.

### 3.3 Classification Module

In the classification module, we concatenate the above two attention vectors as the final representation:

$$\tilde{m} = [m_c; m_f] \tag{11}$$

where $\tilde{m} \in R^{4d}$, and ';' representing the combination of feature vectors. The polarity probability is obtained by applying a Tanh activation, and is normalized with the softmax function as follows:

$$\hat{y} = softmax(w_1 \cdot \tanh(w_2 \cdot \tilde{m} + b_m)) \tag{12}$$

where $\hat{y}$ is the predicted sentiment polarity distribution, $w_2$ and $b_m$ are learned parameters. Suppose that a training corpus contains $N$ training samples $(x_i, y_i)$, we can train the entire networks by minimizing the following loss function:

$$\mathcal{L}(\hat{y}, y) = -\sum_{i=1}^{N}\sum_{j=1}^{C} y_i^j \log\left(\hat{y}_i^j\right) + \lambda\left(\sum_{\theta\in\Theta}\theta^2\right) \tag{13}$$

where $y_i^j$ is the ground truth sentiment polarity, $C$ is the number of sentiment polarity categories, $\hat{y}_i^j$ denotes the predicted sentiment probabilities, $\theta$ represents each parameter to be regularized, $\Theta$ is a collection of all parameters, $\lambda$ is the weight coefficient for $L_2$ regularization.

# 4 Experiments

## 4.1 Datasets and Settings

We conduct experiments on three benchmark datasets as shown in Table 1. Restaurant is the union set of the restaurant domain from SemEval2014 [14], Laptop contains the reviews of the laptop domain from SemEval Challenge 2014 [14]. Twitter is built by [17] consisting of twitter post. They were widely used in previous works [2,10,11,23].

In our experiments, the pre-trained 300-dimensional Glove word vectors are initialized for our experiments, all out-of-vocabulary words are initialized as zero vectors. The dimensions of LSTM hidden layer are set to 300. The dimensions of word embedding, position embedding and binary embedding are set to 150, 150 and 150 respectively. Set the learning rate to 0.01, dropout rate to 0.5 and regularization to 0.001.

**Table 1.** Details of the experimental datasets.

| Dataset | Pos. | Neu. | Neg. | Total |
|---|---|---|---|---|
| Res.-train | 2164 | 637 | 807 | 3608 |
| Res.-test | 728 | 196 | 196 | 1120 |
| Lap.-train | 994 | 464 | 870 | 2328 |
| Lap.-test | 341 | 169 | 128 | 638 |
| Twi.-train | 1561 | 3217 | 1560 | 6338 |
| Twi.-test | 173 | 346 | 173 | 692 |

**Table 2.** The performance (%) comparisons with different methods on three datasets, the results with * are retrieved from [1], the results with † are retrieved from [5] and the NA means the results are not available, the best performance are marked in bold.

| Methods | Laptop | | Restaurant | | Twitter | |
|---|---|---|---|---|---|---|
| | ACC | Macro-F1 | ACC | Macro-F1 | ACC | Macro-F1 |
| SVM-feature [9] | 70.5 | NA | 80.2 | NA | 63.4* | 63.3* |
| TD-LSTM [19] | 68.1 | 63.9 | 75.6 | 64.5 | 66.6* | 64.0* |
| RAM [1] | 72.1† | 68.4† | 78.5† | 68.5† | 69.4 | 67.3 |
| ATAE-LSTM [22] | 68.7 | 64.2 | 77.6 | 65.3 | 67.7 | 65.8 |
| IAN [13] | 72.1 | NA | 78.6 | NA | NA | NA |
| MGAN [4] | 75.4 | 72.5 | 81.3 | 71.9 | 72.5 | 70.8 |
| MemNet [20] | 68.9† | 62.8† | 76.9† | 66.4† | 68.5 | 66.9 |
| HSCN [10] | 76.1 | 72.5 | 77.8 | 70.2 | 69.6 | 66.1 |
| Our Model | **76.3** | **72.8** | **82.4** | **74.8** | **74.0** | **72.7** |
| | (74.4±1.22) | (70.0±1.7) | (80.8 ± 0.7) | (72.2±1.3) | (73.6±0.3) | (72.0±0.5) |

## 4.2 Main Results and Discussions

In experiments, the evaluation metrics are classification accuracy and Macro-averaged F1 [15]. We report the best results by running our model ten times. The mean and the standard deviation of our model are also shown in Table 2.

From the experimental results, we can observe that our model achieves the best performance among all these methods, it suggests that the carefully-designed learning network can be more effective than other models. The improvements presumably benefit from two aspects. Firstly, we extracts the opinion approximate representation that contains the descriptive information, which is helpful to directly judge the sentiment polarity of the aspect and simplified the process of polarity judgment. Secondly, we propose the coarse-to-fine attention algorithm to complete the interaction of useful information from sentence and word level, therefore, our model can achieve better performances compared to other methods.

Moreover, we noticed that our model has the most obvious improvement on Twitter, reaching 5% in terms of ACC. One reason is that his training data is very large, and several households are three times the Laptop.

## 4.3 Effects of Opinion Approximate Representation

In our model, we use the k-nearest neighbor words of the aspect to approximate the opinion information. To verify the validity of the opinion approximate representation, we verified seven sets of nearest neighbor k values for each dataset and the results are shown in Fig. 2.

One the one hand, we can be observed that taking aspect information as the attention goal directly which means the k equals zeros fails to achieve the best performance under seven different conditions in each dataset. And our model improves the performance significantly by 0.7% and 1.64% in F1 on Restaurant and Twitter, this indicates that exploiting the opinion information can improve the performances. One the other hand, we also find that with the proximity information increases, the classification performance does not continue to increase, the best performance is achieved within one word before and after the aspect word, it may be because the representation of a word before and after the aspect word contains rich description information about the aspect, and the contextual semantic information is not too complicated.

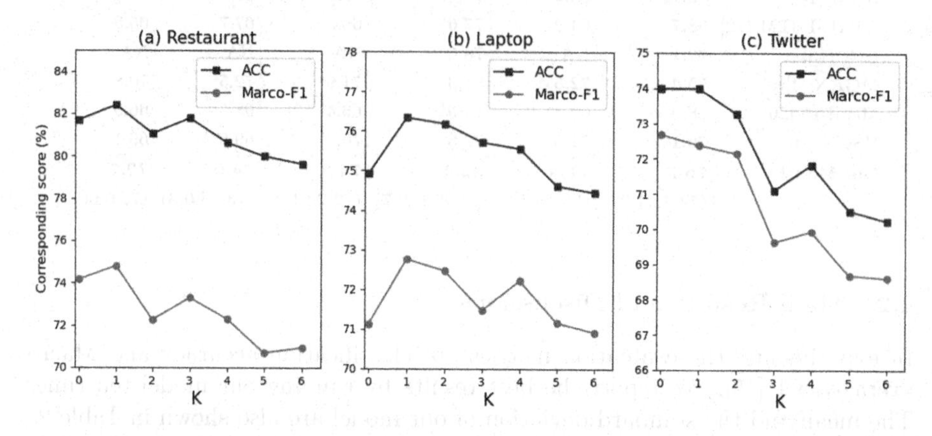


**Fig. 2.** The experimental results for the models with the different K (proximity of aspect words) on three datasets.

### 4.4 Effects of Coarse-to-Fine Attention Algorithm

To fully complete the interaction of useful information, we propose the coarse-to-fine attention n, one to learn the rough representation of the context, while the other learns the precise representation of opinion. To examine the effectiveness of the proposed module, we conduct extensive experiments, and results are shown in Table 3.

The coarse-attention method means that we only utilize the coarse attention module to train the model, and the fine-attention is similar to coarse-attention.

In the Table, we demonstrate that the interaction of coarse and fine attention will affect the classification result. When interacting with the two modules, the accuracy of model will increase 3% and 1% in Restaurant and Twitter respectively. Moreover, we also found that the Fine-attention method is significantly higher than the Coarse-attention method. We think there are the following reasons. Firstly, the coarse-attention learns the rough representation of context by average pooling directly in sentence-level, which may lose some semantic information. Secondly, for the opinion words, each word has different importance for the sentiment polarity judgment, the fine-attention accurately represent the opinion information through word-level interaction.

**Table 3.** The experiment results with different module on Restaurant and Twitter.

| Methods | Restaurant (%) | | Twitter (%) | |
|---|---|---|---|---|
| | ACC | Macro-F1 | ACC | Macro-F1 |
| Coarse-attention | 80.00 | 70.42 | 71.53 | 70.19 |
| Fine-attention | 82.05 | 73.87 | 73.48 | 72.51 |
| Coarse-to-Fine attention | **82.41** | **74.80** | **73.99** | **72.68** |

### 4.5 Case Study

We randomly select the sentence "most importantly, it is reasonably priced." from restaurant-test data to visualize the attention results in Fig. 3. The aspect word is "priced", we take a word before and after it (means k equals 1) as opinion information, so the opinion is "reasonably priced".

We can observe that our model can enforce the model to pay more attention to the important words. In terms of the coarse-attention, which learns the rough representation of the context and obtain the attention vector of context to opinion, it is obvious the word "reasonably" have higher attention weights compared with "priced". When the fine-attention module predicts the aspect "priced", the word "reasonably" gets more than 30% of the attention weights, which helps the sentiment polarity of the aspect "priced" to be correctly predicted to be positive. Hence, our model can automatically capture the attention information to assign the corresponding weights for each word and help it find out the sentiment.

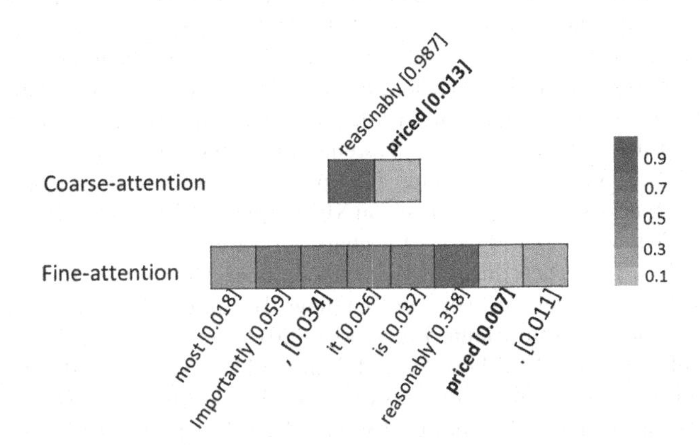


**Fig. 3.** Attention weights of two modules on sentence, and the color depth represents the weight value.

## 5 Conclusion

In this paper, we propose a coarse-to-fine attention network via opinion approximate representation, which first extracts neighboring words of the aspect to approximate the opinion representation which makes full use of the aspect information. Moreover, the coarse-to-fine attention module complete the interaction of context and opinion from the sentence and word level, the former learns the attention vector of the context to opinion, while the latter learns the attention weights of opinion to context. The experimental results on three public datasets show our model achieves the best results.

**Acknowledgement.** This research is supported by Sichuan Science and Technology Program (No. 2020YFS0307, No. 2019YFS0146, No. 2019YFS0155), National Natural Science Foundation of China (No. 61907009), Science and Technology Planning Project of Guangdong Province (No. 2019B010150002), Natural Science Foundation of Guangdong Province (No. 2018A030313802).

## References

1. Chen, P., Sun, Z., Bing, L., Yang, W.: Recurrent attention network on memory for aspect sentiment analysis. In: Proceedings of the 2017 Conference on Empirical Methods in Natural Language Processing, pp. 452–461 (2017)
2. Chen, W., et al.: Target-based attention model for aspect-level sentiment analysis. In: Gedeon, T., Wong, K.W., Lee, M. (eds.) ICONIP 2019. LNCS, vol. 11955, pp. 259–269. Springer, Cham (2019). https://doi.org/10.1007/978-3-030-36718-3_22
3. Cui, Y., Chen, Z., Wei, S., Wang, S., Liu, T., Hu, G.: Attention-over-attention neural networks for reading comprehension. arXiv preprint arXiv:1607.04423 (2016)
4. Fan, F., Feng, Y., Zhao, D.: Multi-grained attention network for aspect-level sentiment classification. In: Proceedings of the 2018 Conference on Empirical Methods in Natural Language Processing, pp. 3433–3442 (2018)
5. He, R., Lee, W.S., Ng, H.T., Dahlmeier, D.: Exploiting document knowledge for aspect-level sentiment classification. arXiv preprint arXiv:1806.04346 (2018)
6. Hermann, K.M., et al.: Teaching machines to read and comprehend. In: Advances in Neural Information Processing Systems, pp. 1693–1701 (2015)
7. Hu, M., Zhao, S., Guo, H., Cheng, R., Su, Z.: Learning to detect opinion snippet for aspect-based sentiment analysis. arXiv preprint arXiv:1909.11297 (2019)
8. Jiang, L., Yu, M., Zhou, M., Liu, X., Zhao, T.: Target-dependent twitter sentiment classification. In: Proceedings of the 49th Annual Meeting of the Association for Computational Linguistics: Human Language Technologies-Volume 1, pp. 151–160. Association for Computational Linguistics (2011)
9. Kiritchenko, S., Zhu, X., Cherry, C., Mohammad, S.: NRC-Canada-2014: detecting aspects and sentiment in customer reviews. In: Proceedings of the 8th International Workshop on Semantic Evaluation (SemEval 2014), pp. 437–442 (2014)
10. Lei, Z., Yang, Y., Yang, M., Zhao, W., Guo, J., Liu, Y.: A human-like semantic cognition network for aspect-level sentiment classification. In: Proceedings of the AAAI Conference on Artificial Intelligence, vol. 33, pp. 6650–6657 (2019)
11. Lin, P., Yang, M., Lai, J.: Deep mask memory network with semantic dependency and context moment for aspect level sentiment classification. In: Proceedings of the 28th International Joint Conference on Artificial Intelligence, pp. 5088–5094. AAAI Press (2019)
12. Luong, M.T., Pham, H., Manning, C.D.: Effective approaches to attention-based neural machine translation. arXiv preprint arXiv:1508.04025 (2015)

13. Ma, D., Li, S., Zhang, X., Wang, H.: Interactive attention networks for aspect-level sentiment classification. arXiv preprint arXiv:1709.00893 (2017)
14. Manandhar, S.: SemEval-2014 task 4: aspect based sentiment analysis. In: Proceedings of the 8th International Workshop on Semantic Evaluation (SemEval 2014) (2014)
15. Manning, C.D., Manning, C.D., Schütze, H.: Foundations of Statistical Natural Language Processing. MIT Press, Cambridge (1999)
16. Mikolov, T., Sutskever, I., Chen, K., Corrado, G.S., Dean, J.: Distributed representations of words and phrases and their compositionality. arXiv: Computation and Language (2013)
17. Mitchell, M., Aguilar, J., Wilson, T., Van Durme, B.: Open domain targeted sentiment. In: Proceedings of the 2013 Conference on Empirical Methods in Natural Language Processing, pp. 1643–1654 (2013)
18. Pennington, J., Socher, R., Manning, C.D.: Glove: Global vectors for word representation. In: Proceedings of the 2014 Conference on Empirical Methods in Natural Language Processing (EMNLP), pp. 1532–1543 (2014)
19. Tang, D., Qin, B., Feng, X., Liu, T.: Target-dependent sentiment classification with long short term memory. arXiv preprint arXiv:1512.01100 (2015)
20. Tang, D., Qin, B., Liu, T.: Aspect level sentiment classification with deep memory network. arXiv preprint arXiv:1605.08900 (2016)
21. Wang, B., Lu, W.: Learning latent opinions for aspect-level sentiment classification. In: Thirty-Second AAAI Conference on Artificial Intelligence (2018)
22. Wang, Y., Huang, M., Zhu, X., Zhao, L.: Attention-based LSTM for aspect-level sentiment classification. In: Proceedings of the 2016 Conference on Empirical Methods in Natural Language Processing, pp. 606–615 (2016)
23. Xue, W., Li, T.: Aspect based sentiment analysis with gated convolutional networks. arXiv preprint arXiv:1805.07043 (2018)
24. Zhang, M., Zhang, Y., Vo, D.T.: Gated neural networks for targeted sentiment analysis. In: Thirtieth AAAI Conference on Artificial Intelligence (2016)

# Deep Cardiovascular Disease Prediction with Risk Factors Powered Bi-attention

Yanlong Qiu, Zhichang Zhang(✉), Xiaohui Qin, and Shengxin Tao

Northwest Normal University, 967 Anning East Road, Lanzhou 730070, China
lankyqiu@163.com, zzc@nwnu.edu.cn, qinxh_qj@163.com, Taoshengxin11@163.com

**Abstract.** Cardiovascular disease (CVD) is one of the serious diseases endangering human life and health. Therefore, using the electronic medical record information to automatically predict CVD has important application value in intelligent auxiliary diagnosis and treatment, and is a hot issue in intelligent medical research. In recent years, attention mechanism utilized in natural language processing has focused heeds on a small part of the context and congregated it using fixed-size vectors, coupling attention in time, and/or often forming a unidirectional attention. In this paper, we propose a CVD risk factor powered bi-directional attention network named as RFPBiA, which is a multi-stage hierarchical architecture, fusing the information at different granularity levels, and employs the bi-directional attention to obtain the text representation of risk factors without early aggregation. The experimental results show that the proposed method can obviously improve the performance of CVD prediction, and the F-score reaches 0.9424, which is better than the existing related methods.

**Keywords:** Chinese electronic medical record · CVD risk factors extraction · CVD prediction · Bi-directional attention · Information fusion

## 1 Introduction

Cardiovascular disease (CVD) continues to be a leading cause of morbidity and mortality in the world [1]. According to data released by the World Health Organization, CVD is the number one cause of death worldwide, with more deaths from CVD each year than any other cause of death. In 2016, an estimated 17.9 million people died of CVD, accounting for 31% of all deaths worldwide. In its 2018 report, China's National Center for CVD noted that CVD mortality remained at the top of 2016, higher than cancer and other diseases, and the number of patients was as high as 290 million. As a chronic disease, CVD will not obviously show the corresponding characteristics in daily life in the hidden stage. However, once the symptoms are manifested, it may affect the life safety of the patient. Therefore, we want to help doctors make rapid diagnosis in time by analyzing the Electronic Medical Record (EMR) of patients during their daily physical examination.


H. Yang et al. (Eds.): ICONIP 2020, LNCS 12532, pp. 716–729, 2020.
https://doi.org/10.1007/978-3-030-63830-6_60

As CVD risk increases in China, interest in strategies to mitigate it is growing. However, information on the prevalence and treatment of CVD in daily life is limited. But in the medical field, many hospitals have been able to systematically accumulate medical records for a large number of patients by introducing an EMR system. Deep learning has been successfully applied to medical field based on accumulated EMR data [2]. In particularly, many studies have been conducted to predict the risk of cardiovascular disease in order to prevent cardiovascular diseases with a high mortality rate globally [3]. Because EMR data is based on patient records, it contains information on the pathogenesis of CVD. However, we found that there is a large amount of irrelevant information in most EMRs. For example, a complete medical record contains only 10 valid information records leading to diseases. The excessively irrelevant information not only reduces model's emphasis on effective disease information. In the field of natural language processing, such problems also exist in text classification tasks. In the article [4], the model proposed by Huang et al. avoids redundant information in the text by skipping the content in the text. Therefore, we intend to use the now steadily developing entity identification model architecture to extract key information. Table 1 is the key information we considered, including 12 risk factors. In response, we propose to extract the risk factors of pathogenesis from EMRs and take the time attributes of these risk factors. In fact, although the training time of the model is reduced based on the extracted CVD risk factors, the experimental results are not optimistic. After experimental analysis, we think that the main reason is the lack of certain contextual information.

In view of this situation, we introduce the Risk Factors Powered Bi-Attention (RFPBiA) network, a hierarchical multi-stage architecture for modeling the representations of the EMR context paragraph at different levels of granularity (Fig. 2). RFPBiA includes character-level and contextual embeddings, and uses bi-directional attention flow to obtain a risk factors representation. In this regard,

**Table 1.** Attributes of CVD.

| No. | Attributes | Description |
|---|---|---|
| 1. | Overweight/Obesity (O2) | A diagnosis of patient overweight or obesity |
| 2. | Hypertension | A diagnosis or history of hypertension |
| 3. | Diabetes | A diagnosis or a history of diabetes |
| 4. | Dyslipidemia | A diagnosis of dyslipidemia, hyperlipidemia or a history of hyperlipidemia |
| 5. | Chronic Kidney Disease (CKD) | A diagnosis of CKD |
| 6. | Atherosis | A diagnosis of atherosclerosis or atherosclerotic plaque |
| 7. | Obstructive Sleep Apnea Syndrome (OSAS) | A diagnosis of OSAS |
| 8. | Smoking | Smoking or a patient history of smoking |
| 9. | Alcohol Abuse (A2) | Alcohol abuse |
| 10. | Family History of CVD (FHCVD) | Patient has a family history of CVD or has a first-degree relative (parents, siblings, or children) who has a history of CVD |
| 11. | Age | The age of the patient |
| 12. | Gender | The gender of patient |

we will predict that the network can take into account both the risk factor information and the context information in EMR. This is very critical for our prediction task. For example, the increased correlation between hypertension risk and controlling blood pressure risk can better predict whether patients suffer from CVD. The experimental results show that the F-score reaches 0.9424, which fully demonstrates the effectiveness of our proposed method and network architecture. To sum up, our contribution is four-fold, with the following conclusions:

- We propose to extract the risk factors leading to CVD by using the existing mature entity recognition technology, which provides a new idea for disease prediction tasks.
- Our attention layer is not used to summarize the context paragraph into a fixed-size vector. Instead, the attention is computed for every time step, and the attended vector at each time step, along with the representations from previous layers, is allowed to flow through to the subsequent modeling layer. This reduces the information loss caused by early summarization.
- We use a memory-*less* attention mechanism. That is, the attention at each time step is a function of only the risk factor and the EMR context paragraph at the current time step and does not directly depend on the attention at the previous time step. We hypothesize that this simplification leads to the division of labor between the attention layer and the modeling layer. It forces the attention layer to focus on learning the attention between the risk factor and the context, and enables the modeling layer to focus on learning the interaction within the factor-aware context representation (the output of the attention layer). It also allows the attention at each time step to be unaffected from incorrect attendances at previous time steps.
- We use attention mechanisms in both directions, factor-to-context and context-to-factor, which provide complimentary information to each other.

## 2 Methodology

The main idea of this paper is to predict whether a patient has CVD by focusing on the risk factors in EMRs. First of all, we need to prepare the data we need. The user enters the appropriate values from his/her EMR report. After this, the historical dataset is uploaded. The fact that most medical dataset may contain missing values makes this accurate prediction difficult. So, for this missing data, we have to transform the missing data into structured data with the help of a data cleaning and data imputation process. After preparing the data, we mainly perform the following two steps. Firstly, the risk factors in the EMRs and their corresponding labels are extracted using the relatively mature entity recognition technology that has been developed. In addition to Age and Gender, the labels for other risk factors include the type of the risk factor and its temporal attributes. We only use the CRF layer to identify the F-score of the extraction result to reach 0.8994. When we use bidirectional LSTM with a CRF layer (BiLSTM-CRF), the F-score identifying the extraction results reached 0.9073. In contrast, the BiLSTM-CRF model has better recognition performance, so we consider

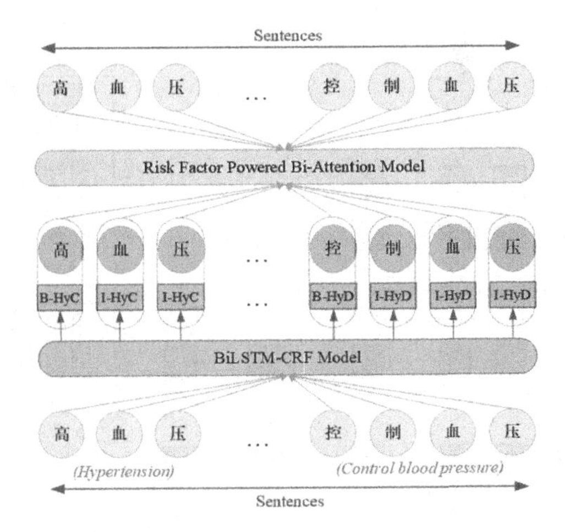


**Fig. 1.** The main process of CVD prediction.

using it to extract the risk factors and corresponding labels in EMRs. These extracted risk factors that carry the corresponding labels serve as the basis for input and predict CVD. In the end, by using the RFPBiA model, we can predict whether a patient has CVD. Figure 1 shows above-mentioned two main processes for predicting CVD.

### 2.1 Technical Details of RFPBiA Model

As shown in Fig. 2, the purpose of our work is to comprehensively model EMRs text by using the characteristics of text content and risk factors in EMRs text, thus further realizing CVD prediction task. Generally speaking, RFPBiA consists of five parts: character embedding layer, contextual embedding layer, bi-attention layer, modeling layer and predicting layer. The details are as follows.

**Character Embedding Layer** maps each character to a vector space using the pre-trained character vectors. Let $X = \{x_1, ..., x_T\}$ and $W = \{word_1, ..., word_J\}$ represent the input EMR context and risk factors in the EMR context, respectively. For risk factors, we add each character-level embedding vector that matches the word, and then average to obtain the embedding vector corresponding to risk factors. In this way, we get a new representation of risk factors $W' = \{w_1, ..., w_J\}$.

**Contextual Embedding Layer** utilizes contextual cues from surrounding characters to refine the embedding of the characters. We use a Long Short-Term Memory Network (LSTM) [5] on top of the embeddings provided by the previous layers to model the temporal interactions between words. We place a LSTM in both directions, and concatenate the outputs of the two LSTMs.

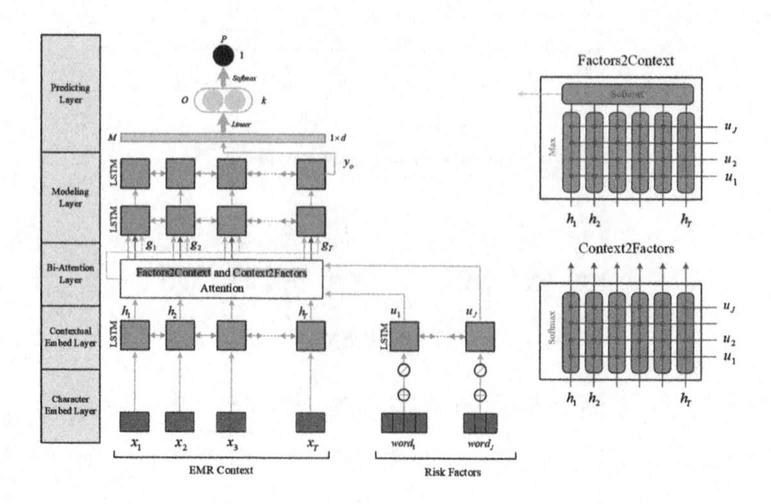


**Fig. 2.** The architecture of RFPBiA model. *(best viewed in color)*

Hence we obtain $H \in \mathbb{R}^{2d \times T}$ from the EMR context character vectors $X$, and $U \in \mathbb{R}^{2d \times J}$ from risk factor vectors $W'$. Note that each column vector of $H$ and $U$ is $2d$-dimensional because of the concatenation of the outputs of the forward and backward LSTMs, each with $d$-dimensional output. The first two layers apply to risk factors and the EMR context.

**Bi-attention Layer** is responsible for linking and fusing information from the EMR context and the risk factors. Unlike previously popular attention mechanisms [6,7], the attention layer is not used to summarize the two kinds of information into single feature vectors. Instead, the attention vector at each time step, along with the embeddings from previous layers, are allowed to flow through to the subsequent modeling layer. This reduces the information loss caused by early summarization.

The inputs to the layer are contextual vector representations of the EMR context $H$ and the risk factors $U$. The outputs of the layer are the factor-aware vector representations of the EMR context characters, $G$, along with the contextual embeddings from the previous layer.

In this layer, we compute attentions in two directions: from EMR context to risk factors as well as from risk factors to EMR context. Both of these attentions, which will be discussed below, are derived from a shared similarity matrix, $S \in \mathbb{R}^{T \times J}$, between the contextual embeddings of the EMR context ($H$) and the risk factors ($U$), where $S_{tj}$ indicates the similarity between $t$-th context character and $j$-th risk factor. The similarity matrix is computed by

$$S_{tj} = \alpha \left( H_{:t} , U_{:j} \right) \in \mathbb{R}, \tag{1}$$

where $\alpha$ is a trainable scalar function that encodes the similarity between its two input vectors, $H_{:t}$ is $t$-th column vector of $H$, and $U_{:j}$ is $j$-th column vector of

$U$. We choose $\alpha(h,u) = w_{(s)}^{\top}[h; u; h \circ u]$, where $w_{(s)}^{\top} \in \mathbb{R}^{6d}$ is a trainable weight vector, $\circ$ is elementwise multiplication, $[;]$ is vector concatenation across row, and implicit multiplication is matrix multiplication. Now we use $S$ to obtain the attentions and the attended vectors in both directions.

*Context-to-Factor Attention.* Context-to-factor (C2F) attention signifies which risk factors are most relevant to each context character. Let $a_t \in \mathbb{R}^J$ represent the attention weights on the risk factors by $t$-th context character, $\sum a_{tj} = 1$ for all $t$. The attention weight is computed by $a_t = softmax(S_{t:}) \in \mathbb{R}^J$, and subsequently each attended risk factor vector is $\widetilde{U} = \sum_j a_{tj} U_{:j}$. Hence $\widetilde{U}$ is a $2d$-by-$T$ matrix containing the attended risk factor vectors for the entire context.

*Factor-to-Context Attention.* Factor-to-context (F2C) attention signifies which context characters have the closest similarity to one of the risk factors and are hence critical for the predicting task. We obtain the attention weights on the context characters by $b = softmax(max_{col}(S)) \in \mathbb{R}^T$, where the maximum function ($max_{col}$) is performed across the column. Then the attended context vector is $\widetilde{h} = \sum_t b_t H_{:t} \in \mathbb{R}^{2d}$. This vector indicates the weighted sum of the most important characters in the context with respect to the risk factors. $\widetilde{h}$ is tiled T times across the column, thus giving $\widetilde{H} \in \mathbb{R}^{2d \times T}$.

Finally, the contextual embeddings and the attention vectors are combined together to yield $G$, where each column vector can be considered as the factor-aware representation of each context character. We define $G$ by

$$G_{:t} = \beta(H_{:t}, \widetilde{U}_{:t}, \widetilde{H}_{:t}) \in \mathbb{R}^{d_G}, \tag{2}$$

where $G_{:t}$ is the $t$-th column vector (corresponding to $t$-th context character), $\beta$ is a trainable vector function that fuses its (three) input vectors, and $d_G$ is the output dimension of the $\beta$ function. While the $\beta$ function can be an arbitrary trainable neural network, such as multi-layer perceptron, a simple concatenation as following still shows good performance in our experiments: $\beta(h, \widetilde{u}, \widetilde{h}) = [h; \widetilde{u}; h \circ \widetilde{u}; h \circ \widetilde{h}] \in \mathbb{R}^{8d \times T}$ (i.e., $d_G = 8d$).

**Modeling Layer** employs a Recurrent Neural Network (RNN) to scan the context. The input to the modeling layer is $G$, which encodes the factor-aware representations of context characters. The output of the modeling layer captures the interaction among the context characters conditioned on the risk factors. This is different from the contextual embedding layer, which captures the interaction among context characters independent of the risk factor. We use two layers of BiLSTM, with the output size of d for each direction. As a result, we take the final hidden layer states of BiLSTM (i.e., $y_o$) as the final output, then we redefine it as $M \in \mathbb{R}^d$, which is passed onto the predicting layer to predict the result. Here, $M$ is exactly the ultimate representation of input EMR context and risk factors.

**Predicting Layer** provides a prediction to the CVD. As a result, we take the output of BiLSTM (i.e., $M$) as the final output. After that, we feed $M$ into a fully-connected neural network to get an output vector $O \in \mathbb{R}^k$ ($K$ is the number of classes):

$$O = sigmoid\left(M \times N\right), \tag{3}$$

where $N \in \mathbb{R}^{d \times k}$ is the weight matrix for dimension transformation, and $sigmoid\left(\cdot\right)$ is a non-linear activation function. Finally, we apply a softmax layer to map each value in $O$ to conditional probability and realize the prediction as follows:

$$P = argmax\left(softmax\left(O\right)\right). \tag{4}$$

**Model Training.** Since what we are trying to solve is a prediction task, we follow the work in [8] to apply the cross-entropy loss function to train our model, and the goal is to minimize the following *Loss*:

$$Loss = -\sum_{T \in Corpus}\sum_{i=1}^{K} p_i\left(T\right) logp_i\left(T\right), \tag{5}$$

where $T$ is the input EMR text, Corpus denotes the training corpus and $K$ is the number of classes. In the training process, we apply *Adagrad* as optimizer to update the parameters of RFPBiA, including $\alpha$, $N$ and all parameters (weights and biases) in each RNN. To avoid the overfitting problem, we apply the dropout mechanism at the end of the embedding layer.

## 3 Results

### 3.1 Dataset and Evaluation Metrics

Our dataset contains two corpora. The first corpus came from a hospital in Gansu Province with 800,000 unlabeled EMRs of internal medicine. The dataset was mainly used to train and generate our character embedding. In Fig. 3, we also added a dictionary of risk factors during the training. In this way, the Skip-gram model in word2vec we use can better make each character in the risk factor more compact. The other one is from the Network Intelligence Research Laboratory of the Language Technology Research Center of the School of Computer Science, Harbin Institute of Technology, which contains 1186 EMRs. In the risk factor identification phase, BiLSTM-CRF model used all EMRs in the experiment, of which 830 were the training dataset, 237 were the test dataset and 119 were the development dataset. Then we will use the EMRs that need to be used for prediction to identify the risk factors through the model. This corpus intends to be used to develop a risk factor information extraction system that, in turn, can be applied as a foundation for the further study of the progress of risk factors and CVD [9]. For the corpus, we divided it into CVD and no CVD according to the clinically diagnosed disease in the electronic medical record. In the corpus we

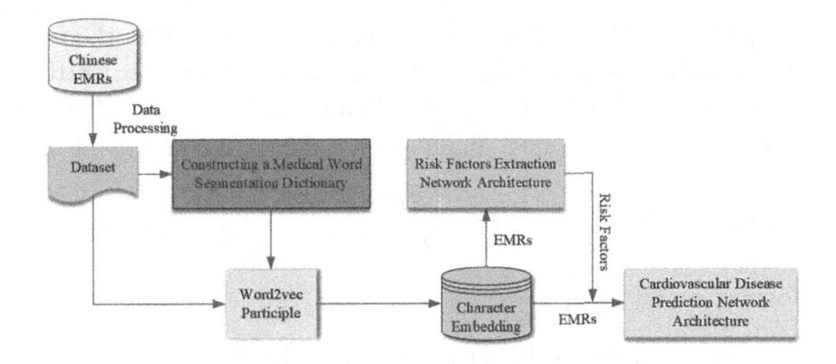


**Fig. 3.** Generate the character embedding for experiments.

used, there were 527 EMRs and 132 EMRs. The basis comes from the following two parts: On the one hand, according to the definition of CVD by the World Health Organization [10]. On the other hand, the first (*Symptoms*) and the third (*Diseases*) in the book of *Clinical Practical Cardiology* [11].

Our experiments involve 12 risk factors which are *Overweight/Obesity* (O2), *Hypertension*, *Diabetes*, *Dyslipidemia*, *Chronic Kidney Disease* (CKD), *Atherosis*, *Obstructive Sleep Apnea Syndrome* (OSAS), *Smoking*, *Age*, *Gender*, *Alcohol Abuse* (A2), *Family History of CVD* (FHCVD), as shown in Table 1. In addition, the dataset also includes 4 temporal attributes which are *Continue*, *During*, *After*, *Before*. Since risk factors of Age and Gender have not consider the temporal attributes, we have added a temporal attribute: *None*. The relevant details of all EMRs are shown in Table 2.

**Table 2.** Distribution of CVD risk factors and their occurrence times.

| Risk factors | Before DHS | During DHS | After DHS | Continuing DHS | Total |
|---|---|---|---|---|---|
| O2 | 0 | 0 | 0 | 18 | 18 |
| Hypertension | 405 | 1909 | 10 | 1405 | 3729 |
| Diabetes | 60 | 57 | 13 | 877 | 1007 |
| Dyslipidemia | 4 | 287 | 6 | 75 | 372 |
| CKD | 0 | 0 | 0 | 26 | 26 |
| Atherosis | 3 | 4 | 0 | 137 | 144 |
| OSAS | 0 | 0 | 0 | 1 | 1 |
| Smoking | 8 | 0 | 0 | 500 | 508 |
| A2 | 9 | 0 | 0 | 86 | 95 |
| FHCVD | 0 | 0 | 0 | 10 | 10 |
| Age | – | – | – | – | 1859 |
| Gender | – | – | – | – | 1909 |

*DHS* duration of hospital stay, "–" denotes not considered.

In the experiment, the training dataset contains 461 EMRs, the test dataset contains 132 EMRs, and the development dataset contains 66 EMRs. Finally, we use *Accuracy* ($A$), *Precision* ($P$), *Recall* ($R$) and *F-score* ($F$) to evaluate the performance [12]:

$$F = \frac{2PR}{P+R}. \tag{6}$$

**Table 3.** Hyper parameters of RFPBiA.

| Parameter | Description | Value |
|---|---|---|
| $d_w$ | Dimension of word embedding | 100 |
| $lr$ | Learning rate | 0.001 |
| $B$ | Batch size | 10 |
| $d_p$ | Each neuron's deactivation rate | 0.5 |
| $dr$ | Decay rate for $lr$ | 0.99 |
| $h$ | Each RNN's hidden unit quantity | 128 |
| $n$ | Number of epochs | 28 |

### 3.2 Models and Parameters

We carry out the experiments to compare the performance of our model with others described in the following.

***CRF:*** This model was used by Mao et al. [13] to recognize the named entity in the electronic medical records based on Conditional Random Field.

***BiLSTM-CRF:*** This model was used by Li et al. In order to realize automatic recognition and extraction of entities in unstructured medical texts, a model combining language model conditional random field algorithm (CRF) and Bi-directional Long Short-term Memory networks (BiLSTM) is proposed [14].

***SVM:*** This model was used by S. Menaria et al. As a traditional machine learning method, the support vector machine algorithm performs well in [15].

***ConvNets:*** This model was used by Xiang et al. [16] offering an empirical exploration on the use of character-level convolutional networks (ConvNets) for text classification.

***LSTM/RFPBiA (no att):*** This model was used by Xin et al. [17], which proposed a LSTM network with fully connected layer and activation layers. We have further built a BiLSTM model for CVD prediction. The model also has a fully connected layer and an activation layer. In fact, this is the RFPBiA model without using the attention mechanism.

***RFPBiA:*** This is the model proposed in this paper. Table 3 gives the chosen hyper-parameters for all experiments. We tune the hyper-parameters on the development set by random search. We try to share as many hyper-parameters as possible in experiments.

### 3.3 Experimental Results

We did a rich comparative experiment on our own model and other models:

In Fig. 4, we performed a comparison of the CRF and BiLSTM-CRF models for the identification of risk factors in EMRs. We did a lot of experiments and summarized why there was such a high F-score based on experimental and EMRs data analysis. Because there are 12 risk factors in the entire data, these risk factors are largely repeated in EMRs. In addition, BiLSTM can take contextual information into account well. This is greater for the system we have proposed.

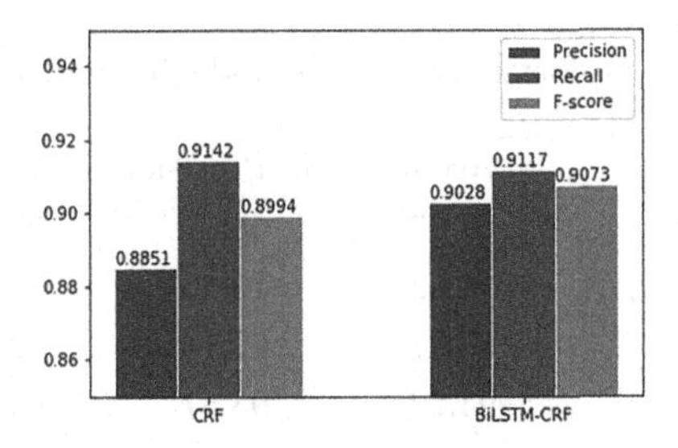


**Fig. 4.** Comparison of CRF and BiLSTM-CRF models.

**Table 4.** The comparison of each model for CVD prediction results.

| Model | Accuracy % | Precision % | Recall % | F-score % |
|---|---|---|---|---|
| $SVM_{(raw)}$ | 90.91 | 90.91 | 90.91 | 90.91 |
| $SVM_{(no\ labels)}$ | 89.39 | 89.03 | 89.39 | 89.21 |
| $ConvNets_{(raw)}$ | 92.83 | 92.64 | 92.83 | 92.73 |
| $LSTM_{(raw)}$ | 92.24 | 93.46 | 92.73 | 93.09 |
| $LSTM_{(risks\ with\ labels)}$ | 82.58 | 81.35 | 83.01 | 82.17 |
| $RFPBiA_{(raw,\ no\ att)}$ | 93.91 | 93.83 | 93.91 | 93.86 |
| $RFPBiA_{(risks\ with\ labels,\ no\ att)}$ | 89.23 | 88.96 | 89.23 | 89.07 |
| $RFPBiA_{(no\ labels)}$ | 93.41 | 92.73 | 93.12 | 92.92 |
| $RFPBiA$ | 94.87 | 93.97 | 94.52 | 94.24 |

"att" denotes the bi-directional attention, "no labels" denotes the risk factors without labels.

In Table 4, we show the comparison between the previous model and our proposed RFPBiA model for *accuracy*, *precision*, *recall* and *F-score*. And the performance of each model when the dataset is the original EMRs, the risk factor with the label, or the risk factor without the label.

In Table 5, we compared four cases: (1) The performance of the ConvNets model in random embedding; (2) The performance of the LSTM model in random embedding; (3) The performance of our model without attention mechanism, that is, the performance of the BiLSTM model; (4) When our model is in random embedding.

In Fig. 5, we have a visual example, which consists of the following three parts: (a) A sample of the case characteristics in EMR is that the risk factors have been marked according to the labeling rules [9]. (b) We translated the case characteristics into English. (c) We have made a visualization of attention matrix. The first point from the case characteristics module in EMR that we considered. It contains 8 risk factors and 61 characters.

**Table 5.** The performance of each model at random embedding.

| Model | Accuracy % | Precision % | Recall % | F-score % |
|---|---|---|---|---|
| $ConvNets_{(risks\ with\ labels)}$ | 91.67 | 90.87 | 91.24 | 91.05 |
| $LSTM_{(risks\ with\ labels)}$ | 81.82 | 79.45 | 82.36 | 80.88 |
| $RFPBiA_{(no\ att)}$ | 92.31 | 92.18 | 92.31 | 91.91 |
| $RFPBiA$ | 94.22 | 93.57 | 94.61 | 94.09 |

## 4 Discussion

Table 4 shows the predictive performance of different models on our evaluation corpus. It can be seen that our RFPBiA model is superior to other methods in various evaluation indexes when pre-trained embedding and labels are used. In addition, the performance of the LSTM and BiLSTM models on the original EMRs text and the risk factor dataset, respectively. We can find that the prediction effect is not optimistic without considering EMR text information. Compared with the ConvNets model proposed by Xiang et al. [16], we find that the sequential model extracts EMR text information better than the model based on convolutional neural networks. Judging from the performance of our model in Table 4 and Table 5, we can use the internal medicine EMRs pre-trained character embedding to help improve the performance of the model. Not only that, judging from the performance of our model in Table 4 without the bi-directional attention mechanism, the importance of the attention mechanism to the performance of our model is shown clearly. This also shows that CVD prediction focusing on risk factors is more important. Moreover, through the presence or

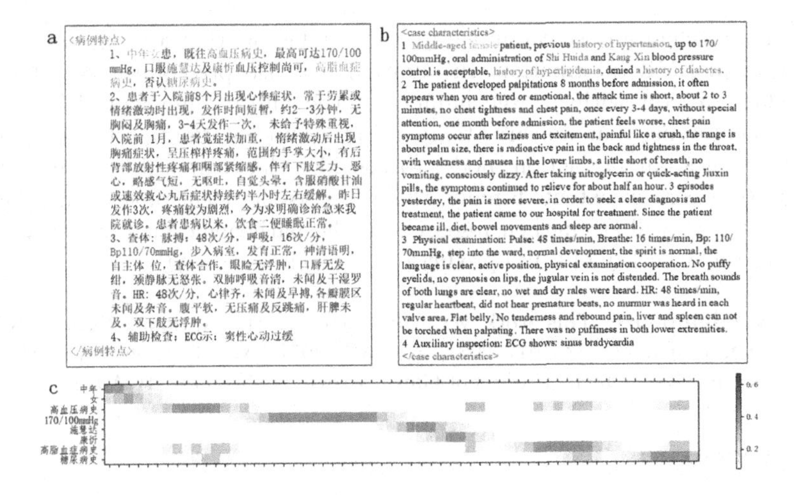


**Fig. 5.** **a** A sample of the case characteristics in EMR with marked CVD risk factors. **b** The sample of the case characteristics (English version). **c** Visualization of attention matrix for the factor-context tuple. Here we only consider the first point of case characteristics. The label on the x-axis in the figure is the characters. The label on the y-axis in the figure is the risk factor.

absence of corresponding labels of risk factors on our dataset, we can verify that the label information is beneficial to our prediction task.

For Fig. 5(a), we can intuitively find that the number of risk factors present in EMR is relatively small. In other words, the key information in an EMR that can determine whether a patient has CVD is limited. For Fig. 5(c), we emphasize that the learned attentions can be very useful to reduce a doctor's reading burden. It also emphasizes the correlation between the characters in EMR and the risk factors therein. For example, in this figure, the risk factor of "(*have a history of hypertension*)" and the risk factor of "(*have a history of hyperlipidemia*)" are clearly related to each other. This is exactly what we need, the risk factors are no longer independent of each other. We believe this can be attributed to our bi-directional attention and pre-trained character embedding. The performance comparison between the RFPBiA model and other models in Table 4 and Table 5 can also prove our analysis.

## 5 Conclusions

In this paper, the disease prediction experiment was carried out on the RFPBiA model using structured data. After a long-term study of EMRs, we focused on the key information leading to CVD, and we defined this key information as risk factors. So, we used BiLSTM-CRF model to identify the risk of CVD and its corresponding risk factors. What's more, we have done the next three aspects in the prediction task. On the one hand, we can use BiLSTM to obtain the context information of EMRs well. On the other hand, with the help of the bi-directional attention mechanism, we can integrate the risk factors that leading to CVD and the original EMR information. In addition, we also pre-trained the character embedding vector with a large number of the internal medicine EMRs. It is worth mentioning that we carry temporal attribute information through labels, which enables our prediction model to consider the temporal attribute of risk factors to a certain extent. Finally, this gives us the ideal prediction effect. In the future, we will strengthen research on the pathogenic factors of CVD and improve the accuracy of CVD prediction as much as possible.

**Acknowledgements.** This work was supported by the National Natural Science Foundation of China (No. 61762081, No. 61662067, No. 61662068) and the Key Research and Development Project of Gansu Province (No. 17YF1GA016).

## References

1. Li, Y., et al.: Systematic review regulatory principles of non-coding RNAs in cardiovascular diseases. Briefi. Bioinformatics **20**(1), 66–76 (2019)
2. Liang, Z., Zhang, G., Huang, J.X., Hu, Q.V.: Deep learning for healthcare decision making with EMRs. 2014 IEEE International Conference on Bioinformatics and Biomedicine, BIBM 2014, Belfast, UK, pp. 556–559. IEEE Computer Society (2014)
3. Wang, J., et al.: Detecting cardiovascular disease from mammograms with deep learning. IEEE Trans. Med. Imaging **36**(5), 1172–1181 (2017)
4. Huang, T., Shen, G., Deng, Z.: Leap-LSTM: enhancing long short-term memory for text categorization. In: Proceedings of the Twenty-Eighth International Joint Conference on Artificial Intelligence, IJCAI 2019, Macao, China, pp. 5017–5023. ijcai.org (2019)
5. Hochreiter, S., Schmidhuber, J.: Long short-term memory. Neural Comput. **9**(8), 1735–1780 (1997)
6. Tao, H., Tong, S., Zhao, H., Xu, T., Jin, B., Liu, Q.: A radical-aware attention-based model for Chinese text classification. In: The Thirty-Third AAAI Conference on Artificial Intelligence, AAAI 2019, Honolulu, Hawaii, USA, p. 5125C5132. AAAI Press (2019)
7. Wang, S., Huang, M., Deng, Z.: Densely connected CNN with multi-scale feature attention for text classification. In: Lang J. (ed.) Proceedings of the Twenty-Seventh International Joint Conference on Artificial Intelligence, IJCAI 2018, Stockholm, Sweden, pp. 4468–4474. ijcai.org (2018)
8. Zhou, Y., Xu, B., Xu, J., Yang, L., Li, C., Xu, B.: Compositional recurrent neural networks for Chinese short text classification. 2016 IEEE/WIC/ACM International Conference on Web Intelligence, WI 2016, Omaha, NE, USA, pp. 137–144. IEEE Computer Society (2016)
9. Su, J., He, B., Guan, Y., Jiang, J., Yang, J.: Developing a cardiovascular disease risk factor annotated corpus of Chinese electronic medical records. BMC Med. Inf. Decis. Making **17**(1), 117:1–117:11 (2017)
10. The details of cardiovascular diseases (CVDs) come from world health organization (who). https://www.who.int/news-room/fact-sheets/detail/cardiovascular-diseases-(cvds). Accessed 6 June 2020
11. Guo, J.: Clinical Practical Cardiology. Peking University Medical Press, New Haven (2015)

12. Hotho, A., Nrnberger, A., Paass, G.: A brief survey of text mining. LDV Forum **20**(1), 19–62 (2005)
13. Mao, X., Li, F., Duan, Y., Wang, H.: Named entity recognition of electronic medical record in ophthalmology based on CRF model. 2017 International Conference on Computer Technology. Electronics and Communication (ICCTEC), Dalian, China, pp. 785–788. IEEE (2017)
14. Li, W., et al.: Drug specification named entity recognition base on biLSTM-CRF model. In: 43rd IEEE Annual Computer Software and Applications Conference, COMPSAC 2019, WI, USA, pp. 429–433. IEEE (2019)
15. Woldemichael, F.G., Menaria, S.: Prediction of diabetes using data mining techniques. In: 2018 2nd International Conference on Trends in Electronics and Informatics (ICOEI), Milwaukee, Tirunelveli, India, pp. 414–418. IEEE (2018)
16. Zhang, X., Zhao, J.J., LeCun, Y.: Character-level convolutional networks for text classification. In: Advances in Neural Information Processing Systems 28: Annual Conference on Neural Information Processing Systems 2015, Montreal, Quebec, Canada, pp. 649–657. Curran Associates (2015)
17. Hong, X., Lin, R., Yang, C., Zeng, N., Cai, C., Gou, J.: Predicting Alzheimer's disease using LSTM. IEEE Access **7**, 80893–80901 (2019)

# Detecting Online Fake Reviews via Hierarchical Neural Networks and Multivariate Features

Chengzhi Jiang, Xianguo Zhang(✉), and Aiyun Jin

College of Computer Science, Inner Mongolia University, Hohhot 010021, China
chengzhi@mail.imu.edu.cn, 2595083628@qq.com, 1723821567@qq.com

**Abstract.** In recent years, as the value and credibility of online reviews tend to influence people's shopping feelings and consumption decisions, various online fake reviews have been constantly emerging. Detecting online fake reviews has attracted widespread attention from both the business and research communities. The existing methods are usually to detect fake reviews on off-the-shelf algorithms using kinds of linguistic and behavioral features respectively. That ignores the fusion of different features and does not take into account that different features have different effects on model performance. In this research, a set of linguistic and non-linguistic features is explored, and an optimal feature subset is selected by using random forest algorithm and the sequential backward selection strategy. Then, a hierarchical neural networks for detecting online fake reviews is empirically proposed, which can learn local and global information from multivariate features. Experimental results show that the model proposed in this paper on multiple datasets is superior to the traditional discrete model and the existing neural network benchmark model, and has a good generalization ability.

**Keywords:** Fake reviews · Hierarchical neural networks · Feature engineering · Word embedding · Multivariate features

## 1 Introduction

With the rapid development and popularization of the Internet and e-commerce website, more and more consumers have changed from offline shopping to online spending. The consumer tends to share and exchange its feelings and opinions by posting reviews on relevant product pages. So many online reviews, which are valuable and informative, will influence people's consumption decision. With 89% of consumers confirming that online reviews of products and services are trustworthy, four out of five consumers would give up on a product based solely on reading a negative review, while 87% would buy a product based on positive reviews [5]. Consumers' reliance on review information has provided some businesses with a chance. To improve their own reputation or derogate and bespatter


H. Yang et al. (Eds.): ICONIP 2020, LNCS 12532, pp. 730–742, 2020.
https://doi.org/10.1007/978-3-030-63830-6_61

other competitors, such malicious businesses have produced abundant synthesized online reviews. These untruth reviews will hurt the consumer's shopping experience, but also not conducive to the stable development of e-commerce platforms. Studies have shown that the artificial recognition accuracy of fake reviews is only slightly higher than 50% [14,27]. Therefore, in order to eliminate the adverse effects caused by fake reviews, it is urgent to improve the fake review detection technology and develop recognition performance.

The detection of fake reviews can be a 2-category problem, and the existing method is usually to train a classifier by features explored from review text and reviewer behavior to predict whether a review is true or fake. Classical classification features, e.g. POS, LIWC and emotional polarity, can denote the semantic and emotional information of plain text. In addition, the characteristic information of reviewers and businesses can also be sought based on review metadata. Previous researches show abundant and effective features enable to give strong performance for detection [1,7,18]. Meanwhile, it can't ignore that different features contribute differently to the detection model. The primary purpose of this paper is to explore a set of effective classification features and propose a robust detection model to integrate different types of features. This study makes several major research contributions. First, we categorize potential clues for fake reviews recognition into linguistic feature and non-linguistic feature by analyzing the source of feature acquisition. The linguistic features mainly include word embedding, n-gram and other discrete linguistic features. Second, based on large-scale corpus pre-training, word embedding features with Labeled-LDA features are generated, which are conducive the acquisition of words with significant recognition. Third, the importance of a single variable in multivariate discrete features is calculated using random forest. According to the ranking results of the importance of features, the feature is trimmed with the sequence backward selection strategy, and the feature subset with the best recognition performance is selected. Forth, we propose a hierarchical neural network model based on word embedding features to mine deep semantic information. In this neural network structure, the attention mechanism can be used to capture the words or sentences with classification significance, so as to improve the recognition performance of fake reviews. At the same time, multivariate discrete features are mapped to low-dimensional, continuous real-valued vector space using Embedding, and then multi-core convolutional neural networks with different convolution kernel sizes are used to extract local and global information through the multivariate discrete features. And finally the semantic features are concatenated with multivariate discrete feature vectors, which work together on the fake review detection model.

## 2 Related Work

### 2.1 Opinion Spam Detection

With the increase in online transactions, various types of spam have caused a lot of trouble for people. Researchers have done a lot of work in the identification of spam in the field of web pages and e-mail, and obtained good detection results [2,3,9,17]. Jindal and Liu first proposed the concept of fake reviews, and employed the characteristics of reviews, reviewers, and products to carry out research on the identification of fake reviews by using logistic regression algorithm modeling [8]. Since then, more and more scholars have begun to work in the field of fake review recognition.

Yoo and Gretzel collected a small hotel review data set that included 40 true reviews and 42 fake reviews, and manually analyzed their language differences [28]. Ott et al. hired professional writers on Amazon's Mechanical Turk to write fake reviews and created a gold-standard dataset of fake reviews. They extracted POS, LIWC and n-gram features and employed two typical classification algorithms , i.e. Naive Bayes and Support Vector Machine, to conduct fake review recognition research [18,19]. After then, a series of research work has been carried out based on this gold data set [11,25]. Li et al. collected the Chinese labeled dataset from dianping.com, and proposed a collective classification algorithm and then extend it to collective positive and unlabeled learning, which achieved good results [11]. Ji and Kang applied the tri-training algorithm to a large amount of unlabeled data for semisupervised learning. The experimental results show that the model is superior to general models in the recognition of fake review [4].

### 2.2 Neural Networks for Representation Learning

An important method for solving natural language processing problems is deep learning. Xu and Rudnicky first proposed that neural networks can be used to train language models [27]. Mikolov et al. proposed two kinds of word embedding architecture, i.e. Continuous Bag-of-Words and Skip-gram, and tried to use negative sampling and hierarchical softmax to improve the calculation speed of the model [10,15]. Researchers have also proposed many methods for sentence and document-level representation learning. Mikolov et al. employ paragraph vectors to learn document representation [15]. For recurrent neural networks and long-short-term memory networks, these sequence models are also applied to semantic composition. Researchers have proposed a variety of recurrent neural network models to learn document semantic information [6,13,23]. Jiang and Zhang used the hierarchical neural network model to merge semantic and non-semantic features to carry out opinion spam detection research [7]. Ren et al. leveraged a convolutional neural network and a bidirectional gated recurrent neural network to construct a hierarchical neural network model for fake review recognition [21,22]. Attention mechanism model refers to recurrent neural network with attention mechanism [2,20,24], which is suitable for many tasks, such as computer vision and natural language processing. The attention mechanism can capture potentially important features from the training data, including local and global semantic information, and has achieved good research results in the field of natural language processing [7,21,25].

# 3 Methodology

The exploration and development of effective feature engineering and detection model can greatly promote the research of fake review detection. This section will introduce the feature extraction and detection model construction of this paper.

## 3.1 Feature Extraction

Feature engineering plays an important role in machine learning tasks. And the richness of feature information has a great impact on the performance of classification model. Therefore, the definition and extraction of effective feature sets is the key to improve the recognition effect of fake reviews. Furthermore, for the fake review detection, it can also extract richer features from the metadata of the review besides extracting feature information directly from the review text. This research divides the potential clues of fake review detection into linguistic features and non-linguistic features.

**Linguistic Features.** The most important source of feature information is the review itself for review spam detection. The review text contains rich linguistic information, such as semantic information and emotional information. Therefore, in this study, the features extracted only from the review text are defined as linguistic features, which are introduced in three parts, i.e. word embedding, n-gram and other linguistic features. Word embedding feature is a low-dimensional, continuous real value vector, n-gram feature is a set of word fragment sequence, and other linguistic features, which include POS, FPP, OPP, RI, RL and ED, are discrete variables.

- **Word Embeddings.** In general, the word vector look-up matrix can be obtained by random initialization or pre-training from large-scale corpus. In this work, we have pre-trained word embeddings via Word2Vec on the open Yelp dataset. Although the word vector can represent the semantic information of the word, that cannot highlight its importance in the specific task. For fake review detection, certain keywords often indicates whether the review is fake or true, but the traditional word vector cannot effectively reflect this importance. Thus, this research introduce Labeled-LDA feature into the traditional word vector to obtain the novel word embedding. In our work, the frequency of word under each topic from labeled data is regarded as Labeled-LDA feature. The product of Labeled-LDA feature and traditional word vector of the word is represented as its new word embedding feature.
- **N-gram.** The N-gram feature is a common feature in NLP tasks. When N takes 1, 2, and 3, it represents unigram, bigram, and trigram features, respectively. Generally, term-frequency or TF-IDF can be used as the weight value of n-gram features.

- **Part-of-Speech (POS).** The ratio of the frequency of part-of-speech words such as nouns, verbs, adjectives, adverbs, and punctuation to the total number of words istaken as the POS feature, which is denoted as POS(*n.*), POS(*v.*), POS(*adj.*), POS(*adv.*) and POS(*punc.*) respectively.
- **First Person Pronouns (FPP).** This feature is the ratio of the number of firstperson pronouns in the review text to the total number of words.
- **Other Personal Pronouns (OPP).** It's the ratio of the number of non-first-person pronouns in the review text to the total number of words.
- **Readability Index (RI).** Researchers have applied the readability index , e.g. CLI and ARI, to the fake review detection and achieved certain results [7, 26]. In the paper, the harmonic average of ARI and CLI values is regarded as the readability index.
- **Review Length (RL).** It's the number of words in the review text.
- **Emothon Diversity (ED).** The feature is the ratio of positive emotion words and negative emotion words in the text to the total number of words, which is denoted as ED(*pos.*) and ED(*ne.g.*)

**Non-linguistic Features.** In addition to extracting features from review text, it also can acquire richer features from review metadata. In this research, the features that are not extracted from text are defined as non-linguistic features.

- **Review Score (RS).** The rating score is given by the reviewer in relation to the content of the review, which is usually an integer between 1 and 5.
- **Rating Deviation (RD).** To measure reviewer's rating deviation, the absolute score bias of a rating score on business from business's rating is computed. Then, the average score bias of all the reviewers' reviews is calculated.
- **Sentiment consistency (SC).** TextBlob can be used to perform sentiment analysis to calculate the sentiment polarity score of the text. Then, the sentiment consistency index of the rating score r and sentiment polarity s is:

$$SC = \begin{cases} 1, if\ rs \geq 0,\ then\ r \geq 3; if\ rs < 0,\ then\ r < 3 \\ 0, \qquad\qquad other \end{cases} \tag{1}$$

- **Average posting rate (APR).** It refers to the average number of reviews that a reviewer has posted in the interval between the first review and the most recent review.
- **Maximum number of review per day (MUR).** This is the maximum number of reviews posted by the same reviewer in a single day.
- **Filtering ratio (FR).** It represents the ratio of the reviews filtered by the business website detection mechanism to the total reviews of the business.
- **Support degree (SD).** When a user posts reviews, others can show their support for review by clicking certain buttons like "Useful", "Funny", "Cool". The total number of behaviors is considered as the support degree of reviews.
- **Maximum content similarity (MCS).** The text representation based on vector space model is used to calculate the similarity between two reviews by cosine distance, and the maximum value is taken as the maximum content similarity of reviewers.

### 3.2 Hierarchical Neural Networks for Classification

The fake review detection model MFNN proposed in this paper mainly employs word embedding to learn the document representation through a hierarchical neural network, and then employs a convolutional neural network to combine multivariate discrete features to generate feature vectors. Finally, the two feature representations are combined and input into the Softmax classifier for classification. The structure of MFNN model is shown in Fig. 1.

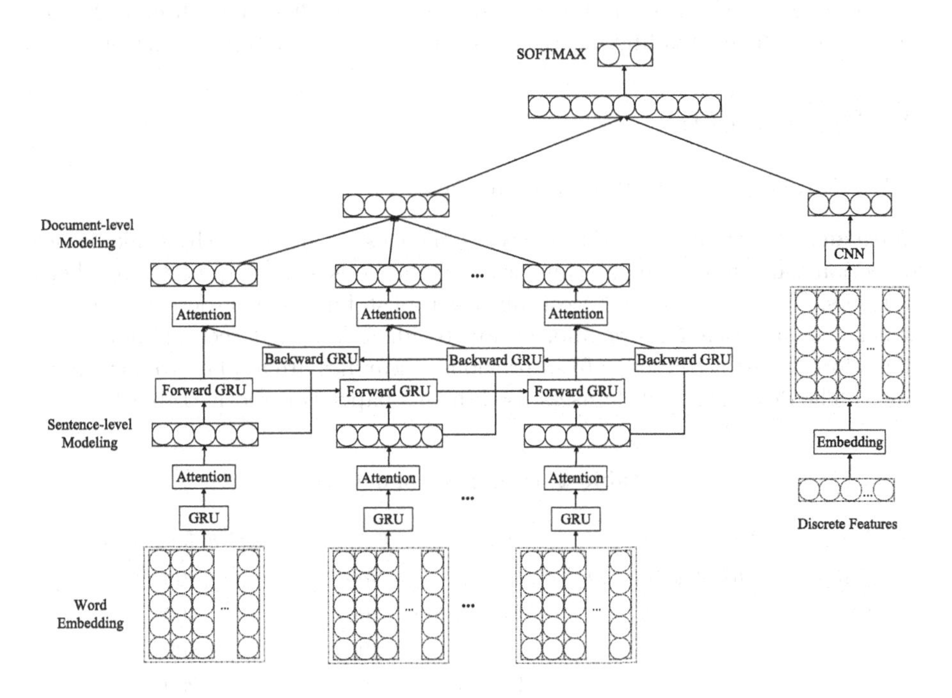


**Fig. 1.** Structure of MFNN.

The review can usually be regarded as a short document with a hierarchical structure, that is, many words form a sentence, and multiple sentences constitute the review text. For this season, a hierarchical neural network model is built using word embedding features to extract the semantic information of text, and the generated semantic representation feature vector can be used for text classification tasks. The document representation learning model based on hierarchical neural network and attention mechanism proposed in this paper is shown in the lower left of Fig. 1. This model mainly consists of two parts: one is to perform sentence-level modeling, which mainly uses the word embedding to learn the sentence representation. And the other is to conduct text-level modeling and generate the document representation through the sentence vector. In addition, since some words or sentences in the review have certain recognition significance, attention mechanisms are added to each presentation learning structure separately to effectively capture this information. The document feature vectors learned by neural network can be directly input into the classifier for fake review detection, and can also be concatenated with other feature vectors for subsequent experiments.

In order to input the discrete multivariate feature and the text semantic feature vector into the classifier for fake review detection, the neural network is also needed to transform the multivariate discrete feature into the low-dimensional continuous feature vector. This research leverage a multi-kernel convolutional neural network with width of 3, 4, and 5 to learn multivariate discrete feature vector from the embedding matrix, as shown in the lower right corner of Fig. 1.

# 4 Experiments

## 4.1 Datasets and Evaluation Metrics

**Dataset.** At present, one of the main challenges in the research of fake review detection using supervised machine learning methods is the acquisition of labeled datasets. This paper aims to establish a robust fake review detection model, which can use rich feature information to identify and detect opinion spam. Therefore, multiple datasets from previous researches are applied to our experiment. The description of datasets used in this paper is listed in Table 1.

**Table 1.** Description of datasets.

| No. | Data source | Domain | # true reviews | # fake reviews |
|---|---|---|---|---|
| Dataset 1 | Ott et al. [18,19] | Hotel | 800 | 800 |
| Dataset 2 | Li et al. [12] | Restaurant | 200 | 400 |
| | | Hotel | 800 | 800 |
| | | Doctor | 200 | 200 |
| Dataset 3 | Mukherjee et al. [16] | Restaurant | 8261 | 58631 |
| | | Hotel | 780 | 5078 |

**Evaluation Metrics.** Considering the unbalanced distribution of positive and negative samples in certain review datasets used in this study, the AUC is introduced to evaluate the performance of detection model.

## 4.2 Feature Selection

The study of fake review detection can be regarded as a two-category problem. For the classification problem, one of the keys to improve the final classification performance is to extract effective features. In this paper, the recognition features of fake reviews are divided into two major feature groups, namely linguistic features and non-linguistic features. However, it is not sufficient to consider only extracting enough features for fake review detection, because not all features contribute equally to the model performance. Therefore, it is very necessary to leverage the importance of features for further feature selection to improve the effectiveness of features and reduce noise interference. This research employs the random forest algorithm to calculate the importance of discrete features and the sequence backward selection strategy is used to trim the features according to the results of importance ranking.

As shown in the Table 2, the importance of multivariate discrete features calculated by random forest based on Dataset 3. When all the features are used for the importance calculation, it is found that the importance of non-linguistic features is greater than that of other linguistic features, and the importance distribution trend of other linguistic features and non-linguistic features is roughly the same as that of the separate calculation The sequential backward selection strategy is a top-down search method. The feature with the smallest importance value is removed from the initial feature set each time. In addition, the classification accuracy of model based on current feature set is calculated and the optimal feature subset with the smallest number of feature variables and the highest classification accuracy is finally obtained by successive iterations. Thus, the feature subset at this time is taken as the result of feature selection.

**Table 2.** Feature importance.

| Category | Feature | Importance | Category | Feature | Importance |
|---|---|---|---|---|---|
| Other linguistic features | RL | 0.1934 | Non-linguistic Features | RP | 0.4842 |
| | RI | 0.1642 | | MPR | 0.1537 |
| | POS(*n.*) | 0.1166 | | APR | 0.1279 |
| | POS(*v.*) | 0.1008 | | MCR | 0.098 |
| | SW(*ne.g.*) | 0.0873 | | RD | 0.0481 |
| | FPP | 0.0641 | | SC | 0.0475 |
| | POS(*adj.*) | 0.0613 | | RS | 0.0243 |
| | SW(*pos.*) | 0.0592 | | FR | 0.0166 |
| | POS(*adv.*) | 0.0581 | | | |
| | OPP | 0.0543 | | | |
| | POS(*punc.*) | 0.0412 | | | |

SVM with ten-fold cross validation is adopted to seek the optimal feature subset. It can be found from the Fig. 2 that the classification accuracy of the model increases continuously with the elimination of the features with the smallest importance measurement, mainly because the features with small contribution and the elimination of feature redundancy improve the performance of the classifier. The accuracy reaches the maximum point twice, and then the classification accuracy showed a downward trend, because the effective features were trimmed, which reduced the performance of the classifier. It can be found that the feature selection method based on random forest can effectively identify and eliminate feature redundancy and improve the performance and simplicity of the model.

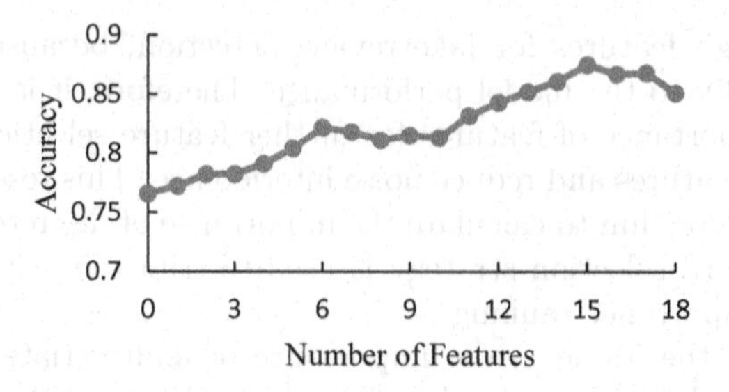


**Fig. 2.** Relationship between model performance and feature number.

# 5 Development Experiment

To better verify the effectiveness of features and detection model proposed in this paper, three kinds of comparison experiments are designed. The detection model with SVM as classifier is trained by 5-fold cross validation. For neural networks, we employ 80% of the samples as the training set, 10% as the validation set, and 10% as the test set. The validation set is mainly to optimize and adjust the super parameters in neural networks to improve the performance of models.

Experiment 1: verify the validity of the proposed text representation model based on hierarchical neural network and attentional mechanism. The following comparative experiments are conducted:

- SVM(Unigram): using word unigram feature in SVM.
- SVM(Bigram): using word bigram feature in SVM.
- SVM(Bigram+): using word unigram feature and bigram feature in SVM.
- SVM(Average): simply using the average of all word vectors as review vector.
- SCNN_1: the model based on a single-kernel CNN.
- MCNN_1: the model based on a multi-kernel CNN with width 1, 2, 3.
- RNN: the model based on a single-directional RNN.
- HRNN: the model based on hierarchical recurrent neural networks.
- HGRU: the model based on hierarchical GRU.
- HANN: the model based on hierarchical neural network and attention mechanism proposed in this paper is used.

Experiment 2: multivariate discrete features, i.e. other linguistic features and non-linguistic features, extracted from review text and metadata are used to verify the validity of the fake review recognition model based on multivariate discrete features. In this experiment, due to the limitation of datasets, only linguistic features can be extracted from Dataset 1 and Dataset 2 mentioned in Sect. 4.1, while Dataset 3 containing multiple metadata can not only extract linguistic feature information, but also rich non-linguistic feature information. Therefore, in this experiment, Dataset 1 and Dataset 2 only use other language features introduced in Sect. 3.1 as their multivariate discrete features, while Dataset 3 uses all feature information. Meanwhile, in addition to all the features extracted from that dataset are used, the optimal feature subset selected by random forest and sequence backward selection strategy is also applied to experiment. Therefore, the following experiments will be carried out:

- SVM: using diverse discrete features in SVM with 5-fold cross validation.
- MCNN_2: the model based on a multi-kernel CNN with width 3, 4, 5.

Experiment 3: in experiment 1 (Exp. 1) and experiment 2 (Exp 2.), it is respectively to verify the proposed text representation model based on word embedding and convolution neural network model based on multivariate discrete features. In experiment 3 (Exp. 3), the fake review detection model based on hierarchical neural network and multivariate features proposed in Sect. 3.2, i.e. MFNN, will be applied directly into experiment.

### 5.1 Results and Analysis

Table 3 shows the results of all development experiments.

**Table 3.** Experiment results.

| Model | | | Dataset 1 | Dataset 2 | | | | Dataset 3 | | |
|---|---|---|---|---|---|---|---|---|---|---|
| | | | | Res. | Hotel | Doc. | All | Res. | Hotel | All |
| Exp. 1 | SVM(Unigram) | | 0.691 | 0.593 | 0.683 | 0.584 | 0.641 | 0.496 | 0.545 | 0.514 |
| | SVM(Bigram) | | 0.675 | 0.591 | 0.691 | 0.571 | 0.665 | 0.517 | 0.529 | 0.511 |
| | SVM(Bigram+) | | 0.693 | 0.579 | 0.687 | 0.582 | 0.693 | 0.509 | 0.551 | 0.534 |
| | SVM(Average) | | 0.743 | 0.631 | 0.759 | 0.629 | 0.762 | 0.653 | 0.635 | 0.651 |
| | SCNN_1 | | 0.773 | 0.658 | 0.771 | 0.651 | 0.754 | 0.694 | 0.651 | 0.665 |
| | MCNN_1 | | 0.781 | 0.664 | 0.779 | 0.674 | 0.785 | 0.716 | 0.673 | 0.691 |
| | RNN | | 0.712 | 0.66 | 0.705 | 0.616 | 0.684 | 0.612 | 0.559 | 0.609 |
| | HRNN | | 0.742 | 0.688 | 0.712 | 0.654 | 0.696 | 0.631 | 0.615 | 0.633 |
| | HGRU | | 0.801 | 0.784 | 0.774 | 0.739 | 0.753 | 0.732 | 0.673 | 0.713 |
| | HANN | | 0.813 | 0.791 | 0.793 | 0.795 | 0.778 | 0.773 | 0.781 | 0.782 |
| Exp. 2 | SVM | All | 0.681 | 0.632 | 0.641 | 0.721 | 0.681 | 0.812 | 0.831 | 0.805 |
| | | Trimmed | 0.716 | 0.667 | 0.665 | 0.755 | 0.730 | 0.843 | 0.867 | 0.841 |
| | MCNN_2 | All | 0.697 | 0.651 | 0.671 | 0.732 | 0.719 | 0.823 | 0.854 | 0.843 |
| | | Trimmed | 0.709 | 0.701 | 0.706 | 0.753 | 0.756 | 0.851 | 0.886 | 0.889 |
| Exp. 3 | MFNN | All | 0.818 | 0.803 | 0.753 | 0.784 | 0.774 | 0.862 | 0.891 | 0.901 |
| | | Trimmed | 0.842 | 0.831 | 0.798 | 0.828 | 0.821 | 0.894 | 0.931 | 0.915 |

According to Exp. 1, it can be found that the performance of SVM using word embedding to express semantics is better than that using n-gram feature, and AUC is improved by more than 4%. Meanwhile, the performance based on n-gram feature on Dataset 1 and Dataset 2 is better than that of Dataset 3. The reason may be that the review data of the first two datasets are mainly collected through crowdsourced websites. Turkers write fake reviews according to the task rules. These fake reviews are quite standard, but real life reviews from Yelp.com are rough and contain noise. In addition, the results of fake reviews labeled by the commercial website filter are not completely correct. From the experimental results of each dataset, the performance of the classification model using neural networks is better than that of the traditional machine learning method. The AUC value of RNN is the lowest among several neural network models. The reason may be that RNN cannot effectively process long sequences, which leads to gradient dispersion and model performance degradation. The HANN model presented in this paper has the best experimental performance in Exp. 1. The AUC obtained by HANN model increased by at least 13.1% compared with SVM and 6.9% compared with ordinary neural networks on Dataset 3 from real business websites.

From the experimental results of Exp. 2, it can be found that the performance results of experiments using discrete features in Dataset 1 and Dataset 2 are lower than that of Exp. 1, because these two datasets can only extract and leverage linguistic features. Multivariate discrete features composed of only other language features contain too little recognition information, which is not conducive to classification prediction. Dataset 3 can not only use other linguistic features, but also extract nonlinguistic feature information, so it can also obtain better classification performance using only multivariate discrete features. By using two feature sets of original feature and optimal feature subset, it is found that the AUC of the trimmed model can increase at least 1.2%, which indicates that the feature selection method proposed in this paper is effective.

The results of Exp. 3 show that MFNN model can take full advantage of both linguistic feature and non-linguistic feature, and also obtain optimal experimental performance in all kinds of datasets.

## 6 Conclusion

We introduce a hierarchical neural networks based on multivariate discrete features and word embedding for fake review detection. The hierarchical neural network with attention mechanism can learn the deep semantic information in the review text. At the same time, CNN can be used to combine the multivariate discrete features and learn the local and global information in the features. The feature vectors learned from the discrete features is concatenated with the semantic feature vectors, and input them into the classifier for training. The experiment results show that the detection model proposed in this paper can obtain the optimal recognition performance. According to the results, it can be found that non-linguistic features obtained from metadata can promote the recognition of fake reviews. And the optimal feature subset is selected by using random forest and sequential backward selection strategy to reduce model complexity and improve simplicity.

## References

1. Arjun, M., et al.: Fake review detection: classification and analysis of real and pseudo reviews. Technical Report (2013)
2. Bahdanau, D., et al.: Neural machine translation by jointly learning to align and translate. ICLR **2015**, 1–15 (2015)

3. Becchetti, L., et al.: Link analysis for Web spam detection. ACM Trans. Web. **2**(1), 1–42 (2008)
4. Chengzhang, J., Kang, D.-K.: Detecting the spam review using tri-training. In: ICACT 2015, pp. 374–377 IEEE (2015)
5. Faville, K., List, A.: Cone releases the 2011 online influence trend tracker. https://www.conecomm.com/news-blog/2011-online-influence-trend-tracker-release
6. Glorot, X., et al.: Domain adaptation for large-scale sentiment classification: a deep learning approach. In: Proceedings of the 28th International Conference on Machine Learning, ICML 2011, pp. 513–520 (2011)
7. Jiang, C., Zhang, X.: Neural networks merging semantic and non-semantic features for opinion spam detection. In: Tang, J., Kan, M.-Y., Zhao, D., Li, S., Zan, H. (eds.) NLPCC 2019. LNCS (LNAI), vol. 11838, pp. 583–595. Springer, Cham (2019). https://doi.org/10.1007/978-3-030-32233-5_45
8. Jindal, N., Liu, B.: Analyzing and Detecting Review Spam. In: ICDM 2007, pp. 547–552 IEEE (2007)
9. Kolcz, A., Alspector, J.: SVM-based filtering of e-mail spam with content-specific misclassification costs. In: Proc. TextDM 2001 Workshop on Text Mining - held 2001 IEEE International Conference Data Mining, pp. 1–6 (2001)
10. Le, Q.V., Mikolov, T.: Distributed Representations of Sentences and Documents. In: ICML 2014 (2014)
11. Li, H., et al.: Spotting fake reviews via collective positive-unlabeled learning. In: 2014 IEEE International Conference on Data Mining, pp. 899–904 IEEE (2014)
12. Li, J., et al.: Towards a general rule for identifying deceptive opinion spam. In: Proceedings of the 52nd Annual Meeting of the Association for Computational Linguistics, pp. 1566–1576 (2014)
13. Li, L., et al.: Document representation and feature combination for deceptive spam review detection. Neurocomputing **254**, 33–41 (2017)
14. Lim, E.P., et al.: Detecting product review spammers using rating behaviors. In: International Conference on Information and Knowledge Management, Proceedings, pp. 939–948 (2010)
15. Mikolov, T., et al.: Distributed representations of words and phrases and their compositionality. NIPS **2013**, 3111–3119 (2013)
16. Mukherjee, A., et al.: What yelp fake review filter might be doing? ICWSM **2013**, 409–418 (2013)
17. Ntoulas, A., et al.: Detecting spam web pages through content analysis. In: Proceedings of the 15th International Conference on World Wide Web, pp. 83–92 (2006)
18. Ott, M., et al.: Finding deceptive opinion spam by any stretch of the imagination. ACL **2011**, 309–319 (2011)
19. Ott, M., et al.: Negative deceptive opinion spam. In: NAACL HLT 2013, pp. 497–501 (2013)
20. Raffel, C., Ellis, D.P.W.: Feed-Forward Networks with Attention Can Solve Some Long- Term Memory Problems (2015)
21. Ren, Y., Ji, D.: Neural networks for deceptive opinion spam detection: an empirical study. Inf. Sci. (Ny) **385–386**, 213–224 (2017)
22. Ren, Y., Zhang, Y.: Deceptive opinion spam detection using neural network. In: COLING 2016–26th International Conference Computer Linguistic Proceedings COLING 2016 Technical Paper 140–150 (2016)
23. Tang, D. et al.: document modeling with gated recurrent neural network for sentiment classification. In: Proceedings of the 2015 Conference on Empirical Methods in Natural Language Processing, pp. 1422–1432

24. Vaswani, A. et al.: Attention is all you need. In: Advances Neural Information Processing System 2017- Decem, Nips, 5999–6009 (2017)
25. Wang, X. et al.: Detecting Deceptive Review Spam via Attention-Based Neural Networks. In: Lecture Notes in Computer Science, pp. 866–876 (2018)
26. Wang, X. et al.: Identification of fake reviews using semantic and behavioral features. In: 2018 4th International Conference on Information Management, ICIM 2018, pp. 92–97 Springer (2018)
27. Xu, W., Rudnicky, A.: Can artificial neural networks learn language models? In: 6th International Conference on Spoken Language Processing, ICSLP 2000 (2000)
28. Yoo, K.-H., Gretzel, U.: Comparison of deceptive and truthful travel reviews. Inf. Commun. Technol. Tourism **2009**, 37–47 (2009)

# Error Heuristic Based Text-Only Error Correction Method for Automatic Speech Recognition

Linhan Zhang(✉), Tieran Zheng(✉), and Jiabin Xue(✉)

School of Computer Science and Technology, Harbin Institute of Technology, Harbin, China
{elenalinhan,zhengtieran,xuejiabin}@hit.edu.cn

**Abstract.** With the fast development of deep learning, automatic speech recognition (ASR) has achieved significant improvement. However, there still exists some errors in ASR transcriptions. They will greatly interfere with the downstream tasks, which take the transcribed text as source data. Obviously, it is necessary to set a corrector to reduce errors in the ASR transcriptions. For various downstream tasks, a text-only based corrector would be more adorable because of its minimal requirements for recognizer. However, the limitation of ASR decoding information exerts considerable influence on the performance of corrector. Correcting a correctly recognized word into a wrong one is one of the most common problems. To relieve this situation, we propose to adopt error knowledge from an error detection model as a heuristic to train a sequence-to-sequence (Seq2seq) correction model by transfer learning. In this way, the corrector can focus on correcting the wrong words with a soft label. The experiment shows our method can effectively correct ASR errors, with a 4.35% word error rate (WER) reduction for the transcription. It outperforms the state of the art Seq2seq baseline with a 1.27% WER reduction.

**Keywords:** ASR error correction · ASR error detection · Error heuristic · Transfer learning

## 1 Introduction

Automatic speech recognition (ASR) has made much progress due to the development of deep learning. Although the accuracy of recognition outcomes is high enough, the recognition result still contains some inevitable errors for the existence of some mismatched factors such as pronunciation of the speaker identity, recording environment or topic domain. In spite of such issue, the transcribed text from ASR is still adopted as data source to be analyzed by many natural language processing tasks. It is necessary to make sure the quality of transcriptions as far as possible. To reduce the WER of transcriptions, an error corrector is always treated as a post-processing module of the recognizer. In general, it


H. Yang et al. (Eds.): ICONIP 2020, LNCS 12532, pp. 743–753, 2020.
https://doi.org/10.1007/978-3-030-63830-6_62

can be separated into two steps: (1) set up a detector to locate the errors in a transcription; (2) build a candidate list from vocabulary and select the best candidate to correct the errors. For instance, Sumita et al. proposed an error correction method based on pattern learning and similarity string algorithm to detect and correct misspellings respectively [3]. In [9], it built a detector and corrector using co-occurrence analysis to correct errors step by step. Nevertheless, above methods still suffer from an issue of error accumulation, which means the errors made by detector can be passed to the corrector. To overcome this problem, a few researches investigated end-to-end corrector models without the help of detector. An external sequence to sequence language model is considered to re-score the output lattices or the N-best candidates. [1] proposed a post-editing ASR errors correction method based on Microsoft N-gram dataset. Guo et al. made use of a LSTM-based sequence to sequence model to correct spelling errors [2]. Besides, Zhang et al. used a Transformer to correct errors especially the substitution errors made by CTC-based Mandarin speech recognition system [13].

Correction strategies above are adopting decoding information like N-best hypotheses made by ASR to train the spelling correction model. However, the information is difficult to be obtained when using the recognizers issued by the third party. This is to say that it might be uneasy to obtain the decoding information of ASR, and most of the time we only have transcribed text, the one-best hypothesis from ASR. In this paper, we conduct a Seq2seq model with attention to correct errors in ASR one-best output for the purpose of addressing error accumulation.

Meanwhile, most of existing end-to-end methods [10] borrow Seq2seq models from NLP tasks like machine translation [6,11]. As a result, these models might neglect the target of ASR and stress too much on the readability of the corrected text. Even though the readability is essential for speech recognition results, there is also a need to keep the corrected transcription as the actual speech content, which may be disfluent or even contain some grammatical errors. Proposed that the spoken sentence in original speech is inversed, or the speaker makes a supplementary speech, then the Seq2Seq model might correct the whole sentence under the assumption it is incorrect grammatically. Some of the correct words may be replaced, which has a detrimental effect on the downstream tasks that rely on specific words. In some extreme cases, the meaning of the whole sentence could also be changed. For the sake of avoiding such circumstances, we put forth an idea that a correction model should learn to correct the "incorrect recognized" part in a transcription while not change the "correct recognized" part.

Based on the previous research, the paper brings up with an innovative ASR error correction method. Seq2seq model is utilized as a correction model as well, while an error detection model is introduced to provide corrector with error prior knowledge. The error prior knowledge is named for error heuristic in this paper. The error detection model, which tagging the transcription with label 0 and 1, tells the corrector a correction suggestion instead of absolute error location and error type information. Different from traditional two-steps methods forcing corrector to generate candidates for incorrect words detected by the detector, the error heuristic can be regarded as a soft label for corrector, only giving advice to corrector without compulsion. To make the corrector receive such knowledge, we transfer the intermediate parameters of detector to the encoder of correction model. We have transcriptions from two ASR systems: Google and Sphinx. However, the data of Google is insufficient to train a Seq2seq model. We propose to train the detector with data from both Google and Sphinx datasets. In this way, the detector not only offers error information but also realizes the data augment for small dataset. Because of the parameters on detector trained on large dataset, richer linguistic and ASR special error pattern information could be delivered to corrector on small dataset. Hence, there is no need to worry about the insufficient information in one-best hypothesis, and our error correction model could also avoid the Seq2seq model making an incorrect correction by manipulating the whole sequence to make predictions.

The main contributions of this paper are summarized as follows:

(1) We present a label-independent ASR error detection model. An unlabeled transcription could be tagged with 0/1 labels automatically. They indicate whether it is recognized correctly or not on word-level. This label-independent sequence tagging model can extract the syntactic and semantic information and locate the error in the transcription, which can be used in a correction model.
(2) We come up with a data augment method for our correction model training. Some datasets we accepted are deficient for Seq2seq based correction model training, so we propose that small dataset is arranged for fine-tuning the correction model pre-trained on the large dataset. Once the correction model for small datasets are provided with sufficient linguistic information, hardly the one-best hypothesis error correction methods would be limited.
(3) We put forward an error-heuristic based error correction method. The heuristic is learned from an error detection model by transferring the network parameters from the detector to the corrector. To be more specific, the parameter from detector is assigned to the encoder of corrector. Therefore, the detector can guide the corrector to be optimized towards correcting the "incorrect recognized" words and paying attention to the "correct recognized" words to make predictions.

## 2 Proposed Methodology

To address the incorrect correction and one-best limitation problem, providing knowledge to corrector as much as possible is a key to improve current Seq2seq based correction method. Actually, no matter what errors might the speech contain should not be corrected by the corrector. Besides, even though there may be a more suitable word to construct a sentence in semantic space, the correct should not make modification if the transcribed word is recognized correctly by ASR. We conduct an error detector to generate prior knowledge to ASR error corrector to predicting corrected transcriptions. This method is an improvement on current Seq2seq-based error correction methods. Figure 1 shows our error heuristic based ASR error correction method.

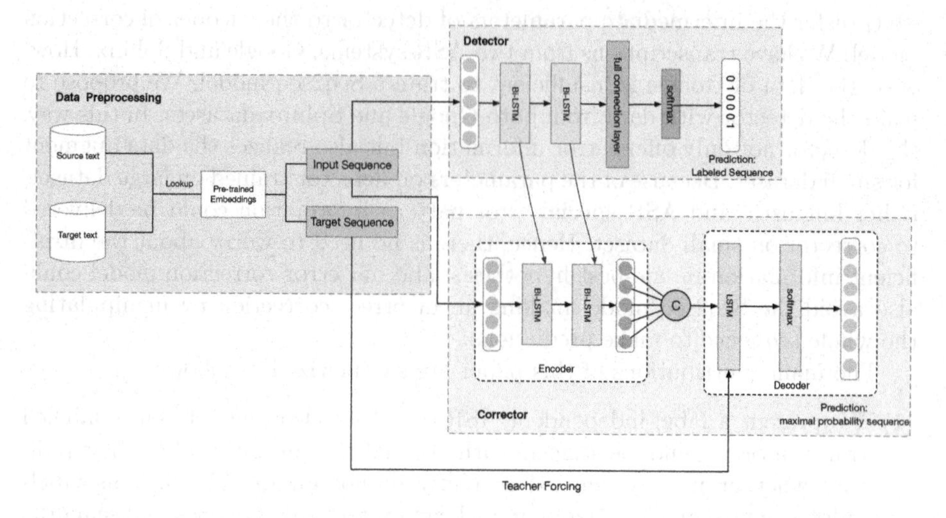


**Fig. 1.** Error Heuristic Based Error Correction Method. The red lines represent the parameters from detector transferred to the encoder of corrector (Color figure online)

## 2.1 Problem Definition

Given a set of transcriptions, $X = \{x_1, x_2, \ldots, x_N\}$, $x_i \in D_T$. $x_i$ is a word in the transcription. $D_T$ is the domain of transcriptions. The ASR reference, $Y = \{y_1, y_2, \ldots, y_M\}, y_i \in D_R$. $y_i$ is a word in the reference. $D_R$ is the domain of references. ASR error correction task is to generate a corrected transcription, $Y' = \{y'_1, y'_2, \ldots, y'_M\}$, $Y' = Y$. We should find out an ASR correction method mapping $X$ from the domain $D_T$ to $D_R$.

## 2.2 Data Processing

Firstly, the transcriptions need to be vectorized before being fed into the deep models. Here we adopt GloVe embeddings pretrained on large scale datasets already to transfer the input text into vectors. Both the transcription and reference should be vectorized since the later one is also necessary for correction model training. The references participated in the procedure of teacher forcing, making sure the decoder of corrector receive the ground truth to make a prediction for next time step.

### 2.3 Detector

An ASR error detection task is as same as a sequence tagging task. Both of them take the text as input and outputs a label sequence. Conditional random field (CRF) algorithm is usually used in sequence tagging tasks like Named Entity Recognition (NER) [14], Part of the Speech (POS) tagging [12]. The labels of tagging tasks are related with each other. In terms of POS tagging task, the POS of the former word has an impact on the label of the latter word. Meanwhile, both the labels and the words should conform to the grammar rules. For example, if a verb is followed by a verb, it is an incorrect annotation. Intuitively, the transfer characteristics of CRF in the outputs can ensure the accuracy of the prediction. Although the ASR error detection task is similar to sequence tagging task, the predicted tags are independent to each other. There is no need to establish the correlation of labels by CRF. Instead, we choose Softmax as the last layer of our detector. The outputs of Softmax are independent of each other, in line with the assumption that the output labels of error detection are independent.

After mapping the transcription $s = \{s_1, s_2, s_3, \ldots s_N\}$ and the reference $z = \{z_1, z_2, z_3, \ldots z_N\}$ to vectors with GloVe word embeddings [8], we could get the input sequence $\mathbf{x} = \{x_1, x_2, \ldots, x_n\}$ and the target sequence $p = \{p_1, p_2, p_3, \ldots\ldots, p_n\}, p \in [0, 1]$. Then the input sequence is fed into two bidirectional long short time memory (Bi-LSTM) layers to obtain contextual information for each time step. The output of the last Bi-LSTM layer on the last time step, a vector concatenating the forward and backward hidden states $\overleftarrow{\mathbf{h}}, \overrightarrow{\mathbf{h}}$, is mapped to the target output dimension by a fully connection layer. After that, a Softmax function is set up to calculate the probability $\mathbf{p}'$ for each time step $i$. Label 0 illustrates the word in transcription is recognized correctly, while 1 represents the opposite.

$$\mathbf{h} = \left[\overleftarrow{\mathbf{h}}, \overrightarrow{\mathbf{h}}\right] \tag{1}$$

$$\mathbf{h}' = \mathbf{W_L h} + b_L \tag{2}$$

$$p' = P(p_i|\mathbf{x}) = softmax(\mathbf{h}') \tag{3}$$

Where $\mathbf{W_L}, b_L$ are trainable parameters. $\mathbf{h}$ is the output of Bi-LSTM on the last layer.

Due to the high performance of ASR, the WER of the transcriptions has been low enough already, so the portion of incorrectly recognized words is smaller than that of the correctly recognized words. It means that the labels for words in a transcription are extremely imbalanced. The number of positive samples is larger than negative samples. Sometimes the error word is closed to the target word in the semantic space, like a synonym, hard to be detected by detector. Because of the large number of positive samples and easy samples, the training for detector is inefficient if we adopt Cross Entropy (CE) as the loss function. Whether the sample is positive or not, the weights for different samples are the same. Hence, focal loss [5] with constraint on easy and negative samples is considered as the loss function to train our detector model.

$$FL(p') = \begin{cases} -\alpha (1-p')^{\gamma} \log(p'), & y = 0 \\ -(1-\alpha) p'^{\gamma} \log(1-p') & , \quad y = 1 \end{cases} \tag{4}$$

$\gamma$ is a tunable focusing parameter, and $\alpha$ is the balance variant used for restraining the learning of positive samples. The larger of $p'$, the easier the sample will be classified. $\gamma$ adjusts the rate at which easy examples are down-weighted. Focal loss function can act as a more effective approach to deal with class imbalance.

## 2.4 Corrector

**The Architecture of Corrector.** As same as the previous work, Seq2seq with attention is taken as our corrector model. The encoder of corrector is a Bi-LSTM, attention mechanism is a linear projection and the decoder is a LSTM followed by a Softmax layer. The architecture of corrector is shown in Fig. 1. The Seq2seq model can capture the deep semantics of sequence via modelling the context. The attention mechanism could filter the effective information for decoder to generate a prediction sequence. The prediction is deemed as the corrected transcription.

**Transfer Learning Based Training.** Despite the Seq2seq model has achieved a respectable performance, it still has some drawbacks when applied in the ASR error correction task. As mentioned before, ASR transcription should consistent with the reference even though the reference is not accurate. In general, Seq2seq predicts the corrected transcriptions according to the grammatical rules and semantics of context. It will trigger a problem that corrector changes the "recognized correctly" words in transcriptions if the context disagrees with these words. To handle the shortcomings of Seq2seq, we introduce error heuristic from detector to guide the training of corrector by transfer learning. Error heuristic refers to the parameters of Bi-LSTM in detector. In advance, the corrector could master the location and error pattern in transcription with heuristic transferred from decoder. The issue that corrector makes wrong predictions can be addressed effectively. The parameters of detector are $\mathbf{W_D}$. The input sequence $\mathbf{x} = \{x_1, x_2, \ldots, x_n\}$ would be encoded by encoder and get the combined hidden state $\mathbf{h} = \{h_1, h_2, \ldots, h_n\}$. With the parameters $\mathbf{W_D} = \{w_i\}, i \in [1, N]$ from detector, $\mathbf{h}$ will be weighted by $\mathbf{W_D}$. $\mathbf{W_D}$ includes the error heuristic which can tell the corrector where and how to correct. Moreover, for time step $j$, the parameter $w_j$ is approximate to $P(p_j|\mathbf{x})$. The context vector $\mathbf{C_i}$ is computed as a weighted sum of $h_j$:

$$\mathbf{C_i} = \sum_{j}^{n} a_{ij}(h_j^T w_j) \tag{5}$$

$j$ is the time step of input, and $i$ is the time step of input. $n$ is the length of input sequence. $a_{ij}$ is the attention distribution.

$$a_{ij} = \frac{exp\left(e_{ij}\right)}{\sum_{k=1}^{n} exp\left(e_{ik}\right)} \tag{6}$$

$e_{ij}$ is an alignment score telling how well the inputs around position $j$ and the output at position $i$ match.

The rest steps for corrector are the same as the Seq2seq model in [12] that the output of decoder will depend on the target word $y_{i-1}$, LSTM hidden state for time $i-1$ and the context vector $C_i$ to predict the LSTM hidden state for time $i$. After training, we can get the predicted corrected transcriptions $y' = \{y'_1, y'_2, \dots y'_m\}$. The loss function for model training is presented as below:

$$Loss_{corrector} = \sum_{i}^{m} CE(y'_i, y_i) = -\sum_{i}^{m} y_i log(y'_i) \tag{7}$$

The pre-trained parameters from detector are not fixed during the training of corrector. The parameters obtained from the detector could also be regarded as a sentence-level embedding extracting the global information from input transcriptions. It contains not only the word-level but also the sentence-level error knowledge that can be utilized to help the error corrector to construct the feature vectors. Compared with the previous Seq2seq error correction methods, the proposed error-heuristic based Seq2seq model has an explicit training direction at the beginning of learning: the corrector needs to focus on the incorrectly recognized words in transcriptions, and makes use of the information from correctly recognized words to make predictions.

# 3 Experiment

## 3.1 Experiment Data

The dataset used in this paper is an open source dataset proposed in [7] intending to provide an ideal evaluation system for the intelligibility of automatic speech recognition system. Google and Sphinx ASR systems are chosen to transcribe five English speech databases (CommonVoice, TIMIT, Accent, UASPEECH, TORGO) into texts. Among these datasets, UASPEECH and TORGO are dysarthric speech databases, whose transcriptions are mostly single words and WER is particularly high compared with other databases. Considering the need for longer and accurate transcriptions to establish context vector for ASR error correction model, the transcriptions from UASPEECH and TORGO would not be employed to train our model. In a word, we have two datasets: Google and Sphinx, containing transcriptions from three database separately. The proportion of training set, validation set and test set in our experiment is 8:1:1. Table 1 demonstrates the specific division of data:

**Table 1.** Division of experiment data

| | Train | Valid | Test | Total |
|---|---|---|---|---|
| Google | 29722 | 3717 | 3716 | 37155 |
| Sphinx | 161153 | 20143 | 20144 | 201440 |
| Overall | 190875 | 23860 | 23860 | 238595 |

**Data Processing.** Our dataset is composed of transcriptions and corresponding references from ASR. But if our error detection task is carried on this dataset, we are short of target sequences. It is essential to tag the transcriptions with labels on word-level firstly and regard the labels as targets to supervise the learning of corrector. Since artificial tagging method is time-consuming and expensive, we propose an automatic tagging method based on editing distance algorithm to implement word-level tagging. Edit distance algorithm is on character-level. When calculate the edit distance between transcriptions and references, the error types should be stored as long as a character in a transcription needs to be edited. The error type includes Insertion ($I$), Deletion ($D$), and Substitution($S$). And then we combine the labels between two space to produce a word-level label. In the end, the letter label is changed into 0 or 1. Figure 2 shows an example of our tagging method.

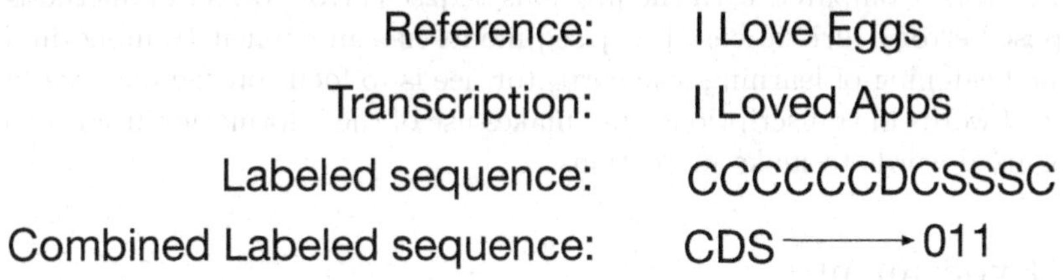

Reference: I Love Eggs
Transcription: I Loved Apps
Labeled sequence: CCCCCCDCSSSC
Combined Labeled sequence: CDS → 011

**Fig. 2.** An example for automatic tagging algorithm

Labeled sequence is on character-level. The red letters in the labeled sequence indicates that the characters of these positions are "space". We combine the letters between spaces by choosing the first error type to represent the error type of the whole word. In this way, the character-level labels could be transformed into word-level. Ultimately, we use 0 or 1 to replace the word-level label showing whether this word is transcribed right or not.

**Data Augment for Training.** The text data are transcribed by Google and Sphinx ASR. For the reason that a transcription is the one-best hypothesis from ASR, the lack of phonemic and semantic knowledge compared with decoding information will greatly influence the corrector performance. In order to solve the insufficient information problem, the detector is trained on the entire transcriptions by aggregating the Google and Sphinx datasets. Then the corrector with the transferred error heuristic will be trained on Google and Sphinx dataset respectively. For errors in Google and Sphinx datasets, they can be corrected more accurately with the abundant knowledge from the detector trained on the large dataset.

### 3.2 Metric

We can see from Table 1 that Google has less transcriptions compared with Sphinx, and the number of the latter is more than five times that of the former.

The evaluation metric is WER defined as the sum of the number of substitutions (S), insertions (I), and deletions (D) divided by the total number of words in the utterance (N).

$$WER = \frac{S + I + D}{N} \tag{8}$$

### 3.3 Model Setting

Clearly, our datasets are not sufficient enough to train a complex Seq2seq model like Transformer for an overfitting issue. To avoid overfitting, we decided to simplify our model as much as possible. Both the detector and encoder of corrector are 2-layers Bi-LSTM, the hidden states of which are 512. The decoder of corrector is a LSTM network with hidden state size 256. The word embeddings for the datasets uses 100-dimensional GloVe embedding. The optimization method is Adam [4]. The batch size is set as 128. Besides, early stopping on the validation data is adopted as a regularization strategy. Plus, we use warmup to make the model learn from a small learning rate at the beginning, gradually speeding up learning.

### 3.4 Result and Discussion

We evaluated our model on the test set. The baseline model is a Seq2Seq model described in [10]. The result of experiments are presented on Table 2.

**Table 2.** Experiment results on Google, Sphinx and overall datasets. The evaluation metric is WER. The bold represents the best result for each dataset.

| Dataset | Original | Seq2seq | Error Heuristic |
|---|---|---|---|
| Google | **11.62%** | 14.33% | 13.75% |
| Sphinx | 35.05% | 33.97% | **32.12%** |
| Overall | 32.97% | 29.89% | **28.62%** |

Original is the WER of transcriptions from ASR without correction. Seq2seq means the WER of transcriptions corrected by a Seq2seq model without error prior knowledge, while Error Heuristic is the WER of transcriptions corrected by our method.

For Google and Sphinx dataset, the WER of their original transcriptions are 11.62% and 35.05%. After Seq2seq based error correction method, the WER of Sphinx reduces to 33.97%, while there are more errors in corrected transcriptions than original for Google dataset. However, corrected by our error heuristic method, the WER for both Google and Sphinx has descended. The WER for Google and Sphinx declines from 14.33% to 13.75% and 33.97% to 32.12%, respectively. This is because the corrector has already "known" errors in the early stage of training. In addition, it is equivalent to carry on data augment for Google or Sphinx dataset by training the detector on the overall dataset. The limitation of one-best hypothesis can be relieved, and the model performance is improved. Despite that the WER for transcriptions corrected by the error heuristic based method is lower than the Seq2seq based corrector on Google dataset, it still cannot surpass the WER of original transcriptions 11.62%. Because the Google dataset with 37155 transcriptions is deficient compared with Sphinx, the corrector suffers from a serious overfitting problem, hard to get a desirable result.

As shown in Table 2, the best performance 28.62% is obtained by error-heuristic based error correction model trained on the overall dataset. The WER of our method is 1.27% lower than the Seq2seq corrector. The number of transcriptions in the overall dataset is the largest. Hence, we can draw a conclusion that the number of data is significant for ASR error correction task with one-beat hypothesis. When training the error correction model on the overall dataset, the large scale of data can make the corrector learn deep error patterns made by ASR. Inspired by the error heuristic, corrector can concentrate more on the correction for error words in contrast with Seq2seq corrector.

According to the experiment, it can be concluded that our error-heuristic based model is superior than the baseline Seq2seq model. In some extend, the limitation of one-best hypothesis and incorrect correction issue can be addressed in some degree.

## 4 Conclusion

In this paper, we present a new error-heuristic based ASR error correction method. Our corrector is a Seq2seq model with attention. It utilizes the error heuristic knowledge transferred from an error detection model to force the corrector only modifying the incorrectly recognized parts in a transcription while not making predictions for the correctly recognized parts. Experiments on three datasets demonstrate our error heuristic corrector achieves a competitive performance compared with the advanced Seq2seq error correction method. Various analyses are conducted to evaluate the corrector model and verify the effectiveness of our error heuristic method.

## References

1. Bassil, Y., Semaan, P.: ASR context-sensitive error correction based on microsoft n-gram dataset. arXiv preprint arXiv:1203.5262 (2012)
2. Guo, J., Sainath, T.N., Weiss, R.J.: A spelling correction model for end-to-end speech recognition. In: ICASSP 2019–2019 IEEE International Conference on Acoustics, Speech and Signal Processing (ICASSP), pp. 5651–5655. IEEE (2019)
3. Kaki, S., Sumita, E., Iida, H.: A method for correcting errors in speech recognition using the statistical features of character co-occurrence. In: COLING 1998 Volume 1: The 17th International Conference on Computational Linguistics (1998)
4. Kingma, D.P., Ba, J.: Adam: a method for stochastic optimization. arXiv preprint arXiv:1412.6980 (2014)
5. Lin, T.Y., Goyal, P., Girshick, R., He, K., Dollár, P.: Focal loss for dense object detection. In: Proceedings of the IEEE international conference on computer vision, pp. 2980–2988 (2017)
6. Mani, A., Palaskar, S., Meripo, N.V., Konam, S., Metze, F.: ASR error correction and domain adaptation using machine translation. In: ICASSP 2020–2020 IEEE International Conference on Acoustics, Speech and Signal Processing (ICASSP), pp. 6344–6348. IEEE (2020)
7. Moore, M., Saxon, M., Venkateswara, H., Berisha, V., Panchanathan, S.: Say what? a dataset for exploring the error patterns that two ASR engines make. In: INTERSPEECH, pp. 2528–2532 (2019)
8. Pennington, J., Socher, R., Manning, C.D.: Glove: global vectors for word representation. In: Proceedings of the 2014 Conference on Empirical Methods in Natural Language Processing (EMNLP), pp. 1532–1543 (2014)
9. Sarma, A., Palmer, D.D.: Context-based speech recognition error detection and correction. In: Proceedings of HLT-NAACL 2004: Short Papers, pp. 85–88 (2004)
10. Tanaka, T., Masumura, R., Masataki, H., Aono, Y.: Neural error corrective language models for automatic speech recognition. In: INTERSPEECH, pp. 401–405 (2018)
11. Vaswani, A., et al.: Attention is all you need. In: Advances in neural information processing systems, pp. 5998–6008 (2017)
12. Zhang, M., Yu, N., Fu, G.: A simple and effective neural model for joint word segmentation and POS tagging. IEEE/ACM Trans. Audio Speech Lang. Process. **26**(9), 1528–1538 (2018)
13. Zhang, S., Lei, M., Yan, Z.: Automatic spelling correction with transformer for ctc-based end-to-end speech recognition. arXiv preprint arXiv:1904.10045 (2019)
14. Zhou, G., Su, J.: Named entity recognition using an HMM-based chunk tagger. In: Proceedings of the 40th Annual Meeting of the Association for Computational Linguistics, pp. 473–480 (2002)

# Exploration on the Generation of Chinese Palindrome Poetry

Liao Chen, Zhichen Lai, Dayiheng Liu, Jiancheng Lv, and Yongsheng Sang(✉)

College of Computer Science, Sichuan University,
Chengdu 610065, People's Republic of China
sangys@scu.edu.cn

**Abstract.** Recently, Chinese poetry generation gains many significant achievement with the development of deep learning. However, existing methods can not generate Chinese palindrome poetry. Besides, there is no public dataset of Chinese palindrome poetry. In this paper, we propose a novel Chinese palindrome poetry generation model, named Chinese Palindrome Poetry Generation Model (CPPGM), based on the universal seq2seq model and language model with specific beam search algorithms. In addition, the proposed model is the first to generate Chinese palindrome poetry automatically, and is applicable to other palindromes, such as palindrome couplets. Compared with several methods we propose, the experimental results demonstrate the superiority of CPPGM with machine evaluation as well as human judgment.

**Keywords:** Palindrome · Poetry · Inference

## 1 Introduction

In the history of Chinese poetry, there were countless talents leaving many outstanding works. Poetry is the carrier of language, due to the different methods of expression and combination of words, various poetry genres and art forms are derived. Palindrome poetry is one of them.

Palindrome poetry is a unique genre of Chinese classical poetry which uses a special rhetorical device. To be specific, a palindrome poetry, by definition, is a group of sentences that read the same forward or backward. Palindrome poems have a long history since the Jin dynasty. In terms of creation technique, the palindrome poetry emphasize the artistic characteristics of poetry's repeated chanting to expressing one's thoughts and telling things. We illustrate an example of famous palindrome quatrains in Fig. 1. It is poems like this that have attracted numerous scholars to devote themselves to the study of palindromes. However, composing a palindrome poem is extremely difficult and poets need

This work is supported by the Key Program of National Natural Science Fund of China (Grant No. 61836006), and the Science and Technology Major Project of Sichuan province (Grant No. 2018GZDZX0028).


H. Yang et al. (Eds.): ICONIP 2020, LNCS 12532, pp. 754–765, 2020.
https://doi.org/10.1007/978-3-030-63830-6_63

not only profound literary accomplishment but also proficient writing skills. It is almost impossible for ordinary people to write poems under the particular setting.

Recently, Chinese poetry generation has achieved tremendous improvement due to the prosperous progress of deep learning and the availability of generative models [7]. Sequence-to-sequence model (Seq2seq) [14] has become the mainstream model to generate Chinese poems. Seq2seq models adopt the Recurrent Neural Network (RNN) Encoder-Decoder model with attention mechanism [2]. Benefiting from the powerful generative capacity of Seq2seq, the models can successfully generate Chinese poems [22].

Girl in the room

春闺

Curtained painting cabinet with the painting curtain droops,

垂帘画阁画帘垂，

Who is longing for others and who is to be long for?

谁系怀思怀系谁？

The shallow bothers flowers while flowers disturb the shallow,

影弄花枝花弄影，

Love links with the willow while the willow reminds her the love.

丝牵柳线柳牵丝。

Corresponding position

**Fig. 1.** An example of 7-character Chinese palindrome poetry. Every line of the poem has the structure of the A-B-C-D-C-B-A (The positions with the same color denote the corresponding positions). (Color figure online)

Most existing methods focus on the generation of Chinese poems with the writing format of traditional Tang poetries. However, there is no work investigating the generation of Chinese palindrome poetry. Different from traditional Chinese poetry generation which has a large dataset for training supervision [8], the generation of Chinese palindrome poetry only has a few sample references. In addition, the format of Chinese palindrome poetry should not only obey the rules of traditional poetries, but also satisfy the constraint of palindrome.

To overcome these difficulties, we propose a novel beam search based algorithm, Chinese Palindrome Poetry Generation Model (CPPGM), to automatically generate Chinese palindrome poems: 1) We first obtain an input word as the middle word of the first line, then we adopt a particular beam search algorithm with a language model trained by tradition Chinese poetry lines to extend the first line. 2) Based on the first line, we adopt another beam search algorithm and two Seq2seq models with attention mechanism, the forward model and the backward model, to generate the second line. 3) After the generation of previous two lines, we repeat step 2 with the previous lines as input to generate the rest lines.

It is worth mentioning that we propose a new beam search algorithm, unlike conventional beam search Algorithm [6], ours picks the most likely words in the corresponding positions in the generation of Chinese palindrome poems.

The main contributions of our work can be summarized as follows:

- To the best of our knowledge, we propose the first Chinese palindrome poetry generation model based on universal models.
- Since we train our model on the traditional poetry dataset, our method can also generate other Chinese palindromes, such as Chinese palindrome couplets.

## 2 Related Works

Poetry generation is a challenging task in Natural Language Processing (NLP). Although the poem is short, its meaning is profound. Thus, the topic of poem generation has attracted many researchers over the past decades. There are several various kinds of approaches to generate poems.

Originally, the traditional approaches rely on templates and rules, they employ templates to construct poems according to series of constraints, e.g., meter, stress, rhyme and word frequency, in combination with corpus-based and lexicographic resources. For instance, [1] use patterns based on parts of speech and WordNet [10]. [19] propose an approach to generate Haiku poem by using rules extracted from a corpus and additional lexical resources to expand user queries. In addition [11] present a Portuguese poem generation platform which utilizes grammar and semantic templates.

The second kind of approach is based on Statistical Machine Translation (SMT). [5] generate Chinese couplets using a phrase-based SMT approach which translates the first line to the second line. And [4] extend this algorithm to generate Chinese quatrains by translating the current line from the previous lines.

Recently, with the rapid development of deep learning on NLP, neural networks have been widely adopted to generate poetry. Different models have been proposed to generate poetry and shown great performance. [24] first propose to generate Chinese quatrains with Recurrent Neural Network (RNN), each generated line is vectorized by a Convolutional Sentence Model (CSM) and then packed into the history vector. To enhance coherence, their model needs to be interpolated with two extra SMT features. Given some input keywords, they use a character-based RNN language model [9] to generate the first line, and then the other lines are generated sequentially by a variant RNN. [17] use Long Short-term Memory (LSTM) based seq2seq model with attention mechanism to generate Song Iambics. Then, [16] extend this system and generate Chinese quatrains successfully. To guarantee that the generated poem is semantically and coherent accordant with the users' intents, [18] propose a two-stage poetry generation method. They first give the sub-topics of the poem and then utilize these sub-topics to generate each line by a revised RNN encoder-decoder model. [22] regard the poetry generation as a sequence-to-sequence learning problem. They construct three generation blocks (word-to-line, line-to-line and context-to-line) based on RNN Encoder-Decoder to generate the whole poem. More recently, [23] saves hundreds of human-authored poems in a static external memory to improve the innovation of generated quatrains and achieve style transfer. [20] employ Conditional Variational AutoEncoder (CVAE) [3,21] to generate Chinese poetry.

To some extent, our approach is related to the works of deep learning mentioned above. Furthermore, we investigate the generation of Chinese palindrome poetry.

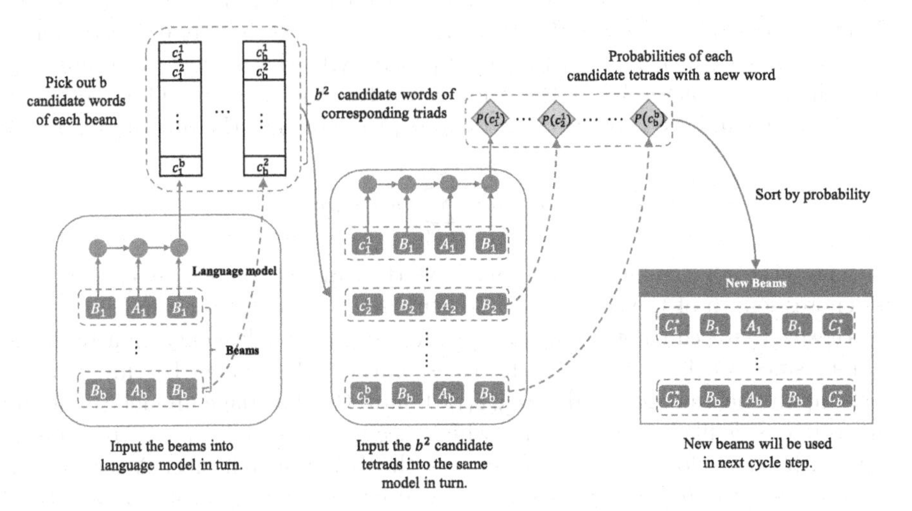


**Fig. 2.** The overall framework of the first line generation and a cyclic step (three-character to five-character) in the inference process.

# 3 Methodology

In this section, we will describe our inference method concretely for Chinese palindrome poetry generation, including: 1) generating the first line with a well-trained language model and a beam search algorithm, 2) generating the rest lines with two Seq2seq models with attention mechanism and another beam search algorithm. As our method bases on probability distribution, it could smoothly tally with other architectures, such as models presented in [15]. Considering the popularity and widespread use of RNNs, we use RNN-based models as an exemplar in this paper. The framework of our models is shown in Fig. 2 and Fig. 3.

## 3.1 Problem Definition

Let $w$ denote the external input, which is then placed in the middle position of the first line. Thus, with the input $w$, we should fill the corresponding positions of the first line with the same character $c$ like $C =$

$[c_1, c_2, ..., w(c_{(n+1)/2}), ..., c_{n-1}, cn]$ (the length of line $n$ is an odd number), and the corresponding positions mean the two characters with the subscripts which sum $n+1$. To compose a whole poem, we have to generate other three lines with the same format as the first line.

### 3.2 The First Line Generation

In order to generate the first line of the poem, we train a neural language model using the first line of traditional Chinese five-character quatrains and seven-character quatrains in our dataset. Different from other Chinese poetry generation methods which input a theme or picture into the models, we input only one Chinese character $W$ as the middle word of the first line. Our method is calculated to find set $\hat{C} = [\hat{c_1}, \hat{c_2}, ..., \hat{c_{n-1}}, \hat{c_n}]$ to maximize the probability $P(\hat{C})$:

$$P(\hat{C}) = \arg\max_{C_i \in C} \sum_{t=1}^{n} logP(c_t | c_{1:t-1}), \tag{1}$$

where the corresponding positions have the same character, i.e. $c_t = c_{n+1-t}$, $n$ denotes length of the sentence to be generated, $t$ denotes the time step.

However, we can not handle it by left-to-right greedy generation or beam search, since we have to meet the rules of the palindrome. To this end, we first input the middle word $w(c_{(n+1)/2})$ into the language model to obtain the top $b \times b$ likely next words, which keeps the search space size consistent with the following steps. $b$ denotes the beam search size. After obtaining the $b \times b$ candidate next words, we compute the probabilities of each triad $[c_{(n-1)/2}, w, c_{(n+3)/2}](c_{(n-1)/2} = c_{(n+3)/2})$ from candidate next words by the language model. Then, we pick out the top $b$ triads with the highest probability. Since we get $b$ triads with high probability, each triad is fed to the model to get $b$ probable next words, so that we could obtain $b \times b$ quintuples $[c_{(n-3)/2}, c_{(n-1)/2}, w, c_{(n+3)/2}, c_{(n+5)/2}]$. Again we compute the probabilities of each quintuple, then we could pick out first $b$ likely sequences. The rest can be done in the same manner to receive $b$ optimum 5-character or 7-character line. Deserved to be mentioned we augment our data by cutting the 5-character or 7-character sentences into 3-character and 5-character phrases so that the candidate triads and quintuples could be more reasonable. The whole algorithm is further summarized in 1, and to make it easier to understand the inference method, we show the approximate steps in Fig. 2.

### 3.3 The Rest Lines Generation

By means of the algorithm above we have now get dozens of candidates first lines, However the sentences selected by probability may not always satisfy the language preference of human-written poetry, as a consequence we filter the generated sentences again by the BiLingual Evaluation Understudy (BLEU) [12]. We use the first lines of human-written poems in our dataset as the references and the generated sentences as the candidates, so that we can generally pick out

**Algorithm 1.** The generation of 7-character first line

**Input:** beam size $b$, a trained language model ($model$) which could output the probabilities of next words with an sequence inputted, an external input $w$, which will be placed in the middle of the first line.
**Output:** $b$ candidate first lines.

Initialize the $beams$ = $[[w]]$, initialize the set produced $\hat{C}_n^l$ where $n$ denotes the position of the line, $l$ denotes the size of the set.
Obtain top $b \times b$ candidate $\hat{C}_5^{b\times b}$ by $model([w])$
for $i$ in $\hat{C}_5^{b\times b}$:
    Compute the $P(i|[i, w])$ by $model([i, w])$
end for
Update $beams$ by the $b$ triads $[\hat{C}_{3_m}^b, w, \hat{C}_{5_m}^b], m \in [1, b]$ with the highest probability
for $t$ in [5,7]:
    for $j$ in $beams$:
        $model(j)$
        Elect top $b$ candidates words of the next position
    end for
    Obtain $b \times b$ $t$-gram sequences sorted by probability
    Update $beams$ by first $b$ $t$-gram sequences
end for
return $beams$

the fluent and reasonable sentences which will be fed as input to two seq2seq models.

For the purpose of generating the second line of the poem, we train two seq2seq models, forward and backward models. We apply the first lines of poems in our dataset as input and the second lines in forward sequence and backward sequence as the target of the forward and backward models, respectively.

Since poetry is an entirety, and every sentence is related to each other, we generate the words from left to right as the common inference method to make the utmost of the information in previous lines. However, in the view of the form of palindrome poem, the probability of a word appearing in its corresponding position should be taken into account when generating a new line. In this case, we come up with another algorithm differs from the method of the first line generation.

Our key idea is to find the set $\hat{C} = [c_0, \hat{c_1}, \hat{c_2}, ..., c_{(n+1)/2}$ to maximize the probability $P(\hat{C}) = P_1(\hat{C}) \times P_2(\hat{C}))$, $P_1(\hat{C})$ and $P_2(\hat{C})$ are calculated in the same way but with different models via:

$$P(\hat{C}) = \underset{C_i \in C}{\arg\max} \sum_{t=1}^{n+1/2} logP(c_t|c_{1:t-1}, X), \tag{2}$$

where the $P_1(\hat{C})$ and $P_2(\hat{C})$ are computed by forward model and backward model, respectively. $c_0$ denotes the start character $< s >$, $X$ denotes the first line generated before, $n$ denotes length of the sentence to be generated, $t$ denotes the time step.

**Algorithm 2.** The generation of second line in 7-character

**Input:**
beam size $b$, two trained Seq2Seq models ($mode_f$,$mode_b$) which could output the probabilities ($P_1$,$P_2$) of next words with inputs fed in, an external input $X$, i.e. the first line.

**Output:** $b$ candidate second lines.

Initialize the $beams = [[[C_0], 1]]$ where $c_0$ denotes the start character $< s >$, initialize set produced $\hat{C}_n^l$ where $n$ denotes the position of the line, $l$ denotes the size of the set.

Obtain top $b \times b$ candidate $\hat{C}_1^{b\times b}$ and their probabilities $P_1(\hat{C}_1^{b\times b}|[C_0], x)$ by $model_f([C_0], X)$

for $i$ in $C_1^{b\times b}$:

    Calculate the $P_2(i|[C_0], X)$ by $model_b([C_0], X)$

    $P = P_1(i|[C_0], X) \times P_2(i|[C_0], X)$

end for

Pick out the first $b$ $[C_0, \hat{C}_{1_m}^b], m \in [1, b]$ with the highest probability

Update $beams$ by these 2-gram sequences and their probabilities

for $t$ in [3,4,5]:

    for $j$ in $beams$:

        $model_f(j[0], X)$

        Obtain top $b$ candidates words of the next position

        Multiply their probabilities with j[1] respectively to get intermediate probabilities

        Update beams by new $t$-gram sequences and their intermediate probabilities

    end for

    for $k$ in $beams$:

        $model_b(k[0][: -1], X)$

        Calculate the $P_2(k[0][-1]|[k[0][: -1], X)$

        $P_k = k[1] \times P_2(k[0][-1]|[k[0][: -1], X)$

    end for

    Obtain $b \times b$ $t$-gram sequences sorted by final probabilities $P$

    Update $beams$ by first $b$ $t$-gram sequences and their probabilities

end for

return $beams$

Take 7-character poems as an example, we first input the start character $< s >$ and the first line $X$ in the forward model to pick up top $b \times b$ candidates of the first position by the probability ($P_1$), and the word that appears in the first position is same as the word in its corresponding position, i.e. the last position. Since the backward model is trained with verses in reverse order, it could calculate the probabilities of the words in the corresponding position with the same inputs as the forward one. So we feed the same inputs into the backward model to obtain the probabilities ($P_2$) of the candidates appearing in their corresponding position. Thus, $P_1 \times P_2$ is the probability of the candidates appearing in both corresponding positions. Based on $P_1 \times P_2$, we could choose final top $b$ candidates of the first position, and save their probabilities

$(P(\hat{c_0}, \hat{c_1}) = P_1(\hat{c_0}, \hat{c_1}) \times P_2(\hat{c_0}, \hat{c_1})))$. Then we input each candidate sequence $[\hat{c_0}, c_1]$ and the first line $X$ respectively in the forward model to get first $b$ candidate next words, and their probabilities $(P_1(\hat{c_2}|\hat{c_0}, \hat{c_1}, X))$, so that we could obtain $b \times b$ candidate sequences. Again we feed the same input to the backward model to calculate the probabilities of the candidate next words in the corresponding position $(P_2(\hat{c_2}|\hat{c_0}, \hat{c_1}, X))$, and then multiply them with their corresponding forward probabilities. In order to pick out $b$ reasonable triads $(\hat{c_0}, \hat{c_1}, \hat{c_2})$, we have to multiply these probabilities with those old $(P(\hat{c_0}, \hat{c_1}))$. The rest can be done in the same rule to receive $b$ optimum $\hat{C} = [\hat{c_0}, \hat{c_1}, \hat{c_2}, \hat{c_3}, \hat{c_4}$. According to the symmetry of the palindrome we could easily obtain the completed lines. The integrated algorithm is specifically summarized in 2, and to make the approximate easier to understand we show the inference process in Fig. 3.

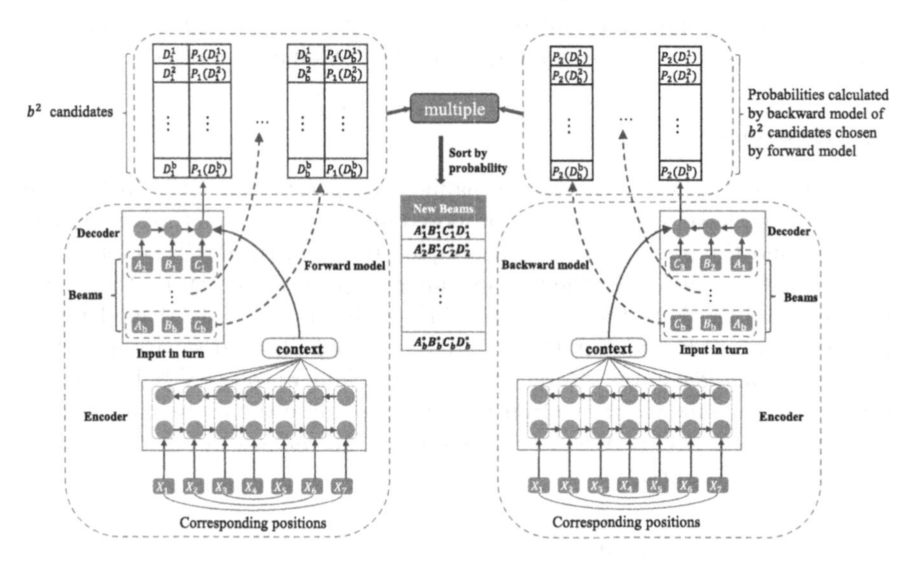


**Fig. 3.** The overall framework and inference process of the rest lines generation. We take a circular reasoning step of the second line generation as an example and the other inference steps are the same.

The generations of the third and fourth lines are almost the same as the second. To be specifically, the Seq2Seq models of the third line generation use the first two lines of the traditional poems as input and the third line in two opposite orders as the targets, respectively. The models of the last line generation use the first three lines of the traditional poems as input and the fourth line in two reverse orders as the targets separately.

# 4 Experiment

In this section, we first introduce the experimental settings, including datasets, data processing and data augmentation. Then, we will briefly introduce the methods that have been eliminated in the exploration process and use them as the baselines. Finally, we evaluate the results by both the qualitative metric and human perception study.

## 4.1 Datasets

We build two palindrome dataset:palindrome poems dataset and palindrome couplets dataset. For palindrome poems, we build a large traditional Chinese poetry corpus which contains 93,474 5-character quatrains and 139,196 7-character quatrain. The dataset we use in this task is built from all the first lines of poems in our corpus. For palindrome couplets, we gathered 478,115 couplets without any punctuation inside from over a million multiform Chinese couplets.

## 4.2 Baselines

Since there was no work on Chinese palindrome poetry generation before ours (CPPGM), we have to compare the proposed method with an intuitive approach (SMCP), and the original method we employed (4-3GM). SMCP simply generates the first four words in the way of universal poetry generation, then copies the first three words to their corresponding positions. For 4-3GM model, since the first four words and last three words are more like two separate parts in each line, it first generates several groups of first four words as SMCP, then uses trained language model or Seq2Seq model to calculate the probability of every candidate A-B-C-D-C-B-A structured sentence and chooses the candidates with the highest probability.

## 4.3 Metrics and Results

In other text generation tasks, following [13], researchers usually compare methods on standard sentence-level metric BLEU scores [12] which generally considers the correspondence between the ground truth and the generated sentences. However, there is no standard reference to evaluate our results, so we can only use part of sentences randomly sampled from the test set as references to evaluate fluency of the generated sentences. Besides BLEU scores, we also train two language models by poems and couplets in our dataset, respectively. Then, we can use the perplexity (PPL) to evaluate the quality of the complete sentences. We use the same models to generate poems and couplets in equal quantity by these three methods. The BLEU (the higher the better) scores and PPL (the lower the better) are shown in Table 1. According to the results, we find the score of proposed model is much better than the other two. It indicates that our method not only meets the structural requirements of the palindromes, but also makes the generated sentences more fluent and closer to those written by human beings. Since the first four words in other two methods are decoded in a regular way, the scores for these two methods are somewhat overrated in the fluency-based evaluation. However, in most cases the proposed method performs better than them. Table 2 shows some examples of palindrome poems and palindrome couplets generated by our model.

**Table 1.** Human evaluation results.

| Dataset | Method | Human evaluation | | | | | | Machine evaluation | |
|---|---|---|---|---|---|---|---|---|---|
| | | Sprachlichkeit | Fluency | Coherence | Poeticness | Meaning | Average | BLEU-2 | PPL |
| *Poetry* | SMCP | 2.34 | 2.22 | 2.36 | 2.02 | 2.12 | 2.21 | 0.38 | 589.27 |
| | 4-3GM | 2.78 | 2.68 | 2.68 | 2.47 | 2.57 | 2.64 | 0.33 | 401.89 |
| | **CPPGM** | **3.35** | **3.28** | **3.33** | **3.13** | **3.27** | **3.27** | **0.56** | **188.76** |
| *Couplet* | SMCP | 2.58 | 2.53 | 2.76 | 2.54 | 2.52 | 2.59 | 0.40 | 602.54 |
| | 4-3GM | 2.85 | 2.81 | 2.90 | 2.66 | 2.84 | 2.81 | 0.37 | 352.16 |
| | **CPPGM** | **3.61** | **3.52** | **3.53** | **3.55** | **3.50** | **3.54** | **0.66** | **167.23** |

### 4.4 Human Evaluation

While the above analysis is useful as a sanity check, it offers limited intuition about how closer the generated samples to man made. The main problem is that we do not have strong-alignment data pairs to evaluate our results. Accordingly, we conduct a human evaluation to compare the perceived sample quality of the different models. Following [4,24], we collect generations of each of the three methods including poetry and couplets on 50 randomly-selected test instances. Then, we launch a crowd-sourcing online study, asking 20 evaluators to rate the generations by **Fluency**, **Coherence**, **Poeticness**, and **Meaning**. In addition, we add a new evaluation metric (**Sprachlichkeit**) to determine whether the palindrome is consistent with human language. The scores range from 1 to 5, the higher the better. The detailed description is listed below: (a) **Sprachlichkeit**: Is the palindrome consistent with human language? (b) **Fluency**: Does the palindrome read smoothly and fluently? (c) **Coherence**: Is the palindrome coherent across lines? (d) **Poeticness**: Does the palindrome follow the rhyme and tone requirements? (e) **Meaning**: Does the palindrome have a certain meaning and artistic conception?

The results are shown in Table 1. Due to the strict requirements of the palindrome, the scores will not be very high, but we can see that CPPGM consistently outperforms all baselines in every aspect.

**Table 2.** Examples of 7-character Chinese palindrome poetry and couplet by CPPGM.

| |
|---|
| 酒饮畅欢畅饮酒，<br>Drinking wine with pleasure further makes me drink more,<br>诗成醉倒醉成诗，<br>Brilliant thoughts come up when I am drunk so that helps me write poems,<br>此生无复无生此。<br>Life comes once, and it won't be the same again.<br>今是仙门仙是今。<br>Right now has been the best time and the best time is now. |
| 舞蝶招花招蝶舞，<br>Butterflies fly around the flowers that attract more butterflies to dance around,<br>飞龙引凤引龙飞。<br>Soar dragon attracts the phoenix to hover toge. |

## 5 Conclusions

In this paper, we propose a model to generate Chinese palindrome poetry. To the best of our knowledge, the method is the first to generate Chinese palindrome poetry and other palindromes such as palindrome couplet based on universal models. We compare the method with several baselines using machine evaluation as well as human judgment. The experimental results prove that the proposed method is an effective and efficient approach for Chinese palindromes generation. Since we only used the basic language model and seq2seq model in the experiment, we can improve the generation effect by improving the models in the future.

## References


1. Agirrezabal, M., Arrieta, B., Astigarraga, A., Hulden, M.: Pos-tag based poetry generation with wordnet. In: Proceedings of the 14th European Workshop on Natural Language Generation. pp. 162–166 (2013)
2. Bahdanau, D., Cho, K., Bengio, Y.: Neural machine translation by jointly learning to align and translate. arXiv preprint arXiv:1409.0473 (2014)
3. Fabius, O., van Amersfoort, J.R.: Variational recurrent auto-encoders. arXiv preprint arXiv:1412.6581 (2014)
4. He, J., Zhou, M., Jiang, L.: Generating chinese classical poems with statistical machine translation models. In: Twenty-Sixth AAAI Conference on Artificial Intelligence (2012)
5. Jiang, L., Zhou, M.: Generating chinese couplets using a statistical mt approach. In: Proceedings of the 22nd International Conference on Computational Linguistics-Volume 1. pp. 377–384. Association for Computational Linguistics (2008)
6. Koehn, P.: Pharaoh: a beam search decoder for phrase-based statistical machine translation models. In: Conference of the Association for Machine Translation in the Americas. pp. 115–124. Springer (2004)
7. Liu, D., Guo, Q., Li, W., Lv, J.: A multi-modal chinese poetry generation model. In: 2018 International Joint Conference on Neural Networks (IJCNN). pp. 1–8. IEEE (2018)

8. Liu, Y., Liu, D., Lv, J.: Deep poetry: A chinese classical poetry generation system. arXiv preprint arXiv:1911.08212 (2019)
9. Mikolov, T., Karafiát, M., Burget, L., Černockỳ, J., Khudanpur, S.: Recurrent neural network based language model. In: Eleventh annual conference of the international speech communication association (2010)
10. Miller, G.A.: WordNet: An electronic lexical database. MIT press (1998)
11. Oliveira, H.G.: Poetryme: a versatile platform for poetry generation. Computational Creativity, Concept Invention, and General Intelligence **1**, 21 (2012)
12. Papineni, K., Roukos, S., Ward, T., Zhu, W.J.: Bleu: a method for automatic evaluation of machine translation. In: Proceedings of the 40th annual meeting on association for computational linguistics. pp. 311–318. Association for Computational Linguistics (2002)
13. Sun, Q., Lee, S., Batra, D.: Bidirectional beam search: Forward-backward inference in neural sequence models for fill-in-the-blank image captioning. In: Proceedings of the IEEE Conference on Computer Vision and Pattern Recognition. pp. 6961–6969 (2017)
14. Sutskever, I., Vinyals, O., Le, Q.V.: Sequence to sequence learning with neural networks. In: Advances in neural information processing systems. pp. 3104–3112 (2014)
15. Vaswani, A., et al.: Attention is all you need. In: Advances in neural information processing systems, pp. 5998–6008 (2017)
16. Wang, Q., Luo, T., Wang, D.: Can machine generate traditional chinese poetry? a feigenbaum test. In: International Conference on Brain Inspired Cognitive Systems. pp. 34–46. Springer (2016)
17. Wang, Q., Luo, T., Wang, D., Xing, C.: Chinese song iambics generation with neural attention-based model. arXiv preprint arXiv:1604.06274 (2016)
18. Wang, Z., et al.: Chinese poetry generation with planning based neural network. arXiv preprint arXiv:1610.09889 (2016)
19. Wu, X., Tosa, N., Nakatsu, R.: New hitch haiku: An interactive renku poem composition supporting tool applied for sightseeing navigation system. In: International Conference on Entertainment Computing. pp. 191–196. Springer (2009)
20. Yan, R., Jiang, H., Lapata, M., Lin, S.D., Lv, X., Li, X.: I, poet: automatic chinese poetry composition through a generative summarization framework under constrained optimization. In: Twenty-Third International Joint Conference on Artificial Intelligence (2013)
21. Yan, X., Yang, J., Sohn, K., Lee, H.: Attribute2image: Conditional image generation from visual attributes. In: European Conference on Computer Vision. pp. 776–791. Springer (2016)
22. Yi, X., Li, R., Sun, M.: Generating chinese classical poems with rnn encoder-decoder. In: Chinese Computational Linguistics and Natural Language Processing Based on Naturally Annotated Big Data, pp. 211–223. Springer (2017)
23. Zhang, J., et al.: Flexible and creative chinese poetry generation using neural memory. arXiv preprint arXiv:1705.03773 (2017)
24. Zhang, X., Lapata, M.: Chinese poetry generation with recurrent neural networks. In: Proceedings of the 2014 Conference on Empirical Methods in Natural Language Processing (EMNLP). pp. 670–680 (2014)

# Improving Mongolian-Chinese Machine Translation with Automatic Post-editing

Shuo Sun, Hongxu Hou(✉), Nier Wu, and Ziyue Guo

College of Computer Science-college of Software, Inner Mongolia University, Hohhot, China
sunshuo07@126.com, cshhx@imu.edu.cn, wunier04@126.com, guoziyue08@126.com

**Abstract.** Fluency and faithfulness are the main criteria to evaluate the quality of machine translation. In order to acquire excellent translation results, the common method is to learn semantic-rich word embeddings by fine-tuning or pre-processing. However, there is no human intervention when generating translation due to the black-box prediction characteristics of neural network, which limits the generation of higher quality translation. Therefore, this paper proposes a translation automatic post-editing method combined with copying-rewriting (CoRe) network and introduces a double-ended attention module to realize the interaction between the source sentences and the machine translation. Meanwhile, we utilize interaction result (copy scores) to determine which fragments in the translation will be copied or be rewritten. Our method is verified on the CWMT2018 Mongolian-Chinese translation task and has obtained significant results.

**Keywords:** Machine translation · Automatic post-editing · Core network

## 1 Introduction

Machine translation is the process of automatically transforming one language into another by computer. Recently, neural machine translation (NMT) [1,2] has improved in leaps and bounds. What's more, numerous translation methods and models have been proposed. Although the quality of machine translation is continually improving, a few translation errors still exist. To correct these errors, automatic post-editing (APE) emerges at a historic moment. It can automatically correct errors in machine translation without starting translation from scratch. Knight el al. [3] first proposed the concept of automatic post-editing in a rule-based approach. Simard el al. [4] trained a phrase-based monolingual statistical machine translation (SMT) to redress machine translations with errors. Pal et al. [5] employed a bidirectional recurrent neural network (RNN) encoder-decoder model to build a monolingual machine translation system for APE of translations. After that, the multi-source neural machine model utilizing machine translation output and its corresponding source language sentences became a


H. Yang et al. (Eds.): ICONIP 2020, LNCS 12532, pp. 766–775, 2020.
https://doi.org/10.1007/978-3-030-63830-6_64

popular method [6,7]. That is, given a source sentence (*src*) and a machine translation (*mt*), APE outputs a post-edited translation (*pe*). Recently, Transformer [8] which has achieved state-of-the-art results on many tasks, has grown into the new favorite of APE tasks. [9,10] proposed multi-encoder APE model based on Transformer to further enhance the quality of translations.

An obvious feature of APE is that many words in *mt* can be copied into *pe*. As shown in Fig. 1, two Chinese words "会议" and "团结胜利" appear in both *mt* and *pe*. It can be seen that the copied words are not necessarily in the same position in *mt* and *pe*. Therefore, major challenges of the APE copy model are not only to determine which words to copy, but also to place the copied words in the correct position. Existing APE models generally have two problems: (1) *src* and *mt* adopt two encoders to learn representations separately, which may lead the translation to be smooth but inadequate. Meanwhile, we have no idea about which words can be copied and which words need to be generated. (2) APE models lack obvious labels to indicate which word in *mt* should be copied.

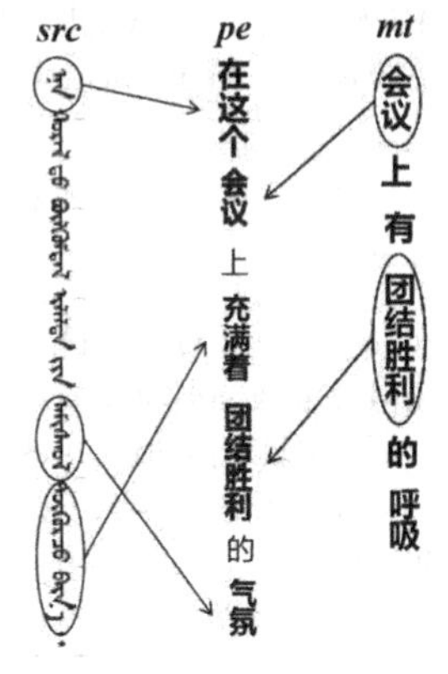


**Fig. 1.** An example of our method. In *pe*, green words such as "会议" are copied from *mt*. The red word such as "气氛" is generated from the corresponding word in the *src*. Note that copying some words directly from *mt* to *pe* is a common phenomenon.

In this paper, we utilize an APE model that incorporates copy network to model Mongolian-Chinese translation tasks. In view of the above two problems, we propose solutions: (1) by learning the interactive representation of *src* and *mt*, we align the representations of the two languages so that they can interact fully to determine when *pe* is generated from *src* or copied from *mt*. (2) label each word in *mt* and employ a prediction module to predict which words in *mt* should be copied. We conduct a several experiments on CWMT2018 Mongolian-Chinese task which fully certificates the effectiveness of our method.

## 2 Background and Related Work

**General Automatic Post-editing Method in NMT.** With the extensive application of deep learning in natural language processing (NLP), many researchers apply it to automatic post-editing (APE) of translations and develop into the mainstream method. It can be roughly divided into two categories: Single-source APE and Multi-source APE.

Pal et al. [5] first integrate APE into NMT. They adopt bidirectional RNN to build a monolingual machine translation system. Recently, many researchers regard APE as a multi-source sequence-to-sequence learning problem [6,7]. They encode the source sentences and machine translation respectively, and then decode it into a post-edited translation. With the advent of Transformer, APE models based on it have emerged endlessly [9,10]. The typical study is Junczys-Dowmunt et al. [10] proposed a Log-linear model that combines monolingual and bilingual neural machine translation model outputs, and has achieved state-of-the-art results on many APE tasks.

Given a source sentence $X = x_1...x_S$ with $S$ words, a machine translation $Y^{mt} = y_1...y_M$ with $M$ words and a post-edited translations $Y = y_1...y_E$ with $E$ words. APE model is given by:

$$P(y|x, y^{mt}; \theta) = \prod_{e=1}^{E} P(y_e|x, y^{mt}, y_{<e}; \theta) \tag{1}$$

where $y_e$ is the $e$-th target word in the post-edit translation, $y_{<e} = y_1...y_{e-1}$ and $\theta$ is a set of parameters of the model. $P(y_e|x, y^{mt}, y_{<e}; \theta)$ is the word-level translation probability, which is proportional to $exp(h_e^{pe} W_g)$ and $W_g$ is a vector matrix. $h_e^{pe}$ is the representation of the $e$-th word in $pe$. The equation is as follows:

$$h_e^{pe} = Decoder(y_{<e}, H^{src}, H^{mt}, \theta) \tag{2}$$

where $H^{src} = Encoder^{src}(x, \theta)$ and $H^{mt} = Encoder^{mt}(y^{mt}, \theta)$ are representations of $src$ and $mt$ respectively. The major disadvantage of this method is the lack of interaction between $src$ and $mt$, resulting in poor APE performance.

# 3 Approach

In this section, we will describe the APE combined with CoRe network. The overall architecture is shown in Fig. 2. It contains four parts: alignment interactive module, prediction module, CoRe and joint training.

## 3.1 Bilingual Alignment Representation

The representation matching between the source language and the target language is the key issue to detect translation faithfulness. According to [11], we adopt Universal Lexical Representation (ULR) to obtain the representations. It is a novel representation for multi-lingual embedding where each word from any language is represented as a probabilistic mixture of universal-space by employing a discrete "universal token space" to make the semantically similar words from different languages have similar representations. Then we directly learn the embedding matrix for these universal tokens in our NMT training. The each word embedding of $src$ and $mt$:

$$X_s = \sum_{i=1}^{M} E^U(u_i) \cdot q(u_i|s) \tag{3}$$

$$Y_m^{mt} = \sum_{i=1}^{M} E^U(u_i) \cdot q(u_i|m) \tag{4}$$

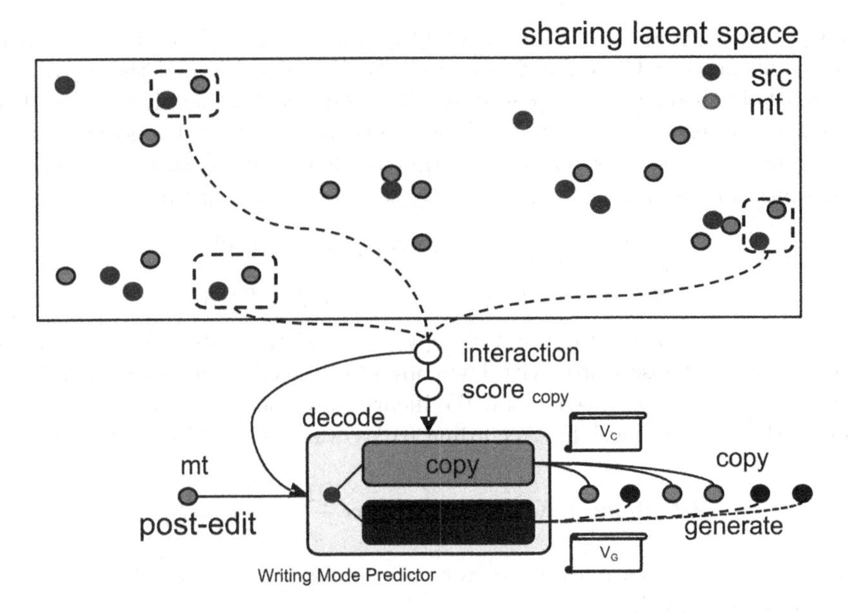


**Fig. 2.** The Illustration of automatic post-editing method combined with CoRe. First, *src* and *mt* are mapped to a share latent space by universal lexical representation alignment information of the two languages, and obtain the copy score of each word in *mt*. After that, the vector of scores is transferred to the CoRe Network to decode *pe*.

where $X_s$ and $Y_m^{mt}$ are the embedding of the $s$-th word at the source sentence and the $m$-th word at the machine translation respectively. $E^U$ and $M$ are respectively the embedding matrix and size for universal tokens. The mapping $q$ projects the multilingual words into the universal space based on their semantic similarity. See [11] for more details.

Following Lample [12], we adopt interactive representation learning method to obtain the representation by adding attention mechanism between shared cross-lingual word embedding. The bilingual alignment representation of *src* and *mt* are as follow:

$$H^{Interact} = Encoder^{Interact}\left(\left[X; Y^{mt}\right]; \theta\right) \tag{5}$$

The method guides model to align the representations of two languages, so that it can learn the alignment relationship between *src* and *mt*. It is useful for discovering which words in *src* is untranslated and which words in *mt* is translated correctly. The interactive representation of the two languages is also used prediction module to get the copy scores of each word. In the following, we will describe in detail how to predict the words that should be copied in *mt*.

### 3.2 Prediction Module

After obtaining the alignment representation of two languages, we still unable to distinguish whether the word needs to be edited. To strengthen the post-editing ability of model, we add a corresponding classification label to the predicted output via Longest Common Sequence(LCS) [13] algorithm. If the word in *mt* also exists in the common sequence, it will be labeled 1, otherwise labeled 0. We label it as sequence $L_1 ... L_M$. Since these labels are discrete and non-differentiable during training, we propose a prediction module that utilizes the interactive representation (Sect. 3.1) to get the copy scores of each word in *mt*:

$$H^{Predictor} = Predictor\left(\left[X; Y^{mt}\right]; \theta\right) \tag{6}$$

$$S = Sigmoid([H_1^{Predictor}; ...; H_M^{Predictor}]W_s) \tag{7}$$

where $S$ is a vector representing the copy scores of $M$ words in $mt$, and $W_s$ is a weight matrix. Those words with low copy scores need to decrease their importance in the attention mechanism. We incorporate the scores into our model to modify the attention weight. According to Vaswani [8], we define attention as:

$$Attention(Q, K, V) = Softmax(\frac{QK^T}{\sqrt{d}} \bigodot [m; S]^T) \tag{8}$$

where $m = \{1.0\}^I$ is a masking vector corresponding to $src$ and $S$ is the vector of copy scores (Eq. 7).

### 3.3 CoRe Network

In Sect. 3.2, we get copy score of each word in *mt*. As shown in Fig. 2, we transfer the score vector to the copy-rewriting (Co-Re) network. Words with high scores are copied from *mt* to *pe* and placed in correct position. Co-Re is a novel Seq2Seq model developed by [14] to fuse a copying decoder and a restricted generative decoder. The copying decoder finds the position to be copied based on attention mechanism form *mt* and the generative decoder produces the words form *src*. At the beginning of decoding, the predictor can predict whether the current translation should be copied or regenerated by using the writing mode predictor. At the same time, the predictor can generate the corresponding vocabularies $V_C$ and $V_G$ according to the results. After that, the model references the corresponding dictionary to generate the token. The probabilities as follow:

$$P(y_e|x, y^{mt}, y_{<e}; \theta) = \mu_e P^C(y_e) + (1 - \mu_e)P^G(y_e) \tag{9}$$

where $P^C(y_e)$ is the copying probability for $y_e$ and $P^G(y_e)$ is the generating probability for $y_e$. $\mu_e$ measures the contributions of the two decoders. It is noted that $\mu_e$ has the following actual supervision in training set: if target word at $e$ exists in the source, $\mu_e = 1$. Otherwise,$\mu_e = 0$. Therefore, we can utilize this supervision to guide the writing mode prediction. See [14] for more details.

### 3.4 Joint Training

Following [15], the training objective of this method consists of three parts: The first part is the log-likelihood loss of APE, which is related to Eq. (1):

$$L_{APE}(\theta) = -LogP\left(y|x, y^{mt}; \theta\right) \tag{10}$$

**Table 1.** Evaluation of translation quality on CWMT2018 tasks.

| Systems | CWMT2018 | | | |
|---|---|---|---|---|
| | MO-ZH | Promote | ZH-MO | Promote |
| Original [8] | 30.63 | – | 27.58 | – |
| POSTECH [9] | 31.02 | 0.39 | 28.13 | 0.55 |
| MS-COPY [9,18] | 31.17 | 0.54 | 28.21 | 0.63 |
| MS-UEDIN [10] | 32.09 | 1.46 | 28.54 | 0.96 |
| Our method | 32.87 | 2.24 | 29.12 | 1.54 |

The second part is the loss function of CoRe:

$$L_{CoRe}(\theta) = -\frac{1}{M}\sum_{m=1}^{M}(L_m - C_m)^2 \tag{11}$$

where $L_m$ is the label of $mt$ (Sect. 3.2), and $C_m = \sum_{e=1}^{E}\mu_e \times P^C(y_m^{mt})$ is used to measure the possibility that the $m$-th word in $mt$ is copied by CoRe.

The third part is the cross-entropy loss of the Prediction module:

$$L_{Predictor}(\theta) = -\sum_{m=1}^{M}[L_m \cdot Log(S_m) + (1 - L_m) \cdot Log(1 - S_m)] \tag{12}$$

where $S_m$ is the copy score of the $m$-th word in $mt$.

Therefore, our overall training objectives are as follows:

$$L(\theta) = (1-\alpha)(L_{APE}(\theta) + \beta L_{CoRe}(\theta)) + \alpha L_{Predictor}(\theta) \tag{13}$$

where $\alpha$ and $\beta$ are hyper-parameters.

## 4 Experiment and Analysis

We evaluate CWMT2018 Mongolian-Chinese tasks to verify the effectiveness of our method.

### 4.1 Datasets and Setting

In this paper, the dataset adopts 261643 sentence pair Mongolian-Chinese bilingual aligned corpus provided by CWMT2018, we utilize 220000 sentence pairs as training set, 20822 as validation set, and the rest as test set. In order to reduce the confusion of the model, the above data deletes sentences with a length greater than 50. We take the target sentence as *pe*, and utilize the translation model with the best performance in MO→ZH and ZH→MO tasks to obtain $mt$. We perform word segmentation processing on the Chinese. On the Mongolian end, due to its own natural separator, we encode it with byte-pair encoding (BPE) [16].

**Table 2.** Ablation study on CWMT2018 MO→ZH tasks. "o" means utilize this module and "×" means not utilize.

| Module | | | | CWMT2018 MO→ZH |
|---|---|---|---|---|
| Alignment | Predictor | CoRe | Joint training | BLEU |
| o | × | × | × | 32.19 |
| × | o | o | o | 31.96 |
| o | × | o | × | 32.34 |
| o | o | × | × | 32.44 |
| o | o | o | × | 32.63 |
| o | o | o | o | **32.87** |

For the original Transformer model, CoRe and the method of this paper, the multi-head attention is set to 6 layers at encoder and decoder. The hidden size and word embedding dimensions of the encoder and decoder are both set to 512 and $dropout = 0.2$. We use the Adam for optimization. A single model was obtained by averaging the last 20 checkpoints and we employ adaptive methods to adjust the learning rate. And the beam search method is used to sample the target token, where the beam size $K = 6$. We utilize the sentence-level evaluation index BLEU [17] as the evaluation standard. All models are implemented in TensorFlow, and the experiment is trained on a single-core NVIDIA TITAN X graphics card. When the model does not improve the evaluation of the development set, the training will be stopped.

### 4.2 Main Results and Analysis

To verify the influence of hyper-parameters (Eq. 13) on the experimental results, we conducted several typical experiments. The best performance is obtained when the values of $\alpha$ and $\beta$ are 0.8 and 1.0, respectively, which indicates that the Prediction module and the CoRe network are necessary in our method. Therefore, in the following experiment, we set $\alpha = 0.8$ and $\beta = 1.0$.

Table 1 shows the MO→ZH and ZH→MO results of different models on the CWMT2018 dataset. Original indicates the original Transformer model without APE. POSTECH [9] is a multi-encoder model based on Transformer. MS-COPY is a multi-source Transformer model that incorporates CopyNet [9,18]. Compared with the original Transformer, both MO→ZH and ZH→MO translation tasks have been improved, which proves that it is effective to integrate the CoRe into APE. And compared with the best-performing MS-UEDIN [10] (a multi-source APE model based on Transformer that shares the encoders of *src* and *mt*), our model is also significantly improved. The experimental results show that the CoRe module, Alignment module and Predictor module are effective for low-resource languages on APE tasks.

| | |
|---|---|
| **Source:** | [illegible] |
| **MT:** | 消息传到这里。 |
| **MS-UEDIN:** | 消息传到我们这了。 |
| **Our Model:** | 远处的消息传来了这里。 |
| **Reference:** | 远方的消息传到我们这里来了。 |
| **Source:** | 在20世纪80年代以前美国没有快餐店。 |
| **MT:** | 20 [illegible] |
| **MS-UEDIN:** | 20 [illegible] |
| **Our Model:** | 20 [illegible] |
| **Reference:** | [illegible] |

**Fig. 3.** An example of our method. In *pe*, green words such as "会议" are copied from *mt*. The red word such as "气氛" is generated from the corresponding word in the *src*. Note that copying some words directly from *mt* to *pe* is a common phenomenon.

### 4.3 Ablation Study

Our method includes four modules: Alignment Interactive module, Prediction module, CoRe module and Joint training. To prove the essential of them, we conducted ablation learning, and the experimental results are shown in Table 2. It can be seen that Alignment module and Prediction module have the greatest influence on the result, which are indispensable parts of the method in this paper. At the same time, the CoRe module also play a critical role, and Joint training is also beneficial for improving APE, but it has less impact than the previous three.

### 4.4 Case Study

As shown in Fig. 3 we list a MO→ZH and a ZH→MO translation cases. Compared with MS-UEDIN, the translation obtained by our method is fluent and faithful. In the first case, our method can accurately copy "消息" and "这里" from *mt* and place them in the correct position. At the same time, the Chinese word "在远方" can also be generated from the *src*. In ZH→MO case, many Mongolian words can be copied directly from *mt* to *pe* and accurately translate "快餐店".

## 5 Conclusion

In this paper, we adopt a novel automatic post-editing method to model Mongolian-Chinese translation tasks. It allows source sentences and machine translation to "interact" through alignment interactive representation learning, so that the model can recognize whether the target word should be copied or rewritten. In addition, through a prediction module to generate copy scores to predict which words should be copied. Experiments at CWMT2018 show that our approach is still improved compared to the previous model with state-of-the-art results, which fully proves the effectiveness of the method.

## References


1. Bahdanau, D., Cho, K., Bengio, Y.: Neural machine translation by jointly learning to align and translate. In: 3rd International Conference on Learning Representations, ICLR 2015, San Diego, CA, USA, 7–9 May 2015, Conference Track Proceedings (2015)
2. He, et al.: Dual learning for machine translation. In: Advances in Neural Information Processing Systems 29: Annual Conference on Neural Information Processing Systems 2016, 5–10 December 2016, Barcelona, Spain, pp. 820–828 (2016)
3. Knight, K., Chander, I.: Automated postediting of documents. In: Proceedings of the 12th National Conference on Artificial Intelligence, Seattle, WA, USA, 31 July–4 August 1994, vol. 1, pp. 779–784 (1994)
4. Simard, M., Goutte, C., Isabelle, P.: Statistical phrase-based post-editing. In: Human Language Technology Conference of the North American Chapter of the Association of Computational Linguistics, Proceedings, 22–27 April 2007, Rochester, New York, USA, pp. 508–515 (2007)
5. Pal, S., Naskar, S.K., Vela, M., van Genabith, J.: A neural network based approach to automatic post-editing. In: Proceedings of the 54th Annual Meeting of the Association for Computational Linguistics, ACL 2016, 7–12 August 2016, Berlin, Germany, Volume 2: Short Papers (2016)
6. Libovicky, J., Helcl, J., Tlusty, M., Bojar, O., Pecina, P.: CUNI system for WMT16 automatic post-editing and multimodal translation tasks. In: Proceedings of the First Conference on Machine Translation, WMT 2016, Colocated with ACL 2016, 11–12 August 2016, Berlin, Germany, pp. 646–654 (2016)
7. Zoph, B., Knight, K.: Multi-source neural translation. In: NAACL HLT 2016, The 2016 Conference of the North American Chapter of the Association for Computational Linguistics: Human Language Technologies, San Diego California, USA, 12–17 2016, pp. 30–34 (2016)
8. Vaswani, A., et al.: Attention is all you need. In: Advances in Neural Information Processing Systems 30: Annual Conference on Neural Information Processing Systems 2017, 4–9 December 2017, Long Beach, CA, USA, pp. 5998–6008 (2017)
9. Shin, J., Lee, J.-H.: Multi-encoder transformer network for automatic post-editing. In: Proceedings of the Third Conference on Machine Translation: Shared Task Papers, WMT 2018, Belgium, Brussels, 31 October–1 November 2018, pp. 840–845 (2018)
10. Junczys-Dowmunt, M., Grundkiewicz, R.: Log-linear combinations of monolingual and bilingual neural machine translation models for automatic post-editing. In: Proceedings of the First Conference on Machine Translation, WMT 2016, Colocated with ACL 2016, 11–12 August 2016, Berlin, Germany, pp. 751–758 (2016)
11. Gu, J., Hassan, H., Devlin, J., Li, V.O.K.: Universal neural machine translation for extremely low resource languages. In: Proceedings of the 2018 Conference of the North American Chapter of the Association for Computational Linguistics: Human Language Technologies, NAACL-HLT 2018, New Orleans, Louisiana, USA, 1–6 June 2018, Volume 1 (Long Papers), pp. 344–354 (2018)

12. Conneau, A., Lample, G.: Cross-lingual language model pretraining. In: Advances in Neural Information Processing Systems 32: Annual Conference on Neural Information Processing Systems 2019, NeurIPS 2019, 8-14 December 2019, Vancouver, BC, Canada, pp. 7057–7067 (2019)
13. Wagner, R.A., Fischer, M.J.: The string-to-string correction problem. J. ACM **21**(1), 168–173 (1974)
14. Cao, Z., Luo, C., Li, W., Li,S.: Joint copying and restricted generation for paraphrase. In: Proceedings of the Thirty-First AAAI Conference on Artificial Intelligence, 4–9 February 2017, San Francisco, California, USA, pp. 3152–3158 (2017)
15. Huang, X., Liu, Y., Luan, H., Xu, J., Sun, M.: Learning to copy for automatic post-editing. In: Proceedings of the 2019 Conference on Empirical Methods in Natural Language Processing and the 9th International Joint Conference on Natural Language Processing, EMNLP-IJCNLP 2019, Hong Kong, China, 3–7 November 2019, pp. 6121–6131 (2019)
16. Sennrich, R., Haddow, B., Birch, A.: Neural machine translation of rare words with subword units. In: Proceedings of the 54th Annual Meeting of the Association for Computational Linguistics, ACL 2016, 7–12 August 2016, Berlin, Germany, Volume 1: Long Papers (2016)
17. Papineni, K., Roukos, S., Ward, T., Zhu, W.-J.: BLEU: a method for automatic evaluation of machine translation. In: Proceedings of the 40th Annual Meeting of the Association for Computational Linguistics, 6–12 July 2002, Philadelphia, PA, USA, pp. 311–318 (2002)
18. Gu, J., Lu, Z., Li, H., Li, V.O.K.: Incorporating copying mechanism in sequence-to-sequence learning. In: Proceedings of the 54th Annual Meeting of the Association for Computational Linguistics, ACL 2016, 7–12 August 2016, Berlin, Germany, Volume 1: Long Papers (2016)

# Improving Personal Health Mention Detection on Twitter Using Permutation Based Word Representation Learning

Pervaiz Iqbal Khan[1,2](✉), Imran Razzak[3], Andreas Dengel[1,2], and Sheraz Ahmed[1]

[1] German Research Center for Artificial Intelligence (DFKI), Kaiserslautern, Germany
{pervaiz.khan,andreas.dengel,sheraz.ahmed}@dfki.de
[2] TU Kaiserslautern, Kaiserslautern, Germany
[3] Deakin University, Geelong, Australia
imran.razzak@deakin.edu.au

**Abstract.** Social media has become a substitute for social interaction, thus the amount of medical and clinical-related information on the web is increasing. Monitoring of Personal Health Mentioning (PHM) on social media is an active area of research that predicts whether a given piece of text contains a health condition or not. To this end, the main idea is to consider the usage of disease or symptom words in the text. However, due to their usage in a figurative sense, disease or symptom words may not always indicate the presence of the health condition. Prior work attempts to address this by considering contextual word representations along with the utilization of the sentiment information. However, these methods are unable to capture the complete context in which symptom word is used. In this work, we incorporate permutation-based contextual word representation for the task of health mention detection which captures the context of disease words efficiently, in the given piece of text, and hence improves the performance of the classifier. To evaluate the integrity of the proposed method, we perform experimentation on the public benchmark dataset that shows an improvement of 5.5% in F-score in comparison to the state of the art health mention detection classifier. (Code is available at https://github.com/pervaizniazi/Figurative-Mention).

## 1 Introduction

The World Health Organization places importance on gathering health-related data to effectively respond to epidemics. Public Health Surveillance (PHS) deals with the collection, analysis, and interpretation of data related to health [1]. PHS systems mainly rely on social media platforms such as Twitter to collect data, which are unstructured in nature and huge in volume. Moreover, data are generally collected from social media platforms based on keyword search using disease words such as headache, fever, depression, etc. Although the retrieved


H. Yang et al. (Eds.): ICONIP 2020, LNCS 12532, pp. 776–785, 2020.
https://doi.org/10.1007/978-3-030-63830-6_65

data contain health-related information, it may not necessarily indicate the presence of disease. For instance the tweets "Did you know that Jeb Bush is used to curing cancer of the human body?"[1], and "*Stimulation of the ear can help manage Parkinson's symptoms* https://t.co/0FRuL1v9Nw" (see footnote 1) contain disease words such as cancer and Parkinson's, but they do not indicate the presence of a disease, instead they suggest treatment for these diseases. Similarly, sometimes, these disease words are used in a figurative sense for exaggeration purposes like "*I made such a good bowl of soup I think I cured my depression*" (see footnote 1) the word, depression, is used figuratively for exaggeration purpose to show that the soup is very delicious.

Personal Health Mentioning Detection (PHMD) is a binary classification problem that predicts whether a given piece of text contains a health condition or not [2]. To classify a text, PHM systems mainly rely on the usage of disease words in the given text. However, due to their usage in a figurative sense, they pose additional challenges to PHM classifiers and hence, do not produce precise results. To deal with these challenges, [2] make use of a contextual word representations. Although their method is able to improve the prediction results, it is still unable to capture the complete context in which disease or symptom word is used, and hence there is a need for the improvement. In this work, we seek to analyze the impact of permutation-based contextual word representation on PHM using XLNet [3]. We investigate to answer the research question, "Can personal health mentioning benefit from contextual word representations if all the words in a sequence are taken into account while learning these representations?" For this purpose, we experiment on the public benchmark dataset provided by [2], and evaluate the results using three evaluation metrics namely precision, recall, and F1 score. Experimental results show that our proposed method outperforms the state-of-the-art in terms of F-score and precision respectively. The key contributions of this work are:

- We present permutation-based contextual word representation in order to detect health mentioning tweets.
- Unlike earlier models such as BERT, we present to use permutation-based representation that capture the complete context of a word.
- We evaluated the performance of proposed approach, which validate the significant gain in performance in comparison to state of the art methods.

The structure of the rest of the paper is as follows: Sect. 2, provides the overview of related work. In Sect. 3, we present our method to improve the PHMD. In Sect. 4 and 5, we present experiment details, and results and discussion respectively. In Sect. 6, we provide a brief error analysis while in Sect. 7, we conclude this paper.

## 2 Related Work

Various unsupervised representation learning tasks have shown success in the Natural Language Processing (NLP) domain [4–11]. These methods train a

[1] This is an original tweet taken from Twitter.

neural network model on unlabeled text data for representation learning and then fine-tune the model on the downstream tasks. Among various other methods, Auto-Regressive language modeling (AR) and Auto-Encoding (AE) also use this idea for representation learning. In the following discussion, we describe the recent work done in the field of PHMD.

Jiang et al. [12] applied general preprocessing on the input tweets and then represented input tokens using pre-trained non-contextual word representations. Then, they used Long Short-Term Memory Networks (LSTMs) [13] to classify a tweet as a health mention or not. Results show that the classifier outperformed Support Vector Machines (SVM), Decision Trees, and K-Nearest Neighbor (KNN). Karisani et al. [14] presented a new method called Word Embedding Space Partitioning and Distortion (WESPAD) to detect personal health mentions on Twitter data. This method learned to partition and distort the word embeddings, which enabled the generalizing capabilities to the network while requiring little training examples containing positive health mentions. Although this method improved the results along with requiring little training data, distorting the original word embedding caused information loss.

Iyer et al. [15] detected whether the usage of disease word in a given tweet was in a figurative or literal sense. Their approach consisted of two steps. The first step detected whether a disease word was used figuratively or literally. Then the second step passed the output of the first step as one of the features to the Convolutional Neural Network (CNN) based classifier. The authors noted the increase in the performance of the classifier if it used the additional information as an input feature from the first step. However, the classifier did not perform well on figurative mentions of diseases, especially for the heart attack, the widely used in a figurative sense. Biddle et al. [2] used the work of [15] as baseline to detect figurative mentions of disease words. Their contribution is twofold: First, they extended the existing dataset by adding 14k new tweets and 4 new disease classes. Then, they improved the performance on health mention classification task. As a preprocessing, they normalized URL, as well as the user mentions in the tweets and converted emojis into string representations using the python library. For word representations, they experimented with both non-contextual representations using word2vec [16] as well as with contextual representations such as ELMO [8] and BERT [5]. To incorporate sentiments in the model, they experimented with WordNet [17], VAD [18], and ULMFit [19]. Word representations were passed to the Bi-LSTM [20] as well as the sentiment module. The last layer concatenated the output of the Bi-LSTM and sentiment module to produce the final binary output. Results showed that the usage of BERT as word representation and VAD as sentiment extractor achieved better performance as compared to ELMO and word2vec used as word representation while ULMFit and WordNet as sentiment extractor. 10-fold cross-validation was used to evaluate the performance of the classifier. Error analysis showed that incongruity and simple linguistic patterns caused misclassifications. Although the use of contextual word representation and incorporation of sentiment information improved the classification results, it could not capture the complete context in which disease words used which resulted in overall performance degradation. In this work, we consider the work of [2] as a baseline, and use permutation-based word representation using XLNet for word representation that improves the performance of PHMD.

# 3 Method

This section provides insights of the proposed method for health mentioning disease classification. We incorporate permutation-based word representation using XLNet to detect personal health mentions in the Twitter tweets. The permutation language modeling through autoregressive language modeling and auto-encoding is able to cover more dependencies thus more informative than its counter pre-trained model that are based on the modeling bidirectional contexts. The pipeline of our method consists of four stages: 1) an embedding layer 2) an encoder block containing 6 XLNet layers, 3) a decoder block containing 6 XLNet layers and 4) the final output layer as shown in Fig. 1. The embedding layer converts input tweets to vector representation and it is pretrained on BooksCorpus [21], Giga5 [22], ClueWeb 2012-B [23], Common Crawl [24], and English Wikipedia datasets. The output of embedding layer goes through encoder block having six XLNet layers, where each layer consists of XLNet relative attention and XLNet feed forward blocks. The output of the encoder block is passed to the decoder block consisting of 6 XLNet layers similar to the encoder block. The output of the decooder block is passed to the final linear layer that produces binary output indicating whether the input tweet contains health mention or not. Figure 1 describes proposed framework for classification of personal health mention tweets.

## 3.1 Language Model Pretraining

For a given text sequence $s = \{s_1, s_2, ..., s_n\}$, AR language model maximizes the likelihood of the each sequence token either in forward or in backward direction as given in the equations below:

$$p_f(s) = \prod_{i=1}^{n} p(s_i|s < i)$$

$$p_b(s) = \prod_{i=n}^{1} p(s_i|s > i)$$

where $p_f(s)$ defines the probability of a sequence 's' considering forward context and $p_b(s)$ defines the probability of a sequence 's' considering backward context.

Since it considers unidirectional context at a time instead of a bidirectional context, AR language modeling is not effective for the real world NLP tasks which need a bidirectional context to capture long term dependencies. On the

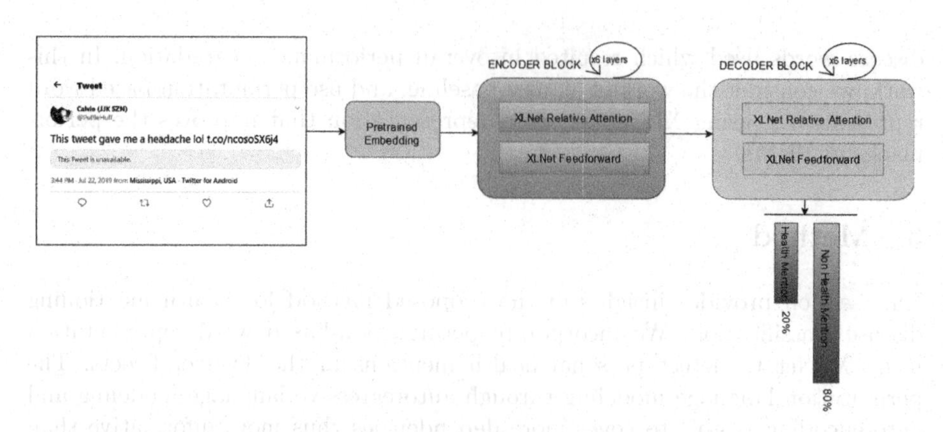


**Fig. 1.** The architecture of the personal health mention detection classifier. The classifier consists of an embedding layer, an encoder block with 6 layers, a decoder block with 6 layers, and the final classification layer.

other hand, instead of estimating density explicitly, AE based models first corrupt the input text and then reconstruct the original input from corrupted input. From the given input sequence, certain tokens are corrupted by replacing them with a special token [MASK], and then the model is trained to reconstruct original token in place of the special token using bidirectional context. However, downstream tasks do not contain this special token in the training data. This results in fine-tuning and pre-training discrepancy. In this way, AE based models such as BERT learn word representations. Literature shows that BERT can improve performance on many downstream tasks by utilizing a bidirectional context, however, it is based on the assumption that predicted tokens are independent of each other which is an oversimplified assumption for the real-world tasks. On the other hand, XLNet is a generalized AR model that combines the best of both AR and AE models while overcoming their limitations. In contrast to the AR models that maximize the likelihood of a sequence by using either forward or backward context of the words, XLNet maximizes the likelihood of a sequence for all the possible permutations of words in a given sequence. In this way, it takes the context from all the words in the input sequence. For example, for tweet "This tweet gave me a headache", XLNet is able to capture the dependency between the (headache, tweet), which BERT can not do. Furthermore, unlike BERT, it does not corrupt the original sequence by masking random words and hence overcomes the fine-tuning and pre-training discrepancy. This also eliminates the assumptions of independence between tokens while learning word representation. XLNet always learns more dependency pairs for the same target and also contains denser effective training signals. This makes XLNet an ideal choice for the task of PHMD.

# 4 Experiments

In this section, we provide some details of dataset used for the evaluation of our proposed method and experimental settings for training our model.

## 4.1 Dataset

To evaluate the performance of our method, we perform experimentation on the extended dataset of English language Tweets from Twitter provided by [2]. The dataset, originally introduced by [14], is extended by adding six new disease categories and 14k new tweets. The extended dataset contains a total of 19,558 tweets with ten disease categories of Alzheimer's, cancer, cough, depression, fever, headache, heart attack, migraine, Parkinson's, and stroke. The extended version of the dataset also contains a new label to identify tweets using disease words in a figurative sense. As the owners of the dataset provide Tweet IDs instead of Tweet text, therefore we firstly downloaded the tweet text from Twitter using its API. At the time of the download, some of the tweets were not available which reduced dataset size from 19,558 to 15,742 tweets. We noted that the most widely occurring disease word in the dataset was Alzheimer's, containing a total of 1715 tweets, whereas headache was the least used disease word with a total of 1429. Alzheimer's was the disease word that was used as non-health mention with a total of 1374 tweets, while fever contained non-health mentioned tweets with a total of 342. Statistics showed that disease word heart attack was used figuratively in 1060 tweets, as a health mention in 209 tweets, and as a non-health mention 349 in tweets. The disease word headache contained minimum non-health mention tweets of 112. There were 4,318, 7,372, and 4,288 tweets in the dataset with health mention, non-health mention, and figurative mention respectively. Table 1 provides the disease level distribution of dataset.

**Table 1.** Disease level statistics in dataset.

| Disease | Tweet count | Health mention | Non-health mention | Figurative mention |
|---|---|---|---|---|
| Alzheimer's | 1,715 | 249 | 1,374 | 92 |
| Cancer | 1,691 | 302 | 1,239 | 150 |
| Cough | 1,452 | 331 | 433 | 688 |
| Depression | 1,579 | 517 | 711 | 351 |
| Fever | 1,484 | 517 | 342 | 625 |
| Headache | 1,429 | 791 | 112 | 526 |
| Heart attack | 1,618 | 209 | 349 | 1,060 |
| Migraine | 1,519 | 904 | 400 | 215 |
| Parkinson's | 1,568 | 153 | 1,362 | 53 |
| Stroke | 1,687 | 255 | 1,000 | 432 |
| Total | 15,742 | 4,228 | 7,322 | 4,192 |

### 4.2 Data Preprocessing

As emojis cannot be directly ingested by the network, we convert all the emojis in the tweets into a string representation, similar to the Biddle et al. [2] data preprocessing. Then, we convert all the input tweets into numeric identifiers using XLNetTokenizer. We take the tweet with the maximum length from the dataset and set its length to maximum sequence length. We pad all the sequences having a length smaller than maximum sequence length with zeroes. Tweets are separated from each other using special tokens <sep> and then <cls>. Along with numeric id's representation of the tweets, we also create attention masks to guide the model where should it pay attention to the input sequences. This preprocessed input is fed into the network.

### 4.3 Experimental Settings

We trained the model for 8 epochs with Adam optimizer, a batch size of 8, and a learning rate of 3e−5. We initialized the weights using XLNet the pretrained model. To classify the tweets as health mention or not, we treat figurative mention tweets as non-health mentions, which reduces the problem to binary classification, which is also consistent with Biddle et al. [2]. Converting figurative mentions to non-health mention results in 11,514 tweets for the non-health mention and 4,288 tweets for the health mention. As it is evident from the stats, there is a class imbalance in the dataset. To make sure the validation fold contains sample data from both classes, we used stratified 10-fold cross-validation.

## 5 Results and Discussion

Table 2 shows that our method outperforms the Biddle et al. [2] by 5.5% in terms of F-score and 13.5% in terms of precision, however, there is a decrease of 3.8% in recall. This shows that our classier is reducing false positives in the prediction that causes a significant increase in precision. On the other hand, our classifier increases the number of false negatives which results in a decrease of overall recall. Statistics of the precision and recall show that our classifier assigns a true class to the sample only when it is highly confident about it. Table 3 presents a disease-wise comparison of the results of our classifier with state of the art methods such as Biddle et al. [2], Jiang et al. [12] and Karisani et al. [14]. We note a maximum performance improvement of 15.4% in F-score for cancer whereas 14.7% improvement of F-score for the heart attack. The improvement in the predictions for the heart attack is encouraging because a heart attack is most commonly used as figurative mentions. We note the minimum performance gain of 4.3% in the F-score for the depression. There is an increase of 14.4%, 11.4%, 7.5%, 6.9%, and 5.3% for the Parkinson's, Alzheimer's, stroke, fever, and cough respectively. There is a decrease of 2.5% and 2.9% in F-score for the headache and migraine respectively. Results show that our method based on permutation based word representation improves the overall performance of the classifier.

**Table 2.** Comparison of experimental results on BERT based methods Biddle et al. [2], Jiang et al. [12] and Karisani et al. [14]

| Method | Precison | Recall | F1 score |
|---|---|---|---|
| Jiang et al. [12] | 0.721 | 0.950 | 0.818 |
| Karisani et al. [14] | 0.752 | 0.896 | 0.818 |
| Biddle et al. [2] | 0.756 | **0.920** | 0.829 |
| Our method | **0.891** | 0.882 | **0.884** |

**Table 3.** Disease-wise F1 score comparison of HMC experiments (here disease refers to disease word)

| Disease | Karisani et al. [14] | Jiang et al. [12] | Biddle et al. [2] | Proposed approach % | Gain/Loss % |
|---|---|---|---|---|---|
| Alzheimer's | 0.716 | 0.731 | 0.735 | **0.849** | △0.114 |
| Cancer | 0.682 | 0.702 | 0.684 | **0.838** | △0.126 |
| Cough | 0.79 | 0.822 | 0.831 | **0.884** | △0.053 |
| Depression | 0.747 | 0.726 | 0.749 | **0.792** | △0.05 |
| Fever | 0.838 | 0.826 | 0.846 | **0.895** | △0.049 |
| Headache | 0.901 | 0.906 | **0.915** | 0.89 | ▽0.025 |
| Heart Attack | 0.686 | 0.713 | 0.705 | **0.852** | △0.139 |
| Migraine | 0.914 | 0.912 | **0.926** | 0.897 | ▽0.029 |
| Stroke | 0.789 | 0.777 | 0.792 | **0.867** | △0.075 |
| Parkinson's | 0.66 | 0.679 | 0.675 | **0.819** | △0.144 |

## 6 Error Analysis

In this section, we provide a brief analysis of misclassified tweets. We take misclassified tweets from the validation folds as described in Sect. 4.

We observe that there are a few factors that cause false classification of figurative mention tweets as health mentions. One of these is the part of the tweet that qualifies a tweet as figurative mention exists in the external link of the image or video. A couple of examples of such tweets are "*A structural Engineer's headache: face_with_head-bandage:* https://t.co/7Zdjl7HrJg" (see footnote 1) and "my mother has a heart attack but who tf cares https://t.co/XwrRKUMKCF" (see footnote 1). Some of the tweets that are misclassified by the classifier as health mention are even hard humans to understand. Examples of such tweets are "*i got a headache from crying so much rip i think that means its time for sleep*" (see footnote 1) and "*My mother been home less than an hour and I already have a headache*" (see footnote 1). For some tweets, the decision of classifying them as health mentions or non-health mentions requires a common sense which the classifier is unable to capture. One such example is, "*My grandma finna have a heart attack over some $100 shoes*" (see footnote 1).

We observe a similar trend for the health mentioning tweets that are misclassified as non-health mentions. Consider the following tweets: "*I drink too much and it gives me a headache*" (see footnote 1) and "*This headache is not the business*" (see footnote 1). These two short tweets are even difficult for humans to distinguish them if they are used in a literal or figurative sense. Another trend we observe is, some tweets contain long text and one disease word that indicates the presence of health mention. Such tweets require common sense to decide about health mention or not. Examples of such tweets are: "*Thailand you have been incredible. Snorkelling at Phi phi Island and visiting an elephant sanctuary. Although I have ended up with Paratyphoid fever I forgive you :winking_face:* https://t.co/SBLpkXGCVd" (see footnote 1) and "I had a mug of hot chocolate with a chai apple spice tea bag in it last night. I was surprised how yummy it was, even when it hit room temperature. Nursing a #migraine that began last night. Heat warning today. Nothing but self care today. #sicknotweak" (see footnote 1).

## 7 Conclusion

In this paper, we targeted the problem of personal health mention detection in social data, using the permutation-based word representation. Thus, the proposed method resulted in performance improvement by leveraging contextual representations of disease as well as sentiment information. Results showed that our method improved the performance of the classifier by 5.5% in terms of F-score and 13.5% in terms of precision as compared to the state-of-the-art. We carried out a brief error analysis which showed that our classifier performed misclassification on the tweets that refer to the content in the image or video. Furthermore, the classifier was unable to capture the commonsense factor. Thus, permutation language modeling through autoregressive language modeling and auto-encoding, was able to cover more dependencies thus more informative than modeling bidirectional contexts. The addition of new disease and symptoms words will introduce new insight and challenges. In the future, we will work on the extension of the dataset and add new challenges in the dataset.

**Acknowledgement.** The authors would like to thank Shoaib Ahmed Siddiqui and Muhammad Nabeel Asim for providing useful feedback during this work.

## References

1. WHO. Epidemic intelligence - systematic event detection (2017)
2. Biddle, R., Joshi, A., Liu, S., Paris, C., Guandong, X.: Leveraging sentiment distributions to distinguish figurative from literal health reports on Twitter. In: Proceedings of The Web Conference 2020, pp. 1217–1227 (2020)
3. Yang, Z., Dai, Z., Yang, Y., Carbonell, J., Salakhutdinov, R.R., Le, Q.V.: XLNet: generalized autoregressive pretraining for language understanding. In: Advances in Neural Information Processing Systems, pp. 5754–5764 (2019)
4. Dai, A.M., Le, Q.V.: Semi-supervised sequence learning. In: Advances in Neural Information Processing Systems, pp. 3079–3087 (2015)
5. Devlin, J., Chang, M.-W., Lee, K., Toutanova, K.: BERT: pre-training of deep bidirectional transformers for language understanding. arXiv preprint arXiv:1810.04805 (2018)

6. Saeed, Z., Ayaz Abbasi, R., Razzak, I.: EveSense: what can you sense from Twitter? In: Jose, J.M., et al. (eds.) ECIR 2020. LNCS, vol. 12036, pp. 491–495. Springer, Cham (2020). https://doi.org/10.1007/978-3-030-45442-5_64
7. McCann, B., Bradbury, J., Xiong, C., Socher, R.: Learned in translation: contextualized word vectors. In: Advances in Neural Information Processing Systems, pp. 6294–6305 (2017)
8. Peters, M.E., et al.: Deep contextualized word representations. arXiv preprint arXiv:1802.05365 (2018)
9. Saeed, Z., et al.: What's happening around the world? A survey and framework on event detection techniques on twitter. J. Grid Comput. **17**(2), 279–312 (2019)
10. Radford, A., Narasimhan, K., Salimans, T., Sutskever, I.: Improving language understanding by generative pre-training (2018)
11. Saeed, Z., Abbasi, R.A., Razzak, I., Maqbool, O., Sadaf, A., Xu, G.: Enhanced heartbeat graph for emerging event detection on twitter using time series networks. Expert Syst. Appl. **136**, 115–132 (2019)
12. Jiang, K., Feng, S., Song, Q., Calix, R.A., Gupta, M., Bernard, G.R.: Identifying tweets of personal health experience through word embedding and LSTM neural network. BMC Bioinf. **19**(8), 210 (2018)
13. Hochreiter, S., Schmidhuber, J.: Long short-term memory. Neural Comput. **9**(8), 1735–1780 (1997)
14. Karisani, P., Agichtein, E.: Did you really just have a heart attack? Towards robust detection of personal health mentions in social media. In: Proceedings of the 2018 World Wide Web Conference, pp. 137–146 (2018)
15. Iyer, A., Joshi, A., Karimi, S., Sparks, R., Paris, C.: Figurative usage detection of symptom words to improve personal health mention detection. arXiv preprint arXiv:1906.05466 (2019)
16. Mikolov, T., Sutskever, I., Chen, K., Corrado, G.S., Dean, J.: Distributed representations of words and phrases and their compositionality. In: Advances in Neural Information Processing Systems, pp. 3111–3119 (2013)
17. Baccianella, S., Esuli, A., Sebastiani, F.: SentiWordNet 3.0: an enhanced lexical resource for sentiment analysis and opinion mining. In: LREC, vol. 10, pp. 2200–2204 (2010)
18. Mohammad, S.: Obtaining reliable human ratings of valence, arousal, and dominance for 20,000 English words. In: Proceedings of the 56th Annual Meeting of the Association for Computational Linguistics (Volume 1: Long Papers), pp. 174–184 (2018)
19. Howard, J., Ruder, S.: Universal language model fine-tuning for text classification. arXiv preprint arXiv:1801.06146 (2018)
20. Graves, A., Mohamed, A., Hinton, G.: Speech recognition with deep recurrent neural networks. In: 2013 IEEE International Conference on Acoustics, Speech and Signal Processing, pp. 6645–6649. IEEE (2013)
21. Zhu, Y., et al.: Aligning books and movies: towards story-like visual explanations by watching movies and reading books. In: Proceedings of the IEEE International Conference on Computer Vision, pp. 19–27 (2015)
22. Parker, R., Graff, D., Kong, J., Chen, K., Maeda, K.: English gigaword fifth edition LDC2011T07 (technical report). Technical report. Linguistic Data Consortium, Philadelphia (2011)
23. Callan, J.: The lemur project and its ClueWeb12 dataset. In: Invited Talk at the SIGIR 2012 Workshop on Open-Source Information Retrieval (2012)
24. Common Crawl. Common crawl corpus (2019). http://commoncrawl.org

# Learning Discrete Sentence Representations via Construction & Decomposition

Haohao Song, Dongsheng Zou[✉], and Weijia Li

College of Computer Science, Chongqing University, Chongqing 400044, China
{songhaohao2018,dszou,liwj}@cqu.edu.cn

**Abstract.** In this paper, we address the problem of learning low-dimensional, discrete representations of real-valued vectors. We propose a new algorithm called similarity matrix construction and decomposition (C&D). In the preparation phase, we constructively generate a set of consistent, unbiased and comprehensive anchor vectors, and obtain their low-dimensional forms with PCA. The C&D algorithm learns the discrete representations of vectors in batches. For a batch of input vectors, we first construct a similarity matrix between them and the anchor vectors, and then learn their discrete representations from the similarity matrix decomposition, where the low-dimensional forms of the anchor vectors are regarded as a fixed factor of the similarity matrix. The matrix decomposition is a mixed-integer optimization problem. We obtain the optimal solution for each bit with mathematical derivation, and then use the discrete coordinate descent method to solve it. The C&D algorithm does not learn directly discrete representations from the input vectors, which distinguishes it from other discrete learning algorithms. We evaluate the C&D algorithm on sentence embedding compression tasks. Extensively experimental results reveal the C&D algorithm outperforms the latest 4 methods and reaches state-of-the-art. Detailed analysis and ablation study further validate the rationality of the C&D algorithm.

**Keywords:** Natural language processing · Sentence representation · Binarization · Dimensionality reduction

## 1 Introduction

In computer science, the learning for discrete representations has never stopped. Related research can be traced back to that using limited precision represents real-valued numbers (e.g., 32-bit float), where the use of fixed bits can be regarded as the beginning of discrete learning. Nowadays, applying resource-intensive deep learning models to low-resource devices becomes a problem, which attracts a large number of researchers to study model compression. Among the many methods that have the potential to reduce the resource requirements of the models, learning discrete representations for real-valued vectors becomes a


H. Yang et al. (Eds.): ICONIP 2020, LNCS 12532, pp. 786–798, 2020.
https://doi.org/10.1007/978-3-030-63830-6_66

key fundamental technology. The learned discrete representations can reduce the memory requirements, improve the calculation speed and lower the hardware threshold for running these models. As of now, the technology of learning discrete representations has been successfully applied to sentence representations [15], word embeddings [19], social recommendations [11] and other fields.

In life, we can use a ruler to measure a box and get its sizes, and then we can build a box of the same sizes according to the ruler and the obtained sizes. Inspired by this approach, we propose a new algorithm for learning discrete representations, called construction and decomposition (C&D). The C&D is a batch learning algorithm, which includes 3 steps. In step 1, we first obtain a ruler that can be used to measure the nature of the input vectors. Intuitively, this ruler should meet some conditions: a unified ruler is used to measure different vectors; this ruler cannot be biased; this ruler has the ability to measure all vectors. We constructively generate a set of vectors that meet these characteristics as our ruler. Since these vectors are consistent in batches, they are called anchor vectors. Then in step 2, we use the anchor vectors to measure the nature of the input vectors. We construct a similarity matrix between a batch of original vectors and anchor vectors. Finally, we learn **binary** representations of this batch of original vectors from matrix decomposition, where we use the anchor vectors processed by PCA as a fixed factor of the similarity matrix. The anchor vectors in this step can be seen as a mini version of the original ruler. Step 3 is a mixed-integer optimization problem. We first obtain the optimal solution for each bit with mathematical derivation, and then solve it with discrete coordinate descent method [17]. In the C&D algorithm, step 1 only requires to be initialized once, and then iterate step 2 and 3 repeatedly until all discrete representations of the original vectors are learned.

We evaluate the C&D algorithm on compressing sentence representation tasks. Shen et al. [15] systematically proposed 4 different binarization strategies and applied them to a pre-trained sentence representation encoder called InferSent [5]. SentEval [4] is a tool for evaluating the quality of sentence embeddings, which was used to evaluate the 4 strategies [15]. The experimental results demonstrated that the 4 strategies established very strong baselines. Under the evaluation environment is consistent with Shen et al. [15], the average performances of the C&D algorithm outperform all the strategies proposed by Shen et al. [15].

Compared with the InferSent real-valued sentence representations, the binary codes learned by C&D algorithm are very advantageous in terms of memory and calculation speed. With the C&D algorithm, the relatedness between sentence pairs can be calculated with the Hamming distance, which is quite efficient. On nearest neighbor retrieval tasks, the calculation of binary codes based on Hamming distance is 11 times faster than the real-valued; the learned binary codes also maintain the semantic information. The experimental results on SentEval sentence pair tasks reveal that the compression ratio is 0.8%, while the performances drop by only 0.6%. On some tasks, the performances of the C&D

algorithm even outperform the raw representations. The code is available at github.com/songs18/cd_algorithm. In summary, our contributions are as follows:

1. We propose a new algorithm for learning discrete representations. It is an indirect learning algorithm: it learns discrete representations from matrix decomposition instead of the input vectors. Under the guidance of high-quality anchor vectors, the decomposed discrete representations are globally consistent, and almost retain the semantic information without loss.
2. Experimental results on sentence pair tasks reveal that the C&D algorithm's average performances outperform the latest 4 strategies proposed by Shen et al. [15] and reach state-of-the-art.

## 2 Related Work

Learning discrete representations contains two goals: dimensionality reduction and discretization. We also introduce related work from the two aspects.

Dimensionality reduction is an important branch of machine learning. Related technologies can be divided into two parts: linear and non-linear. The well-known techniques for linear dimensionality reduction are PCA [9] and LDA [6]; There are many methods available for nonlinear dimensionality reduction, e.g., LLE [13]. Our method belongs to the latter.

The other goal is to learn discrete representations [3,15,16,18,19]. Among these discrete codes, binary codes are the most memory-efficient and have the advantage of using Hamming distance. They attracted much attention. There are [3,15,18,19] methods in this regard. However, most of them are two-stage methods (relax the binarization restriction first and then use threshold cutoff). Semantic hashing is also related to our work. It is a method for learning binary codes designed for information retrieval tasks. Related methods are [14,16,20].

We evaluate the performances of our algorithm on the sentence embedding compression tasks. The related work of learning sentence representations can be roughly divided into two categories: pre-training a model and then fine-tuning on specific tasks [2]; training a general-purpose sentence embedding encoder that can be used directly in downstream tasks [5]. This paper is related to the latter. It should be additionally noted that Kiros et al. [10] and Shen et al. [15] report that to train a binary sentence representation encoder from scratch produces inferior empirical results than from pre-trained sentence representation encoders. Therefore, learning binary codes from pre-trained sentence representation encoders is a task worth paying attention to.

## 3 Methodology

We use bold uppercase and lowercase letters as matrices and vectors, respectively. In particular, we use $\mathbf{a}_i$ as the $i$-th column vector of matrix $\mathbf{A}$. We denote $\|\cdot\|_2$ as the $\ell_2$ norm for a vector, $\|\cdot\|_{\mathrm{F}}$ as the Frobenius norm for a matrix and $\mathrm{tr}(\cdot)$

as the matrix trace. We denote $\text{sgn}(\cdot) : \mathbb{R} \to \{\pm 1\}$ as the round-off function, *i.e.*, $\text{sgn}(x) = +1$ if $x \geq 0$ and $\text{sgn}(x) = -1$ otherwise.

Suppose we have $t$ sentences, each of which is represented with an $g$-dimensional real-valued vector, then these vectors can be written as a matrix $\mathbf{A} \in \mathbb{R}^{g \times t}$. In this paper, We aim to learn a matrix $\mathbf{A}_\mathrm{b} \in \{\pm 1\}^{r \times t}$ without semantic information loss, where $r$ is fewer than $g$. Figure 1 depicts our algorithm flow. Next, we first introduce the generation of anchor vectors in Sect. 3.1, and then present C&D algorithm in Sect. 3.2 and 3.3, finally we analyze the algorithm in Sect. 3.4.

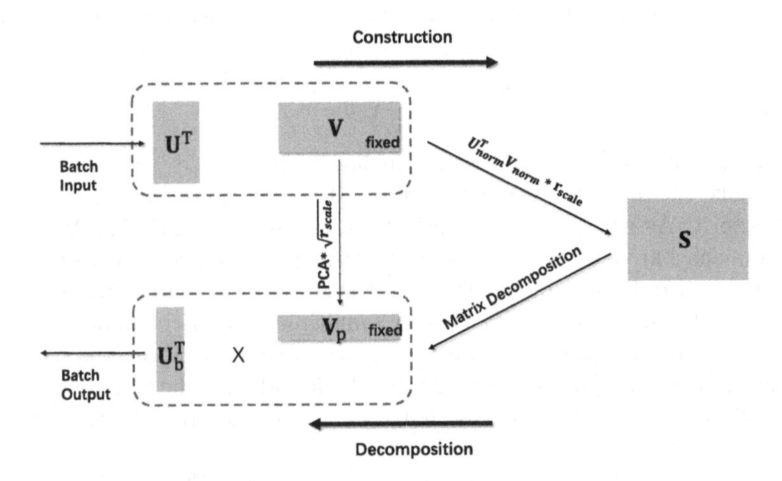


**Fig. 1.** Algorithm flow of learning binary representations for real-valued vectors.

### 3.1 Preparation: Generating Anchor Vectors

In our algorithm, constructing the similarity matrix with the anchor vectors can be regarded as measuring the characteristics of the input vectors. To ensure the quality of the learned binary vectors, the anchor vectors must have some quality. Here, we make the following 3 intuitive assumptions:

1. Consistency. Since the number of real-valued vectors is often very large, constructing a similarity matrix between anchor vectors and all input vectors will result in huge memory consumption. A common way to solve this problem is to discretize the vectors in batches. Obviously, anchor vectors participating in the matrix construction must be consistent. Otherwise, there will be a problem of vector drift: the discrete vectors decomposed from the matrix cannot maintain the relative spatial relationships among the original vectors;
2. Unbiasedness. We also think that good anchor vectors should be unbiased, that is to say, the anchor vectors should be objective to any vectors, which requires their spatial distribution to be as balanced as possible;

3. Comprehensiveness. For any vector, the anchor vectors must have the ability to measure it from all dimensions. This requires that the number of anchor vectors must be greater than or equal to the number of dimensions of the vector, and be as linearly irrelevant as possible.

In 3 assumptions, unbiasedness constrains anchor vectors from the perspective of spatial distribution balance, which can be regulated by zero mean; the comprehensiveness can be considered as a requirement for the singularity of the anchor vector matrix, and we can measure it from the rank of matrices.

Generating $n$ $g$-dimensional anchor vectors is to initialize a matrix $\mathbf{V} \in \mathbb{R}^{g \times n}$. We use GlorotNormal [7] to realize it, and the initialization can be written as

$$\mathbf{V} \sim \mathcal{N}\left(0, \sqrt{\frac{2}{g+n}}\right), \tag{1}$$

where each column of matrix $\mathbf{V}$ is an anchor vector.

To make the matrix $\mathbf{V}$ satisfy the above 3 assumptions, we make some slight changes to it. As will be introduced in Sect. 3.2, we only initialize $\mathbf{V}$ once, and use it in each batch, so the $\mathbf{V}$ is consistent; In practice, we observe that the $\mathbf{V}$ initialized by Eq. 1 does not strictly meet the zero mean, so we subtract its mean value for each anchor vector; We choose a multiple of $g$ for the number of anchor vectors, e.g., $4g$; The comprehensiveness of the matrix $\mathbf{V}$ is easy to satisfy, and this comes from observations in practice. We found that the rank of the matrix $\mathbf{V}$ can often reach $g$, which means that there are $g$ linearly independent vectors in the matrix $\mathbf{V}$.

We obtain a $r$-dimensional matrix $\mathbf{V}_\mathrm{p}$ by applying PCA on $\mathbf{V}$. Then scale the $\mathbf{V}_\mathrm{p}$ by $\sqrt{r}$ times. The $\mathbf{V}_\mathrm{p}$ will be used later in the matrix decomposition.

### 3.2 Construction

We learn the binary representations in batches. We use $\mathbf{U} \in \mathbb{R}^{g \times m}$ to denote a batch of sentence vectors, where $m$ is the batch size. Given the matrix $\mathbf{V}$ obtained in Sect. 3.1, we can obtain a similarity matrix $\mathbf{S} \in \mathbb{R}^{m \times n}$ between the anchor vectors and this batch of vectors based on cosine similarity function. Each element in $\mathbf{S}$ can be calculated by

$$\mathbf{S}_{ij} = \frac{\mathbf{u}_i^{\mathrm{T}} \mathbf{v}_j}{\|\mathbf{u}_i\|_2 \|\mathbf{v}_j\|_2}, \qquad i \in [1, m], j \in [1, n], \tag{2}$$

where $\mathbf{S}_{ij}$ is the value of the $i^{th}$ row and the $j^{th}$ column of $\mathbf{S}$. According to the definition of cosine similarity function, the value range of $\mathbf{S}_{ij}$ is $[-1, 1]$. We scale the matrix $\mathbf{S}$ to $[-r, r]$ by multiplying $r$.

### 3.3 Decomposition

We denote the binarized $\mathbf{U}$ as $\mathbf{U}_\mathrm{b}$. We use $\ell_2$ loss for matrix factorization, and the problem of learning binary representations is formulated as

$$\begin{aligned} &\arg\min_{\mathbf{u}_{\mathrm{b},i}} \sum_{i=1}^{m}\sum_{j=1}^{n}\left(\mathbf{S}_{ij}-\mathbf{u}_{\mathrm{b},i}^{\mathrm{T}}\mathbf{v}_{\mathrm{p},j}\right)^2, \\ &\text{s.t. } \mathbf{u}_{\mathrm{b},i} \in \{\pm 1\}^{r\times 1}, \end{aligned} \tag{3}$$

where $\mathbf{u}_{\mathrm{b},i}$ is the $i^{th}$ column vector of the matrix $\mathbf{U}_\mathrm{b}$, and $\mathbf{v}_{\mathrm{p},j}$ is the $j^{th}$ column vector in $\mathbf{V}_\mathrm{p}$. The problem 3 is a generally NP-hard and difficult to solve with the discrete variables in matrix $\mathbf{U}_\mathrm{b}$. Specifically, finding the global optimum solution needs to involve $\mathcal{O}(2^{rm})$ combinatorial search [8]. We simplify problem 3 by mathematical derivation. First, we rewrite problem 3 as

$$\begin{aligned} &\arg\min_{\mathbf{U}_\mathrm{b}} \|\mathbf{S}-\mathbf{U}_\mathrm{b}^{\mathrm{T}}\mathbf{V}_\mathrm{p}\|_\mathrm{F}^2, \\ &\text{s.t. } \mathbf{U}_\mathrm{b} \in \{\pm 1\}^{r\times m}. \end{aligned} \tag{4}$$

Problem 4 is:

$$\begin{aligned} &\arg\min_{\mathbf{U}_\mathrm{b}} \|\mathbf{S}\|_\mathrm{F}^2 - 2\mathrm{tr}\left(\mathbf{S}^{\mathrm{T}}\mathbf{U}_\mathrm{b}^{\mathrm{T}}\mathbf{V}_\mathrm{p}\right) + \|\mathbf{U}_\mathrm{b}^{\mathrm{T}}\mathbf{V}_\mathrm{p}\|_\mathrm{F}^2, \\ &\text{s.t. } \mathbf{U}_\mathrm{b} \in \{\pm 1\}^{r\times m}, \end{aligned} \tag{5}$$

where $\|\mathbf{S}\|_\mathrm{F}^2$ is a constant term. We remove it and problem 5 is equivalent to

$$\begin{aligned} &\arg\min_{\mathbf{U}_\mathrm{b}} \|\mathbf{U}_\mathrm{b}^{\mathrm{T}}\mathbf{V}_\mathrm{p}\|_\mathrm{F}^2 - 2\mathrm{tr}(\mathbf{U}_\mathrm{b}^{\mathrm{T}}\mathbf{Q}), \\ &\text{s.t. } \mathbf{U}_\mathrm{b} \in \{\pm 1\}^{r\times m}, \end{aligned} \tag{6}$$

where $\mathbf{Q} = \mathbf{V}_p\mathbf{S}^{\mathrm{T}}$. Let $\mathbf{u}^{\mathrm{T}}$ be the $l^{th}$ row of $\mathbf{U}_\mathrm{b}$ and $\mathbf{U}_\mathrm{b}'$ the matrix $\mathbf{U}_\mathrm{b}$ excluding $\mathbf{u}^{\mathrm{T}}$, where $l$ is from 1 to $r$. Similarly, let $\mathbf{v}^{\mathrm{T}}$ be the $l^{th}$ row of $\mathbf{V}_\mathrm{p}$ and $\mathbf{V}_\mathrm{p}'$ the matrix $\mathbf{V}_\mathrm{p}$ excluding $\mathbf{v}^{\mathrm{T}}$, $\mathbf{q}^{\mathrm{T}}$ the $l^{th}$ row of matrix of $\mathbf{Q}$ and $\mathbf{Q}'$ the matrix $\mathbf{Q}$ excluding $\mathbf{q}^{\mathrm{T}}$. We simplify the two items in problem 6 separately:

$$\begin{aligned} \left\|\mathbf{U}_\mathrm{b}^{\mathrm{T}}\mathbf{V}_\mathrm{p}\right\|_\mathrm{F}^2 &= \left\|\begin{bmatrix}\mathbf{U}_\mathrm{b}' \\ \mathbf{u}^{\mathrm{T}}\end{bmatrix}^{\mathrm{T}}\begin{bmatrix}\mathbf{V}_\mathrm{p}' \\ \mathbf{v}^{\mathrm{T}}\end{bmatrix}\right\|_\mathrm{F}^2 \\ &= \left\|\mathbf{U}_\mathrm{b}'^{\mathrm{T}}\mathbf{V}_\mathrm{p}' + \mathbf{u}\mathbf{v}^{\mathrm{T}}\right\|_\mathrm{F}^2 \\ &= \left\|\mathbf{U}_\mathrm{b}'^{\mathrm{T}}\mathbf{V}_\mathrm{p}'\right\|_\mathrm{F}^2 + \left\|\mathbf{u}\mathbf{v}^{\mathrm{T}}\right\|_\mathrm{F}^2 + 2\mathrm{tr}\left(\mathbf{v}\mathbf{u}^{\mathrm{T}}\mathbf{U}_\mathrm{b}'^{\mathrm{T}}\mathbf{V}_\mathrm{p}'\right) \\ &= \mathit{const} + 2\mathbf{u}^{\mathrm{T}}\mathbf{U}_\mathrm{b}'^{\mathrm{T}}\mathbf{V}_\mathrm{p}'\mathbf{v}. \end{aligned} \tag{7}$$

The first two items in the third line are constants: $\left\|\mathbf{U}_\mathrm{b}'^{\mathrm{T}}\mathbf{V}_\mathrm{p}'\right\|_\mathrm{F}^2$ is constant obviously; $\left\|\mathbf{u}\mathbf{v}^{\mathrm{T}}\right\|_\mathrm{F}^2 = \mathrm{tr}\left(\mathbf{v}\mathbf{u}^{\mathrm{T}}\mathbf{u}\mathbf{v}^{\mathrm{T}}\right) = m\mathrm{tr}\left(\mathbf{v}\mathbf{v}^{\mathrm{T}}\right) = \mathit{const}$. Similarly, we have

$$\mathrm{tr}\left(\mathbf{U}_\mathrm{b}^{\mathrm{T}}\mathbf{Q}\right) = \mathit{const} + \mathbf{q}^{\mathrm{T}}\mathbf{u}. \tag{8}$$

**Algorithm 1:** C&D algorithm

**Input:** $\mathbf{A} \in \mathbb{R}^{g \times t}$: $t$ $g$-dimensional real-valued vectors; $r$: code length; $n$: anchor vector number
**Output:** $\mathbf{A}_\mathrm{b} \in \{\pm 1\}^{r \times t}$: $t$ $r$-dimensional binarized vectors

```
1  Initialization: V ∈ R^{g×n} (Eq. 1), V_p ∈ R^{r×n} (PCA) and V_p^{-1};
2  for one batch of U ∈ R^{g×m} in A do
3  |   Calculation: S (Eq. 2) and Q ← V_p S^T;
4  |   Initialization: U_b^T ← S V_p^{-1} ;
5  |   do
6  |   |   for one row u^T in U_b do
7  |   |   |   u ← sgn(q − U_b'^T V_p' v);
8  |   |   end
9  |   while U_b changes;
10 end
11 Concatenate all batch output U_b into A_b;
12 return A_b;
```

Putting Eq. 6, 7 and 8 together, we can rewrite problem 6 as

$$\begin{aligned} &\arg\min_{\mathbf{u}} \left(\mathbf{v}^\mathrm{T}\mathbf{V}_\mathrm{p}^{'\mathrm{T}}\mathbf{U}_\mathrm{b}^{'} - \mathbf{q}^\mathrm{T}\right)\mathbf{u}, \\ &\text{s.t. } \mathbf{u}^\mathrm{T} \in \{\pm 1\}^{1 \times m}. \end{aligned} \tag{9}$$

Problem 9 has the optimal solution:

$$\mathbf{u} = \mathrm{sgn}\left(\mathbf{q} - \mathbf{U}_\mathrm{b}^{'\mathrm{T}}\mathbf{V}_\mathrm{p}^{'}\mathbf{v}\right). \tag{10}$$

We use the discrete coordinate descent method [17] to update $\mathbf{U}_b$. Specifically, we update one bit of all vectors in $\mathbf{U}_b$ at a time. This bit is learned from the pre-learned $r$-1 bits $\mathbf{B}^{'}$, so our algorithm can learn a better $\mathbf{U}_b$ after each iteration. The C&D algorithm is summarized in Algorithm 1. Note that the C&D algorithm does not use any task related information, and it is task-independent. Therefore, it can be applied to other fields without modifications.

### 3.4 Algorithmic Analysis

*Complexity.* In Algorithm 1, the matrices ($\mathbf{V}$, $\mathbf{V}_\mathrm{p}$ and $\mathbf{V}_\mathrm{p}^{-1}$) related to the anchor vectors are constants and do not change with the increase of problem size, so the complexity has nothing to do with them. For space complexity, Algorithm 1 requires $\mathcal{O}(mn)$ to store $\mathbf{S}$, $\mathbf{Q}$ and $\mathbf{U}_\mathrm{b}^{'\mathrm{T}}\mathbf{V}_\mathrm{p}^{'}\mathbf{v}$. The time complexity of Algorithm 1 is $\mathcal{O}(T_s t n r^3)$, where $T_s$ is the number of iterations. In fact, due to the optimization of the matrix multiplication and the parallelization of hardware, the actual time complexity of Algorithm 1 is lower than the given here.

*Initialization.* Since C&D solves a hard mixed-integer optimization problem, the initial value of $\mathbf{U}_{\mathrm{b}}$ has a great influence on the convergence and binary codes learned. Here we introduce a heuristic strategy. We relax the binarization restriction in problem 4 and use the closed solution of the matrix $\mathbf{U}_{\mathrm{b}}$ as its initial value. Specifically, we first calculate the pseudoinverse $\mathbf{V}_{\mathrm{p}}^{-1}$ of the matrix $\mathbf{V}_{\mathrm{p}}$, and then obtain the initial value from $\mathbf{V}_{\mathrm{p}}^{-1}$ right multiply $\mathbf{S}$. This strategy is successful: in practice, we found that it takes only 4 to 6 iterations for all bits to converge. Note that the initial value for $\mathbf{U}_{\mathrm{b}}$ does not need to be binarized.

*Out-of-Sample Extension.* Above we gave the C&D algorithm in the case of off-line vectors. When a new vector comes, we can perform the same operations as above with the anchor vectors. This situation can be seen as a special case of batch size is set to be 1. This is all thanks to the consistency of the anchor vectors.

# 4 Experiments

## 4.1 Experimental Setup

*Pre-trained Continuous Embeddings.* To make an objective comparison with the latest 4 methods [15], we use the same experimental setup as Shen et al. [15]. Specifically, we use a pre-trained InferSent [5] in SNLI [1] as sentence representation encoder. InferSent [5] learns a sentence vector via a bidirectional LSTM architecture along with a max-pooling operation over the hidden units, which is effective and widely used [15]. The sentence vectors produced with InferSent are 4096-dimensional and 32-bit float.

*Hyperparameter Settings.* The number of anchor vectors $n$ is 16384 ($4g$). The dimensions $r$ of binary codes are 512, 1024 and 2048. The batch size $m$ is 64.

*Evaluation.* SentEval [4] is used to evaluate the binary sentence representations we learned. The C&D algorithm is based on the similarity matrix, and we also select 7 sentence pair tasks to evaluate it: STS13-16, STSB and SICK-R. These tasks include pairs of sentences taken from news articles, forum discussions, news conversations, headlines, image, video descriptions and etc. labeled with a similarity score. The goal is to calculate the Spearman coefficients between human scores and the scores computed with sentence vectors. We use cosine similarity to compute scores for real-valued vectors, and Sokal & Michener similarity function [12] for binary vectors. The latter can be written as: $sim\left(\mathbf{b}_1, \mathbf{b}_2\right) = \left(n_{11} + n_{00}\right)/n$, where $n_{11}$ and $n_{00}$ are the numbers of corresponding positions in $\mathbf{b}_1$ and $\mathbf{b}_2$ that are 1 or 0 at the same time, respectively. $n$ is the length of vector $\mathbf{b}_1$ or $\mathbf{b}_2$. To compare these methods synthetically, we also calculate Pearson coefficients of each method in all tasks, and they are aggregated with the Fisher's transform average and are displayed in the average column.

*Comparative Methods.* We use 4 very strong baselines in Shen et al. [15] as our comparative methods, and they are *i*) Hard Threshold. This strategy requires setting a threshold in advance, and then binarize the vector by comparing each dimension of the vector with the threshold. This method does not require training and can only generate binary vectors with the same dimensions as the original vectors. *ii*) Random Projection. Shen et al. [15] proposes to first randomly initialize a project matrix, and then use this matrix to reduce the dimensions of the original sentence embeddings. Finally, use the hard threshold method to binarize them. This method overcomes the problem of dimensional invariance of the hard threshold method. *iii*) Principal Component Analysis. This strategy is similar to random projection. The only difference is that the dimensionality reduction method used is singular value decomposition. *iv*) Autoencoder. The autoencoder proposed by Shen et al. [15] includes a linear encoder and a linear decoder. With a reconstruction loss term and a semantic preservation term, the model is guided to learn an intermediate latent variable rich in semantic information. Note that the C&D and method *i*, *ii* do not need training, and method *iii*, *iv* are trained on SNLI, which are consistent with Shen et al. [15]. Moreover, autoencoder used here is fine-tuned and can achieve better performances than Shen et al. report in [15], which is quite challenging.

## 4.2 Experimental Results and Analysis

**Table 1.** Performances for InferSent, C&D and 4 comparative methods on 7 downstream tasks. Spearman coefficients are reported, and the Fisher's average scores of Pearson coefficients are also reported in the average column.

| Model | Dim | STS 2012 | STS 2013 | STS 2014 | STS 2015 | STS 2016 | STS B | SICK-R | Average |
|---|---|---|---|---|---|---|---|---|---|
| InferSent | 4096 | 62.6 | 67.4 | 68.8 | 74.2 | 73.5 | 70.8 | 64.9 | 70.7 |
| HT-binary | 4096 | 57.2 | 54.8 | 59.8 | 66.0 | 66.5 | 53.6 | 61.9 | 60.6 |
| Rand-binary | 2048 | 60.7 | 62.7 | 65.1 | 72.4 | 71.5 | 64.3 | 66.4 | 67.3 |
| PCA-binary | 2048 | 61.4 | 66.0 | 67.9 | 74.4 | 73.7 | 64.5 | 61.3 | 66.3 |
| AE-binary-SP | 2048 | **62.8** | 66.5 | 68.0 | **74.6** | 73.3 | **71.2** | 66.6 | 70.2 |
| C&D | 2048 | 61.9 | **66.7** | **68.5** | **74.6** | **73.8** | 71.1 | **66.7** | **70.4** |

Table 1 reports the experimental results of C&D algorithm and 4 comparative methods on 7 downstream tasks. Compared with raw InferSent, both C&D and autoencoder methods achieve very good performances. Specifically, when both are 2048-dimensional (~1.6% compression ratio), the absolute loss on 7 tasks is within 1%, and the absolute average loss does not exceed 0.5%. The C&D algorithm outperforms the autoencoder in 5 out of 7 tasks, with advantage on the average scores. It achieves state-of-the-art, demonstrating the effectiveness of the proposed algorithm. In comparison, the performances of random projection and PCA method are weaker than C&D and autoencoder. PCA performs better than random projection in 6 out of 7 tasks. On the dataset (SICK-R) lost by

PCA, the codes learned by random projection achieve the only performances beyond the original sentence embeddings, which shows that the performances of PCA are smoother and better than the random projection. The hard threshold method achieves the worst performances among all methods, and compared with the best performances (C&D achieves), the absolute loss is 9.8%. Not only that, in the experiments, but we also find that the hard threshold method is very sensitive to the threshold. If the threshold is set to 0, the performances are very poor. Table 1 reports the performances when the threshold is 0.07.

### 4.3 Nearest Neighbor Retrieval

*Case Study.* To get more intuitions about the semantic information encoded in learned binary codes, we also convert all the sentences in SNLI to continuous vectors (4096-dimensional InferSent) and binary vectors (2048-dimensional C&D). We randomly select a sentence as a query to retrieve in SNLI, and the top-3 results are shown in Table 2. It can be observed that the learned binary codes cover very similar semantic meanings, and even give more reasonable responses to some queries than the original vectors, e.g., for the first query, 'five' appears in the retrieval results of the binary codes, while the continuous codes fail.

**Table 2.** Nearest neighbor retrieval results on the SNLI dataset. Given a query sentence, the top-3 retrieved samples based on the Hamming distance with all sentences' binary representations are shown in the left column, while the right column exhibits the samples according to the cosine similarity of their continuous embeddings.

| Hamming Distance (binary embeddings) | Cosine Similarity (continuous embeddings) |
|---|---|
| **Query: The five people are best friends** | |
| The three people are best friends | The three people are best friends |
| The two people are best friends | The two people are best friends |
| The five people are all friends | Four people are best friends |
| **Query: A group of people are sitting around a table, conversing** | |
| A group of people sitting around a table talking | A group of people sitting around a table talking |
| A group of people are sitting around a table waiting and talking | A group of people sitting around a table |
| A group of people sitting around a table | A group of people are sitting around a table waiting and talking |
| **Query: A person is baking a cake in a kitchen** | |
| A guy is baking a cake in the kitchen | A guy is baking a cake in the kitchen |
| The person is baking a cake | A woman baking a cake in the kitchen |
| A person working baking bread in a kitchen | A person working baking bread in a kitchen |

*Retrieval Speed.* The learned binary codes can use Hamming distance, which is very computationally efficient. We made a simple exploration of the speed improvement. We randomly select a sentence as a query and a sentence database composed of 10000 sentences in SNLI, and then use the InferSent and C&D

algorithm to encode them, respectively. Under the Python implementation, the binary codes with Hamming distance took 14 ms, and the continuous representations with cosine distance took 154 ms. Binary codes can perform 11 times faster than the real-valued. Our implementation is not optimal. The implementation in [19] reports that the calculation can be 37.5 times faster.

### 4.4 Ablation Study

*The Influence of the Assumptions of Anchor Vectors.* In Sect. 3.1, we make 3 assumptions on the anchor vectors: consistency, unbiasedness, and comprehensiveness. If lack of consistency in batch learning, the results are obvious and there is no necessary to explore further. We conduct an experimental analysis on the impacts of unbiasedness and comprehensiveness. In the case that the experimental settings are consistent with Sect. 4.1, if the unbiasedness is removed, the average score of C&D is 69.3; if the comprehensiveness is removed, the average score is 64.2; if both are removed, the performance of C&D will drop sharply and the average score is 9.1. Compared with the C&D with 3 qualities, removing any property will result in performance loss, and removing both at the same time will cause C&D to work abnormally. This reveals that *the assumptions of unbiasedness and comprehensiveness are reasonable.*

**Table 3.** Performances for C&D and 2 competitive methods on 7 downstream tasks when low-dimensional. The cell settings are the same as Table 1.

| Model | Dim | STS 2012 | STS 2013 | STS 2014 | STS 2015 | STS 2016 | STS B | SICK-R | Average |
|---|---|---|---|---|---|---|---|---|---|
| Rand-binary | 512 | 59.5 | 58.4 | 64.2 | 70.6 | 70.3 | 63.1 | **65.3** | 65.8 |
| PCA-binary | 512 | 60.1 | **65.8** | **67.9** | **72.5** | 72.5 | 65.4 | 63.3 | 67.3 |
| C& D | 512 | **60.2** | 64.2 | 65.3 | 71.4 | 70.6 | **67.2** | **65.3** | **67.7** |
| Rand-binary | 1024 | 60.6 | 60.7 | 65.2 | 71.0 | 71.3 | 65.0 | **66.2** | 66.8 |
| PCA-binary | 1024 | **61.5** | 65.9 | **68.4** | **73.8** | **73.4** | 64.6 | 62.2 | 67.0 |
| C& D | 1024 | **61.5** | **66.2** | 66.9 | 72.5 | 72.6 | **69.5** | 65.9 | **70.1** |

*The Effect of Embedding Dimension.* Shen et al. [15] report that autoencoder is more sensitive to dimensions than random projection and PCA. On some tasks, autoencoder and linear projection (random projection and PCA) achieve similar performances when 2048-dimensional, but when the dimension is reduced to 512, autoencoder is more than 5% worse than linear projection methods. We also conducted similar dimension transformation experiments on the dimensions of 512 and 1024. Table 3 reports the comparative results of C&D and linear projection methods on 7 tasks. We can see that with the decrease of available dimensions, the performances of C&D method decrease, even weaker than the linear projection methods on some tasks. However, C&D still outperforms the two competitive methods on the average scores. It is closely related to our mathematical derivation that C&D can achieve such performances.

## 5 Conclusion

This paper presents a new algorithm for learning discrete representations of continuous vectors. Extensive experiments verify the effectiveness of the algorithm. From the academic point of view, the C&D algorithm is a supplement to the research in the field of lightweight models. It does not directly learn the discretized representations from the original vectors, but measures the nature and restores them with a set of universal anchor vectors. The binary codes are learned from the similarity matrix decomposition. This method not only does not cause a huge loss of information, but also outperforms the latest strategies, and reaches the state-of-the-art. From the application perspective, the binarized representations learned with C&D can greatly reduce memory requirements and improve calculation speed. The C&D has huge application value in reality.

## References

1. Bowman, S.R., Angeli, G., Potts, C., Manning, C.D.: A large annotated corpus for learning natural language inference. In: Proceedings of EMNLP, pp. 632–642. ACL (2015)
2. Cer, D., et al.: Universal sentence encoder for English. In: Proceedings of EMNLP, pp. 169–174. ACL (2018)
3. Chen, T., Min, M.R., Sun, Y.: Learning K-way D-dimensional discrete codes for compact embedding representations. In: Proceeding of ICML, vol. 80, pp. 853–862. PMLR (2018)
4. Conneau, A., Kiela, D.: SentEval: an evaluation toolkit for universal sentence representations. In: Proceedings of LREC. European Language Resources Association (ELRA) (2018)
5. Conneau, A., Kiela, D., Schwenk, H., Barrault, L., Bordes, A.: Supervised learning of universal sentence representations from natural language inference data. In: Proceedings of EMNLP, pp. 670–680. ACL (2017)
6. Fisher, R.A.: The use of multiple measurements in taxonomic problems. Ann. Hum. Genet. **7**(2), 179–188 (1936)
7. Glorot, X., Bengio, Y.: Understanding the difficulty of training deep feedforward neural networks. In: Proceedings of AISTATS, vol. 9, pp. 249–256. JMLR.org (2010)
8. Håstad, J.: Some optimal inapproximability results. J. ACM **48**(4), 798–859 (2001)
9. Hotelling, H.: Analysis of a complex of statistical variables into principal components. J. Educ. Psychol. **24**(6), 498–520 (1933)
10. Kiros, J.R., Chan, W.: InferLite: simple universal sentence representations from natural language inference data. In: Proceedings of EMNLP, pp. 4868–4874. ACL (2018)
11. Liu, C., Wang, X., Lu, T., Zhu, W., Sun, J., Hoi, S.C.H.: Discrete social recommendation. In: Proceedings of AAAI, pp. 208–215. AAAI Press (2019)
12. Sokal, R., Michener, C.: A statistical method for evaluating systematic relationships. Univ. Kansas Sci. Bull. **38**, 1409–1438 (1958)
13. Roweis, S.T., Saul, L.K.: Nonlinear dimensionality reduction by locally linear embedding. Science **290**(5500), 2323–2326 (2000)

14. Salakhutdinov, R., Hinton, G.E.: Semantic hashing. Int. J. Approx. Reason. **50**(7), 969–978 (2009)
15. Shen, D., et al.: Learning compressed sentence representations for on-device text processing. In: Proceeding of ACL, pp. 107–116. ACL (2019)
16. Shen, D., et al.: NASH: toward end-to-end neural architecture for generative semantic hashing. In: Proceeding of ACL, pp. 2041–2050. ACL (2018)
17. Shen, F., Shen, C., Liu, W., Shen, H.T.: Supervised discrete hashing. In: Proceeding of CVPR, pp. 37–45. IEEE Computer Society (2015)
18. Shu, R., Nakayama, H.: Compressing word embeddings via deep compositional code learning. In: Proceeding of ICLR. OpenReview.net (2018)
19. Tissier, J., Gravier, C., Habrard, A.: Near-lossless binarization of word embeddings. In: Proceeding of AAAI, pp. 7104–7111. AAAI Press (2019)
20. Xu, J., et al.: Convolutional neural networks for text hashing. In: Proceedings of IJCAI, pp. 1369–1375. AAAI Press (2015)

# Sparse Hierarchical Modeling of Deep Contextual Attention for Document-Level Neural Machine Translation

Jianshen Zhang[1], Yong Liao[1], YongAn Li[2], and Gongshen Liu[1](✉)

[1] School of Electronic Information and Electrical Engineering, Shanghai Jiao Tong University, Shanghai, China
{zjs_007,liaoyong,lgshen}@sjtu.edu.cn

[2] Research and Development Department, Shanghai Oriental Webcasting Co. Ltd., Shanghai, China
foxl.studio@outlook.com

**Abstract.** Document-level machine translation has shown its advantages and importance, but we still have to face some challenges due to the difficulty in efficiently using document context for translation. In this work, we propose a model based on the Transformer to translate the whole paragraph or document. We extend the Transformer model with a new context encoder, which can be incorporated into the original sentence encoder. Then we use a sparsity gate on the context encoder to extract document-level context attention, which is then incorporated into the hierarchical decoder so that we can feedback the document-level inter-sentence consistency and coherence to each word to distinguish different translations of a word according to its specific surrounding context. In addition, we use the pre-training strategy instead of the two-step strategy to take advantage of large-scale parallel sentence pairs and a small-scale corpus with in-domain parallel document pairs to achieve the domain adaptability. The results of experiment on three language-pair corpora have shown that our proposed model can achieve better translation performance quality than baselines.

**Keywords:** Neural machine translation · Deep sentential attention · Sparse attention gate · Natural language processing

## 1 Introduction

Neural Machine Translation (NMT) [1] has drawn more and more attention in academy and industry [8]. Compared with traditional Statistical Machine Translation (SMT), NMT achieves similar or even better translation results in an end-to-end framework. To dynamically and generate the sentence as long as possible, the attention mechanism [10] which enables the model to focus on the relevant previous words in the source side sentence is usually developed and applied. Recently, Transformer [15], the first sequence-to-sequence model entirely


H. Yang et al. (Eds.): ICONIP 2020, LNCS 12532, pp. 799–810, 2020.
https://doi.org/10.1007/978-3-030-63830-6_67

based on attention, seems to gradually replace the traditional RNN in many cases of NLP tasks due to the state-of-the-art performance. Despite the success of Transformer, it still suffers from a major drawback: when it comes to document-level translation, even the Transformer model yields a low performance as it translates each sentence in the document independently and suffers from the problem of ignoring document context, which is actually the work that human usually do in the real world, and has been proven to be beneficial for improving translation performance [6,14,16].

Most of the works try to make the machine predict the words more like human. What these works consider is that the machine should also learn the information from the context before the sentence needed to be translated, which upgrades the traditional sentence-level translation to the document-level [9,13,16,19,22]. They extract previous context (pre-context) and then process the context information as the additional attention of the encoder by traditional neural network or Transformer. However, when there exists a huge gap between the pre-context and the context after the current sentence, the guidance from pre-context is not sufficient for the NMT model to fully disambiguate the sentence. On the other hand, the translation of the current sentence may be inaccurate due to the one-sidedness of partial context. To avoid the issue of translation bias propagation caused by improper pre-context, previous work [13] extracts global context from all sentences of a document with a novel method to feed back the extracted global document context to each word in a top-to-down manner to clarify the translation of words in specific surrounding contexts.

However, the distribution of context attention extraction to the encoding phase is too complicated, and the context attention is calculated mainly based on the sentence-level encoder. Therefore, in this paper, we design two encoder on sentence-level and document-level which can influence each other, and inspired by the idea of Sparse Hierarchical Attention (SHA) mechanism [3] and deep sentential context [17], we can extract the global context more efficiently and then equip the global context to the exact words. Then the both global context and sentence attention can be incorporated into the decoder by a hierarchical gate structure. Furthermore, motivated by [22] and [19], who exploit a large amount of sentence-level parallel pairs to improve the performance of document-level translation, we employ a two-step training strategy in taking advantage of a large scale corpus of out-of-domain sentence-level parallel pairs to pre-train the model and a small-scale corpus of in-domain document-level parallel pairs to fine-tune the pre-trained model.

We conduct experiments on the Transformer model with three parallel language-pair tasks. Experimental results on Japanese-Chinese, Chinese-English and French-English translation show that our proposed model can achieve the state-of-the art performance due to its ability in well capturing global document context.

## 2 Related Work

Developing document-level models for machine translation has been an important research direction, both or conventional SMT [5] and NMT [8]. Most existing work on document-level NMT has focused on integrating document-level context into the RNNsearch model [1,12]. These approaches can be roughly divided into two broad categories: computing the representation of the full document-level context [9,13,22] and using a cache to memorize most relevant information in the document-level context [14,16]. Our approach belongs to the first category by using multi-head attention to represent and integrate document-level context to the sentence encoder and decoder for the whole paragraph translation.

Although these approaches have achieved some progress in document-level machine translation, they still suffer from incomplete document context, since the context encoder chooses a specific number of words before the source sentence, it may not be an entire paragraph which means the content will probably loses some related information. Further more, most of previous works are based on the RNNSearch model [1,12], and only few exceptions [19,22] are on top of the state-of-the-art Transformer model.

## 3 Our Model

In this section, we describe the architecture of the proposed document-level. In the traditional NMT model, the goal is to maximize the likelihood of a set of sentences in a target language represented as sequences of words $y = (y_1, y_2, ..., y_t)$ when given a sentence in the source language as a sequence $x = (x_1, x_2, ...x_n)$, therefore our ultimate goal on document-level translation is to

$$\max \frac{1}{N} \sum_{n=1}^{N} \log \left(P_\theta(y^n | x^n, X_{<n}, Y_{<n})\right) \tag{1}$$

where $X_{<n}$ and $Y_{<n}$ denote the previous sentences before the $n-th$ sentence from source and target sides respectively. The whole architecture of the model is described in Fig. 1.

### 3.1 Sentence-Level Encoder

Given the document with $N$ sentences $X = (X_1, X_2, ...X_N)$, we can use $x_j^{(i)} \in \mathbb{R}^{d \times 1}$ to denote the word vector representation of the $j$-th source word in the $i$-th source sentence. Then the sentence-level encoder maps each word into the corresponding state by:

$$\widetilde{S}_i = \text{MultiHead}(X_i, X_i, X_i) \tag{2}$$

where $S_i$ is the $i-th$ sentence in the document. MultiHead is the self-attention following the transformer [15], and $\widetilde{S}_i = (s_{i,1}, \cdots, s_{i,n}) \in \mathbb{R}^{d \times n}$ is the output of sentence-level encoder.

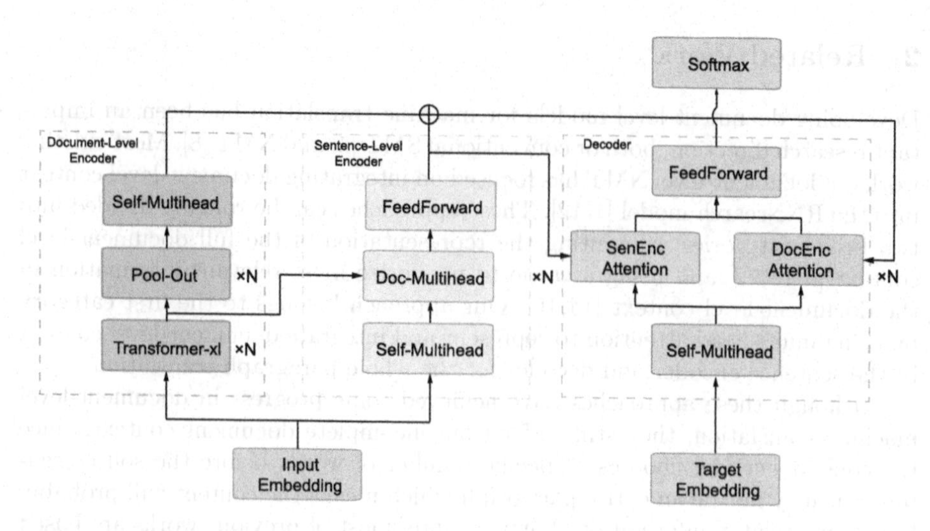


**Fig. 1.** This is whole structure of the document-level Transformer, where the purple blocks are the document encoder for contextual attention while the green blocks are the original sentence encoder of Transformer. The decoder uses a gate structure with blue blocks of original Transformer decoder and the red block for the document-level contextual attention. (Color figure online)

### 3.2 Document-Level Encoder

**Contextual Attention.** In our method, we suppose the context encoder should put the emphasis on the attention between each sentence in the same document. To get the related context attention for the document, we need to let the multi-head attention calculate on the sentence level. However, because the large length of $X_c$, we use transformer-xl structure with two-stream self-attention from xlnet [21] to calculate the bi-directional contexts as the attention. Therefore, we have:

$$H^{(i)} = \text{DocEncoder}\left(H^{(i-1)}, H^{(i-1)}, H^{(i-1)}\right) \quad (3)$$

where ConEncoder is the transformer-xl structure with two-stream self-attention. $H^{(i)}$ is the output of the $i-th$ Context Encoder layer, and $H^{(0)} = X_c$.

**Pool-Out Layer for Deep Sentential Contextual Attention.** No matter how the transformer is set, the output dimension of sequence length is the same as the input query, so here we define a pool-out layer which randomly initializes a trainable vector $Q$ with the same dimension of the last dimension by normal distribution as the input query. In this layer, a multi-head attention will calculate the attention between the input query and the output of the context encoder transformer by

$$C^{(i)} = \text{MultiHead}\left(Q, H^{(i)}, H^{(i)}\right) \quad (4)$$

Then following the idea of deep sentential context [17], we use a new pool-out layer to get the extracted feature of the paragraph, with a multi-head layer to get the attention between the sentences by:

$$\widetilde{C} = \text{MultiHead}\Big(\mathbf{Pool-out}\Big(\Big[C^{(1)}, \cdots, C^{(N)}\Big]\Big)\Big) \tag{5}$$

where the input of the pool-out layer is the concatenation of the output of the $N$ layers on the sequence length dimension. The pool-out function is defined in Eq. 4.

**Sparse Attention Gate Block for Contextual Attention.** Through the context encoder, we can get the attention vector for each sentence in the paragraph. Because we use a randomly initialized vector to calculate the contextual attention vector, therefore we can consider the calculation of queries $Q$ and keys $K$ in the multi-head of the pool-out step as the influential weights of the tokens in the sentences. These tokens with higher weights should contain more information which can represent the sentence. Here we use a Sparse Attention Gate (SAG), which is realized by a novel Sparse Hierarchical Attention (SHA) mechanism, to get these emphasized sequences and tokens. The Fig. 2a shows that the block contains two steps. First, we need to get the attention matrix $M$ by getting the mean value of $M^{(i)}$ in each pool-out layer. Then we use a sparse gate to enforce sparsity via a learned, input-dependent threshold $\tau$ by:

$$\begin{aligned} M &= \frac{1}{N}\sum_{i=1}^{N} M^{(i)} \\ \hat{M} &= \mathbf{Boolean}(\text{ReLu}(M - \tau_M)) \end{aligned} \tag{6}$$

Here Boolean is an element-wise operation and is 1 if the corresponding element in $\hat{M}$ is non-zero. Because the $M$ is a matrix with the last dimension of sequence length, so as Fig. 2b, the contextual attention can be multiplied by the sparse gate matrix by:

$$\hat{C} = \hat{M} \cdot \widetilde{C} \tag{7}$$

The Hierarchical Attention Block (HAB) can propagate the connections across network layers. With multiple layers stacked together, Sparse hierarchical attention (SHA) simulates a hierarchical process that combines tokens into trees using learned attention connections.

### 3.3 Integration with Sentence-Level Encoder

Consider that the document context is first extracted during the encoding phase and then distributed to each word in the document. On this basis, following the previous work [22], we first incorporate the extracted context self-attention into the word representation as:

$$S = \text{MultiHead}\Big(H, \widetilde{S}, \widetilde{S}\Big) \tag{8}$$

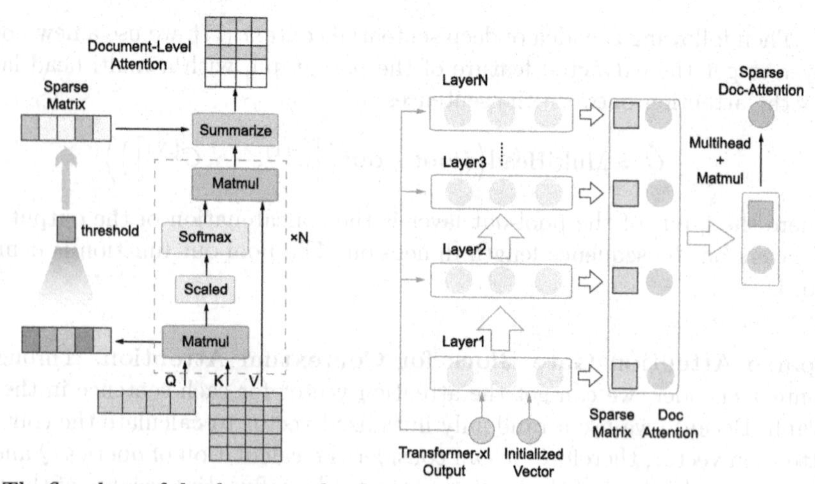


(a) The flowchart of the document-level attention calculation in the document-level encoder and the details of the sparse attention gate.

(b) The flowchart of deep sentential extraction in the document-level encoder.

**Fig. 2.** Main changes in the model due to the document-level.

where $H$ is the contextual attention output of context-level encoder. $S$ is the integrated representation of the words.

We propose to employ the residual connection function [7] to incorporate the extracted global context information into the word representation by:

$$C = \hat{C} + \widetilde{S} \tag{9}$$

where $\widetilde{S}$ is the self attention output of sentence-level encoder. $C$ is the global contextual attention.

### 3.4 Document-Level Decoder

The decoder can be separated into two parts, one for sentence-level attention calculation and document-level attention calculation. At the $N-th$ level in the whole decoder, the first step is to do the self-attention by:

$$U^{(N)} = \text{MultiHead}\Big(T^{(N-1)}, T^{(N-1)}, T^{(N-1)}\Big) \tag{10}$$

Because we need to integrate context attention of sentences to the decoder so as to generate document or paragraph, so the sentence-level and document-level Encoder-Decoder attention sub-layer is formulated by:

$$\begin{aligned} E^{(N)} &= \text{MultiHead}\Big(U^{(N)}, S, S\Big) \\ G^{(N)} &= \text{MultiHead}\Big(U^{(N)}, C, C\Big) \end{aligned} \tag{11}$$

where $S$ and $C$ are the output of sentence-level and document-level encoders. With sentence-level result $E$ and document-level result $G$, we use a gate structure to get the result and the distribution of the corresponding translated words by softmax, which can be formulated by:

$$\begin{aligned} T^{(N)} &= \mathrm{FFN}(\alpha \cdot E^{(N)} + (1-\alpha) \cdot G^{(N)}) \\ P(y_j^k|X) &\propto \exp(W_o T_{\cdot,j}^{N_c}) \end{aligned} \tag{12}$$

### 3.5 Training Strategy of Model

The goal of most NMT models is to maximize the log-likelihood of the decoder output. Considered that the model still needs to be evaluated on sentences and the the size limitation of document-level parallel corpora, the two-step training strategy [22] is also applied in the method. The difference is that we first optimize the sentence-level parameters $\theta_s$ as pre-train in the whole training phase by:

$$\hat{\theta}_s = \underset{\theta_s}{\mathrm{argmax}} \sum_{(x,y) \in S} \log(P(y|X;\theta_s)) \tag{13}$$

Next we start to train the whole parameters in the model by:

$$\hat{\theta} = \underset{\theta}{\mathrm{argmax}} \sum_{(X,Y) \in D} \log(P(Y|X;\theta)) \tag{14}$$

In this way, we find that we can avoid the model dropped into over-fitting to some degrees. The model will perform well in both the sentence-level translation and document-level coherence.

## 4 Experiments

To further prove the effect of our method, we conduct experiments on Japanese-Chinese, Chinese-English and French-English translation tasks.

### 4.1 Dataset

For Japanese-Chinese translation, we conduct our experiments on Asian Scientific Paper Excerpt Corpus (ASPEC) [11]. ASPEC corpus consists of a Chinese-Japanese scientific paper excerpt corpus of approximately 0.68 million parallel sentences. It is a parallel corpus consisting of Japanese scientific papers from the reference database and electronic journal site J-STAGE of the Japan Science and Technology Agency (JST) that have been translated to Chinese.

For Chinese-English and French-English translation task, we use the Ted talks corpus from the IWSLT [2] as the document-level parallel corpus, including 1,906 documents with 226K sentence pairs on zh-en task and 1,824 documents with 220K sentence pairs on de-en task. We use dev2010 which contains 8 documents with 879 sentence pairs on zh-en and 8 documents with 887 sentence pairs on de-en for development. We use tst2012-tst2015 which contain 62 documents with 5566 sentence pairs on zh-en task and tst2010-2014 which 61 documents with 5,937 sentence pairs on de-en task for testing respectively.

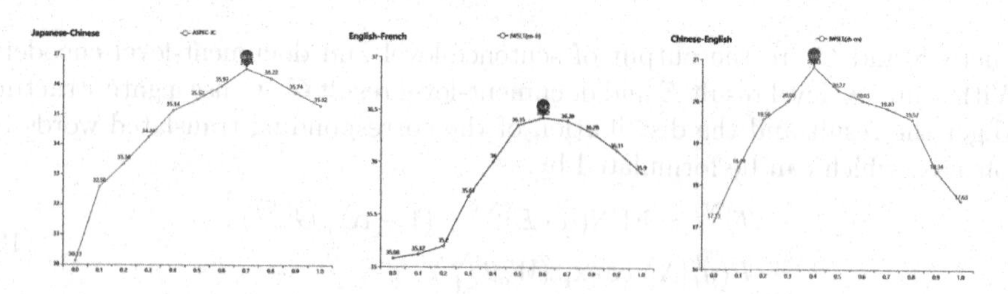

**Fig. 3.** BLEU score of different portion $\alpha$

## 4.2 Model Setup

We use the case insensitive 4-gram BLEU score to evaluate the results and the script from Moses to test the BLEU scores. The vocabulary size of English and French is 50,000 and the vocabulary size of Japanese and Chinese is 32000. In addition, according to the previous work [22], we also take the number of sentences in the document into account which will make influence on the performance. If the sentence is too long, we will divide the sentences into two shorter sentences.

The encoder and decoder are each made up of 1 to 6 hidden layers. All hidden states have a dimension of 512, dropout of 0.1 and heads of 8 for Multi-Head attention. For the efficiency, we set the layer number of all the models to 1 in our experiments. The optimization and regularization methods were the same as proposed by [20].

# 5 Analysis and Discussion

To further demonstrate the effectiveness of our proposed hierarchical model with sparse context attention, we illustrate several experimental results in this section and give our analysis on them.

## 5.1 Effect of Gate Portion $\alpha$

Because we want to see how the parameter $\alpha$ in the gate structure effects the model. As in Fig. 3, we can see that the $\alpha$ between 0.7 to 0.8 can have a better result on BLEU on ASPEC Japanese-Chinese dataset while about 0.4 to 0.5 on English-Chinese dataset. That means the document-level attention based on sentences can actually improve the model in decoding, but we still need to optimize the sentence-level decoding in the decoder to make the output of the model have higher accuracy on words. And we can find that different language-pair datasets need to consider the balance of attention between the two levels. Also, we will try to explain the trend of the change of ratio $\alpha$ in the future work.

**Table 1.** BLEU score of the experiments on three datasets.

| Models | ASPEC | IWSLT | |
|---|---|---|---|
| | ja->zh | fr->en | zh->en |
| Doc-TRANS [22](Baseline) | 35.32 | 34.00 | 17.63 |
| Bi-LSTM (attention) | 31.65 | 33.20 | 15.87 |
| ConvS2S [4] | 31.90 | 33.85 | 17.15 |
| Memory Network [9] | 31.85 | 32.00 | 17.24 |
| H-RNN Search [16] | 33.42 | 32.24 | 16.18 |
| HAN encoder-decoder [19] | 35.22 | 34.24 | 17.87 |
| Our Model | **36.54** | **34.42** | **20.83** |

### 5.2 Overall Performance

Here we use Doc-Transformer [22] as the baseline model, and we also make some comparison tests with other models. The overall experimental results of different models are evaluated by the BLEU score, based on the paragraph-parallel corpus made from ASPEC Chinese-Japanese parallel corpus and the document-level corpus from IWSLT 2010. Our model will use the entire paragraph or document as the input, and the output will be the translated paragraph or document. While in the decoding phase, we use the greedy search to get the result in all the comparison experiments because of the efficiency.

As shown in Table 1, using the same data for Japanese-Chinese translations, our approach achieves significant improvements over the original Transformer model [15], which gains on the ASPEC-JC is 1.22 BLEU points. On IWSLT, it also outperforms the Transformer and Doc-Transformer, but smaller than that on JC set, only 0.42 points and the Chinese-English task got a significant improvement of 3.2 points. Besides the comparison with previous works, the results also show that our model also outperforms the hierarchical method [16] adapted for Transformer, which also uses the two-step training strategy. By this experiment we can infer that sparsity of context attention in the encoder can help the decoder to put more emphasis on the specific tokens and sentences in the paragraph, which eliminates the back propagation of error in the decoder. This combination further improves translation performance, showing that they contribute complementary information.

### 5.3 Accuracy of Pronoun/Noun Translations

We use the reference-based metric: accuracy of pronoun translation [18] to evaluate co-reference. From the results, our proposed sparsity hierarchical model can well improve the performance of pronoun translation in both corpora due to the well captured global document context assigned to each word. Correspondingly, we display a translation example in Table 3 to further illustrate this. From the example, given the surrounding context, our proposed model can well infer the latent pronoun it and thus improve the translation performance of the Transformer model (Table 2).

**Table 2.** Evaluation on pronoun translation of document-level translation on different language pairs.

| Models | Japanese Chinese | French English | | | | | Chinese English | | | |
|---|---|---|---|---|---|---|---|---|---|---|
| | ASPEC | tst11 | tst12 | tst13 | tst14 | Avg | tst13 | tst14 | tst15 | Avg |
| Baseline | 65.23 | 86.22 | 85.01 | 86.47 | 84.78 | 85.63 | 55.10 | 51.71 | 52.48 | 54.34 |
| Our Model | **68.25** | 87.03 | 86.06 | 87.36 | 85.64 | **86.50** | 55.68 | 52.75 | 53.26 | **55.22** |

## 5.4 Examples of Translation

We use an example to illustrate how document-level context helps translation. To better understand the advantages of our model, we compared the translation results of different models on our corpus, as shown in Table 3. In Doc-Trans model of this case, we also find some incorrect translations and some incoherently translated sentences. While our model can not only translate the paragraph well in the meaning, but also very brief and coherent, especially when it comes to some conjunctions or turning words, the traditional methods can not exactly translate them, our model can find out these words and perform well and output the words that referred in the previous context. Meanwhile, through the output sentences, we can also find that the same source sentence will sometimes have different

**Table 3.** Examples of the output from different NMT models.

| | |
|---|---|
| Source | 環境マネジメントシステムの🄵🄵性を担保するには，運用システムを🄵独で構築🄵運用するのではなく，ほかのシステムの要素とのつながりを明確にする必要がある。\| その上で，組織の形態や🄵態に即した運用システムを構築する必要がある。すなわち，本業における環境マネジメントを展開することが重要である。このような視点から，本稿では，ＩＳＯ１４００１規格の４．４項「🄵施及び運用」について，教育訓練と運用管理を取り上げて解説した。 |
| Reference | 为了确保环境管理系统的实效性，有必要明确与其他系统之间的关连，而不是单独地构筑及运用该系统。在此基础上，须构筑适合组织的形式与实际情况的运用系统。即，重要的是在本行业开展环境管理系统。从这样的角度出发，本文中列举了ＩＳＯ１４０１规格的４．４项“实施及运用”，进行解说。 |
| Doc-TRANS | 为了确保运用环境管理体系的实效性，不是单独构建和运用系统，而是需要明确与其他系统要素的联系。在此基础上，有必要构筑符合组织的形态和实际状态的运用系统。也就是说，开展环境管理是非常重要的。从这样的观点出发，在本稿中，关于ＩＳＯ１４００１规格的４．４项“实施及运用”的教育训练和运用管理进行了解说。 |
| Our Model | 为确保环境管理系统的实效性，不仅仅需要单独构筑运用系统，还需要明确其他的系统要素的联系。在此基础上，必须构筑符合组织形态和实际的运用系统。即，在本行业中展开环境管理系统是很重要的。从这样的角度出发，本文中列举了ＩＳＯ１４０１规格的４．４项“实施及运用”，进行解说。 |

translated output due to the position in the document, which also demonstrates that our model captures the sequence order in the document.

## 6 Conclusion

In this paper, we introduced two modifies in the original NMT models with transformer. First, we have presented a method for exploiting paragraph translation by using deep sentential context extraction to get document-level attention. Then, we using sparse attention gate with hierarchical deep sentential context attention model to integrate both sentence-level and document-level attention to encoder-decoder system inside the state-of-the-art neural translation model Transformer. Second, to improve the performance of NMT in paragraph-level translation, we try to make the balance between document-level and sentence-level parts in the gate structure of decoder. Also, we use the pre-training strategy to alleviate the imbalance between sentence-level corpus and document-level corpus. Experiments on different language-pair translation tasks show that our method is able to translate long paragraphs and improve over Transformer. We can also find that the paragraph translated by the model will put more emphasis on the related words in the sentences so as to make the paragraph be more reasonable. In the future, we plan to further validate the effectiveness of our approach on more language pairs.

**Acknowledgement.** This research work has been funded by the National Natural Science Foundation of China (No. 61772337), and the Eastday-SJTU Artificial Intelligence Media Joint Lab.

## References

1. Bahdanau, D., Cho, K., Bengio, Y.: Neural machine translation by jointly learning to align and translate. In: ICLR (2015)
2. Cettolo, M., Girardi, C., Federico, M.: WIT3: web inventory of transcribed and translated talks. In: Conference of European Association for Machine Translation, pp. 261–268 (2012)
3. Correia, G.M., Niculae, V., Martins, A.F.T.: Adaptively sparse transformers. In: EMNLP/IJCNLP (1), pp. 2174–2184. Association for Computational Linguistics (2019)
4. Gehring, J., Auli, M., Grangier, D., Yarats, D., Dauphin, Y.N.: Convolutional sequence to sequence learning. In: ICML. Proceedings of Machine Learning Research, vol. 70, pp. 1243–1252. PMLR (2017)
5. Gong, Z., Zhang, M., Zhou, G.: Cache-based document-level statistical machine translation. In: EMNLP, pp. 909–919. ACL (2011)
6. Hardmeier, C., Nivre, J., Tiedemann, J.: Document-wide decoding for phrase-based statistical machine translation. In: EMNLP-CoNLL, pp. 1179–1190. ACL (2012)
7. He, K., Zhang, X., Ren, S., Sun, J.: Deep residual learning for image recognition. In: CVPR, pp. 770–778. IEEE Computer Society (2016)

8. Jean, S., Cho, K., Memisevic, R., Bengio, Y.: On using very large target vocabulary for neural machine translation. In: ACL (1), pp. 1–10. The Association for Computer Linguistics (2015)
9. Maruf, S., Haffari, G.: Document context neural machine translation with memory networks. In: ACL (1), pp. 1275–1284. Association for Computational Linguistics (2018)
10. Mnih, V., Heess, N., Graves, A., Kavukcuoglu, K.: Recurrent models of visual attention. In: NIPS, pp. 2204–2212 (2014)
11. Nakazawa, T., et al.: ASPEC: asian scientific paper excerpt corpus. In: LREC. European Language Resources Association (ELRA) (2016)
12. Sutskever, I., Vinyals, O., Le, Q.V.: Sequence to sequence learning with neural networks. In: NIPS, pp. 3104–3112 (2014)
13. Tan, X., Zhang, L., Xiong, D., Zhou, G.: Hierarchical modeling of global context for document-level neural machine translation. In: EMNLP/IJCNLP (1), pp. 1576–1585. Association for Computational Linguistics (2019)
14. Tu, Z., Liu, Y., Shi, S., Zhang, T.: Learning to remember translation history with a continuous cache. TACL **6**, 407–420 (2018)
15. Vaswani, A., et al.: Attention is all you need. In: NIPS, pp. 5998–6008 (2017)
16. Wang, L., Tu, Z., Way, A., Liu, Q.: Exploiting cross-sentence context for neural machine translation. In: EMNLP, pp. 2826–2831. Association for Computational Linguistics (2017)
17. Wang, X., Tu, Z., Wang, L., Shi, S.: Exploiting sentential context for neural machine translation. In: ACL (1), pp. 6197–6203. Association for Computational Linguistics (2019)
18. Werlen, L.M., Popescu-Belis, A.: Validation of an automatic metric for the accuracy of pronoun translation (APT). In: DiscoMT@EMNLP, pp. 17–25. Association for Computational Linguistics (2017)
19. Werlen, L.M., Ram, D., Pappas, N., Henderson, J.: Document-level neural machine translation with hierarchical attention networks, pp. 2947–2954 (2018)
20. Yang, Z., Chen, W., Wang, F., Xu, B.: Improving neural machine translation with conditional sequence generative adversarial nets. In: NAACL-HLT, pp. 1346–1355. Association for Computational Linguistics (2018)
21. Yang, Z., Dai, Z., Yang, Y., Carbonell, J.G., Salakhutdinov, R., Le, Q.V.: XLNet: generalized autoregressive pretraining for language understanding. In: NeurIPS, pp. 5754–5764 (2019)
22. Zhang, J., et al.: Improving the transformer translation model with document-level context. In: EMNLP, pp. 533–542. Association for Computational Linguistics (2018)

# Author Index

Printed in the United States
By Bookmasters